AF147428

KIAS Springer Series in Mathematics

Volume 5

Founding Editor
Jaigyoung Choe

Editor-in-Chief

Nam-Gyu Kang, School of Mathematics, Korea Institute for Advanced Study, Seoul, Korea (Republic of)

Series Editors

Kyeongsu Choi, School of Mathematics, Korea Institute for Advanced Study, Seoul, Korea (Republic of)

Sang-hyun Kim, School of Mathematics, Korea Institute for Advanced Study, Seoul, Korea (Republic of)

Young-Hoon Kiem, School of Mathematics, Korea Institute for Advanced Study, Seoul, Korea (Republic of)

The KIAS Springer Series in Mathematics publishes original content in the form of high level research monographs, lecture notes, proceedings and contributed volumes as well as advanced textbooks in English language only, in any field of Pure and Applied Mathematics. The books in the Series are connected to the research activities carried out by the Korea Institute for Advanced Study (KIAS), and will discuss recent results and analyze new trends in mathematics and its applications. The Series is aimed at providing useful reference material to academics and researchers at an international level.

Nam-Gyu Kang • Ionut Ciocan-Fontanine •
David Favero • Yuan-Pin Lee • Hyenho Lho •
Jeongseok Oh
Editors

Categorical and Enumerative Aspects of Mirror Symmetry

A Tribute to the Life and Work of Professor Bumsig Kim

Springer

Editors
Nam-Gyu Kang
School of Mathematics and June E Huh Center for Mathematical Challenges
Korea Institute for Advanced Study
Seoul, Korea (Republic of)

David Favero
School of Mathematics
University of Minnesota
Minneapolis, MN, USA

Hyenho Lho
Department of Mathematics
Chungnam National University
Daejeon, Taejon-jikhalsi, Korea (Republic of)

Ionut Ciocan-Fontanine
Institute of Mathematics
Academia Sinica
Taipei, Taiwan

Yuan-Pin Lee
Institute of Mathematics
Academia Sinica
Taipei, Taiwan

Jeongseok Oh
Department of Mathematical Sciences and Research Institute of Mathematics
Seoul National University
Seoul, Korea (Republic of)

International Conference on Categorical and Enumerative Aspects of Mirror Symmetry in Memory of Professor Bumsig Kim 2022 KIAS Seoul, Korea (Republic of)

ISSN 2731-5142 ISSN 2731-5150 (electronic)
KIAS Springer Series in Mathematics
ISBN 978-981-95-0384-1 ISBN 978-981-95-0385-8 (eBook)
https://doi.org/10.1007/978-981-95-0385-8

Mathematics Subject Classification: 14C17, 14D20, 14A22, 14F10, 14J32, 14L24, 14N35, 53D42

This work was supported by Korea Institute for Advanced Study.

© The Editor(s) (if applicable) and The Author(s) 2026. This book is an open access publication.

Open Access This book is licensed under the terms of the Creative Commons Attribution-NonCommercial-NoDerivatives 4.0 International License (http://creativecommons.org/licenses/by-nc-nd/4.0/), which permits any noncommercial use, sharing, distribution and reproduction in any medium or format, as long as you give appropriate credit to the original author(s) and the source, provide a link to the Creative Commons license and indicate if you modified the licensed material. You do not have permission under this license to share adapted material derived from this book or parts of it.
The images or other third party material in this book are included in the book's Creative Commons license, unless indicated otherwise in a credit line to the material. If material is not included in the book's Creative Commons license and your intended use is not permitted by statutory regulation or exceeds the permitted use, you will need to obtain permission directly from the copyright holder.
This work is subject to copyright. All commercial rights are reserved by the author(s), whether the whole or part of the material is concerned, specifically the rights of translation, reprinting, reuse of illustrations, recitation, broadcasting, reproduction on microfilms or in any other physical way, and transmission or information storage and retrieval, electronic adaptation, computer software, or by similar or dissimilar methodology now known or hereafter developed. Regarding these commercial rights a non-exclusive license has been granted to the publisher.
The use of general descriptive names, registered names, trademarks, service marks, etc. in this publication does not imply, even in the absence of a specific statement, that such names are exempt from the relevant protective laws and regulations and therefore free for general use.
The publisher, the authors and the editors are safe to assume that the advice and information in this book are believed to be true and accurate at the date of publication. Neither the publisher nor the authors or the editors give a warranty, expressed or implied, with respect to the material contained herein or for any errors or omissions that may have been made. The publisher remains neutral with regard to jurisdictional claims in published maps and institutional affiliations.

This Springer imprint is published by the registered company Springer Nature Singapore Pte Ltd.
The registered company address is: 152 Beach Road, #21-01/04 Gateway East, Singapore 189721, Singapore

If disposing of this product, please recycle the paper.

Foreword

This volume is based on the memorial conference, "A Tribute to the Life and Work of Professor Bumsig Kim: Categorical and Enumerative Aspects of Mirror Symmetry," held on September 19–23, 2022, at KIAS (http://events.kias.re.kr/h/SEP22/). The conference brought together more than 80 participants to honor Professor Bumsig Kim, his life, and his contributions to Mathematics. Nineteen of his friends, including former collaborators, postdocs, and students, were invited as speakers. Additionally, eight speakers, along with their collaborators, and three teams who were unable to attend the conference have graciously contributed to this volume.

"Gopakumar–Vafa invariants = Quantum K-invariants on Calabi–Yau threefolds" by Y.-C. Chou and Y.-P. Lee is an expository article exploring the equivalence between genus-zero Gopakumar–Vafa invariants and genus-zero Quantum K-invariants on Calabi–Yau threefolds. The authors establish this equivalence for the local projective line proving the multiple cover formula.

"Twisted wild character varieties and Gromov–Witten invariants" by D.-E. Diaconescu explores the construction of a Calabi–Yau threefold with torus action from a given twisted wild character variety. Then the author shows the residual Gromov–Witten invariants of this threefold reproduce the E-polynomial of the character variety, reflecting its weight filtration.

"Holomorphic anomaly equations for $[\mathbb{C}^5/\mathbb{Z}_5]$" by D. Genlik and H.-H. Tseng proves holomorphic anomaly equations for the equivariant Gromov–Witten invariants of $[\mathbb{C}^5/\mathbb{Z}_5]$, as indicated by the title.

"Permutation–Equivariant Quantum K-theory IX: Quantum Hirzebruch–Riemann-Roch in all genera" by A. Givental establishes a higher-genus version of the adelic characterization, which expresses the descendant potential encoding K-theoretic Gromov–Witten invariants in terms of cohomological Gromov–Witten invariants. This result is achieved using the Kawasaki–Riemann–Roch formula.

"Calabi–Yau complete intersections in exceptional Grassmannians" by A. Ito, M. Miura, A. Okawa, and K. Ueda classifies all equivariant vector bundles on G/P, where G is an exceptional group, with rank $= \dim(G/P) - 3$ and the first Chern class $= c_1(G/P)$. For Grassmannians with Picard number one, the authors classify all families of Calabi–Yau threefolds that can be realized as complete intersections of these bundles and compute their Hodge and Chern numbers.

"Virtual fundamental classes of the vanishing loci of cosections" by Y.-H. Kiem and H. Park is a tutorial paper that investigates the equivalence between the cosection-localized virtual cycle and the (-2)-shifted symplectic virtual cycle, achieved by imposing a (-2)-shifted symplectic structure on the degeneracy locus of the cosection.

"Globalization of Chern Characters and canonical pairings" by T. Kim is an article surveying work on Chern characters of matrix factorizations and Riemann–Roch type theorems.

"Bordered contact instantons and their Fredholm theory and generic transversalities" by Y.-G. Oh develops the Fredholm theory for the moduli space of bordered contact instantons, with a primary focus on proving generic mapping transversality and evaluation transversality.

"Duality for Landau–Ginzburg models" by C. Sabbah compares two hypercohomologies associated with a Landau–Ginzburg model: the twisted de Rham cohomology and the cohomology without de Rham differentials. These are shown to arise as distinct specializations within the hypercohomology of a family of complexes under suitable conditions. The paper also provides results comparing the pairings with compactly supported hypercohomologies.

"A Kleiman criterion for GIT stack quotients" by M. Shoemaker proposes an analogue of Kleiman's numerical criterion of ampleness in the context of GIT quotients. This analogous criterion is proven by introducing an equivariant lift of the cone of curves generated by the numerical classes of quasimaps.

"Derived categories of Quot schemes of zero-dimensional quotients on curves" by Y. Toda examines a semiorthogonal decomposition of the derived category of a relative Quot scheme parametrizing zero-dimensional quotients of a sheaf with homological dimension at most 1 on a family of smooth, projective curves.

Seoul, Korea (Republic of)
February 2025

Ionut Ciocan-Fontanine
David Favero
Nam-Gyu Kang
Yuan-Pin Lee
Hyenho Lho
Jeongseok Oh

Memories of Bumsig Kim (1968–2021)

I think Bumsig was Sasha Givental's first student. Between Bumsig (1990–96 Berkeley math) and I (1994–99), there were 4 years and three other Sasha's PhD students: an Indian girl, a Taiwanese, and a Japanese boy. It turned out that all three dropped out of the PhD program. For a year (1995–96), Bumsig and I were the only PhD students Sasha had. After he left for Stockholm (Mittag-Leffler Institute), I became Sasha's only PhD student until nearing my graduation when Tom Coates joined. I did not get to know Bumsig as much as I should have, but I did get to know him a little, and he was the one I interacted with the most in that year. For one thing, I decided to study with Sasha in Spring 1995, and at that time Bumsig already had a paper with Sasha. At that time, Sasha had only three papers on quantum cohomology and mirror symmetry proper (although he had many on closely related Floer homology and symplectic topology.) So I was naturally very keen at understanding their joint paper Quantum Cohomology of Flag Manifolds and Toda Lattices. However, that was near the end of the Spring 1995 semester, and I was busy preparing for my oral qualifying exam. Furthermore, neither of us possessed any English language skills to speak about at that point (and Sasha's English wasn't great either!) The deciding factor was that my wife was in her third-trimester of pregnancy, and I soon was completely occupied, indeed overwhelmed, by the birth of our son.

Fortunately, in Fall 1995, Sasha gave a topics course on the symplectic aspect of mirror symmetry and Floer homology and what have you. Most of the material flew over my head, so I spoke with Bumsig more than a couple of times after class to consult with him about what was taught in class. (It wasn't really discussions, but mostly a one-way street from him to me.) Two events stand out during the semester. The first was that in one lecture, Sasha proposed a "Lefschetz hyperplane theorem" for Floer homology, or something like that. As with most other things in that class, I was intrigued but understood little. Then Sasha gave his manuscript of Equivariant Gromov–Witten Invariants to us and encouraged us to generalize it to products of the projective spaces. Foolish as I was, I thought that was not really exciting and did not pursue it seriously. Another obstacle was of course the difficulty in understanding Sasha's paper. Then Sasha went on sabbatical in Spring 1996, not before he gave me a problem to work on. That became my first paper and a lucky one to boot. That one overlapped with one of Rahul's small papers and I got to know him because of that. That was before the Internet browser and I remember using the terminal command and "gopher" to connect to other servers, etc., but I have sidetracked . . .

I got to know Bumsig more after he left for Mittag-Leffler. He told me that there was a learning seminar going on there, reading Fulton–Pandharipande's intro paper. At some point, he sent me a draft entitled Givental's Lefschetz Principle, or something like that. (The title was later revised as Quantum Lefschetz Hyperplane Principle.) You see, while I was busy changing my son's diapers, Bumsig really understood the potential of Givental's philosophy and was able to combine what he learned from Sasha's lectures and the paper on mirror theorem to produce a nice result.

After I managed to slightly generalize Bumsig's result on quantum Lefschetz, I was invited to visit him in Pohang University of Science and Technology. It was a very nice visit, and I have since visited Bumsig more than a couple of times, first in Pohang and then in Seoul. Unfortunately, our discussions remained essentially a one-way street: while he received very little from me, I have learned a great deal from him. For example, when I heard his (and Ionut's) idea of abelian vs. non-abelian quotients, I was intrigued. However, I did not manage to pursue it further. It had to wait for more qualified Ionut and Aaron to get the job done.

It was a great shock when I heard from Jeongseok Oh that Bumsig passed away while having dinner with him and others. That night I could barely sleep, even though an international conference call ended past 2 AM and I was exhausted. A lot of random memories came back to me. I sat up and wrote most of the above.

For an online meeting reminiscing on our friend Bumsig, 2021/12/20, 0:00 Korean Standard Time

Y.-P. Lee

I met Bumsig Kim in 2011 at the Kavli Institute for the Physics and Mathematics of the Universe. I remember it fondly. He approached me after I gave a talk on variation of geometric invariant theory quotients and derived categories, a topic I had been working on with Matthew Ballard and Ludmil Katzarkov as a Postdoctoral Fellow at the University of Vienna. Already on our first encounter, he was kind, humble, and earnest in his approach to mathematics and I suppose . . . most everything.

I'm sure he had wonderful questions and am also sure I had no idea where they were coming from. Of course, Bumsig was an absolute expert on geometric invariant theory, yet, he asked his questions simply and with incredible politeness. I did my best to answer and I doubt he corrected any mistakes I made. We were soon joined by another conference speaker: Ionut Ciocan-Fontanine. The two then were truly engaged (though again I had no idea why as I knew nothing about quasi-maps or, more generally, enumerative geometry) and the three of us spoke for some time, first in the hallway outside the lecture room, then at the blackboards, then as we rode the elevator to the final conference tea, and more at the tea itself.

After the conference, I returned to Vienna and to my work with Ludmil. That could have been the end of my relationship with Bumsig but fortunately, two years later in 2013, Ludmil and I would organize a 3-month thematic period at the Erwin Schrodinger Institute and Bumsig would spend two weeks there as a visiting scholar. This time, I approached Bumsig after his talk on quasi-maps and we spoke for some time again. His 2-week visit and the conference events gave us a better opportunity to get to know each other. We ate dinner and drank wine together at the conference dinner which was at one of the Huerigers north of the city (these are traditional Austrian taverns located in the vineyards).

The next part is also a memory I often return to. We were walking down the hallway together at the Erwin Schrodinger Institute and he asked me if I'd like to visit him at the Korea Institute for Advanced Study (KIAS), in Seoul. Though I was very excited about it, it was a tricky question. This was the Summer of 2013 and among my last days in Austria. I had just started a new position at the University of Alberta and would be immigrating to Canada in August. Nevertheless, it was a very exciting opportunity and I decided to go. Two months later, I spent my first month at KIAS and subsequently immigrated to Canada directly from South Korea.

This first experience at KIAS was wonderful. I believe Bumsig always put a lot of thought and effort into making his visitors welcome and comfortable. On my first visit, he put my wife and I in the largest KIAS apartment during my one month stay. Seoul was exciting, the KIAS staff were so kind and helpful, and discussing math with Bumsig was fascinating and new. It was an amazing visit.

After that, Bumsig asked me to return as a KIAS scholar. This time, it was very easy to say yes and, in fact, I would return for the next 6 years. We became close friends, working continuously during numerous visits. Nevertheless, every time I visited, Bumsig continued to treat me as an honored guest. He would always insist I decided where to go for lunch and often order the same thing as me despite the fact that I am a vegetarian which leads to very limited food options in Korea. I can't say precisely what it was, but there was something very kind and humble about this simple gesture. He would also negotiate the KIAS apartments for me, even though they were sometimes fully booked, and would always make sure I had everything I needed (e.g., after I told him I had no way to boil eggs at a KIAS apartment, he brought me a microwave egg boiler the next day—without my asking).

The last time I visited Bumsig was in December of 2019 right before the Covid pandemic. My daughter was one year old and she came with me on her first plane ride. She got to meet Bumsig and his family on December 25, 2019, at the KIAS Christmas party. We all laughed and smiled and celebrated together. That was the last time I saw him.

David Favero 2024/08/27

When Prof. Bumsig Kim and I visited Berkeley in Spring 2018, we did many things together. Discussions on lunch menus was the number one among them. We also played tennis together quite often, almost once every other day. Being a beginner, most of my shots were line-out. Instead of pointing out my inadequacy, he constantly encouraged me and told me that my skills had been improving during the two months we played together.

The moment I asked him to be my PhD advisor was at the conference in Osaka, July 2012. It was an unforgettable memory. I felt like I just asked Roger Federer to be my double partner. Anyways that Roger Federer in math kindly answered me yes and I became his student.

Again, in Berkeley, I heard from him that his wife warned him that a young person usually does not like to be with an older one. I worried maybe he told me this because I bothered him too much, but he was likely worried that he was boring me. The truth is I treasured every moment with him—not just our precious time in Berkeley. Back in Bonn 2012, he was surprised how I loved beer—the truth was that I drank one or two glasses of beer in every dinner, a canonical minimal pair in Germany. Another memory came from Guanajuato in 2013 when Bumsig, Hyenho, and myself became like K-pop stars—thanks to Korean singer PSY with the song Gangnam style, Mexican students wanted to take a photo with quasi-PSYs in that lovely city. Also in Kobe 2016, we and Kyusik ran to escape a rain on the mountain Rokko, but I led them in a wrong direction. When we finished running the rain already stopped. I also remembered the first time he ever said "no" to me. I was much relieved and paradoxically very happy when he said "no."

On 1st December 2021, we spent a whole day together. At lunch, he gave me an (indirect but clear) compliment that he was proud of me to become a postdoc of Richard, another Roger Federer. Bumsig was a very kind and very humble person. He has never judged other people mathematically because he did not think too highly of himself. Therefore the compliment on that day was incredibly special to me. It was at dinner of that day, Bumsig had a heart attack and tragically passed away.

Sometimes my friends ask me what a mathematician's life is like. Whether or not I answer them I always think it is a perfect life and the mathematics community is one of the warmest communities. I am not sure whether this is due to the fact that my mathematical life has been greatly influenced by Bumsig and other friends. Bumsig has been a giant in my math life and I have been greatly influenced by what I have learned from him. To be very honest I started tennis because I wanted to spend more time with him. It was my greatest honor and privilege to be a student of Bumsig, my hero.

From Jeongseok Oh's website

Contents

Chapter 1
Gopakumar–Vafa Invariants = Quantum K-Invariants on Calabi–Yau Threefolds

Y.-C. Chou and Y.-P. Lee

In memory of Bumsig Kim

Abstract The main purpose of this article is to discuss a project relating Gopakumar–Vafa invariants to quantum K-invariants on Calabi–Yau threefolds. Results in genus zero, including recent and forthcoming works, are reported.

1.1 Introduction

This is an expository article based on Part I [1] and Part II [2] of our ongoing project about quantum K-theory of Calabi–Yau threefolds (CY3). In this project, we hope to explore various special properties of *quantum K-theory on Calabi–Yau threefolds*, which have not yet been extensively studied in the mathematical literature. We chose to start with its relation with the Gopakumar–Vafa invariants partly because we feel that the *integrality* of quantum K-theory has not received much attention so far.

Gromov–Witten theory of Calabi–Yau threefolds enjoys various beautiful properties, among them the *Kodaira–Spencer theory of gravity* and *holomorphic anomaly equation* of Bershadsky, Cecotti, Ooguri and Vafa, the appearance of (quasi-)modular forms etc.. Some of these remarkable phenomena should have counterparts in K-theory. We hope to explore these properties in the context of quantum K-theory in the future.

Y.-C. Chou
Institute of Mathematics, Academia Sinica, Taipei, Taiwan

Y.-P. Lee (✉)
Department of Mathematics, Institute of Mathematics, Academia Sinica, Taipei, Taiwan

© The Author(s) 2026

N.-G. Kang et al. (eds.), *Categorical and Enumerative Aspects of Mirror Symmetry*, KIAS Springer Series in Mathematics 5,
https://doi.org/10.1007/978-981-95-0385-8_1

1.1.1 Gopakumar–Vafa Invariants and Quantum K-Theory

Among the *integral* (virtual) enumerative invariants on Calabi–Yau threefolds (CY3), two will be the foci of this paper. The first, called the *Gopakumar–Vafa invariants* (GV) [11, 12], was introduced in theoretical physics as "new topological invariants on *Calabi–Yau threefolds*", counting the "numbers of BPS states". There have been various attempts at giving these BPS invariants rigorous mathematical definitions. At the moment, there are still unresolved issues of these definitions. We refer the readers to [19] and references therein for mathematical definitions of various degrees of generalities of these invariants. Gopakumar and Vafa also argue that these BPS invariants and the Gromov–Witten invariants are intimately related by the following formula

$$\begin{aligned}&\sum_{g=0}^{\infty}\sum_{\beta\in H_2(M,\mathbb{Z})} \mathrm{GW}_{g,\beta}q^{\beta}\lambda^{2g-2}\\ =&\sum_{g=0}^{\infty}\sum_{k=1}^{\infty}\sum_{\beta\in H_2(M,\mathbb{Z})} \mathrm{GV}_{g,\beta}\frac{1}{k}\left(2\sin\left(\frac{k\lambda}{2}\right)\right)^{2g-2} q^{k\beta},\end{aligned} \tag{1.1.1}$$

which all viable mathematical definitions must satisfy. In fact, this formula is ultimately the most important test for any geometric definition. The detailed definitions and discussions can be found in Sect. 1.2.2. In this paper, we will use the above (invertible) relation as the *definition* of the Gopakumar–Vafa invariants (in terms of Gromov–Witten invariants). As defined, it is not at all obvious these invariants are integral. The integrality of this ad hoc definition was proven by E. Ionel and T. Parker using symplectic techniques in a remarkable work [14] .

The second is the *quantum K-invaraints* (QK) [6, 17], a K-theoretic variant of the (cohomological) Gromov–Witten invariants (GW). The definition and references for quantum K-theory will be recalled in Sect. 1.2.3. Whereas Gromov–Witten invariants are *rational* numbers, quantum K-invariants, counting alternating sum of ranks of sheaf cohomology (Euler characteristic), produces *integral* invariants by definition. As an application of this fact, we prove the integrality of Gopakumar–Vafa invariants by relating them to quantum K-invariants.

1.1.2 GV = QK on CY3?

The question is: Are there any relations between the Gopakumar–Vafa invariants and quantum K-invariants on Calabi–Yau threefolds? We think this is the case. In fact, we think that one should be able to *define* the Gopakumar–Vafa invariants (GV) in terms of the quantum K-invariants, similar to (1.1.1). While this definition would still be *ad hoc*, it would have the benefit of being integral by definition. It would also give an alternative proof of Ionel–Parker's integrality theorem within the algebraic category.

Why do we believe such relationship exist? On the conceptual level, there is a clear link between GV and QK through GW. On the one hand, as pointed out above, Gopakumar–Vafa invariants and Gromov–Witten invariants are reconstructible from each other by the Gopakumar–Vafa formula (1.1.1). On the other hand, Givental and his collaborators has furnished a clear link between QK and GW via a virtual orbifold Hirzebruch–Riemann–Roch theorem [8, 10].

The close relationship between the quantum K-theory and quantum cohomology was understood since the early phase of quantum K-theory. In fact, many K-theoretic results were inspired by their cohomological counterparts. Many formulas and results were first guessed based on their cohomological counterparts. See, e.g., [9, 17]. In [10] and subsequent works [8], A. Givental and his collaborators completely characterized genus zero quantum K-theory in terms of quantum cohomology. Subsequently, Givental generalizes these results to permutation equivariant setting and to higher genera [8].

Therefore, there is a relation between quantum K-invariants and Gopakumar–Vafa invariants for Calabi–Yau threefolds via Gromov–Witten invariants. In this series of papers, we seek to explore these links and to "codify" the relations between QK and GV. Our starting point is the conjecture by Jockers and Mayr [15] and Garoufalidis and Scheidegger [5] in genus zero. We will formulate a precise version of the conjecture for all Calabi–Yau threefolds in genus zero based on their works.

The Gopakumar–Vafa formula (1.1.1), together with results in Sect. 1.3.1, can be interpreted as the following statement: *in genus zero*, the collection of all Gopakumar–Vafa invariants of a fixed Calabi–Yau threefold contains exactly the same information as the collection of all Gromov–Witten invariants. We hope that in the future to study the higher genus counterpart.

1.2 GW, GV and QK

1.2.1 Gromov–Witten Invariants

Let X be a smooth complex projective variety, and $\overline{\mathcal{M}}_{g,n}(X, \beta)$ be the M. Kontsevich's moduli space of n-pointed, genus g, degree β stable maps. Given $i \in \{1, \ldots, n\}$, there is an evaluation map

$$\begin{aligned} \mathrm{ev}_i : \overline{\mathcal{M}}_{g,n}(X, \beta) &\to X \\ [f : (C; x_1, \ldots, x_n)] &\mapsto f(x_i), \end{aligned}$$

and a line bundle $L_i := x_i^* w_{\mathcal{C}/\overline{\mathcal{M}}}$ on $\overline{\mathcal{M}}_{g,n}(X, \beta)$, where $w_{\mathcal{C}/\overline{\mathcal{M}}}$ is the relative dualizing sheaf of the universal curve $\mathcal{C} \to \overline{\mathcal{M}}_{g,n}(X, \beta)$ and $x_i : \overline{\mathcal{M}}_{g,n}(X, \beta) \to \mathcal{C}$ is the i-th mark point.

(Cohomological) Gromov-Witten invariants with descendants on X are defined to be

$$\langle \tau_{k_1}(\phi_1) \ldots \tau_{k_n}(\phi_n) \rangle_{g,n,\beta}^{X,H} := \pi_*^H \left(\cup_{i=1}^n \mathrm{ev}_i^*(\phi_i) c_1(L_i)^{k_i} \cap [\overline{\mathcal{M}}_{g,n}(X,\beta)]^{\mathrm{vir}} \right) \in \mathbb{Q},$$

where

$$\pi : \overline{\mathcal{M}}_{g,n}(X,\beta) \to pt := \mathrm{Spec}(\mathbb{C})$$

is the structural map and π_*^H is the (cohomological) pushforward to the point. Here $\phi_1, \ldots, \phi_n \in H(X)$, $k_1, \ldots, k_n \in \mathbb{Z}_{\geq 0}$, and $[\overline{\mathcal{M}}_{g,n}(X,\beta)]^{\mathrm{vir}}$ are the (cohomological) virtual fundamental classes.

The genus g Gromov–Witten invariants of X can be encoded in a generating function, called genus-g descendant potential

$$F_g^H(t) = \sum_{n \geq 0} \sum_{\beta} \frac{Q^\beta}{n!} \langle t(L), \ldots, t(L) \rangle_{g,n,\beta}^{X,H}.$$

Here the sum is over all curve class $\beta \in H_2(X)_{\geq 0}$ and Q^β are formal variables, called the Novikov variables, which keep track of the curve classes. $t(q)$ stands for any polynomial of one variable with coefficients in $H(X)$. That is

$$t(q) = \sum_{k \in \mathbb{Z}_{\geq 0}} \sum_{\alpha=1}^{N} t_k^\alpha \phi_\alpha q^k$$

with $\{\phi_\alpha\}_{\alpha=1}^N$ a basis of $H(X)$.

1.2.2 Gopakumar–Vafa Invariants

In theoretical physics, R. Gopakumar and C. Vafa in [11, 12] introduced new topological invariants on *Calabi–Yau threefolds* (CY3) X, which are now commonly called *Gopakumar–Vafa invariants*. These invariants represent the counts of "numbers of BPS states" on X. Unlike the Gromov–Witten invariants, which are defined for any symplectic manifolds or orbifolds, Gopakumar–Vafa invariants only make sense for Calabi–Yau threefolds (or their variants).

The virtual dimensions for moduli spaces (of stable maps) to CY3 are always equal to the number of marked points. That is,

$$\mathrm{vdim}\, \overline{\mathcal{M}}_{g,n}(X,\beta) = n.$$

By general properties of the moduli spaces, more precisely the string equation, the divisor equation and the dilaton equation, all Gromov–Witten invariants can be easily reconstructed from 0-pointed invariants

$$\mathrm{GW}_{g,\beta} := \pi_*^H\left([\overline{\mathcal{M}}_{g,0}(X,\beta)]^{\mathrm{vir}}\right) = \int_{[\overline{\mathcal{M}}_{g,0}(X,\beta)]^{\mathrm{vir}}} 1.$$

In fact, a *closed formula* was obtained in [4] in terms of generating functions. We will therefore focus on 0-pointed invariants $\mathrm{GW}_{g,\beta}$.

There have been various attempts at defining Gopakumar–Vafa invariants mathematically. We refer the readers to [19] and references therein. If one is interested in defining Gopakumar–Vafa invariants to allow for insertions, the proposed generalizations are expected to be compatible with the dilaton and divisor equations. Therefore, the counting of the BPS states is again reduced to similarly defined $\mathrm{GV}_{g,\beta}$.

A remarkable relation between GV and GW in [11, 12] can be expressed in terms of generating functions:

$$\begin{aligned}&\sum_{g=0}^{\infty}\sum_{\beta\in H_2(X,\mathbb{Z})_{\geq 0}} \mathrm{GW}_{g,\beta}q^{\beta}\lambda^{2g-2}\\ =&\sum_{g=0}^{\infty}\sum_{k=1}^{\infty}\sum_{\beta\in H_2(X,\mathbb{Z})_{\geq 0}} \mathrm{GV}_{g,\beta}\frac{1}{k}\left(2\sin\left(\frac{k\lambda}{2}\right)\right)^{2g-2}q^{k\beta}.\end{aligned} \tag{1.2.1}$$

We note that this formula gives an invertible relation between GV and GW, filtered by genus. Namely, one can obtain all $\{\mathrm{GV}_{g,\beta}\}_{g\leq g_0,\beta}$ from $\{\mathrm{GW}_{g,\beta}\}_{g\leq g_0,\beta}$ and vice versa. In this paper, we use this formula to *define* the Gopakumar–Vafa invariants.

Example 1.2.1 For the quintic CY3, the above relation can be written as

$$\begin{aligned}\mathrm{GW}_{g=0,\beta=d[\mathrm{line}]} &=: \mathrm{GW}_d = \sum_{e|d}\frac{1}{e^3}\mathrm{GV}_{d/e},\\ \mathrm{GV}_{g=0,\beta=d[\mathrm{line}]} &=: \mathrm{GV}_d := \sum_{e|d}\frac{\mu(e)}{e^3}\mathrm{GW}_{d/e},\end{aligned} \tag{1.2.2}$$

where we have used the Möbius inversion and $\mu(e)$ is the Möbius function. For our purpose, we will *use* (1.2.2) *as the definition of* GV_d. The first 4 terms are listed for readers' convenience.

d	1	2	3	4
GW_d	2875	4876875/8	8564575000/27	15517926796875/64
GV_d	2875	609250	317206375	242467530000

This *ad hoc* definition has among other things one difficulty. Namely, the Gopakumar–Vafa invariants are to be intrinsically *integers*, while the Gromov–Witten invariants are generally *rational* numbers, as the above table demonstrates. Fortunately, the integrality of this definition has been shown in [14].

There is a variant of the Gromov–Witten theory which also produces *integral* invariants, namely, the *quantum K-theory* [6, 17]. This leads to the possibility of relating quantum K-invariants with Gopakumar–Vafa invariants for Calabi–Yau threefolds.

1.2.3 Quantum K-Invariants

The formulation of *quantum K-theory* is similar to that of quantum cohomology, i.e., Gromov–Witten theory. The *quantum K-invariants*, or K-theoretic Gromov-Witten invariants, of X are defined to be

$$\langle \tau_{d_1}(\Phi_1)\dots\tau_{d_n}(\Phi_n)\rangle^{X,K}_{g,n,\beta} := \chi\left(\overline{\mathcal{M}}_{g,n}(X,\beta); \left(\otimes_{i=1}^n \mathrm{ev}_i^*(\Phi_i)L_i^{d_i}\right)\otimes \mathcal{O}^{\mathrm{vir}}\right) \in \mathbb{Z}.$$

Here $\Phi_1,\dots,\Phi_n \in K^0(X)$, $d_1,\dots,d_n \in \mathbb{Z}$, and $\mathcal{O}^{\mathrm{vir}}$ is the virtual structure sheaf on $\overline{\mathcal{M}}_{g,n}(X,\beta)$ defined in [17]. Roughly, let 0 be the zero section of the vector bundle stack $\mathfrak{E} \to \overline{\mathcal{M}}(X)$. The intrinsic normal cone $\mathfrak{C}$ of $\overline{\mathcal{M}}(X)$ is a closed substack of $\mathfrak{E}$ by the perfect obstruction theory. Then the K-theoretic pullback $0_K^*(\mathscr{O}_{\mathfrak{C}})$ is the virtual structure sheaf of $\overline{\mathcal{M}}(X)$. Alternatively $\mathcal{O}^{\mathrm{vir}}$ is the classical truncation of the structure sheaf of the underlying derived moduli stack. As in Gromov–Witten theory, all genus g invariants can be packed into a formal power series, called genus-g descendant potential of X:

$$F_g^K(t) = \sum_{n\geq 0}\sum_{\beta} \frac{Q^\beta}{n!}\langle t(L),\dots,t(L)\rangle^{X,K}_{g,n,\beta}.$$

Here the sum is over all curve classes $\beta \in H_2(X)_{\geq 0}$. $t(q)$ stands for any Laurent polynomial of one variable q with coefficients in $K^0(X)$

$$t(q) = \sum_{k\in\mathbb{Z}}\sum_{\alpha=1}^{N} t_k^\alpha \Phi_\alpha q^k$$

with $\{\Phi_\alpha\}_{\alpha=1}^N$ a basis in $K^0(X)$. We may rewrite $F_g^K(t)$ as

$$F_g^K(t) = \sum_{n\geq 0}\sum_{\beta} \frac{Q^\beta}{n!} \sum_{\substack{k_1,\ldots,k_n\in\mathbb{Z}\\ \alpha_1,\ldots,\alpha_n\in\{1,\ldots,N\}}} t_{k_1}^{\alpha_1}\cdots t_{k_n}^{\alpha_n}\langle\tau_{k_1}(\Phi_{\alpha_1}),\ldots,\tau_{k_n}(\Phi_{\alpha_n})\rangle_{g,n,\beta}^{X,K}.$$

Reamrk 1.2.2 In general, we will allow more general insertions in $K^0(X)[q,q^{-1}]\otimes\mathbb{Q}[\![Q]\!]$. More precisely, for any curve class β, the coefficient of Q^β of a insertion will lie in $K^0(X)[q,q^{-1}]$.

1.3 Quantum K-Theory on Calabi–Yau Threefolds

1.3.1 GW on CY3

Before we proceed to quantum K-theory for the Calabi–Yau threefolds, we first discuss some relevant statements in Gromov–Witten theory. In the following, we use $\deg_{\mathbb{C}}$ for the Chow degree, i.e., one half of the usual degree in cohomology.

Proposition 1.3.1 *For any Calabi-Yau threefold X, if* $\deg_{\mathbb{C}}\phi_1\geq 2$ *then*

$$\langle\tau_{k_1}(\phi_1),\ldots,\tau_{k_n}(\phi_n)\rangle_{g,n,\beta\neq 0}^{X,H,\mathrm{tw}}=0,$$

where tw *denotes cohomological GW invariants with twistings.*[1]

Proof Let

$$\pi_1:\overline{\mathcal{M}}_{g,n}(X,\beta)\to\overline{\mathcal{M}}_{g,1}(X,\beta)$$

be the forgetful map forgetting the last $n-1$ marked points and $T\in H^*(\overline{\mathcal{M}}_{g,n}(X,\beta))$ be the twisting class. By projection formula

$$\begin{aligned}\int_{[\overline{\mathcal{M}}_{g,n}(X,\beta)]^{\mathrm{vir}}} & T\prod_{i=1}^n\left(\psi_i^{k_i}\mathrm{ev}_i^*\phi_i\right)\\ &=\int_{[\overline{\mathcal{M}}_{g,1}(X,\beta)]^{\mathrm{vir}}}(\mathrm{ev}_1^*\phi_1)\,(\pi_1)_*\left(T\,\psi_1^{k_1}\prod_{i=2}^n\psi_i^{k_i}\mathrm{ev}_i^*\phi_i\right).\end{aligned}\tag{1.3.1}$$

[1] That is, an integration over virtual classes of the usual Gromov–Witten insertions *and* characteristic classes of the K-theoretic classes of the form $R\pi_*f^*E$, where $E\in K(X)$, $\pi:\mathcal{C}\to\overline{\mathcal{M}}(X)$ the universal curve and $f:\mathcal{C}\to X$ the universal stable map.

Since $[\overline{\mathcal{M}}_{g,1}(X,\beta)]^{\mathrm{vir}}$ has virtual dimension 1, while $\deg_{\mathbb{C}}(\phi_i) \geq 2$. This completes the proof. □

Proposition 1.3.2 *All descendant Gromov–Witten invariants on a Calabi–Yau threefold X can be reconstructed from* 0*-pointed invariants* $\{\langle\cdot\rangle^H_{g,0,\beta}\}_{g,\beta}$.

Proof This follows from a combination of virtual dimension counts, the string equation, dilaton equation and divisor equation. A closed formula is available in [4, Proposition 1.6]. □

1.3.2 QK on CY3

In quantum K-theory, things are somewhat different, mostly due to the fact that K-theory is more sensitive to the stack structures. For example, let G be a finite group acting on X and $\pi : X \to [X/G]$ the G-torsor. In cohomology or Chow,

$$H([X/G],\mathbb{Q}) = H(X,\mathbb{Q})$$

and

$$\int_{[X/G]} \alpha = \frac{1}{|G|}\int_X \pi^*\alpha.$$

However, in K-theory,

$$K([X/G]) = K_G(X)$$

is the G-equivariant K-theory, which is much richer than $K(X)$. Furthermore, the pushforward of a vector bundle to a point for X is the Euler characteristic of the bundle, while the same operation for $[X/G]$ extracts the G-invariant part of the sheaf cohomologies, which is much more intricate.

Nevertheless, one can apply the virtual Hirzebruch–Riemann–Roch theorem on Deligne–Mumford stacks, [16, 20] which we briefly recall.

Let M be a quasi-smooth DM stack, i.e., a virtual orbifold,

$$IM = \sqcup_i M_i$$

its inertia stack and N_i^{vir} the virtual normal bundle. Let $\lambda_{-1}(N_i^{\mathrm{vir}})^*$ be the "K-theoretic Euler class" of $(N_i^{\mathrm{vir}})^*$. For example, if $(N_i^{\mathrm{vir}})^*$ decomposes into a direct sum of line bundles L_α, then

$$\lambda_{-1}(N_i^{\mathrm{vir}})^* = \prod_\alpha (1 - L_\alpha).$$

Let g_i be the generic automorphism element of M_i. For every vector bundle E on M_i, g_i acts on E. Let $E = \oplus_j E_j$ be an eigenbundle decomposition with eigenvalues ϵ_j on E_j. Denote

$$\mathrm{Tr}(E) := \oplus_j \epsilon_j E_j.$$

Let F be a vector bundle on M, and $\tau(F) \in H(IM)$ defined by

$$\tau(F)|_{M_i} := \frac{\mathrm{ch}(\mathrm{Tr}\, F)}{\mathrm{ch}\,\mathrm{Tr}(\lambda_{-1}(N_i^{\mathrm{vir}})^*)} Td(T_{M_i}^{\mathrm{vir}}).$$

The virtual HRR states that

$$\chi(M, F) = \sum_i \frac{1}{m_i} \int_{[M_i]^{\mathrm{vir}}} \tau(F)|_{M_i}, \tag{1.3.2}$$

where m_i is the order of the generic automorphism associated to M_i. We now prove a K-theoretic version of Proposition 1.3.1.

Proposition 1.3.3 ([2]) *Let $E_1 \in K(X)$ be any element such that* $\deg_{\mathbb{C}} \mathrm{ch}(E_1) \geq 2$. *Then*

$$\langle \tau_{k_1}(E_1), \ldots, \tau_{k_n}(E_n) \rangle^K_{g,n,\beta} = 0.$$

Proof The proof is a combination of the virtual orbifold Hirzebruch–Riemann–Roch theorem (1.3.2) and the arguments in the proof of Proposition 1.3.1, including the virtual dimension count and the projection formula. Here two key facts are used. First, $\mathrm{ev}^* E_1$ is pulled back from a scheme and

$$\mathrm{Tr}(\mathrm{ev}^* E_1) = \mathrm{ev}^* E_1. \tag{1.3.3}$$

Second, the virtual dimension of any non-identity component of the intertial stack is no greater than the virtual dimension of the identity component. More precisely,

$$\begin{aligned}
&\chi\left(\overline{\mathcal{M}}_{g,n}(X,\beta), \prod_{j=1}^n L_j^{\otimes k_j} \mathrm{ev}_j^* E_j\right) \\
=&\sum_i \frac{1}{m_i} \int_{[(\overline{\mathcal{M}}_{g,n}(X,\beta))_i]^{\mathrm{vir}}} \tau\Big(\prod_{j=1}^n L_j^{\otimes k_j} \mathrm{ev}_j^* E_j\Big) \\
=&\sum_i \frac{1}{m_i} \int_{[(\overline{\mathcal{M}}_{g,1}(X,\beta))_i]^{\mathrm{vir}}} \mathrm{ch}(\mathrm{ev}_1^* E_1)(\pi_1)_* \left(\tau\Big(L_1^{k_1} \prod_{j=2}^n L_j^{k_j} \mathrm{ev}_j^* E_j\Big)\right) \\
=&0,
\end{aligned}$$

where we have used the fact that the virtual dimensions of $(\overline{\mathcal{M}}_{g,1}(X,\beta))_i$ are less than or equal to 1 and $\deg_{\mathbb{C}} \mathrm{ch}(\mathrm{ev}_1^* E_1) \geq 2$. □

At this moment, we do not have a general result in K-theory corresponding to Proposition 1.3.2. Nevertheless, in genus zero there is a reconstruction theorem from the set of all one-pointed descendants (i.e., including the cotangent line bundles) to all quantum K-invariants with descendants, in the spirit of the reconstruction theorem in [18].

The small J-function is a generating function of one-pointed descendants

$$J(q, Q) := (1-q) + \sum_{\alpha} \sum_{\beta \neq 0} \Phi_\alpha \langle \frac{\Phi^\alpha}{1-qL} \rangle^K_{0,1,\beta} Q^\beta$$

where $\{\Phi_\alpha, \Phi^\alpha\}$ are dual classes with respect to the K-theoretic Poincaré pairing

$$(\Phi_\alpha, \Phi^{\alpha'}) := \chi(X, \Phi_\alpha \otimes \Phi^{\alpha'}) = \delta_\alpha^{\alpha'}.$$

Proposition 1.3.4 ([2]) *The small J-function in quantum K-theory determines all genus zero (descendant) quantum K-invariants for Calabi–Yau threefolds.*

Proof This can be shown by a combination of the string equation in the quantum K-theory [17], a reconstruction theorem in quantum K-theory [17, (22) (23)], a version of Riemann–Roch for virtually smooth stacks, together with virtual dimension counts.

By the string equation, if any insertion is $1 = \mathcal{O}$, then it can be reduced to K-invariants of fewer points. One can assume that there is no insertion of 1 by induction. We can therefore assume that the insertions at all marked point look like

$$(\mathrm{ev}_i^*(E_i) - r_i 1)$$

where E_i are vector bundles of rank r_i. The Chern character of the i-th insertion starts at $\deg_{\mathbb{C}} \geq 1$, otherwise the insertion is 1, contradictory to the assumption.

One can now apply the Hirzebruch–Riemann–Roch for the virtually smooth stacks. Since the K-classes pulled back by evaluations are acted trivially by automorphisms, we conclude that only $\prod_i c_1(\mathrm{ev}_i^*(E_i))$ contribute to stacky HRR by virtual dimension counting. Therefore, the quantum K-invariants remain unchanged if we replace $(\mathrm{ev}_i^*(E_i) - r_i 1)$ by the corresponding *line bundles* $(\mathrm{ev}_i^*(\bigwedge^{\mathrm{top}} E_i) - 1)$. That is, one can assume that the insertions are all linear combinations of line bundles.

Then the reconstruction theorem in quantum K-theory in [18] applies: any descendant quantum K-invariants of insertions by line bundles can be reconstructed from the small J-function in the quantum K-theory. More precisely, all cotangent lines can be moved to the first marked points by induction. Assuming that there are no cotangent line bundles at any but the first marked point, we may further assume that there are no 1's at all other points after applying the string equation. Therefore the quantum K-invariants look like

$$\mathrm{ev}_1^*(E) \otimes L_1^k \otimes \prod_{j=2}^{n} (\mathrm{ev}_j^*(E_j) - r_j 1),$$

where E can be arbitrary, e.g., 1. By Proposition 1.3.3, $\mathrm{ev}_1^*(E)$ can only contribute through its ch_0 or ch_1 to the virtual HRR formula. ch_0 can be absorbed by the string equation and

$$c_1(\mathrm{ev}_i^*(E_i) - r_i 1) = c_1(\bigwedge^{\mathrm{top}} E_i) - 1).$$

Thus, vector bundle can be replaced by line bundles without changing the quantum K-invariants. The reconstruction theorem in [18] applies. □

1.4 Multiple Cover Formula and the JMGS Conjecture

1.4.1 GV = QK in Genus Zero

In [15] and [5] H. Jockers, P. Mayr and S. Garoufalidis, E. Scheidegger formulate a conjectural relation between the Gopakumar–Vafa invariants and quantum K-invariants for the Calabi–Yau threefolds. The conjecture is formulated in terms of *small J-functions*, a generating function in quantum K-theory as well as quantum cohomology.

We fix the following notation:

$$\{\Phi_\alpha\}_{\alpha=1}^N = \bigsqcup_{i=0}^{3} \{\Phi_{ij}\}_{j=1}^{n_i},$$

where $\{\mathrm{ch}(\Phi_{ij})\}_{j=1}^{n_i}$ forms a basis in $H^{2i}(X)$. In particular,

$$\{\Phi_{0j}\}_{j=1}^{n_0} = \{\Phi_{01} = \mathcal{O}\}.$$

Let $\{\Phi^{ij}\}$ be the dual basis of $\{\Phi_{ij}\}$ with respect to the K-theoretic Poincaré pairing:

$$(\Phi_a, \Phi_b)^K := \chi(X, \Phi_a \Phi_b).$$

The following fact will be used later.

Lemma 1.4.1 *For $i = 0$ and 1,*

$$\mathrm{ch}(\Phi^{ij}) \in H^{2(3-i)}(X).$$

For $i = 2$ and 3,

$$\mathrm{ch}(\Phi^{ij}) \in H^{\geq 2(3-i)}(X).$$

Proof Φ^{ij} can be written as:[2]

$$\Phi^{ij} := \mathrm{ch}^{-1}\Big(\mathrm{td}(TX)^{-1} PD(\mathrm{ch}(\Phi_{ij}))\Big),$$

where PD denotes the Poincaré dual. The lemma follows from the definition of Φ_{ij} and that $\mathrm{td}(TX)^{-1} \in 1 + H^{\geq 2}(X)$. □

We now generalize the JMGS conjecture to general Calabi–Yau threefolds.

Conjecture 1.4.2 (cf. [5, 15])

$$\begin{aligned}\frac{1}{1-q} J(q, Q) :=& \frac{1}{1-q}\left[(1-q) + \sum_{\alpha}\sum_{\beta\neq 0} \Phi_\alpha \langle \frac{\Phi^\alpha}{1-qL}\rangle^K_{0,1,\beta} Q^\beta\right] \\ =& 1 + \sum_{\vec{d}\in H_2(X,\mathbb{Z})} \sum_{r=1}^{\infty}\Bigg[\sum_{j=1}^{n_1} \Phi^{1j}\Big(\int_{\vec{d}} \mathrm{ch}(\Phi_{1j})\Big)\, a(r, q^r)\, \mathrm{GV}_{\vec{d}}\, Q^{r\vec{d}} \\ &+ \Phi^{01} b(r, q^r)\, \mathrm{GV}_{\vec{d}}\, Q^{r\vec{d}}\Bigg],\end{aligned}$$

where

$$\begin{aligned} a(r, q^r) &= \frac{(r-1)}{1-q^r} + \frac{1}{(1-q^r)^2}, \\ b(r, q^r) &= \frac{r^2-1}{1-q^r} + \frac{3}{(1-q^r)^2} - \frac{2}{(1-q^r)^3}. \end{aligned} \tag{1.4.1}$$

The main result of Part II is a proof of this conjecture.

Theorem 1.4.3 ([2]) *Conjecture 1.4.2 holds.*

Corollary 1.4.4 $\mathrm{GV}_{\vec{d}} \in \mathbb{Z}$.

---

[2] Since the chern character isomorphism only works over $\mathbb{Q}$, we may work on $K^0_{\mathbb{Q}}(X)$ if necessary. It wouldn't change any discussion in this paper.

Proof By Proposition 1.3.4, Theorem 1.4.3 implies that the quantum K-theory and Gopakumar–Vafa theory are equivalent for all Calabi–Yau threefolds in genus zero. In particular, $\mathrm{GV}_{\vec{d}}$ can be written as a finite *integral* combination of quantum K-invariants

$$\mathrm{GV}_{\vec{d}} = \sum_{r|\vec{d}} \mu(r) r^2 \langle 1 \rangle^K_{0,1,\vec{d}/r},$$

where μ is the Möbius function. This shows that $\mathrm{GV}_{\vec{d}} \in \mathbb{Z}$. □

1.4.2 Multiple Cover Formula

Lemma 1.4.5 *For the total space* $X_{-1,-1}$ *of* $\mathcal{O}(-1) \oplus \mathcal{O}(-1) \to \mathbb{P}^1$*, the Gopakumar–Vafa invariants* $\mathrm{GV}_{0,1} = 1$ *(genus zero and degree* 1*) and* $\mathrm{GV}_{g,d} = 0$ *otherwise*

Proof This follows from the Gopakumar–Vafa equation (1.2.1), and the results in Gromov–Witten theory: for $g = 0$ the Voisin–Aspinwall–Morrison formula [21]

$$\mathrm{GW}_{0,d} = \frac{1}{d^3},$$

for $g = 1$, the BCOV and Graber–Pandharipande formula [13]

$$\mathrm{GW}_{1,d} = \frac{1}{12d},$$

and for $g \geq 2$ the Faber–Pandharipande formula [3]

$$\mathrm{GW}_{g,d} = \frac{|B_{2g}| d^{2g-3}}{2g \cdot (2g-2)!}.$$

Since $\{\mathrm{GW}_{g,d}\}_{g,d}$ uniquely determine $\{\mathrm{GV}_{g,d}\}_{g,d}$ by Eq. (1.2.1) and vice versa, it suffices to check the validity of Eq. (1.2.1) by plugging in values of $\mathrm{GW}_{g,d}$ mentioned above and the claimed values of $\mathrm{GV}_{g,d}$. It reduces to prove the following identity

$$\left(2 \sin \frac{x}{2}\right)^{-2} = \frac{1}{x^2} + \frac{1}{12} + \sum_{g=2}^{\infty} \frac{|B_{2g}| x^{2g-2}}{2g \cdot (2g-2)!}.$$

It follows from the two identities:

$$\left(2\sin\frac{x}{2}\right)^{-2} = \sum_{k=-\infty}^{\infty} \frac{1}{(-2k\pi + x)^2} = \frac{1}{x^2} + \sum_{k=1}^{\infty} \frac{2(2k-1)}{(2\pi)^{2k}} \zeta(2k) x^{2k-2},$$

and

$$\zeta(2k) = \frac{|B_{2k}|(2\pi)^{2k}}{2(2k)!}.$$

This completes the proof. □

In order to consider the small J-function of $X_{-1,-1}$ in quantum K-theory, we consider its compatification by the "infinity divisor". Let

$$Y_{-1,-1} := \mathbb{P}_{\mathbb{P}^1}(\mathcal{O}(-1) \oplus \mathcal{O}(-1) \oplus \mathcal{O}).$$

Let $P = \pi^* \mathcal{O}(-1)$ with $\pi : Y_{-1,-1} \to \mathbb{P}^1$, and $t = \mathcal{O}(D_\infty)$ with $D_\infty \subset Y_{-1,-1}$, the infinity divisor. We denote $Q^r := Q^{r\ell}$, where

$$\ell := [\mathbb{P}^1] \overset{0}{\hookrightarrow} Y_{-1,-1}$$

is the line class in the "zero section" of the projective bundle. In the following, we consider the *specialized* small I-function and J-function of $Y_{-1,-1}$ only for the curve classes in the zero section, i.e., multiples of ℓ.

Lemma 1.4.6 *The small I-function for $Y_{-1,-1}$, with curve classes in the zero section, is*

$$I^{Y_{-1,-1}}(q, Q) = (1-q)\left[1 + \sum_{r=1}^{\infty} Q^r \frac{(1-Pt)^2 \prod_{m=1}^{r-1}(1-Ptq^m)^2}{(Pt)^{2r} q^{r(r-1)} \prod_{m=1}^{r}(1-Pq^m)^2}\right].$$

Proof This lemma follows from the computations of the I-functions for toric manifolds via fixed point localization by A. Givental and collaborators. See [8, 10]. Their formula for small I-function gives

$$I^{Y_{-1,-1}}(q, Q) = (1-q)\left[1 + \sum_{r=1}^{\infty} Q^r \frac{\prod_{m=-r+1}^{0}(1-P^{-1}t^{-1}q^m)^2}{\prod_{m=1}^{r}(1-Pq^m)^2}\right].$$

A simple manipulation gives the above presentation. □

We note that in this case the small I-function includes a factor of $(1-Pt)^2$, the K-theoretic normal bundle of $\mathbb{P}^1$ embedded in $Y_{-1,-1}$. One may think of this as an I-function of a toric completion $\mathbb{P}_{\mathbb{P}^1}(\mathcal{O}(-1) \oplus \mathcal{O}(-1) \oplus \mathcal{O})$, with curve class in the base $\mathbb{P}^1$. This small I-function is different from the small J-function, as it has poles at $q = 0$. (However, it satisfies $\lim_{q\to\infty} I^{X_{-1,-1}}(q) = 0$.) Using this I-function one can obtain the small J-function via the "*generalized* mirror transform" (also known as the explicit reconstruction, or Birkhoff factorization).

Proposition 1.4.7 *The small J-function for $Y_{-1,-1}$, with curve classes in the zero section, is*

$$\begin{aligned}\frac{1}{1-q}J^{Y_{-1,-1}}(q,Q) = 1 + (1-Pt)^2\Big(1+(1-P)\Big)\sum_{r\geq 1} Q^r a(r,q^r)\\ + (1-Pt)^2(1-P)\sum_{r\geq 1} Q^r\, b(r,q^r),\end{aligned} \tag{1.4.2}$$

where $a(r,q^r)$, $b(r,q^r)$ are defined in (1.4.1).

Proof The K-ring of $Y_{-1,-1}$ has the following presentation

$$K(Y_{-1,-1}) = \frac{\mathbb{Z}[P,t]}{\big((1-P)^2, (1-Pt)^2(1-t)\big)}.$$

We use the relations of $K(Y_{-1,-1})$ to rewrite $I^{Y_{-1,-1}}$ as follows. Since $(1-Pt)^2(1-t) = 0$, we have

$$(1-Pt)^2(1-Pq^m t) = (1-Pt)^2(1-Pq^m). \tag{1.4.3}$$

Therefore,

$$\begin{aligned}&I^{Y_{-1,-1}} - (1-q)\\ =&(1-q)\sum_{r=1}^{\infty} Q^r \frac{(1-Pt)^2}{(Pt)^{2r} q^{r(r-1)}(1-Pq^r)^2}\\ =&(1-q)\sum_{r\geq 1} Q^r\left[\Big(\sum_{i=1}^{r-1}\frac{i}{q^{r(r-i)}}\Big)(1-Pt)^2 + \Big(\sum_{i=1}^{r-1}\frac{i(2r-i+1)}{q^{r(r-i)}}\Big)(1-Pt)^3\right]\\ &+ (1-q)\Big((1-Pt)^2 + (1-Pt)^3\Big)\sum_{r\geq 1} Q^r\Big(\frac{r-1}{1-q^r} + \frac{1}{(1-q^r)^2}\Big)\\ &+ (1-q)(1-Pt)^3\sum_{r\geq 1} Q^r\Big(\frac{r^2-1}{1-q^r} + \frac{3}{(1-q^r)^2} - \frac{2}{(1-q^r)^3}\Big).\end{aligned}$$

In the first equality, the factor $\prod_{m=1}^{r-1}(1-Ptq^m)^2$ in the numerator and $\prod_{m=1}^{r-1}(1-Pq^m)^2$ in the denominator cancel each other due to the presence of $(1-Pt)^2$ and (1.4.3). The second equality follows from an explicit computation.

The first line after the second equal sign lies in $\mathcal{K}_+$ and the rest lies in $\mathcal{K}_-$. Consider the reconstruction theorem [7, Theorem 2]

$$
\begin{aligned}
J^{Y_{-1,-1}}(q,Q) = \sum_{d\geq 0} I_d^{Y_{-1,-1}} Q^d.\\
\exp\left(\left(\sum_{k>0}\frac{\Psi^k}{k}\right)\left(\frac{\delta(Q)(1-Pq^d)+\sum_{i=0}^{3}\epsilon_i(Q)(1-Ptq^d)^i}{(1-q)}\right)\right)\cdot\\
\left(s(q,Q)(1-Pq^d)+\sum_{i=0}^{3} r_i(q,Q)(1-Ptq^d)^i\right),
\end{aligned}
$$

for some uniquely determined $\epsilon_i(Q)$, $\delta(Q)$, $s(q,Q)$ and $r_i(q,Q)$, where

$$
\begin{aligned}
\epsilon_i(Q) &= \sum_{j\geq 1}\epsilon_{ij}Q^j \in \mathbb{Q}[\![Q]\!],\\
\delta(Q) &= \sum_{j\geq 1}\delta_j Q^j \in \mathbb{Q}[\![Q]\!],\\
r_i(q,Q) &= \sum_{j\geq 0} r_{ij}(q)Q^j \in \mathbb{Q}[q,q^{-1}][\![Q]\!],\\
s(q,Q) &= \sum_{j\geq 0} s_j(q)Q^j \in \mathbb{Q}[q,q^{-1}][\![Q]\!].
\end{aligned}
$$

A direct computation by induction on the degree of the Novikov variable shows that

$$
r_0(q,Q)=1, \quad \epsilon_1(Q)=\epsilon_2(Q)=\epsilon_3(Q)=\delta(Q)=r_1(q,Q)=s(q,Q)=0,
$$

and that $r_2(q,Q)$ and $r_3(q,Q)$ will not change the $\mathcal{K}_-$ part. This concludes the proof. □

Corollary 1.4.8 (Multiple Cover Formula [2]) *The small J-function for $X_{-1,-1}$ is*

$$
\begin{aligned}
\frac{1}{1-q}J^{X_{-1,-1}}(q) = 1 + \Big(1+(1-P)\Big)\sum_{r\geq 1} Q^r a(r,q^r)\\
+ (1-P)\sum_{r\geq 1} Q^r\, b(r,q^r),
\end{aligned}
$$

where $a(r,q^r)$, $b(r,q^r)$ are defined in (1.4.1).

Proof Since the zero section $\mathbb{P}^1$ has normal bundle $\mathcal{O}(-1)\oplus\mathcal{O}(-1)$, the quantum K-invariants of $r\ell$ in $X_{-1,-1}$ are exactly the same as those in $Y_{-1,-1}$. The only difference in J-functions comes from different bases of the K-groups and the Poincaré pairings. The net result is the removal of the factor $(1 - Pt)^2$ from the specialized $J^{Y_{-1,-1}}(q, Q)$ for non-zero degree terms. □

1.4.3 Virtual Clemens' Conjecture

We now give a heuristic derivation of Conjecture 1.4.2 and a heuristic interpretation of the relationship between Gopakumar–Vafa invariants and quantum K-invariants at genus zero by a multiple cover formula. This has served to guide us in our search for the current formulation of Conjecture 1.4.2, even though the actually proof follows a completely different approach. Of course, the original formulations of Jockers and Mayr [15] and Garoufalidis and Scheidegger [5] have been enormous help.

Assume that we are given an "ideal" Calabi–Yau threefold X satisfying a "virtual Clemens' conjecture". That is, there are, up to deformations, finitely many isolated rational curves $\{C_i\}$. Furthermore, they are all smooth $(-1, -1)$ curves.

By Lemma 1.4.5, each isolated $(-1, -1)$-curve (in any degree $\vec{d}$) contributes 1 to the Gopakumar–Vafa invariants, independently of $\vec{d}$. Therefore, there are $\mathrm{GV}_{0,\vec{d}}$ isolated $(-1, -1)$-curves in degree $\vec{d}$. For each of these isolated curves, quantum K-theory allows multiple r-covers of the isolated $(-1, -1)$-curve. The coefficients $a(r, q^r)$ and $b(r, q^r)$ of the r-covers come from the J-function of $X_{-1,-1}$. The only addition is the factor of $\int_{\vec{d}} D_j$, where the divisor $D_j = \mathrm{ch}(\Phi_{1j})$ comes from the divisor axiom.

In summary, the "virtual Clemens conjecture" implies that GV = QK for all Calabi–Yau threefolds in genus zero via the multiple cover contributions.

This line of thoughts lead us to believe that, in order to generalize this to higher genera, the most important ingredient is the higher genus multiple cover formula in quantum K-theory. It is entirely possible that the higher genus multiple cover formulas will serve as universal coefficients, similar to the genus zero case. We intend to pursue this in future works.

Acknowledgments We wish to thank A. Givental, R. Pandharipande, H. Tseng and Y. Wen for discussions about this work. We also wish to thank anonymous referees for corrections and suggestions which have improved the exposition of this article. The research is partially supported by the Simons Foundation, the NSTC, University of Utah and Academia Sinica.

References

1. Chou, Y.C., Lee, Y.P.: Quantum K-theory and Gopakumar–Vafa invariants I. The quintic threefolds. Amer. J. Math., to appear (2026), arXiv:2211.00788
2. Chou, Y.C., Lee, Y.P.: Quantum K-theory and Gopakumar–Vafa invariants II. Calabi–Yau threefolds. Geom. Topol. **29**(9), 4665–4693 (2025)

3. Faber, C., Pandharipande, R.: Hodge integrals and Gromov-Witten theory. Invent. Math. **139**(1), 173–199 (2000)
4. Fan, H., Lee, Y.P.: Towards a quantum Lefschetz hyperplane theorem in all genera. Geom. Topol. **23**(1), 493–512 (2019)
5. Garoufalidis, S., Scheidegger, E.: On the quantum K-theory of the quintic. SIGMA Symmetry Integrabil. Geom. Methods Appl. **18**, 021 (2022)
6. Givental, A.: On the WDVV Equation in Quantum K-Theory, vol. 48, pp. 295–304 (2000). Dedicated to William Fulton on the occasion of his 60th birthday, Michigan Math. J. **48**, 295–304 (2000)
7. Givental, A.: Permutation-equivariant quantum quantum K-theory VIII. Explicit reconstruction (2015). arXiv:1510.06116
8. Givental, A.: Permutation-equivariant quantum quantum K-theory I-XI (2015–2017). arXiv:Algebraic Geometry
9. Givental, A., Lee, Y.P.: Quantum K-theory on flag manifolds, finite-difference Toda lattices and quantum groups. Invent. Math. **151**(1), 193–219 (2003)
10. Givental, A., Tonita, V.: The Hirzebruch-Riemann-Roch theorem in true genus-0 quantum K-theory. In: Symplectic, Poisson, and Noncommutative Geometry. Mathematical Sciences Research Institute Publications, vol. 62, pp. 43–91. Cambridge University Press, New York (2014)
11. Gopakumar, R., Vafa, C.: M-theory and topological strings–I, pp. 1–14 (1998). arXiv:hep-th/9809187
12. Gopakumar, R., Vafa, C.: M-theory and topological strings–II, pp. 1–19 (1998). arXiv:hep-th/9812127
13. Graber, T., Pandharipande, R.: Localization of virtual classes. Invent. Math. **135**(2), 487–518 (1999)
14. Ionel, E.N., Parker, T.H.: The Gopakumar-Vafa formula for symplectic manifolds. Ann. Math. **187**(1), 1–64 (2018)
15. Jockers, H., Mayr, P.: Quantum K-theory of Calabi-Yau manifolds. J. High Energy Phys. **11**, 011, 20 (2019)
16. Kawasaki, T.: The Riemann-Roch theorem for complex V-manifolds. Osaka Math. J. **16**(1), 151–159 (1979)
17. Lee, Y.P.: Quantum K-theory. I. Foundations. Duke Math. J. **121**(3), 389–424 (2004). ArXiV.org math.AG/0105014
18. Lee, Y.P., Pandharipande, R.: A reconstruction theorem in quantum cohomology and quantum K-theory. Am. J. Math. **126**(6), 1367–1379 (2004)
19. Maulik, D., Toda, Y.: Gopakumar-Vafa invariants via vanishing cycles. Invent. Math. **213**(3), 1017–1097 (2018)
20. Tonita, V.: A virtual Kawasaki-Riemann-Roch formula. Pac. J. Math. **268**(1), 249–255 (2014)
21. Voisin, C.: A mathematical proof of a formula of Aspinwall and Morrison. Compos. Math. **104**(2), 135–151 (1996)

Open Access This chapter is licensed under the terms of the Creative Commons Attribution-NonCommercial-NoDerivatives 4.0 International License (http://creativecommons.org/licenses/by-nc-nd/4.0/), which permits any noncommercial use, sharing, distribution and reproduction in any medium or format, as long as you give appropriate credit to the original author(s) and the source, provide a link to the Creative Commons license and indicate if you modified the licensed material. You do not have permission under this license to share adapted material derived from this chapter or parts of it.

The images or other third party material in this chapter are included in the chapter's Creative Commons license, unless indicated otherwise in a credit line to the material. If material is not included in the chapter's Creative Commons license and your intended use is not permitted by statutory regulation or exceeds the permitted use, you will need to obtain permission directly from the copyright holder.

Chapter 2
Twisted Wild Character Varieties and Gromov-Witten Invariants

D.-E. Diaconescu

Dedicated to the memory of Bumsig Kim

Abstract Conjectural results for E-polynomials of twisted wild character varieties are obtained from the Gromov-Witten theory of normal crossing degenerations.

2.1 Introduction

Given a smooth projective curve C over $\mathbb{C}$, a character variety is defined as a moduli space of monodromy and Stokes data associated via Riemann-Hilbert correspondence to non-abelian flat connections on C, possibly with prescribed analytic singularities. First introduced in [43] in the context of geometric Langlands duality, these moduli spaces were rigorously constructed in [4]. The connection between character varieties and the enumerative geometry of Calabi-Yau threefolds stems from non-abelian Hodge correspondence and the spectral cover construction for Higgs bundles on curves. As a real manifold, a given character variety is identified through non-abelian Hodge correspondence [2, 3, 17, 31, 39, 40] to a moduli space of Higgs bundles on C. The resulting Higgs bundle moduli space is further identified in the algebraic category to a moduli space of purely one dimensional sheaves on a spectral surface S, which is a holomorphic symplectic surface S equipped with a natural projection $\pi : S \to C$. The Higgs bundles obtained through this process may be regular or singular, depending on the nature of the given character variety.

Using the $P = W$ conjecture [13], now a theorem [14, 23, 33] in the regular case, this yields concrete conjectural relations between the enumerative invariants of the K-trivial threefold $Y = S \times \mathbb{A}^1$ and the cohomological invariants of the character

D.-E. Diaconescu (✉)
Department of Physics, Rutgers University, Piscataway, NJ, USA
e-mail: duiliu@physics.rutgers.edu

© The Author(s) 2026

N.-G. Kang et al. (eds.), *Categorical and Enumerative Aspects of Mirror Symmetry*, KIAS Springer Series in Mathematics 5,
https://doi.org/10.1007/978-981-95-0385-8_2

variety. In particular the Gromov-Witten theory of Y is identified with a generating function for E-polynomials character varieties via the BPS expansion conjectured in [18]. In the context of stable pair theory [38], the BPS expansion is known as the strong rationality conjecture and it has been shown to follow from wallcrossing in the derived category in [41, 42]. The connection between Gromov-Witten invariants and cohomological invariants of moduli spaces of one dimensional sheaves follows from the mathematical definition of BPS numbers formulated in [24, 34].

The relation between cohomological invariants of character varieties and enumerative geometry has been progressively made explicit first for the unramified case [9, 10], followed by the simply ramified case [11], and for certain wildly ramified cases in [12, 16]. The resulting conjectural formulas are in agreement with the rigorous mathematical results [20–22] obtained by counting rational points on character varieties in all cases where the latter are available in the literature. It should be noted however, that in many ramified cases this approach yields completely new conjectural results. Furthermore, a new facet of this correspondence, employing the Gromov-Witten theory of normal crossing degenerations, was found in [15].

The goal of the present paper is to obtain explicit results for the degenerate Gromov-Witten theory associated to the twisted wild character varieties studied in [12]. These are character varieties associated to flat $GL(N, \mathbb{C})$ connections on a curve C of analytic type

$$\frac{z^{1-n}}{1-n} J_{\ell,r} + \frac{z^{2-n}}{2-n} E_{\ell,r} + \text{regular}, \qquad r, \ell \geq 1,\ n \geq 3$$

in the formal neighborhood of a marked point $p \in C$, where

$$J_{\ell,r} = \begin{pmatrix} 0 & 0 & \cdots & 0 & 0 \\ \mathbf{1}_r & 0 & \cdots & 0 & 0 \\ \vdots & \vdots & & \vdots & \vdots \\ 0 & 0 & \cdots & \mathbf{1}_r & 0 \end{pmatrix}, \qquad E_{\ell,r} = \begin{pmatrix} 0 & 0 & \cdots & 0 & \mathbf{1}_r \\ 0 & 0 & \cdots & 0 & 0 \\ \vdots & \vdots & & \vdots & \vdots \\ 0 & 0 & \cdots & 0 & 0 \end{pmatrix}, \tag{2.1.1}$$

are square matrices of size $N = r\ell$ and all entries are $r \times r$ blocks. This is a subclass of the twisted wild character varieties constructed in [7]. For simplicity only a single marked point will be considered in this paper. The generalization to an arbitrary number of marked points is straightforward as long as all of them are characterized by the same analytic type. The case of several marked points of different analytic types remains an open problem.

As shown in [4, 7], for any n, r, ℓ, the resulting character variety $\mathcal{C}_{n,\ell,r}$ is a smooth holomorphic symplectic variety of dimension $d(g, n, \ell, r) = 2 + r^2(n\ell(\ell-1) + 2(g-1)\ell^2)$, where g is the genus of C. As usual, the E-polynomial of $\mathcal{C}_{n,\ell,r}$ encodes the graded dimension of the Deligne's weight filtration,

$$E_{n,\ell,r}(u) = \sum_{i,j} u^{i/2}(-1)^j \dim Gr_i^W H^j(\mathcal{C}_{n,\ell,r}).$$

An explicit formula was conjectured in [12, Section 1.5] for the following generating function

$$F_{n,\ell}(q,\mathsf{x}) = \sum_{k\geq 1}\sum_{r\geq 1} \frac{\mathsf{x}^{kr}}{k} \frac{q^{kd(g,n,\ell,r)/2}E_{n,\ell,r}(q^{-k})}{(1-q^k)(1-q^{-k})} \in \mathbb{C}(q^{1/2})[[\mathsf{x}]].$$

The expression conjectured in loc. cit. is written as a formal sum over partitions, or, equivalently Young diagrams. In order to fix the notation, a partition $\mu = (\mu_1 \geq \cdots \geq \mu_{l(\mu)} \geq 1)$ will be identified with a Young diagram consisting $l(\mu)$ rows of lengths $\mu_1, \ldots, \mu_j$. The size of μ will be denoted by $|\mu|$ and the conjugate partition will be denoted by μ^t. For the i-th box on the j-th row, the arm and the leg lengths are defined by

$$a(\square) = \mu_j - i, \qquad l(\square) = \mu^t_i - j$$

The content $c(\mu)$ will de defined by

$$c(\mu) = \sum_{\square\in\mu} (a(\square) - l(\square))$$

and the Schur function associated to μ will be denoted by $s_\mu(x_1, x_2, \ldots)$. For any pair of partitions μ, ρ and any integer $k \geq 1$ the coefficients $c^{(k)}_{\mu,\rho}$ are defined by the symmetric function identity

$$s_\mu(x_1^k, x_2^k, \ldots) = \sum_\rho c^{(k)}_{\mu,\rho} s_\rho(x_1, x_2, \ldots). \tag{2.1.2}$$

For fixed n, ℓ, let

$$Z_{n,\ell}(q,\mathsf{x}) = \sum_\mu q^{-\mathbf{n}c(\mu)/\ell} s_{\mu^t}(\underline{q})^{1-2g} \sum_\sigma c^{(\ell)}_{\sigma,\mu} q^{\mathbf{n}c(\sigma)} s_{\sigma^t}(\underline{q})\mathsf{x}^{|\sigma|} \in \mathbb{C}(q^{1/2}, q^{1/\ell})[[\mathsf{x}]],$$

where $\mathbf{n} = (\ell-1)n - 1$ and $\underline{q} = (q^{1/2}, q^{3/2}, \ldots)$. Then the conjecture formulated in [7] reads:

Conjecture 2.1.1 The following relation holds in $\mathbb{C}(q^{1/2}, q^{1/\ell})[[\mathsf{x}]]$

$$\ln Z_{n,\ell}(q,\mathsf{x}) = -F_{n,\ell}(q,\mathsf{x}) \tag{2.1.3}$$

for any $n \geq 3$, $\ell \geq 1$.

Given the data (C, p), n, ℓ, the present paper will construct a quasi-projective K-trivial normal crossing threefold $\overline{Y}$ equipped with a torus action $\mathbf{T} \times \overline{Y} \to \overline{Y}$ so that the generating function of residual equivariant Gromov-Witten invariants of $\overline{Y}$

coincides with $Z_{n,\ell}$. Building on the results of [12], this is a two-step construction carried out in Sects. 2.2.1 and 2.3.

Let K denote the field of fraction of the complex cohomology ring of the classifying space of $\mathbf{T}$. Since the second homology of $\overline{Y}$ is freely generated by two curve classes, one obtains a generating function

$$Z_{GW}(\lambda, Q_0, Q) = 1 + \sum_{h\in\mathbb{Z}} \sum_{\substack{d,r\geq 0\\ (d,r)\neq(0,0)}} GW_h(r,d)\, \lambda^{2h-2}\, Q_0^d\, Q^r \in K[\lambda^{-1}][[\lambda, Q_0, Q]]$$

for equivariant residual Gromov-Witten invariants, where Q_0, Q are formal counting variables associated to the two generators and λ is the genus counting variable. Here the stable maps are allowed to have disconnected domains as long as no connected component is contracted to a point. The main theorem in this paper, proven in Sect. 2.3 Corollary 2.3.1, reads:

Theorem 2.1.2 *The following identity holds in* $K[\lambda^{-1}][[\lambda, Q_0, Q]]$

$$Z_{GW}(\lambda, Q_0, Q) = Z_{n,\ell}(e^{i\lambda}, Q_0^\ell Q) \tag{2.1.4}$$

In conclusion, Conjecture 2.1.1 and Theorem 2.1.2 provide an enumerative geometric realization for the cohomological invariants of the twisted wild character varieties $\mathcal{C}_{n,\ell,r}$. This is the second occurrence of degenerate Gromov-Witten theory in this context, the first being considered in [15]. Since in positive characteristic the E polynomial encodes the number of rational points on the character variety, this framework points towards a possible TQFT formalism for rational point counting. While natural in Gromov-Witten theory [5, 6], such a formalism is not equally natural for rational point counting problems. This is an open problem which deserves further investigation.

2.2 The Twisted Wild Cap

2.2.1 Spectral Surfaces by Iterated Blow-Ups

In this section C will be the projective line $\mathbf{Proj}\,\mathbb{C}[z_1, z_2]$ and $q_0, \infty \in C$ will denote the closed points $z_1 = 0$ and z_2 respectively. Let M_0 be the total space of the line bundle $L = \mathcal{O}_{\mathbb{P}^1}((n-2)q_0 + \infty)$ for some integer $n \geq 3$. Let $p : M_0 \to C$ denote the canonical projection, and let $\zeta \in H^0(p^*L)$ be the tautological section. For ease of exposition, the curve C will be implicitly identified with the zero section $\zeta = 0$, and no notational distinction will be made between the homogeneous coordinates z_1, z_2 and their pull-backs to M_0. This convention will also apply to closed points in C, which will be implicitly identified with their images in M_0 via the zero section.

The distinction will be clear from the context. Note also that M_0 is a toric surface with homogeneous toric coordinates (z_1, z_2, ζ).

Let $\ell \geq 1$ and let $\Sigma_0 \subset M_0$ be the reduced irreducible curve defined by

$$\zeta^\ell - z_1 z_2^{(n-1)\ell-1} = 0.$$

Note that Σ_0 intersects the zero section $\zeta = 0$ at the points q_0, ∞ determined by $\zeta = 0,\ z_1 = 0$ and $\zeta = 0,\ z_2 = 0$ respectively. Let $U_0 = \{z_2 \neq 0\}$ and $U_\infty = \{z_1 \neq 0\}$ be the natural affine coordinate charts on M centered at q_0, ∞ respectively. More explicitly, $U_0 = \operatorname{Spec} \mathbb{C}[u, x]$ and $U_\infty = \operatorname{Spec} \mathbb{C}[v, y]$, where

$$x = z_1 z_2^{-1},\ u = \zeta z_2^{-(n-1)}, \qquad y = z_1^{-1} z_2,\ v = \zeta z_1^{-(n-1)}$$

on the intersection. Then note that the local equations of Σ_0 are

$$\Sigma_0 \cap U_0 : \quad x = u^\ell, \qquad \Sigma_0 \cap U_\infty : \quad v^\ell = y^{(n-1)\ell-1}.$$

Therefore Σ_0 passes through q_0, ∞ and has a unique singular point at ∞.

Let

$$M_{n\ell} \xrightarrow{\beta_{n\ell}} M_{n\ell-1} \to \cdots \to M_1 \xrightarrow{\beta_1} M_0 \tag{2.2.1}$$

be a sequence of one point-blowups with centers $q_i \in M_i$, $0 \leq i \leq n\ell - 1$, constructed recursively as follows. Let $E_i \subset M_i$ be the exceptional divisor of the i-th blow-up and let $\Sigma_i \subset M_i$ be the strict transform of $\Sigma_0 \subset M$. Then:

- $\beta_1 : M_1 \to M_0$ is the blow-up of M_0 at q_0, which is a smooth point of Σ_0, and
- q_{i+1} is the unique set theoretic intersection point of Σ_i and E_i in M_i for all $1 \leq i \leq n\ell - 2$.

Let $\pi_i : M_i \to M$ be the resulting natural projection, $1 \leq i \leq n\ell$. Then the following holds in complete analogy to [12, Section 3.1].

Lemma 2.2.1 *Let Δ_0, Δ_∞ denote the scheme theoretic inverse images of the points $q_0, \infty \in C$ under the composition $p \circ \pi_{n\ell} : M_{n\ell} \to C$. Let $q_{n\ell}$ denote the unique set theoretic intersection point of $E_{n\ell}$ and $\Sigma_{n\ell}$. Then*

(i) *Δ_0 is a scheme theoretic union $\Delta_0 = E_{n\ell} \cap \Delta_0'$ where Δ_0' is disjoint from $\Sigma_{n\ell}$ and intersects $E_{n\ell}$ at a single point.*
(ii) *$E_{n\ell}$ and $\Sigma_{n\ell}$ intersect transversely at $q_{n\ell}$, which is a smooth point of $\Sigma_{n\ell}$.*
(ii) *One has an identity $-K_{M_{n\ell}} = -K'_{M_{n\ell}} + \Delta_\infty$ in the Picard group of $M_{n\ell}$ where $-K'_{M_{n\ell}}$ is an effective divisor supported on $(\Delta_0')_{\mathrm{red}}$.*

Remark 2.2.2 Using the notation in Lemma 2.2.1, let $S = M_{n\ell} \setminus \Delta_0'$. By construction, S contains $\Sigma_{n\ell}$ as a closed subscheme, as well as the inverse image $\pi_{n\ell}^{-1}(U_\infty)$. Since the latter is mapped isomorphically to $U_\infty = \operatorname{Spec} \mathbb{C}[v, y]$ by the projection $\pi_{n\ell}$, they will be identified in the following. The distinction being clear from the context. Similarly, the divisor $\Delta_\infty \subset U_\infty$ will be identified with its inverse image in S.

Again, in complete analogy to [12, Section 3.1], the main properties of S are the following.

Lemma 2.2.3 *S is a smooth quasi-projective surface containing $\Sigma_{n\ell}$ as a reduced irreducible effective divisor. Furthermore its dualizing sheaf is $\omega_S \simeq \mathcal{O}_S(-\Delta_\infty)$ and the second homology group $H_2(S)$ is freely generated the curve class $[\Sigma_{n\ell}]$.*

In order to simplify the notation, let $\Sigma = \Sigma_{n\ell}$ and $q = q_{n\ell}$ as well as $\pi = \pi_{n\ell}|_S$. Let $\mathbf{T}\times M \to M$, where $\mathbf{T} = \mathbb{C}^\times$, be the torus action with weights $(0, \ell, (n-1)\ell-1)$ on (z_1, z_2, ζ) respectively. Then, a straightforward computation shows the following.

Lemma 2.2.4

(*i*) *The above torus action leaves Σ_0 invariant and its fixed locus $M_0^{\mathbf{T}}$ coincides with the fixed locus $\Sigma_0^{\mathbf{T}} = \{q_0, \infty\}$. Moreover, Σ_0 is the only torus invariant reduced proper closed subscheme of M_0.*

(*ii*) *There is a unique lift $\mathbf{T} \times M_{n\ell} \to M_{n\ell}$ preserving S, Δ_∞, and Σ so that the projection $\pi_{n\ell} : M_{n\ell} \to M$ is equivariant. Furthermore the fixed locus $S^{\mathbf{T}}$ coincides with the fixed locus $\Sigma^{\mathbf{T}} = \{q, \infty\}$ and Σ is the only torus invariant reduced proper closed subscheme of S.*

For further reference, note the following basic result.

Lemma 2.2.5 *Let X be a smooth quasi-projective surface, let $\beta : X' \to X$ be a one-point pull-back of X, and let $\iota : E \hookrightarrow X'$ be the exceptional divisor. Let $\Omega^1_{X'} \to \iota_*\Omega^1_E$ the composition of the natural morphisms*

$$\Omega^1_{X'} \to \iota_*\iota^*\Omega^1_{X'} \to \iota_*\Omega^1_E \,.$$

Then one has an exact sequence

$$0 \to \beta^*\Omega^1_X \to \Omega^1_{X'} \to \iota_*\Omega^1_E \to 0. \tag{2.2.2}$$

Proof First note that the natural morphism $\pi^*\Omega^1_X \to \Omega^1_{X'}$ is an isomorphism on the complement of the exceptional divisor E. Therefore it is injective, and its cokernel Q is a purely one dimensional sheaf with set theoretic support on E. The isomorphism $Q \simeq \iota_*\Omega^1_E$ follows by a simple local computation. □

Corollary 2.2.6 *Under the assumptions of Lemma 2.2.5 let $\Gamma \subset X$ be a reduced irreducible curve, containing the blow-up center as a smooth point, and let $\Gamma' \subset X'$ be its strict transform in X'. Let $\gamma : \Gamma' \to \Gamma$ be the morphism induced by π. Then Γ' intersects the exceptional divisor transversely E at a unique point p and one has an exact sequence*

$$0 \to \gamma^*\,\Omega^1_X|_\Gamma \to \Omega^1_{X'}|_{\Gamma'} \to \iota_*\Omega^1_E|_{\Gamma'} \to 0 \tag{2.2.3}$$

where $\iota_*\Omega^1_E\big|_{\Gamma'}$ *is a length one zero dimensional sheaf supported at* p.

Proof Under the stated conditions, γ is an isomorphism, and the intersection $\Gamma' \cap E$ consists indeed of a unique transverse intersection point. Moreover, since Γ' is purely one dimensional, one has $\mathrm{Tor}_1^{X'}(\mathcal{O}_E, \mathcal{O}_{\Gamma'}) = 0$. Then the claim follows from Lemma 2.2.5. □

2.2.2 Relative Gromov-Witten Invariants

Let $Y = S \times \mathbb{A}^1$ and let $D_\infty \subset Y$ be the inverse image of $\Delta_\infty \subset S$ under the natural projection $p_S : Y \to S$. Then $\omega_Y \simeq \mathcal{O}_Y(-D_\infty)$ and, as a consequence of Lemma 2.2.3, the second homology group $H_2(Y)$ is freely generated by the curve class $[\Sigma]$.

Here $\Sigma \subset S$ is naturally viewed as a curve on Y using the zero section $S \to Y$. Writing $\mathbb{A}^1 = \mathrm{Spec}\,\mathbb{C}[z]$, the local equations of Σ in the open chart $V_\infty = p_S^{-1}(U_\infty) \subset Y$ are

$$v^\ell = y^{(n-1)\ell-1}, \qquad z = 0. \tag{2.2.4}$$

Moreover, the divisor $D_\infty \subset Y$ coincides with the divisor $y = 0$ in V_∞. In particular, $\Sigma \cdot D_\infty = \ell$ in the intersection ring of Y. A map $f : X \to Y$, with X a projective curve, will be said to be of degree d if and only if $f_*[X] = d[\Sigma]$.

Let $\mathbf{T} \times Y \to Y$ be the unique lift of the given torus action to X so that the isomorphism $\omega_Y \simeq \mathcal{O}_Y(-D_\infty)$ holds equivariantly. Note that the fixed locus $Y^{\mathbf{T}}$ coincides again with $\Sigma^{\mathbf{T}} = \{q, \infty\}$.

For any $h, d \in \mathbb{Z}$, $d > 0$ and any partition $\mu = (\mu_1 \geq \cdots \geq \mu_{l(\mu)} \geq 0)$ of $d\ell$ let $\overline{M}_{h,d}(\mu)$ denote the moduli stack of arithmetic genus h, degree d, relative stable maps to (Y, D_∞) with contact conditions along D_∞ specified by μ. This paper will use the same flavor of relative stable maps as in [5, 6, 36, 37], where the points in the inverse image of D_∞ are not marked on the domain. As shown in [25, 27], this is a Deligne-Mumford stack which has a perfect obstruction theory. Moreover, it is also proper for proper targets, which is not the case in the present situation since Y is non-compact. However, as shown below, in the present case the fixed locus of the induced torus action the moduli stack $\overline{M}_{h,d}(\mu)$ is proper. Therefore residual relative invariants will be defined by equivariant localization as in [6, 29, 30, 35–37]. Foundational results for equivariant localization in the context of relative stable maps were proven in [19, 32].

In order to facilitate the exposition, recall that the moduli stack $\overline{M}_{h,d}(\mu)$ contains a canonical open substack $\mathcal{S}_{h,d}(\mu)$ consisting of simple relative stable maps. These are stable maps $f : X \to Y$ so that $f_*[X] = d[\Sigma]$ and $f^{-1}(D_\infty)$ is an effective divisor of the form

$$f^{-1}(D_\infty) = \sum_{i=1}^{l(\mu)} \mu_i p_i \tag{2.2.5}$$

with p_i contained in the smooth locus of X. In particular, the points p_i are unmarked. In addition, X is allowed to have several connected components, as long as no connected component is mapped to a point in Y.

The complement of $\mathcal{S}_{h,d}(\mu)$ consists of relative stable maps to expanded degenerations of (Y, D_∞). A length $n \geq 1$ expanded degeneration is a normal crossing threefold Y' with irreducible components $Y, Y_1, \ldots, Y_n$, each Y_i being isomorphic to the total space of the projective bundle

$$P = \mathbb{P}(N \oplus \mathcal{O}_{D_\infty}),$$

where N denotes the normal bundle to D_∞ in Y. This projective bundle has two disjoint canonical sections Z_0, Z_∞ over D_∞ with normal bundles N^{-1} and N respectively. The irreducible components are glued by identifying $D_\infty \subset Y$ to $Z_{0,1} \subset Y_1$ and, successively, $Z_{\infty,i} \subset Y_i$ to $Z_{0,i+1} \subset Y_{i+1}$ for $1 \leq i \leq n-1$. The Weil divisor obtained by identifying $Z_{\infty,i}$ and $Z_{0,i+1}$ will be denoted by D_i, $1 \leq i \leq n-1$. Note that no gluing occurs along the section $Z_{\infty,n} \subset Y_n$, which will be denoted by D'_∞. For notational purposes, such an expanded degeneration will be denoted by

$$Y' = Y_0 \cup_{D_\infty} Y_1 \cup_{D_1} \cdots \cup_{D_{n-1}} Y_n \ . \tag{2.2.6}$$

where by convention $Y_0 = Y$ and $D_0 = D_\infty$.

The domain X of a relative stable map $f : X \to Y'$ has a similar normal crossing structure

$$X = X_0 \cup_{\nu_0} X_1 \cup_{\nu_1} \cdots \cup_{\nu_{n-1}} X_n$$

where the gluing occurs along reduced divisors ν_i so that both X_i and X_{i+1} are smooth at ν_i. Each divisor ν_i in the domain is mapped to the divisor D_i in the expanded target, so that the restrictions $f_i : X_i \to Y'$ satisfy natural gluing conditions. Namely, each closed point ν_i occurs with the same multiplicity in the inverse images $f_i^{-1}(D_i)$ and $f_{i+1}^{-1}(D_{i+1})$, which are Cartier divisors on X_i and X_{i+1} respectively. Furthermore, $f^{-1}(D'_\infty)$ is a divisor of the form (2.2.5) contained in the smooth locus of X_n.

In the present case, $D_\infty \subset Y$ has trivial normal bundle, hence $P \simeq D_\infty \times \mathbb{P}^1$. Moreover, $D_\infty \simeq \mathbb{A}^2$ and the fixed locus of the induced torus action on D_∞ consists only of the origin in $\mathbb{A}^2$. This lifts to a torus action on P with fixed locus is $\{0\} \times \mathbb{P}^1$.

Since the torus action preserves both canonical sections Z_0 and Z_∞, it will lift to a torus action on any expanded degeneration Y'. Then Lemmas 2.2.3 and 2.2.4 yield

Lemma 2.2.7 *The fixed locus of the induced* $\mathbf{T}$*-action on* $\overline{M}_{h,d}(\mu)$ *is proper for any* h, d, μ.

The twisted wild cap is then defined as the generating function of residual relative Gromov-Witten invariants

$$GW_{h,d}(\lambda) = \int_{[\overline{M}_{h,d}(\lambda)^{\mathbf{T}}]^{vir}} \frac{1}{e_{\mathbf{T}}(N^{vir})}$$

where $[\overline{M}_{h,d}(\lambda)^{\mathbf{T}}]^{vir}$ is the induced virtual cycle on the fixed locus and N^{vir} denotes the equivariant virtual normal bundle to the fixed locus. The goal of this section is to derive an explicit expression for this generating function. The first step is to evaluate the residual contributions of the $\mathbf{T}$-fixed components contained in the open substack $\mathcal{S}_{h,d}(\mu)$, which will be called simple fixed loci in the following.

2.2.3 Simple Fixed Loci

Let normalization $\widetilde{\Sigma} \simeq \mathbb{P}^1$ be the normalization of Σ and let $\nu : \widetilde{\Sigma} \to S$ be the composition of the normalization map with the natural closed immersion $\Sigma \to Y$. Let $\mathbf{T} \times \widetilde{\Sigma} \to \widetilde{\Sigma}$ be the unique torus action which makes ν equivariant.

Let $\Xi \simeq \mathbb{P}^1$ and let $\xi : \Xi \to \widetilde{\Sigma}$ be a $\mathbf{T}$-equivariant $r : 1$ cover branched at the fixed locus $\widetilde{\Sigma}^{\mathbf{T}}$. Let $\phi : \Xi \to S$ be the map obtained by composition with ν. Let $\mathbf{T}^{1/r} \to \mathbf{T}$, where $\mathbf{T}^{1/r} = \mathbb{C}^\times$, be the natural $r : 1$ cover. By construction, f is clearly equivariant with respect to the unique torus action $\mathbf{T}^{1/r} \times \Xi \to \Xi$ which makes the $r : 1$-cover ξ equivariant. Let $T^{1/r}$ denote the canonical generator of the representation ring of $\mathbf{T}^{1/r}$ and set $T^{m/r} = (T^{1/r})^m$ for any $m \in \mathbb{Z}$. Let $g : \Xi \to M_0$ denote the composition $\gamma = \pi_{n\ell} \circ \phi$ and note that γ is $\mathbf{T}^{1/r}$-equivariant. Let $\mathbf{o}$ denote the unique fixed point of Ξ mapped to the fixed point $q_0 \in M_0$.

Lemma 2.2.8 *Let* $\Omega^1_{M_0}(\log \Delta_\infty)$ *denote the sheaf of Kähler differentials on* M_0 *with logarithmic singularities along* Δ_∞*. Then one has the following identities*

$$\mathrm{Ext}^0_\Xi(\gamma^*\Omega^1_{M_0}(\log \Delta_\infty), \mathcal{O}_\Xi) = T^{1-(n-1)\ell} \sum_{k=0}^{r(n-1)\ell} T^{k/r} + \sum_{k=0}^{r\ell} T^{k/r}, \tag{2.2.7}$$

in the representation ring of $\mathbf{T}^{1/r}$*. Moreover*

$$\mathrm{Ext}^1_\Xi(\gamma^*\Omega^1_{M_0}(\log \Delta_\infty), \mathcal{O}_\Xi) = 0. \tag{2.2.8}$$

Proof Let $\Xi_{\mathbf{0}} = \mathrm{Spec}\, \mathbb{C}[s]$ and $\Xi_{\infty} = \mathrm{Spec}\, \mathbb{C}[t]$ be a standard affine open cover of $\Xi \simeq \mathbb{P}^1$ so that the map g is locally given by

$$x = s^{r\ell},\ u = s^r, \qquad \text{respectively} \qquad y = t^{r\ell},\ v = t^{r(n-1)\ell - r}.$$

Note that $\Omega^1_{M_0}(\log \Delta_\infty)$ is locally free and let $\Theta = \gamma^*\Omega^1_{M_0}(\log \Delta_\infty)^\vee$. Then Θ is a is a subsheaf of the pull-back $\gamma^* T_{M_0}$ of the Zariski tangent sheaf T_{M_0} and its local sections are given by

$$\Gamma(\Theta|_{\Xi_{\mathbf{0}}}) \simeq \mathbb{C}[s]\, \gamma^*\partial_x \oplus \mathbb{C}[s]\, \gamma^*\partial_u\,, \qquad \Gamma(\Theta|_{\Xi_\infty}) \simeq (t^{r\ell})\, \gamma^*\partial_y \oplus \mathbb{C}[t]\gamma^*\partial_v$$

where $(t^{r\ell}) \subset \mathbb{C}[t]$ is the ideal generated by $t^{r\ell}$. The transition functions on the intersection $\Xi_{\mathbf{0}} \cap \Xi_\infty$ are

$$\gamma^*\partial_x = -t^{2r\ell}\, \gamma^*\partial_y - (n-1)t^{r(n\ell-1)}\, \gamma^*\partial_v\,, \qquad \gamma^*\partial_u = t^{r(n-1)\ell}\, \gamma^*\partial_v$$

on the intersection. Then the claim follows from the associated Čech cohomology complex. □

Lemma 2.2.9 *The following identity holds in the representation ring of* $\mathbf{T}^{1/r}$.

$$\begin{aligned} &\mathrm{Ext}^0(\phi^*\Omega^1_S(\log \Delta_\infty), \mathcal{O}_\Xi) - \mathrm{Ext}^1(\phi^*\Omega^1_S(\log \Delta_\infty), \mathcal{O}_\Xi) = \\ &\mathrm{Ext}^0(\gamma^*\Omega^1_{M_0}(\log \Delta_\infty), \mathcal{O}_\Xi) - T^{-(n-1)\ell} \sum_{k=1}^{rn\ell} T^{k/r}. \end{aligned} \tag{2.2.9}$$

Proof For each $0 \le i \le n\ell$ let $\phi_i : \Xi \to M_i$ denote the composition

$$\Xi \xrightarrow{\phi} M_{n\ell} \to M_i$$

where $M_{n\ell} \to M_i$ is the natural projection induced by the sequence of blowups (2.2.2). In particular, $\phi_{n\ell} = \phi$ and $\phi_0 = \gamma$. Then Corollary 2.2.6 yields an exact sequence

$$0 \to \phi_i^*\Omega^1_{M_i}(\log \Delta_\infty) \to \phi_{i+1}^*\Omega^1_{M_{i+1}}(\log \Delta_\infty) \to \phi_{i+1}^*\Omega^1_{E_{i+1}} \otimes \mathcal{O}_{\Sigma_{i+1}} \to 0$$

for each $0 \le i \le n\ell - 1$, where $\Omega^1_{E_{i+1}}$ is extended by zero to M_{i+1}. Since Σ_{i+1} and E_{i+1} have a unique transverse intersection point, one has an isomorphism

$$\phi_{i+1}^*\Omega^1_{E_{i+1}} \otimes \mathcal{O}_{\Sigma_{i+1}} \simeq \phi_{i+1}^*\Omega^1_{E_{i+1}} \otimes \mathcal{O}_{r\mathbf{o}}\,.$$

In particular, this yields

$$\mathcal{E}xt^1_\Xi(\phi_{i+1}^*\Omega^1_{E_{i+1}} \otimes \mathcal{O}_{\Sigma_{i+1}}, \mathcal{O}_\Xi) \simeq \phi_{i+1}^* T_{E_{i+1}} \otimes \mathcal{O}_{r\mathbf{o}}(r\mathbf{o})$$

where $T_{E_{i+1}}$ denotes the extension by zero of the Zariski tangent sheaf to the exceptional divisor $E_{i+1} \subset M_{i+1}$. Then the local to global sequence yields the isomorphism

$$\mathrm{Ext}^1(\phi_{i+1}^*\Omega^1_{E_{i+1}} \otimes \mathcal{O}_{\Sigma_{i+1}}, \mathcal{O}_\Xi) \simeq H^0(\phi_{i+1}^* T_{E_{i+1}} \otimes \mathcal{O}_{r\mathbf{o}}(r\mathbf{o})).$$

Hence one obtains the long exact sequence

$$\begin{aligned} 0 &\to \mathrm{Ext}^0(\phi_{i+1}^*\Omega^1_{M_{i+1}}(\log \Delta_\infty), \mathcal{O}_\Xi) \\ &\to \mathrm{Ext}^0(\phi_i^*\Omega^1_{M_i}(\log \Delta_\infty)), \mathcal{O}_\Xi) \to H^0(\phi_{i+1}^* T_{E_{i+1}} \otimes \mathcal{O}_{\mathbf{o}}(\mathbf{o})) \\ &\to \mathrm{Ext}^1(\phi_{i+1}^*\Omega^1_{M_{i+1}}(\log \Delta_\infty), \mathcal{O}_\Xi) \\ &\to \mathrm{Ext}^1(\phi_i^*\Omega^1_{M_i}(\log \Delta_\infty), \mathcal{O}_\Xi) \to 0. \end{aligned} \tag{2.2.10}$$

Furthermore, a local computation shows that

$$H^0(\phi_i^* T_{E_i} \otimes \mathcal{O}_{r\mathbf{o}}(r\mathbf{o})) = T^{\ell-i}\sum_{k=1}^{r} T^{k/r}$$

for $1 \le i \le n\ell$. Then identity (2.2.9) follows from the the exact sequences (2.2.10) and Lemma 2.2.8. □

As a consequence of Lemmas 2.2.8 and 2.2.9, one has:

Corollary 2.2.10 *The following identity holds in the representation ring of* $\mathbf{T}^{1/r}$*:*

$$\mathrm{Ext}^0(\phi^*\Omega^1_Y(\log D_\infty), \mathcal{O}_\Xi) - \mathrm{Ext}^1(\phi^*\Omega^1_Y(\log D_\infty), \mathcal{O}_\Xi) =$$
$$T^{(n-1)\ell-1} + \sum_{k=0}^{r} T^{k/r} - T^{-(n-1)\ell}\sum_{k=1}^{r-1} T^{k/r}. \tag{2.2.11}$$

Let $\mu = (r)$ be the one part partition of r. Let $f : \Xi \to Y$ be the composition of $\phi : \Xi \to S$ with the zero section $S \to Y$ and note that f determines a closed $\mathbf{T}$-fixed point in the simple locus $\mathcal{S}_{1,r}(\mu)$.

Lemma 2.2.11 *Then the restriction of the perfect obstruction complex to the fixed point determined by* f *is given by*

$$T^{(n-1)\ell-1} + \sum_{k=2}^{r} T^{k/r} - T^{-(n-1)\ell}\sum_{k=1}^{r-1} T^{k/r} \tag{2.2.12}$$

in the representation ring of $\mathbf{T}^{1/r}$.

Proof Let $\mathbf{o}, \boldsymbol{\xi}$ denote the fixed points of the $\mathbf{T}^{1/r}$-action on Ξ, which are mapped respectively to $q, \infty \in Y$. Let $f^*\Omega^1_Y(\log D_\infty) \to \Omega^1_\Xi(\log \boldsymbol{\xi})$ be the natural morphism, and let C denote its cone. The restriction of the perfect obstruction complex to the closed point determined by f in $\mathcal{S}_{1,r}(\mu)$ is given by

$$\mathrm{Ext}^0(\mathsf{C}, \mathcal{O}_\Xi) - \mathrm{Ext}^1(\mathsf{C}, \mathcal{O}_\Xi).$$

in the representation ring of $\mathbf{T}^{1/r}$. Then identity (2.2.12) follows from the long exact sequence associated to the exact triangle

$$f^*\Omega^1_Y(\log D_\infty) \to \Omega^1_\Xi(\log \boldsymbol{\xi}) \to \mathsf{C}$$

and Corollary 2.2.10. □

The conventions for partitions, or Young diagrams, used in the following are the same as in Sect. 2.1, above Eq. (2.1.2). In addition, for any partition ν let χ^ν denote the associated character of the symmetric group $\mathcal{S}_{|\nu|}$ and, for any partition μ with $|\mu| = |\nu|$, let $\chi^\nu(\mu)$ denote the value of χ^ν on the conjugacy class determined by μ. Let also $\mathcal{S}_\nu \subset \mathcal{S}_{|\nu|}$ denote the associated Young subgroup. If ν consists of m_j parts equal to j, for any $j \geq 1$, one has

$$S_\nu = \prod_{j\geq 1} \mathcal{S}_{m_j} .$$

Furthermore, the size of the associated conjugacy class in $\mathcal{S}_{|\nu|}$ is

$$\zeta(\nu) = \prod_{j\geq 1} m_j! j^{m_j}$$

Recall the coefficients $c^{(k)}_{\mu,\rho}$ defined in Eq. (2.1.2) through the Schur function identity

$$s_\mu(x_1^k, x_2^k, \ldots) = \sum_\rho c^{(k)}_{\mu,\rho} s_\rho(x_1, x_2, \ldots) \tag{2.2.13}$$

Given a partition ν consisting of n_j parts equal to j, with $j \geq 1$, let $k\nu$ be the partition of $k|\nu|$ consisting of n_j parts equal to kj for any $j \geq 1$. In partiocular, $l(k\nu) = l(\nu)$. Then note the following basic identity.

Lemma 2.2.12 *The coefficients $c^{(k)}_{\mu,\rho}$ are given by*

$$c^{(k)}_{\mu,\rho} = \sum_{|\nu|=|\mu|} \frac{\chi^\mu(\nu)\chi^\rho(k\nu)}{\zeta(\nu)} \tag{2.2.14}$$

for any partition ρ *with* $|\rho| = k|\mu|$, *while* $c^{(k)}_{\mu,\rho} = 0$ *for all partitions* ρ *of size* $|\rho| \neq k|\mu|$.

Proof This follows by a straightforward computation from the identities

$$s_\mu(x) = \sum_{|\nu|=|\mu|} \frac{\chi^\mu(\nu)}{\zeta(\nu)} p_\nu(x), \qquad p_\sigma(x) = \sum_{|\rho|=|\sigma|} \chi^\rho(\sigma) s_\rho(x),$$

where $p_\nu(x)$, $p_\sigma(x)$ are the power symmetric functions. It suffices to note that

$$p_\nu(x_1^k, x_2^k, \ldots) = p_{k\nu}(x_1, x_2, \ldots).$$

□

Let $H_{\mathbf{T}}(*)$ denote the equivariant cohomology ring of the classifying space of $\mathbf{T}$ and let $\alpha = e_{\mathbf{T}}(T)$ denote its canonical generator. For a fixed nonempty partition μ let

$$W^{(s)}_\mu(\lambda) = \sum_{h\in\mathbb{Z}} GW^{(s)}_{h,\mu} \lambda^{2h-2} \in \mathbb{C}(\alpha)[\lambda^{-1}][[\lambda]]$$

denote the generating function for the equivariant residual contributions associated to the simple fixed loci. Let $q = e^{i\lambda}$.

Lemma 2.2.13 *The following identity holds*

$$W^{(s)}_\mu = \sum_\nu \delta_{\mu,\ell\nu} (i\mathbf{n}\alpha)^{-l(\nu)} \sum_{|\sigma|=|\nu|} \frac{\chi^\sigma(\nu)}{\zeta(\nu)} q^{\mathbf{n}c(\sigma)} s_{\sigma^t}(\underline{q}), \qquad (2.2.15)$$

where $\mathbf{n} = (n-1)\ell - 1$ *and* $\underline{q} = (q^{1/2}, q^{3/2}, \ldots)$.

Proof Generically, the domain of a stable map in the simple fixed locus is a normal crossing curve

$$X = \Gamma_0 \cup \bigcup_{i=1}^{l(\mu)} \Xi_i$$

where

- Γ_0 is a finite union of connected components, and
- Ξ_i, $1 \leq 1 \leq l(\mu)$, are smooth projective lines mapped equivariantly to Y with degrees r_i, as in Lemma 2.2.11 such that
- each Ξ_i intersects Γ_0 transversely at the fixed point $\mathbf{o}_i \in \Xi^{\mathbf{T}^{1/r_i}}$, which is mapped to $q \in S^{\mathbf{T}}$,
- each component of Γ_0 is a stable marked curve, where the marked points are the nodes $\mathbf{o}_i$ belonging to that particular component, and

- no connected component of X is mapped to a point in Y.

Furthermore, the partition μ encoding the contact conditions along D_∞ has parts ℓr_i. This follows from the local equations (2.2.4) of Σ in Y. Hence, using the notation in Lemma 2.2.12, one has $\mu = \ell\nu$, where ν be the partition with parts r_i, $1 \leq i \leq l(\mu)$. In particular, $l(\nu) = l(\mu)$ and $\mathcal{S}_\nu = \mathcal{S}_\mu$. Then the associated connected component of the simple fixed locus is isomorphic to the stack quotient

$$\overline{M}_{h,l(\nu)}/\mathcal{S}_\nu$$

where $\overline{M}_{h,l(\nu)}$ denotes the moduli stack of stable, not necessarily connected, stable curves of arithmetic genus h curves with $l(\nu)$ marked points.

In addition, one has rigid fixed loci where X is either a projective line mapped equivariantly to Σ, i.e. $X = \Xi$, or a nodal curve $X = \Xi_1 \cup \Xi_2$ where Ξ_1, Ξ_2 are projective lines and the unique nodal point coincides simultaneously with the fixed points $\mathbf{o}_1, \mathbf{o}_2$.

In order to evaluate the residual contributions of the simple fixed loci one then has to compute the equivariant K-theory class of the perfect obstruction complex restricted to the fixed locus. This follows by a standard computation from Lemma 2.2.11 using normalization exact sequences. Since this is a well trodden path, the details will be omitted. The final expression for the equivariant residues is written as follows in terms of the tautological classes ψ_j associated to the marked points and the Chern classes of the Hodge bundle $\mathbb{E}$ on $\overline{M}_{h,l(\nu)}$:

$$\frac{(-1)^{l(\nu)-1}}{\mathrm{Aut}(\nu)}((n-1)\ell)^{l(\nu)-1}((n-1)\ell-1)^{l(\nu)-1}((n-1)\ell-1)^{-l(\nu)}\,\alpha^{2l(\nu)-3}$$

$$\prod_{j=1}^{l(\nu)} \frac{\prod_{k=1}^{\nu_j-1}(k-\nu_j(n-1)\ell)}{(\nu_j-1)!}$$

$$\int_{\overline{M}_{h,l(\nu)}} \frac{\Lambda_h^\vee(\alpha)\Lambda_g^\vee(-(n-1)\ell\alpha)\Lambda_g^\vee(((n-1)\ell-1)\alpha)}{\alpha-\nu_j\psi_j}$$

where

$$\Lambda_h^\vee(\mathsf{x}) = \sum_{i=0}^{h}(-1)^{h-i}\mathsf{x}^i(-1)^{h-i}c_{h-i}(\mathbb{E}).$$

The residual contributions of the rigid fixed loci are completely analogous. Then identity (2.2.15) follows from the Marino-Vafa formula proven in [29, 35]. □

2.2.4 The Cap Formula

Using the same notation as in Lemma 2.2.12, one has:

Theorem 2.2.14 *For a fixed nonempty partition* μ*, the twisted wild cap is given by*

$$W_\mu = (i\mathbf{n}\alpha)^{-l(\mu)}\zeta(\mu)^{-1}\sum_{|\eta|=|\mu|}\chi^\eta(\mu)q^{-\mathbf{n}c(\eta)/\ell}\sum_\sigma c^{(\ell)}_{\sigma,\eta}q^{\mathbf{n}c(\sigma)}s_{\sigma^t}(\underline{q}) \tag{2.2.16}$$

Proof A detailed analysis of torus fixed loci and residual relative invariants can be found for example in [5, 29, 30, 35]. A summary is also provided in Appendix A of [15] where the context is similar to the present paper.

First note that $D_\infty \simeq \mathbb{A}^2$ has normal bundle $N \simeq \mathcal{O}_{D_\infty}$ in Y and the origin in $\mathbb{A}^2$ is identified with the fixed point $\infty \in Y^{\mathbf{T}}$. Moreover, one has following identities in the equivariant ring of $\mathbf{T}$.

$$T_\infty D_\infty = T^{-\ell} + T^{-\mathbf{n}}, \qquad N_\infty = T^{\mathbf{n}}, \tag{2.2.17}$$

where $\mathbf{n} = (n-1)\ell - 1$. Then, proceeding by analogy with loc. cit. the twisted wild cap has a presentation

$$W_\mu = W^{(s)}_\mu + \sum_{|\rho|=|\mu|}(i\mathbf{n}\alpha)^{2l(\rho)}\zeta(\rho)W^{(s)}_\rho R_{\rho,\mu} \tag{2.2.18}$$

where $R_{\rho,\mu}$ is the rubber theory encoding the contributions of the expanded degenerations. As shown in [30, 36, 37], the rubber theory of an affine plane equipped with a torus action with weights $(1, k)$ is given by

$$(ik\alpha)^{-l(\rho)-l(\mu)}\sum_{|\eta|=|\mu|}\left(q^{-kc(\eta)}-1\right)\frac{\chi^\eta(\rho)}{\zeta(\rho)}\frac{\chi^\eta(\mu)}{\zeta(\mu)},$$

A succinct summary of the main steps in this computation can be found for example in [15, Appendix A]. In complete analogy to [30, 36, 37], given the equivariant weights (2.2.17), in the present case, the rubber theory reads

$$R_{\rho,\mu} = (i\mathbf{n}\alpha)^{-l(\rho)-l(\mu)}\sum_{|\eta|=|\mu|}\left(q^{-\mathbf{n}c(\eta)/\ell}-1\right)\frac{\chi^\eta(\rho)}{\zeta(\rho)}\frac{\chi^\eta(\mu)}{\zeta(\mu)},$$

where $|\rho| = |\mu|$. Moreover, using the orthogonality relations

$$\sum_{|\eta|=|\mu|}\chi^\eta(\rho)\chi^\eta(\mu) = \delta_{\rho,\mu}\zeta(\mu),$$

the rubber theory is rewritten as

$$R_{\rho,\mu} = (i\mathbf{n}\alpha)^{-l(\rho)-l(\mu)} \sum_{|\eta|=|\mu|} q^{-\mathbf{n}c(\eta)/\ell} \frac{\chi^{\eta}(\rho)}{\zeta(\rho)} \frac{\chi^{\eta}(\mu)}{\zeta(\mu)} - (i\mathbf{n}\alpha)^{-2l(\rho)}\zeta(\rho)\delta_{\rho,\mu}.$$

This implies the identity

$$W_{\mu} = (i\mathbf{n}\alpha)^{-l(\mu)} \sum_{\substack{\eta,\rho \\ |\eta|=|\rho|=|\mu|}} q^{-\mathbf{n}c(\eta)/\ell} \frac{\chi^{\eta}(\rho)\chi^{\eta}(\mu)}{\zeta(\mu)} (i\mathbf{n}\alpha)^{l(\rho)} W^{(s)}_{\rho},$$

As shown in Lemma 2.2.13,

$$W^{(s)}_{\rho} = \sum_{\nu} \delta_{\rho,\ell\nu}(i\mathbf{n}\alpha)^{-l(\nu)} \sum_{|\sigma|=|\nu|} \frac{\chi^{\sigma}(\nu)}{\zeta(\nu)} q^{\mathbf{n}c(\sigma)} s_{\sigma^t}(\underline{q}).$$

Furthermore identity (2.2.14) yields

$$\sum_{\nu,\rho} \delta_{\rho,\ell\nu} \frac{\chi^{\sigma}(\nu)\chi^{\eta}(\rho)}{\zeta(\nu)} = c^{(\ell)}_{\sigma,\eta}.$$

Then Eq. (2.2.16) follows by direct substitution. □

2.3 Twisted Wild Local Curves

Let C be a smooth connected rational curve as in Sect. 2.2.1 and let $\pi : S \to C$ be the spectral surface constructed in Sect. 2.2.1. Note that the generic fiber of the projection π is isomorphic to the affine line. Let (Y, D_∞) be the log Calabi-Yau threefold constructed in Sect. 2.2.2, where $Y = S \times \mathbb{A}^1$ and $D_\infty = \Delta_\infty \times \mathbb{A}^1$, with $\Delta_\infty \simeq \mathbb{A}^1$ a fiber of π. Let $\mathbf{T} \times Y \to Y$ be the torus action on Y constructed in loc. cit. and recall that its fixed point set consists of two points $q, \infty \in Y$, where ∞ coincides with the origin in $D_\infty \simeq \mathbb{A}^2$.

Let C_0 be a smooth projective genus g curve over $\mathbb{C}$ and let $\overline{C}$ be a nodal curve with irreducible components C_0, C intersecting at a nodal point $\infty \in \overline{C}$, which is the unique singular point of $\overline{C}$. Let $\omega_{\overline{C}}$ be the dualizing line bundle of $\overline{C}$ and let $\overline{M}$ be the total space of the line bundle $\omega_{\overline{C}} \otimes \mathcal{O}_{\overline{C}}(q)$, where q a smooth point of $\overline{C}$ contained in C. The points $q, \infty \in C$ will be identified with their images in $\overline{M}$ via the zero section $\overline{C} \to \overline{M}$.

As in Sect. 2.2.1 one then constructs a normal crossing spectral surface $\overline{S}$ as an open subscheme in an iterated blowup of $\overline{M}$, starting with a blowup at q. Let $\bar{\pi} : \overline{S} \to \overline{C}$ be the natural projection to $\overline{C}$, and note that $\overline{S}$ has two smooth irreducible components, the spectral surface $S = \bar{\pi}^{-1}(C)$ and a second component

$S_0 = \bar{\pi}^{-1}(C_0)$. The latter is isomorphic to the total space of the twisted canonical bundle $L = K_{C_0}(\infty)$ and intersects S along the fiber of L at ∞, which coincides with the affine line $\Delta_\infty \in S$.

Finally, let $\overline{Y} = \overline{S} \times \mathbb{A}$ and note that $\overline{Y}$ is a normal crossing Calabi-Yau threefold with irreducible components $Y_0 = S_0 \times \mathbb{A}^1$ and $Y = S \times \mathbb{A}^1$. By construction, the former is naturally isomorphic to the total space of the rank two bundle $V = K_{C_0}(\infty) \oplus \mathcal{O}_{C_0}$, and the two components meet along an affine plane which coincides simultaneously with the fiber $V_\infty \subset Y_0$ and with the divisor $D_\infty \in Y$. Furthermore, the torus action $\mathbf{T} \times Y \to Y$ used in Sect. 2.2.2 extends to a unique torus action $\mathbf{T} \times \overline{Y} \to \overline{Y}$ whose restriction to Y_0 coincides with the anti-diagonal torus action on V with weights $(-\mathbf{n}, \mathbf{n})$ on the two direct summands $K_{C_0}(\infty) \oplus \mathcal{O}_{C_0}$ respectively.

Since $H_2(D_\infty) = 0$ the Mayer-Vietoris sequence yields a natural isomorphism

$$H_2(\overline{Y}) \simeq H_2(Y_0) \oplus H_2(Y),$$

where $H_2(Y_0)$ is freely generated by the zero section class $[C_0]$ while $H_2(Y)$ is freely generated by $[\Sigma]$. For any $h, d, r \in \mathbb{Z}$, $r, d \geq 0$, let $\overline{M}_h(\overline{Y}, d, r)$ denote the moduli stack of genus h stable maps $f : X \to \overline{Y}$ with $f_*[X] = d[C_0] + r[\Sigma]$. Again, disconnected domains are allowed provided that no connected component is contracted to a point in $\overline{Y}$. Moduli stacks of stable maps to normal crossing targets were constructed in [1, 8, 26, 28], where it shown that they have a perfect obstruction theory. Moreover, the moduli stacks are proper for proper targets. Although this is not the case here, by analogy to Lemma 2.2.7, the torus fixed locus $\overline{M}_h(\overline{Y}, d, r)^{\mathbf{T}}$ is proper. Let

$$\begin{aligned} &Z_{GW}(\lambda, Q_0, Q) \\ &= 1 + \sum_{h\in\mathbb{Z}} \sum_{\substack{d,r\geq 0 \\ (d,r)\neq(0,0)}} GW_h(r, d)\, \lambda^{2h-2}\, Q_0^d\, Q^r \in \mathbb{C}(\alpha)[\lambda^{-1}][[\lambda, Q_0, Q]] \end{aligned}$$

be the resulting generating function of equivariant residual Gromov-Witten invariants. Here $\alpha \in H_{\mathbf{T}}(*)$ denotes again the canonical generator of the cohomology ring of the classifying space of $\mathbf{T}$.

Using the degeneration formula proven in [1, 8, 26, 28] and the TQFT formalism of [6] by analogy to [15, Section 3.3], Theorem 2.2.14 yields:

Corollary 2.3.1 *The following identity holds in* $\mathbb{C}(q^{1/2}, q^{1/\ell})[\lambda^{-1}][[\lambda, Q_0, Q]]$*:*

$$Z_{GW}(\lambda, Q_0, Q) = \sum_{\mu} q^{-\mathbf{n}c(\mu)/\ell} s_{\mu^t}(\underline{q})^{1-2g} \sum_{\sigma} c^{(\ell)}_{\sigma,\mu} q^{\mathbf{n}} c(\sigma) s_{\sigma^t}(\underline{q}) \mathsf{x}^{|\sigma|},$$

where $\mathsf{x} = Q_0^\ell Q$.

Acknowledgments This note is based on a talk given at the memorial conference *A tribute to the life and work of Professor Bumsig Kim: categorical and enumerative aspects of mirror symmetry*, KIAS, September 19-23, 2022. The author acknowledges the honor and the privilege to pay tribute to the mathematical legacy of a wonderful friend and an inspirational role model.

References

1. Abramovich, D., Fantechi, B.: Orbifold techniques in degeneration formulas. Ann. Sc. Norm. Super. Pisa Cl. Sci. **16**(2), 519–579 (2016). MR 3559610
2. Biquard, O., Boalch, P.: Wild non-abelian Hodge theory on curves. Compos. Math. **140**(1), 179–204 (2004). MR 2004129
3. Boalch, P.: Hyperkahler manifolds and nonabelian Hodge theory of (irregular) curves (2012). Arxiv:1203.6607
4. Boalch, P.P.: Geometry and braiding of Stokes data; fission and wild character varieties. Ann. Math. **179**(1), 301–365 (2014). MR 3126570
5. Bryan, J., Pandharipande, R.: Curves in Calabi-Yau threefolds and topological quantum field theory. Duke Math. J. **126**(2), 369–396 (2005). MR 2115262
6. Bryan, J., Pandharipande, R.: The local Gromov-Witten theory of curves. J. Am. Math. Soc. **21**(1), 101–136 (2008). With an appendix by Bryan, C. Faber, A. Okounkov and Pandharipande. MR 2350052
7. Boalch, P., Yamakawa, D.: Twisted wild character varieties (2015). arXiv:1512.08091
8. Chen, Q.: The degeneration formula for logarithmic expanded degenerations. J. Algebraic Geom. **23**(2), 341–392 (2014). MR 3166394
9. Chuang, W.Y., Diaconescu, D.E., Pan, G.: Wallcrossing and cohomology of the moduli space of Hitchin Pairs. Commun. Number Theory Phys. **5**, 1–56 (2011)
10. Chuang, W.Y., Diaconescu, D.E., Pan, G.: BPS States and the $P = W$ Conjecture. Moduli Spaces, London Mathematical Society Lecture Note Series, vol. 411, pp. 132–150. Cambridge University Press, Cambridge (2014). MR 3221294
11. Chuang, W.Y., Diaconescu, D.E., Donagi, R., Pantev, T.: Parabolic refined invariants and Macdonald polynomials. Commun. Math. Phys. **335**(3), 1323–1379 (2015). arXiv:1311.3624
12. Chuang, W.Y., Diaconescu, D.E., Donagi, R., Nawata, S., Pantev, T.: Twisted spectral correspondence and torus knots. J. Knot Theory Ramifications **29**(6), 2050040, 65 (2020). MR 4125177
13. de Cataldo, M.A.A., Hausel, T., Migliorini, L.: Topology of Hitchin Systems and Hodge Theory of Character Varieties: The Case A_1. Ann. Math. **175**(3), 1329–1407 (2012). MR 2912707
14. de Cataldo, M.A.A., Maulik, D., Shen, J.: Hitchin fibrations, abelian surfaces, and the $P = W$ conjecture. J. Am. Math. Soc. **35**(3), 911–953 (2022). MR 4433080
15. Diaconescu, D.E.: Local curves, wild character varieties, and degenerations. Commun. Number Theory Phys. **12**(3), 491–542 (2018). MR 3862071
16. Diaconescu, D.E., Donagi, R., Pantev, T.: BPS states, torus links and wild character varieties. Commun. Math. Phys. **359**(3), 1027–1078 (2018). MR 3784539
17. Donaldson, S.K.: Twisted harmonic maps and the self-duality equations. Proc. London Math. Soc. **55**(1), 127–131 (1987). MR 887285 (88g:58040)
18. Gopakumar, R., Vafa, C.: M theory and topological strings II (1998). arXiv:9812127
19. Graber, T., Vakil, R.: Relative virtual localization and vanishing of tautological classes on moduli spaces of curves. Duke Math. J. **130**(1), 1–37 (2005). MR 2176546
20. Hausel, T., Rodriguez-Villegas, F.: Mixed Hodge polynomials of character varieties. Invent. Math. **174**(3), 555–624 (2008). With an appendix by Nicholas M. Katz. MR MR2453601 (2010b:14094)
21. Hausel, T., Letellier, E., Rodriguez-Villegas, F.: Arithmetic harmonic analysis on character and quiver varieties. Duke Math. J. **160**(2), 323–400 (2011). MR 2852119

22. Hausel, T., Mereb, M., Wong, M.L.: Arithmetic and representation theory of wild character varieties. J. Eur. Math. Soc. **21**(10), 2995–3052 (2019). MR 3994099
23. Hausel, T., Mellit, A., Minets, A., Schiffmann, O.: $P = W$ via H_2 (2022). arXiv:2209.05429
24. Hosono, S., Saito, M.H., Takahashi, A.: Relative Lefschetz action and BPS state counting. Int. Math. Res. Not. **15**, 783–816 (2001). MR 1849482
25. Kim, B.: Logarithmic stable maps. In: New Developments in Algebraic Geometry, Integrable Systems and Mirror Symmetry (RIMS, Kyoto, 2008). Advanced Studies in Pure Mathematics, vol. 59. The Mathematical Society of Japan, Tokyo, pp. 167–200 (2010). MR 2683209
26. Kim, B., Lho, H., Ruddat, H.: The degeneration formula for stable log maps. Manuscripta Mathematica (2021)
27. Li, J.: Stable morphisms to singular schemes and relative stable morphisms. J. Differ. Geom. **57**(3), 509–578 (2001). MR 1882667
28. Li, J.: A degeneration formula of GW-invariants. J. Differ. Geom. **60**(2), 199–293 (2002). MR1938113 (2004k:14096)
29. Liu, C.C.M., Liu, K., Zhou, J.: A proof of a conjecture of Mariño-Vafa on Hodge integrals. J. Differ. Geom. **65**(2), 289–340 (2003). MR 2058264
30. Liu, C.C.M., Liu, K., Zhou, J.: A formula of two-partition Hodge integrals. J. Am. Math. Soc. **20**(1), 149–184 (2007). MR 2257399
31. Mochizuki, T.: Wild harmonic bundles and wild pure twistor D-modules. Astérisque **340**, x+607 (2011). MR 2919903
32. Molcho, S., Routis, E.: Localization for logarithmic stable maps. Trans. Am. Math. Soc. Ser. B **6**, 80–113 (2019). MR 3905962
33. Maulik, D., Shen, J.: The $P = W$ conjecture for GL_n (2022). arXiv:2209.02568
34. Maulik, D., Toda, Y.: Gopakumar-Vafa invariants via vanishing cycles. Invent. Math. **213**(3), 1017–1097 (2018). MR 3842061
35. Okounkov, A., Pandharipande, R.: Hodge integrals and invariants of the unknot. Geom. Topol. **8**, 675–699 (2004). MR 2057777
36. Okounkov, A., Pandharipande, R.: Gromov-Witten theory, Hurwitz theory, and completed cycles. Ann. Math. **163**(2), 517–560 (2006). MR 2199225
37. Okounkov, A., Pandharipande, R.: Virasoro constraints for target curves. Invent. Math. **163**(1), 47–108 (2006). MR 2208418
38. Pandharipande, R., Thomas, R.P.: Curve counting via stable pairs in the derived category. Invent. Math. **178**(2), 407–447 (2009). MR MR2545686
39. Sabbah, C.: Harmonic metrics and connections with irregular singularities. Ann. Inst. Fourier **49**(4), 1265–1291 (1999)
40. Simpson, C.T.: Harmonic bundles on noncompact curves. J. Am. Math. Soc. **3**(3), 713–770 (1990). MR 1040197
41. Toda, Y.: Generating functions of stable pair invariants via wall-crossings in derived categories. In: New Developments in Algebraic Geometry, Integrable Systems and Mirror Symmetry (RIMS, Kyoto, 2008). Advanced Studies in Pure Mathematics, vol. 59, pp. 389–434. The Mathematical Society of Japan, Tokyo (2010). MR 2683216
42. Toda, Y.: Curve counting theories on Calabi-Yau 3-folds: approach via stability conditions on derived categories [translation of MR3244103]. Sugaku Expos. **31**(2), 199–229 (2018). MR 3863903
43. Witten, E.: Gauge theory and wild ramification. Anal. Appl. **6**(4), 429–501 (2008). MR 2459120

Open Access This chapter is licensed under the terms of the Creative Commons Attribution-NonCommercial-NoDerivatives 4.0 International License (http://creativecommons.org/licenses/by-nc-nd/4.0/), which permits any noncommercial use, sharing, distribution and reproduction in any medium or format, as long as you give appropriate credit to the original author(s) and the source, provide a link to the Creative Commons license and indicate if you modified the licensed material. You do not have permission under this license to share adapted material derived from this chapter or parts of it.

The images or other third party material in this chapter are included in the chapter's Creative Commons license, unless indicated otherwise in a credit line to the material. If material is not included in the chapter's Creative Commons license and your intended use is not permitted by statutory regulation or exceeds the permitted use, you will need to obtain permission directly from the copyright holder.

Chapter 3
Holomorphic Anomaly Equations for $[\mathbb{C}^5/\mathbb{Z}_5]$

Deniz Genlik and Hsian-Hua Tseng

To the memory of Bumsig Kim

Abstract We prove holomorphic anomaly equations for $[\mathbb{C}^5/\mathbb{Z}_5]$.

3.1 Introduction

The cyclic group $\mathbb{Z}_5$ acts naturally on $\mathbb{C}^5$ by letting its generator $1 \in \mathbb{Z}_5$ act by multiplication by the fifth root of unity

$$e^{\frac{2\pi\sqrt{-1}}{5}}.$$

This action commutes with the diagonal action of the torus $\mathrm{T} = (\mathbb{C}^*)^5$ on $\mathbb{C}^5$ and induces a T-action on $[\mathbb{C}^5/\mathbb{Z}_5]$. Consequently $[\mathbb{C}^5/\mathbb{Z}_5]$ is a toric Deligne-Mumford stack.

This paper is concerned with T-equivariant Gromov-Witten invariants of $[\mathbb{C}^5/\mathbb{Z}_5]$. By definition, these are the following integrals

$$\int_{\left[\overline{M}^{\mathrm{orb}}_{g,n}([\mathbb{C}^5/\mathbb{Z}_5],0)\right]^{vir}} \prod_{i=1}^{n} \mathrm{ev}_i^*(\gamma_i)\, \psi_i^{k_i}. \tag{3.1.1}$$

D. Genlik (✉)
Department of Mathematics, University of Illinois Urbana-Champaign, IL, USA
e-mail: genlik@illinois.edu

H.-H. Tseng
Department of Mathematics, Ohio State University, Columbus, OH, USA
e-mail: hhtseng@math.ohio-state.edu

© The Author(s) 2026

N.-G. Kang et al. (eds.), *Categorical and Enumerative Aspects of Mirror Symmetry*,
KIAS Springer Series in Mathematics 5,
https://doi.org/10.1007/978-981-95-0385-8_3

Here, $[\overline{M}^{\mathrm{orb}}_{g,n}([\mathbb{C}^5/\mathbb{Z}_5],0)]^{vir}$ is the (T-equivariant) virtual fundamental class of the moduli space $\overline{M}^{\mathrm{orb}}_{g,n}([\mathbb{C}^5/\mathbb{Z}_5],0)$ of stable maps to $[\mathbb{C}^5/\mathbb{Z}_5]$. The classes $\psi_i \in H^2(\overline{M}^{\mathrm{orb}}_{g,n}([\mathbb{C}^5/\mathbb{Z}_5],0),\mathbb{Q})$ are descendant classes, for a detailed treatment see [8]. The evaluation maps

$$\mathrm{ev}_i : \overline{M}^{\mathrm{orb}}_{g,n}\left(\left[\mathbb{C}^5/\mathbb{Z}_5\right],0\right) \to I[\mathbb{C}^5/\mathbb{Z}_5]$$

take values in the inertia stack $I[\mathbb{C}^5/\mathbb{Z}_5]$ of $[\mathbb{C}^5/\mathbb{Z}_5]$. The classes $\gamma_i \in H^*_{\mathrm{T,Orb}}([\mathbb{C}^5/\mathbb{Z}_5]) := H^*_{\mathrm{T}}(I[\mathbb{C}^5/\mathbb{Z}_5])$ are classes in the Chen-Ruan cohomology of $[\mathbb{C}^5/\mathbb{Z}_5]$. Standard references for the Chen-Ruan cohomology are [2] and [3].

Let

$$\lambda_0, \lambda_1, \lambda_2, \lambda_3, \lambda_4 \in H^*_{\mathrm{T}}(\mathrm{pt}) = H^*(B\mathrm{T})$$

be the first Chern classes of the tautological line bundles of $B\mathrm{T} = (B\mathbb{C}^*)^5$. Then (3.1.1) takes value in $\mathbb{Q}(\lambda_0, \lambda_1, \lambda_2, \lambda_3, \lambda_4)$.

Foundational treatments of orbifold Gromov-Witten theory can be found in many references. For compact target stacks, the original reference is [1]. For non-compact target stacks admitting torus actions, such as $[\mathbb{C}^5/\mathbb{Z}_5]$, one can define Gromov-Witten theory for them using virtual localization formula [7]. In this case, their Gromov-Witten theory should be understood as certain *twisted* Gromov-Witten theory of stacks. Generalities on twisted Gromov-Witten theory of stacks can be found in [4] and [18].

The main results of this paper concern structures of Gromov-Witten invariants (3.1.1), formulated in terms of generating functions. The definition of inertia stacks implies that

$$I[\mathbb{C}^5/\mathbb{Z}_5] = [\mathbb{C}^5/\mathbb{Z}_5] \cup \bigcup_{k=1}^{4} B\mathbb{Z}_5.$$

Let

$$\phi_0 = 1 \in H^0_{\mathrm{T}}([\mathbb{C}^5/\mathbb{Z}_5]), \phi_k = 1 \in H^0_{\mathrm{T}}(B\mathbb{Z}_5), 1 \le k \le 4.$$

Then $\{\phi_0, \phi_1, \phi_2, \phi_3, \phi_4\}$ is an additive basis of $H^*_{\mathrm{T,Orb}}([\mathbb{C}^5/\mathbb{Z}_5])$. The orbifold Poincaré dual $\{\phi^0, \phi^1, \phi^2, \phi^3, \phi^4\}$ of this basis is given by

$$\phi^0 = 5\lambda_0\lambda_1\lambda_2\lambda_3\lambda_4\phi_0, \quad \phi^1 = 5\phi_4, \quad \phi^2 = 5\phi_3, \quad \phi^3 = 5\phi_2, \quad \phi^4 = 5\phi_1.$$

To simplify notation, in what follows we set

$$\phi_i := \phi_j \quad \text{if } j \equiv i \mod 5 \quad \text{and} \quad \phi^i := \phi^j \quad \text{if } j \equiv i \mod 5,$$

for all $i \ge 0$ and $0 \le j \le 4$.

For $\phi_{c_1}, \ldots, \phi_{c_n} \in H^*_{\mathrm{T,Orb}}\left([\mathbb{C}^5/\mathbb{Z}_5]\right)$, define the Gromov-Witten potential by

$$\mathcal{F}_{g,n}^{[\mathbb{C}^5/\mathbb{Z}_5]}\left(\phi_{c_1}, \ldots, \phi_{c_n}\right) = \sum_{d=0}^{\infty} \frac{\Theta^d}{d!} \int_{\left[\overline{M}^{\mathrm{orb}}_{g,n+d}([\mathbb{C}^5/\mathbb{Z}_5],0)\right]^{vir}} \prod_{i=1}^{n} \mathrm{ev}_i^*\left(\phi_{c_i}\right) \prod_{i=n+1}^{n+d} \mathrm{ev}_i^*\left(\phi_1\right) \tag{3.1.2}$$

where Θ is a formal variable keeping track of the insertion ϕ_1. The potential $\mathcal{F}_{g,n}^{[\mathbb{C}^5/\mathbb{Z}_5]}$ is regarded as a formal power series in Θ. We also use the standard double bracket notation for the Gromov-Witten potential throughout the paper:

$$\left\langle\left\langle \phi_{c_1}, \ldots, \phi_{c_n} \right\rangle\right\rangle_{g,n}^{[\mathbb{C}^5/\mathbb{Z}_5]} = \mathcal{F}_{g,n}^{[\mathbb{C}^5/\mathbb{Z}_5]}\left(\phi_{c_1}, \ldots, \phi_{c_n}\right).$$

The main results of this paper are differential equations for these generating functions $\mathcal{F}_g^{[\mathbb{C}^5/\mathbb{Z}_5]}$ for $g \geq 2$ after the following specializations of equivariant parameters:

$$\lambda_i = e^{\frac{2\pi\sqrt{-1}i}{5}}, \quad 0 \leq i \leq 4. \tag{3.1.3}$$

There are two differential equations, given precisely in Theorems 3.5.6 and 3.5.7 below. Theorem 3.5.6 is an analogue of main results of [12] and [13]. To the best of our knowledge, Theorem 3.5.7 does not have analogue in previous studies. Borrowing terminology from String Theory, we call these two differential equations *holomorphic anomaly equations* for $[\mathbb{C}^5/\mathbb{Z}_5]$.

Our approach to proving holomorphic anomaly equations is the same as that of [13] and is based on the fact that genus 0 Gromov-Witten theory of $[\mathbb{C}^5/\mathbb{Z}_5]$ yields a *semisimple* Frobenius structure on $H^*_{\mathrm{T,Orb}}([\mathbb{C}^5/\mathbb{Z}_5])$. Consequently, the *cohomological field theory* (in the sense of [9]) associated to the Gromov-Witten theory of $[\mathbb{C}^5/\mathbb{Z}_5]$ is semisimple. The Givental-Teleman classification [6, 17] of semisimple cohomological field theories can then be applied to yield an explicit formula for $\mathcal{F}_g^{[\mathbb{C}^5/\mathbb{Z}_5]}$, which can be used to prove holomorphic anomaly equations.

The rest of the paper is organized as follows. In Sect. 3.2, we state the mirror theorem for $[\mathbb{C}^5/\mathbb{Z}_5]$ and study certain power series arising from the I-function. In Sect. 3.3, we describe necessary ingredients of the Frobenius structure from Gromov-Witten theory of $[\mathbb{C}^5/\mathbb{Z}_5]$. In Sect. 3.4, we study lifting to certain ring of functions of an important ingredient called the R-matrix. Section 3.5 contains the main results of this paper. In Sect. 3.5.1, we give the formula for Gromov-Witten potentials of $[\mathbb{C}^5/\mathbb{Z}_5]$ arising from Givental-Teleman classification of semisimple

CohFTs. In Sect. 3.5.2, we state the two holomorphic anomaly equations and use the formula in Sect. 3.5.1 to prove them.

3.2 On Mirror Theorem

In this Section we discuss mirror theorem for Gromov-Witten theory of $[\mathbb{C}^5/\mathbb{Z}_5]$.

The I-function of $\left[\mathbb{C}^5/\mathbb{Z}_5\right]$ is defined[1] to be

$$I(x,z) = \sum_{k=0}^{\infty} \frac{x^k}{z^k k!} \prod_{\substack{b:0\le b<\frac{k}{5} \\ \langle b\rangle=\langle\frac{k}{5}\rangle}} \left(1-(bz)^5\right)\phi_k. \tag{3.2.1}$$

It is easy to see that I-function (3.2.1) of $[\mathbb{C}^5/\mathbb{Z}_5]$ is of the form

$$I(x,z) = \sum_{k=0}^{\infty} \frac{I_k(x)}{z^k}\phi_k. \tag{3.2.2}$$

The small J-function for $[\mathbb{C}^5/\mathbb{Z}_5]$ is defined by

$$J(\Theta,z) = \phi_0 + \frac{\Theta\phi_1}{z} + \sum_{i=0}^{n-1}\phi^i \left\langle\left\langle \frac{\phi_i}{z(z-\psi)} \right\rangle\right\rangle_{0,1}^{[\mathbb{C}^5/\mathbb{Z}_5]}.$$

The following is a consequence of the main result of [4].

Proposition 3.2.1 *We have the following mirror identity,*

$$J(T(x),z) = I(x,z), \tag{3.2.3}$$

with the mirror transformation[2]

$$T(x) = I_1(x) = \sum_{k\ge 0} \frac{(-1)^{5k}x^{5k+1}}{(5k+1)!}\left(\frac{\Gamma\left(k+\frac{1}{5}\right)}{\Gamma\left(\frac{1}{5}\right)}\right)^5. \tag{3.2.4}$$

[1] Here, $\langle\bullet\rangle$ is the fractional part of $\bullet$.

[2] The gamma function is defined by $\Gamma(z) = \int_0^\infty t^{z-1}e^{-t}dt$ where $\Re(z) > 0$. One of its fundamental properties is $\Gamma(z+1) = z\Gamma(z)$. We mainly use the gamma function in the expression of mirror transformation (3.2.3) to give a compact formula.

Define the operator

$$D : \mathbb{C}[[x]] \to x\mathbb{C}[[x]]$$

by

$$Df(x) = x\frac{df(x)}{dx}.$$

Next, we consider the following series[3] in $\mathbb{C}[[x]]$ arising from the I-function, which will be useful later:

$$\begin{aligned} L =& x\left(1+\left(\frac{x}{5}\right)^5\right)^{-\frac{1}{5}}, \\ C_1 =& DI_1, \\ C_2 =& D\left(\frac{DI_2}{C_1}\right), \\ C_3 =& D\left(\frac{D\left(\frac{DI_3}{C_1}\right)}{C_2}\right). \end{aligned} \tag{3.2.5}$$

It is easy to verify that

$$\frac{DL}{L} = 1 - \frac{L^5}{5^5}. \tag{3.2.6}$$

In [11, Proposition 4], the following identity is given:

$$C_1^2 C_2^2 C_3 = L^5. \tag{3.2.7}$$

The following lemma is a direct result of the definition (3.1.2) of Gromov-Witten potential and the mirror map $\Theta = T(x)$.

Lemma 3.2.2 *For $k \geq 1$, we have*

$$\frac{\partial^k \mathcal{F}_{g,n}^{[\mathbb{C}^5/\mathbb{Z}_5]}(\phi_{c_1}, \ldots, \phi_{c_n})}{\partial T^k} = \mathcal{F}_{g,n+k}^{[\mathbb{C}^5/\mathbb{Z}_5]}(\phi_{c_1}, \ldots, \phi_{c_n}, \underbrace{\phi_1, \ldots, \phi_1}_{k-many}).$$

---

[3] Here, our L differs from L defined in [11] by a sign. Although the definitions of C_1 and C_2 look different from those in [11], it is easy to check that these definitions match with those in [11].

We further define the following series:

$$
\begin{aligned}
X_1 &= \frac{DC_1}{C_1}, \\
X_2 &= \frac{DC_2}{C_2}, \\
A_1 &= \frac{1}{L}\left(\frac{DL}{L} - X_1\right), \\
A_2 &= \frac{1}{L}\left(2\frac{DL}{L} - X_1 - X_2\right), \\
B_i &= \frac{1}{5^i}(D + X_1)^{i-1}X_1 \quad \text{for} \quad 1 \leq i \leq 4.
\end{aligned}
\tag{3.2.8}
$$

In [11, Section 3], the following equations are given:

$$
B_4 = \left(1 - \frac{L^5}{5^5}\right)\left(2B_3 - \frac{7}{5}B_2 + \frac{2}{5}B_1 - \frac{24}{625}\right), \tag{3.2.9}
$$

$$
\begin{aligned}
DX_2 = &-10\left(1 - \frac{L^5}{5^5}\right) + 10\left(1 - \frac{L^5}{5^5}\right)X_1 \\
&+ 5\left(1 - \frac{L^5}{5^5}\right)X_2 - 2X_1^2 - 4DX_1 - 2X_1X_2 - X_2^2.
\end{aligned}
\tag{3.2.10}
$$

Since there is a linear relation between $\{A_1, A_2\}$ and $\{X_1, X_2\}$ with coefficients from the ring $\mathbb{C}[L^{\pm 1}]$, we can rewrite these two equations in terms of A_i's and their D derivatives. For example, Eq. (3.2.10) can be rewritten as

$$
DA_2 = LA_1^2 + LA_2^2 - DA_1 - 15\left(1 - \frac{L^5}{5^5}\right)\frac{L^5}{5^5}. \tag{3.2.11}
$$

Moreover, these linear relations show that the differential ring

$$
\mathbb{C}[L^{\pm 1}][A_1, A_2, DA_1, DA_2, D^2A_1, D^2A_2, \ldots]
$$

is a quotient of the free polynomial ring

$$
\mathbb{F} := \mathbb{C}[L^{\pm 1}][A_1, DA_1, D^2A_1, A_2].
$$

3.3 Frobenius Structures

In this Section, we spell out details of the Frobenius structure on $H^*_{\mathrm{T,Orb}}\left([\mathbb{C}^5/\mathbb{Z}_5]\right)$ defined using genus 0 Gromov-Witten theory of $[\mathbb{C}^5/\mathbb{Z}_5]$. We refer to [10] for generalities of Frobenius structures.

Let $\gamma = \sum_{i=0}^4 t_i\phi_i \in H^*_{\mathrm{T,Orb}}\left([\mathbb{C}^5/\mathbb{Z}_5]\right)$. The full genus 0 Gromov-Witten potential is defined to be

$$\mathcal{F}_0^{[\mathbb{C}^5/\mathbb{Z}_5]}(t,\Theta) = \sum_{n=0}^{\infty}\sum_{d=0}^{\infty}\frac{1}{n!d!}\int_{\left[\overline{M}^{\mathrm{orb}}_{0,n+d}([\mathbb{C}^5/\mathbb{Z}_5],0)\right]^{vir}}\prod_{i=1}^{n}\mathrm{ev}_i^*(\gamma)\prod_{i=n+1}^{n+d}\mathrm{ev}_i^*(\Theta\phi_1). \tag{3.3.1}$$

In the basis $\{\phi_0,\phi_1,\phi_2,\phi_3,\phi_4\}$ and under the specialization (3.1.3), the orbifold Poincaré pairing

$$g(-,-): H^*_{\mathrm{T,Orb}}\left(\left[\mathbb{C}^5/\mathbb{Z}_5\right]\right)\times H^*_{\mathrm{T,Orb}}\left(\left[\mathbb{C}^5/\mathbb{Z}_5\right]\right)\to\mathbb{Q}(\lambda_0,\lambda_1,\lambda_2,\lambda_3,\lambda_4)$$

has the matrix representation

$$G = \frac{1}{5}\begin{bmatrix} 1&0&0&0&0\\ 0&0&0&0&1\\ 0&0&0&1&0\\ 0&0&1&0&0\\ 0&1&0&0&0 \end{bmatrix}. \tag{3.3.2}$$

The quantum product $\bullet_\gamma$ at $\gamma \in H^*_{\mathrm{T,Orb}}\left([\mathbb{C}^5/\mathbb{Z}_5]\right)$ is a product structure on $H^*_{\mathrm{T,Orb}}\left([\mathbb{C}^5/\mathbb{Z}_5]\right)$. It can be defined as follows:

$$g(\phi_i\bullet_\gamma\phi_j,\phi_k) := \frac{\partial^3}{\partial t_i\partial t_j\partial t_k}\mathcal{F}_0^{[\mathbb{C}^5/\mathbb{Z}_5]}(t,\Theta).$$

In what follows, we focus on the quantum product $\bullet_{\gamma=0}$ at $\gamma = 0 \in H^*_{\mathrm{T,Orb}}\left([\mathbb{C}^5/\mathbb{Z}_5]\right)$, which we denote by $\bullet$. Note that $\bullet$ still depends on Θ.

It is proved in [11, Section 5] that the quantum product at $0 \in H^*_{\mathrm{T,Orb}}\left([\mathbb{C}^5/\mathbb{Z}_5]\right)$ is semisimple with the idempotent basis $\{e_0,e_1,e_2,e_3,e_4\}$, that is,

$$e_i\bullet e_j = \delta_{i,j}e_j.$$

The corresponding normalized idempotent basis $\{\widetilde{e}_0,\widetilde{e}_1,\widetilde{e}_2,\widetilde{e}_3,\widetilde{e}_4\}$ is given by

$$\widetilde{e}_i = 5e_i \quad \text{for} \quad 0\le i\le 4.$$

In [11, Section 5], the transition matrix given by $\Psi_{ij} = g(\widetilde{e}_i, \phi_j)$ is calculated to be

$$\Psi = \frac{1}{5}\begin{bmatrix} 1 & \frac{L}{C_1} & \frac{L^2}{C_1C_2} & \frac{C_1C_2}{L^2} & \frac{C_1}{L} \\ 1 & \zeta\frac{L}{C_1} & \zeta^2\frac{L^2}{C_1C_2} & \zeta^3\frac{C_1C_2}{L^2} & \zeta^4\frac{C_1}{L} \\ 1 & \zeta^2\frac{L}{C_1} & \zeta^4\frac{L^2}{C_1C_2} & \zeta\frac{C_1C_2}{L^2} & \zeta^3\frac{C_1}{L} \\ 1 & \zeta^3\frac{L}{C_1} & \zeta\frac{L^2}{C_1C_2} & \zeta^4\frac{C_1C_2}{L^2} & \zeta^2\frac{C_1}{L} \\ 1 & \zeta^4\frac{L}{C_1} & \zeta^3\frac{L^2}{C_1C_2} & \zeta^2\frac{C_1C_2}{L^2} & \zeta\frac{C_1}{L} \end{bmatrix}.$$

Let $\left\{u^0, u^1, u^2, u^3, u^4\right\}$ be canonical coordinates associated to the idempotent basis $\{e_0, e_1, e_2, e_3, e_4\}$ which satisfy

$$u^\alpha\,(t_i = 0, \Theta = 0) = 0.$$

By [11, Lemma 6], we have

$$\frac{du^\alpha}{dx} = \zeta^\alpha L \frac{1}{x} \tag{3.3.3}$$

at $t = 0$, for $0 \le \alpha \le 4$.

The full genus 0 Gromov-Witten potential (3.3.1) satisfies the following property

$$\mathcal{F}_0^{[\mathbb{C}^5/\mathbb{Z}_5]}(t, \Theta) = \mathcal{F}_0^{[\mathbb{C}^5/\mathbb{Z}_5]}(t|_{t_1=0}, \Theta + t_1).$$

that is, $\mathcal{F}_0^{[\mathbb{C}^5/\mathbb{Z}_5]}(t, \Theta)$ depends on t_1 and Θ through $\Theta + t_1$. In particular, the operator

$$\frac{\partial}{\partial t_1} - \frac{\partial}{\partial \Theta} \tag{3.3.4}$$

annihilates $\mathcal{F}_0^{[\mathbb{C}^5/\mathbb{Z}_5]}(t, \Theta)$.

Denote by

$$R(z) = \mathrm{Id} + \sum_{k \ge 1} R_k z^k \in \mathrm{End}(H^*_{\mathrm{T,Orb}}([\mathbb{C}^5/\mathbb{Z}_5]))[[z]]$$

the R-matrix of the Frobenius structure associated to the (T-equivariant) Gromov-Witten theory of $[\mathbb{C}^5/\mathbb{Z}_5]$ near the semisimple point 0. The R-matrix plays a central role in the Givental-Teleman classification of semisimple cohomological field theories. By definition of R, the symplectic condition

$$R(z) \cdot R(-z)^* = \mathrm{Id}$$

holds. The following flatness equation

$$z(d\Psi^{-1})R + z\Psi^{-1}(dR) + \Psi^{-1}R(dU) - \Psi^{-1}(dU)R = 0 \tag{3.3.5}$$

also holds, see [10, Section 4.6] and [5, Proposition 1.1]. Here $d = \frac{d}{dt}$.

Since $\mathcal{F}_0^{[\mathbb{C}^5/\mathbb{Z}_5]}(t, \Theta)$ depends on t_1 and Θ through $\Theta + t_1$, it follows that Ψ and $R(z)$ also depend on t_1 and Θ through $\Theta + t_1$. So we have[4]

$$\frac{\partial}{\partial t_1}\Psi = \frac{\partial}{\partial \Theta}\Psi, \quad \frac{\partial}{\partial t_1}R(z) = \frac{\partial}{\partial \Theta}R(z).$$

In Eq. (3.3.5), we set $t_{\neq 1} = 0$ and only consider $\frac{d}{dt_1}$. It follows that (3.3.5) can be written as

$$z(\frac{d}{d\Theta}\Psi^{-1})R + z\Psi^{-1}(\frac{d}{d\Theta}R) + \Psi^{-1}R(\frac{d}{d\Theta}U) - \Psi^{-1}(\frac{d}{d\Theta}U)R = 0.$$

Since

$$\frac{d}{d\Theta} = \frac{dx}{d\Theta}\frac{d}{dx},$$

after cancelling $\frac{dx}{d\Theta}$ and multiplying by x, we rewrite the above equation as

$$z(x\frac{d}{dx}\Psi^{-1})R + z\Psi^{-1}(x\frac{d}{dx}R) + \Psi^{-1}R(x\frac{d}{dx}U) - \Psi^{-1}(x\frac{d}{dx}U)R = 0.$$

By equating coefficients of z^k, we see that

$$\Psi\left(D\Psi^{-1}\right)R_{k-1} + DR_{k-1} + R_k\,(DU) - (DU)\,R_k = 0 \tag{3.3.6}$$

or equivalently,

$$D\left(\Psi^{-1}R_{k-1}\right) + \left(\Psi^{-1}R_k\right)DU - \Psi^{-1}\,(DU)\,\Psi\left(\Psi^{-1}R_k\right) = 0 \tag{3.3.7}$$

where $D = x\frac{d}{dx}$ as before.

Now set $t_1 = 0$. By Eq. (3.3.3), we have

$$DU = \mathrm{diag}(L, L\zeta, L\zeta^2, L\zeta^3, L\zeta^4). \tag{3.3.8}$$

[4] An argument for this (written for a different target space) from the CohFT viewpoint can be found in [15, Section 3.3].

Let $P^k_{i,j}$ denote the (i, j)-entry of the matrix defined by $P_k = \Psi^{-1} R_k$ after being restricted to the semisimple point $0 \in H^*_{\mathrm{T,Orb}}\left([\mathbb{C}^5/\mathbb{Z}_5]\right)$. Set

$$\widetilde{P}^k_{i,j} = \frac{L^i}{K_i} P^k_{i,j} \zeta^{(k+i)j}$$

where $0 \leq i, j \leq 4$ and $k \geq 0$. Then, the flatness equation (3.3.7) reads as

$$\begin{aligned}
\widetilde{P}^k_{4,j} &= \widetilde{P}^k_{0,j} + \frac{1}{L} D \widetilde{P}^{k-1}_{0,j}, \\
\widetilde{P}^k_{3,j} &= \widetilde{P}^k_{4,j} + \frac{1}{L} D \widetilde{P}^{k-1}_{4,j} + A_1 \widetilde{P}^{k-1}_{4,j}, \\
\widetilde{P}^k_{2,j} &= \widetilde{P}^k_{3,j} + \frac{1}{L} D \widetilde{P}^{k-1}_{3,j} + A_2 \widetilde{P}^{k-1}_{3,j}, \\
\widetilde{P}^k_{1,j} &= \widetilde{P}^k_{2,j} + \frac{1}{L} D \widetilde{P}^{k-1}_{2,j} - A_2 \widetilde{P}^{k-1}_{2,j}, \\
\widetilde{P}^k_{0,j} &= \widetilde{P}^k_{1,j} + \frac{1}{L} D \widetilde{P}^{k-1}_{1,j} - A_1 \widetilde{P}^{k-1}_{1,j}.
\end{aligned} \tag{3.3.9}$$

We call Eq. (3.3.9) the *modified flatness equations*.

In [19], properties of some hypergeometric series associated with mirror symmetry are analyzed, and certain formal power series are associated to these hypergeometric series. They obtained equations like (3.2.7), and showed that some parts of the asymptotic expansions of those hypergeometric series lie in polynomial rings in those formal power series. By the methodology of [19], we obtain the following result.[5]

Lemma 3.3.1 *We have* $\widetilde{P}^k_{0,j} \in \mathbb{C}[L^{\pm 1}]$ *for all* $0 \leq j \leq 4$ *and* $k \geq 0$.

3.4 Lift of Modified R-Matrix

3.4.1 Canonical Lift

The functions $\widetilde{P}^k_{i,j} \in \mathbb{C}[[x]]$ in modified flatness equations have canonical lifts to the the ring

$$\mathbb{F} = \mathbb{C}[L^{\pm 1}][A_1, DA_1, D^2 A_1, A_2]$$

[5] The same result for the total space $K\mathbb{P}^4$ of the canonical bundle of $\mathbb{P}^4$ can be inferred from [11, Proposition 14]. Hence, Lemma 3.3.1 can also be deducted from the main result of [11].

via Eqs. (3.2.6), (3.2.11) and the first four rows of the flatness equations (3.3.9) in the descending order. More precisely, we start with Lemma 3.3.1, that is

$$\widetilde{P}^k_{0,j} \in \mathbb{C}[L^{\pm 1}] \subset \mathbb{F}.$$

Then, by Eq. (3.2.6), we obtain

$$\widetilde{P}^k_{4,j} = \widetilde{P}^k_{0,j} + \frac{1}{L} D\widetilde{P}^{k-1}_{0,j} \in \mathbb{C}[L^{\pm 1}] \subset \mathbb{F}.$$

By proceeding in a similar manner, we see that

$$\widetilde{P}^k_{3,j} = \widetilde{P}^k_{4,j} + \frac{1}{L} D\widetilde{P}^{k-1}_{4,j} + A_1 \widetilde{P}^{k-1}_{4,j} \in \mathbb{C}[L^{\pm 1}][A_1] \subset \mathbb{F},$$

$$\widetilde{P}^k_{2,j} = \widetilde{P}^k_{3,j} + \frac{1}{L} D\widetilde{P}^{k-1}_{3,j} + A_2 \widetilde{P}^{k-1}_{3,j} \in \mathbb{C}[L^{\pm 1}][A_1, DA_1, A_2] \subset \mathbb{F}.$$

Lastly, using Eq. (3.2.11) we get

$$\widetilde{P}^k_{1,j} = \widetilde{P}^k_{2,j} + \frac{1}{L} D\widetilde{P}^{k-1}_{2,j} - A_2 \widetilde{P}^{k-1}_{2,j} \in \mathbb{C}[L^{\pm 1}][A_1, DA_1, D^2A_1, A_2] = \mathbb{F}.$$

This procedure gives us a canonical lift of $\widetilde{P}^k_{i,j} \in \mathbb{C}[[x]]$ to the free polynomial ring $\mathbb{F}$, which we denote again as $\widetilde{P}^k_{i,j}$. We state this result in the following way.

Lemma 3.4.1 *We have* $\widetilde{P}^k_{i,j} \in \mathbb{F}$ *for all* $0 \le i, j \le 4$ *and* $k \ge 0$.

3.4.2 Preparations

In this subsection, we use the lift $\widetilde{P}^k_{i,j} \in \mathbb{F}$ and prove two lemmas which will be used for the proof of holomorphic anomaly equations.

Lemma 3.4.2 *The following identity holds*

$$\frac{\partial \widetilde{P}^k_{i,j}}{\partial A_2} = \delta_{i,2} \widetilde{P}^{k-1}_{3,j}.$$

Proof It is clear that the degrees of $\widetilde{P}^k_{0,j}$, $\widetilde{P}^k_{4,j}$ and $\widetilde{P}^k_{3,j}$ in A_2 are all zero. Hence, we get

$$\frac{\partial \widetilde{P}^k_{0,j}}{\partial A_2} = \frac{\partial \widetilde{P}^k_{4,j}}{\partial A_2} = \frac{\partial \widetilde{P}^k_{3,j}}{\partial A_2} = 0. \tag{3.4.1}$$

The place where we see A_2 for the first time are the next two equations in (3.3.9),

$$\widetilde{P}^k_{2,j} = \widetilde{P}^k_{3,j} + \frac{1}{L} D \widetilde{P}^{k-1}_{3,j} + A_2 \widetilde{P}^{k-1}_{3,j}, \tag{3.4.2}$$

$$\widetilde{P}^k_{1,j} = \widetilde{P}^k_{2,j} + \frac{1}{L} D \widetilde{P}^{k-1}_{2,j} - A_2 \widetilde{P}^{k-1}_{2,j}, \tag{3.4.3}$$

From the first Eq. (3.4.2), we see that

$$\frac{\partial \widetilde{P}^k_{2,j}}{\partial A_2} = \widetilde{P}^{k-1}_{3,j}.$$

Note that Eq. (3.2.11) gives

$$\frac{\partial (D A_2)}{\partial A_2} = 2 L A_2.$$

Now, we compute the last derivative. By flatness equations (3.3.9) and Eq. (3.4.2), we obtain

$$\begin{aligned} \frac{\partial \widetilde{P}^k_{1,j}}{\partial A_2} &= \frac{\partial \widetilde{P}^k_{2,j}}{\partial A_2} + \frac{1}{L} \frac{\partial \left(D \widetilde{P}^{k-1}_{2,j} \right)}{\partial A_2} - \widetilde{P}^{k-1}_{2,j} - A_2 \frac{\partial \widetilde{P}^{k-1}_{2,j}}{\partial A_2} \\ &= \frac{\partial \widetilde{P}^k_{2,j}}{\partial A_2} + \frac{1}{L} \left(2 L A_2 \widetilde{P}^{k-2}_{3,j} + D \widetilde{P}^{k-2}_{3,j} \right) - \widetilde{P}^{k-1}_{2,j} - A_2 \frac{\partial \widetilde{P}^{k-1}_{2,j}}{\partial A_2}. \end{aligned} \tag{3.4.4}$$

Then, by Eq. (3.4.2) and again by flatness equations (3.3.9), we get

$$\begin{aligned} \frac{\partial \widetilde{P}^k_{1,j}}{\partial A_2} &= \widetilde{P}^{k-1}_{3,j} + 2 A_2 \widetilde{P}^{k-2}_{3,j} + \frac{1}{L} D \widetilde{P}^{k-2}_{3,j} - \widetilde{P}^{k-1}_{2,j} - A_2 \widetilde{P}^{k-2}_{3,j} \\ &= \widetilde{P}^{k-1}_{3,j} + A_2 \widetilde{P}^{k-2}_{3,j} + \frac{1}{L} D \widetilde{P}^{k-2}_{3,j} - \widetilde{P}^{k-1}_{2,j} = 0. \end{aligned} \tag{3.4.5}$$

This completes the proof. □

Lemma 3.4.3 *The following identity holds*

$$\frac{\partial \widetilde{P}^k_{i,j}}{\partial (D^2 A_1)} = \delta_{i,1} \frac{1}{L^2} \widetilde{P}^{k-3}_{4,j}.$$

Proof It is clear that the only $\widetilde{P}^k_{i,j} \in \mathbb{F}$ that has non-zero degree in D^2A_1 is $\widetilde{P}^k_{1,j}$, and the degree of $\widetilde{P}^k_{1,j}$ in D^2A_1 is 1. So, we obtain

$$\frac{\partial \widetilde{P}^k_{0,j}}{\partial(D^2A_1)} = \frac{\partial \widetilde{P}^k_{4,j}}{\partial(D^2A_1)} = \frac{\partial \widetilde{P}^k_{3,j}}{\partial(D^2A_1)} = \frac{\partial \widetilde{P}^k_{2,j}}{\partial(D^2A_1)} = 0. \tag{3.4.6}$$

The coefficient of D^2A_1 in $\widetilde{P}^k_{1,j}$ descends from the coefficient of A_1 in $\widetilde{P}^{k-2}_{3,j}$, which is $\widetilde{P}^{k-3}_{4,j}$. Keeping track of this term in the procedure of canonical lifting, we see that the coefficient of D^2A_1 in $\widetilde{P}^k_{1,j}$ is

$$\frac{1}{L^2}\widetilde{P}^{k-3}_{4,j}.$$

This completes the proof. □

3.5 Holomorphic Anomaly Equations

3.5.1 *Formula for Potentials*

By general considerations, Gromov-Witten theory of $[\mathbb{C}^5/\mathbb{Z}_5]$ has the structure of a cohomological field theory (CohFT). We refer to [9] and [14] for discussions on CohFTs.

Graphs

We describe the graphs needed in the formula for Gromov-Witten potentials.

Recall that a *stable graph* Γ is a tuple

$$\Gamma = \big(\mathrm{V}_\Gamma, \mathrm{H}_\Gamma, \mathrm{L}_\Gamma, \mathrm{g} : \mathrm{V}_\Gamma \to \mathbb{Z}_{\geq 0},$$
$$\nu : \mathrm{H}_\Gamma \cup \mathrm{L}_\Gamma \to \mathrm{V}_\Gamma, \iota : \mathrm{H}_\Gamma \to \mathrm{H}_\Gamma, \ell : \mathrm{L}_\Gamma \to \{1, \ldots, n\}\big)$$

satisfying:

(1) V_Γ is the vertex set with a genus assignment $\mathrm{g} : \mathrm{V}_\Gamma \to \mathbb{Z}_{\geq 0}$,
(2) H_Γ is the half-edge set equipped with an involution $\iota : \mathrm{H}_\Gamma \to \mathrm{H}_\Gamma$,
(3) E_Γ is the edge set defined by the orbits of $\iota : \mathrm{H}_\Gamma \to \mathrm{H}_\Gamma$ in H_Γ (self-edges are allowed at the vertices) and the tuple $(\mathrm{V}_\Gamma, \mathrm{E}_\Gamma)$ defines a connected graph,
(4) L_Γ is the set of legs, the subset of H_Γ fixed by the involution $\iota : \mathrm{H}_\Gamma \to \mathrm{H}_\Gamma$ and the map $\ell : \mathrm{L}_\Gamma \to \{1, \ldots, m\}$ is an isomorphism labeling legs,
(5) The map $\nu : \mathrm{H}_\Gamma \cup \mathrm{L}_\Gamma \to \mathrm{V}_\Gamma$ is a vertex assignment,

(6) For each vertex v, let $\mathrm{n}(\mathfrak{v}) = \mathrm{l}(\mathfrak{v}) + \mathrm{h}(\mathfrak{v})$ be the valence of the vertex (where $\mathrm{l}(\mathfrak{v})$ and $\mathrm{h}(\mathfrak{v})$ are the number of legs and the number of edges attached to the vertex $\mathfrak{v}$ respectively). Then, the following stability condition holds:

$$2\mathrm{g}(\mathfrak{v}) - 2 + \mathrm{n}(\mathfrak{v}) > 0.$$

The *genus* of a stable graph Γ is defined by

$$\mathrm{g}(\Gamma) = h^1(\Gamma) + \sum_{\mathfrak{v} \in \mathrm{V}_\Gamma} \mathrm{g}(\mathfrak{v}).$$

Let $\mathrm{G}_{g,m}$ be the isomorphism classes of stable graphs of genus g with n legs. A *decorated stable graph*

$$\Gamma \in \mathrm{G}_{g,n}^{\mathrm{Dec}}(5)$$

of order 5 is a stable graph $\Gamma \in \mathrm{G}_{g,n}$ with an extra assignment $\mathrm{p} : \mathrm{V}_\Gamma \to \{0, 1, 2, 3, 4\}$ to each vertex $\mathfrak{v} \in \mathrm{V}_\Gamma$. For a decorated stable graph $\Gamma \in \mathrm{G}_{g,n}^{\mathrm{Dec}}(5)$ we denote its underlying stable graph by

$$\Gamma^{\mathrm{St}} \in \mathrm{G}_{g,n}$$

after forgetting the decoration.

Formula for $\mathcal{F}_g$

By the results stated in Sect. 3.3, the CohFT of Gromov-Witten theory of $[\mathbb{C}^5/\mathbb{Z}_5]$ is semisimple. By Givental-Teleman classification of semisimple CohFTs (see e.g. [14] for a survey), we can write Gromov-Witten potential as a sum over decorated stable graphs,

$$\mathcal{F}_{g,n}^{[\mathbb{C}^5/\mathbb{Z}_5]}\left(\phi_{c_1}, \ldots, \phi_{c_n}\right) = \sum_{\Gamma \in \mathrm{G}_{g,m}^{\mathrm{Dec}}(5)} \mathrm{Cont}_\Gamma\left(\phi_{c_1}, \ldots, \phi_{c_n}\right). \tag{3.5.1}$$

Details about how this formula works in general can be found in e.g. [16] and [13].

In order to state the contributions of graphs to $\mathcal{F}_{g,n}^{[\mathbb{C}^5/\mathbb{Z}_5]}\left(\phi_{c_1}, \ldots, \phi_{c_n}\right)$, we need to introduce the following series in $\mathbb{C}[[x]]$:

$$K_0, \quad K_1 = C_1, \quad K_2 = C_1C_2, \quad K_3 = C_1C_2C_3, \quad \text{and} \quad K_4 = C_1C_2^2C_3,$$

and the following involution

$$\mathrm{Inv} : \{0, 1, 2, 3, 4\} \to \{0, 1, 2, 3, 4\},$$

with $\mathrm{Inv}(0) = 0$ and $\mathrm{Inv}(i) = 5 - i$ for $1 \le i \le 4$.

Proposition 3.5.1 *The contribution associated to a decorated stable graph* $\Gamma \in \mathrm{G}^{Dec}_{g,n}(5)$ *is*

$$\begin{aligned}\mathrm{Cont}_\Gamma\left(\phi_{c_1}, \dots, \phi_{c_n}\right) \\ = \frac{1}{|\mathrm{Aut}(\Gamma^{\mathrm{St}})|} \sum_{\mathrm{A}\in\mathbb{Z}_{\ge 0}^{\mathrm{F}(\Gamma)}} \prod_{\mathfrak{v}\in\mathrm{V}_\Gamma} \mathrm{Cont}^{\mathrm{A}}_\Gamma(\mathfrak{v}) \prod_{\mathfrak{e}\in\mathrm{E}_\Gamma} \mathrm{Cont}^{\mathrm{A}}_\Gamma(\mathfrak{e}) \prod_{\mathfrak{l}\in\mathrm{L}_\Gamma} \mathrm{Cont}^{\mathrm{A}}_\Gamma(\mathfrak{l})\end{aligned}$$

where $\mathrm{F}(\Gamma) = |\mathrm{H}_\Gamma|$. *Here,* $\mathrm{Cont}^{\mathrm{A}}_\Gamma(\mathfrak{v})$, $\mathrm{Cont}^{\mathrm{A}}_\Gamma(\mathfrak{e})$, *and* $\mathrm{Cont}^{\mathrm{A}}_\Gamma(\mathfrak{l})$ *are the* vertex, edge *and* leg *contributions with flag* A*–values*[6] $(a_1, \dots, a_m, b_{m+1}, \dots, b_{|\mathrm{H}_\Gamma|})$ *respectively, and they are given by*

$$\begin{aligned}\mathrm{Cont}^{\mathrm{A}}_\Gamma(\mathfrak{v}) &= \sum_{k\ge 0} \frac{g(e_{\mathrm{p}(\mathfrak{v})}, e_{\mathrm{p}(\mathfrak{v})})^{-\frac{2\mathrm{g}(\mathfrak{v})-2+\mathrm{n}(\mathfrak{v})+k}{2}}}{k!} \\ &\times \int_{\overline{M}_{\mathrm{g}(\mathfrak{v}),\mathrm{n}(\mathfrak{v})+k}} \psi_1^{a_{\mathfrak{v}1}} \cdots \psi_{\mathrm{l}(v)}^{a_{\mathfrak{v}\mathrm{l}(\mathfrak{v})}} \psi_{\mathrm{l}(\mathfrak{v})+1}^{b_{\mathfrak{v}1}} \\ &\cdots \psi_{\mathrm{n}(\mathfrak{v})}^{b_{\mathfrak{v}\mathrm{h}(\mathfrak{v})}} t_{\mathrm{p}(\mathfrak{v})}(\psi_{\mathrm{n}(\mathfrak{v})+1}) \cdots t_{\mathrm{p}(\mathfrak{v})}(\psi_{\mathrm{n}(\mathfrak{v})+k}), \\ \mathrm{Cont}^{\mathrm{A}}_\Gamma(\mathfrak{e}) &= \frac{(-1)^{b_{\mathfrak{e}1}+b_{\mathfrak{e}2}}}{5} \sum_{m=0}^{b_{\mathfrak{e}2}} (-1)^m \sum_{r=0}^{4} \frac{\widetilde{P}^{b_{\mathfrak{e}1}+m+1}_{\mathrm{Inv}(r),\mathrm{p}(\mathfrak{v}_1)} \widetilde{P}^{b_{\mathfrak{e}2}-m}_{r,\mathrm{p}(\mathfrak{v}_2)}}{\zeta^{(b_{\mathfrak{e}1}+m+1+\mathrm{Inv}(r))\mathrm{p}(\mathfrak{v}_1)} \zeta^{(b_{\mathfrak{e}2}-m+r)\mathrm{p}(\mathfrak{v}_2)}}, \\ \mathrm{Cont}^{\mathrm{A}}_\Gamma(\mathfrak{l}) &= \frac{(-1)^{a_{\ell(\mathfrak{l})}}}{5} \frac{K_{\mathrm{Inv}(c_{\ell(\mathfrak{l})})}}{L^{\mathrm{Inv}(c_{\ell(\mathfrak{l})})}} \frac{\widetilde{P}^{a_{\ell(\mathfrak{l})}}_{\mathrm{Inv}(c_{\ell(\mathfrak{l})}),\mathrm{p}(\nu(\mathfrak{l}))}}{\zeta^{(a_{\ell(\mathfrak{l})}+\mathrm{Inv}(c_{\ell(\mathfrak{l})}))\mathrm{p}(\nu(\mathfrak{l}))}},\end{aligned}$$

where

$$t_{\mathrm{p}(\mathfrak{v})}(z) = \sum_{j\ge 2} \mathrm{T}_{\mathrm{p}(\mathfrak{v})j} z^j \quad \textit{with} \quad \mathrm{T}_{\mathrm{p}(\mathfrak{v})j} = \frac{(-1)^j}{n} \widetilde{P}^{j-1}_{0,\mathrm{p}(\mathfrak{v})} \zeta^{-(j-1)\mathrm{p}(\mathfrak{v})}.$$

We should emphasize that Proposition 3.5.1 holds in $\mathbb{C}[[x]]$. Using Proposition 3.5.1 and lifting procedure in Sect. 3.4, we obtain the following lift of Gromov-Witten potential to certain polynomial rings.

---

[6] Notation: The values $b_{\mathfrak{v}1}, \dots, b_{\mathfrak{v}\mathrm{h}(\mathfrak{v})}$ and $b_{\mathfrak{e}1}, b_{\mathfrak{e}2}$ are the entries of $(a_1, \dots, a_m, b_{m+1}, \dots, b_{|\mathrm{H}_\Gamma|})$ corresponding to $\mathrm{Cont}^{\mathrm{A}}_\Gamma(\mathfrak{v})$ and $\mathrm{Cont}^{\mathrm{A}}_\Gamma(\mathfrak{e})$ respectively.

Theorem 3.5.2 (Finite Generation Property) *Let* $\mathrm{Cont}_\Gamma^{\mathrm{A}}(\mathfrak{v})$, $\mathrm{Cont}_\Gamma^{\mathrm{A}}(\mathfrak{e})$, *and* $\mathrm{Cont}_\Gamma^{\mathrm{A}}(\mathfrak{l})$ *be as in Proposition 3.5.1. We have* $\mathrm{Cont}_\Gamma^{\mathrm{A}}(\mathfrak{v}) \in \mathbb{C}[L^{\pm 1}]$, $\mathrm{Cont}_\Gamma^{\mathrm{A}}(\mathfrak{e}) \in \mathbb{F}$, *and* $\mathrm{Cont}_\Gamma^{\mathrm{A}}(\mathfrak{l}) \in \mathbb{F}[C_1, C_2, C_3]$. *Hence, we have*

$$\mathcal{F}_{g,n}^{[\mathbb{C}^5/\mathbb{Z}_5]}\left(\phi_{c_1}, \ldots, \phi_{c_n}\right) \in \mathbb{F}[C_1, C_2, C_3]$$

and when there is no insertions, we have

$$\mathcal{F}_g^{[\mathbb{C}^5/\mathbb{Z}_5]} \in \mathbb{F}$$

where $\mathbb{F} = \mathbb{C}[L^{\pm 1}][A_1, DA_1, D^2A_1, A_2]$ *as before.*

Proof By Lemma 3.3.1, we have $\mathrm{Cont}_\Gamma^{\mathrm{A}}(\mathfrak{v}) \in \mathbb{C}[L^{\pm 1}]$ since its expression involves only $\widetilde{P}_{0,j}^k$'s. By Lemma 3.4.1, and definitions of K_i's, we see that $\mathrm{Cont}_\Gamma^{\mathrm{A}}(\mathfrak{e}) \in \mathbb{F}$ and $\mathrm{Cont}_\Gamma^{\mathrm{A}}(\mathfrak{l}) \in \mathbb{F}[C_1, C_2, C_3]$. Hence, results for Gromov-Witten potentials follow. □

Depending on the insertions, we can give a better description of the polynomial ring that contains Gromov-Witten potentials. For example, by Proposition 3.5.1 we have

$$\mathcal{F}_{g,n}^{[\mathbb{C}^5/\mathbb{Z}_5]}\left(\phi_{c_1}, \ldots, \phi_{c_1}\right) \in \mathbb{F}[C_1^{-1}] = \mathbb{C}[L^{\pm 1}][A_1, DA_1, D^2A_1, A_2, C_1^{-1}] \tag{3.5.2}$$

and the degree of C_1^{-1} in $\mathcal{F}_{g,n}^{[\mathbb{C}^5/\mathbb{Z}_5]}\left(\phi_{c_1}, \ldots, \phi_{c_1}\right)$ is n. Then, we obtain the following result by Lemma 3.2.2.

Corollary 3.5.3 *For all,* $k \geq 1$, *we have*

$$\frac{\partial^k \mathcal{F}_g^{[\mathbb{C}^5/\mathbb{Z}_5]}}{\partial T^k} \in \mathbb{F}[C_1^{-1}] = \mathbb{C}[L^{\pm 1}][A_1, DA_1, D^2A_1, A_2, C_1^{-1}]$$

and the degree of C_1^{-1} *in* $\frac{\partial^k \mathcal{F}_g^{[\mathbb{C}^5/\mathbb{Z}_5]}}{\partial T^k}$ *is* k.

3.5.2 Proof of Holomorphic Anomaly Equations

Lemma 3.5.4 *We have*

$$\frac{\partial}{\partial A_2}\mathrm{Cont}_\Gamma^{\mathrm{A}}(\mathfrak{e}) = \frac{(-1)^{b_{\mathfrak{e}1}+b_{\mathfrak{e}2}}}{5}\frac{\widetilde{P}_{3,\mathrm{p}(\mathfrak{v}_1)}^{b_{\mathfrak{e}1}}\widetilde{P}_{3,\mathrm{p}(\mathfrak{v}_2)}^{b_{\mathfrak{e}2}}}{\zeta^{(b_{\mathfrak{e}1}+3)\mathrm{p}(\mathfrak{v}_1)}\zeta^{(b_{\mathfrak{e}2}+3)\mathrm{p}(\mathfrak{v}_2)}}.$$

Proof By Proposition 3.5.1 and Lemma 3.4.2, we obtain

$$\begin{aligned}\frac{\partial}{\partial A_2}\mathrm{Cont}^{\mathrm{A}}_{\Gamma}(\mathfrak{e}) &= \frac{(-1)^{b_{\mathfrak{e}1}+b_{\mathfrak{e}2}}}{5}\sum_{m=0}^{b_{\mathfrak{e}2}}(-1)^m\sum_{r=0}^{4}\frac{\frac{\partial}{\partial A_2}\left(\widetilde{P}^{b_{\mathfrak{e}1}+m+1}_{\mathrm{Inv}(r),\mathrm{p}(\mathfrak{v}_1)}\widetilde{P}^{b_{\mathfrak{e}2}-m}_{r,\mathrm{p}(\mathfrak{v}_2)}\right)}{\zeta^{(b_{\mathfrak{e}1}+m+1+\mathrm{Inv}(r))\mathrm{p}(\mathfrak{v}_1)}\zeta^{(b_{\mathfrak{e}2}-m+r)\mathrm{p}(\mathfrak{v}_2)}}\\
&= \frac{(-1)^{b_{\mathfrak{e}1}+b_{\mathfrak{e}2}}}{5}\sum_{m=0}^{b_{\mathfrak{e}2}}(-1)^m\frac{\widetilde{P}^{b_{\mathfrak{e}1}+m}_{3,\mathrm{p}(\mathfrak{v}_1)}\widetilde{P}^{b_{\mathfrak{e}2}-m}_{3,\mathrm{p}(\mathfrak{v}_2)}}{\zeta^{(b_{\mathfrak{e}1}+m+3)\mathrm{p}(\mathfrak{v}_1)}\zeta^{(b_{\mathfrak{e}2}-m+3)\mathrm{p}(\mathfrak{v}_2)}}\\
&+\frac{(-1)^{b_{\mathfrak{e}1}+b_{\mathfrak{e}2}}}{5}\sum_{m=0}^{b_{\mathfrak{e}2}}(-1)^m\frac{\widetilde{P}^{b_{\mathfrak{e}1}+m+1}_{3,\mathrm{p}(\mathfrak{v}_1)}\widetilde{P}^{b_{\mathfrak{e}2}-m-1}_{3,\mathrm{p}(\mathfrak{v}_2)}}{\zeta^{(b_{\mathfrak{e}1}+m+4)\mathrm{p}(\mathfrak{v}_1)}\zeta^{(b_{\mathfrak{e}2}-m+2)\mathrm{p}(\mathfrak{v}_2)}}.\end{aligned}$$

Since $\widetilde{P}^k_{i,j}$ is defined to be 0 for $k < 0$, the second summation ends actually at $m = b_{\mathfrak{e}2} - 1$. Then, by shifting the second summmation by 1, and cancelling out terms in total expression, we get

$$\begin{aligned}\frac{\partial}{\partial A_2}\mathrm{Cont}^{\mathrm{A}}_{\Gamma}(\mathfrak{e}) =&\frac{(-1)^{b_{\mathfrak{e}1}+b_{\mathfrak{e}2}}}{5}\sum_{m=0}^{b_{\mathfrak{e}2}}(-1)^m\frac{\widetilde{P}^{b_{\mathfrak{e}1}+m}_{3,\mathrm{p}(\mathfrak{v}_1)}\widetilde{P}^{b_{\mathfrak{e}2}-m}_{3,\mathrm{p}(\mathfrak{v}_2)}}{\zeta^{(b_{\mathfrak{e}1}+m+3)\mathrm{p}(\mathfrak{v}_1)}\zeta^{(b_{\mathfrak{e}2}-m+3)\mathrm{p}(\mathfrak{v}_2)}}\\
&+\frac{(-1)^{b_{\mathfrak{e}1}+b_{\mathfrak{e}2}}}{5}\sum_{m=1}^{b_{\mathfrak{e}2}}(-1)^{m-1}\frac{\widetilde{P}^{b_{\mathfrak{e}1}+m}_{3,\mathrm{p}(\mathfrak{v}_1)}\widetilde{P}^{b_{\mathfrak{e}2}-m}_{3,\mathrm{p}(\mathfrak{v}_2)}}{\zeta^{(b_{\mathfrak{e}1}+m+3)\mathrm{p}(\mathfrak{v}_1)}\zeta^{(b_{\mathfrak{e}2}-m+3)\mathrm{p}(\mathfrak{v}_2)}}\\
=&\frac{(-1)^{b_{\mathfrak{e}1}+b_{\mathfrak{e}2}}}{5}\frac{\widetilde{P}^{b_{\mathfrak{e}1}}_{3,\mathrm{p}(\mathfrak{v}_1)}\widetilde{P}^{b_{\mathfrak{e}2}}_{3,\mathrm{p}(\mathfrak{v}_2)}}{\zeta^{(b_{\mathfrak{e}1}+3)\mathrm{p}(\mathfrak{v}_1)}\zeta^{(b_{\mathfrak{e}2}+3)\mathrm{p}(\mathfrak{v}_2)}}.\end{aligned}$$

□

Lemma 3.5.5 *We have*

$$\frac{\partial}{\partial(D^2A_1)}\mathrm{Cont}^{\mathrm{A}}_{\Gamma}(\mathfrak{e}) = \frac{(-1)^{b_{\mathfrak{e}1}+b_{\mathfrak{e}2}}}{5L^2}\sum_{m=0}^{2}(-1)^m\frac{\widetilde{P}^{b_{\mathfrak{e}1}+m-2}_{4,\mathrm{p}(\mathfrak{v}_1)}\widetilde{P}^{b_{\mathfrak{e}2}-m}_{4,\mathrm{p}(\mathfrak{v}_2)}}{\zeta^{(b_{\mathfrak{e}1}+m+2)\mathrm{p}(\mathfrak{e}_1)}\zeta^{(b_{\mathfrak{e}2}-m+4)\mathrm{p}(\mathfrak{v}_2)}}.$$

Proof The strategy of proof is similar to that of Lemma 3.4.2. The only difference is that we use Lemma 3.4.3 instead of Lemma 3.4.2 and shift one of the summations by 3 rather than by 1. □

Theorem 3.5.6 (The First Holomorphic Anomaly Equation) *For $g \geq 2$, we have*

$$\frac{C_3}{5L}\frac{\partial \mathcal{F}_g^{[\mathbb{C}^5/\mathbb{Z}_5]}}{\partial A_2} = \frac{1}{2}\mathcal{F}_{g-1,2}^{[\mathbb{C}^5/\mathbb{Z}_5]}(\phi_2,\phi_2) + \frac{1}{2}\sum_{i=1}^{g-1}\mathcal{F}_{g-i,1}^{[\mathbb{C}^5/\mathbb{Z}_5]}(\phi_2)\,\mathcal{F}_{i,1}^{[\mathbb{C}^5/\mathbb{Z}_5]}(\phi_2)$$

in $\mathbb{F}[C_1, C_2, C_3]$.

Proof Let $\Gamma \in \mathrm{G}_{g,0}^{\mathrm{Dec}}(5)$ be a decorated graph and $\tilde{\mathfrak{e}} \in \mathrm{E}_\Gamma$ be an edge of Γ connecting two vertices $\mathfrak{v}_1$ and $\mathfrak{v}_2$. After deleting the edge $\tilde{\mathfrak{e}}$, we obtain a new graph. (By deleting, we mean breaking the edge $\tilde{\mathfrak{e}}$ into two legs $\mathfrak{l}_{\tilde{\mathfrak{e}}}$ and $\mathfrak{l}'_{\tilde{\mathfrak{e}}}$.) There are two possibilities for the resulting graph after deletion of edge $\mathfrak{e}$:

(i) If it is connected, then we obtain an element of $\mathrm{G}_{g-1,2}^{\mathrm{Dec}}(5)$, which we denote as $\Gamma_{\tilde{\mathfrak{e}}}^0$.
(ii) If it is disconnected, then the resulting graph has two connected components, which we denote as $\Gamma_{\tilde{\mathfrak{e}}}^1 \in \mathrm{G}_{g_1,1}^{\mathrm{Dec}}(5)$ and $\Gamma_{\tilde{\mathfrak{e}}}^2 \in \mathrm{G}_{g_2,1}^{\mathrm{Dec}}(5)$ where we have $g = g_1 + g_2$.

By Proposition 3.5.1 and Lemma 3.5.4, we observe that

$$\begin{aligned}\frac{\partial \mathrm{Cont}_\Gamma^{\mathrm{A}}(\tilde{\mathfrak{e}})}{\partial A_2} =& \frac{(-1)^{b_{\tilde{\mathfrak{e}}1}+b_{\tilde{\mathfrak{e}}2}}}{5}\frac{\widetilde{P}_{3,\mathrm{p}(\mathfrak{v}_1)}^{b_{\tilde{\mathfrak{e}}1}}\widetilde{P}_{3,\mathrm{p}(\mathfrak{v}_2)}^{b_{\tilde{\mathfrak{e}}2}}}{\zeta^{(b_{\tilde{\mathfrak{e}}1}+3)\mathrm{p}(\mathfrak{v}_1)}\zeta^{(b_{\tilde{\mathfrak{e}}2}+3)\mathrm{p}(\mathfrak{v}_2)}}\\ =&5\left(\frac{L^3}{K_3}\right)^2\begin{cases}\mathrm{Cont}_{\Gamma_{\tilde{\mathfrak{e}}}^0}^{\mathrm{A}}(\mathfrak{l}_{\tilde{\mathfrak{e}}})\mathrm{Cont}_{\Gamma_{\tilde{\mathfrak{e}}}^0}^{\mathrm{A}}(\mathfrak{l}'_{\tilde{\mathfrak{e}}}) & \text{for the case (i),}\\ \mathrm{Cont}_{\Gamma_{\tilde{\mathfrak{e}}}^1}^{\mathrm{A}}(\mathfrak{l}_{\tilde{\mathfrak{e}}})\mathrm{Cont}_{\Gamma_{\tilde{\mathfrak{e}}}^2}^{\mathrm{A}}(\mathfrak{l}'_{\tilde{\mathfrak{e}}}) & \text{for the case (ii)}\end{cases}\end{aligned}$$

with $\ell(\mathfrak{l}_{\tilde{\mathfrak{e}}}) = \ell(\mathfrak{l}'_{\tilde{\mathfrak{e}}}) = 2$, i.e., with insertions ϕ_2 .

By definition of K_3 and Eq. (3.2.7), we also note that

$$\left(\frac{L^3}{K_3}\right)^2 = \frac{L}{C_3}.$$

Then, for case (i), we easily see that we have

$$\begin{aligned}&\mathrm{Cont}_{\Gamma_{\tilde{\mathfrak{e}}}^0}(\phi_2,\phi_2)\\ &= \frac{1}{|\mathrm{Aut}(\Gamma_{\tilde{\mathfrak{e}}}^{0,\mathrm{St}})|}\sum_{\mathrm{A}\in\mathbb{Z}_{\geq 0}^{\mathrm{F}(\Gamma_{\tilde{\mathfrak{e}}}^0)}}\prod_{\mathfrak{v}\in\mathrm{V}_{\Gamma_{\tilde{\mathfrak{e}}}^0}}\mathrm{Cont}_{\Gamma_{\tilde{\mathfrak{e}}}^0}^{\mathrm{A}}(\mathfrak{v})\prod_{\mathfrak{e}\in\mathrm{E}_{\Gamma_{\tilde{\mathfrak{e}}}^0}}\mathrm{Cont}_{\Gamma_{\tilde{\mathfrak{e}}}^0}^{\mathrm{A}}(\mathfrak{e})\prod_{\mathfrak{l}\in\mathrm{L}_{\Gamma_{\tilde{\mathfrak{e}}}^0}}\mathrm{Cont}_{\Gamma_{\tilde{\mathfrak{e}}}^0}^{\mathrm{A}}(\mathfrak{l})\\ &= \frac{1}{|\mathrm{Aut}(\Gamma_{\tilde{\mathfrak{e}}}^{0,\mathrm{St}})|}\sum_{\mathrm{A}\in\mathbb{Z}_{\geq 0}^{\mathrm{F}(\Gamma)}}\frac{C_3}{5L}\frac{\partial \mathrm{Cont}_\Gamma^{\mathrm{A}}(\tilde{\mathfrak{e}})}{\partial A_2}\prod_{\mathfrak{v}\in\mathrm{V}_\Gamma}\mathrm{Cont}_\Gamma^{\mathrm{A}}(\mathfrak{v})\prod_{\substack{\mathfrak{e}\in\mathrm{E}_\Gamma\\ \mathfrak{e}\neq\tilde{\mathfrak{e}}}}\mathrm{Cont}_\Gamma^{\mathrm{A}}(\mathfrak{e}).\end{aligned} \tag{3.5.3}$$

Similarly, for case (ii), we observe the following

$$\begin{aligned}
&\mathrm{Cont}_{\Gamma^1_{\tilde{\mathfrak{e}}}}(\phi_2)\,\mathrm{Cont}_{\Gamma^2_{\tilde{\mathfrak{e}}}}(\phi_2)\\
&=\frac{1}{|\mathrm{Aut}(\Gamma^{1,\mathrm{St}}_{\tilde{\mathfrak{e}}})|}\sum_{\mathrm{A}\in\mathbb{Z}^{\mathrm{F}(\Gamma^1_{\tilde{\mathfrak{e}}})}_{\geq 0}}\mathrm{Cont}^{\mathrm{A}}_{\Gamma^1_{\tilde{\mathfrak{e}}}}(\mathfrak{l}_{\tilde{\mathfrak{e}}})\prod_{\mathfrak{v}\in\mathrm{V}_{\Gamma^1_{\tilde{\mathfrak{e}}}}}\mathrm{Cont}^{\mathrm{A}}_{\Gamma^1_{\tilde{\mathfrak{e}}}}(\mathfrak{v})\prod_{\mathfrak{e}\in\mathrm{E}_{\Gamma^1_{\tilde{\mathfrak{e}}}}}\mathrm{Cont}^{\mathrm{A}}_{\Gamma^1_{\tilde{\mathfrak{e}}}}(\mathfrak{e})\\
&\times\frac{1}{|\mathrm{Aut}(\Gamma^{2,\mathrm{St}}_{\tilde{\mathfrak{e}}})|}\sum_{\mathrm{A}\in\mathbb{Z}^{\mathrm{F}(\Gamma^2_{\tilde{\mathfrak{e}}})}_{\geq 0}}\mathrm{Cont}^{\mathrm{A}}_{\Gamma^2_{\tilde{\mathfrak{e}}}}(\mathfrak{l}'_{\tilde{\mathfrak{e}}})\prod_{\upsilon\in\mathrm{V}_{\Gamma^2_{\tilde{\mathfrak{e}}}}}\mathrm{Cont}^{\mathrm{A}}_{\Gamma^2_{\tilde{\mathfrak{e}}}}(\mathfrak{v})\prod_{\mathfrak{e}\in\mathrm{E}_{\Gamma^2_{\tilde{\mathfrak{e}}}}}\mathrm{Cont}^{\mathrm{A}}_{\Gamma^2_{\tilde{\mathfrak{e}}}}(\mathfrak{e})\\
&=\frac{1}{|\mathrm{Aut}(\Gamma^{1,\mathrm{St}}_{\tilde{\mathfrak{e}}})||\mathrm{Aut}(\Gamma^{,\mathrm{St}}_{\tilde{\mathfrak{e}}})|}\sum_{\mathrm{A}\in\mathbb{Z}^{\mathrm{F}(\Gamma)}_{\geq 0}}\frac{C_3}{5L}\frac{\partial\mathrm{Cont}^{\mathrm{A}}_{\Gamma}(\tilde{\mathfrak{e}})}{\partial A_2}\prod_{\mathfrak{v}\in\mathrm{V}_{\Gamma}}\mathrm{Cont}^{\mathrm{A}}_{\Gamma}(\mathfrak{v})\prod_{\substack{\mathfrak{e}\in\mathrm{E}_{\Gamma}\\ \mathfrak{e}\neq\tilde{\mathfrak{e}}}}\mathrm{Cont}^{\mathrm{A}}_{\Gamma}(\mathfrak{e}).
\end{aligned}\tag{3.5.4}$$

By Lemma 3.3.1 and Theorem 3.5.2, we have the following vanishing result:

$$\frac{\partial\mathrm{Cont}^{\mathrm{A}}_{\Gamma}(\mathfrak{v})}{\partial A_2}=0$$

for any vertex $\mathfrak{v}\in\mathrm{V}_{\Gamma}$.

Then, this vanishing result gives us the following

$$\begin{aligned}
\frac{\partial\mathrm{Cont}_{\Gamma}}{\partial A_2}&=\frac{1}{|\mathrm{Aut}(\Gamma^{\mathrm{St}})|}\sum_{\mathrm{A}\in\mathbb{Z}^{\mathrm{F}(\Gamma)}_{\geq 0}}\prod_{\mathfrak{v}\in\mathrm{V}_{\Gamma}}\mathrm{Cont}^{\mathrm{A}}_{\Gamma}(\mathfrak{v})\frac{\partial}{\partial A_2}\left(\prod_{\mathfrak{e}\in\mathrm{E}_{\Gamma}}\mathrm{Cont}^{\mathrm{A}}_{\Gamma}(\mathfrak{e})\right)\\
&=\sum_{\tilde{\mathfrak{e}}\in\mathrm{E}_{\Gamma}}\frac{1}{|\mathrm{Aut}(\Gamma^{\mathrm{St}})|}\sum_{\mathrm{A}\in\mathbb{Z}^{\mathrm{F}(\Gamma)}_{\geq 0}}\frac{\partial\mathrm{Cont}^{\mathrm{A}}_{\Gamma}(\tilde{\mathfrak{e}})}{\partial A_2}\prod_{\mathfrak{v}\in\mathrm{V}_{\Gamma}}\mathrm{Cont}^{\mathrm{A}}_{\Gamma}(\mathfrak{v})\prod_{\substack{\mathfrak{e}\in\mathrm{E}_{\Gamma}\\ \mathfrak{e}\neq\tilde{\mathfrak{e}}}}\mathrm{Cont}^{\mathrm{A}}_{\Gamma}(\mathfrak{e}).
\end{aligned}$$

So, we have

$$\frac{C_3}{5L}\frac{\partial\mathrm{Cont}_{\Gamma}}{\partial A_2}=\sum_{\tilde{\mathfrak{e}}\in\mathrm{E}_{\Gamma}}\frac{1}{|\mathrm{Aut}(\Gamma^{\mathrm{St}})|}\sum_{\mathrm{A}\in\mathbb{Z}^{\mathrm{F}(\Gamma)}_{\geq 0}}\frac{C_3}{5L}\frac{\partial\mathrm{Cont}^{\mathrm{A}}_{\Gamma}(\tilde{\mathfrak{e}})}{\partial A_2}\prod_{\mathfrak{v}\in\mathrm{V}_{\Gamma}}\mathrm{Cont}^{\mathrm{A}}_{\Gamma}(\mathfrak{v})\prod_{\substack{\mathfrak{e}\in\mathrm{E}_{\Gamma}\\ \mathfrak{e}\neq\tilde{\mathfrak{e}}}}\mathrm{Cont}^{\mathrm{A}}_{\Gamma}(\mathfrak{e}).\tag{3.5.5}$$

Then, summing Eqs. (3.5.3) and (3.5.4) over all decorated stable graphs $\Gamma^0_{\tilde{\mathfrak{e}}}$ and $(\Gamma^1_{\tilde{\mathfrak{e}}},\Gamma^2_{\tilde{\mathfrak{e}}})$ we obtain

$$\langle\langle\phi_2,\phi_2\rangle\rangle^{[\mathbb{C}^5/\mathbb{Z}_5]}_{g-1,2}\quad\text{and}\quad\sum_{i=1}^{g-1}\langle\langle\phi_2\rangle\rangle^{[\mathbb{C}^5/\mathbb{Z}_5]}_{g-i,1}\langle\langle\phi_2\rangle\rangle^{[\mathbb{C}^5/\mathbb{Z}_5]}_{i,1}$$

respectively. Then, by Eq. (3.5.5) and an analysis of automorphism factors appearing in Eqs. (3.5.3) and (3.5.4), we conclude that

$$2\frac{C_3}{5L}\frac{\partial}{\partial A_2}\langle\langle\rangle\rangle_g^{[\mathbb{C}^5/\mathbb{Z}_5]} = \langle\langle\phi_2,\phi_2\rangle\rangle_{g-1,2}^{[\mathbb{C}^5/\mathbb{Z}_5]} + \sum_{i=1}^{g-1}\langle\langle\phi_2\rangle\rangle_{g-i,1}^{[\mathbb{C}^5/\mathbb{Z}_5]}\langle\langle\phi_2\rangle\rangle_{i,1}^{[\mathbb{C}^5/\mathbb{Z}_5]}$$

after summing over all decorated stable graphs Γ. The reason we have 2 in front of the left hand side is due to not having a canonical order of labelings of each of the legs $\mathfrak{l}_{\tilde{\mathfrak{e}}}$ and $\mathfrak{l}'_{\tilde{\mathfrak{e}}}$ for case (i) and double counting of different genera of connected components for case (ii). This completes the proof. □

Let π be the morphism to the moduli space of stable curves determined by the domain,

$$\pi : \overline{M}^{\mathrm{orb}}_{g,k}\left(\left[\mathbb{C}^5/\mathbb{Z}_5\right],0\right) \to \overline{M}_{g,k}.$$

We define the Gromov-Witten potentials with certain types of ancestor insertions by

$$\begin{aligned}&\mathcal{F}_{g,n}^{[\mathbb{C}^5/\mathbb{Z}_5]}\left(\phi_{c_1}\psi_1^{a_1},\ldots,\phi_{c_n}\psi_n^{a_n}\right)\\&=\sum_{d=0}^{\infty}\frac{\Theta^d}{d!}\int_{\left[\overline{M}^{\mathrm{orb}}_{g,n+d}([\mathbb{C}^5/\mathbb{Z}_5],0)\right]^{vir}}\prod_{i=1}^{n}\mathrm{ev}_i^*(\phi_{c_i})\pi^*(\psi_i^{a_i})\prod_{i=n+1}^{n+d}\mathrm{ev}_i^*(\phi_1).\end{aligned}$$

These Gromov-Witten potentials can also be written as a sum over decorated stables graphs as in Proposition 3.5.1. The only difference with Proposition 3.5.1 happens at vertex contributions and it is the shift in the powers of ψ classes according to the insertions. The rest is the same. Hence, we again conclude that

$$\mathcal{F}_{g,n}^{[\mathbb{C}^5/\mathbb{Z}_5]}\left(\phi_{c_1}\psi_1^{a_1},\ldots,\phi_{c_n}\psi_n^{a_n}\right) \in \mathbb{F}[C_1,C_2,C_3].$$

Now, we are ready to state another holomorphic anomaly equation.

Theorem 3.5.7 (The Second Holomorphic Anomaly Equation) *For $g \geq 2$, we have*

$$\begin{aligned}&\frac{C_2^2C_3}{5L^3}\frac{\partial\mathcal{F}_g^{[\mathbb{C}^5/\mathbb{Z}_5]}}{\partial(D^2A_1)}\\&=\mathcal{F}_{g-1,2}^{[\mathbb{C}^5/\mathbb{Z}_5]}\left(\phi_1\psi_1^2,\phi_1\right)+\sum_{i=1}^{g-1}\mathcal{F}_{g-i,1}^{[\mathbb{C}^5/\mathbb{Z}_5]}\left(\phi_1\psi_1^2\right)\mathcal{F}_{i,1}^{[\mathbb{C}^5/\mathbb{Z}_5]}(\phi_1)\\&\quad-\frac{1}{2}\mathcal{F}_{g-1,2}^{[\mathbb{C}^5/\mathbb{Z}_5]}(\phi_1\psi_1,\phi_1\psi_2)-\frac{1}{2}\sum_{i=1}^{g-1}\mathcal{F}_{g-i,1}^{[\mathbb{C}^5/\mathbb{Z}_5]}(\phi_1\psi_1)\mathcal{F}_{i,1}^{[\mathbb{C}^5/\mathbb{Z}_5]}(\phi_1\psi_1)\end{aligned}$$

in $\mathbb{F}[C_1,C_2,C_3]$.

Proof The proof is similar to that of Theorem 3.5.6 with some technical difference. Instead of giving full details, this time we point out these different technicalities. Throughout the proof, let Γ, $\tilde{\mathfrak{e}}$, $\mathfrak{v}_1$, $\mathfrak{v}_2$, $\mathfrak{l}_{\tilde{\mathfrak{e}}}$, $\mathfrak{l}'_{\tilde{\mathfrak{e}}}$, $\Gamma^0_{\tilde{\mathfrak{e}}}$, $\Gamma^1_{\tilde{\mathfrak{e}}}$, $\Gamma^2_{\tilde{\mathfrak{e}}}$, "case (i)" and "case (ii)" be as in the proof of Theorem 3.5.6.

By Proposition 3.5.1 and Lemma 3.5.5, we have

$$\begin{aligned}\frac{\partial}{\partial(D^2A_1)}\mathrm{Cont}^{\mathrm{A}}_{\Gamma}(\mathfrak{e}) =&\frac{(-1)^{b_{\tilde{\mathfrak{e}}1}+b_{\tilde{\mathfrak{e}}2}}}{5L^2}\frac{\widetilde{P}^{b_{\tilde{\mathfrak{e}}1}-2}_{4,\mathrm{p}(\mathfrak{v}_1)}\widetilde{P}^{b_{\tilde{\mathfrak{e}}2}}_{4,\mathrm{p}(\mathfrak{v}_2)}}{\zeta^{(b_{\tilde{\mathfrak{e}}1}+2)\mathrm{p}(\tilde{\mathfrak{e}}_1)}\zeta^{(b_{\tilde{\mathfrak{e}}2}+4)\mathrm{p}(\mathfrak{v}_2)}}\\ &-\frac{(-1)^{b_{\tilde{\mathfrak{e}}1}+b_{\tilde{\mathfrak{e}}2}}}{5L^2}\frac{\widetilde{P}^{b_{\tilde{\mathfrak{e}}1}-1}_{4,\mathrm{p}(\mathfrak{v}_1)}\widetilde{P}^{b_{\tilde{\mathfrak{e}}2}-1}_{4,\mathrm{p}(\mathfrak{v}_2)}}{\zeta^{(b_{\tilde{\mathfrak{e}}1}+3)\mathrm{p}(\tilde{\mathfrak{e}}_1)}\zeta^{(b_{\tilde{\mathfrak{e}}2}+3)\mathrm{p}(\mathfrak{v}_2)}}\\ &+\frac{(-1)^{b_{\tilde{\mathfrak{e}}1}+b_{\tilde{\mathfrak{e}}2}}}{5L^2}\frac{\widetilde{P}^{b_{\tilde{\mathfrak{e}}1}}_{4,\mathrm{p}(\mathfrak{v}_1)}\widetilde{P}^{b_{\tilde{\mathfrak{e}}2}-2}_{4,\mathrm{p}(\mathfrak{v}_2)}}{\zeta^{(b_{\tilde{\mathfrak{e}}1}+4)\mathrm{p}(\tilde{\mathfrak{e}}_1)}\zeta^{(b_{\tilde{\mathfrak{e}}2}+2)\mathrm{p}(\mathfrak{v}_2)}},\end{aligned} \tag{3.5.6}$$

where right hand side of this equation is equal to

$$\begin{aligned}&5\left(\frac{L^4}{K_4}\right)^2\Big(\mathrm{Cont}^{\mathrm{A}_{b_{\tilde{\mathfrak{e}}1}-2}}_{\Gamma^0_{\tilde{\mathfrak{e}}}}(\mathfrak{l}_{\tilde{\mathfrak{e}}})\mathrm{Cont}^{\mathrm{A}}_{\Gamma^0_{\tilde{\mathfrak{e}}}}(\mathfrak{l}'_{\tilde{\mathfrak{e}}})-\mathrm{Cont}^{\mathrm{A}_{b_{\tilde{\mathfrak{e}}1}-1}}_{\Gamma^0_{\tilde{\mathfrak{e}}}}(\mathfrak{l}_{\tilde{\mathfrak{e}}})\mathrm{Cont}^{\mathrm{A}_{b_{\tilde{\mathfrak{e}}2}-1}}_{\Gamma^0_{\tilde{\mathfrak{e}}}}(\mathfrak{l}'_{\tilde{\mathfrak{e}}})\\ &\quad+\mathrm{Cont}^{\mathrm{A}}_{\Gamma^0_{\tilde{\mathfrak{e}}}}(\mathfrak{l}_{\tilde{\mathfrak{e}}})\mathrm{Cont}^{\mathrm{A}_{b_{\tilde{\mathfrak{e}}2}-2}}_{\Gamma^0_{\tilde{\mathfrak{e}}}}(\mathfrak{l}'_{\tilde{\mathfrak{e}}})\Big)\end{aligned}$$

for the case (i), and it is equal to

$$\begin{aligned}&5\left(\frac{L^4}{K_4}\right)^2\Big(\mathrm{Cont}^{\mathrm{A}_{b_{\tilde{\mathfrak{e}}1}-2}}_{\Gamma^1_{\tilde{\mathfrak{e}}}}(\mathfrak{l}_{\tilde{\mathfrak{e}}})\mathrm{Cont}^{\mathrm{A}}_{\Gamma^2_{\tilde{\mathfrak{e}}}}(\mathfrak{l}'_{\tilde{\mathfrak{e}}})-\mathrm{Cont}^{\mathrm{A}_{b_{\tilde{\mathfrak{e}}1}-1}}_{\Gamma^1_{\tilde{\mathfrak{e}}}}(\mathfrak{l}_{\tilde{\mathfrak{e}}})\mathrm{Cont}^{\mathrm{A}_{b_{\tilde{\mathfrak{e}}2}-1}}_{\Gamma^2_{\tilde{\mathfrak{e}}}}(\mathfrak{l}'_{\tilde{\mathfrak{e}}})\\ &\quad+\mathrm{Cont}^{\mathrm{A}}_{\Gamma^1_{\tilde{\mathfrak{e}}}}(\mathfrak{l}_{\tilde{\mathfrak{e}}})\mathrm{Cont}^{\mathrm{A}_{b_{\tilde{\mathfrak{e}}2}-2}}_{\Gamma^2_{\tilde{\mathfrak{e}}}}(\mathfrak{l}'_{\tilde{\mathfrak{e}}})\Big)\end{aligned}$$

for the case (ii), with $\ell(\mathfrak{l}_{\tilde{\mathfrak{e}}})=\ell(\mathfrak{l}'_{\tilde{\mathfrak{e}}})=1$. Here, by $\mathrm{A}_{b_{\tilde{\mathfrak{e}}s}-r}$ we mean the flag value $b_{\tilde{\mathfrak{e}}s}$ is shifted by r in $\mathrm{A}\in\mathbb{Z}^{\mathrm{F}(\bullet)}_{\geq 0}$ where $\bullet$ is $\Gamma^0_{\tilde{\mathfrak{e}}}$, $\Gamma^1_{\tilde{\mathfrak{e}}}$ or $\Gamma^2_{\tilde{\mathfrak{e}}}$. Also note that, by definition of K_4 and Eq. (3.2.7), we have

$$\left(\frac{L^4}{K_4}\right)^2=\frac{L^3}{C_2^2C_3}.$$

By shifting the flag values by r in Eq. (3.5.6), we get

$$\begin{aligned}&\frac{(-1)^{b_{\tilde{\mathfrak{e}}1}+b_{\tilde{\mathfrak{e}}2}}}{5L^2}\left(\frac{\widetilde{P}^{b_{\tilde{\mathfrak{e}}1}}_{4,\mathrm{p}(\mathfrak{v}_1)}\widetilde{P}^{b_{\tilde{\mathfrak{e}}2}}_{4,\mathrm{p}(\mathfrak{v}_2)}}{\zeta^{(b_{\tilde{\mathfrak{e}}1}+4)\mathrm{p}(\tilde{\mathfrak{e}}_1)}\zeta^{(b_{\tilde{\mathfrak{e}}2}+4)\mathrm{p}(\mathfrak{v}_2)}}-\frac{\widetilde{P}^{b_{\tilde{\mathfrak{e}}1}}_{4,\mathrm{p}(\mathfrak{v}_1)}\widetilde{P}^{b_{\tilde{\mathfrak{e}}2}}_{4,\mathrm{p}(\mathfrak{v}_2)}}{\zeta^{(b_{\tilde{\mathfrak{e}}1}+4)\mathrm{p}(\tilde{\mathfrak{e}}_1)}\zeta^{(b_{\tilde{\mathfrak{e}}2}+4)\mathrm{p}(\mathfrak{v}_2)}}\right.\\ &\quad\left.+\frac{\widetilde{P}^{b_{\tilde{\mathfrak{e}}1}}_{4,\mathrm{p}(\mathfrak{v}_1)}\widetilde{P}^{b_{\tilde{\mathfrak{e}}2}}_{4,\mathrm{p}(\mathfrak{v}_2)}}{\zeta^{(b_{\tilde{\mathfrak{e}}1}+4)\mathrm{p}(\tilde{\mathfrak{e}}_1)}\zeta^{(b_{\tilde{\mathfrak{e}}2}+4)\mathrm{p}(\mathfrak{v}_2)}}\right),\end{aligned} \tag{3.5.7}$$

where r is 2 in the first term for $b_{\tilde{\mathfrak{e}}1}$, r is 1 in the second term for $b_{\tilde{\mathfrak{e}}1}$, $b_{\tilde{\mathfrak{e}}2}$, and r is 2 in the last term for $b_{\tilde{\mathfrak{e}}2}$. If we make these shifts after taking the derivative of a graph contribution with respect to D^2A_1, they will also effect the contributions of vertices that $\tilde{\mathfrak{e}}$ is attached to. More precisely, they will also shift the powers of the corresponding ψ classes appearing in the contributions of the vertices that $\tilde{\mathfrak{e}}$ is attached to.

If we adapt of the proof of Theorem 3.5.6 at this point, then the first and last terms in Eq. (3.5.7) explain the terms

$$\mathcal{F}_{g-1,2}^{[\mathbb{C}^5/\mathbb{Z}_5]}\left(\phi_1\psi_1^2,\phi_1\right)+\sum_{i=1}^{g-1}\mathcal{F}_{g-i,1}^{[\mathbb{C}^5/\mathbb{Z}_5]}\left(\phi_1\psi_1^2\right)\mathcal{F}_{i,1}^{[\mathbb{C}^5/\mathbb{Z}_5]}(\phi_1)$$

in the holomorhic anomaly equation and the middle term in Eq. (3.5.7) explains the terms

$$-\frac{1}{2}\mathcal{F}_{g-1,2}^{[\mathbb{C}^5/\mathbb{Z}_5]}(\phi_1\psi_1,\phi_1\psi_2)-\frac{1}{2}\sum_{i=1}^{g-1}\mathcal{F}_{g-i,1}^{[\mathbb{C}^5/\mathbb{Z}_5]}(\phi_1\psi_1)\mathcal{F}_{i,1}^{[\mathbb{C}^5/\mathbb{Z}_5]}(\phi_1\psi_1)$$

in the holomorphic anomaly equation. □

Acknowledgments D. G. is supported in part by a Special Graduate Assignment fellowship by Ohio State University's Department of Mathematics and H.-H. T. is supported in part by Simons foundation collaboration grant.

References

1. Abramovich, D., Graber, T., Vistoli, A.: Gromov-Witten theory of Deligne-Mumford stacks. Am. J. Math. **130**(5), 1337–1398 (2008)
2. Adem, A., Leida, J., Ruan, Y.: Orbifolds and Stringy Topology. Cambridge Tracts in Mathematics. Cambridge University Press, Cambridge (2007)
3. Chen, W., Ruan, Y.: A new cohomology theory of orbifold. Commun. Math. Phys. **248**, 1–31 (2004)
4. Coates, T., Corti, A., Iritani, H., Tseng, H.H.: Computing Genus-Zero twisted Gromov-Witten invariants. Duke Math. J. **147**(3), 377–438 (2009)
5. Givental, A.: Elliptic Gromov-Witten invariants and the generalized mirror conjecture. In: Integrable Systems and Algebraic Geometry (Kobe/Kyoto, 1997), pp. 107–155. World Scientific, River Edge (1998)
6. Givental, A.: Symplectic geometry of Frobenius structures. In: Frobenius manifolds. Aspects of Mathematics, vol. E36, pp. 91–112. Friedr. Vieweg, Wiesbaden (2004)
7. Graber, T., Pandharipande, R.: Localization of virtual classes. Invent. Math. **135**, 487–518 (1999)
8. Kock, J.L Notes on PSI clases. https://mat.uab.cat/~kock/GW/notes/psi-notes.pdf
9. Kontsevich, M., Manin, Y.: Gromov-Witten classes, quantum cohomology, and enumerative geometry. Commun. Math. Phys. **164**, 525–562 (1994)

10. Lee, Y.P., Pandharipande, R.: Frobenius manifolds, Gromov-Witten theory, and Virasoro constraints. Manuscript available from the authors' websites
11. Lho, H.: Crepant resolution conjecture for $\mathbb{C}^5/\mathbb{Z}_5$ (2017). arXiv:1707.02910
12. Lho, H., Pandharipande, R.: Stable quotients and the holomorphic anomaly equation. Adv. Math. **332**, 349–402 (2018)
13. Lho, H., Pandharipande, R.: Crepant resolution and the holomorphic anomaly equation for $[\mathbb{C}^3/\mathbb{Z}_3]$. Proc. Lond. Math. Soc. **119**, 781–813 (2019)
14. Pandharipande, R.: Cohomological field theory calculations. In: Proceedings of the ICM (Rio de Janeiro 2018), vol. 1, pp. 869–898. World Scientific, Hackensack (2018)
15. Pandharipande, R., Tseng, H.H.: Higher genus Gromov-Witten theory of $\mathsf{Hilb}^n(\mathbb{C}^2)$ and CohFTs associated to local curves. In: Forum of Mathematics, Pi, vol. 7, e4, 63pp. Cambridge University Press (2019). arXiv:1707.01406
16. Pandharipande, R., Pixton, A., Zvonkine, D.: Relations on $\overline{M}_{g,n}$ via 3-spin structures. J. Am. Math. Soc. **28**, 279–309 (2015)
17. Teleman, C.: The structure of 2D semi-simple field theories. Invent. Math. **188**, 525–588 (2012)
18. Tseng, H.H.: Orbifold quantum Riemann-Roch, Lefschetz and Serre. Geom. Topol. **14**, 1–81 (2010)
19. Zagier, D., Zinger, A.: Some properties of hypergeometric series associated with mirror symmetry. In: Modular Forms and String Duality. Fields Institute Communications, vol. 54, pp. 163–177. AMS, Providence (2008)

Open Access This chapter is licensed under the terms of the Creative Commons Attribution-NonCommercial-NoDerivatives 4.0 International License (http://creativecommons.org/licenses/by-nc-nd/4.0/), which permits any noncommercial use, sharing, distribution and reproduction in any medium or format, as long as you give appropriate credit to the original author(s) and the source, provide a link to the Creative Commons license and indicate if you modified the licensed material. You do not have permission under this license to share adapted material derived from this chapter or parts of it.

The images or other third party material in this chapter are included in the chapter's Creative Commons license, unless indicated otherwise in a credit line to the material. If material is not included in the chapter's Creative Commons license and your intended use is not permitted by statutory regulation or exceeds the permitted use, you will need to obtain permission directly from the copyright holder.

Chapter 4
Permutation-Equivariant Quantum K-Theory IX: Quantum Hirzebruch-Riemann-Roch in All Genera

Alexander Givental

To the memory of my student Bumsig Kim

Abstract We introduce the most general to date version of the permutation-equivariant quantum K-theory, and express its total descendant potential in terms of cohomological Gromov-Witten invariants. This is the higher-genus analogue of adelic characterization (Givental–Tonita The Hirzebruch-Riemann–Roch Theorem in True Genus-0 Quantum K-Theory. Noncommutative Geometry, vol. 62, pp. 43–91. MSRI Publication, Berkeley (2014). arXiv:1106.3136), and is based on the application of the Kawasaki-Riemann-Roch formula (Kawasaki, Osaka J. Math. **16**(1), 151–159 (1979)) to moduli spaces of stable maps.

4.1 Introduction

Cohomological Gromov-Witten invariants of a compact Kähler manifold X are defined as various intersection numbers in moduli spaces of stable maps, denoted here $X_{g,n,d}$ with g, n, d standing for the genus, number of marked points, and degree of the maps. The K-theoretic counterpart of GW-theory studies holomorphic Euler characteristics of appropriate vector bundles over the moduli spaces. The action of permutations of the marked points on the sheaf cohomology of such bundles leads to the refined version of the theory, which we call permutation-equivariant. In genus 0, a complete description of K-theoretic GW-invariants in terms of cohomological ones was obtained in [7], and then applied to the permutation-equivariant theory in the previous papers of the present series (see Part III [8] or Part VII [9].

Conceptually the cohomological description of K-theoretic invariants is based on Kawasaki's version of Hirzebruch–Riemann–Roch formula [12] (or more precisely,

A. Givental (✉)
University of California, Berkeley, CA, USA
e-mail: givental@math.berkeley.edu

© The Author(s) 2026

N.-G. Kang et al. (eds.), *Categorical and Enumerative Aspects of Mirror Symmetry*,
KIAS Springer Series in Mathematics 5,
https://doi.org/10.1007/978-981-95-0385-8_4

its virtual variant [16]) applied on the moduli spaces $X_{g,n,d}$. An early version of this approach to the higher genus problem is used in the preprint [18] by V. Tonita. I am thankful to him for numerous discussions and corrections.[1]

As it was found in [7], in genus 0 the solution can be described in the form of *adelic characterization*. Roughly speaking, genus-0 K-theoretic GW-invariants of X are encoded by a certain Lagrangian cone in a symplectic space whose elements are rational functions in one complex variable, q, with vector values in $K^0(X)$. The adelic characterization says that a rational function lies in the cone if and only if the Laurent series expansion of it at each root of unity $q = \zeta$ passes a certain test. Namely, the expansion (as an element in the symplectic space of Laurent series with coefficients in $K^0(X)$) should represent certain cohomological GW-invariants of the *orbifold* target space $X/\mathbb{Z}_M$, where M is the order of ζ as a root of unity.

This paper establishes the higher genus version of adelic characterization. It involves quantization of the aforementioned symplectic formalism. In this Introduction, we don't give a complete formulation of the ultimate theorem (because it requires so many poorly motivated ingredients and notations, that the resulting formula, we fear, would become incomprehensible), but merely outline the quantum-mechanical structure of the adelic formula relating K-theoretic GW-invariants with cohomological ones.

A thorough definition of the permutation-equivariant GW-invariants and of the appropriate generating functions will be given in Sect. 4.2. In Sect. 4.3, we sketch the geometric machinery which shows, in principle, how to reduce the computation of K-theoretic to cohomological GW-invariants. In Sects. 4.4 and 4.5, we describe the language of symplectic loop spaces and their Fock spaces where various generating functions for GW-invariants live. Using this language, we will accurately build the ingredients of the ultimate formula starting from cohomological GW-invariants. The remaining details of the proof will be provided in Sects. 4.6–4.10.

By definition, permutation-equivariant K-theoretic GW-invariants take values in a ground coefficient ring, Λ, which is a λ-algebra, i.e. is equipped with the action of Adams operations $\Psi^r : \Lambda \to \Lambda$, $r = 1, 2, 3, \dots$, which are ring homomorphisms from Λ to itself, and satisfy $\Psi^1 = \text{id}$, $\Psi^r \Psi^s = \Psi^{rs}$.

The *total descendant potential* $\mathcal{D}_X$ for permutation-equivariant GW-invariants of X is defined (in Sect. 4.2) as a Λ-valued function of a sequence $\mathbf{t} = (\mathbf{t}_1, \mathbf{t}_2, \dots, \mathbf{t}_r, \dots)$ of Laurent polynomials[2] in q with vector coefficients in $K^0(X) \otimes \Lambda$. It also depends on the "Planck constant" $\hbar$, and can be interpreted as an element of the Fock space associated with a certain symplectic space $(\mathcal{K}^\infty, \Omega^\infty)$.

[1] I am also thankful to Irit Huq-Kuruvilla for helping me fix several errors, and to Jeongseok Oh for correcting numerous typos.

[2] Foreshadowing the definition let us mention here that $\mathbf{t}_r$ will be used as the input in the correlators of permutation-equivariant quantum K-theory at those marked points which belong to cycles of length r in the cycle structure of the permutation.

Namely, put $K := K^0(X) \otimes \Lambda$, and consider the space $\mathcal{K}$ or rational K-valued functions of q which are allowed to have poles only at $q = 0, \infty$, or at roots of unity. Equip $\mathcal{K}$ with the Λ-valued symplectic form

$$\Omega(\mathbf{f}, \mathbf{g}) := -\left[\mathrm{Res}_{q=0} + \mathrm{Res}_{q=\infty}\right] (\mathbf{f}(q^{-1}), \mathbf{g}(q)) \frac{dq}{q},$$

where $(a, b) := \chi(X; a \otimes b)$ is the K-theoretic Poincaré pairing on K, and with the Lagrangian polarization $\mathcal{K} = \mathcal{K}_+ \oplus \mathcal{K}_-$, where

$$\mathcal{K}_+ := K[q, q^{-1}], \quad \mathcal{K}_- := \{\mathbf{f} \in \mathcal{K} \mid \mathbf{f}(\infty) = 0, \ \mathbf{f}(0) \neq \infty\}.$$

By definition, $\mathcal{K}^\infty$ consists of sequences $\mathbf{f} = (\mathbf{f}_1, \mathbf{f}_2, \ldots, \mathbf{f}_r, \ldots)$ of elements of $\mathcal{K}$. It is equipped with the symplectic form

$$\Omega^\infty(\mathbf{f}, \mathbf{g}) := \sum_{r=1}^{\infty} \frac{\Psi^r}{r} \Omega(\mathbf{f}_r, \mathbf{g}_r),$$

and Lagrangian polarization $\mathcal{K}_\pm^\infty = \{\mathbf{f} = (\mathbf{f}_1, \mathbf{f}_2, \ldots) \mid \forall r, \mathbf{f}_r \in \mathcal{K}_\pm\}$. The total descendant potential $\mathcal{D}_X$, which is naturally a function of $\mathbf{t} = (\mathbf{t}_1, \mathbf{t}_2, \ldots) \in \mathcal{K}_+^\infty$ (depending on the parameter $\hbar$), is considered as a function on $\mathcal{K}^\infty$ constant in the direction of $\mathcal{K}_-^\infty$, and in this capacity is interpreted as a "quantum state", $\langle \mathcal{D}_X \rangle$, an element of the Fock space associated with $(\mathcal{K}^\infty, \Omega^\infty)$.

On the cohomological side, for each $M = 1, 2, 3, \ldots$, let $\mathbb{Z}_M = \mathbb{Z}/M\mathbb{Z}$ denote the cyclic group of order M, and $\mathbb{C}^{M-1} = \mathbb{C}[\mathbb{Z}_M]/\mathbb{C}$ be the quotient of the regular representation of $\mathbb{Z}_M$ by the trivial one. Over the global quotient orbifold $X/\mathbb{Z}_M$ (where the action of $\mathbb{Z}_M$ is trivial), introduce the orbibundle $T_X \otimes \mathbb{C}^{M-1}$, and denote by E_M its total (orbi)space. What we need is a certain twisted cohomological GW-theory of $X/\mathbb{Z}_M$, which can be interpreted as the fake quantum K-theory[3] of the non-compact orbifold E_M. Denote by $\mathcal{D}^{tw}_{X/\mathbb{Z}_M}$ the total descendant potential of such a theory. Using a series of "quantum Riemann-Roch theorems" available in the literature (see [3, 4, 11, 17, 19, 20]), it will be shown in Sects. 4.7 and 4.8 how to link this generating function directly to the total descendant potential $\mathcal{D}^H_X$ of the ordinary cohomological GW-theory of X. So we will assume here that all the functions $\mathcal{D}^{tw}_{X/\mathbb{Z}_M}$ are given.

Each $\mathcal{D}^{tw}_{X/\mathbb{Z}_M}$ can be considered as a quantum state, $\langle \mathcal{D}^{tw}_{X/\mathbb{Z}_M} \rangle$, an element of the Fock space associated with the appropriate symplectic space, $(\mathcal{K}^{tw}_{(M)}, \Omega^{tw}_{(M)})$. This

[3] In fake quantum K-theory, genuine holomorphic Euler characteristics of orbibundles over moduli spaces of stable maps are replaced with their fake versions: $\chi^{fake}(\mathcal{M}; V) := \int_{\mathcal{M}} \mathrm{ch}(V)\,\mathrm{td}(T_{\mathcal{M}})$, and are therefore cohomological in nature.

space is a direct sum of M *sectors* corresponding to Mth roots of unity ζ. Each sector is represented by the space $\mathcal{K}^{(\zeta)}$ isomorphic to the space $K((q-1))$ of vector-valued Laurent series in $q-1$. The symplectic form $\Omega^{tw}_{(M)}$ pairs $\mathcal{K}^{(\zeta)}$ with $\mathcal{K}^{(\zeta^{-1})}$ by the non-degenerate pairing

$$(f, g) \mapsto \frac{1}{M} \operatorname{Res}_{q=1}(f(q^{-1}), g(q))^{(r)} \frac{dq}{q}.$$

It is based on the twisted Poincaré pairing on K characterized by

$$(\Psi^r a, \Psi^r b)^{(r)} = r\Psi^r(a, b),$$

where $r = r(\zeta)$ equals the index of the subgroup generated by ζ in the multiplicative group of all Mth roots of unity.

Note that when M runs through all positive integers, each root of unity ζ of *primitive* order $m = m(\zeta)$ occurs among Mth roots of unity infinitely many times distinguished by the values of the index $r(\zeta) = M/m(\zeta) = 1, 2, 3 \ldots$. Consequently the direct sum $\oplus_{M=1}^{\infty} \mathcal{K}^{tw}_{(M)}$ can be rearranged according to the indices r into the *adelic space*

$$\underline{\mathcal{K}}^{\infty} := \oplus_{\text{roots of unity } \zeta} \oplus_{r=1}^{\infty} \mathcal{K}_r^{(\zeta)}$$

(here $\mathcal{K}_r^{(\zeta)}$ is the rth copy of $\mathcal{K}^{(\zeta)}$) with the symplectic form

$$\underline{\Omega}^{\infty}(\mathbf{f}, \mathbf{g}) = \sum_{\zeta} \frac{1}{m(\zeta)} \sum_{r=1}^{\infty} \frac{1}{r} \operatorname{Res}_{q=1}(f_r^{(\zeta)}(q^{-1}), g_r^{(\zeta^{-1})}(q))^{(r)} \frac{dq}{q}.$$

Thus, the *adelic tensor product*

$$\underline{\mathcal{D}}_X := \otimes_{M=1}^{\infty} \mathcal{D}^{tw}_{X/\mathbb{Z}_M}$$

can be considered as an element $\langle \underline{\mathcal{D}}_X \rangle$ in the Fock space associated with the adelic symplectic space.

We define the *adelic map* $\underline{\ } : \mathcal{K}^{\infty} \to \underline{\mathcal{K}}^{\infty}$ by

$$\mathbf{f} = (\mathbf{f}_1, \mathbf{f}_2, \ldots, \mathbf{f}_r, \ldots) \mapsto \underline{\mathbf{f}} = \{f_r^{(\zeta)}\} : \ f_r^{(\zeta)} := \Psi^r(\mathbf{f}_r(q^{1/m}/\zeta)),$$

where the last expression is to be expanded into a Laurent series near $q = 1$ after applying Adams' operations Ψ^r, acting naturally on $K = K^0(X) \otimes \Lambda$, and by

$\Psi^r(q) = q^r$ on functions of q. The residue theorem implies that the adelic map is symplectic:

$$\Omega^\infty(\mathbf{f}, \mathbf{g}) = \sum_{r=1}^{\infty} \frac{\Psi^r}{r} \sum_{\zeta} \mathrm{Res}_{q=\zeta} (\mathbf{f}_r(q^{-1}), \mathbf{g}_r(q)) \frac{dq}{q}$$

$$= \sum_{r=1}^{\infty} \frac{1}{r^2} \sum_{\zeta} \mathrm{Res}_{q=1} (\Psi^r(\mathbf{f}_r(q^{-1/m}\zeta)), \Psi^r(\mathbf{g}_r(q^{1/m}/\zeta))^{(r)} \frac{dq^{r/m}}{q^{r/m}}$$

$$= \sum_{\zeta} \frac{1}{m(\zeta)} \sum_{r=1}^{\infty} \frac{1}{r} \mathrm{Res}_{q=1} \, (\underline{\mathbf{f}}_r^{(\zeta)}(q^{-1}), \underline{\mathbf{g}}_r^{(\zeta^{-1})}(q))^{(r)} \frac{dq}{q}.$$

Our "higher genus quantum RR formula" can be stated this way.

Main Theorem *The adelic map* $\underline{\ \ } : (\mathcal{K}^\infty, \Omega^\infty) \to (\underline{\mathcal{K}}^\infty, \underline{\Omega}^\infty)$ *between the symplectic loop spaces transforms the adelic quantum state* $\langle \underline{\mathcal{D}}_X \rangle$ *into the total descendant potential* $\langle \mathcal{D}_X \rangle$ *of permutation-equivariant quantum K-theory of the target Kähler manifold X.*

How does a map between symplectic spaces map respective Fock spaces? Elements of the Fock space are functions on the symplectic space constant in the direction of the negative space of a chosen Lagrangian polarization. A map between symplectic spaces respecting the negative spaces of the chosen polarizations induces a map between the quotients, and hence maps the Fock spaces naturally (in the reverse direction). When the given polarizations disagree, one needs first to change one of them to identify the models of the Fock space based on different polarizations by the construction of Stone-von Neumann's theorem, and only after that apply the natural pull-back.

In the situation of our theorem, the polarizations disagree, and the precursory change of polarization in the adelic space is one of the key ingredients of the relation between $\mathcal{D}_X$ and $\underline{\mathcal{D}}_X$ as generating functions.

The space $\mathcal{K}_-^\infty$ consists of sequences $\mathbf{f} = (\mathbf{f}_1, \mathbf{f}_2, \dots, \mathbf{f}_r, \dots)$ of vector-values rational functions of q with poles at roots of unity ζ, but vanishing at $q = \infty$ and having no pole at $q = 0$. Such rational functions uniquely decompose into the sums of their partial fractions, $\mathbf{f}_r = \sum_\zeta \mathbf{f}_r^{(\zeta)}$, i.e. reduced rational functions of q with only one pole $q = \zeta$. In fact the negative space of polarization $\underline{\mathcal{K}}_-^\infty$ in the adelic space (we've neglected to describe it so far, but it is involved in the interpretation of the infinite product $\underline{\mathcal{D}}_X$ as an element of the Fock space) is exactly the direct sum of subspaces $\{\Psi^r(\mathbf{f}_r^{(\zeta)}(q^{1/m(\zeta)}/\zeta))\} \subset \mathcal{K}_r^{(\zeta)} = K((q-1))$ obtained from such partial fractions.

By the way, we encounter here an interesting phenomenon impossible in finite-dimensional symplectic geometry. The adelic map $\underline{\ \ } : \mathcal{K}^\infty \to \underline{\mathcal{K}}^\infty$ is a symplectic injection which embeds the Lagrangian subspace $\mathcal{K}_+^\infty$ into the much bigger Lagrangian subspace $\underline{\mathcal{K}}_+^\infty$, but it identifies the Lagrangian subspaces $\mathcal{K}_-^\infty$ and $\underline{\mathcal{K}}_-^\infty$ considered as quotient spaces $\mathcal{K}^\infty / \mathcal{K}_+^\infty$ and $\underline{\mathcal{K}}^\infty / \underline{\mathcal{K}}_+^\infty$.

At the same time, the image of $\mathcal{K}_-^{\infty}$ under the adelic map does not coincide with $\underline{\mathcal{K}}_-^{\infty}$, and it is now easy to understand why: the image of $\mathbf{f}_r^{(\zeta)}$ consists of the expansions of $\Psi^r(\mathbf{f}_r^{(\zeta)}(q^{1/m(\eta)}/\eta))$ for all roots of unity η, and not only for $\eta = \zeta$ where the partial fraction $\mathbf{f}_r^{(\zeta)}$ has its pole. Consequently, the relation between the quantum states $\langle \mathcal{D}_X \rangle$ and $\langle \underline{\mathcal{D}}_X \rangle$ described in the theorem actually means that the total descendant potential $\mathcal{D}_X$ is obtained from the infinite product $\underline{\mathcal{D}}_X$ as

$$\mathcal{D}_X = \text{pull-back by } \underline{\ \ } : \mathcal{K}_+^{\infty} \subset \underline{\mathcal{K}}_+^{\infty} \text{ of } e^{\frac{1}{2}\sum_r r\Psi^r(\hbar \sum \nabla_{\eta,\zeta})} \otimes_{M=1}^{\infty} \mathcal{D}^{tw}_{X/\mathbb{Z}_M}.$$

Here $\nabla_{\eta,\zeta}$ are certain 2nd order differential operators whose coefficients are tautologically determined by expansions of partial fractions with poles at roots of unity ζ into power series near all other roots of unity, while the embedding $\underline{\ \ } : \mathcal{K}_+^{\infty} \to \underline{\mathcal{K}}_+^{\infty}$ maps sequences $\mathbf{t} = (\mathbf{t}_1, \mathbf{t}_2, \ldots, \mathbf{t}_r \ldots)$ of Laurent polynomial $\mathbf{t}_r \in K[q, q^{-1}]$ into the collection of power series expansions $\Psi^r(\mathbf{t}_r(q^{1/m(\zeta)}/\zeta))$ of the Laurent polynomials at the roots of unity.

The above description of our main formula is neither complete not totally accurate, and should be supplemented with further clarifications.

1. The quantum state $\langle \mathcal{D}_X \rangle$ differs from the total descendant potential $\mathcal{D}_X$ (though both are functions on $\mathcal{K}^{\infty}/\mathcal{K}_-^{\infty} = \mathcal{K}_+^{\infty}$) by the translation of the origin called the *dilaton shift*: $\langle \mathcal{D}_X \rangle(\mathbf{v} + \mathbf{t}) = \mathcal{D}_X(\mathbf{t})$, where $\mathbf{v} = ((1-q)\mathbf{1}, (1-q)\mathbf{1}, \ldots)$, and $\mathbf{1}$ stands for the unit element in $K^0(X)$. Likewise, $\langle \mathcal{D}^{tw}_{X/\mathbb{Z}_M} \rangle((1-q)\mathbf{1} + t) = \mathcal{D}^{tw}_{X/\mathbb{Z}_M}(t)$. Here $\mathbf{1}$ belongs to the unit sector, i.e. among the components $t^{(\zeta)} \in K[[q-1]]$ labeled by the Mth roots of unity ζ only the component with $\zeta = 1$ is dilaton-shifted.
2. In the generating functions for GW-invariants, one weighs contributions of degree-d stable maps by the binomials Q^d in Novikov's variables $Q = (Q_1, \ldots, Q_r)$, where $r = \operatorname{rk} H_2(X, \mathbb{Z})$. Novikov's variables are adjoined to the ground λ-ring Λ so that $\Psi^r Q^d := Q^{rd}$. Furthermore, the expression "rational functions" ("Laurent series," "power series", etc.) of q should be understood as formal Q-series whose coefficients are rational functions (formal Laurent series, power series etc.) of q, and the notations like $K[q, q^{-1}]$, $K((q-1))$, etc. have to be understood in the sense of such a Q-adic completion.
3. To avoid some divergences, we require that Λ is a local algebra with the maximal ideal Λ_+, that Adams' operations respect the filtration by its powers: $\Psi^r \Lambda_+ \subset \Lambda_+^r$, and assume that the components of the variables in generating functions lie in Λ_+. In particular, the quantum states $\langle \mathcal{D}_X \rangle$, $\langle \underline{\mathcal{D}}_X \rangle$, etc. are functions on $\mathcal{K}_+^{\infty}$, $\underline{\mathcal{K}}_+^{\infty}$, etc. defined in a Λ_+-neighborhood of the dilaton shift.
4. A peculiar phenomenon overlooked in the previous discussion is that the symplectic structure Ω^{∞}, the adelic map, and other ingredient of our formalism are not Λ-linear in the usual sense. For instance, for $\nu \in \Lambda$ and $\mathbf{f} = (\mathbf{f}_1, \mathbf{f}_2, \ldots, \mathbf{f}_r, \ldots) \in \mathcal{K}^{\infty}$, the adelic image $\underline{\nu \mathbf{f}} =$

$(\nu\underline{\mathbf{f}}_1, \Psi^2(\nu)\underline{\mathbf{f}}_2, \ldots, \Psi^r(\nu)\underline{\mathbf{f}}_r, \ldots)$, i.e. the map between the rth components is *linear relative to* the scalar transformation Ψ^r.[4]

5. The previous feature manifests in the quantization formalism as well. Namely the Planck constant, which needs to be adjoint to the ground ring Λ, is acted upon by Adams' operations as $\Psi^r\hbar := \hbar^r$. Respectively, $\hbar^r$ plays the role of the Planck constant in the quantization formalism on the rth component of the adelic space $\underline{\mathcal{K}}^\infty$. This is manifest in our formula $\sum_r r\hbar^r\Psi^r \sum \nabla_{\eta,\zeta}$ for the propagator, where $\nabla_{\eta,\zeta}$ are 2nd order differential operators.
6. This brings up the question about the status of the Planck constant in the adelic product $\otimes_{M=1}^\infty \mathcal{D}^{tw}_{X/\mathbb{Z}_M}$ since each factor mixes up sectors with different values of the index r. In fact the quantum state $\langle\mathcal{D}^{tw}_{X/\mathbb{Z}_M}\rangle(t,\hbar,Q))$ (i.e. the generating function for twisted fake K-theoretic GW-invariants of the orbifold E_M *after* the dilaton shift) is homogeneous (due to the so-called dilaton equation):

$$\langle\mathcal{D}^{tw}_{X/\mathbb{Z}_M}\rangle(t,\hbar,Q)) = \hbar^{\frac{M \dim K^0(X)}{48}} \langle\mathcal{D}^{tw}_{X/\mathbb{Z}_M}\rangle(\frac{t}{\sqrt{\hbar}}, 1, Q).$$

By the rules of quantum mechanics, scalar factors don't affect "quantum states." The accurate definition of the infinite tensor product in our main theorem is

$$\langle\underline{\mathcal{D}}_X\rangle\ (\{t_r^{(\zeta)}\},\hbar,Q) = \otimes_{M=1}^\infty\langle\mathcal{D}^{tw}_{X/\mathbb{Z}_M}\rangle\left(\frac{\{t^{(\zeta)}_{M/m(\zeta)}\}}{\sqrt{\hbar}}, 1, Q^M\right).$$

Note the change of Q into Q^M in the Mth factor.
7. Our final remark here is about equivariant generalizations of the theorem. In applications of GW-theory, the target space is often equipped with an action of a torus T, and all holomorphic Euler characteristics are replaced with the characters of the T-action on the sheaf cohomology. In particular, Lefschetz' fixed point localization technique, when combined with the formalism of symplectic loop spaces, leads to dealing with fractions of the form $1/(1-q^m\tau)$, where τ is a coordinate on T, and the poles in q are at roots of $1/\tau$ rather than roots of unity. Nevertheless our theory carries over *verbatim* to the equivariant case. Namely, the homotopy theory construction of equivariant K-theory yields $K^0_T(pt) = K^0(BT)$ which is not the character ring of T, but its completion into functions on T defined in the formal neighborhood of the identity. Our ground λ-algebra Λ should be changed into $\Lambda \otimes K^0_T(pt)$. To make sense, the

[4] Perhaps one can rectify this by noticing that *de facto* $\mathcal{D}_X$ depends not on $\mathbf{t}_r$, $r = 1, 2, 3, \ldots$, but on $\Psi^r\mathbf{t}_r$.

above fractions must be expanded into series in $\tau - 1$ with coefficients in rational fractions of q having poles at roots of unity only:

$$\frac{1}{1 - q^m \tau} = \frac{1}{1 - q^m - q^m(\tau - 1)} = \sum_{n=0}^{\infty} \frac{q^{mn}(\tau - 1)^n}{(1 - q^m)^{n+1}}.$$

Thus, in the homotopy theory interpretation of T-equivariant K-theory, localization to fixed points of T makes no sense, but our "quantum RR formula" holds unchanged for T-equivariant GW-invariants, which take values in $\Lambda \otimes K_T^0(pt)$.

4.2 Redefining the Invariants

Let us recall and generalize the definition of permutation-equivariant K-theoretic GW-invariants given in Part I [10], and of the mixed genus-g potential given in Part VII [9].

Let X be a compact Kähler manifold, $K := K^0(X) \otimes \Lambda$, where Λ is a local λ-algebra that contains Novikov's ring as it was explained in Introduction.

Let $X_{g,n,d}$ be the moduli space of degree-d stable maps to X of complex curved of arithmetic genus g with n marked points, and let $h \in S_n$ be a permutation, acting on the moduli space by renumbering the marked points. Let V be a holomorphic vector bundle over $X_{g,n,d}$ equivariant with respect to the action of the permutation h. Then the sheaf cohomology $\pi_*(V) := H^*(X_{g,n,d}; V \otimes \mathcal{O}_{g,n,d})$, where $\mathcal{O}_{g,n,d}$ is the (S_n-invariant) virtual structure sheaf introduced by Y.-P. Lee [14], inherits the action of h. Therefore the supertrace $\operatorname{str}_h \pi_*(V)$ is defined.

Denote $l_k = l_k(h)$ the number of cycles of length k in the cycle structure of h, and by $\mathbf{l} = (l_1, l_2, l_3, \dots)$ the corresponding partition of $n = \sum r l_r$. Our current goal is to define *correlators*

$$\langle \mathbf{u}_1, \dots, \mathbf{u}_{l_1}; \mathbf{v}_1, \dots, \mathbf{v}_{l_2}; \dots; \mathbf{w}_1, \dots, \mathbf{w}_{l_r}; \dots \rangle_{g,\mathbf{l},d},$$

where the *inputs* $\mathbf{u}_i, \mathbf{v}_j, \mathbf{w}_k, \dots$ are elements of $K^0(X) \otimes \Lambda[q, q^{-1}]$. Note that groups of the seats in the correlator have lengths l_1, l_2 etc., and the total number $\sum l_r$ of the seats is equal to the number of *non-empty* cycles.

Let $\sigma_1, \dots, \sigma_r$ be indices of the marked points cyclically permuted by h, and let out of all the l_r cycles of length r, this be the kth cycle. We take the h-equivariant bundle W_k on $X_{g,n,d}$ determined by the input $\mathbf{w}_k = \sum_m \phi_m q^m$ ($\phi_m \in K^0(X)$) in the form

$$W_k := \bigotimes_{\alpha=1}^{r} \sum_m (\mathrm{ev}_{\sigma_\alpha}^* \phi_m) L_{\sigma_\alpha}^m,$$

where $\mathrm{ev}_{\sigma_\alpha} : X_{g,n,d} \to X$ is the evaluation map, and L_{σ_α} is the universal cotangent line bundle at the marked point with the index σ_α. This way, for each cycle of length 1, 2, etc. we associate the inputs $\mathbf{u}_i$, $\mathbf{v}_j$, etc. and define respectively the bundles U_i, V_j, etc. We define the above correlator as

$$\prod_{r=1,2,\ldots} r^{-l_r} \,\mathrm{str}_h\, H^*\left(X_{g,n,d}; \mathcal{O}_{g,n,d} \bigotimes_{i=1}^{l_1} U_i \bigotimes_{j=1}^{l_2} V_j \cdots \bigotimes_{k=1}^{l_r} W_k \cdots \right).$$

The factor in front of the supertrace is motivated by the number $n!/\prod_r r^{l_r} l_r!$ of permutations with the cycle structure described by the partition $\mathbf{l}$.

Note that the correlator is poly-additive with respect to each input. Namely, if $\mathbf{w}_k = \mathbf{w}_k' + \mathbf{w}_k''$, then

$$\bigotimes_{\alpha=1}^{r} \mathbf{w}_k(L_\alpha) = \sum_{I\subset\{1,\ldots,r\}} \bigotimes_{\alpha\in I} \mathbf{w}_k'(L_\alpha) \bigotimes_{\beta\notin I} \mathbf{w}''(L_\beta).$$

The sheaf cohomology splits into 2^r summands accordingly, but the summands with $I \neq \emptyset$ or $\{1, \ldots, n\}$ are permuted by h non-trivially, and hence don't contribute to str_h. Therefore

$$\langle \ldots, \mathbf{w}_k, \ldots \rangle_{g,\mathbf{l},d} = \langle \ldots, \mathbf{w}_k', \ldots \rangle_{g,\mathbf{l},d} + \langle \ldots, \mathbf{w}_k'', \ldots \rangle_{g,\mathbf{l},d}.$$

We extend the correlator to inputs from $\mathcal{K}_+ := K^0(X) \otimes \Lambda[q, q^{-1}]$ in the way *linear relative to* Ψ^r on each input corresponding to the cycles of length r, i.e.

$$\langle \ldots, \nu\mathbf{w}_k, \ldots \rangle_{g,\mathbf{l},d} = \Psi^r(\nu)\langle \ldots, \mathbf{w}_k, \ldots \rangle_{g,\mathbf{l},d}.$$

This is motivated by the fact that if $\Lambda = K^0(Y)$, then for a vector bundle ν on Y, the trace bundle of the cyclic permutation of the factors in $\nu^{\otimes r}$ coincides with $\Psi^r(\nu)$.

Now, we define the genus-g potential of permutation-equivariant quantum K-theory of X as the sum over degrees and partitions $\mathbf{l}$ of all $n = 0, 1, 2, \ldots$:

$$\mathcal{F}_g(\mathbf{t}) = \sum_d Q^d \sum_{\mathbf{l}} \frac{1}{\prod_r l_r!} \langle \ldots \mathbf{t}_1 \ldots; \ldots, \mathbf{t}_2, \ldots; \ldots \rangle_{g,\mathbf{l},d}.$$

Here $\mathbf{t} = (\mathbf{t}_1, \mathbf{t}_2, \ldots, \mathbf{t}_r, \ldots)$, each $\mathbf{t}_r \in \mathcal{K}_+$, and all the inputs in the correlator corresponding to the cycles of length r are taken to be the same and equal $\mathbf{t}_r$.

In fact we need to make one more adjustment to these definitions, and redefine $\mathcal{F}_0(\mathbf{t})$ as $\mathcal{F}_0(\mathbf{t}) + (\Psi^2\mathbf{t}_2(1), 1)/2$, i.e. add this linear function to the above formula for $\mathcal{F}_0$. The reasons will become clear in Sect. 4.6 (see Remark (d) therein) on Kawasaki strata. In a way, this anomalous term represents contributions of degree 0 curves with 2 marked points carrying $\mathbb{Z}_2$-symmetry: they are unstable as curves in X, but represent stable maps in the Chen–Ruan theory of $X/\mathbb{Z}_2$.

Remark The correlators $\langle \mathbf{u}, \ldots, \mathbf{u}\rangle^{S_n}_{g,n,d}$ defined in Part I [10] by taking averages over S_n can be expressed in terms of the above correlators via re-summation over the conjugacy classes labeled by partitions $\mathbf{l}$ of n:

$$\langle \mathbf{u}, \ldots, \mathbf{u}\rangle^{S_n}_{g,n,d} = \frac{1}{n!}\sum_{h\in S_n} \mathrm{str}_h[\mathbf{u}, \ldots, \mathbf{u}]_{g,n,d} = \sum_{\mathbf{l}} \frac{1}{\prod_r l_r!}\langle \mathbf{u}; \ldots; \mathbf{u}\rangle_{g,\mathbf{l},d}.$$

Respectively the mixed genus-g potential of Part VII [9]

$$\sum_{m,n\geq 0,d} Q^d \langle \mathbf{x}, \ldots, \mathbf{x}; \mathbf{t}, \ldots, \mathbf{t}\rangle^{S_n}_{g,m+n,d} = \sum_d Q^d \sum_{\mathbf{l}} \frac{1}{\prod_r l_r!}\langle \mathbf{x}+\mathbf{t}; \mathbf{t}; \mathbf{t}; \ldots\rangle_{g,\mathbf{l},d}$$

coincides with the specialization of $\mathcal{F}_g$ to the inputs $\mathbf{t}_1 = \mathbf{x}+\mathbf{t}$, $\mathbf{t}_2 = \mathbf{t}$, $\mathbf{t}_3 = \mathbf{t}, \ldots$.

While moduli spaces $X_{g,n,d}$ parameterize stable maps of *connected* curves, the *total* descendant potential is to account for contributions of possibly disconnected curves, as well as for symmetries of such curves caused by permutations of identical connected components.

Abstractly speaking, if $\nu \in \Lambda$ represents the contribution of "connected" objects, then the sum over n of contributions of objects with n components is given by

$$\sum_{n\geq 0}\frac{1}{n!}\sum_{h\in S_n}\prod_{k>0}\Psi^k(\nu)^{l_k(h)} = \sum_{\mathbf{l}}\prod_{k>0}\frac{(\Psi^k(\nu)/k)^{l_k}}{l_k!} = e^{\sum_{k>0}\Psi^k(\nu)/k}.$$

This motivates the following definition of the *total descendant potential* of the permutation-equivariant quantum K-theory on X:

$$\mathcal{D}_X := e^{\sum_{g\geq 0}\left[\sum_{k>0}\hbar^{k(g-1)}\Psi^k(R_k\mathcal{F}_g)/k\right]},$$

where $(R_k\mathcal{F})(\mathbf{t}_1, \mathbf{t}_2, \ldots, \mathbf{t}_r, \ldots) := \mathcal{F}(\mathbf{t}_k, \mathbf{t}_{2k}, \ldots, \mathbf{t}_{rk}, \ldots)$.

In order to explain the rescaling R_k of the indices in the variables $\mathbf{t}_r$, note that automorphisms of $\bigsqcup\prod_{\alpha=1}^k X_{g_\alpha,n_\alpha,d_\alpha}$ induced by cyclic permutations of k connected components of a disconnected curve accompanied by a renumbering h of marked points, generate traceless operators on the sheaf cohomology unless $g_\alpha, n_\alpha, d_\alpha$ don't depend on α, and h^k renumbers the marked points of all components separately in consistent ways. In this case, we have an automorphism of $X^k_{g,n,d}$ whose kth power is the automorphism of each factor $X_{g,n,d}$ induced by the renumbering h^k. If the orbit of one of the marked points under the renumbering h^k has order r, then the orbit under the renumbering h has order rk. Therefore the input corresponding to this cycle of marked points must be $\mathbf{t}_{rk}$.

Finally, the factor $\hbar^{k(g-1)}$, whose exponent is $-1/2$ times the Euler characteristic of k copies of a genus-g Riemann surface, can be interpreted as $\Psi^k(\hbar^{g-1})$ by adjoining $\hbar$ to Λ and setting $\Psi^k(\hbar) = \hbar^k$.

Note that all $\mathcal{F}_g$ can be recovered from $\mathcal{G} := \log \mathcal{D}_X$ by Möbius' exclusion-inclusion formula

$$\sum_g \hbar^{g-1} \mathcal{F}_g = \prod_{p \text{ prime}} \left(1 - \frac{\Psi^p}{p} R_p\right) \mathcal{G}.$$

4.3 Kawasaki's Riemann–Roch Formula

The expression of K-theoretic GW-invariants in terms of cohomological ones is based on the use of the virtual variant [16] of Kawasaki's Riemann–Roch formula [12].

Let $\mathcal{M}$ be a compact complex orbifold, and V be a holomorphic orbibundle on $\mathcal{M}$. The holomorphic Euler characteristic of V, defined in terms of Čech cohomology as $\chi(\mathcal{M}; V) := \sum_i (-1)^i \dim H^i(\mathcal{M}; V)$, is expressed by Kawasaki's RR formula in cohomological terms of the *inertia orbifold* $I\mathcal{M}$:

$$\chi(\mathcal{M}; V) = \chi^{fake}\left(I\mathcal{M}; \frac{\operatorname{tr}_h V}{\operatorname{str}_h \wedge^\bullet N^*_{I\mathcal{M}}}\right).$$

Recall that a point in $I\mathcal{M}$ is represented by a pair (x, h) where $x \in \mathcal{M}$, and $h \in \Gamma(x)$ is an element of the *inertia group* of $x \in \mathcal{M}$ (i.e. the group of local symmetries of x in the orbifold structure). In the formula, $N^*_{I\mathcal{M}}$ denotes the conormal bundle to the stratum of fixed points of the symmetry h. The bundle V can be restricted to the stratum and decomposed into eigenbundles V_λ of h corresponding to the eigenvalues λ. The trace operation $\operatorname{tr}_h V$ denotes the virtual bundle $\sum_\lambda \lambda V_\lambda$, and the supertrace str_h in the denominator denotes the similar operation on the $\mathbb{Z}_2$-graded bundle $\wedge^\bullet N^*_{I\mathcal{M}}$. Finally, the notation χ^{fake} stands for the *fake* holomorphic Euler characteristic of an orbibundle over an orbifold:

$$\chi^{fake}(M; W) := \int_M \operatorname{ch}(W) \operatorname{td}(T_M),$$

where $\operatorname{ch}(W)$ is the Chern character of the orbibundle W, and $\operatorname{td}(T_M)$ is the Todd class of tangent orbibundle T_M (both defined over $\mathbb{Q}$).

In effect, the RHS of Kawasaki's RR formula is the sum of certain fake holomorphic Euler characteristics, i.e. of certain integrals over the strata of the inertia orbifold, which are rational numbers adding up to the integer defined by the LHS.

It is no accident that Kawasaki's RR formula resembles Lefschetz' holomorphic fixed point formula. To make the connection, let h be an automorphism of a holomorphic bundle V over a compact complex manifold $\widetilde{\mathcal{M}}$. For our goals it suffices to assume that h belongs to a finite group G of such automorphisms (although abstractly speaking this restriction can be relaxed). Lefschetz' fixed point formula computes the supertrace of h on the sheaf cohomology as an integral over the fixed point submanifold $\widetilde{\mathcal{M}}^h$:

$$\mathrm{str}_h\, H^*(\widetilde{\mathcal{M}};\, V) = \chi^{fake}\left(\widetilde{\mathcal{M}}^h;\, \frac{\mathrm{tr}_h\, V}{\mathrm{str}_h \wedge^\bullet N^*_{\widetilde{\mathcal{M}}^h}}\right).$$

On the other hand, V can be considered as an orbibundle over the quotient orbifold $\mathcal{M} := \widetilde{\mathcal{M}}/G$, and the holomorphic Euler characteristic $\chi(\mathcal{M};\, V)$ of the orbibundle can be found as the average over G:

$$\frac{1}{|G|}\sum_{h\in G} \mathrm{str}_h\, H^*(\widetilde{\mathcal{M}};\, V) = \frac{1}{|G|}\sum_{h\in G} \chi^{fake}\left(\widetilde{\mathcal{M}}^h;\, \frac{\mathrm{tr}_h\, V}{\mathrm{str}_h \wedge^\bullet N^*_{\widetilde{\mathcal{M}}^h}}\right).$$

The last sum coincides with the right hand side of Kawasaki's RR formula on $\mathcal{M} = \widetilde{\mathcal{M}}/G$ since in the global quotient case

$$I\mathcal{M} = \left[\bigsqcup_{h\in G} \widetilde{\mathcal{M}}^h\right]/G.$$

In fact, we need a combination of Kawasaki's RR with Lefschetz' fixed point formula, computing $\mathrm{str}_h\, H^*(\mathcal{M};\, V)$ where h is a finite order automorphism of an *orbi*bundle V over an *orbi*fold $\mathcal{M}$:

$$\mathrm{str}_h\, H^*(\mathcal{M};\, V) = \chi^{fake}\left(I\mathcal{M}^h;\, \frac{\mathrm{tr}_{\tilde{h}}\, V}{\mathrm{str}_{\tilde{h}} \wedge^\bullet N^*_{I\mathcal{M}^h}}\right),$$

where the "fixed point inertia orbifold" $I\mathcal{M}^h$ can be described as follows. Let $x \in \mathcal{M}$ be a fixed point of h, and $U_x \to U_x/\Gamma(x) \subset \mathcal{M}$ be its orbifold chart. The transformation h can be lifted to automorphisms $\tilde{h}$ of the chart (and of the bundle V over the chart) in $|\Gamma(x)|$ possible ways. Each transformation $\tilde{h}$ has a fixed point submanifold $U_x^{\tilde{h}} \subset U_x$ whose union is $\Gamma(x)$ invariant. The quotient $\left[\bigcup_{\tilde{h}} U_x^{\tilde{h}}\right]/\Gamma(x)$ provides the local description of the orbifold $I\mathcal{M}^h$ near $x \in \mathcal{M}$. The ingredients $\mathrm{tr}_{\tilde{h}}\, V$ and $N_{I\mathcal{M}^h}$ are obtained from the fibers of V over $U_x^{\tilde{h}}$ and from the normal space to $U_x^{\tilde{h}}$ in U_x respectively.

A justification of Lefschetz-Kawasaki's RR formula can be obtained formally from Kawasaki's RR formula applied to the orbifold $\mathcal{M}/G$ where G is the cyclic group generated by h. Indeed, let $\mathbb{C}_\lambda$ denotes the 1-dimensional representation of G where h acts by a root of unity λ. Then

$$\mathrm{str}_h\, H^*(\mathcal{M};\, V) := \sum_\lambda \lambda\, H^*(\mathcal{M};\, V)_\lambda = \sum_\lambda \lambda\, \chi(\mathcal{M}/G;\, V \otimes \mathbb{C}_{\lambda^{-1}}).$$

The last sum can be computed on the inertia orbifold $I(\mathcal{M}/G)$ using Kawasaki's RR. However $\sum_\lambda \lambda\, \mathbb{C}_{\lambda^{-1}}$ is a virtual representation of G whose character equals $|G|$ on h and equals 0 on all other elements of G. Therefore only the strata of $I(\mathcal{M}/G)$ made of fixed points of $\tilde{h}$ will contribute. Note that the factor $|G|$ from the character is compensated by the factor $1/|G|$ arising from the comparison between the fundamental classes of strata in $I\mathcal{M}^h$ with those in $I(\mathcal{M}/G)$.

In applications to quantum K-theory, the orbifold $\mathcal{M}$ is replaced with moduli space $X_{g,n,d}$ of stable maps to X, which are virtual orbifolds, or with products of such spaces (since the curves are allowed to be disconnected). An automorphism h of such a product is induced by a renumbering of the marked points on the curve. A fixed point of h is represented by a stable map $\phi : \Sigma \to X$ for which there exists a symmetry accomplishing the required permutation h, i.e. there exists an isomorphism $\tilde{h} : \Sigma \to \Sigma$ which permutes the marked points by h, and such that $\phi\tilde{h} = \phi$. It is the result of [16] which justifies the application of Kawasaki's RR to *virtual* orbifolds.[5] Respectively, our generating function $\mathcal{D}_X$ (which incorporates contributions of all stable maps and all renumberings of the markings) can be described in terms of suitable fake holomorphic Euler characteristics on the strata of the inertia orbifold $I\mathcal{M}^h$. We will call them *Kawasaki strata*. They parameterize *stable maps with prescribed symmetries*, i.e. equivalence classes of pairs $(\phi, \tilde{h})$, where ϕ is a stable map of a (possibly disconnected) curve to X, and $\tilde{h}$ is a symmetry of the map, accomplishing a (possibly non-trivial) permutation of the marked points.

How does a Kawasaki stratum look like? Given a stable map $\phi : \Sigma \to X$ with a symmetry h (note that now on we omit the tilde), it defines the map of the quotient $\widehat{\Sigma}$ of the curve Σ by the cyclic group generated by h. On Fig. 4.1, we attempt to show a typical picture of a (connected) quotient curve. The quotient map $\Sigma \to \widehat{\Sigma}$ may have different number of branches (shown as the multiplicity of lines) over different irreducible components of $\widehat{\Sigma}$. This shows that the summation over Kawasaki strata will have the structure of Wick's formula of *summation over graphs*. The vertices of the graphs represent contribution of Kawasaki strata parameterizing irreducible quotient maps, while the edges correspond to the nodes connecting the irreducible components.

Furthermore, an M-fold quotient map $\Sigma \to \widehat{\Sigma}$ over an irreducible curve $\widehat{\Sigma}$ can be described as the principal $\mathbb{Z}_M$-bundle over the complement to marked and nodal points, possibly ramified at such points. Consequently, Kawasaki strata representing

[5] The set-up of the virtual Kawasaki RR is axiomatic, but it eventually employs Kawasaki's RR theorem for (ambient) compact orbifolds. For moduli spaces of stable maps, the existence of such ambient orbifolds is easily obtained in genus 0 by projective embedding of X (since $\mathcal{M}_{0,n}(\mathbb{C}P^n, d)$ are orbifolds). In higher genus, the existence of such *compact* ambient orbifolds is a result of A. Kresch [13]. Of course, one expects Kawasaki's RR formula to remain true for compactly supported orbisheaves on non-compact orbifolds (which would settle this technical issue in a more natural way). For compactly suppotred sheaves on manifolds, this was proved in [15] some quarter of a century later than Hirzebruch's celebrated result for compact manifolds. The orbifold story develops slower, and almost 40 years after Kawasaki's result [12], its vision for compactly supported orbisheaves seems still missing in the literature. The most promising approximations we could find were [6] and [5].

Fig. 4.1 Stable maps with prescribed symmetries

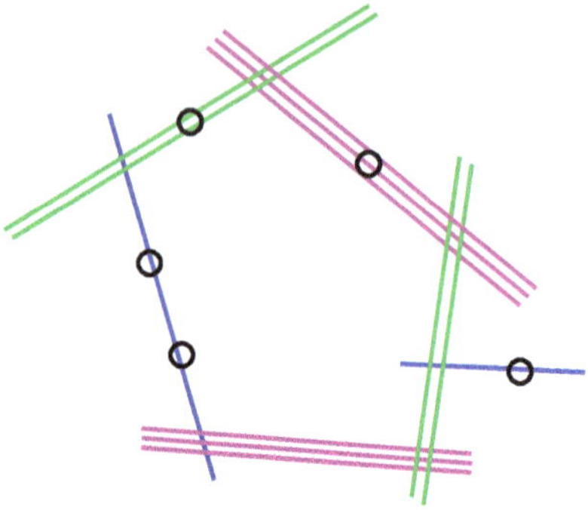

the vertices can be identified with *moduli spaces of stable maps to the orbifold target spaces* $X/\mathbb{Z}_M$ ($= X \times B\mathbb{Z}_M$ in the notation of [11], i.e. assuming that $\mathbb{Z}_M$ acts trivially on X).

We will denote by $\mathcal{D}^{fake}_{X/\mathbb{Z}_M}$ the total descendant potential of the *fake* quantum K-theory of the orbifold $X/\mathbb{Z}_M$. Using the results [19], one can obtain the K-theoretic counterpart to the theorem of Jarvis and Kimura [11] and express $\mathcal{D}^{fake}_{X/\mathbb{Z}_M}$ in terms of $\mathcal{D}^{fake}_X$, the total descendant potential of quantum K-theory of X. The latter can be, in its turn, expressed in terms of the cohomological total GW-potential $\mathcal{D}^H_X$, using the quantum Hirzebruch-Riemann-Roch formula [3, 4] for fake GW-invariants with values in complex cobordisms, specialized to the case of complex K-theory. However, the vertex contributions in our Wick's formula are not $\mathcal{D}^{fake}_{X/\mathbb{Z}_M}$, but some *twisted* fake K-theoretic GW-invariants of these orbifolds. This means that the virtual fundamental classes of moduli spaces of stable maps to $X/\mathbb{Z}_M$ need to be systematically modified—-in fact by the factors accounting for the denominators in the Kawasaki-RR formula. The total descendant potential $\mathcal{D}^{tw}_{X/\mathbb{Z}_M}$ for suitably twisted fake quantum K-theory of $X/\mathbb{Z}_M$ can be expressed in terms of $\mathcal{D}^{fake}_{X/\mathbb{Z}_M}$ using the results of Tseng [20] and Tonita [17].

In the next two sections, we first explain (or recall) how to pass from $\mathcal{D}^H_X$ to $\mathcal{D}^{fake}_X$, and then to $\mathcal{D}^{fake}_{X/\mathbb{Z}_M}$. Then we will formulate the twisting result relating $\mathcal{D}^{fake}_{X/\mathbb{Z}_M}$ with $\mathcal{D}^{tw}_{X/\mathbb{Z}_M}$. Then the vertex contributions of our graph summation formula will be described, roughly speaking, as the product $\bigotimes_{M=1}^{\infty} \mathcal{D}^{tw}_{X/\mathbb{Z}_M}$ over all $M = 1, 2, 3, \ldots$, leading to the concise quantum-mechanical description of $\mathcal{D}_X$ given in Introduction.

4.4 Symplectic Loop Spaces and Quantization

The formalism of symplectic loop spaces and their quantizations starts with the data: a vector *space* H (or a module over a ground ring Λ), a symmetric Λ-valued *Poincaré pairing* $(\cdot, \cdot)$ on H, and a nonzero vector $v \in H$. Using this datum, one cooks up a *loop space* $\mathcal{H}$, equipped with a symplectic Λ-valued form Ω, a Lagrangian *polarization* $\mathcal{H} := \mathcal{H}_+ \oplus \mathcal{H}_-$, and a vector $\mathbf{v} \in \mathcal{H}_+$ called the *dilaton shift*.

Given a sequence of functions $\mathcal{F}_g : \mathcal{H}_+ \to \Lambda$, one combines them into the *total descendant potential* $\mathcal{D} := e^{\sum \hbar^{g-1}\mathcal{F}_g}$, and interprets the latter as an "asymptotical element" in the Fock space associated with $(\mathcal{H}, \Omega)$ by lifting from $\mathcal{H}_+$ to $\mathcal{H}$ the dilaton-shifted function $\mathbf{t} \mapsto \mathcal{D}(\mathbf{t} - \mathbf{v})$ from $\mathcal{H}_+$ so that it stays constant along the Lagrangian subspaces parallel to $\mathcal{H}_-$.

According to the ideology of quantum mechanics, the Heisenberg Lie algebra of the symplectic space acts irreducibly in the Fock space (of functions constant in the direction of $\mathcal{H}_-$), which by Schur's lemma, projectively identifies Fock spaces defined using different polarizations. Furthermore, the symplectic group moves the polarizations around, which therefore defines a projective action of the Lie algebra of quadratic hamiltonians on the Fock space. Explicit formulas for this action provide the standard quantization of quadratic hamiltonians. Namely, let $\{ q_\alpha\}$ be coordinates on $\mathcal{H}_+$, and $\{p_\alpha\}$ the Darboux-dual coordinates on $\mathcal{H}_-$. Then the quantization $\widehat{\ }$ of Darboux monomials is given by the multiplication and differentiation operators on functions of $\{q_\alpha\}$:

$$\widehat{q_\alpha q_\beta} := \hbar^{-1} q_\alpha q_\beta, \quad \widehat{q_\alpha p_\beta} := q_\alpha \partial_{q_\beta}, \quad \widehat{p_\alpha p_\beta} := \hbar\, \partial_{q_\alpha} \partial_{q_\beta}.$$

Finally, given a linear symplectic transformation $\square$ on $(\mathcal{H}, \Omega)$, the *Stone—von Neumann quantization* of it acts on the Fock space by the operator $\widehat{\square} := e^{\widehat{\log \square}}$.

A typical application of this formalism in GW-theory relates generating functions for two kinds of GW-invariants as follows. The functions $\mathcal{D}^i$, $i = 1, 2$, are lifted to asymptotical elements $\langle \mathcal{D}^i \rangle$ of the respective Fock spaces associated with symplectic loop spaces $(\mathcal{H}^i, \Omega^i)$ using Lagrangian polarizations $\mathcal{H}^i_\pm$ and dilaton shifts $\mathbf{v}_i$. The respective quantum states are related by

$$\langle \mathcal{D}^1 \rangle = \widehat{\mathrm{qch}}\, \widehat{\square}\, \langle \mathcal{D}^2 \rangle,$$

where $\square$ is a suitable symplectic automorphism of $(\mathcal{H}^2, \Omega^2)$, while the "quantum Chern character" $\mathrm{qch} : \mathcal{H}^1 \to \mathcal{H}^2$ is a symplectic isomorphism (i.e. $\mathrm{qch}^* \Omega^2 = \Omega^1$), and hence identifies the respective Fock spaces. Note that the isomorphism qch may not respect the polarizations (in practice, qch respects $\mathcal{H}^i_+$, but not $\mathcal{H}^i_-$), nor the dilaton shifts ($\mathrm{qch}\, \mathbf{v}_1 \neq \mathbf{v}_2$). Consequently, the generating functions $\mathcal{D}^1$ and $\mathcal{D}^2$ are obtained from each other by three consecutive transformations: the quantized operator $\square$, the change of polarization, and the correction for the discrepancy in the dilaton shifts.

To begin with cohomological GW-invariants of X, we set

$$H := H^{even}(X; \Lambda), \quad (a, b)^H := \int_X ab, \quad v = \mathbf{1},$$

take $\mathcal{H}$ to be the space $H((z))$ of Laurent series in one indeterminate z with vector coefficients from H. We assume that the ground ring Λ contains Novikov's variables, Q, and the Laurent series are *Q-adically convergent* for $z \neq 0$, i.e. that modulo any fixed power of (Q), the series in question contain finitely many negative powers of z. We equip $\mathcal{H}$ with the symplectic form

$$\Omega^H(\mathbf{f}, \mathbf{g}) := \mathrm{Res}_{z=0}(\mathbf{f}(-z), \mathbf{g}(z))^H\, dz,$$

and Lagrangian polarization $\mathcal{H} = \mathcal{H}_+ \oplus \mathcal{H}_-$, where $\mathcal{H}_+$ consists of the power series part of the Laurent series, and $\mathcal{H}_-$ of their principal parts.

Recall that genus-g generating functions for GW-invariants of X are defined by

$$\mathcal{F}_g^H(\mathbf{t}) := \sum_{d,n} \frac{Q^d}{n!} \int_{[X_{g,n,d}]} \prod_{i=1}^{n} \sum_{k=0}^{\infty} \sum_{\alpha} t_{k,\alpha}\, \mathrm{ev}_i^*(\phi_\alpha) \psi_i^k,$$

where $[X_{g,n,d}]$ is the virtual fundamental classes of the moduli spaces of stable maps to X, $\psi_i := c_1(L_i)$ is the 1st Chern class of universal cotangent line bundle at the ith marked point, and $\{\phi_\alpha\}$ is a basis in $H^{even}(X, \Lambda)$. They are functions of $\mathbf{t} = \sum_{k,\alpha} t_{k,\alpha} \phi_\alpha z^k$, which lie in $\mathcal{H}_+$. Respectively, the total descendant potential of the cohomological GW-theory of X is defined as $\mathcal{D}_X^H = e^{\sum_g \hbar^{g-1} \mathcal{F}_g^H(\mathbf{t})}$, subject to the dilaton shift $\mathbf{v} = -z\mathbf{1}$, i.e. $\langle \mathcal{D}_X^H \rangle(\mathbf{t} - z\mathbf{1}) = \mathcal{D}_X^H(\mathbf{t})$.

In the fake quantum K-theory of X, one puts

$$H := K = K^0(X) \otimes \Lambda, \;\; (a, b) := \chi(X; a \otimes b) = \int_X \mathrm{ch}(a)\, \mathrm{ch}(b)\, \mathrm{td}(T_X),$$

uses $\mathcal{K}^{fake} = K((q-1))$, i.e. the space of Q-adically convergent Laurent series in $q-1$ with vector coefficients in K, and equips it with the symplectic form

$$\Omega^{fake}(\mathbf{f}, \mathbf{g}) := \mathrm{Res}_{q=1}(\mathbf{f}(q^{-1}), \mathbf{g}(q))\, \frac{dq}{q},$$

and Lagrangian polarization $\mathcal{K} = \mathcal{K}_+ \oplus \mathcal{K}_-$, taking $\mathcal{K}_+$ to consist of power series, and $\mathcal{K}_-$ of the principal parts of Laurent series in $q-1$.

The genus-g generating functions $\mathcal{F}_g^{fake}$ are defined on $\mathcal{K}_+$ by

$$\mathcal{F}_g^{fake}(\mathbf{t}) = \sum_{d,n} \frac{Q^d}{n!} \chi^{fake}\left(X_{g,n,d}; \bigotimes_{i=1}^{n} \sum_{k=0}^{\infty} \sum_{\alpha} t_{k,\alpha}\, \mathrm{ev}_i^*(\phi_\alpha)(L_i - 1)^k \right),$$

where $\{\phi_\alpha\}$ form a basis in $K^0(X)$, and the fake holomorphic Euler characteristic of a bundle V on $X_{g,n,d}$ is defined using the virtual fundamental cycle $[X_{g,n,d}]$ and the virtual tangent bundle $T_{X_{g,n,d}}$:

$$\chi^{fake}(X_{g,n,d}; V) := \int_{[X_{g,n,d}]} \mathrm{ch}(V)\,\mathrm{td}(T_{X_{g,n,d}}).$$

The total descendant potential of fake quantum K-theory is defined by $\mathcal{D}_X^{fake} = e^{\sum \hbar^{g-1}\mathcal{F}_g^{fake}(\mathbf{t})}$ as a function on $\mathcal{K}_+$ subject to the dilaton shift by $\mathbf{v} = (1-q)\mathbf{1}$, i.e. $\langle \mathcal{D}_X^{fake}\rangle((1-q)\mathbf{1}+\mathbf{t}) = \mathcal{D}_X^{fake}(\mathbf{t})$. It is expressed in terms of $\mathcal{D}_X^H$ following [3, 4].

Namely, introduce the *quantum Chern character* $\mathrm{qch} : \mathcal{K} \to \mathcal{H}$ by

$$\mathcal{K} \ni \mathbf{f} = \sum_k f_k(q-1)^k \mapsto \sqrt{\mathrm{td}(T_X)} \sum_k \mathrm{ch}(f_k)(e^z - 1)^k \in \mathcal{H}.$$

It is symplectic: $\mathrm{qch}^*\,\Omega^H = \Omega^{fake}$. Then

$$\langle \mathcal{D}_X^{fake}\rangle = \widehat{\mathrm{qch}}^*\, \widehat{\triangle}\, \langle \mathcal{D}_X^H\rangle,$$

where $\triangle$ is the Euler–Maclaurin asymptotics of the infinite product $\prod_{r=1}^{\infty} \mathrm{td}((T_X - 1)\otimes q^{-r})$. The equality holds up to a scalar factor explicitly described in [3]. Recall that the Euler–Maclaurin asymptotics of the product $\sqrt{S(E)}\prod_{r=1}^{\infty} S(E\otimes q^{-r})$, where E is a vector bundle over X, q is the universal line bundle (so that $c_1(q) = z$), and $S(\cdot) = e^{\sum_k s_k\, \mathrm{ch}_k(\cdot)/k!}$ is an invertible multiplicaive characteristic class, is

$$e^{\sum_{m\geq 0}\sum_{l\geq 0} s_{2m-1+l} \frac{B_{2m}}{(2m)!}\mathrm{ch}_l(E) z^{2m-1}},$$

where B_{2m} are Bernoulli numbers, and $\mathrm{ch}_l(E)$ in the exponent are understood as operators of classical multiplication in the cohomology algebra of X by the components of the Chern character.

Our next step is to describe in terms of $\mathcal{D}_X^{fake}$ the total descendant potential $\mathcal{D}_{X/\mathbb{Z}_M}^{fake}$ of the fake quantum K-theory of the orbifold $X/\mathbb{Z}_M$. The Grothendieck group $K^0(X/\mathbb{Z}_M)$ of orbibundles on $X/\mathbb{Z}_M$ is identified with $K^0(X)\otimes \mathrm{Repr}(\mathbb{Z}_M)$. Respectively, the total descendant potential $\mathcal{D}_{X/\mathbb{Z}_M}^{fake}$ in the *fake* quantum K-theory of $X/\mathbb{Z}_M$ is a function on the space of vector power series

$$\mathbf{t} := \sum_{\chi\in \mathrm{Repr}(\mathbb{Z}_M)} \mathbf{t}_\chi\, \chi,$$

where each $\mathbf{t}_\chi$ is a power series in $q-1$ with coefficients in $K^0(X)\otimes\Lambda$. In down-to-earth terms we have:

$$\mathcal{D}^{fake}_{X/\mathbb{Z}_M}(\mathbf{t}) = \prod_{\chi\in \mathrm{Repr}(\mathbb{Z}_M)} \mathcal{D}^{fake}_X(\mathbf{t}_\chi).$$

This follows from the analogous cohomological result of Jarvis and Kimura [11] by application of twisting theorems of Tseng [20] and Tonita [17] (combined with a description of the virtual tangent bundles to the moduli spaces of stable maps to $X/\mathbb{Z}_M$). Alternatively, this result can be extracted from Sect. 4.4 of their joint paper [19].

To go on, we need to describe the element of the Fock space defined by $\mathcal{D}^{fake}_{X/\mathbb{Z}_M}$, and the respective symplectic loop space. We have

$$H := K\otimes \mathrm{Repr}(\mathbb{Z}_M),\;\; (a,b) := \frac{1}{M^2}\sum_\chi (a_\chi, b_\chi)^{fake},\;\; v=\sum_\chi 1\chi.$$

Respectively the loop space

$$\mathcal{K}^{fake}_{X/\mathbb{Z}_M} = \mathcal{K}^{fake}_X \otimes \mathrm{Repr}(\mathbb{Z}_M),$$

is equipped with the symplectic form

$$\Omega^{fake}_{X/\mathbb{Z}_M}(\mathbf{f},\mathbf{g}) = \frac{1}{M^2}\sum_{\chi\in\mathrm{Repr}(\mathbb{Z}_M)} \Omega^{fake}(\mathbf{f}_\chi,\mathbf{g}_\chi).$$

The Lagrangian polarization is given by $\mathcal{K}^{fake}_\pm\otimes \mathrm{Repr}(\mathbb{Z}_M)$, and the dilaton shift by $\mathbf{v}=(1-q)v$.

The specifics of the orbifold situation, however, is that the evaluation maps involved in the construction of the invariants take values in the *inertia orbifold* IX, in the case of the orbifold $X/\mathbb{Z}_M$ consisting of M disjoint copies of X, which are labeled not by representations of $\mathbb{Z}_M$, but by its elements $h\in\mathbb{Z}_M$ (referred to as *sectors*). In sector notation

$$\mathbf{f} = \sum_{\chi\in\mathrm{Repr}(\mathbb{Z}_M)} \mathbf{f}_\chi \chi = \sum_{h\in\mathbb{Z}_M}\mathbf{f}^{(h)} h,$$

where (by Fourier transform)

$$\mathbf{f}^{(h)} = \frac{1}{M}\sum_\chi \mathbf{f}_\chi \chi(h),\;\; \mathbf{f}_\chi = \sum_h \mathbf{f}^{(h)}\chi(h^{-1}).$$

Consequently,

$$(a,b) = \frac{1}{M}\sum_h (a^{(h)}, b^{(h^{-1})})^{fake},$$

the symplectic form decomposes as

$$\Omega^{fake}_{X/\mathbb{Z}_M}(\mathbf{f},\mathbf{g}) = \frac{1}{M}\sum_{h\in\mathbb{Z}_M} \Omega^{fake}(\mathbf{f}^{(h)}, \mathbf{g}^{(h^{-1})}),$$

the polarization spaces have the form $\oplus_{h\in\mathbb{Z}_M} \mathcal{K}^{fake}_{\pm} h$, where $\mathcal{K}^{fake}_{+} h$ is Darboux-dual to $\mathcal{K}^{fake}_{-} h^{-1}$, while the dilaton shift $\mathbf{v} = (1-q)\mathbf{1}$ belongs to the sector of the unit element $\mathbf{1} \in \mathbb{Z}_M$.

We will label the sectors by Mth roots of unity ζ (primitive or not) as follows. To the element $h = h_0^{rs}$, where h_0 is the standard generator of $\mathbb{Z}_M$, $M = rm$, and $(s,m) = 1$, we assign $\zeta(h)$ to be the *primitive* root of unity of order m such that $\zeta^s = e^{2\pi i/m}$. Conversely, to $\zeta = e^{2\pi i t/m}$, where $m|M$, and $(t,m) = 1$, we assign $h_{(\zeta)} \in \mathbb{Z}_M$ to be h_0^{rs}, where $r = M/m$, and s is the multiplicative inverse to t modulo m.

4.5 Formulation of the Results

We describe $\mathcal{D}^{tw}_{X/\mathbb{Z}_M}$ in terms of $\mathcal{D}^{fake}_{X/\mathbb{Z}_M}$.

The Fock space where $\mathcal{D}^{tw}_{X/\mathbb{Z}_M}$ lies quantizes the loop space

$$\mathcal{K}^{tw}_{(M)} := \oplus_{\zeta:\zeta^M=1}\mathcal{K}^{(\zeta)}$$

equipped with the symplectic form $\Omega^{tw}_{(M)}$ as follows. Let $m = m(\zeta)$ denote the order of ζ as a *primitive* root of unity, and let $M = mr$. On the space $K = K^0(X)\otimes\Lambda$, introduce a new Λ-valued pairing

$$(a,b)^{(r)} := \chi\left(X; a\otimes b\otimes \frac{\mathrm{Eu}(T_X - 1)}{\mathrm{Eu}(\Psi^r(T_X-1))}\right).$$

Here Eu is the K-theoretic Euler class defined by $\mathrm{Eu}\,L = (1-L^{-1}) = e^{-\sum_{k>0} L^{-k}/k}$ on line bundles, and extended to arbitrary complex vector bundles by multiplicativity using the splitting principle. The pairing satisfies

$$(\Psi^r a, \Psi^r b)^{(r)} = r\Psi^r(a,b),$$

which is simply the abstract Grothendieck-RR formula (called also *Adams-RR*) for the operation Ψ^r from K-theory to itself, while the factor r comes from

$$\frac{\mathrm{Eu}(\Psi^r 1)}{\mathrm{Eu}(1)} = \lim_{L\to 1} \frac{1-L^{-r}}{1-L^{-1}} = r.$$

Introduce the symplectic form on $\mathcal{K}^{tw}_{(M)}$:

$$\Omega^{tw}_{(M)}(\mathbf{f},\mathbf{g}) := \frac{1}{M} \sum_{\zeta:\,\zeta^M=1} \mathrm{Res}_{q=1}(\mathbf{f}^{(\zeta)}(q^{-1}), \mathbf{g}^{(\zeta^{-1})}(q))^{(r(\zeta))} \frac{dq}{q}.$$

To describe the polarization in $\mathcal{K}^{tw}_{(M)}$, introduce basis in $\mathcal{K}^{(\zeta)}$:

$$\mathbf{f}^{(\zeta)}_{k,\alpha} := \Psi^r\left(\phi^\alpha (q^{1/m}-1)^k\right), \quad \mathbf{g}^{(\zeta)}_{k,\alpha} := r\Psi^r\left(\phi_\alpha \frac{q^{k/m}}{(1-q^{1/m})^{k+1}}\right),$$

where $m = m(\zeta)$, $r = r(\zeta)$, ϕ^α runs through a basis in $K^0(X)$ Poincaré-dual to ϕ_α, and k run through non-negative integers. Then $\mathbf{f}^{\zeta}_{k,\alpha}$ run through a basis in the positive space of polarization, while $\mathbf{g}^{(\zeta^{-1})}_{k,\alpha}$ run through the Darboux-dual basis in the negative space of the polarization in question. The generating function $\mathcal{D}^{tw}_{X/\mathbb{Z}_M}$ is represented by an element $\langle \mathcal{D}^{tw}_{X/\mathbb{Z}_M}\rangle$ in the Fock space of the symplectic loop space $(\mathcal{K}^{tw}_{(M)}, \Omega^{tw}_{(M)})$, using this polarization, and the dilaton shift $\mathbf{v} = (1-q^M)\mathbf{1} = \Psi^M(1-q)\mathbf{1}$ (in the unit sector):

$$\langle D^{tw}_{X/\mathbb{Z}_M}\rangle((1-q^M)\mathbf{1}+\mathbf{t}) = \mathcal{D}^{tw}_{X/\mathbb{Z}_M}(\mathbf{t}).$$

We will also assume that a quantum state does not change when the function is multiplied by a non-zero constant (so that $\langle\mathcal{D}\rangle$ actually denotes the 1-dimensional subspace spanned by $\mathcal{D}$.)

To state the quantum Riemann-Roch formula relating $\langle\mathcal{D}^{tw}_{X/\mathbb{Z}_M}\rangle$ with $\langle\mathcal{D}^{fake}_{X/\mathbb{Z}_M}\rangle$, define operator $\square_{(M)} : \mathcal{K}^{tw}_{(M)} \to \mathcal{K}^{fake}_{X/\mathbb{Z}_M}$ acting block-diagonally by sectors:

$$(\square_{(M)}\mathbf{f})^{(\zeta)} = \square_{\zeta,r(\zeta)}(\mathbf{f}^{(\zeta)}),$$

where for a primitive mth root of unity η and $r = 1, 2, 3, \ldots,$

$$\square_{\eta,r} := e^{\sum_{k>0}\left(\frac{\Psi^{kr}(T^*_X-1)}{k(1-\eta^{-k}q^{kr/m})} - \frac{\Psi^k(T^*_X-1)}{k(1-q^k)}\right)}.$$

We claim that $\square_{(M)}$ is symplectic, i.e.

$$\Omega^{fake}_{X/\mathbb{Z}_M}(\square_{(M)}\mathbf{f}, \square_{(M)}\mathbf{g}) = \Omega^{tw}_{(M)}(\mathbf{f}, \mathbf{g}).$$

This follows from the identity

$$\square_{\eta,r}(q^{-1})\square_{\eta^{-1},r}(q) = e^{\sum_{k>0} \frac{\Psi^{kr}(T_X^*-1)-\Psi^k(T_X^*-1)}{k}} = \frac{\mathrm{Eu}(T_X - 1)}{\mathrm{Eu}(\Psi^r(T_X - 1))}.$$

Note that $\square_{(M)}$ respects positive spaces of our polarizations in its source and target loop spaces, but does not respect the negative ones, nor the dilaton shifts.

Proposition 4.5.1 $\langle \mathcal{D}^{tw}_{X/\mathbb{Z}_M}\rangle = \widehat{\square}_{(M)}\, \langle \mathcal{D}^{fake}_{X/\mathbb{Z}_M}\rangle.$

Let us now recall the dilaton equation, which says that in the expression $\mathcal{D}^H_X = e^{\sum_g \hbar^{g-1}\mathcal{F}_g}$, *after* the dilaton shift, the functions $\mathcal{F}_g$ are homogeneous of degree $2-2g$ (with some anomaly for $g=1$). Namely,

$$(\mathbf{t}\partial_{\mathbf{t}} + 2\hbar\partial_\hbar)\langle \mathcal{D}^H_X\rangle(\mathbf{t}, \hbar) = -\frac{\mathrm{eu}(X)}{24}\langle \mathcal{D}^H_X\rangle(\mathbf{t}, \hbar).$$

In the transition from $\langle \mathcal{D}^H_X\rangle$ to $\langle \mathcal{D}^{tw}_{X/\mathbb{Z}_M}\rangle$, the homogeneity property is preserved, because our quantization formulas (from Sect. 4.4) for quadratic Darboux monomials are homogeneous of zero degree. This allows one to recast the dependence of $\hbar$ (omitting the scalar factors such as $\hbar^{\mathrm{eu}(X)/48}$) this way:

$$\langle \mathcal{D}^{tw}_{X/\mathbb{Z}_M}\rangle(\mathbf{t}, \hbar, Q) = \langle \mathcal{D}^{tw}_{X/\mathbb{Z}_M}\rangle(\frac{\mathbf{t}}{\sqrt{\hbar}}, 1, Q).$$

Note that $\mathbf{t}$ can be rewritten by sectors as $\sum_{\zeta:\, \zeta^M=1} t^{(\zeta)}h_{(\zeta)}$, where each $t_{(\zeta)} \in \mathcal{K}^{fake}_+$.

Now, for each *primitive* mth root of unity ζ, introduce a sequence of variables $t^{(\zeta)}_r \in \mathcal{K}^{fake}_+$, where $r = 1, 2, 3, \ldots$, and define the *adelic* tensor product

$$\langle \underline{\mathcal{D}}_X\rangle\,(\{t^{(\zeta)}_r\}, \hbar, Q) := \bigotimes_{M=1}^{\infty} \langle \mathcal{D}^{tw}_{X/\mathbb{Z}_M}\rangle\,(\sum_{\zeta:\, \zeta^M=1} \frac{\mathbf{t}^{(\zeta)}_{r(\zeta)}}{\sqrt{\hbar^{r(\zeta)}}}h_{(\zeta)}, 1, Q^M),$$

where for ζ of primitive order $m|M$, we put $r(\zeta) = M/m$.

Proposition 4.5.2 *The contribution to Wick's formula for $\langle \mathcal{D}_X\rangle$ of the one-vertex graph (i.e. by the moduli spaces of connected quotient curves $\widehat{\Sigma}$ in the notation of Sect. 4.3) is given by the logarithm* $\log\langle \underline{\mathcal{D}}_X\rangle$ *of adelic tensor product.*

The technical point in this proposition is that the dependence of the formula on $\hbar$ and Q correctly accounts for the Euler characteristics and degrees of the *covering* curves $\Sigma \to \widehat{\Sigma}$.

As we have already explained in Introduction, the adelic tensor product belongs to the Fock space associated with the symplectic loop space $(\underline{\mathcal{K}}^{\infty}, \underline{\Omega}^{\infty})$, which is obtained by rearranging sectors in the direct sum of the spaces $(\mathcal{K}^{tw}_{(M)}, \Omega^{tw}_{(M)})$. This direct sum comes with a Lagrangian polarization inherited from those of the summands. Let us call this polarization *standard*.

Recall now that adelic map $\underline{\ }$: $(\mathcal{K}^{\infty}, \Omega^{\infty}) \to (\underline{\mathcal{K}}^{\infty}, \underline{\Omega}^{\infty})$, defined in Introduction, is symplectic but does not respect polarizations. More precisely, the adelic image of $\mathcal{K}^{\infty}_{+}$ is a proper subspace in the positive space $\underline{\mathcal{K}}^{\infty}_{+}$ of the standard polarization, while the adelic image of $\mathcal{K}^{\infty}_{-}$ is Lagrangian in $\underline{\mathcal{K}}^{\infty}$, but does not coincide with the negative space of the standard polarization. Let us call *uniform* the polarization of the adelic loop space formed by the positive space of the standard polarization and by the adelic image of $\mathcal{K}^{\infty}_{-}$.

Proposition 4.5.3 *The change from the standard to the uniform polarization accounts for the edges (propagators) of Wick's summation over graphs.*

Sections 4.6–4.10 will be dedicated to the proof of Propositions 4.5.1–4.5.3. Also, in Sect. 4.8 we will see that the adelic embedding $\underline{\ }$: $\mathcal{K}^{\infty}_{+} \subset \underline{\mathcal{K}}^{\infty}_{+}$ of the positive spaces of our polarizations correctly transforms the inputs $\mathbf{t}_r$ of $\langle \mathcal{D}_X \rangle$ into the inputs of the adelic tensor product (they occur in the numerators of the fake holomorphic Euler characteristics in Kawasaki's RR formula). Altogether these results imply our Main Theorem:

The adelic map transforms the quantum state $\langle \underline{\mathcal{D}}_X \rangle$ *into* $\langle \mathcal{D}_X \rangle$.

4.6 Kawasaki Strata

We begin here with a detailed description of Kawasaki strata of moduli spaces of stable maps to X in terms moduli spaces of stable maps to *orbifolds* $X/\mathbb{Z}_M$.

Let $\phi : \Sigma \to X$ be a stable map of a compact nodal curve (not necessarily connected) with n non-singular marked points, and let $h : \Sigma \to \Sigma$ be a symmetry of this stable map (i.e. $\phi \circ h = \phi$) which is allowed to permute the marked points. Due to the stability condition, the symmetry has finite order, and therefore induces the quotient map $\widehat{\phi} : \widehat{\Sigma} \to X$ of the quotient curve $\widehat{\Sigma} := \Sigma/(h)$. Our nearest goal is to represent the combinatorial structure of the quotient map by a certain decorated graph Γ.

Let $p : \Sigma \to \widehat{\Sigma}$ denote the projection of factorization.

The *edges* of Γ correspond to *unbalanced nodes* of $\widehat{\Sigma}$. For a node $\widehat{\sigma} \in \widehat{\Sigma}$, denote by $r = r(\widehat{\sigma})$ the cardinality of its inverse image $p^{-1}(\widehat{\sigma})$ in Σ. The inverse image is an orbit of the action of (h) on Σ, and each point σ in it is a node of Σ fixed by h^r. In the case (let's call it *typical*) when h^r preserves each of the two branches of Σ at σ, it acts on the tangent lines to these branches at σ by eigenvalues $\zeta_{\pm}$.

The node is *unbalanced* if $\zeta_+\zeta_- \neq 1$. The same becomes true in the *atypical* case when h^r interchanges the two branches of Σ at σ, but only after a certain stabilysing modification of the curves gS and $\widehat{gS}$ which will be discussed later.

Normalizing the quotient curve $\widehat{\Sigma}$ at all unbalanced nodes, we obtain a collection of connected curves $\widehat{\Sigma}_v$ which by definition correspond to *vertices* v of graph Γ, and the maps $\widehat{\phi}_v : \widehat{\Sigma}_v \to X$, obtained by the restrictions of $\widehat{\phi}$. Moreover, each vertex comes with the ramified (h)-cover $\Sigma_v := p^{-1}\widehat{\Sigma}_v \to \widehat{\Sigma}_v$. More precisely, let $M = M_v$ be the order of h on Σ_v. Then outside the ramification locus, $p : \Sigma_v \to \widehat{\Sigma}_v$ is a principal $\mathbb{Z}_M$-bundle. This allows one to identify $\widehat{\phi}_v$ with a stable map in the sense of [1, 2, 11] to the *orbifold* target space $X/\mathbb{Z}_M$, the quotient of X by the trivial action of the cyclic group $(h)/(h^M)$.

The moduli space of stable maps to $X/\mathbb{Z}_M$ is characterized by certain discrete invariants, which we now describe in terms of $\widehat{\phi}_v$. First, it is the arithmetical *genus* $\widehat{g}_v$ of $\widehat{\Sigma}_v$. Next, it is the *degree* $\widehat{d}_v$, i.e. the homology class in $H_2(X;\mathbb{Z})$ represented by the map $\widehat{\phi}_v$. Furthermore, the vertex carries marked points, which represent in $\widehat{\Sigma}_v$ the orbits of marked points in Σ_v, ramification points which are not marked in Σ_v, and (the remnants in $\widehat{\Sigma}_v$ of) the unbalanced nodes. At each such marked point $\widehat{\sigma} \in \widehat{\Sigma}_v$, the *order* $r = r(\widehat{\sigma})$ of the inverse image of $\widehat{\sigma}$ in Σ_v is defined, as well as the eigenvalue $\zeta = \zeta(\widehat{\sigma})$ by which the symmetry h^r of Σ_v acts on the tangent line at any $\sigma \in p^{-1}\widehat{\sigma}$. Note that ζ is a primitive mth root of unity for some $m = M_v/r$. Therefore for some s (unique $\mod m$), we have $\zeta^s = e^{2\pi i/m}$. This determines the *sector* of the marked point, i.e. the element, h^{rs}, of the cyclic group $\mathbb{Z}_{M_v}$ which acts on $T_\sigma\Sigma_v$ by the generator $e^{2\pi i/m}$ of the isotropy group of $\widehat{\sigma}$ in the orbifold curve $\widehat{\Sigma}$.

Thus, the Kawasaki stratum in question is characterized by the graph Γ whose vertices correspond to moduli spaces of genus $\widehat{g}_v$ degree $\widehat{d}_v$ stable maps to $X/\mathbb{Z}_{M_v}$ with certain numbers $\widehat{n}_v$ of marked points. The marked points (which are usually depicted as flags sticking out of the vertices) are decorated by the sectors (or, equivalently, primitive mth roots of unity ζ with $m|M_v$), while the edges pair the *unbalanced* flags ($\zeta_+\zeta_- \neq 1$) of *the same* order: $r_+ = M_{v_+}/m_+ = M_{v_-}/m_- = r_-$.

Conversely, given such a decorated graph Γ, one can form the corresponding Kawasaki stratum by *gluing* stable maps to $X/\mathbb{Z}_{M_v}$ corresponding to the vertices of Γ over the diagonal constraints ($\mathrm{ev}_+ = \mathrm{ev}_-$) corresponding to the edges. More precisely, each stable map to $X/\mathbb{Z}_{M_v}$ comes equipped with a principle $\mathbb{Z}_{M_v}$-bundle, possibly ramified at the markings. The generators of the groups $\mathbb{Z}_{M_v}$ define a symmetry h of the total map to X from the union of the covers. Since the glued marked points have the same order r, the covers can be glued h-equivariantly, resulting in stable maps to X (possibly disconnected), equipped with prescribed symmetries h (of order equal to the least common multiple of all M_v).

By applying this construction to all (possibly disconnected) decorated graphs Γ, one obtains all Kawasaki strata of all moduli spaces of (possibly disconnected) stable maps to X.

Remarks

(a) When a node $\widehat{\sigma}$ of the curve $\widehat{\Sigma}$ is *balanced*, i.e. h^r fixes a node $\sigma \in p^{-1}(\widehat{\sigma})$ but acts on the branches of Σ at the node by inverse primitive mth roots of unity

$(\zeta_+\zeta_- = 1)$, the stable map is deformable, at least in the virtual sense, to a non-nodal curve within the same Kawasaki stratum. The local model of h^r near σ is given by

$$xy = \epsilon, h^r(x, y) = (\zeta_+ x, \zeta_- y),$$

where $\epsilon = 0$ corresponds to the nodal curve. The requirement above that the nodes corresponding to the edges of the graph are unbalanced prevents such deformations and guarantees that the stratum of symmetric maps glued according to a given graph is maximal (e.g. in the sense that 1 does not occur as an eigenvalue of the symmetry on the virtual normal bundle to the stratum).

(b) One more type of deformable nodes of Σ occurs when h^r fixes a node σ but interchanges the branches of Σ. The local model of this phenomenon can be described by the formulas:

$$xy = \epsilon, \ h^r(x, y) = (y, x), \ \phi(x, y) = x + y,$$

so that at $\epsilon = 0$, the quotient curve doesn't seem to have a node. Here is how this situation is captured in terms of orbifold stable maps. For $\epsilon \neq 0$, the map $\phi = x + y$ restricted to $xy = \epsilon$ has two ramification points: $(x, y) = \pm(\sqrt{\epsilon}, \sqrt{\epsilon})$. Thus, the quotient curve has two marked points $\pm 2\sqrt{\epsilon}$ with inertia groups $\mathbb{Z}_2$. When ϵ tends to 0, the quotient curve becomes reducible, with a new component $\mathbb{C}P^1$ mapped with degree 0, and carrying both marked points with the inertia group $\mathbb{Z}_2$ (as well as the node with the trivial isotropy group, see Fig. 4.2). The covering curve has now 3 components: two branches interchanged by the symmetry and connected by $\mathbb{C}P^1$, which carries two marked points (say, at $z = 0, \infty$), and two nodes (at $z = \pm 1$). The symmetry acts on this component by $z \mapsto -z$, so that the quotient has the node at $z^2 = 1$, and two marked points $z^2 = 0, \infty$. Thus, the quotient map, properly understood in terms of stable maps to $X/\mathbb{Z}_2$, has a balanced node of order $r = 2$ with the eigenvalues $\zeta_\pm = 1$.

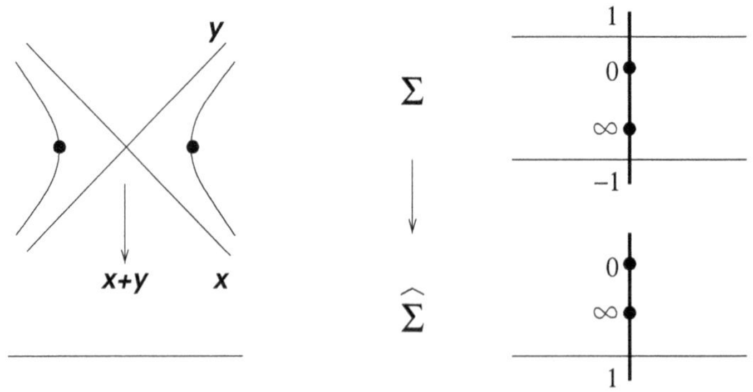

Fig. 4.2 $\mathbb{Z}_2$-invariant nodes with interchanged branches

(c) The constantly mapped curves $\mathbb{C}P^1$ with the $\mathbb{Z}_2$-symmetry just described also play the key role in the assessment of the mentioned earlier atypical case of nodes $\sigma \in \Sigma$ fixed by h^r which interchanges the branches of gS at the node. The projection $p : \Sigma \to \widehat{gS}$ in this case does not qualify on the role of the principal bundle defining the orbifold structure of $\widehat{gS}$ in the Chen-Ruan theory. Instead, the curve Σ needs to be modified by gluing to it at $\widehat{\sigma}$ a new (constantly mapped) component $\mathbb{C}P^1/\mathbb{Z}_2$ by the marked point $z^2 = 1$ (in the above notation) as shown on the bottom right side of Fig. 4.2. At the same time, the curve gS needs to be modified by normalizing it at σ and other nodes from its h-orbits and gluing in r copies of $\mathbb{C}P^1$ attached by its marked points $z = \pm 1$ as shown on the top right of Fig. 4.2. The symmetry h extends to the modified curve Σ in the obvious way by cyclically permutting the r copies of $\mathbb{C}P^1$ in such a way that h^r, though preserves each $\mathbb{C}P^1$, acts on in it as the $\mathbb{Z}_2$-symmetry $z \mapsto -z$. Consequently h^{2r} now fixes each of the $2r$ inverse images of the node $\widehat{\sigma} \in \widehat{gS}$ and preserves each of the branches of the modified covering curve, acting on them with the eigenvalues 1 (on the incerted $\mathbb{C}P^1$) and ζ. The balanced case $\zeta = 1$ is deformable and was discussed in Remark (b) above, while the case $\zeta \neq 1$ is unbalanced. The order of the node $\widehat{\sigma}$ in this case equals $2r$.

(d) The newly attached component $\mathbb{C}P^1$ of the quotient curve $\widehat{gS}$, taken on its own, is a degree-0 stable map to $X/\mathbb{Z}_2$ with 3 marked points, of which two ($z^2 = 0, \infty$) are ramification points of the covering map $z \mapsto z^2$. On the covering r copies $\mathbb{C}P^1$ considered as constantly mapped curves with a $\mathbb{Z}_2$-symmetry in X, both points $z = 0, \infty$ are unmarked, making these covering $\mathbb{C}P^1$ (carrying only the 2-cycle of nodal points $z = \pm 1$) unstable. As far as stable maps to X are concerned, these component have to be contracted, and thus the original stable map $\phi : \Sigma \to X$ is recovered.

Note, however, that such constantly mapped $\mathbb{C}P^1/\mathbb{Z}_2$ (and r-fold collections of them) not glued to anything are legitimate stable maps to $X/\mathbb{Z}_2$. Their contributions to the respective descendant potentials $\mathcal{D}^{tw}_{X/\mathbb{Z}_2}$ are present in the adelic tensor product and should therefore be removed from it. Instead, we chose to include these contributions into $\mathcal{D}_X$ in the form of the anomalous term added to $\mathcal{F}_0$.

4.7 Twistings

The denominators $\mathrm{str}_h \wedge^\bullet N^*_{I\mathcal{M}}$ in Kawasaki's RR formula can be interpreted as certain *twistings* of the fake quantum K-theory of $X/\mathbb{Z}_M$, in fact a combination of several types of twistings, corresponding to different ingredients of the virtual conormal bundles.

Let $\mathcal{M}$ denote a Kawasaki stratum, i.e. (a component of) a moduli space $(X/\mathbb{Z}_M)_{\widehat{g},\widehat{n},\widehat{d}}$. Let $\mathrm{ft} : \mathcal{C} \to \mathcal{M}$ be the corresponding universal curve, and $\mathrm{ev} : \mathcal{C} \to X/\mathbb{Z}_M$ the universal stable map, while $\widetilde{\mathrm{ft}} : \widetilde{\mathcal{C}} \to \mathcal{M}$ and $\widetilde{\mathrm{ev}} : \widetilde{\mathcal{C}} \to X$ denote $\mathbb{Z}_M$-equivariant lifts of ft and ev to the family of ramified $\mathbb{Z}_M$-covers.

The Kawasaki stratum $\mathcal{M}$ carries (the restriction to $\mathcal{M}$ of) the virtual tangent bundle (let's call it $\mathcal{T}$) to the ambient moduli space of stable maps to X (say, $X_{g,n,d}$). Following [3] (see p. 99), we describe it in terms of the universal curve $\mathrm{ft} : \mathcal{C} \to \mathcal{M}$:

$$\mathcal{T} = \widetilde{\mathrm{ft}}_* \widetilde{\mathrm{ev}}^* (T_X - 1) + \widetilde{\mathrm{ft}}_* (1 - \widetilde{L}^{-1}) - (\widetilde{\mathrm{ft}}_* \widetilde{j}_* \mathcal{O}_{\widetilde{\mathcal{Z}}})^\vee .$$

Here $\widetilde{L}$ is the universal cotangent line bundle to the fibers of $\widetilde{\mathrm{ft}}$ (i.e. the cotangent line bundle L_{n+1} at the marked point forgotten by $\widetilde{\mathrm{ft}} : \mathcal{C} \subset X_{g,n+1,d} \to X_{g,n,d}$), and $\widetilde{j}$ is the embedding of the nodal locus $\widetilde{\mathcal{Z}}$ into $\widetilde{\mathcal{C}}$. Loosely speaking, the three summands correspond to: (A) deformations of the maps of curves with a fixed complex structure, (B) deformations of the complex structure of curves with fixed combinatorics, and (C) the smoothing of the nodes.

The summands carry the action of $\mathbb{Z}_M$, and can be decomposed into the eigenbundles corresponding to the eigenvalues $\lambda = e^{2\pi i k/M}$ of the generator. The normal bundle $N_{I\mathcal{M}}$, featuring in the denominator of Kawasaki's RR formula, consists of the eigenbundles corresponding to $\lambda \neq 1$.

To decompose $\mathcal{T}$ into the eigenbundles, introduce the 1-dimensional representation $\mathbb{C}_\lambda$ of $\mathbb{Z}_M$ where the generator acts by λ. Then the eigenbundles have the form

$$\begin{aligned} \mathcal{T}_{\lambda^{-1}} = (\mathcal{T} \otimes \mathbb{C}_\lambda)^{\mathbb{Z}_M} &= \mathrm{ft}_* \, \mathrm{ev}^* [(T_X - 1) \otimes \mathbb{C}_\lambda] \\ &+ \mathrm{ft}_* [(1 - \widetilde{L}^{-1}) \otimes \mathrm{ev}^* \, \mathbb{C}_\lambda] - (\mathrm{ft}_* [j_* \mathcal{O}_{\widetilde{\mathcal{Z}}} \otimes \mathrm{ev}^* \, \mathbb{C}_\lambda])^\vee , \end{aligned}$$

where j is the embedding of $\mathcal{Z} = \widetilde{\mathcal{Z}}/\mathbb{Z}_M$ into $\mathcal{C}$. The terms on the right are interpreted as K-theoretic push-forwards by $\mathrm{ft} : \mathcal{C} \to \mathcal{M}$ of orbibundles on the global quotient $\mathcal{C} = \widetilde{\mathcal{C}}/\mathbb{Z}_M$. By the very definition, such push-forward automatically extracts from the sheaf cohomology its $\mathbb{Z}_M$-invariant part.

Now we use the three twisting results of [17] to express the effect of the denominator in Kawasaki's RR formula in terms of twisted GW-invariants of orbifolds $X/\mathbb{Z}_M$.

The answer consists in the application of three operations:

(A) Transformation

$$\mathcal{D}^{fake}_{X/\mathbb{Z}_M} \mapsto \mathcal{D}^{tw}_{X/\mathbb{Z}_M} = \widehat{\square}_M \mathcal{D}^{fake}_{X/\mathbb{Z}_M}$$

by some quantized symplectic operator (to be described and calculated later) acting block-diagonally by $\square_M^{(h^s)}$ in the decomposition into sectors $h^s \in \mathbb{Z}_M$ of the appropriate symplectic loop spaces.

(B) Change in the dilaton shift: $(1 - q)\mathbf{1} \mapsto (1 - q^M)\mathbf{1} = \Psi^M (1 - q)\mathbf{1}$.

(C) Change of polarization, different on each sector (to be described later).

In fact the three twisting theorems of [17] are stated in terms of cohomological GW-invariants of the orbifold target ($X/\mathbb{Z}_M$ in our case). In order to relate the fake K-theory of $\mathcal{M}$ in Kawasaki's formula with cohomology theory, one needs to apply the three twistings with the same bundles as above, but with $\lambda = 1$, and the Todd characteristic class, $\mathrm{td}(x) = x/(1-e^{-x})$. This results in the respective three operations described in the previous section and transforming $\mathcal{D}^H_{X/\mathbb{Z}_m}$ to $\mathcal{D}^{fake}_{X/\mathbb{Z}_M}$: by (A) application of $\widehat{\mathrm{qch}}^{-1}\widehat{\triangle}$ (the same in each sector), (B) change of the dilaton shift $-z \mapsto 1-e^z = 1-q$ (in sector **1**), and (C) change of polarization from $\mathcal{H}_-$ to $\mathcal{K}^{fake}_-$ (the same in each sector). Such operations result in expressing $\mathcal{D}^{fake}_{X/\mathbb{Z}_M}$ in terms of $\mathcal{D}^H_X$ as it was explained in Sect. 4.4. The twistings A,B,C with $\lambda \neq 1$ come on the top of these, which makes it easy to phrase their outcomes directly in terms of fake quantum K-theory of $X/\mathbb{Z}_M$.

(A) The first twisting result goes back to Tseng's "orbifold quantum RR Theorem" [20]. It allows us to expresses cohomological GW-invariants of $X/\mathbb{Z}_M$ twisted by the orbibundle $E = (T_X - 1)\otimes \mathbb{C}_\lambda$ and by the multiplicative characteristic class td_λ defined by its value $1/(1-\lambda e^{-x})$ on a line bundle with the 1st Chern class x. Namely,

$$\langle \mathcal{D}^{tw}_{X/\mathbb{Z}_M}\rangle = \left[\prod_{k=1}^{M-1} \widehat{\triangle}_{e^{2\pi i k/M}}\right] \langle \mathcal{D}^{fake}_{X/\mathbb{Z}_M}\rangle,$$

where $\triangle_{e^{2\pi i k/M}}$ is the operator $\mathcal{K}^{fake}_{X/\mathbb{Z}_M} \to \mathcal{K}^{tw}_{X/\mathbb{Z}_M}$ which on the copy of $\mathcal{K}^{fake}_X$ corresponding to the *sector* $h^s \in \mathbb{Z}_M$ acts as the multiplication by the Euler–Maclaurin asymptotics of the following infinite product:

$$\triangle_{e^{2\pi i k/M}} \sim \frac{\prod_{l=1}^{\infty}(1-e^{2\pi i k/M}q^l q^{-\{ks/M\}})}{\prod_{l=1}^{\infty}\prod_{i=1}^{\dim_{\mathbb{C}} X}(1-e^{2\pi i k/M}e^{-x_i}q^l q^{-\{ks/M\}})}.$$

Here x_i are Chern roots of T_X, and $\{ks/M\}$ denotes the fractional part of ks/M.

We rearrange the product $\prod_{k=1}^{M-1}\triangle_{e^{2\pi i k/M}}$. Let $r = (s, M)$ be the greatest common divisor of s and M, so that $s = rs'$, $M = rm$, $(s', m) = 1$, and $ks/M = ks'/m$. Let t' be inverse to s' modulo m. Write $k = k't' + mu$ with $0 \le k' < m$. Then $\{ks'/m\} = k'/m$ for any u. Since $\prod_{u=1}^{r}(1-Ye^{2\pi i u/r}) = 1-Y^r$ for any Y, we have

$$\prod_{k=1}^{M-1}\prod_{l=1}^{\infty}(1-e^{2\pi i k/M}Yq^l q^{-\{ks/M\}}) = \prod_{k'}\prod_{l=1}^{\infty}(1-e^{2\pi i k't'/m}Y^r q^{rl}q^{-k'r/m})$$

$$= \prod_{l=0}^{\infty}(1-Y^r q^{lr/m}\eta^{-l})/\prod_{l=0}^{\infty}(1-Yq^l).$$

Here $\eta := e^{2\pi i t'/m}$ satisfies $\eta^{s'} = e^{2\pi i/m}$, i.e. η is the eigenvalue by which the symmetry h^r acts on the tangent lines to the curves at the marked point of order r and sector $h^s = h^{s'r}$. Also note that the Euler–Maclaurin asymptotics of the infinite product near $q = 1$ is written as

$$\prod_{l=0}^{\infty}(1 - Yq^l) \sim e^{-\sum_{k>0} Y^k/k(1-q^k)}.$$

Using this, and the abbreviation $\sum_j e^{-kx_j} - 1 = \mathrm{ch}(\Psi^k(T_X^* - 1))$, we can summarize the above computation this way:

$$\square_M^{(h^s)} = e^{\sum_{k>0}\left(\frac{\Psi^{kr}(T_X^*-1)}{k(1-\eta^{-k}q^{kr/m})} - \frac{\Psi^k(T_X^*-1)}{k(1-q^k)}\right)}.$$

The answer for $\square_M^{(h^s)}$ coincides with what was denoted by $\square_{\eta,r}$ in Sect. 4.5, where η is a primitive mth root of unity, $M = mr$, $s = rs'$, and $\eta^{s'} = e^{2pii/m}$.

(B) The effect of the twisting by $\widetilde{\mathrm{ft}}_*\left[(1 - \widetilde{L}^{-1}) \otimes \widetilde{\mathrm{ev}}^*\mathbb{C}_\lambda\right]$ is described by Corollary 6.1 in [17]. That paper, instead of the bundle $\widetilde{L}$ on the covering universal curve $\widetilde{\mathcal{C}}$, deals with the universal cotangent line bundle $L = L_{n+1}$ on $\mathcal{C} = \widetilde{\mathcal{C}}/\mathbb{Z}_M$. To apply the result of that paper, it is important to realize that $\widetilde{L} = p^*L$ where $p : \widetilde{\mathcal{C}} \to \mathcal{C}$ is the projection of factorization. Indeed, $\widetilde{L}$ is the canonical bundle of the covering curve twisted by the marked points. In local coordinates, it has a local section $x^{-1}dx$ near a marked point $x = 0$, and $dx \wedge dy/d(xy)$ on the curves $xy = \epsilon$ near a node. The formulas

$$\frac{dx^m}{x^m} = m\frac{dx}{x}, \text{ and } \frac{dx^m \wedge dy^m}{d(x^m y^m)} = m\frac{dx \wedge dy}{d(xy)}$$

identify p^*L with $\widetilde{L}$ near a ramified m-fold marked point and a balanced m-fold node respectively. The answer, as we've already said, is the change of the dilaton shift: $(1 - q)\mathbf{1} \mapsto \Psi^M(1 - q)\mathbf{1} = (1 - q^M)\mathbf{1}$.

Remark The result does not depend on the character λ of $\mathbb{Z}_M$. To understand why, the reader is invited to examine the details of the proof in [17], namely formula (4.2). The explanation is that the bundle L is trivialized at the marked points and at the nodes (as the above local coordinate sections indicate). Consequently, Kawasaki's Chern character of $1 - L^{-1}$ vanishes on all twisted sectors of the inertia orbifold $I\Sigma$ of the orbi-curve Σ. On the unit sector, however, all $\mathbb{C}_\lambda$ coincide.

(C) To describe the change of polarization caused by the twistings of type C, consider the expression $(1 - L_+^{1/m} \otimes L_-^{1/m})^{-1}$. It comes from the inverse to the K-theoretic Euler class $1 - L_+^{1/m}L_-^{1/m}$ of the virtual normal line bundle to the nodal stratum in $\mathcal{M}$ at the nodes of order $r = M/m$, assuming that $L_\pm$ represent the universal cotangent lines to the branches of quotient curve at the node. We expand the expression in powers of $L_-^{1/m} - 1$:

$$\frac{1}{1 - L_+^{1/m} \otimes L_-^{1/m}} = \frac{1}{1 - L_-^{1/m} - L_-^{1/m} \otimes (L_+^{1/m} - 1)}$$
$$= \sum_{k \geq 0} \frac{L_-^{k/m}}{(1 - L_-^{1/m})^{k+1}} \otimes (L_+^{1/m} - 1)^k.$$

Let $\{\phi_\alpha\}$ and $\{\phi^\alpha\}$ denote bases in $K^0(X)$ dual with respect to the K-theoretic Poincaré pairing. In the subspace $\mathcal{K}_+^{fake} h^{-s} \subset \mathcal{K}_{X/\mathbb{Z}_M}^{fake}$ (here h^{-s} indicates the sector, and $r = (s, M)$ is assumed), we have a topological basis in $\mathcal{K}_+^{fake}$ $(k \geq 0, \alpha = 1, \ldots, \dim K^0(X))$:

$$\Psi^r \left(\phi^\alpha (q^{1/m} - 1)^k \right) = \Psi^r (\phi^\alpha)(q^{r/m} - 1)^k.$$

Then the following rational functions

$$r\Psi^r \left(\phi_\alpha \frac{q^{k/m}}{(1 - q^{1/m})^{k+1}} \right) = r\Psi^r (\phi_\alpha) \frac{q^{kr/m}}{(1 - q^{r/m})^{k+1}},$$

expanded into Laurent series near $q = 1$, span the negative space of the polarization in question in the sector h^{-s} of $\mathcal{K}_{X/\mathbb{Z}_M}^{fake}$. Moreover, the indicated vectors altogether form a Darboux basis in $\mathcal{K}_{X/\mathbb{Z}_M}^{tw}$ with respect to the symplectic form based on the following twisted pairing:

$$(ah^s, bh^t)^{(r)} = \frac{\delta_{h^s h^t, 1}}{M} \int_X \mathrm{td}(\Psi^r (T_X - 1) \,\mathrm{ch}(a) \,\mathrm{ch}(b).$$

The result just described can be derived from a general theorem in [17] (see Corollary 6.3 therein). It can be justified in a more direct way as well. Namely, in the non-orbifold situation, the effect of the nodal twisting leads, as it was found in the thesis [3] of T. Coates, to the change of polarization based (as it has just been described) on the "inverse Euler class" $(1 - L_+ \otimes L_-)^{-1}$. In our situation of the target $X/\mathbb{Z}_M$, the smoothing of the nodes of order $r = 1$ contributes into the virtual tangent bundle to Kawasaki's stratum $\mathcal{M}$ the same 1-dimensional summand, $\widetilde{L}_+^{-1} \widetilde{L}_-^{-1} = L_+^{-1/M} L_-^{-1/M}$, as into the virtual tangent bundle $\mathcal{T}$ to the ambient moduli space of stable maps to X. This means that Coates' computation still applies, with the only change that the "inverse Euler class" has the form $(1 - L_+^{1/M} \otimes L_-^{1/M})^{-1}$. In the case of nodes of order $r > 1$, the covering curves contain the $\mathbb{Z}_M$-orbit consisting of r copies $\mathbb{Z}_m$-invariant nodes ($mr = M$), each contributing into $\mathcal{T}$ a copy of $L_+^{-1/m} L_-^{-1/m}$, cyclically permuted by $\mathbb{Z}_r = \mathbb{Z}_M / \mathbb{Z}_m$. The "inverse Euler class" of their sum is $\Psi^r (1 - L_+^{1/m} \otimes L_-^{1/m})^{-1} = (1 - L_+^{r/m} \otimes L_-^{r/m})^{-1}$ due to the following fact that $\Psi^r(V) = \mathrm{tr}_h\, V^{\otimes r}$, where h acts on the tensor product by the cyclic permutation of the r factors.

This completes the proof of Proposition 4.5.1.

Remark We should revisit the phenomenon of $\mathbb{Z}_2$-invariant nodes with interchanged branches to examine their contribution to the type C twistings. The cotangent line bundles $L_\pm$ to the branches at the node are identified by the $\mathbb{Z}_2$-symmetry: $L_+ = L_- =: L$. Respectively the smoothing of the node contributes $L_+^{-1}L_-^{-1} = L^{-2}$ to the tangent bundle $\mathcal{T}$, and the corresponding Euler factor in the denominator of Kawasaki's formula is $1 - L^2$. It turns out that the interpretation of the situation in terms of maps to $X/\mathbb{Z}_2$ leads to the same contribution of the nodal locus. The line bundles $L_\pm$ are now identified with the cotangent lines to the interchanged branches at the two nodes ± 1 of the resolved curve (top right on Fig. 4.2). Since the configuration of $0, \infty, 1, -1$ on the exceptional $\mathbb{C}P^1$ (vertical line at the top right) is standard, the tangent lines to this $\mathbb{C}P^1$ at ± 1 are trivialized. Consequently the smoothing deformation modes of the curve add up to $L_+^{-1} \oplus L_-^{-1}$ with the $\mathbb{Z}_2$-action interchanging the summands. Therefore the Euler factors representing the $\mathbb{Z}_2$-invariant and anti-invariant modes in the denominator of Kawasaki's formula are $1 - L$ and $1 + L$, and their product is $1 - L^2$, i.e. the same as above.

4.8 Inputs

We denote by $\mathbf{t}_r(q) = \sum_{k\in\mathbb{Z}} t_{r,k} q^k$ the inputs in the total descendant potential $\mathcal{D}_X$ of quantum K-theory on X, corresponding to the cycles of length $r = 1, 2, 3, \ldots,$ and examine how they contribute to the numerators in Kawasaki's RR formula on the stratum $\mathcal{M}$ (still assuming that the decorated graph of the stratum consists of one vertex).

The numerators have the form of the *trace* tr_h of the tensor product of contributions which come from the marked points.

Let M be the degree of the covers associated with a given vertex. Let $L = L_i$ denote the universal cotangent line at a marked point on the *quotient* curve $\widehat{\Sigma} = \Sigma/\mathbb{Z}_M$, $r = r_i$ the order of the marked point, $m = M/r$ the ramification index (of the r copies in Σ) of this marked point, and ζ the primitive mth root of unity by which the symmetry h^r acts on the *tangent* line to Σ at each of the r copies of the marked point. Omitting the index i and the pull-back by the evaluation map ev_i, we can express the resulting input of $\mathcal{D}^{tw}_{X/\mathbb{Z}_M}$ in the sector determined by ζ this way:

$$\mathrm{tr}_h[\mathbf{t}_r(\zeta^{-1}L^{1/m})]^{\otimes r} = \Psi^r[\mathbf{t}_r(\zeta^{-1}L^{1/m})] = \sum_{k\in\mathbb{Z}} \Psi^r(t_{r,k})\zeta^{-k}L^{kr/m}.$$

Note the presence of the weight factors $1/\prod r_i^{l_i}$ in front of the supertraces str_h in the definition of the correlators involved in $\mathcal{D}_X$. In the expression of the correlators in terms of Kawasaki strata, these factors are compensated in the following way. Given a stable map $\widehat{\Sigma} \to X/\mathbb{Z}_M$, a marked point $\widehat{\sigma} \in \widehat{\Sigma}$ with the ramification index $m = M/r$ represents stable maps $\Sigma \to X$ with a prescribed symmetry, which in particular cyclically permutes r marked points $\sigma_1, \dots, \sigma_r$ over $\widehat{\sigma}$. Even when the indices of the r marked points are already decided (e.g. 1 goes to 2 etc. goes to r goes to 1), there still remain r choices for deciding which of the marked points $\sigma_i \in \Sigma$ is numbered by 1. Thus totally for each map $\widehat{\Sigma} \to X/\mathbb{Z}_M$ in the Kawasaki stratum there are $\prod r_i^{l_i}$ symmetric maps $\Sigma \to X$, and this compensates the weight factor.

It is now time to realize that not all marked points of $\widehat{\Sigma}$ come from marked points of $\phi : \Sigma \to X$. Namely, in the theory of stable maps to $X/\mathbb{Z}_M$, all ramification points are declared marked, even if they are unmarked for the covering stable map to X. Consequently, the virtual cotangent bundle $\mathcal{T}^*$ which was analyzed in the previous section, and whose Euler class occurs in the denominator of Kawasaki's formula, is in fact the cotangent bundle to the ambient moduli space of stable maps to X with extra marked points introduced at the ramifications. To compensate for these modes of deformation of stable maps, we thus need to multiply the numerator by the appropriate Euler class. Namely, if our marked point is such a ramification point, the correction has the form (one factor $1 - \zeta^{-1}L^{1/m}$ per each of the r copies of the ramification points):

$$\mathrm{tr}_h(1 - \zeta^{-1}L^{1/m})^{\otimes r} = \Psi^r(1 - \zeta^{-1}L^{1/m}) = 1 - \zeta^{-1}L^{r/m}.$$

There is an exception: the unramified marked points ($m = 1$, $r = M$) of $\widehat{\Sigma}$ can come only from the orbits of marked points on Σ. Note however, that in this case the same formula yields

$$\Psi^M(1 - L) = 1 - L^M,$$

which agrees with the dilaton shift in $\mathcal{D}^{tw}_{X/\mathbb{Z}_M}$ in the unramified sector.

To summarize our observations, let us assume that the generating function $\mathcal{D}^{tw}_{X/\mathbb{Z}_M}$ is already dilaton-shifted by $1 - q^M$ in the unit sector, and denote by $t_r^{(\zeta)} \in \mathcal{K}_+^{fake}$ the input of it through the sector indicated by the primitive mth root of unity ζ, where $r = M/m$. Then the substitution

$$t_r^{(\zeta)}(q) = \Psi^r\left[1 - \zeta^{-1}q^{1/m} + \mathbf{t}_r(\zeta^{-1}q^{1/m})\right],$$

factors correctly into the numerators (and denominators) of Kawasaki's RR formula. In other words, the inputs $t_r^{(\zeta)} \in \mathcal{K}_+^{fake}$ of $\mathcal{D}^{tw}_{X/\mathbb{Z}_M}$ (dilaton-shifted by $1 - q^M$ when $\zeta = 1$) are obtained from the inputs $\mathbf{t}_r \in K[q, q^{-1}]$ of $\mathcal{D}_X$ *dilaton shifted by* $(1-q)\mathbf{1}$ *for each* $r = 1, 2, 3, \dots$ by expanding $\Psi^r\mathbf{t}_r(\zeta^{-1}q^{1/m})$ into $q - 1$-series.

This is what we claimed at the end of Sect. 4.5.

4.9 Hurwitz' Formula

Here we determine the discrete characteristics of the *covering* of the map $\phi : \Sigma \to X$, given the decorated graph Γ of the quotient map $\widehat{\phi} : \widehat{\Sigma} \to X$.

The degree of ϕ is given by $d = \sum_v \widehat{d}_v M_v$, where $\widehat{d}_v \in H_2(X;\mathbb{Z})$ is the degree of the vertex v, and M_v is the degree of the covering $\Sigma_v \to \widehat{\Sigma}_v$.

Let us find the topological Euler characteristic eu of typical curves from the moduli spaces to which $\phi : \Sigma \to X$ belongs. The computation is similar to that in Hurwitz' genus formula. The vertex curves $\widehat{\Sigma}_v$ with all the $\widehat{n}_v$ ramification (i.e. marked or special) points removed have the Euler characteristics $2-2\widehat{g}_v-\widehat{n}_v$, which need to be multiplied by the degrees M_v of the coverings. Gluing in the orbits of the ramification points of order r_i, $i = 1, \dots, \widehat{n}_v$, adds r_i units for each respective orbit. Each edge e of order r_e corresponds to an $\mathbb{Z}_{r_e}$-orbit of (unbalanced) nodes. This subtracts $\sum_e r_e$ units from $\mathrm{eu}(\Sigma)$, but the smoothing of all nodes subtracts $\sum_e r_r$ once more. We get

$$\mathrm{eu} = \sum_v M_v(2-2\widehat{g}_v-\widehat{n}_v) + \sum_v \sum_{i=1}^{\widehat{n}_v} r_i - 2\sum_e r_e.$$

Recall that our eventual goal is to represent the total descendant potential of X by Kawasaki's RR formula as the sum over decorated graphs of the contributions of the respective Kawasaki strata. The contribution of the stratum represented by a given Γ should be somehow obtained, starting from the product of the twisted fake potentials $\mathcal{D}^{tw}_{X/\mathbb{Z}_{M_v}}$, corresponding to the vertices of Γ, and then "marrying" them appropriately by the edges. What we want to discuss now is how to dispose of the Planck constant variable $\hbar$ and Novikov's variables Q in the vertex factors in order to achieve the correct overall occurrence of $\hbar$ and Q in the total descendant potential of X.

Recall that contributions to $\mathcal{D}_X$ are weighted by the powers $\hbar^{-\mathrm{eu}/2}$, where eu is the Euler characteristic of the curve, connected or not, mapped to X. We have:

$$-\frac{\mathrm{eu}}{2} = \sum_v M_v(\widehat{g}_v-1) + \sum_v M_v\frac{\widehat{n}_v}{2} - \sum_v \sum_{i=1}^{\widehat{n}_v} \frac{r_i}{2} + \sum_e r_e.$$

The four terms of the sum lead to the following strategy.

(i) In each factor $\langle \mathcal{D}^{tw}_{X/\mathbb{Z}_{M_v}} \rangle$, replace $\hbar$ with $\hbar^{M_v}$.
(ii) Replace the (dilaton-shifted) input $\mathbf{t}$ of the marked points with $\hbar^{M_v/2}\mathbf{t}$.
(iii) At each marked point of order r divide the input by (another) factor $\hbar^{r/2}$.

(iv) Each "marriage" by an edge of order r should be accompanied by the factor $\hbar^r$.

(v) Each monomial $Q^{\widehat{d}_v}$ representing in $\langle \mathcal{D}^{tw}_{X/\mathbb{Z}_{M_v}} \rangle$ the contributions of degree $\widehat{d}$ orbicurves in $X/\mathbb{Z}_{M_v}$ should be replaced with $Q^{M_v \widehat{d}_v}$.

Now, the point is that *due to the homogeneity of* $\langle \mathcal{D}^{tw}_{X/\mathbb{Z}_M} \rangle$, the steps (i) and (ii) of our strategy cancel each other, and so the steps (iii), (iv), and (v) suffice.

In particular, referring to (iii) and (v), together with the results of the previous section, we find that the vertex contribution into Wick's formula can be described in terms of $\langle \mathcal{D}^{tw}_{X/\mathbb{Z}_M}(\mathbf{t}, \hbar, Q) \rangle$ as the adelic product:

$$\prod_{M=1}^{\infty} \langle \mathcal{D}^{tw}_{X/\mathbb{Z}_M} \rangle \left(\sum_{\zeta : \zeta^M = 1} \Psi^{r(\zeta)} \left[\frac{\mathbf{t}_{r(\zeta)}(\zeta^{-1} q^{1/m(\zeta)})}{\sqrt{\hbar}} \right] h_{(\zeta)}, 1, Q^M \right),$$

where $\mathbf{t}_r \in K[q, q^{-1}]$ are the arguments of $\langle \mathcal{D}_X \rangle$.

This completes the proof of Proposition 4.5.2.

4.10 Propagators

Recall that the edges of decorated graphs Γ correspond to unbalanced nodes of the quotient curves $\widehat{\Sigma}$. Such a node *of order* r represents r-tuples of nodes of the covering curve Σ cyclically permuted by the symmetry h. On the two branches of the curve Σ at such a node, h^r acts with the eigenvalues $\eta_\pm$, which are primitive roots of unity of certain orders $m_\pm$. The node is *unbalanced* if $\eta_+ \eta_- \neq 1$ (regardless of whether $m_\pm$ coincide or not). The effect of the unbalanced node on the contribution of the stratum $\mathcal{M}$ (determined by Γ) to Kawasaki's RR formula can be described as follows.

Let $L_\pm$ denote the cotangent lines to the two branches of the quotient curve $\widehat{\Sigma}$ at the node, so that $L_\pm^{1/\mathrm{m}_\pm}$ denote such cotangent lines to the covering curves. The following expression

$$\Psi^r \nabla_{\eta_+, \eta_-} = \Psi^r \frac{\sum_\alpha \phi_\alpha \otimes \phi^\alpha}{1 - \eta_+^{-1} L_+^{1/m_+} \otimes \eta_-^{-1} L_-^{1/m_-}} = \frac{\sum_\alpha \Psi^r \phi_\alpha \otimes \Psi^r \phi^\alpha}{1 - \eta_+^{-1} L_+^{r/m_+} \otimes \eta_-^{-1} L_-^{r/m_-}}$$

can be considered as an element of $K[[L_+^{r/m_+} - 1]] \otimes K[[L_-^{r/m_-} - 1]]$, where $K = K^0(X) \otimes \Lambda$, and $\{\phi_\alpha\}$ and $\{\phi^\alpha\}$ are Poincaré-dual bases in $K^0(X)$. In this capacity, $\Psi^r \nabla_{\eta_+, \eta_-}$ act as biderivations in the variables $\mathbf{t}_r^{(\eta_\pm)}$ of the factors $\langle \mathcal{D}^{tw}_{X/\mathbb{Z}_{M_\pm}} \rangle$ (with $M_\pm = r m_\pm$) in the adelic tensor product $\langle \underline{\mathcal{D}}_X \rangle$. With this notation,

Wick's summation over all graphs consists in the application to $\langle \underline{\mathcal{D}}_X \rangle$ (i.e. to the contribution of one-vertex graphs) of the following "propagator" (edge) operator:

$$\langle \underline{\mathcal{D}}_X \rangle \mapsto \exp\left[\bigoplus_{r>0} \frac{r}{2} \hbar^r \Psi^r \left(\sum_{\eta_+\eta_- \neq 1} \nabla_{\eta_+,\eta_-} \right)\right] \langle \underline{\mathcal{D}}_X \rangle.$$

The summation sign $\oplus$ is to emphasize that the operator is block-diagonal, namely the sums with different values of r act on different groups of variables, $\mathbf{t}_r$.

The justification of this description is quite standard. The ingredient $\sum_\alpha \phi_\alpha \otimes \phi^\alpha$ is responsible for the "ungluing" of the diagonal constraint $\Delta \subset X \times X$ at the node. The denominator $1 - \eta_+^{-1} L_+^{1/m_+} \otimes \eta_-^{-1} L_-^{1/m_-}$ represents the trace str (from the denominator of the Kawasaki-RR formula) of the smoothing deformation of the curve at the node, which is normal to the Kawasaki stratum of stable maps with the prescribed symmetry h. The Adams operation Ψ^r occurs at the nodes of order r due to the general fact: $\mathrm{tr}_h(V^{\otimes r}) = \Psi^r(V)$, assuming that h acts on the tensor product by the cyclic permutation of the r factors. The factor r accounts for the number of $\mathbb{Z}_r$-equivariant ways of gluing the components of the covering curve Σ over a node of order r on the quotient curve $\widehat{\Sigma}$. The factors $\hbar^r$ comes from the item (iv) in our strategy of the previous section to account for the change of the Euler characteristic of the covering curves Σ under gluing at the r nodes. The factor $1/2$ is due to the symmetry between η_+ and η_-.

Our goal is to show that the application of the operator

$$e^{\bigoplus_{r>0} r \Psi^r \hbar \sum_{\eta_+\eta_- \neq 1} \nabla_{\eta_+,\eta_-}/2}$$

to a function on $\underline{\mathcal{K}}_+$, considered as a quantum state in the standard polarization on the adelic loop space $(\underline{\mathcal{K}}^\infty, \underline{\Omega}^\infty)$, is equivalent to representing the same quantum state in the uniform polarization.

In traditional Darboux coordinate notation $\mathbf{p} = \{p_\alpha\}, \mathbf{q} = \{q_\alpha\}$ a second order differential operator $\hbar\nabla/2 = (\hbar/2)\sum_{\alpha\beta} s_{\alpha\beta} \partial_{q_\alpha} \partial_{q_\beta}$ quantizes the quadratic hamiltonian $(\mathbf{p}, S\mathbf{p})/2$. The time-one map generated by the corresponding hamiltonian system $\dot{\mathbf{q}} = S\mathbf{p}, \dot{\mathbf{p}} = 0$ transforms the negative polarization space $\mathbf{q} = 0$ into $\mathbf{q} = S\mathbf{p}$. According to Stone-von Neumann' theorem, the operator $\exp \hbar\nabla/2$ intertwines the representations of the Heisenberg Lie algebra in the Fock spaces corresponding to these polarizations. Thus, we need to compute the operator S in our situation, and check that the space $\mathbf{q} = S\mathbf{p}$ is the adelic image of $\mathcal{K}_-^\infty$. In invariant terms, the operator $S : \underline{\mathcal{K}}_- \to \underline{\mathcal{K}}_+$ is computed by contracting the symmetric tensor $S \in \underline{\mathcal{K}}_+ \otimes \underline{\mathcal{K}}_+$ using the symplectic pairing $\underline{\mathcal{K}}_- \otimes \underline{\mathcal{K}}_+ \to \Lambda$.

Since our operator is block-diagonal, let us first do the computation for the block $r = 1$. Here we have the adelic space $\underline{\mathcal{K}} = \oplus_\zeta \mathcal{K}^{(\zeta)}$, where each sector $\mathcal{K}^{(\zeta)}$ is isomorphic to $\mathcal{K}^{fake} = K((q-1))$. It is equipped with the symplectic form

$$\underline{\Omega}(\mathbf{f}, \mathbf{g}) := \sum_\zeta \frac{1}{m(\zeta)} \mathrm{Res}_{q=1}(f^{(\zeta)}(q^{-1}), g^{(\zeta^{-1})}(q)) \frac{dq}{q}.$$

The spaces $\underline{\mathcal{K}}_+$ and $\underline{\mathcal{K}}_-$ of the standard polarization are spanned respectively by (the superscript indicates the only non-zero component):

$$\mathbf{f}_{k,\alpha}^{(\zeta)} = \phi^\alpha (q^{1/m} - 1)^k \ \text{ and } \ \mathbf{g}_{k,\alpha}^{(\zeta^{-1})} = \phi_\alpha \frac{q^{k/m}}{(1 - q^{1/m})^{k+1}},$$

which form a Darboux basis as k,α and ζ run through their ranges. Namely $\underline{\Omega}(\mathbf{f}_{k,\alpha}^{(\zeta)}, \mathbf{g}_{k,\alpha}^{(\zeta^{-1})}) = -1$, and $= 0$ in all the cases when the indices mismatch.

As it was discussed earlier,

$$\nabla_{\eta,\zeta} = \frac{\sum_\alpha \phi_\alpha \otimes \phi^\alpha}{1 - \eta^{-1} x^{1/m} \otimes \zeta^{-1} y^{1/n}} \in K[[x^{1/m} - 1]] \otimes K[[y^{1/n} - 1]]$$

defines a biderivation on the space of functions on $\mathcal{K}_+^{(\eta)} \oplus \mathcal{K}_+^{(\zeta)}$. Here η and ζ are primitive roots of unity of orders m and n respectively with $\eta\zeta \neq 1$ (and we write x, y instead of $L_\pm$ used earlier). Equivalently, $\nabla_{\eta,\zeta}$ can be considered as a bilinear form on $\mathcal{K}_-^{(\eta^{-1})} \oplus \mathcal{K}_-^{(\zeta^{-1})}$ (the symbol of the biderivation), or as a linear map ∇_η^ζ : $\mathcal{K}_-^{(\zeta^{-1})} \to \mathcal{K}_+^{(\eta)}$, which is what we want to compute.

In explicit form, the linear map ∇_η^ζ is described by[6]

$$\mathcal{K}_-^{(\zeta^{-1})} \ni f = \sum_\alpha f^\alpha(q)\phi_\alpha \mapsto -\operatorname{Res}_{y=1} \frac{\sum_\alpha \phi_\alpha f^\alpha(y)}{(1 - \eta^{-1}\zeta^{-1} q^{1/m} y^{-1/n})} \frac{dy^{1/n}}{y^{1/n}}.$$

Take $f = \phi_\alpha q^{k/n}/(1 - q^{1/n})^{k+1}$, and put $x = y^{1/n}$. Then

$$\begin{aligned}
\nabla_\eta^\zeta f &= -\phi_a \operatorname{Res}_{x=1} \frac{x^k}{(1-x)^{k+1}} \frac{1}{(1 - \eta^{-1}\zeta^{-1} q^{1/m} x^{-1})} \frac{dx}{x} \\
&= \phi_\alpha \operatorname{Res}_{x=\eta^{-1}\zeta^{-1}q^{1/m}} \frac{x^k}{(1-x)^{k+1}} \frac{dx}{(x - \eta^{-1}\zeta^{-1} q^{1/m})} \\
&= \phi_\alpha \frac{(\eta^{-1}\zeta^{-1} q^{1/m})^k}{(1 - \eta^{-1}\zeta^{-1} q^{1/m})^{k+1}}.
\end{aligned}$$

The last expression is interpreted as an element of $\mathcal{K}_+^{(\eta)}$ by expanding it as a power series in $q^{1/m} - 1$.

Note that when η runs through all roots of unity, k runs through all non-negative integers, and ϕ_α runs through a basis of $K^0(X)$, the vector monomials $\mathbf{f} = \mathbf{f}_{\alpha,k,\zeta} :=$

[6] The negative sign comes from $\underline{\Omega}(\mathbf{f}_{k,\alpha}^{(\zeta)}, \mathbf{g}_{k,\alpha}^{(\zeta^{-1})}) = -1$.

$\phi_\alpha(\zeta^{-1}q)^k/(1-\zeta^{-1}q)^{k+1}$ run through a basis in $\mathcal{K}_-$. The adelic map is defined so that

$$\underline{\mathbf{f}}^{(\zeta^{-1})} = \phi_\alpha \frac{q^{k/n}}{(1-q^{1/n})^{k+1}}, \quad \underline{\mathbf{f}}^{(\eta)} = \phi_\alpha \frac{(\eta^{-1}\zeta^{-1}q^{1/m})^k}{(1-\eta^{-1}\zeta^{-1}q^{1/m})^{k+1}},$$

where n and m are the orders of ζ and $\eta \neq \zeta^{-1}$, and the expressions have to be expanded into Laurent series near $q = 1$. The above computation shows that $\underline{\mathbf{f}}^{(\eta)} \oplus \underline{\mathbf{f}}^{(\zeta^{-1})} \in \mathcal{K}_+^{(\eta)} \oplus \mathcal{K}_-^{(\zeta^{-1})}$ lies in the graph of ∇_η^ζ. Since $\mathbf{f}_{\alpha,k,\zeta}^{(\zeta^{-1})}$ form a basis in the domain $\mathcal{K}_-^{(\zeta^{-1})}$ of ∇_η^ζ when α and k run through their ranges, we find that $\underline{\mathbf{f}}_{\alpha,k,\zeta}$ form a basis in the graph of

$$\oplus_{\eta\neq\zeta^{-1}} \nabla_\eta^\zeta : \mathcal{K}_-^{(\zeta^{-1})} \to \underline{\mathcal{K}}_+,$$

and altogether form a basis in the direct sum of the graphs over ζ.

For general block $r \geq 1$, we have the adelic map: $\mathcal{K}^{(r)} = \mathcal{K} \to \underline{\mathcal{K}}^{(r)}$, which maps $\mathbf{f} \in \mathcal{K}$ to $\Psi^r\underline{\mathbf{f}}$. It satisfies

$$\underline{\Omega}^{(r)}(\Psi^r\underline{\mathbf{f}}, \Psi^r\underline{\mathbf{g}}) = \frac{\Psi^r}{r}\underline{\Omega}(\underline{\mathbf{f}}, \underline{\mathbf{g}}) = \frac{\Psi^r}{r}\Omega(\mathbf{f}, \mathbf{g}),$$

where $\Psi^r\Omega/r$ is the restriction of Ω^∞ to $\mathcal{K}^{(r)}$, and $\underline{\Omega}^{(r)}$ is the restriction of $\underline{\Omega}^\infty$ to the block $\underline{\mathcal{K}}^{(r)}$ in the total adelic space $\underline{\mathcal{K}}^\infty$. It is equal to

$$\underline{\Omega}^{(r)}(\mathbf{f}, \mathbf{g}) := \frac{1}{r}\sum_\zeta \frac{1}{m(\zeta)} \mathrm{Res}_{q=1}(f^{(\zeta)}(q^{-1}), g^{(\zeta^{-1})}(q))^{(r)}\frac{dq}{q}.$$

In fact the factor $1/r$ interacts with the factor r in the biderivation $r\Psi^r\nabla_{\eta,\zeta}$ in such a way that the operator from $\Psi^r(\mathcal{K}_-^{(\eta)})$ to $\underline{\mathcal{K}}_+^{(r)}$ generated by it (or by the corresponding bilinear form on $\Psi^r(\mathcal{K}_-^{(\eta)} \oplus \mathcal{K}_-^{(\zeta)})$) acts as

$$\Psi^r\left(\phi_\alpha \frac{q^{k/n}}{(1-q^{1/n})^{k+1}}\right) \mapsto \Psi^r\left(\phi_\alpha \frac{(\eta^{-1}\zeta^{-1}q^{1/m})^k}{(1-\eta^{-1}\zeta^{-1}q^{1/m})^{k+1}}\right).$$

Therefore the graph of the map (defined by all $r\Psi^r\nabla_{\eta,\zeta}$) from the negative space $\underline{\mathcal{K}}_-^{(r)}$ of the standard polarization to $\underline{\mathcal{K}}_+^{(r)}$ indeed coincides with the negative space of the uniform polarization on $\underline{\mathcal{K}}^{(r)}$, defined as the adelic image of $\mathcal{K}_-^{(r)} = \mathcal{K}_-$.

This completes the proof of Proposition 4.5.3, and our Main Theorem follows.

Acknowledgments This material is based upon work supported by the National Science Foundation under Grants DMS-1611839 and DMS-1906326, and by the IBS Center for Geometry and Physics, POSTECH, Korea.

References

1. Abramovich, D., Graber, T., Vistoli, A.: Algebraic orbifold quantum products. In: Orbifolds in Mathematics and Physics (Madison, WI, 2001). Contemporary Mathematics, vol. 310, p. 124. American Mathematical Society, Providence (2002)
2. Chen, W., Ruan, Y.: Orbifold Gromov–Witten Theory. Orbifolds in Mathematics and Physics (Madison, WI, 2001). Contemporary Mathematics, vol. 310, 2585 pp. American Mathematical Society, Providence (2002)
3. Coates, T.: Riemann–Roch theorems in Gromov–Witten theory. Ph.D. Thesis (2003)
4. Coates, T., Givental, A.: Quantum cobordisms and formal group laws. In: The Unity of Mathematics. Progress in Mathematics, vol. 244, pp. 155–171. Birkhäuser, Boston (2006)
5. Edidin, D.: Riemann-Roch for Deligne-Mumford Stacks. In: A Celebration of Algebraic Geometry. Clay Mathematics Institute, vol. 18, pp. 241–266. American Mathematical Society, Providence (2013)
6. Farsi, C.: An orbifold relative index theorem. J. Geom. Phys. **57**(8), 1653–1668 (2007)
7. Givental, A., Tonita, V.: The Hirzebruch-Riemann–Roch Theorem in True Genus-0 Quantum K-Theory. Noncommutative Geometry, vol. 62, pp. 43–91. MSRI Publication, Berkeley (2014). arXiv:1106.3136
8. Givental, A.: Permutation-equivariant quantum K-theory III. Lefschetz' formula on $\overline{M}_{0,n}/S_n$ and adelic characterization (2015). arXiv:1508.06697
9. Givental, A.: Permutation-equivariant quantum K-theory VII. General theory (2015). arXiv:1510.03076
10. Givental, A.: Permutation-equivariant quantum K-theory I. Definitions. Elementary K-theory on $\overline{M}_{0,n}/S_n$. Moscow Math. J. **17**(4), 691–698 (2017). arXiv:1508.02690
11. Jarvis, T., Kimura, T.: Orbifold quantum cohomology of the classifying space of a finite group. Orbifolds in Mathematics and Physics. Contemporary Mathematics, vol. 310, pp. 123–134. American Mathematical Society, Providence (2002)
12. Kawasaki, T.: The Riemann-Roch theorem for complex V-manifolds. Osaka J. Math. **16**(1), 151–159 (1979)
13. Kresch, A.: On the geometry of Deligne-Mumford stacks. Algebraic Geometry, Proceedings of Symposia in Pure Mathematics, vol. 80, Part 1, pp. 259–271. American Mathematical Society, Providence (2009)
14. Lee, Y.P. Quantum K-Theory I. Foundations. Duke Math. J. **121**(3), 389–424 (2004)
15. O'Brien, N., Toledo, D., Tong, Y.L.: Hirzebruch-Riemann-Roch for coherent sheaves. Am. J. Math. **103**(2), 253–271 (1981)
16. Tonita, V.: A virtual Kawasaki Riemann–Roch formula. Pac. J. Math. **268**(1), 249–255 (2014). arXiv:1110.3916.
17. Tonita, V.: Twisted orbifold Gromov–Witten invariants. Nagoya Math. J. **213**, 141–187 (2014). arXiv:1202.4778
18. Tonita, V.: A formula for the total permutation-equivariant K-theoretic Gromov-Witten potential, 13pp. (2016, preprint). arXiv:1603.09562
19. Tonita, V., Tseng, H.H.: Quantum orbifold Hirzebruch-Riemann-Roch theorem in genus zero, 26pp. (2013, preprint). arXiv:1307.0262
20. Tseng, H.H.: Orbifold quantum Riemann-Roch, Lefschetz and Serre. Geom. Top. **14**, 1–81 (2010)

Open Access This chapter is licensed under the terms of the Creative Commons Attribution-NonCommercial-NoDerivatives 4.0 International License (http://creativecommons.org/licenses/by-nc-nd/4.0/), which permits any noncommercial use, sharing, distribution and reproduction in any medium or format, as long as you give appropriate credit to the original author(s) and the source, provide a link to the Creative Commons license and indicate if you modified the licensed material. You do not have permission under this license to share adapted material derived from this chapter or parts of it.

The images or other third party material in this chapter are included in the chapter's Creative Commons license, unless indicated otherwise in a credit line to the material. If material is not included in the chapter's Creative Commons license and your intended use is not permitted by statutory regulation or exceeds the permitted use, you will need to obtain permission directly from the copyright holder.

Chapter 5
Calabi–Yau Complete Intersections in Exceptional Grassmannians

Atsushi Ito, Makoto Miura, Shinnosuke Okawa, and Kazushi Ueda

Dedicated with admiration to the memory of Bumsig Kim

Abstract We classify completely reducible equivariant vector bundles on Grassmannians of exceptional Lie groups which give Calabi–Yau 3-folds as complete intersections. We also calculate Hodge numbers for those Calabi–Yau 3-folds.

5.1 Introduction

A smooth projective manifold is said to be *Calabi–Yau* if the canonical bundle is trivial. Calabi–Yau manifolds have attracted attentions from both mathematicians and string theorists, not only because of their importance in the classification of algebraic varieties, but also because of their relation with string theory and mirror symmetry. A Calabi–Yau manifold in dimensions at most two are either an elliptic curve, an abelian surface, or a K3 surface. In dimensions greater than two, it is not known whether the number of deformation equivalence classes (or even homeomorphism types) of Calabi–Yau manifolds is finite or not.

In this paper, we study Calabi–Yau 3-folds in rational homogeneous spaces of exceptional types. The main result is the following:

A. Ito
Department of Mathematics, Institute of Pure and Applied Sciences, University of Tsukuba, Tsukuba, Japan
e-mail: ito-atsushi@math.tsukuba.ac.jp

M. Miura · S. Okawa
Department of Mathematics, Graduate School of Science, The University of Osaka, Osaka, Japan
e-mail: miurror.jp@gmail.com; okawa@math.sci.osaka-u.ac.jp

K. Ueda (✉)
Graduate School of Mathematical Sciences, The University of Tokyo, Tokyo, Japan
e-mail: kazushi@ms.u-tokyo.ac.jp

© The Author(s) 2026

N.-G. Kang et al. (eds.), *Categorical and Enumerative Aspects of Mirror Symmetry*, KIAS Springer Series in Mathematics 5,
https://doi.org/10.1007/978-981-95-0385-8_5

Table 5.1 Complete intersection Calabi–Yau 3-folds in exceptional Grassmannians

No.	G/P	$\mathcal{E}$	$h^{1,1}$	$h^{1,2}$	deg	c_2
1	×⊞▪	$(1,1)$	1	50	42	84
2	⊞×	$(1,1)$	1	50	14	56
3	•–×–•–•–•	$(1,0,0,0,0;0) \oplus (0,0,0,0,1;0)^{\oplus 4}$	1	31	192	132
4	×⊞▪	$(1,0) \oplus (2,0)$	1	61	36	84

Theorem 5.1.1 *A complete intersection Calabi–Yau 3-fold of a globally generated completely reducible equivariant vector bundle $\mathcal{E}$ on an exceptional Grassmannian G/P, which is not a complete intersection of line bundles on a projective space, is one of those appearing in Table 5.1.*

In particular, there is no such Calabi–Yau 3-fold in exceptional Grassmannians of types E_7, E_8, and F_4.

In Table 5.1, we label the simple roots of E_6 as

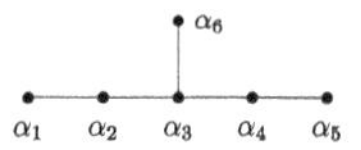

and write the coordinates of a weight $\lambda = \lambda_1 \omega_1 + \cdots + \lambda_6 \omega_6$ with respect to the corresponding fundamental weights $\omega_1, \ldots, \omega_6$ as $(\lambda_1, \ldots, \lambda_5; \lambda_6)$. Similarly, we label the simple roots as

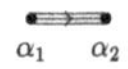

and write $\lambda = \lambda_1 \omega_1 + \lambda_2 \omega_2 = (\lambda_1, \lambda_2)$ for G_2. A weight λ is identified with the equivariant vector bundle associated with the irreducible representation of the Levi subgroup with highest weight λ. The degree and the second Chern number are with respect to the restriction of the ample generator of the Picard group of the ambient space.

Note that the classification itself in Theorem 5.1.1 was already obtained in [2]. The Hodge numbers are newly calculated in this work, and are also reverified in [11, Table 2] shortly after that.

The G_2-Grassmannian ⊞⊟ is the zero locus of the section $s \in H^0(\mathcal{Q}^{\vee\vee}(1)) \cong \bigwedge^3 \mathbb{C}^7$ corresponding to the G_2-invariant 3-form, and $\mathcal{E}_{(1,1)}$ is the restriction of $\mathcal{S}^\vee(1)$, where $\mathcal{S}$ and $\mathcal{Q}$ are the universal subbundle and the universal quotient bundle on $\mathrm{Gr}(2, 7)$. Hence No. 1 in Table 5.1 is the same as No. 16 in [5, Table 1], which is known by Inoue et al. [5, Proposition 5.1] to be deformation-equivalent to the intersection of the image of $\mathrm{Gr}(2, 7)$ and a linear subspace of codimension 7 in $\mathbb{P}\left(\bigwedge^2 \mathbb{C}^7\right)$.

The G_2-Grassmannian ⊞× is a smooth quadric hypersurface in $\mathbb{P}^6$. Calabi–Yau 3-folds contained in a (not necessarily smooth) quadric 5-fold are classified in [9,

Section 5], and it is shown in [9, Theorem 7.1] that No. 2 in Table 5.1 is deformation-equivalent to the Pfaffian Calabi–Yau 3-fold appearing in [13].

Although the families No. 1 and No. 2 were known, their relation with G_2-Grassmannians were new, and led to the discoveries of an L-equivalence [8], a derived equivalence [10], and a 7-fold flop [14].

To the best of our knowledge, no known Calabi–Yau 3-fold has the same topological invariants as No. 3 in Table 5.1. The restriction $\mathcal{O}_X(1)$ of the ample generator of the ambient space is primitive since

$$\chi(\mathcal{O}_X(t)) = 32t^3 + 11t \tag{5.1.1}$$

by the Hirzebruch–Riemann–Roch theorem (5.4.8).

Families of Calabi–Yau 3-folds described as complete intersections of line bundles on the G_2-Grassmannian are omitted in Theorem 5.1.1, since they are complete intersections of line bundles in $\mathbb{P}^6$.

Calabi–Yau complete intersection 3-folds of completely reducible equivariant bundles on the Cayley plane containing $\mathcal{E}_{\omega_5}$ as a direct summand turn out to be the empty set, since the zero locus of a general section of $\mathcal{E}_{\omega_5}$ on the Cayley plane is the empty set because $h^0(\mathcal{O}_X) = 0$.

The proof of Theorem 5.1.1 also shows the following:

Theorem 5.1.2 *There is no complete intersection Fano 4-fold of a globally generated completely reducible equivariant vector bundle on an exceptional Grassmannian, which is neither a complete intersection of line bundles on a projective space nor a hypersurface of the G_2-Grassmannian .*

We also classify a class of Calabi–Yau 3-folds in flag varieties of exceptional types.

Theorem 5.1.3 *A globally generated completely reducible equivariant vector bundle $\mathcal{E}$ on an exceptional flag variety G/P of Picard number greater than one satisfying* $\operatorname{rank}\mathcal{E} = \dim G/P - 3$ *and $c_1(\mathcal{E}) = c_1(G/P)$ is one of those appearing in Table 5.2.*

The condition $c_1(\mathcal{E}) = c_1(G/P)$ is sufficient for the zero locus X to be Calabi–Yau. This condition will be necessary if the restriction $\operatorname{Pic} G/P \to \operatorname{Pic} X$ is injective, and it is an interesting problem to decide when this is the case.

Table 5.2 Complete intersection Calabi–Yau 3-folds in exceptional flag manifolds

No.	G/P	$\mathcal{E}$	$h^{1,1}$	$h^{1,2}$
5		$(1,0)^{\oplus 2} \oplus (0,2)$	2	38
6		$(1,0) \oplus (0,1) \oplus (1,1)$	2	48
7		$(2,0) \oplus (0,1)^{\oplus 2}$	2	58

5.2 Equivariant Vector Bundles Over G/P

Let $\mathfrak{g}$ be a complex semisimple Lie algebra of rank r. The corresponding simply-connected Lie group is denoted by G. Fix a Cartan subgroup $H \subset G$ with the associated Cartan subalgebra $\mathfrak{h} \subset \mathfrak{g}$ and set

$$\mathfrak{g}_\alpha := \{v \in \mathfrak{g} \mid [h, v] = \alpha(h)v \text{ for any } h \in \mathfrak{h}\} \tag{5.2.1}$$

for $\alpha \in \mathfrak{h}^\vee := \mathrm{Hom}_{\mathbb{C}}(\mathfrak{h}, \mathbb{C})$. One has the root decomposition

$$\mathfrak{g} = \mathfrak{h} \oplus \bigoplus_{\alpha \in \Delta} \mathfrak{g}_\alpha, \tag{5.2.2}$$

where

$$\Delta := \left\{\alpha \in \mathfrak{h}^\vee \;\middle|\; \mathfrak{g}_\alpha \neq \{0\}, \alpha \neq 0\right\}. \tag{5.2.3}$$

We choose a system of simple roots $\mathcal{S} := \{\alpha_1, \ldots, \alpha_r\} \subset \Delta$. This choice is equivalent to the choice of the sets Δ^+ and Δ^- of positive and negative roots.

The Dynkin diagram of $\mathfrak{g}$ is a graph whose nodes correspond to the simple roots $\alpha_i \in \mathcal{S}$ and whose edges represent the Cartan integers $\langle \alpha_i, \alpha_j^\vee \rangle$, where $\langle -, - \rangle$ is the Killing form on $\mathfrak{h}^\vee$ and $\alpha^\vee = 2\alpha/\langle \alpha, \alpha \rangle$. One has $\langle \alpha_i, \alpha_i^\vee \rangle = 2$ for all $\alpha_i \in \mathcal{S}$, and the correspondence between edges and the Cartan integers is given by

$$\iff \quad \langle \alpha, \beta^\vee \rangle = \langle \beta, \alpha^\vee \rangle = 0, \tag{5.2.4}$$

$$\iff \quad \langle \alpha, \beta^\vee \rangle = \langle \beta, \alpha^\vee \rangle = -1, \tag{5.2.5}$$

$$\iff \quad \langle \alpha, \beta^\vee \rangle = -2, \quad \langle \beta, \alpha^\vee \rangle = -1, \tag{5.2.6}$$

$$\iff \quad \langle \alpha, \beta^\vee \rangle = -3, \quad \langle \beta, \alpha^\vee \rangle = -1. \tag{5.2.7}$$

A subgroup P of G is said to be *parabolic* if G/P is a projective variety. Conjugacy classes of parabolic subgroups are in one-to-one correspondence with subsets $\mathcal{S}_{\mathfrak{p}} \subset \mathcal{S}$ of the set of simple roots in such a way that the corresponding subalgebra $\mathfrak{p}$ is given by

$$\mathfrak{p} := \mathfrak{l} \oplus \mathfrak{n}, \tag{5.2.8}$$

where the *Levi part* $\mathfrak{l}$ is

$$\mathfrak{l} := \mathfrak{h} \oplus \bigoplus_{\alpha \in (\operatorname{span} \mathcal{S}_{\mathfrak{p}}) \cap \Delta} \mathfrak{g}_\alpha, \tag{5.2.9}$$

and the *nilpotent part* $\mathfrak{n}$ is

$$\mathfrak{n} := \bigoplus_{\alpha \in \Delta^+ \setminus \operatorname{span} \mathcal{S}_\mathfrak{p}} \mathfrak{g}_\alpha. \tag{5.2.10}$$

Here, $\operatorname{span} \mathcal{S}_\mathfrak{p} \subset \mathfrak{h}^\vee$ is the linear subspace spanned by $\mathcal{S}_\mathfrak{p}$. The subset $\mathcal{S}_\mathfrak{p} \subset \mathcal{S}$ can be described by a *crossed Dynkin diagram*, where elements not in $\mathcal{S}_\mathfrak{p}$ are crossed out (i.e., elements of $\mathcal{S}_\mathfrak{p}$ correspond to uncrossed nodes). The inclusion relation of $\mathcal{S}_\mathfrak{p}$ corresponds to the inclusion relation of P. For example, the Borel subgroup is the minimal parabolic subgroup, so that all the nodes are crossed out in the corresponding crossed Dynkin diagram. We write the Weyl group of $\mathfrak{l}$ as W_P, which is the subgroup of $W = W_G$ generated by simple reflections associated with elements of $\mathcal{S}_\mathfrak{p}$. One has

$$\dim G/P = \#\left(\Delta^- \setminus \operatorname{span} \mathcal{S}_\mathfrak{p}\right) = \dim G/B - \dim G'/B', \tag{5.2.11}$$

where G'/B' is the full flag variety corresponding to the full subgraph of the Dynkin diagram of G consisting of uncrossed nodes.

A finite-dimensional representation V of the parabolic subalgebra $\mathfrak{p}$ naturally carries a filtration

$$V = F^1 V \supset \cdots \supset F^j V \supset F^{j+1} V \supset \cdots \supset F^s V \supset F^{s+1} = F^{s+2} = \cdots = 0 \tag{5.2.12}$$

for some s, where we set

$$F^j V = \mathfrak{n}^{j-1} \cdot V \tag{5.2.13}$$

for all j. The nilpotent part $\mathfrak{n}$ acts trivially on each quotient

$$V_j := F^j V / F^{j+1} V, \tag{5.2.14}$$

so that one may regard V_j as a representation of the Levi subalgebra $\mathfrak{l}$. Since $\mathfrak{l}$ is reductive, any finite-dimensional representation of $\mathfrak{l}$ is completely reducible, i.e., the direct sum of irreducible representations. We write

$$V = \sum_j V_j = V_1 + V_2 + \cdots + V_s, \tag{5.2.15}$$

where $F^j V = V_j + \cdots + V_s$ is a sub-representation. Let $W = \sum_j W_j$ be another finite-dimensional representation of $\mathfrak{p}$. It is easy to see

$$V \oplus W = \sum_{j \geq 1} \left(V_j \oplus W_j\right), \tag{5.2.16}$$

$$V \otimes W = \sum_{l \geq 2} \bigoplus_{j+k=l} \left(V_j \otimes W_k\right), \tag{5.2.17}$$

$$\bigwedge^p V = \sum_{k \geq p} \bigoplus_{h(\mathbf{p})=k} \bigotimes_{j \geq 1} \bigwedge^{p_j} V_j, \tag{5.2.18}$$

$$\operatorname{Sym}^p V = \sum_{k \geq p} \bigoplus_{h(\mathbf{p})=k} \bigotimes_{j \geq 1} \operatorname{Sym}^{p_j} V_j, \tag{5.2.19}$$

for any positive integer p, where $\mathbf{p} = (p_1, \ldots, p_r)$ runs over partitions of p and $h(\mathbf{p}) = \sum_j j p_j$.

Irreducible representations of the parabolic subalgebra $\mathfrak{p}$ correspond bijectively to those of the Levi subalgebra $\mathfrak{l}$, which are highest weight representations. Since $\mathfrak{g}$ and $\mathfrak{l}$ share the same Cartan subalgebra, weights of $\mathfrak{l}$ can be regarded naturally as weights of $\mathfrak{g}$. Since G and L share the same Cartan subgroup, the notion of integrality of weights is the same for both $\mathfrak{g}$ and $\mathfrak{l}$. The fundamental weight associated with the simple root $\alpha_i \in \mathcal{S}$ is denoted by ω_i;

$$\langle \omega_i, \alpha_j^\vee \rangle = \delta_{ij}, \quad i, j = 1, \ldots, r. \tag{5.2.20}$$

A weight $\lambda = \sum_{i=1}^r \lambda_i \omega_i$ is

- *integral* if $\lambda_i \in \mathbb{Z}$ for any $i = 1, \ldots, r$,
- $\mathfrak{p}$-*dominant* if $\lambda_i \in \mathbb{N}$ for any i such that $\alpha_i \in \mathcal{S}_\mathfrak{p}$, and
- $\mathfrak{g}$-*dominant* if $\lambda_i \in \mathbb{N}$ for any $i = 1, \ldots, r$.

A highest weight representation of $\mathfrak{l}$ integrates to a representation of P if and only if the highest weight is integral and $\mathfrak{p}$-dominant (see e.g. [1, Remark 3.1.6]). The irreducible representations of P and G with highest weight λ are denoted by V_λ^P and V_λ^G respectively.

The category of equivariant vector bundles on G/P is equivalent to that of representations of P. For a representation V of P, the corresponding equivariant vector bundle on G/P is denoted by

$$\mathcal{E}_V := G \times_P V. \tag{5.2.21}$$

We write the filtration of an equivariant bundle corresponding to the filtration (5.2.15) as

$$\mathcal{E} = \sum_{j \geq 1} \mathcal{E}_j = \mathcal{E}_1 + \mathcal{E}_2 + \cdots + \mathcal{E}_s. \tag{5.2.22}$$

For an irreducible equivariant vector bundle, we set

$$\mathcal{E}_\lambda := \mathcal{E}^\vee_{V_\lambda^P}, \tag{5.2.23}$$

which is globally generated if and only if λ is $\mathfrak{g}$-dominant. The Borel–Weil–Bott theorem gives an isomorphism

$$H^{\ell(w)}(G/P, \mathcal{E}_\lambda) \cong \left(V_{w.\lambda}^G\right)^\vee \tag{5.2.24}$$

of G-vector spaces, where $\rho := 1/2 \sum_{\alpha \in \Delta^+} \alpha$ is the Weyl vector,

$$w.\lambda := w(\lambda + \rho) - \rho \tag{5.2.25}$$

is the affine Weyl action, $w \in W_P$ is the unique element such that $w.\lambda$ is $\mathfrak{g}$-dominant (the left hand side of (5.2.24) is zero if there is no such w), and $\ell(w)$ is the length of (the minimal representative in W of) $w \in W_P$. The filtration (5.2.22) gives a spectral sequence

$$E_1^{p,q} = H^{p+q}(G/P, \mathcal{E}_p) \Rightarrow H^{p+q}(G/P, \mathcal{E}), \tag{5.2.26}$$

which allows us to compute the cohomology of $\mathcal{E}$ using the Borel–Weil–Bott theorem.

The Picard group $\operatorname{Pic} G/P$ is isomorphic to the group $\operatorname{Hom}(P, \mathbb{C}^\times)$ of characters of P. The set of weights of a representation V of P is denoted by $\Delta(V)$. One has

$$\operatorname{rank} \mathcal{E}_V = \dim V = |\Delta(V)|, \ \det \mathcal{E}_V \cong \mathcal{E}_{\det V}, \text{ and } \Delta(\det V) = \left\{ \sum_{\lambda \in \Delta(V)} \lambda \right\}. \tag{5.2.27}$$

The tangent bundle $T_{G/P}$ corresponds to the P-vector space $\mathfrak{g}/\mathfrak{p}$ with respect to the adjoint action;

$$T_{G/P} \cong \mathcal{E}_{\mathfrak{g}/\mathfrak{p}}. \tag{5.2.28}$$

Since the weights of the adjoint action are roots, one has

$$\Delta(\mathfrak{g}/\mathfrak{p}) = \Delta(\mathfrak{g}) \setminus \Delta(\mathfrak{p}) = \Delta^- \setminus \operatorname{span} \mathcal{S}_{\mathfrak{p}}. \tag{5.2.29}$$

Although the tangent bundle $T_{G/P}$ and hence the cotangent bundle $\Omega^1_{G/P}$ are indecomposable if G is simple, they are not irreducible unless G/P is a Hermitian symmetric space [6, Theorem 6]. Instead, it carries the following filtration. Recall that the *height* of a positive root is the sum of the coefficients of the simple roots.

Lemma 5.2.1 (cf. e.g. [1, Section 9.9]) *For a complex semisimple simply-connected Lie group G and a parabolic subgroup $P \subset G$, the cotangent bundle has the filtration*

$$\Omega^1_{G/P} = \sum_{j \geq 0} \bigoplus_{|\mathbf{n}|=j} \mathcal{E}_{-\alpha(\mathbf{n})}, \tag{5.2.30}$$

where $\mathbf{n} = (n_\alpha)_\alpha$ runs over the image of the map

$$\pi : \Delta\left((\mathfrak{g}/\mathfrak{p})^\vee\right) = \Delta^+ \setminus \operatorname{span} \mathcal{S}_\mathfrak{p} \to \mathbb{Z}_{\geq 0}^{\mathcal{S} \setminus \mathcal{S}_\mathfrak{p}} \tag{5.2.31}$$

taking the coefficients of the roots in $\mathcal{S} \setminus \mathcal{S}_\mathfrak{p}$, $|\mathbf{n}| := \sum_{\alpha \in \mathcal{S} \setminus \mathcal{S}_\mathfrak{p}} n_\alpha$, and $\alpha(\mathbf{n}) \in \Delta^+ \setminus \operatorname{span} \mathcal{S}_\mathfrak{p}$ is the unique element of minimal height such that $\pi(\alpha) = \mathbf{n}$.

Sketch of Proof Since the coadjoint action of $\mathfrak{n}$ on $(\mathfrak{g}/\mathfrak{p})^\vee \subset \mathfrak{g}^\vee$ increases $|\mathbf{n}|$, it suffices to show the decomposition for each graded component. For each $\mathbf{n} \in \operatorname{Im} \pi$, the root $\alpha(\mathbf{n})$ is uniquely determined, since the existence of distinct positive roots β_1, β_2 with the same height contradicts the fact that the difference $\beta_1 - \beta_2$ must be a (positive or negative) root [4, Lemma 9.4]. For any $\beta \in \pi^{-1}(\mathbf{n})$, the difference $\alpha(\mathbf{n}) - \beta$ is a root in $\operatorname{span} \mathcal{S}_\mathfrak{p}$, and the ladder operator of the corresponding $\mathfrak{sl}_2$-triple sends $\mathbb{C}(-\alpha(\mathbf{n})) \subset (\mathfrak{g}/\mathfrak{p})^\vee \subset \mathfrak{g}^\vee$ onto $\mathbb{C}(-\beta)$ by the coadjoint action [4, Proposition 8.4]. This implies the existence of the filtration (5.2.30). □

5.3 Complete Intersections of Equivariant Vector Bundles

Let $\mathcal{E} := \mathcal{E}_V$ be the equivariant vector bundle on $F := G/P$ associated with a representation V of P. Assume that $\mathcal{E}$ is globally generated. For a general section s of $\mathcal{E}$, the zero locus $X := s^{-1}(0)$ is a smooth complete intersection by a generalization of the theorem of Bertini [12, Theorem 1.10].

Since X is a complete intersection, the differential ds of the section s induces an isomorphism

$$N_{X/F} \cong \mathcal{E}|_X. \tag{5.3.1}$$

By taking the determinant of the exact sequence

$$0 \to T_X \to T_F|_X \to N_{X/F} \to 0, \tag{5.3.2}$$

one obtains an isomorphism

$$\det T_X \cong \det T_F|_X \otimes \det^{-1} \mathcal{E}|_X. \tag{5.3.3}$$

Hence $\det V \cong \det(\mathfrak{g}/\mathfrak{p})$ is a sufficient condition for $\det T_X \cong \mathcal{O}_X$, which is necessary if the restriction map $\operatorname{Pic} F \to \operatorname{Pic} X$ is injective.

The exact sequence

$$0 \to \operatorname{Sym}^j \mathcal{E}^\vee|_X \to \cdots \to \operatorname{Sym}^{j-k} \mathcal{E}^\vee \otimes \Omega_F^k|_X \to \cdots \to \Omega_F^j|_X \to \Omega_X^j \to 0 \tag{5.3.4}$$

obtained as the j-th exterior power of the exact sequence

$$0 \to \mathcal{E}^\vee|_X \to \Omega_F^1|_X \to \Omega_X^1 \to 0 \tag{5.3.5}$$

dual to (5.3.2) gives the spectral sequence

$$E_1^{-q,p} = H^p\left(\operatorname{Sym}^q \mathcal{E}^\vee \otimes \Omega_F^{j-q}|_X\right) \Rightarrow H^{p-q}\left(\Omega_X^j\right). \tag{5.3.6}$$

The Koszul resolution

$$0 \to \wedge^{\operatorname{rank} \mathcal{E}} \mathcal{E}^\vee \to \cdots \to \mathcal{E}^\vee \to \mathcal{O}_F \to \mathcal{O}_X \to 0 \tag{5.3.7}$$

gives the spectral sequence

$$E_1^{-q,p} = H^p\left(\wedge^q \mathcal{E}^\vee \otimes \mathcal{G}\right) \Rightarrow H^{p-q}\left(\mathcal{G}|_X\right) \tag{5.3.8}$$

for any coherent sheaf $\mathcal{G}$ on F.

Together with the Hodge symmetry $h^{p,q}(X) = h^{q,p}(X)$ and the obvious fact that $h^{p,q}(X) = 0$ unless $0 \le p \le \dim X$, the Hodge numbers are often determined only from dimensions of the cohomology groups on the E_1-page, although there are cases where one should look at morphisms more carefully.

The topological Euler number $\chi(X)$ can be computed by

$$\chi(X) = \int_X c(T_X) = \int_F \frac{c(T_F)}{c(\mathcal{E})} c_{\mathrm{top}}(\mathcal{E}), \tag{5.3.9}$$

where $c(\mathcal{G})$ and $c_{\mathrm{top}}(\mathcal{G})$ denote the total and the top Chern classes. The first equality in (5.3.9) is the Chern–Gauss–Bonnet theorem, and the second equality comes from (5.3.1) and (5.3.2). From the splitting principle, it follows

$$\chi(X) = \int_F \frac{\prod_{\mu \in \Delta(\mathfrak{g}/\mathfrak{p}^\vee)} \left(1 + c_1(\mathcal{L}_\mu)\right)}{\prod_{\nu \in \Delta(V^\vee)} \left(1 + c_1(\mathcal{L}_\nu)\right)} \prod_{\lambda \in \Delta(V^\vee)} c_1(\mathcal{L}_\lambda), \tag{5.3.10}$$

where $\mathcal{L}_\lambda = \mathcal{E}_{V_\lambda^B}^\vee$ is a line bundle on G/B for any weight λ, and the integrand is an element of $H^*(F, \mathbb{Z})$ considered as a subgroup of $H^*(G/B, \mathbb{Z})$ by the pull-back along the natural projection $G/B \to F$; an element in $H^*(F, \mathbb{Z})$ is described as a polynomial in $x_i := c_1(\mathcal{L}_{\omega_i}) \in H^2(G/B, \mathbb{Z})$ for $i = 1, \ldots, r$. Note that $c_1(\mathcal{L}_\lambda) = \sum_{i=1}^{r} \lambda_i x_i$. One can perform the integral in terms of representation theory by using the following Lemma 5.3.1.

Lemma 5.3.1 (cf. e.g. [1, Lemma 6.3.2]) *For a monomial* $x_i x_j \ldots x_k \in H^{2l}(G/B, \mathbb{Z})$ *of degree* l *and a Schubert cycle* $[X_w] := [\overline{BwB/B}] \in H_{2l}(G/B, \mathbb{Z})$ *associated with* $w \in W$ *of length* l, *one has*

$$x_i x_j \ldots x_k [X_w] = \sum \langle \lambda_i, \beta_{(1}^{\vee} \rangle \langle \lambda_j, \beta_2^{\vee} \rangle \cdots \langle \lambda_k, \beta_{l)}^{\vee} \rangle, \tag{5.3.11}$$

where the sum runs over all collections $\beta_1, \ldots, \beta_l \in \Delta^+$ *such that*

$$w = \sigma_{\beta_1} \sigma_{\beta_2} \ldots \sigma_{\beta_l} \tag{5.3.12}$$

is a reduced expression, and the parentheses on the right-hand side denote the symmetrized product.

5.4 Rational Homogeneous Spaces of Exceptional Types

Let G be the complex simple Lie group of exceptional type. We call a rational homogeneous space G/P an *exceptional flag variety*, or an *exceptional Grassmannian* if P is maximal.

For type G_2, there are three homogeneous spaces G/P_1, G/P_2 and G/B associated with the crossed Dynkin diagrams ×⇚•, •⇚× and ×⇚× respectively. The sets of weights of the irreducible representations $V^P_{(a,b)}$ with the highest weight $(a, b) := a\omega_1 + b\omega_2$ are given by

$$\Delta\left(V^{P_1}_{(a,b)}\right) = \{(a + j, b - 2j) \mid j = 0, 1, \ldots, b\},$$

$$\Delta\left(V^{P_2}_{(a,b)}\right) = \{(a - 2j, b + 3j) \mid j = 0, 1, \ldots, a\},$$

$$\Delta\left(V^{B}_{(a,b)}\right) = \{(a, b)\}.$$

In particular, the dimensions and the determinants of these representations are given as follows:

representation	dimension	determinant
$V^{P_1}_{(a,b)}$	$b + 1$	$(a(b + 1) + b(b + 1)/2, 0)$
$V^{P_2}_{(a,b)}$	$a + 1$	$(0, (a + 1)b + 3a(a + 1)/2)$
$V^{B}_{(a,b)}$	1	(a, b)

(5.4.1)

By considering the action of the nilpotent parts on the roots, one can directly observe that the representations $(\mathfrak{g}/\mathfrak{p})^\vee$ are given by

$$(\mathfrak{g}/\mathfrak{p}_1)^\vee \cong V^{P_1}_{(-1,3)} + V^{P_1}_{(1,0)}, \tag{5.4.2}$$

$$(\mathfrak{g}/\mathfrak{p}_2)^\vee \cong V^{P_2}_{(1,-1)} + V^{P_2}_{(0,1)} + V^{P_2}_{(1,0)}, \tag{5.4.3}$$

$$(\mathfrak{g}/\mathfrak{b})^\vee \cong V^{B}_{(2,-3)} \oplus V^{B}_{(-1,2)} + V^{B}_{(1,-1)} + V^{B}_{(0,1)} + V^{B}_{(-1,3)} + V^{B}_{(1,0)}, \tag{5.4.4}$$

which agree with the formula for the cotangent bundles in Lemma 5.2.1. The determinants are given by

$$\det(\mathfrak{g}/\mathfrak{p}_1)^\vee \cong V^{P_1}_{(3,0)}, \tag{5.4.5}$$

$$\det(\mathfrak{g}/\mathfrak{p}_2)^\vee \cong V^{P_2}_{(0,5)}, \tag{5.4.6}$$

$$\det(\mathfrak{g}/\mathfrak{b})^\vee \cong V^{B}_{(2,2)}. \tag{5.4.7}$$

Theorem 5.1.1 for G_2-Grassmannians is an easy consequence of these facts. For example, to obtain a Calabi–Yau 3-fold from $(G/P_1, \mathcal{E}_V^\vee)$, the completely reducible representation V must satisfy $\dim V = 2$ and $\det V = V^{P_1}_{(3,0)}$ since $\operatorname{Pic} G/P_1 \cong \mathbb{Z}$ must inject to $\operatorname{Pic} X$. If V is decomposable, then $V \cong V^{P_1}_{(1,0)} \oplus V^{P_1}_{(2,0)}$ is the only choice, and if V is indecomposable, then $V \cong V^{P_1}_{(1,1)}$ is the only choice.

For the Calabi–Yau 3-fold X associated with $\left(G/P_1, \mathcal{E}_{(1,1)}\right)$, one can use the spectral sequences (5.2.26), (5.3.6), and (5.3.8) to prove $h^{0,1} = h^{0,2} = 0$, $h^{1,1} = 1$, and $h^{1,2} = 50$ by hand. The Koszul resolution (5.3.7) allows us to compute the cohomology of $\mathcal{O}_X(i)$, which together with the Hirzebruch–Riemann–Roch theorem

$$\chi(\mathcal{O}_X(i)) = \frac{1}{6}\deg X \cdot i^3 + \frac{1}{12} c_2(X) \cdot i \tag{5.4.8}$$

implies $\deg X = 42$ and $c_2(X) = 84$.

Similar calculations aided by a Mathematica package [3] give Theorems 5.1.1, 5.1.2, and 5.1.3. The conditions $\operatorname{rank} \mathcal{E} = \dim G/P - 3$ and $c_1(\mathcal{E}) = c_1(G/P)$ are strong, and many of 439 exceptional flag varieties are eliminated quickly. The topological invariants are calculated by using the spectral sequences (5.2.26), (5.3.6), (5.3.8), and the Chern–Gauss–Bonnet theorem (5.3.10) in one case. All calculations are recorded in [7].

Acknowledgments We thank Daisuke Inoue for helpful discussions on Pfaffian Calabi–Yau 3-folds, Grzegorz Kapustka and Michał Kapustka for pointing out the reference [9], Pieter Belmans and Maxim Smirnov for correcting an error about representation theory of parabolic subalgebra and for informing us of the reference [6], and Kyeong-Dong Park for pointing out that $\mathcal{E}_{\omega_5}$ defines the empty set in the Cayley plane. A. I. was supported by Grant-in-Aid for Scientific Research (14J01881, 17K14162). M. M. was supported by Frontiers of Mathematical Sciences and Physics at University of Tokyo, Korea Institute for Advanced Study, and Grant-in-Aid for Scientific Research (21K03156). S. O. was partially supported by Grants-in-Aid for Scientific Research (16H05994, 16K13746, 16H02141, 16K13743, 16K13755, 16H06337) and the Inamori Foundation. K. U. was partially supported by Grants-in-Aid for Scientific Research (24740043, 15KT0105, 16K13743, 16H03930).

References

1. Baston, R.J., Eastwood, M.G.: The Penrose Transform. Oxford Mathematical Monographs. The Clarendon Press/Oxford University Press, New York (1989). Its interaction with representation theory, Oxford Science Publications. MR 1038279 (92j:32112)
2. Benedetti, V.: Sous-variétés spéciales des espaces homogénes (special subvarieties of homogeneous spaces). Ph.D. Thesis, Aix-Marseille Université
3. Feger, R., Kephart, T.W., Saskowski, R.J.: LieART 2.0 – a mathematica application for lie algebras and representation theory. Comput. Phys. Commun. **257**, 107490 (2020)
4. Humphreys, J.E.: Introduction to Lie Algebras and Representation Theory. Graduate Texts in Mathematics, vol. 9. Springer, New York (1972). MR 0323842
5. Inoue, D., Ito, A., Miura, M.: Complete intersection Calabi-Yau manifolds with respect to homogeneous vector bundles on Grassmannians. Math. Z. **292**(1–2), 677–703 (2019). MR 3968921
6. Ise, M.: Some properties of complex analytic vector bundles over compact complex homogeneous spaces. Osaka Math. J. **12**, 217–252 (1960). MR 124919
7. Ito, A., Miura, M., Okawa, S., Ueda, K.: The Mathematica notebook file *calculation.nb* in the ancillary files to the preprint version of this paper (2016). arXiv:1606.04076
8. Ito, A., Miura, M., Okawa, S., Ueda, K.: The class of the affine line is a zero divisor in the Grothendieck ring: via G_2-Grassmannians. J. Algebraic Geom. **28**(2), 245–250 (2019). MR 3912058
9. Kapustka, G., Kapustka, M.: Calabi–Yau threefolds in $\mathbb{P}^6$. Ann. Mat. Pura Appl. **195**(2), 529–556 (2016). MR 3476687
10. Kuznetsov, A.: Derived equivalence of Ito-Miura-Okawa-Ueda Calabi-Yau 3-folds. J. Math. Soc. Japan **70**(3), 1007–1013 (2018). MR 3830796
11. Lee, E., Park, K.D.: Complete intersection hyperkähler fourfolds with respect to equivariant vector bundles over rational homogeneous varieties of Picard number one (2021). https://doi.org/10.1016/j.geomphys.2024.105348
12. Mukai, S.: Polarized $K3$ surfaces of genus 18 and 20. In: Complex Projective Geometry (Trieste, 1989/Bergen, 1989). London Mathematical Society Lecture Note Series, vol. 179. Cambridge University Press, Cambridge, pp. 264–276 (1992). MR 1201388 (94a:14039)
13. Rødland, E.A.: The Pfaffian Calabi-Yau, its mirror, and their link to the Grassmannian $G(2, 7)$. Compos. Math. **122**(2), 135–149 (2000). MR 1775415 (2001h:14051)
14. Ueda, K.: G_2-Grassmannians and derived equivalences. Manuscript. Math. **159**(3–4), 549–559 (2019). MR 3959275

Open Access This chapter is licensed under the terms of the Creative Commons Attribution-NonCommercial-NoDerivatives 4.0 International License (http://creativecommons.org/licenses/by-nc-nd/4.0/), which permits any noncommercial use, sharing, distribution and reproduction in any medium or format, as long as you give appropriate credit to the original author(s) and the source, provide a link to the Creative Commons license and indicate if you modified the licensed material. You do not have permission under this license to share adapted material derived from this chapter or parts of it.

The images or other third party material in this chapter are included in the chapter's Creative Commons license, unless indicated otherwise in a credit line to the material. If material is not included in the chapter's Creative Commons license and your intended use is not permitted by statutory regulation or exceeds the permitted use, you will need to obtain permission directly from the copyright holder.

Chapter 6
Virtual Fundamental Classes of the Vanishing Loci of Cosections

Young-Hoon Kiem and Hyeonjun Park

Dedicated to the memory of Bumsig Kim.

Abstract Let X be a Deligne-Mumford stack equipped with a perfect obstruction theory $\phi : \mathbb{E} \to \mathbb{L}_X$. By Kiem and Li (J. Am. Math. Soc. **26**(4), 1025–1050 (2013)), if the obstruction sheaf $Ob_X = h^1(\mathbb{E}^\vee)$ admits a cosection $\sigma : Ob_X \to \mathcal{O}_X$, the virtual fundamental class $[X]^{\mathrm{vir}}$ of X is localized to a class $[X]^{\mathrm{vir}}_\sigma$ supported in the zero locus $X(\sigma)$ of $\sigma^\vee$. In many natural examples, $X(\sigma)$ is an interesting space on its own and we may ask if there is a natural induced perfect obstruction theory of $X(\sigma)$ whose virtual fundamental class equals $[X]^{\mathrm{vir}}_\sigma$. In this expository article, we discuss two possible approaches, by global complex Kuranishi chart and by derived algebraic geometry.

Keywords Virtual fundamental class · Cosection localization · Derived scheme

6.1 Introduction

A fundamental problem in mathematics is to find the number of geometric objects with certain required properties. In algebraic geometry, such a problem is solved by first constructing the moduli space X of all geometric objects and then by integrating cohomology classes ξ on X representing the required properties.

Unfortunately the moduli space is often very singular of wrong dimension and hence it is hard to make sense of the integrals $\int_X \xi$. On the other hand, the moduli space often comes with a refined structure such as a perfect obstruction theory or a derived scheme structure. Using a perfect obstruction theory, one may construct a homology class $[X]^{\mathrm{vir}}$, called the virtual fundamental class (cf. [1, 27]), against,

Y.-H. Kiem (✉) · H. Park
School of Mathematics, Korea Institute for Advanced Study, Seoul, South Korea
e-mail: kiem@snu.ac.kr; kiem@kias.re.kr; hyeonjunpark@kias.re.kr

© The Author(s) 2026

N.-G. Kang et al. (eds.), *Categorical and Enumerative Aspects of Mirror Symmetry*, KIAS Springer Series in Mathematics 5,
https://doi.org/10.1007/978-981-95-0385-8_6

against which cohomology classes can be integrated to define enumerative invariants

$$\int_{[X]^{\mathrm{vir}}} \xi.$$

A perfect obstruction theory on a Deligne-Mumford stack X provides us with an obstruction sheaf Ob_X which contains obstruction classes to square zero infinitesimal extensions of morphisms to X. When there is a cosection $\sigma : Ob_X \to \mathcal{O}_X$, the virtual fundamental class $[X]^{\mathrm{vir}}$ is represented by a cycle supported in the zero locus $X(\sigma)$ of $\sigma^\vee$ by Kiem and Li [17]. In fact, we can define the cosection localized virtual fundamental class

$$[X]^{\mathrm{vir}}_\sigma \in H_*(X(\sigma))$$

which coincides with $[X]^{\mathrm{vir}}$ when pushed forward to X.

The cosection localization turned out to be an effective technique to compute enumerative invariants. Moreover in many problems, the vanishing locus $X(\sigma)$ in X of the cosection σ is an interesting space on its own. For instance, when X is the moduli space of stable maps to a projective surface S equipped with a smooth canonical curve C, there is a cosection of the obstruction sheaf on X whose vanishing locus is the moduli space of stable maps to C [15, 16]. So it seems natural to ask the following.

Question 6.1.1 Is there a natural induced perfect obstruction theory on $X(\sigma)$ whose virtual fundamental class $[X(\sigma)]^{\mathrm{vir}} \in H_*(X(\sigma))$ satisfies

$$[X(\sigma)]^{\mathrm{vir}} = [X]^{\mathrm{vir}}_\sigma? \tag{6.1.1}$$

In the single chart case where X is the zero locus $s^{-1}(0)$ of a section of a vector bundle E over a smooth Deligne-Mumford stack V, we have a standard perfect obstruction theory

$$E^\vee|_X \xrightarrow{ds^\vee} \Omega_V|_X$$

and a virtual cycle $[X]^{\mathrm{vir}}$ of dimension

$$\dim V - \operatorname{rank} E.$$

Suppose we have a homomorphism $\sigma : E \to \mathcal{O}_V$ such that $\sigma \circ s = 0$. The vanishing locus $X(\sigma)$ of $\sigma|_X$ is the zero locus $\nu^{-1}(0)$ of $E \oplus E^\vee$ where $\nu = (s, \sigma^\vee)$. Then we have the obvious perfect obstruction theory

$$E^\vee \oplus E|_{X(\sigma)} \xrightarrow{(ds^\vee, d\sigma)} \Omega_V|_{X(\sigma)} \tag{6.1.2}$$

which cannot be a correct answer to Question 6.1.1 because its virtual dimension

$$\dim V - 2\,\mathrm{rank}\, E$$

is smaller than the virtual dimension $\dim V - \mathrm{rank}\, E$ of X. The way out is to enlarge (6.1.2) to a 3-term symmetric obstruction theory

$$T_V|_{X(\sigma)} \xrightarrow{(d\sigma^\vee, ds)} E^\vee \oplus E|_X \xrightarrow{(ds^\vee, d\sigma)} \Omega_V|_{X(\sigma)} \tag{6.1.3}$$

and then we can apply the Oh-Thomas construction [28] to obtain a virtual fundamental class $[X(\sigma)]^{\mathrm{vir}}$ of $X(\sigma)$. In this single chart case, it is not hard to prove (6.1.1) and hence (6.1.3) gives a correct answer to Question 6.1.1.

In general, we may find an open cover $\{X_\alpha\}$ of X such that X_α is the zero locus $s_\alpha^{-1}(0)$ of a section s_α of a vector bundle E_α over a smooth Deligne-Mumford stack V_α. So we have 3-term symmetric obstruction theories

$$T_{V_\alpha}|_{X_\alpha(\sigma)} \xrightarrow{(d\sigma_\alpha^\vee, ds_\alpha)} E_\alpha^\vee \oplus E_\alpha|_{X_\alpha} \xrightarrow{(ds_\alpha^\vee, d\sigma_\alpha)} \Omega_{V_\alpha}|_{X_\alpha(\sigma)} \tag{6.1.4}$$

for X_α but we do not know how to glue (6.1.4) to a global derived category object on X.

In this paper, we discuss two possible approaches for constructing a 3-term symmetric obstruction theory on $X(\sigma)$ and giving a correct answer to Question 6.1.1. The first approach is topological. One may avoid the gluing issue by using the global complex Kuranishi chart in [29]. A projective scheme X equipped with a perfect obstruction theory is the zero locus of a C^∞-section s of a complex vector bundle E over a C^∞ manifold V, which are holomorphic in the first or second infinitesimal neighborhood of X. One may try to extend the constructions in [17] and in [28] to this setting and prove (6.1.1).

The second approach is derived. We assume that X is the classical truncation of a derived scheme and σ comes from a (-1)-shifted closed 1-form. Then the gluing issue is handled by derived algebraic geometry and we have a 3-term symmetric obstruction theory for $X(\sigma)$. The equality (6.1.1) then follows from a deformation to the normal cone in derived sense and the virtual pullback in [31]. See [22] for more details.

Most of the results discussed in this paper hold for any homology theory satisfying certain reasonable conditions such as Chow groups and algebraic K-groups [20] by using [14, 18, 25, 26]. For simplicity, we only consider Chow groups and Borel-Moore homology in this paper.

The layout of this paper is as follows. In Sect. 6.2, we recall perfect obstruction theories and virtual fundamental classes. In Sect. 6.3, we recall the cosection localization and discuss Question 6.1.1. In Sect. 6.4, we recall the Oh-Thomas construction of a virtual fundamental class for a 3-term symmetric obstruction theory and discuss the topological approach. In Sect. 6.5, we discuss the derived algebraic geometry approach to Question 6.1.1.

In this paper, every scheme or stack is defined over the complex number field $\mathbb{C}$. We denote by $A_*(X)$ (resp. $H_*(X)$) the Chow group (resp. Borel-Moore homology) of a scheme X with rational coefficients.

6.2 Virtual Fundamental Classes

The goal of this section is to provide a quick introduction to virtual fundamental classes and set up the notation to be used below.

Finding the number of objects of interest is a fundamental problem in most branches of mathematics. In algebraic geometry, we often want to know whether a variety, a subvariety or a vector bundle with desired properties exists or not. And if it does, we want to know how many. This is the goal of enumerative geometry and we can often calculate the number of desired objects even though we cannot explicitly find them.

For instance, given a polynomial equation in one variable of degree d, we often do not know how to find the roots explicitly but we know that there are d roots. Similarly, we know there are exactly 27 lines on a smooth cubic surface S in three dimensional projective space $\mathbb{P}^3$ although we often cannot write down the equations of the lines.

A line in the projective space $\mathbb{P}^3$ is given by a two dimensional subspace in $\mathbb{C}^4$ and hence the set of all lines in $\mathbb{P}^3$ is parameterized by the Grassmannian variety

$$Gr(2,4) = M_2^{4\times 2}/\mathrm{GL}(2)$$

which is the space of GL(2)-orbits in the set $M_2^{4\times 2}$ of 4×2 matrices of rank 2, where GL(2) acts by matrix multiplication on the right. The set of lines in a smooth cubic surface $S = \mathrm{zero}(g) \subset \mathbb{P}^3$, defined by a homogeneous cubic polynomial $g \in \mathbb{C}[x_0, x_1, x_2, x_3]$, is the closed subvariety of $Gr(2,4)$ defined locally by 4 equations. More precisely, the universal rank 2 vector bundle $\mathcal{U}$ over $Gr(2,4)$ gives us a diagram

$$\begin{array}{ccc} \mathbb{P}\mathcal{U} & \xrightarrow{f} & \mathbb{P}^3 \\ {\scriptstyle \pi}\downarrow & & \\ Gr(2,4) & & \end{array}$$

where f restricted to each fiber is the inclusion of the line corresponding to the base point. Then g gives us the section $s = \pi_* f^*(g)$ of the vector bundle $E = \pi_* f^* \mathcal{O}_{\mathbb{P}^3}(3)$ of rank 4. Hence the set of lines in the smooth cubic surface S is the zero locus zero$(s) = s^{-1}(0)$ of the section $s \in H^0(E)$ and the number of lines in S is

$$\deg\,[Gr(2,4)] \cap e(E) \in \mathbb{Q} \tag{6.2.1}$$

where $e(E)$ denotes the Euler class of E, $[Gr(2,4)]$ is the fundamental class of the Grassmannian and

$$\deg : H_0(Gr(2,4)) \longrightarrow \mathbb{Q}$$

is the degree map counting the points in a zero cycle.

As we have seen above, enumerative problems are often solved in two steps. We first construct a nice compact moduli space X that parameterizes all the objects, like the Grassmannian $Gr(2,4)$ for lines in $\mathbb{P}^3$ above. Then we look for the subvariety Z of desired objects in the moduli space, like zero(s) in $Gr(2,4)$ above. If we are lucky, the moduli space X is smooth and s is transveral to the zero section so that the homology class of the subvariety X is written as the cap product

$$[X] \cap \xi$$

where ξ is a suitable cohomology class like $e(E)$ above and $[X]$ is the fundamental class of X. Then

$$\int_X \xi := \deg\,[X] \cap \xi \tag{6.2.2}$$

is the number we are looking for. In enumerative geometry, we expect our answer to stay the same even if we deform our requirements for desired objects. For instance, we expect our count of lines in a cubic surface $S = \text{zero}(g)$ to be independent of the choice of g.

Unfortunately, the moduli space X is usually highly singular and the integral (6.2.2) gives us a "wrong" answer as the number may not be *deformation invariant*. So we have to find a homology class $[X]^{\mathrm{vir}}$ which plays the role of the fundamental class $[X]$ such that

$$\int_{[X]^{\mathrm{vir}}} \xi := \deg\,[X]^{\mathrm{vir}} \cap \xi \tag{6.2.3}$$

is invariant under deformation.

Suppose the moduli space X fits into the diagram

$$\begin{array}{ccc} & & E \\ & & \big\downarrow\uparrow\, s \\ X = s^{-1}(0) & \hookrightarrow & V \end{array} \tag{6.2.4}$$

where E is a natural vector bundle of rank r over a smooth variety V of dimension n and

$$X = s^{-1}(0) = \Gamma_s \cap \Gamma_0$$

is the zero locus of a section s of E where Γ_s and Γ_0 denote the graph of the sections s and 0 respectively. If Γ_s is transversal to Γ_0, then X is smooth and the fundamental class

$$[X] = [\Gamma_s \cap \Gamma_0] = [V] \cap e(E)$$

is the Euler class of E capped with the fundamental class of V. If Γ_s is not transversal to Γ_0, the graph $\Gamma_{s'}$ of a general topological deformation s' of s is transversal to Γ_0 and (the pushforward to X of) their intersection represents the homology class

$$[\Gamma_{s'} \cap \Gamma_0] = [V] \cap e(E, s) \in H_{2n-2r}(X) \tag{6.2.5}$$

where $e(E, s) \in H^{2r}(V, V - X)$ is the Euler class of E localized by s, which is the pullback $s^* or_E$ of the orientation class

$$or_E \in H^{2r}(E, E - \Gamma_0)$$

by $s : (V, V - X) \to (E, E - \Gamma_0)$. Hence the correct way to integrate a cohomology class ξ on X is to calculate (6.2.3) with

$$[X]^{\text{vir}} := [V] \cap e(E, s) \in H_{2n-2r}(X)$$

in this single chart case. An astute reader probably has observed that in (6.2.5), we could also perturb the zero section Γ_0 as well as Γ_s. This is what we are going to do for the cosection localization that we will review in the subsequent section.

If we want to find an *algebraic* cycle representing $[X]^{\text{vir}}$, by Fulton [9, §5.1], we deform V to the normal cone

$$C = C_{X/V} = \text{Spec}_X(\bigoplus_n I^n/I^{n+1})$$

of X in V where I is the defining ideal of X in V. Since X is defined by the vanishing of s, I is the image of $s^\vee : E^\vee \to \mathcal{O}_V$ and hence we have a surjection

$$\text{Sym}(E^\vee|_X) \longrightarrow \bigoplus_n I^n/I^{n+1}$$

which gives us the canonical inclusion $C \subset E|_X$. In fact, the normal cone $C_{X/V}$ is the limit of the graph Γ_{ts} as $t \to \infty$.

By Fulton [9, §1.9], there is a cycle η on X such that C is rationally equivalent to $\pi^{-1}(\eta)$ where $\pi : E|_X \to X$ denotes the bundle projection. As $\pi^{-1}(\eta)$ is a deformation of C which is transversal to the zero section Γ_0, we find that

$$\eta = \pi^{-1}(\eta) \cap \Gamma_0 = [V] \cap e(E, s) = [X]^{\mathrm{vir}}$$

is an algebraic cycle representing the virtual fundamental class $[X]^{\mathrm{vir}}$. We call η a *virtual fundamental cycle* of X. In summary, the graph Γ_s is deformed to the normal cone $C = C_{X/V}$ and then to $\pi^{-1}(\eta)$ which is transversal to Γ_0. Then it leads us to

$$[X]^{\mathrm{vir}} = \Gamma_s \cdot \Gamma_0 = C_{X/V} \cdot \Gamma_0 = \pi^{-1}(\eta) \cdot \Gamma_0 = \eta \in A_{n-r}(X).$$

As we will see below, if we further deform Γ_0, we can localize the virtual fundamental class $[X]^{\mathrm{vir}}$ to a smaller subset.

In general, we may not be able to write X as the zero locus of a section for a globally defined bundle E. But we can find an open cover $\{X_\alpha \to X\}$ of X such that

$$\begin{array}{ccc} & & E_\alpha \\ & & \big\downarrow\big\uparrow s_\alpha \\ X_\alpha = s_\alpha^{-1}(0) & \hookrightarrow & V_\alpha \end{array} \tag{6.2.6}$$

where V_α are smooth varieties of dimension n_α and E_α are vector bundles of rank r_α with section s_α. Usually, V_α and E_α arise intrinsically from the moduli problem by deformation theory. Then we say (6.2.6) is a *Kuranishi chart* for X_α. On each X_α, we have a virtual fundamental class

$$[X_\alpha]^{\mathrm{vir}} \in A_{n_\alpha - r_\alpha}(X_\alpha)$$

by the recipe above and we want to glue $[X_\alpha]^{\mathrm{vir}}$ on X_α to a cycle class $[X]^{\mathrm{vir}}$ on X.

The minimum requirement for such a gluing is that the degrees $n_\alpha - r_\alpha$ of the cycle classes should be indepedent of α. Moreover, the cotangent sheaves Ω_{X_α} of X_α should also glue to the cotangent sheaf Ω_X of X. By Hartshorne [11], the cotangent sheaves Ω_{X_α} are the zeroth cohomology of the truncated cotangent complex

$$\mathbb{L}_{X_\alpha} = [I_\alpha/I_\alpha^2 \xrightarrow{d} \Omega_{V_\alpha}|_{X_\alpha}] \tag{6.2.7}$$

where

$$I_\alpha = \mathrm{im}(s_\alpha^\vee : E_\alpha^\vee \to \mathcal{O}_{V_\alpha}) \tag{6.2.8}$$

is the defining ideal of X_α in V_α. By (6.2.8), we have a morphism of complexes

$$\begin{array}{ccccc} \mathbb{E}_\alpha & & E_\alpha^\vee|_{X_\alpha} & \xrightarrow{ds_\alpha^\vee} & \Omega_{V_\alpha}|_{X_\alpha} \\ \phi_\alpha \downarrow & & s_\alpha^\vee \downarrow & & \| \\ \mathbb{L}_{X_\alpha} & & I_\alpha/I_\alpha^2 & \xrightarrow{d} & \Omega_{V_\alpha}|_{X_\alpha}. \end{array} \tag{6.2.9}$$

Although we may not be able to glue the local charts (6.2.6) to a global chart (6.2.4), often the derived category morphisms (6.2.9) glue to a morphism, called a *perfect obstruction theory*,

$$\phi : \mathbb{E} \longrightarrow \mathbb{L}_X \tag{6.2.10}$$

in the derived category $D^b(\mathrm{Coh}X)$ of coherent sheaves on X. Conversely, a morphism (6.2.10) with $\mathbb{E}$ perfect of amplitude $[-1,0]$ such that $h^{-1}(\phi)$ is surjective and $h^0(\phi)$ is an isomorphism comes from gluing (6.2.9) for some local charts (6.2.6). This is enough to define the virtual fundamental class because we can glue the normal cones, after normalizing by taking quotients by $T_{V_\alpha}|_{X_\alpha}$, and then intersect it with the zero section as follows.

A key observation in [1] is that the quotient stack

$$\mathfrak{C}_\alpha := C_\alpha / T_{V_\alpha}|_{X_\alpha} \tag{6.2.11}$$

of the normal cone $C_\alpha = C_{X_\alpha/V_\alpha}$ by the tangent bundle T_{V_α} of V_α restricted to X_α is independent of α. Hence (6.2.11) glue to define a cone stack $\mathfrak{C}_X$ over X, called the *intrinsic normal cone* of X, whose restriction $\mathfrak{C}_X \times_X X_\alpha$ to X_α is isomorphic to $\mathfrak{C}_\alpha$. The canonical inclusion $C_\alpha \hookrightarrow E_\alpha|_{X_\alpha}$ induces the inclusion

$$\mathfrak{C}_\alpha \hookrightarrow E_\alpha|_{X_\alpha}/T_{V_\alpha}|_{X_\alpha} = \mathcal{E}_\alpha.$$

As (6.2.9) glue to (6.2.10), $\mathcal{E}_\alpha$ glue to a vector bundle stack $\mathcal{E} =: h^1/h^0(\mathbb{E}^\vee)$ over X and hence we have an inclusion

$$\mathfrak{C}_X \hookrightarrow \mathcal{E}$$

of the intrinsic normal cone into the vector bundle stack $\mathcal{E}$. By Kresch [24], the intersection $\mathfrak{C}_X \cdot \Gamma_0$ of $\mathfrak{C}$ with the zero section makes sense and gives us the virtual fundamental class

$$[X]^{\mathrm{vir}} = \mathfrak{C}_X \cdot \Gamma_0 = 0^!_{\mathcal{E}}[\mathfrak{C}_X] \in A_{\mathrm{rank}\,\mathbb{E}}(X). \tag{6.2.12}$$

Once again, note that there is a chance that we may further localize $[X]^{\mathrm{vir}}$ to a smaller subset if we can perturb Γ_0 in a controlled manner.

Virtual invariants are now defined as integrals (6.2.3) of cohomology classes on X against $[X]^{\mathrm{vir}}$ which is provided by a perfect obstruction theory (6.2.10).

6.3 Cosection Localization

In this section, we recall the localization of the virtual fundamental class $[X]^{\mathrm{vir}}$ by a cosection of the obstruction sheaf $Ob_X = h^1(\mathbb{E}^\vee)$.

Let X be a Deligne-Mumford stack and $\phi : \mathbb{E} \to \mathbb{L}_X$ denote a perfect obstruction theory (6.2.10). The derived category object $\mathbb{E}$ is a complex of coherent sheaves, perfect of amplitude $[-1, 0]$, i.e. locally a two-term complex of locally free sheaves concentrated in degrees $-1, 0$.

Definition 6.3.1 A *cosection* of the perfect obstruction theory ϕ is a morphism

$$\sigma : \mathbb{E}^\vee[1] \longrightarrow \mathcal{O}_X \tag{6.3.1}$$

in the derived category $D^b(\mathrm{Coh}X)$.

Equivalently, by taking its dual, a cosection can be thought of as a section

$$\sigma^\vee \in \mathrm{Hom}_{D^b(\mathrm{Coh}X)}(\mathcal{O}_X, \mathbb{E}[-1]) = \mathbb{H}^0(\mathbb{E}[-1]). \tag{6.3.2}$$

By assumption, the dual $\mathbb{E}^\vee$ of $\mathbb{E}$ is a perfect complex of amplitude $[0, 1]$ whose zeroth cohomology $H^0(\mathbb{E}^\vee)$ is the tangent sheaf $T_X = \Omega_X^\vee$. The higher cohomology $H^{>1}(\mathbb{E}^\vee)$ vanishes and the first cohomology

$$H^1(\mathbb{E}^\vee) =: Ob_X$$

is called the obstruction sheaf of X. Therefore, under our assumptions, we have a canonical morphism $\mathbb{E}^\vee[1] \to Ob_X$ and σ is uniquely determined by the sheaf homomorphism

$$Ob_X \longrightarrow \mathcal{O}_X$$

which we also denote by σ by abuse of notation. Let $X(\sigma)$ be the closed substack of X defined by the image of Ob_X.

If we restrict Ob_X to X_α in (6.2.6), we have

$$Ob_X|_{X_\alpha} = \mathrm{coker}\left((ds_\alpha^\vee)^\vee : T_{V_\alpha}|_{X_\alpha} \to E_\alpha|_{X_\alpha}\right)$$

by (6.2.9) and $\sigma|_{X_\alpha}$ gives us a homomorphism $\sigma_\alpha : E_\alpha|_{X_\alpha} \to \mathcal{O}_{X_\alpha}$. Let

$$E(\sigma)_\alpha = E_\alpha|_{X(\sigma)_\alpha} \cup \ker(E_\alpha|_{X_\alpha - X(\sigma)} \xrightarrow{\sigma_\alpha} \mathcal{O}_{X_\alpha - X(\sigma)})$$

denote the *kernel* of σ. As $\sigma_\alpha \circ (ds_\alpha^\vee)^\vee = 0$, we have a substack

$$\mathcal{E}(\sigma)_\alpha = E(\sigma)_\alpha / T_{V_\alpha}|_{X_\alpha} \subset E_\alpha|_{X_\alpha} / T_{V_\alpha}|_{X_\alpha} = \mathcal{E}_\alpha$$

and these glue to a substack

$$\mathcal{E}(\sigma) \subset \mathcal{E} = h^1/h^0(\mathbb{E}^\vee).$$

Now the main observations in [17] are as follows.

Theorem 6.3.2

(1) *The intrinsic normal cone* $\mathfrak{C}_X$ *has support in* $\mathcal{E}(\sigma)$.
(2) *There is a homomorphism*

$$0^!_{\mathcal{E},\sigma} : A_*(\mathcal{E}(\sigma)) \to A_{*+\operatorname{rank}\mathbb{E}}(X(\sigma)) \tag{6.3.3}$$

such that

$$\imath_* \circ 0^!_{\mathcal{E},\sigma} = 0^!_{\mathcal{E}} \circ \jmath_* \tag{6.3.4}$$

where $\imath : X(\sigma) \hookrightarrow X$ *and* $\jmath : \mathcal{E}(\sigma) \hookrightarrow \mathcal{E}$ *denote the inclusion maps.*

Definition 6.3.3 The cosection localized virtual fundamental class is

$$[X]^{\mathrm{vir}}_\sigma := 0^!_{\mathcal{E},\sigma}[\mathfrak{C}_X] \in A_{\operatorname{rank}\mathbb{E}}(X(\sigma)).$$

By (6.3.4), we obviously have $\imath_*[X]^{\mathrm{vir}}_\sigma = [X]^{\mathrm{vir}}$.

To see Theorem 6.3.2 (1), we only need to look into the local charts (6.2.6), because the issue is local. Under the stronger assumption that σ_α extends to a homomorphism $E_\alpha \to \mathcal{O}_{V_\alpha}$ and that

$$\sigma_\alpha \circ s_\alpha = 0, \tag{6.3.5}$$

the graph Γ_{ts_α} lies entirely in the kernel of σ_α and so is the normal cone

$$C_{X_\alpha/V_\alpha} = \lim_{t\to\infty} \Gamma_{ts_\alpha} \subset E(\sigma)_\alpha.$$

Without (6.3.5), we need more refined arguments in [17].

For Theorem 6.3.2 (2), let $\rho : \widetilde{X} \to X$ denote the blowup along $X(\sigma)$. Let D denote the exceptional divisor so that σ induces a surjection $\mathcal{E}|_{\widetilde{X}} \to \mathcal{O}_{\widetilde{X}}(-D)$ whose kernel is denoted by $\mathcal{E}'$. For any cycle η in $\mathcal{E}(\sigma)$, we can easily find cycles ζ in $\mathcal{E}'$ and μ in $\mathcal{E}|_{X(\sigma)}$ such that

$$\eta = \tilde{\imath}_*\mu + \tilde{\rho}_*\zeta$$

where $\tilde{\imath} : \mathcal{E}|_{X(\sigma)} \hookrightarrow \mathcal{E}(\sigma)$ is the inclusion and $\tilde{\rho} : E' \hookrightarrow \mathcal{E}(\sigma)|_{\widetilde{X}} \to \mathcal{E}(\sigma)$ is the inclusion followed by the natural map induced by ρ. Then we may define

$$0^!_{\mathcal{E},\sigma}\eta := 0^!_{\mathcal{E}|_{X(\sigma)}}\mu - (\rho|_{X(\sigma)})_*(D \cdot 0^!_{\mathcal{E}'}\zeta) \in A_*(X(\sigma))$$

and check all the desired properties.

In fact, the construction above gives us a homomorphism

$$0^!_{E,\sigma} : A_*(E(\sigma)) \longrightarrow A_{*-r}(X(\sigma)) \tag{6.3.6}$$

when E is a vector bundle of rank r over a Deligne-Mumford stack X equipped with a cosection $\sigma : E \to \mathcal{O}_X$, $X(\sigma)$ is the zero locus of $\sigma^\vee$ and $E(\sigma)$ is the kernel of σ. Like (6.3.4), (6.3.6) is related to the ordinary Gysin map $0^!_E$ by

$$\imath_* \circ 0^!_{E,\sigma} = 0^!_E \circ \jmath_*$$

where $\imath : X(\sigma) \hookrightarrow X$ and $\jmath : E(\sigma) \hookrightarrow E$ denote the inclusion maps. In the single chart case of (6.2.4), if we have a cosection $\sigma : E \to \mathcal{O}_V$ satisfying $\sigma \circ ds = 0$, then the virtual fundamental class $[X]^{\mathrm{vir}}$ of X is localized to

$$[X]^{\mathrm{vir}}_\sigma = 0^!_{E|_X,\sigma}[C_{X/V}] \in A_{n-r}(X(\sigma)). \tag{6.3.7}$$

There is also a topological explanation of Theorem 6.3.2 in [17, Appendix] as follows. Let us first consider the single chart case of (6.2.4). Suppose we have a homorphism $\sigma : E \to \mathcal{O}_V$ satisfying $\sigma \circ s = 0$. We want to perturb the zero section to a continuous section s'' such that

$$[X]^{\mathrm{vir}} = \Gamma_s \cdot \Gamma_0 = [\Gamma_s \cap \Gamma_{s''}] \in H_*(X(\sigma)).$$

Pick a tubular neighborhood U of $X(\sigma)$ in V. Let E' be the kernel of the surjection $\sigma|_{V-U} : E|_{V-U} \to \mathcal{O}_{V-U}$. Pick a hermitian metric on $E|_{V-U}$ so that we have a splitting

$$E' \oplus \mathcal{O}_{V-U} \cong E|_{V-U} \xrightarrow{\sigma} \mathcal{O}_{V-U}, \quad (v,t) \mapsto t.$$

By standard fiber bundle theory, the section $(0,1)$ of $E' \oplus \mathcal{O}_{V-U}$ extends to a continuous section s'' of E which is transversal to Γ_s. Because $\Gamma_s|_{V-U}$ lies in E', $\Gamma_s \cap \Gamma_{s''}$ is a cycle lying over U whose projection to $X(\sigma)$ represents the virtual fundamental class $[X]^{\mathrm{vir}}_\sigma$.

We can make the above arguments work in general, assuming that we have a global resolution

$$\mathbb{E} \cong [E^{-1} \to E^0]$$

of $\mathbb{E}$ where E^{-1} and E^0 are locally free sheaves on X. Let E_i denote the dual of E^{-i} for $i = 0, 1$. Then the intrinsic normal cone

$$\mathfrak{C}_X \subset h^1/h^0(\mathbb{E}^\vee) = E_1/E_0$$

lifts to a cone $C_X = \mathfrak{C}_X \times_{E_1/E_0} E_1$ whose support lies in $E_1(\sigma)$ where $E_1(\sigma)$ denotes the kernel of the homomorphism

$$E_1 \longrightarrow h^1(\mathbb{E}^\vee) = Ob_X \longrightarrow \mathcal{O}_X.$$

The cosection localized virtual fundamental class is now

$$[X]^{\mathrm{vir}}_\sigma = 0^!_{E_1,\sigma}[C_X] = [C_X \cap \Gamma_{s''}] \in H_*(X(\sigma))$$

where s'' is a perturbed zero section of E_1 by σ as in the previous paragraph.

The cosection localization turned out to be quite useful as in [4–8, 10, 13, 15, 16, 19, 23] to name a few. In many of these examples, the vanishing locus $X(\sigma)$ of the cosection σ is an interesting space on its own. So we may ask the following.

Question 6.3.4 Is there a natural induced perfect obstruction theory on $X(\sigma)$ whose virtual fundamental class $[X(\sigma)]^{\mathrm{vir}}$ coincides with the cosection localized virtual fundamental class $[X]^{\mathrm{vir}}_\sigma$ of X?

As a start, let us consider the single chart case of (6.2.4) equipped with a cosection $\sigma : E \to \mathcal{O}_V$. In this case, $X(\sigma)$ is the zero locus of the section $(s, \sigma^\vee)$ of the vector bundle $E \oplus E^\vee$ and fits into the diagram

$$\begin{array}{ccc} & & E \oplus E^\vee \\ & & \downarrow\uparrow (s,\sigma^\vee) \\ X(\sigma) & \hookrightarrow & V. \end{array} \tag{6.3.8}$$

So we have a natural perfect obstruction theory

$$\begin{array}{ccccc} \mathbb{E}_{X(\sigma)} & & E^\vee \oplus E|_{X(\sigma)} & \xrightarrow{ds^\vee + d\sigma} & \Omega_V|_{X(\sigma)} \\ \phi \downarrow & & s^\vee+\sigma \downarrow & & \| \\ \mathbb{L}_{X(\sigma)} & & J/J^2 & \xrightarrow{d} & \Omega_V|_{X(\sigma)} \end{array} \tag{6.3.9}$$

for $X(\sigma)$ where J is the image of $s^\vee + \sigma : E^\vee \oplus E \to \mathcal{O}_V$ that defines $X(\sigma)$. This cannot be a correct answer to Question 6.3.4 because the rank of $\mathbb{E}_{X(\sigma)}$ is

$$\dim V - 2\,\mathrm{rank}\, E = n - 2r$$

while the rank of the perfect obstruction theory

$$E^\vee|_X \xrightarrow{ds^\vee} \Omega_V|_X$$

for X is $\dim V - \operatorname{rank} E = n - r$. Unless E and σ are trivial, the equality

$$[X(\sigma)]^{\mathrm{vir}} = [X]^{\mathrm{vir}}_\sigma \tag{6.3.10}$$

can never hold if we use (6.3.9) as our perfect obstruction theory for $X(\sigma)$. To get a correct perfect obstruction theory, we have to modify (6.3.9) into a 3-term symmetric obstruction theory

$$\begin{array}{ccccccc} \mathbb{E}_{X(\sigma)} & & T_V|_{X(\sigma)} & \xrightarrow{\delta} & E^\vee \oplus E|_{X(\sigma)} & \xrightarrow{\delta^\vee} & \Omega_V|_{X(\sigma)} \\ \phi\downarrow & & & & \downarrow s^\vee+\sigma & & \| \\ \mathbb{L}_{X(\sigma)} & & & & J/J^2 & \xrightarrow{d} & \Omega_V|_{X(\sigma)} \end{array} \tag{6.3.11}$$

where $\delta^\vee = ds^\vee + d\sigma$ and use the Oh-Thomas construction in the subsequent section.

6.4 Virtual Fundamental Class for a 3-Term Symmetric Obstruction Theory

In this section, we recall the Oh-Thomas construction of virtual fundmental classes for 3-term symmetric obstruction theories in [28].

Definition 6.4.1 Let Y be a Deligne-Mumford stack. A *3-term symmetric obstruction theory* for Y is a morphism

$$\phi : \mathbb{E} \longrightarrow \mathbb{L}_Y$$

in the derived category $D^b(\mathrm{Coh}Y)$ of coherent sheaves on Y where $\mathbb{E}$ is

(1) perfect of amplitude $[-2, 0]$, and
(2) equipped with an isomorphism $\theta : \mathbb{E}^\vee[2] \to \mathbb{E}$ satisfying $\theta^\vee[2] = \theta$ and an isomorphism $o : \mathcal{O}_Y \to \det(\mathbb{E})$ satisfying $\det(\theta) = o \circ o^\vee$

such that $h^{-1}(\phi)$ is surjective and $h^0(\phi)$ is an isomorphism.

By Oh and Thomas [28], if Y is a quasi-projective scheme, we may assume that the 3-term symmetric complex $\mathbb{E}$ is given by an orthogonal bundle as follows.

Proposition 6.4.2 ([28]) *Let Y be a quasi-projective scheme and J be the ideal of a closed embedding $Y \hookrightarrow P$ into a smooth scheme P. Then a 3-term symmetric obstruction theory $\phi : \mathbb{E} \to \mathbb{L}_Y$ for Y admits a global resolution*

$$\begin{array}{ccccccc} \mathbb{E} & & [B & \xrightarrow{\delta} & F \cong F^\vee & \xrightarrow{\delta^\vee} & B^\vee] \\ \phi\downarrow & & & & \downarrow & & \downarrow \\ \mathbb{L}_Y & & & & [J/J^2 & \xrightarrow{d} & \Omega_P|_Y] \end{array} \tag{6.4.1}$$

with surjective vertical arrows, where B is a locally free sheaf of rank n on Y and F is an $SO(2r)$-bundle on Y. The isomorphism $q : F \cong F^\vee$ in the middle of the top row of (6.4.1) *comes from the orthogonal structure and satisfies $q^\vee = q$.*

From the definition, it is obvious that the stupid truncation

$$\begin{array}{ccc} [F^\vee & \xrightarrow{\delta^\vee} & B^\vee] \\ \downarrow & & \downarrow \\ [J/J^2 & \xrightarrow{d} & \Omega_P|_Y] \end{array}$$

of (6.4.1) is a 2-term perfect obstruction theory and the intrinsic normal cone $\mathfrak{C}_Y$ of Y embeds into the vector bundle stack F/B which gives us a cone

$$C_Y = \mathfrak{C}_Y \times_{F/B} F \subset F. \tag{6.4.2}$$

By the Darboux theorem in [2], if Y comes from a derived scheme equipped with a (-2)-shifted symplectic structure, then we may assume the following.

Assumption 6.4.3 The cone C_Y in (6.4.2) is isotropic in F, i.e. C_Y is contained in the quadratic cone

$$Q = \{v \in F \mid q(v, v) = 0\} \subset F \tag{6.4.3}$$

of isotropic vectors in the fibers of F.

For instance, by Pantev et al. [30], a moduli scheme Y of stable sheaves on a Calabi-Yau 4-fold is the classical truncation of a derived scheme equipped with a (-2)-shifted symplectic structure and thus Assumption 6.4.3 holds.

The key construction in [28] is the following.

Proposition 6.4.4 ([28]) *Let F be an $SO(2r)$-bundle over a scheme Y. Then the Gysin map $0^!_F$ admits a square root*

$$\sqrt{0^!_F} : A_*(Q) \longrightarrow A_{*-r}(Y). \tag{6.4.4}$$

Here is an outline of the construction. Firstly by lifting to the isotropic flag variety over Y, we may assume that F admits a maximal isotropic subbundle Λ thats fits into an exact sequence

$$0 \longrightarrow \Lambda \longrightarrow F \longrightarrow \Lambda^\vee \longrightarrow 0.$$

The inclusion of Q into F induces a tautological section τ of $F|_Q$ and a section τ_+ of $\Lambda^\vee$ in the commutative diagram

$$\begin{array}{ccccccccc} 0 & \longrightarrow & \Lambda & \longrightarrow & F|_Q & \longrightarrow & \Lambda^\vee & \longrightarrow & 0 \\ & & & & {\scriptstyle\tau}\uparrow & \nearrow_{\tau_+} & & & \\ & & & & \mathcal{O}_Q. & & & & \end{array}$$

Secondly, we deform Q to the normal cone

$$C_{Z/Q} \subset \Lambda^\vee|_Z$$

to the zero locus $Z = \tau_+^{-1}(0)$ of τ_+. As $\tau_+ = 0$ over Z, τ induces a section τ_- of $\Lambda|_Z$ and hence a cosection

$$\tau_-^\vee : \Lambda^\vee|_Z \longrightarrow \mathcal{O}_Z.$$

Note that $Z(\tau_-^\vee) = Y$ and we have a canonical inclusion

$$C_{Z/Q} \hookrightarrow \Lambda^\vee|_Z(\tau_-) \tag{6.4.5}$$

because Q is isotropic. Then we use (6.3.6) to define

$$\sqrt{0_F^!} : A_*(Q) \longrightarrow A_*(C_{Z/Q}) \longrightarrow A_*(\Lambda^\vee|_Z(\tau_-)) \xrightarrow{0^!_{\Lambda^\vee|_Z,\tau_-}} A_{*-r}(Y)$$

where the first arrow is the usual specialization map in [9] and the second arrow is the pushforward by the inclusion (6.4.5). The composition of $\sqrt{0_F^!}$ with the pushforward $\imath_*$ by the inclusion $\imath : V \hookrightarrow Q \subset F$ of the zero section is the square root Euler class

$$\sqrt{e}(F) = \sqrt{0_F^!} \circ \imath_* : A_*(Y) \longrightarrow A_{*-r}(Y).$$

By Kiem and Park [21], if there is an isotropic section s of F, the square root Euler class can be localized to the zero locus $Y(s)$ of s as

$$\sqrt{e}(F, s) : A_*(Y) \longrightarrow A_{*-r}(Y(s))$$

so that we have $\jmath_* \circ \sqrt{e}(F, s) = \sqrt{e}(F)$ where $\jmath$ denotes the inclusion of $Y(s)$ into Y.

Now the virtual fundamental class for a 3-term symmetric obstruction theory is defined as follows.

Definition 6.4.5 ([28]) Let Y be a quasi-projective scheme equipped with a 3-term symmetric obstruction theory (6.4.1). Under Assumption 6.4.3, the virtual fundamental class of Y is defined as

$$[Y]^{\mathrm{vir}} = \sqrt{0^!_F}[C_Y] \in A_*(Y). \tag{6.4.6}$$

In particular, if a projective scheme Y is the classical truncation of a derived scheme equipped with a (-2)-shifted symplectic structure, then we have the virtual fundamental class

$$[Y]^{\mathrm{vir}} \in A_{\mathrm{rank}\,\mathbb{E}/2}(Y)$$

where $\phi : \mathbb{E} \to \mathbb{L}_Y$ is the 3-term symmetric obstruction theory induced from the cotangent complex of the derived scheme.

With Proposition 6.4.4 and Definition 6.4.5 at hand, let us look back at Question 6.3.4. As a start, let us consider the single chart case of (6.3.8). It is obvious that (6.3.11) is a 3-term symmetric obstruction theory for $Y = X(\sigma)$ with the natural pairing

$$q : E \oplus E^\vee \longrightarrow \mathcal{O}_Y, \quad (v, w) \mapsto w \circ v$$

as the orthogonal structure on $F = E \oplus E^\vee$. For Assumption 6.4.3, we assume

$$\sigma \circ s = 0. \tag{6.4.7}$$

Then we find that the graph $\Gamma_{s+\sigma^\vee}$ is contained in the isotropic cone

$$Q = \{(v, w) \in E \oplus E^\vee \mid w \circ v = 0\}$$

of the orthogonal bundle $E \oplus E^\vee$. Hence the normal cone

$$C_{X(\sigma)/V} = \lim_{t \to \infty} \Gamma_{t(s+\sigma^\vee)}$$

is contained in Q. Hence the virtual fundamental class

$$[X(\sigma)]^{\mathrm{vir}} = \sqrt{0^!_{E \oplus E^\vee}}[C_{X(\sigma)/V}] \tag{6.4.8}$$

is a cycle class of correct degree $n - r$. We are suppressing the sign issue by the choice of an orientation on F. So any statement on $[X(\sigma)]^{\mathrm{vir}}$ from now on should be understood up to sign $\pm$.

In this case, (6.3.11) and (6.4.8) turn out to give a correct answer to Question 6.3.4 by the following.

Proposition 6.4.6 *Let X be the zero locus of a section s of a vector bundle E over a smooth Deligne-Mumford stack V as in* (6.2.4) *so that we have a perfect obstruction theory*

$$\begin{array}{ccccc} \mathbb{E} & & E^\vee|_X & \xrightarrow{ds^\vee} & \Omega_V|_X \\ \phi\downarrow & & s^\vee\downarrow & & \| \\ \mathbb{L}_X & & I/I^2 & \xrightarrow{d} & \Omega_V|_X \end{array} \tag{6.4.9}$$

where I is the image of $s^\vee : E^\vee \to \mathcal{O}_V$. Let $\sigma : E \to \mathcal{O}_V$ be a cosection satisfying $\sigma \circ s = 0$. Then the cosection localized virtual fundamental class $[X]^{\mathrm{vir}}_\sigma$ in (6.3.7) *with respect to* (6.4.9) *coincides with the virtual fundamental class $[X(\sigma)]^{\mathrm{vir}}$ in* (6.4.8) *induced by the 3-term symmetric obstruction theory* (6.3.11).

Proof By deforming V to the normal cone $C_{X(\sigma)/V}$ of $X(\sigma)$, we find that

$$[X(\sigma)]^{\mathrm{vir}} = \sqrt{0^!_{E\oplus E^\vee|_{X(\sigma)}}}[C_{X(\sigma)/V}] \tag{6.4.10}$$

$$= \sqrt{e}(E \oplus E^\vee|_{C_{X(\sigma)/V}}, \tau_{X(\sigma)}) \cap [C_{X(\sigma)/V}] \tag{6.4.11}$$

$$= \sqrt{e}(E \oplus E^\vee, (s, \sigma^\vee)) \cap [V] \tag{6.4.12}$$

$$= 0^!_{E|_X,\sigma}[C_{X/V}] = [X]^{\mathrm{vir}}_\sigma \tag{6.4.13}$$

where $\tau_{X(\sigma)}$ is the tautological section. □

Therefore Proposition 6.4.6 is a positive answer to Question 6.3.4 in the single chart case. Comparing (6.4.9) with (6.3.11), it seems natural to expect the following.

Conjecture 6.4.7 Let X be a Deligne-Mumford stack with a perfect obstruction theory $\phi_X : \mathbb{E} \to \mathbb{L}_X$ and a cosection $\sigma : \mathbb{E}^\vee[1] \to \mathcal{O}_X$. Under suitable assumptions, there is a 3-term symmetric obstruction theory $\phi_{X(\sigma)} : \mathbb{E}_{X(\sigma)} \to \mathbb{L}_{X(\sigma)}$ that fits into an exact triangle

$$\mathbb{E}|_{X(\sigma)} \longrightarrow \mathbb{E}_{X(\sigma)} \longrightarrow \mathbb{E}|^\vee_{X(\sigma)}[2] \longrightarrow$$

whose virtual fundamental class $[X(\sigma)]^{\mathrm{vir}}$ by (6.4.6) equals the cosection localized virtual fundamental class $[X]^{\mathrm{vir}}_{\sigma}$ of X, i.e. we have

$$[X(\sigma)]^{\mathrm{vir}} = [X]^{\mathrm{vir}}_{\sigma}.$$

We have two possible approaches for Conjecture 6.4.7, topological and derived. Let us briefly outline the topological approach. The details will appear elsewhere. In this approach, we may use the complex Kuranish chart in [29] and extend all the above constructions to make them applicable to complex Kuranishi charts. In fact, a main result in [29] tells us that if X is a projective scheme equipped with a perfect obstruction theory, then there are a smooth manifold V, a complex vector bundle E over V and a smooth section s of E such that $X = s^{-1}(0)$. So we are in the single chart case of Proposition 6.4.6 except that V, E and s are holomorphic only in the first or second order infinitesimal neighborhood of X. Now we may extend the constructions in Sects. 6.3 and 6.4 such that (6.3.7) and (6.4.8) make sense for a complex Kuranishi chart. Then the proof of Proposition 6.4.6 should give us a proof of Conjecture 6.4.7.

In the subsequent section, we will discuss a derived algebraic geometry approach to Conjecture 6.4.7.

6.5 Derived Cosection Localization

The cosection localization in Sect. 6.3 looks more natural in derived algebraic geometry. The cosection σ in (6.3.1) should be thought of as a (-1)-shifted 1-form and if it is closed, its vanishing locus $X(\sigma)$ admits a natural (-2)-shifted symplectic structure, which gives us a 3-term symmetric obstruction theory on the classical truncation of X and the virtual fundamental class (6.4.6). In this section, we outline this derived approach for the cosection localization and a proof of Conjecture 6.4.7 in this setting. The details will appear in [22].

Let us set up the notation for this section. Let X be a derived quasi-smooth Deligne-Mumford stack and $\mathbb{E} = \mathbb{L}_X$ denote the cotangent complex of X. Let $\underline{X}$ denote the classical truncation of X. Then the natural morphism

$$\mathbb{E}|_{\underline{X}} = \mathbb{L}_X|_{\underline{X}} \longrightarrow \mathbb{L}_{\underline{X}}$$

of cotangent complexes is a perfect obstruction theory of $\underline{X}$ with amplitude $[-1,0]$.

In derived algebraic geometry, it is more natural to consider the dual of a cosection (6.3.1).

Definition 6.5.1 A *cosection* of X is a (-1)-shifted 1-form

$$\sigma^{\vee} : \mathcal{O}_X \longrightarrow \mathbb{L}_X[-1] = \mathbb{E}[-1] \tag{6.5.1}$$

whose vanishing locus is the fiber product

$$X(\sigma) = X \times_{0,\mathbb{T}^*_X[-1],\sigma^\vee} X.$$

Here $\mathbb{T}^*_X[-1] = \mathrm{Tot}(\mathbb{E}[-1])$ denotes the total space of the (-1)-shifted cotangent complex of X and we have the fiber diagram

$$\begin{array}{ccc} X(\sigma) & \longrightarrow & X \\ \downarrow & & \downarrow \sigma^\vee \\ X & \xrightarrow{0} & \mathbb{T}^*_X[-1]. \end{array} \tag{6.5.2}$$

By definition, the classical truncation of $X(\sigma)$ is the classical vanishing locus of the cosection discussed in Sect. 6.3 and the dual of (6.5.1) gives us a cosection of the obstruction sheaf $Ob_{\underline{X}} = h^1(\mathbb{E}|_{\underline{X}})$ which defines the cosection localized virtual fundamental class

$$[\underline{X}]^{\mathrm{vir}}_\sigma \tag{6.5.3}$$

by Definition 6.3.3.

The following is fundamental for our derived approach for Conjecture 6.4.7.

Theorem 6.5.2 ([3, 30])

(1) $\mathbb{T}^*_X[-1]$ *has a* (-1)*-shifted symplectic structure and the cotangent complex* $\mathbb{E}_{\mathbb{T}^*_X[-1]}$ *of* $\mathbb{T}^*_X[-1]$ *fits into an exact triangle*

$$\mathbb{E}|_{\mathbb{T}^*_X[-1]} \longrightarrow \mathbb{E}_{\mathbb{T}^*_X[-1]} \longrightarrow \mathbb{E}|^\vee_{\mathbb{T}^*_X[-1]}[1] \longrightarrow .$$

(2) *If the* (-1)*-shifted 1-form* $\sigma^\vee$ *is closed, then* $\sigma^\vee : X \to \mathbb{T}^*_X[-1]$ *is Lagrangian and the Lagrangian intersection* $X(\sigma)$ *in* (6.5.2) *admits a natural* (-2)*-shifted symplectic structure.*

In fact, the cotangent complex $\mathbb{E}_{X(\sigma)}$ of the Lagrangian intersection $X(\sigma)$ is the cone

$$\begin{aligned} &\mathrm{cone}\left(\mathbb{E}_{\mathbb{T}^*_X[-1]}|_{X(\sigma)} \xrightarrow{(0^*,\sigma^{\vee *})} \mathbb{E}|_{X(\sigma)} \oplus \mathbb{E}|_{X(\sigma)}\right) \\ \cong\ &\mathrm{cone}\left(\mathbb{E}^\vee[1]|_{X(\sigma)} \xrightarrow{\sigma^{\vee *}} \mathbb{E}|_{X(\sigma)}\right) \end{aligned}$$

and fits into an exact triangle

$$\mathbb{E}|_{X(\sigma)} \longrightarrow \mathbb{E}_{X(\sigma)} \cong \mathbb{E}^\vee_{X(\sigma)}[2] \longrightarrow \mathbb{E}|^\vee_{X(\sigma)}[2] \longrightarrow . \tag{6.5.4}$$

By Pantev et al. [30] and Brav et al. [2], we then have a 3-term symmetric obstruction theory on the classical truncation $\underline{X}(\sigma)$ of $X(\sigma)$ and Assumption 6.4.3 holds for $\underline{X}(\sigma)$. We therefore have the virtual fundamental class

$$[\underline{X}(\sigma)]^{\mathrm{vir}} \tag{6.5.5}$$

by Definition 6.4.5.

The two virtual fundamental classes (6.5.3) and (6.5.5) coincide as expected by Conjecture 6.4.7.

Theorem 6.5.3 ([22]) *Let X be a derived quasi-smooth Deligne-Mumford stack equipped with a* closed *(-1)-shifted 1-form $\sigma^\vee : \mathcal{O}_X \to \mathbb{E}[-1]$ where $\mathbb{E} = \mathbb{L}_X$ denotes the cotangent complex of X. Then we have the equality*

$$[\underline{X}]^{\mathrm{vir}}_{\sigma} = [\underline{X}(\sigma)]^{\mathrm{vir}}. \tag{6.5.6}$$

Here is an outline of the proof. Firstly, we note that the inclusion map

$$\imath : X(\sigma) \longrightarrow X$$

has a canonical Lagrangian fibration structure compatible with (6.5.4). Secondly, we construct a canonical relative (-2)-shifted non-degenerate 2-form on the composition

$$X(\sigma) \times \mathbb{A}^1 \longrightarrow X \times \mathbb{A}^1 \longrightarrow \mathbf{D}_X$$

where $\mathbf{D}_X$ denotes the derived deformation space in [12]. Then we apply the square root virtual pullback in [31] to deduce (6.5.6).

Unfortunately, the details of the proof are quite involved and long. Hence we refer to [22] for a detailed proof.

Acknowledgments This work was partially supported by Korea NRF grant 2021R1F1A1046556.

References

1. Behrend, K., Fantechi, B.: The intrinsic normal cone. Invent. Math. **128**(1), 45–88 (1997)
2. Brav, C., Bussi, V., Joyce, D.: A Darboux theorem for derived schemes with shifted symplectic structure. J. Am. Math. Soc. **32**(2), 399–443 (2019)
3. Calaque, D.: Shifted cotangent stacks are shifted symplectic. Ann. Faculté Sci Toulouse Math. **28**(1), 67–90 (2019)
4. Chang, H.L., Kiem, Y.H.: Poincaré invariants are Seiberg-Witten invariants. Geom. Topol. **17**(2), 1149–1163 (2013)
5. Chang, H.L., Li, J.: Gromov-Witten invariants of stable maps with fields. Int. Math. Res. Not. IMRN **2012**(18), 4163–4217 (2012)

6. Chang, H.L., Li, J., Li, W.P.: Witten's top Chern class via cosection localization. Invent. Math. **200**(3), 1015–1063 (2015)
7. Chang, H.L., Li, J., Li, W.P., Liu, M.C.C.: Mixed-Spin-P fields of Fermat quintic polynomials. Cambridge J. Math. **7**(3), 319–364 (2019)
8. Clader, E.: Landau-Ginzburg/Calabi-Yau correspondence for the complete intersections $X_{3,3}$ and $X_{2,2,2,2}$. Adv. Math. **307**, 1–52 (2017)
9. Fulton, W.: Intersection Theory. Ergebnisse der Mathematik und ihrer Grenzgebiete, vol. 3, Folge. 2. Springer, Berlin (1998)
10. Gholampour, A., Sheshmani, A.: Donaldson-Thomas invariants of 2-dimensional sheaves inside threefolds and modular forms. Adv. Math. **326**, 79–107 (2018)
11. Hartshorne, R.: Algebraic Geometry. Graduate Texts in Mathematics, vol. 52, xvi+496 pp. Springer, New York (1977)
12. Hekking, J., Khan, A., Rydh, D.: Deformation to the normal cone and blowups via derived Weil restrictions (2025). arXiv:2511.19412
13. Hu, J., Li, W.P., Qin, Z.: The Gromov-Witten invariants of the Hilbert schemes of points on surfaces with $p_g > 0$. Int. J. Math. **26**(1), 1550009, 26pp. (2015)
14. Kiem, Y.H.: Localizing virtual fundamental cycles for semi-perfect obstruction theories. Int. J. Math. **29**(4), 1850032, 30pp. (2018)
15. Kiem, Y.-H., Li, J.: Low degree GW invariants of spin surfaces. Pure Appl. Math. Q. **7**(4), 1449–1475 (2011). Special Issue: In memory of Eckart Viehweg
16. Kiem, Y.H., Li, J.: Low Degree GW Invariants of Surfaces II. Sci. China Math. **54**(8), 1679–1706 (2011)
17. Kiem, Y.H., Li, J.: Localizing virtual cycles by cosections. J. Am. Math. Soc. **26**(4), 1025–1050 (2013)
18. Kiem, Y.H., Li, J.: Localizing virtual structure sheaves by cosections. Int. Math. Res. Not. **22**, 8387–8417 (2020)
19. Kiem, Y.H., Li, J.: Quantum singularity theory via cosection localization. J. Reine Angew. Math. **766**, 73–107 (2020)
20. Kiem, Y.H., Park, H.: Virtual intersection theories. Adv. Math. **388**, 107858, 51pp. (2021)
21. Kiem, Y.H., Park, H.: Localizing virtual cycles for Donaldson-Thomas invariants of Calabi-Yau 4-folds. J. Algebraic Geom. **32**, 585–639 (2023)
22. Kiem, Y.H., Park, H.: Cosection Localization via Shifted Symplectic Geometry (preprint) arXiv:2504.19542
23. Kool, M., Thomas, R.: Reduced classes and curve counting on surfaces I: theory. Algebr. Geom. **1**(3), 334–383 (2014)
24. Kresch, A.: Cycle groups for Artin stacks. Invent. Math. **138**(3), 495–536 (1999)
25. Lee, Y.P.: Quantum K-theory. I. Foundations. Duke Math. J. **121**(3), 389–424 (2004)
26. Levine, M., Morel, F.: Algebraic Cobordism. Springer Monographs in Mathematics. Springer, Berlin (2007)
27. Li, J., Tian, G.: Virtual moduli cycles and Gromov-Witten invariants of algebraic varieties. J. Am. Math. Soc. **11**(1), 119–174 (1998)
28. Oh, J., Thomas, R.P.: Counting sheaves on Calabi-Yau 4-folds, I. Duke. Math. J. **172**(7), 1333–1409
29. Oh, J., Thomas, R.P.: Complex Kuranishi structures and counting sheaves on Calabi-Yau 4-folds, II (preprint) arXiv:2305.16441
30. Pantev, T., Toën, B., Vaquié, M., Vezzosi, G.: Shifted symplectic structures. Publ. Math. IHES **117**, 271–328 (2013)
31. Park, H.: Virtual pullbacks in Donaldson-Thomas theory of Calabi-Yau 4-folds (preprint) arXiv:2110.03631

Open Access This chapter is licensed under the terms of the Creative Commons Attribution-NonCommercial-NoDerivatives 4.0 International License (http://creativecommons.org/licenses/by-nc-nd/4.0/), which permits any noncommercial use, sharing, distribution and reproduction in any medium or format, as long as you give appropriate credit to the original author(s) and the source, provide a link to the Creative Commons license and indicate if you modified the licensed material. You do not have permission under this license to share adapted material derived from this chapter or parts of it.

The images or other third party material in this chapter are included in the chapter's Creative Commons license, unless indicated otherwise in a credit line to the material. If material is not included in the chapter's Creative Commons license and your intended use is not permitted by statutory regulation or exceeds the permitted use, you will need to obtain permission directly from the copyright holder.

Chapter 7
Globalization of Chern Characters and Canonical Pairings

Taejung Kim

Abstract We survey how one can derive an explicit realization of the Hirzebruch-Riemann-Roch theorem for global matrix factorizations by unfolding elaborate investigations evolving from the local to global cases by several researchers during the last decade.

Keywords Matrix factorizations · Hochschild homology · Negative cyclic homology · Periodic cyclic homology · Chern characters · Canonical pairings · Hirzebruch-Riemann-Roch theorem

2020 Mathematics Subject Classification Primary 14A22; Secondary 16E40, 18G80

7.1 Introduction

The Hirzebruch-Riemann-Roch (for short, HRR) formula in algebraic geometry holds a special place due to its usefulness in applications and beauties in the various generalizations. It states that if E is a vector bundle of rank r on a non-singular complete variety X of dimension n over $\mathbb{C}$, then

$$\chi(X, E) = \int_X \mathsf{ch}(E)\mathrm{td}(T_X)$$

T. Kim was supported by NRF-2018R1D1A3B07043346.

T. Kim (✉)
Department of Mathematics, Inha University, Incheon, Republic of Korea
e-mail: tjkim@kias.re.kr

© The Author(s) 2026

N.-G. Kang et al. (eds.), *Categorical and Enumerative Aspects of Mirror Symmetry*, KIAS Springer Series in Mathematics 5,
https://doi.org/10.1007/978-981-95-0385-8_7

where $\mathsf{ch}(E) = \sum_{k=1}^{r} \exp(\alpha_i)$, $\mathrm{td}(T_X) = \prod_{k=1}^{n} \beta_k/(1 - \exp(-\beta_k))$, $\chi(X, E) = \sum_{k=0}^{n} (-1)^k \dim_{\mathbb{C}} H^i(X, E)$, $\alpha_1, \ldots, \alpha_r$ are the Chern roots of E, and $\beta_1, \ldots, \beta_n$ are the Chern roots of the tangent bundle T_X; see [7, Chapter 15] for details.

Adopting a viewpoint of categorical noncommutative geometry, one can generalize it in an abstract and tautological way; see formula (7.2.6). However, such a tautological abstraction tends to suffer from a lack of practical computational formulae. During the past decade, researchers in the field have developed methods to make this abstraction explicitly computable in several cases. The goals of this expository paper are to describe how the previous works are connected with an explicit computable realization of the HRR formula for the category of global matrix factorizations and to provide the reader with a road map and links to the original references for detailed verification.

The structure of this article is as follows: In Sect. 7.2.1, after recalling definitions and some preparatory material about a curved differential algebra and its modules, we introduce various homology theories associated with mixed complexes and discuss some of their properties. In Sect. 7.2.2, we construct categorical Chern characters and canonical pairings, which lead to an abstract HRR formula. In Sect. 7.3, the works for local explicit formulae of the Chern characters by A. Polishchuk and A. Vaintrob [15], D. Shklyarov [17], E. Segal [16], and M. Brown and M. Walker [1] are explained. In Sect. 7.4, the explicit descriptions of the canonical pairings by A. Polishchuk and A. Vaintrob [15] and D. Shklyarov [17, 18] are given. In Sect. 7.5, after global formulae for the Chern characters in [5, 12] are characterized, the main idea of globalization of the HRR formula in Bumsig Kim's works [4, 8] is outlined. Moreover, its generalization and possible application are discussed.

7.2 Abstract Hirzebruch-Riemann-Roch Formula

7.2.1 *Curved Differential Graded Algebra and Mixed Complexes*

Let k be a field of characteristic zero and $\mathbb{G}$ be either $\mathbb{Z}$ or $\mathbb{Z}/2$. We call (A, d_A, h) a *curved differential graded* (for short, cdg) *k-algebra* if A is a unital $\mathbb{G}$-graded k-algebra with a differential d_A, a degree one k-linear endomorphism of A, and a curvature $h \in A$ with $|h| = 2$ such that

(i) $d_A(a_1 a_2) = d_A(a_1)a_2 + (-1)^{|a_1|} a_1 d_A(a_2)$ for $a_1, a_2 \in A$;
(ii) $d_A^2(a) = ha - ah$ for $a \in A$;
(iii) $d_A(h) = 0$.

Moreover, if $d_A = 0$, we call $(A, 0, h)$, for short (A, h), a *curved algebra* and if $h = 0$, we call $(A, d_A, 0)$, for short (A, d_A), a *dg k-algebra*.

A right cdg module (E, δ_E) over (A, d_A, h) is a right $\mathbb{G}$-graded A-module E with a k-linear map $\delta_E : E \to E$ with $|\delta_E| = 1$ such that $\delta_E(ma) = \delta_E(m)a + (-1)^{|m|}md_A(a)$ for every $m \in E$, $a \in A$, and $\delta_E^2 = \rho_{-h}$ where ρ_{-h} is the right multiplication by $-h$. One can check that $(\mathrm{End}_A(E), [\delta_E, -], 0)$ becomes a dg algebra. Moreover, if we do not require $\delta_E^2 = \rho_{-h}$, then we call it a right cdg *quasi-module* (E, δ_E) over (A, d_A, h). In this case, $(\mathrm{End}_A(E), [\delta_E, -], \delta_E^2 + \rho_h)$ becomes a cdg algebra. When A is a commutative k-algebra concentrated in degree 0 and E is a finitely generated projective A-module, a right cdg module (E, δ_E) over $(A, 0, h)$ is called a *matrix factorization* of $-h$. That is, the category $\mathrm{MF}(A, h)$ of matrix factorizations of h over A is the category $\mathrm{Perf}_{\mathbb{Z}/2}(A, 0, -h)$ of perfect right cdg $\mathbb{Z}/2$-graded modules over $(A, 0, -h)$.

We say that (M, b, B) is a mixed complex if M is a $\mathbb{G}$-graded vector space, $b^2 = 0 = B^2$, and $bB + Bb = 0$ with $|b| = 1$ and $|B| = -1$. A morphism $\phi : (M, b, B) \to (M', b', B')$ between mixed complexes is a k-linear map preserving degrees and both differentials. We call ϕ a *quasi-isomorphism* if it is a quasi-isomorphism between $(M, b) \to (M', b')$ such that $\phi \circ B = B' \circ \phi$.

Noting that an algebra is a category with just one object, a cdg k-category can be defined as a generalization of a cdg k-algebra; see [1, Section 2.1] for information about a cdg k-category. Let $\mathcal{A}$ be a cdg k-category with curvature h and $\mathcal{A}(x, y)[1]$ denote the degree shifted by 1 where $\mathcal{A}(x, y)$ means the Hom space from an object y to an object x in $\mathcal{A}$. The Hochschild complex $C(\mathcal{A})$ of $\mathcal{A}$ is defined by

$$C(\mathcal{A}) := \bigoplus_{x \in \mathcal{A}} \mathcal{A}(x, x) \oplus \bigoplus_{n \geq 1} \Big(\bigoplus_{x_i \in \mathcal{A}} \mathcal{A}(x_1, x_0) \otimes_k \underbrace{\mathcal{A}(x_2, x_1)[1] \otimes_k \ldots \otimes_k \mathcal{A}(x_0, x_n)[1]}_{n} \Big),$$

with differential $b := b_2 + b_1 + b_0$ defined as follows:

$$\begin{aligned} b_2(a_0[a_1|\ldots|a_n]) &:= (-1)^{|a_0|}a_0a_1[a_2|\ldots|a_n] \\ &\quad + \sum_{j=1}^{n-1}(-1)^{\sum_{i=0}^{j}|a_i|-j}a_0[a_1|\ldots|a_ja_{j+1}|\ldots|a_n] \\ &\quad - (-1)^{(|a_n|+1)(\sum_{i=0}^{n-1}|a_i|-(n-1))}a_na_0[a_1|\ldots|a_{n-1}]; \\ b_1(a_0[a_1|\ldots|a_n]) &:= d(a_0)[a_1|\ldots|a_n] \\ &\quad + \sum_{j=1}^{n}(-1)^{\sum_{i=0}^{j-1}|a_i|-j}a_0[a_1|\ldots|d(a_j)|\ldots|a_n]; \\ b_0(a_0[a_1|\ldots|a_n]) &:= (-1)^{|a_0|}a_0[h|a_1|\ldots|a_n] + \cdots \\ &\quad + (-1)^{\sum_{i=0}^{n}|a_i|-n}a_0[a_1|\ldots|a_n|h]. \end{aligned}$$

The Hochschild homology of $\mathcal{A}$ is defined as the cohomology of the complex $(C(\mathcal{A}), b)$:

$$HH_*(\mathcal{A}) := H^{-*}(C(\mathcal{A}), b).$$

When $\mathcal{A}$ is a cdg algebra (A, d_A, h), we write $C(A, d_A, h)$ for $C(\mathcal{A})$. It is known from [3, 14] that if the curvature $h \neq 0$, then

$$HH_*(A, d_A, h) \cong 0,$$

which justifies the introduction of another complex, the *Hochschild complex of the second kind*. It is defined by

$$\begin{aligned} C^{II}(\mathcal{A}) &:= \bigoplus_{x\in\mathcal{A}} \mathcal{A}(x,x) \\ &\oplus \prod_{n\geq 1}\Big(\bigoplus_{x_i\in\mathcal{A}} \mathcal{A}(x_1,x_0)\otimes_k \underbrace{\mathcal{A}(x_2,x_1)[1]\otimes_k \ldots \otimes_k \mathcal{A}(x_0,x_n)[1]}_{n}\Big) \end{aligned}$$

with the differentials b_i, $i = 0, 1, 2$. Its homology is denoted by $HH_*^{II}(\mathcal{A})$.

On the graded k-module $C(\mathcal{A})$, the Connes boundary map is given by $B := (1 - t^{-1})sN$ where

$$\begin{aligned} t(a_0[a_1|\ldots|a_n]) &:= (-1)^{(|a_0|-1)\sum_{i=1}^{n}(|a_i|-1)} a_1[a_2|\ldots|a_n|a_0]; \\ s(a_0[a_1|\ldots|a_n]) &:= 1[a_0|a_1|\ldots|a_n]; \\ N(a_0[a_1|\ldots|a_n]) &:= \sum_{i=0}^{n} t^i(a_0[a_1|\ldots|a_n]). \end{aligned}$$

Since $bB + Bb = 0$, $\mathrm{MC}(\mathcal{A}) := (C(\mathcal{A}), b, B)$ is a mixed complex, sometimes, called the *mixed Hochschild complex* of $\mathcal{A}$. We also write $\mathrm{MC}(A, d_A, h)$ for $\mathrm{MC}(\mathcal{A})$ when one studies the mixed complex of a cdg algebra.

Let D be the subcomplex of $C(\mathcal{A})$ generated by elements $a_0[a_1|\ldots|a_n]$ for which $a_i = c \cdot \mathrm{id}_x$ for $x \in \mathcal{A}$, $c \in k$ and some $i \geq 1$. We define the normalized Hochschild complex to be

$$\overline{C}(\mathcal{A}) := (C(\mathcal{A}), b)/D =: \bigoplus_{n\geq 0} \overline{C}_n(\mathcal{A}),$$

where n denotes the tensor degree. The Connes operator descends to an operator on $\overline{C}(\mathcal{A})$, which will be given by a simpler formula

$$B(a_0[\overline{a}_1|\cdots|\overline{a}_n]) \\ = \sum_{l=0}^{n}(-1)^{(|a_l|+\cdots+|a_n|-(n-l+1))(|a_0|+\cdots+|a_{l-1}|-l)}1[\overline{a}_l|\cdots|\overline{a}_n|\overline{a}_0|\cdots|\overline{a}_{l-1}]. \tag{7.2.1}$$

We let $\overline{\mathrm{MC}}(\mathcal{A}) := (\overline{C}(\mathcal{A}), b, B)$ and $\overline{\mathrm{MC}}^{II}(\mathcal{A}) := (\overline{C}^{II}(\mathcal{A}), b, B)$. For a cdg category $\mathcal{A}$ the quotient map

$$quot^{II} : \mathrm{MC}^{II}(\mathcal{A}) \overset{\sim}{\to} \overline{\mathrm{MC}}^{II}(\mathcal{A}) \tag{7.2.2}$$

is a quasi-isomorphism; see [1, Proposition 3.15] for details.

From a mixed complex, we may construct the negative cyclic homology and the periodic cyclic homology in $\mathbb{Z}/2$-grading as follows:

$$HN_*(\mathcal{A}) := H^{-*}(C(\mathcal{A})[\![u]\!], b + uB);$$
$$HP_*(\mathcal{A}) := H^{-*}(C(\mathcal{A})(\!(u)\!), b + uB)$$

where u is a formal variable with degree 2; see [13] for details.

7.2.2 Categorical Chern Character and Canonical Pairing

Let $E \in \mathrm{Perf}(A, d_A, 0)$ where $(A, d_A, 0)$ is a dg algebra over k. Consider a dg functor $T_E : (k, 0, 0) \to \mathrm{Perf}(A, d_A, 0)$ given by $T_E(\star) = E$ where $(k, 0, 0)$ is a dg category with a trivial differential and a zero curvature with one object denoted by $\star$. The dg functor induces a linear map

$$HH(T_E) : HH_*(k, 0, 0) \to HH_*(\mathrm{Perf}(A, d_A, 0)).$$

Since $HH_*(k, 0, 0) \cong HH_0(k, 0, 0) \cong k$, we let 1_k be the unit of $HH_0(k, 0, 0)$. A Hochschild homology valued Chern character of E is defined by

$$\mathrm{Ch}_{\mathrm{HH}}(E) := HH(T_E)(1_k) \in HH_0(\mathrm{Perf}(A, d_A, 0)).$$

Note that $\mathrm{Ch}_{\mathrm{HH}}(E)$ is the class of id_E in $HH_0(\mathrm{Perf}(A, d_A, 0))$.

Let $E \in \mathrm{Perf}(A, d_A, h)$. Using the dg functor $T_E : (k, 0, 0) \to \mathrm{Perf}(A, d_A, h)$, which can be seen as a strict cdg functor, we define a normalized negative cyclic homology valued Chern character of the second kind to be

$$\mathrm{Ch}_{\overline{\mathrm{HN}}^{II}}(E) := \overline{HN}^{II}(T_E)(1_k) \in \overline{HN}_0^{II}(\mathrm{Perf}(A, d_A, h)).$$

Note that $\mathrm{Ch}_{\overline{\mathrm{HN}}^{II}}(E)$ is the class of id_E in $\overline{HN}_0^{II}(\mathrm{Perf}(A, d_A, h))$. We observe that 1_k is not a closed element in the "unnormalized" negative cyclic complex $(C(\mathrm{Perf}(k, 0, 0)[[u]], b+uB)$. There are two ways to find a negative cyclic homology valued Chern character of the second kind from the normalized one $\mathrm{Ch}_{\overline{\mathrm{HN}}^{II}}(E)$. One way is to find an element in $HN_0^{II}(k, 0, 0)$ corresponding to $1_k \in \overline{HN}_0^{II}(k, 0, 0)$ through the isomorphism $HN^{II}(k, 0, 0) \cong \overline{HN}^{II}(k, 0, 0) \cong k[[u]]$. It is given by a degree zero element

$$\gamma_k := 1_k + \sum_{i=1}^{\infty}(-1)^i \frac{(2i)!}{2(i!)} 2 \cdot 1_k[\underbrace{1_k|\cdots|1_k}_{2i}]u^i; \text{ (see [1, Remark 4.4].)}$$

It means that we may define

$$\mathrm{Ch}_{\mathrm{HN}^{II}}(E) := HN^{II}(T_E)(\gamma_k) \in HN_0^{II}(\mathrm{Perf}(A, d_A, h)).$$

On the other hand, Kuerak Chung, Bumsig Kim, and the author find another representative for the Chern character of E by two different methods. That is, $\mathrm{Ch}_{\mathrm{HN}^{II}}(E)$ is represented by a degree zero element

$$\eta_\pi := \pi + \sum_{i=1}^{\infty}(-1)^i \frac{(2i)!}{2(i!)}(2\pi - 1_N)[\underbrace{\pi|\cdots|\pi}_{2i}]u^i \in HN_0^{II}(\mathrm{Perf}(A, d_A, h))$$

where E is a direct summand of a free module N in a dg category $\mathrm{Perf}(A, d_A, h)$ such that degree 0 closed homomorphisms $i : E \to N$ and $j : N \to E$ satisfy $1_E = j \circ i$ and $\pi^2 = \pi$ for $\pi := i \circ j$; see [5, 6] for details.

A question one can ask is what the explicit characterization of the categorical Chern character is. The answer awaits further analyses, which will be given in Sect. 7.3 for a couple of cases. Before going into that, we study a canonical pairing. For convenience's sake, we take the following notational conventions. Let $\mathcal{A}$ be $\mathrm{Perf}(A, d_A, 0)$ or $\mathrm{Perf}(A, d_A, h)$. $\mathbf{H}(\mathcal{A})$ will be either $HH(\mathrm{Perf}(A, d_A, 0))$ or $HN^{II}(\mathrm{Perf}(A, d_A, h))$ and $\mathbf{H}(\mathrm{Perf}k)$ will be $HH(\mathrm{Perf}(k, 0, 0))$ or $HN^{II}(\mathrm{Perf}(k, 0, 0))$. A categorical canonical pairing on $\mathbf{H}(\mathcal{A}) \times \mathbf{H}(\mathcal{A}^{\mathrm{op}})$ is by definition a composition of the following maps:

$$\langle -, - \rangle_{\mathsf{can}} := \mathbf{H}(\mathbf{Hom}_{\mathcal{A}}) \circ \mathsf{kun}. \tag{7.2.3}$$

kun is a Künneth map

$$\mathbf{H}(\mathcal{A} \otimes 1) \otimes_{\mathbf{H}(\mathrm{Perf}k)} \mathbf{H}(1 \otimes \mathcal{A}^{\mathrm{op}}) \simeq \mathbf{H}(\mathcal{A} \otimes \mathcal{A}^{\mathrm{op}})$$

and we are assuming that $\mathbf{Hom}_{\mathcal{A}} : \mathcal{A} \otimes \mathcal{A}^{\mathrm{op}} \to \mathrm{Perf}k$ given by $N \otimes M \mapsto \mathrm{Hom}_{\mathcal{A}}(M, N)$ is well-defined, which is a definition of properness of dg or cdg algebras. It induces a linear map $\mathbf{H}(\mathbf{Hom}_{\mathcal{A}}) : \mathbf{H}(\mathcal{A} \otimes \mathcal{A}^{\mathrm{op}}) \to \mathbf{H}(\mathrm{Perf}k)$. Again we may ask how explicitly we can describe the pairing. We will answer it in Sect. 7.4.

An observation by M. Kontsevich and Y. Soibelman manifests the importance of the canonical pairing, which says the pairing should be inverse to the Chern character of A as A-bimodule, denoted by Δ_A, (see [17, page 25]). Let us assume that $\Delta_A \in \mathcal{A}^{\mathrm{op}} \otimes \mathcal{A}$, which means that A is smooth. More specifically, noting that $\mathrm{Ch}_{\mathbf{H}}(\Delta_A) \in \mathbf{H}(\mathcal{A}^{\mathrm{op}} \otimes \mathcal{A})$, we have

$$\big(\langle -, - \rangle_{\mathsf{can}} \otimes \mathrm{id}_{\mathbf{H}(\mathcal{A})}\big) \circ \big(\gamma \otimes \mathrm{Ch}_{\mathbf{H}}(\Delta_A)\big) = \gamma \text{ for all } \gamma \in \mathbf{H}(\mathcal{A}); \tag{7.2.4}$$

see [15, page 1875]. More elaborately, via the Künneth isomorphism, after writing

$$\mathrm{Ch}_{\mathbf{H}}(\Delta_A) = \sum_i T^i \otimes T_i \text{ for some } T^i \in \mathbf{H}(\mathcal{A}^{op}),\ T_i \in \mathbf{H}(\mathcal{A}),$$

we can see that $\langle -, - \rangle_{\mathsf{can}}$ is a unique non-degenerate bilinear pairing $\langle -, - \rangle$ satisfying

$$\sum_i \langle \gamma, T^i \rangle \langle T_i, \gamma' \rangle = \langle \gamma, \gamma' \rangle \text{ for every } \gamma \in \mathbf{H}(\mathcal{A}), \gamma' \in \mathbf{H}(\mathcal{A}^{op}); \tag{7.2.5}$$

see [17, Proposition 4.2] and [8, Section 2.5]. We will call Eq. (7.2.5) the characteristic equation of the canonical pairing, which is often conveniently used to find an explicit formula for the canonical pairing.

Now we have all the categorical ingredients to formulate the categorical HRR formula in a noncommutative setting. Let $\xi := 1_k$ when $\mathbf{H}(\mathcal{A})$ is $HH(\mathrm{Perf}(A, d_A, 0))$ and $\xi := \gamma_k$ when $\mathbf{H}(\mathcal{A})$ is $HN^{II}(\mathrm{Perf}(A, d_A, h))$. For any $E, F \in \mathcal{A}$, we have

$$\begin{aligned}
\langle \mathrm{Ch}_{\mathbf{H}(\mathcal{A})}(E), \mathrm{Ch}_{\mathbf{H}(\mathcal{A})}(F)^{\vee} \rangle_{\mathsf{can}} &= \mathbf{H}(\mathbf{Hom}_{\mathcal{A}}) \big(\mathbf{H}(T_E)(\xi) \otimes (\mathbf{H}(T_F)(\xi))^{\vee}\big) \\
&= \mathbf{H}(\mathbf{Hom}_{\mathcal{A}}) \left(\mathbf{H}(T_E)(\xi) \otimes \mathbf{H}(T_{F^{\mathrm{op}}})(\xi)\right) \\
&= \mathbf{H}(\mathbf{Hom}_{\mathcal{A}}) \left(\mathbf{H}(T_{E \otimes F^{\mathrm{op}}})(\xi)\right) \\
&= \mathbf{H}(\mathbf{Hom}_{\mathcal{A}} \circ T_{E \otimes F^{\mathrm{op}}})(\xi) \\
&= \mathbf{H}(\mathrm{Hom}_{\mathcal{A}}(F, E))(\xi) \\
&= \mathrm{Ch}_{\mathbf{H}(\mathrm{Perf}k)}(\mathrm{Hom}_{\mathcal{A}}(F, E))
\end{aligned} \tag{7.2.6}$$

where $^\vee : \mathbf{H}(\mathcal{A}) \simeq \mathbf{H}(\mathcal{A}^{op})$ is a functorial isomorphism mapping ξ to ξ; see [17]. To make this abstract formula a concrete form has been a topic of many research works and it will be explained in the following sections.

7.3 Local Computable Formulae for Chern Characters

7.3.1 Morita Invariance

It is known that there are isomorphisms

$$HH(\mathrm{Perf}(A, d_A, 0)) \cong HH(A, d_A, 0);$$
$$HN^{II}(\mathrm{Perf}(A, d_A, h)) \cong HN^{II}(A, d_A, h),$$

which are induced by so-called Morita equivalence and pseudo-equivalence respectively; see [1, 5, 14] for details. For example, for a unital k-algebra A, the isomorphism can be explicitly realized by a generalized trace map

$$\mathrm{tr} : HH(\mathsf{M}_{n\times n}(A)) \to HH(A)$$

where $\mathsf{M}_{n\times n}(A)$ is the space of $n \times n$ matrices over A with $n \geq 1$; see [13, Chapter 1] for details. However, the trace map usually does not give a chain map if A has a dg structure. The key to make a Chern character explicit in a dg or cdg category is to seek out how to modify the trace map to give a chain map representing the Morita isomorphism.

Let us mention Segal's work [16] for a cdg algebra, which gives an explicit Morita isomorphism for a cdg case by constructing a chain map. Let $(A, 0, h)$ be a curved algebra and $\mathcal{P} \subset \mathrm{Perf}(A, 0, h)$ be a full sub-category of the category of perfect $(A, 0, h)$-modules satisfying that there is a module $N \in \mathcal{P}$ that contains A as a direct summand. Segal's translation map T_d is defined as follows. Let $(E_0, \delta_0), \ldots, (E_k, \delta_k)$ be cdg-modules in $\mathcal{P}$ and let

$$E_0 \overset{\alpha_0}{\to} E_1 \overset{\alpha_1}{\to} \cdots \overset{\alpha_{k-1}}{\to} E_k \overset{\alpha_k}{\to} E_0$$

be morphisms. Then the translation map is defined by

$$T_d : \alpha_0 \otimes \cdots \otimes \alpha_k \mapsto \sum_{s_0, \ldots, s_k \geq 0} \alpha_0 \otimes (\delta_1)^{\otimes s_1} \otimes \alpha_1 \otimes (\delta_2)^{\otimes s_2} \otimes \cdots \otimes \alpha_k \otimes (\delta_0)^{\otimes s_0}.$$

Then assuming the Hochschild homology of A is bounded with respect to the tensor grading, we have

$$\mathrm{str} \circ T_d : \big(C^{II}(\mathcal{P}), b\big) \to \big(C^{II}(A, 0, -h), b\big) \tag{7.3.1}$$

is a quasi isomorphism; see [16, Theorem 2.14].

7.3.2 Dg Algebra Case

Recall that $(E, \delta_E) \in \mathrm{Perf}(A, d_A, 0)$ can be regard as a homotopy direct summand of a finitely generated semi-free module $(N_\alpha, d_F + \alpha)$ where N_α is a finitely generated dg free A-module with the induced differential d_F from d_A and α is strictly upper triangular and satisfies the Maurer-Cartan equation

$$d_F(\alpha) + \alpha \cdot \alpha = 0.$$

In particular, there are degree 0 closed homomorphisms $i : E \to N_\alpha$ and $j : N_\alpha \to E$ satisfying $1_E = j \circ i$ and $\pi^2 = \pi$ for $\pi := i \circ j$ in the homotopy category. D. Shklyarov constructs an explicit formula of the Chern character taking values in $HH(A, d_A)$, which is equivalent to constructing an explicit isomorphism $\mathsf{morita}_A : HH(\mathrm{Perf}(A, d_A)) \to HH(A, d_A)$ for specific elements in $HH(\mathrm{Perf}(A, d_A, 0))$.

Theorem 7.3.1 ([17]) *Let* $N_\alpha = (\bigoplus_{j=1}^{n} A[r_j], d_F + \alpha)$ *and* E *be a homotopy direct summand of* N_α *corresponding to a homotopy idempotent* $\pi : N_\alpha \to N_\alpha$. *Then*

$$\mathrm{ch}_{\mathrm{HH}}(E) = \sum_{l=0}^{n-1} (-1)^l \mathrm{str}(\pi \underbrace{[\alpha| \dots |\alpha]}_{l})$$

where $\mathrm{ch}_{\mathrm{HH}}(E) := \mathsf{morita}_A \circ \mathrm{Ch}_{\mathrm{HH}}(E)$ *and* $\mathsf{morita}_A : HH(\mathrm{Perf}(A, d_A)) \to HH(A, d_A)$ *is an isomorphism.*

Segal's result can reproduce Shklyarov's result. That is, if we let $d_A = 0$, we see that $d_F = 0$. Then since α is strictly upper triangular, map (7.3.1) also gives

$$\mathrm{str} \circ T_d(\pi) = \sum_{l=0}^{n-1} (-1)^l \mathrm{str}(\pi \underbrace{[\alpha| \dots |\alpha]}_{l}).$$

The generalization of both cases to the negative cyclic homology valued Chern character can be found in [6].

7.3.3 Hochschild-Kostant-Rosenberg Type Maps

Let A be a smooth commutative algebra over k. It is well-known that the Hochschild-Kostant-Rosenberg (for short, HKR) map gives an isomorphism

$$\epsilon : HH(A) \to (\Omega^\bullet_{A/k}, 0)$$

where $\Omega^\bullet_{A/k}$ is the space of Kähler differentials; see [13]. This can be extended to a category of perfect complexes over a cdg algebra. That is, we have

$$HH(\mathrm{MF}(R, h)) \overset{\epsilon}{\cong} (\Omega^\bullet_{R/k}, -dh\wedge) \text{ where } R = k[\![x_1, \ldots, x_n]\!]; \tag{7.3.2}$$

see [3, 15] for details. Let us recall that the HKR type isomorphism is nothing but a kind of an explicit Morita isomorphism. A. Polishchuk and A. Vaintrob construct an explicit formula of the boundary-bulk map for a category of matrix factorization in this viewpoint.

Let $(E, \delta_E) \in \mathrm{MF}(R, h)$ and $\mathsf{Hom}^*_{\mathrm{MF}}(E, E)$ be the homology of the dg structure, i.e., b_1-differential, of the hom complex. Since $\alpha \in \mathsf{Hom}^*_{\mathrm{MF}}(E, E)$ is closed under $b = b_1 + b_2$, there is a well-defined natural map

$$\tau^E : \mathsf{Hom}^*_{\mathrm{MF}}(E, E) \to HH(\mathrm{MF}(R, h)) \cong (\Omega^\bullet_{R/k}, -dh\wedge).$$

After composing the HKR isomorphism, τ^E is called the boundary-bulk map associated with (E, δ_E). A. Polishchuk and A. Vaintrob give its explicit formula.

Theorem 7.3.2 ([15]) *Let $h \in R = k[\![x_1, \ldots, x_n]\!]$ have an isolated singularity and let $\mathcal{J}_h = (\partial_1 h, \ldots, \partial_n h)$. For $\alpha \in \mathsf{Hom}^*_{\mathrm{MF}}(E, E)$ of $(E, \delta_E) \in \mathrm{MF}(R, h)$,*

$$\tau^E(\alpha) = (-1)^n \cdot \frac{1}{n!} \cdot \mathrm{str}((d\delta_E)^n \circ \alpha) \mod \mathcal{J}_h \cdot dx_1 \wedge \ldots \wedge dx_n. \tag{7.3.3}$$

In particular, $\mathrm{ch}(E) = \tau^E(\mathrm{id}_E)$ *where* $\mathrm{ch}(E) = \epsilon \circ \mathrm{ch}_{\mathrm{HH}}(E)$.

This result is extended by M. Brown and M. Walker [1] using an algebraic connection. First, using the same symbol ϵ for the HKR type maps by abuse of notation, let us mention that there is a HKR type isomorphism for an essentially smooth curved algebra $(A, 0, h)$

$$\epsilon : HN^{II}_q(A, 0, h) \overset{\cong}{\to} H^{-q}(\Omega^\bullet_{A/k}[\![u]\!], ud + dh) \tag{7.3.4}$$

of graded $k[\![u]\!]$-modules for all $q \in \mathbb{Z}$; see [1, Theorem 3.31]. Let $(E, \delta_E) \in q\mathrm{Perf}(A, 0, h)$, $\mathsf{morita}_A : HN^{II}(q\mathrm{Perf}(A, 0, h)) \to HN^{II}(A, 0, h)$ be an isomorphism by abuse of notation, and

$$\mathsf{morita}_A(\mathrm{Ch}_{\mathrm{HN}^{II}}(E)) := \mathrm{ch}_{\mathrm{HN}^{II}}(E).$$

Similar to Segal's translation map, for any $(E, \delta_E) \in q\mathrm{Perf}(A, d_A, h)$

$$(\mathrm{id}, \delta_E)_* : (\overline{C}^{II}(\mathrm{End}(E, \delta_E)), b, B) \to (\overline{C}^{II}(\mathrm{End}(E, 0)), b, B) \tag{7.3.5}$$

gives a quasi-isomorphism; see [1, Formula (3.6)] for the formula of $(\mathrm{id}, \delta_E)_*$. Let $\mathcal{D}^\natural$ be the full subcategory of $q\mathrm{Perf}(A, 0, h)$ consisting of objects with trivial differentials where A is an essentially smooth algebra over k. Consider a connection

$$\nabla_E : E \to E \otimes_A \Omega^1_{A/k}$$

for an object $(E, 0) \in \mathcal{D}^\natural$. Define $\mathrm{tr}_\nabla : \overline{C}^{II}(\mathcal{D}^\natural)[\![u]\!] \to \Omega^\bullet_{A/k}[\![u]\!]$ to be the $k[\![u]\!]$-linear map given by

$$\begin{aligned} &\mathrm{tr}_\nabla\, (\alpha_0[\overline{\alpha_1}|\cdots|\overline{\alpha_n}]) \\ &= \sum_{J=0}^{\infty} \sum_{j_0+\cdots+j_n=J} (-1)^J \frac{\mathrm{str}\left(\alpha_0 \nabla_1^{2j_0} \alpha_1' \nabla_2^{2j_1} \alpha_2' \cdots \nabla_n^{2j_{n-1}} \alpha_n' \nabla_0^{2j_n}\right)}{(J+n)!} u^J \end{aligned} \tag{7.3.6}$$

where $\alpha_i : E_{i-1} \to E_i$, $\nabla_i := \nabla_{E_i}$ (with $\nabla_{n+1} = \nabla_0$), $\alpha_i' := \nabla_i \circ \alpha_i - (-1)^{|\alpha_i|}\alpha_i \circ \nabla_{i-1}$, and the inner sum ranges over $(n+1)$-tuples of non-negative integers that sum to J. M. Brown and M. Walker prove that it is a chain map; see [1, Theorem 5.19]. From it, a Chern character can be explicitly given as follows.

Theorem 7.3.3 ([1, Theorem 5.7]) *Let k be a field of characteristic 0. Assume that $(A, 0, h)$ is a $\mathbb{G}$-graded essentially smooth curved k-algebra and let $(E, \delta_E) \in \mathrm{Perf}(A, 0, h)$ be a perfect right $\mathcal{A}$-module. For any connection ∇ on E, we have*

$$\epsilon \circ \mathrm{ch}_{\mathrm{HN}^{II}}(E) = \mathrm{tr}_\nabla \circ (\mathrm{id}, \delta_E)_*(\mathrm{id}_E).$$

That is,

$$\epsilon \circ \mathrm{ch}_{\mathrm{HN}^{II}}(E) = \mathrm{str}(\exp(-\mathrm{R})) \in H_0(\Omega^\bullet_{A/k}[\![u]\!], ud + dh) \tag{7.3.7}$$

where $\mathrm{R} = u\nabla^2 + [\nabla, \delta_E]$.

If E is a free A-module of finite rank, then we can take the de Rham differential d as a connection ∇. Note that it is flat, i.e., $d^2 = 0$. Then formula (7.3.7) becomes Polishchuk-Vaintrob's formula (7.3.3) when $\alpha = \mathrm{id}_E$.

7.4 Explicit Realizations of the Canonical Pairings

To achieve the goal of making the categorical HRR formula (7.2.6) explicit, we need an explicit formula for the canonical pairing defined in (7.2.3) on the Hochschild homology. We introduce two different methods to find the formula in this section.

7.4.1 *Polishchuk and Vaintrob's Method*

Let

$$\langle f \otimes d\mathbf{x}, g \otimes d\mathbf{x}\rangle_{\mathsf{PV}} := (-1)^{\binom{n}{2}} \mathrm{Res}_{k[\mathbf{x}]/k} \begin{bmatrix} f(\mathbf{x})g(\mathbf{x})dx_1 \wedge \ldots \wedge dx_n \\ \partial_1 h, \ldots, \partial_n h \end{bmatrix} \tag{7.4.1}$$

where Res is the Grothendieck residue and $f, g \in R := k[\![x_1, \ldots, x_n]\!]$. Polishchuk and Vaintrob's idea is to use Eq. (7.2.4) to find a formula for the canonical pairing. More precisely, the argument goes as follows. Let $h \in R$ have an isolated singularity. Then Theorem 7.3.2 and some calculation give

$$\mathrm{ch}(\Delta_R) = (-1)^{\binom{n}{2}} \cdot \det(\Delta_j(\partial_i h)) \in \mathcal{R}^e_{\widetilde{h}} \cdot dx_1 \wedge \cdots \wedge dx_n \wedge dy_1 \wedge \cdots \wedge dy_n$$

where $\mathcal{R}^e_{\widetilde{h}} = k[\![x_1, \ldots, x_n, y_1, \ldots, y_n]\!]/(\partial_{x_1}\widetilde{h}, \ldots, \partial_{x_n}\widetilde{h}, \partial_{y_1}\widetilde{h}, \ldots, \partial_{y_n}\widetilde{h})$ with $\widetilde{h} = h(\mathbf{y}) - h(\mathbf{x})$ and

$$\Delta_j h := \frac{h(x_1, \ldots, x_{j-1}, y_j, y_{j+1} \ldots, y_n) - h(x_1, \ldots, x_{j-1}, x_j, y_{j+1}, \ldots, y_n)}{y_j - x_j} \in R \otimes R.$$

They show that the pairing in (7.4.1) satisfies Eq. (7.2.4) after identification (7.3.2), i.e.,

$$\big(\langle -, -\rangle_{\mathsf{PV}} \otimes \mathrm{id}\big) \circ \big(\gamma \otimes \mathrm{ch}(\Delta_R)\big) = \gamma \text{ for all } \gamma \in (\Omega^\bullet_{R/k}, -dh\wedge); \tag{7.4.2}$$

see [15, Corollary 4.1.3]. That is, (7.4.1) is indeed an explicit formula for the canonical paring.

7.4.2 *Shklyarov's Works*

In [17], D. Shklyarov characterizes a canonical pairing on $HH(A, d_A) \times HH((A, d_A)^{\mathrm{op}})$. His method consists of unwinding the definition of the canonical pairing in (7.2.3) and explicitly expressing a Morita isomorphism between

$$\mathsf{morita}_k : HH(\mathrm{Perf}(k,0)) \cong HH(k,0) = k. \tag{7.4.3}$$

Let (A, d_A) be a dg algebra over k. Consider two cycles

$$a := \sum a_0[a_1|\ldots|a_l] \in C(A), \quad b := \sum b_0[b_1|\ldots|b_m] \in C(A^{\mathrm{op}}).$$

Recall that the shuffle product on $C(A) \times C(A^{\mathrm{op}})$ is given by the following formula:

$$\mathsf{sh}(a_0[a_1|\ldots|a_l] \otimes b_0[b_1|\ldots|b_m]) := (-1)^{\checkmark} \cdot a_0 b_0\, \mathsf{sh}_{lm}[a_1|\ldots|a_l|b_1|\ldots|b_m]$$
$$\checkmark := |b_0|(|sb_1| + \ldots + |sb_n|)$$
$$\mathsf{sh}_{lm}[x_1|\ldots|x_l|x_{l+1}|\ldots|x_{l+m}] := \sum_{\sigma} \pm[x_{\sigma^{-1}(1)}|\ldots|x_{\sigma^{-1}(l)}|x_{\sigma^{-1}(l+1)}|\ldots|x_{\sigma^{-1}(l+m)}]$$

where the sum is taken over all permutations that don't shuffle the first l and the last m elements, $|sx| = |x| - 1$, and the transposition $[\ldots|x|y|\ldots] \to [\ldots|y|x|\ldots]$ contributes $(-1)^{|sx||sy|}$ to the sign in the third line. Then by the definition of each map in (7.2.3), one can see that

$$HH(\mathbf{Hom}_A) \circ \mathsf{kun}(a \otimes b) = a \wedge b \text{ where}$$
$$a \wedge b := \sum \pm L(a_0)R(b_0)\mathsf{sh}_{lm}[L(a_1)|\ldots|L(a_l)|R(b_1)|\ldots|R(b_m)].$$

Here $L(a) \in \mathrm{End}_k(A)$ is a left multiplication and $R(b) \in \mathrm{End}_k(A)$ is a right multiplication, i.e., $R(b) : c \mapsto (-1)^{|c||b|}cb$. Since there is homotopy equivalence between the complex A and its cohomology $H^*(A)$, we have a pair of degree 0 maps $p : A \to H^*(A)$ and $i : H^*(A) \to A$ such that

$$p \circ i = 1_{H^*(A)}, \quad i \circ p = 1_A - [d_A, H]$$

where $H : A \to A$ is a degree -1 map. D. Shklyarov proves that

$$\mathsf{morita}_k \circ \langle a, b\rangle_{\mathsf{can}} = \sum_{j=0}^{l+m} \mathrm{str}_{H^*(A)}(\mathcal{F}_{l+m+1}(\tau^j(a \wedge b))) \text{ where}$$
$$\tau(T_1[T_2|\ldots|T_n]) := (-1)^{|sT_n|(|sT_1|+\ldots+|sT_{n-1}|)} T_n[T_1|\ldots|T_{n-1}]$$
$$\mathcal{F}_n(T_1[T_2|\ldots|T_n]) := p \circ T_1 \circ H \circ T_2 \circ H \circ \ldots \circ H \circ T_n \circ i;$$

see [17, Theorem 1.2] for details.

Remark 7.4.1 In the negative cyclic homology case, using a u-connection D. Shklyarov proves that the canonical pairing on the negative cyclic homology of the category of matrix factorizations of an isolated singularity can be identified with Kyoji Saito's higher residue pairing on the twisted de Rham cohomology of the singularity; see [18] for details. We also remark that in [11] Hoil Kim and the author provide another proof for the equivalence between the canonical pairing and Saito's higher residue pairing on the twisted de Rham cohomology using a different method from [18].

7.5 Globalization of Hirzebruch-Riemann-Roch Theorem

7.5.1 Global Chern Character Formula

A work for globalizing the Chern character formula appears in [12] in terms of the globalization of the boundary-bulk map. Using the Atiyah class of a matrix factorization, Bumsig Kim and Alexander Polishchuk characterize it as follows. Let X be a smooth separated scheme of finite type over k. Recall that a *global matrix factorization* (E, δ_E) over (X, h) is a cdg $\mathscr{A}_h$-module such that E is a locally free and coherent as an $\mathcal{O}_X$-module and $\delta_E^2 = \rho_h$ where $\mathscr{A}_h := (\mathcal{O}_X, 0, -h)$ and $\mathrm{Hom}_{\mathrm{MF}}(E, F)$ is defined to be $\mathrm{Hom}_{\mathcal{O}_X}(I_E, I_F)$ using quasi-coherent curved injective replacements I_E, I_F of E, F, respectively. One can consider a model using Čech resolutions of matrix factorizations for injective resolutions as follows.

Let us fix a finite open affine covering $\mathfrak{U} = \{U_i\}_{i\in I}$ of X with a total ordering of I. For an $\mathcal{O}_X$-sheaf $\mathcal{F}$, let $\check{\mathrm{C}}(\mathcal{F})$ the sheafified version of the (ordered) Čech complex of $\mathcal{F}$ with respect to the covering $\mathfrak{U}$:

$$\check{\mathrm{C}}^p(\mathcal{F}) := \prod_{i_0<\ldots<i_p} f_*(\mathcal{F}|_{U_{i_0,\ldots,i_p}}) = \mathcal{F} \otimes_{\mathcal{O}_X} \prod_{i_0<\ldots<i_p} f_*\mathcal{O}_{U_{i_0,\ldots,i_p}} \tag{7.5.1}$$

where f denotes the immersions $U_{i_0,\ldots,i_p} := U_{i_0} \cap \ldots \cap U_{i_p} \to X$. Here the second equality follows from the projection formula. The Čech differential $1_{\mathcal{F}} \otimes d_{\check{\mathrm{C}}\mathrm{ech}}$ for

$$\check{\mathrm{C}}(\mathcal{F}) = \mathcal{F} \otimes \check{\mathrm{C}}(\mathcal{O}_X) \tag{7.5.2}$$

will be written simply $d_{\check{\mathrm{C}}\mathrm{ech}}$ by abuse of notation. In this setting, a global matrix factorization (E, δ_E) becomes a Čech model $(\check{\mathrm{C}}(E), \delta_E + d_{\check{\mathrm{C}}\mathrm{ech}})$ which is also a cdg module over $\mathscr{A}_{\check{\mathrm{C}},h} := (\check{\mathrm{C}}(\mathcal{O}_X), d_{\check{\mathrm{C}}\mathrm{ech}}, -h)$ with the Alexander-Čech-Whitney product which is an $\mathcal{O}_X$-homomorphism

$$\cdot : \check{\mathrm{C}}(\mathcal{O}_X) \otimes_{\mathcal{O}_X} \check{\mathrm{C}}(\mathcal{O}_X) \to \check{\mathrm{C}}(\mathcal{O}_X)$$

defined by setting

$$(a \cdot b)_{i_0,\ldots,i_{p+q}} := a_{i_0,\ldots,i_p}|_{U_{i_0,\ldots,i_{p+q}}} b_{i_p,\ldots,i_{p+q}}|_{U_{i_0,\ldots,i_{p+q}}} \tag{7.5.3}$$

for $a \in \check{C}^p(\mathcal{O}_X), b \in \check{C}^q(\mathcal{O}_X)$; see [5, Section 4] for more information about the Čech model of a global matrix factorization.

Let ∇_i be a connection of $E|_{U_i}$ which certainly exists for each affine open set U_i. The Čech hyper-cohomology $\check{\mathbb{H}}^*(\mathfrak{U}, (\Omega^\bullet_X, (-1)^i dh\wedge))$-valued Chern character via a HKR type isomorphism is given by

$$\begin{aligned} \mathrm{ch}^{\mathsf{global}}(E) &= \mathrm{str}\exp(\mathrm{at}(E)) \\ \mathrm{at}(E) &= -\Big(\prod_{i\in I}[\nabla_i, \delta_E] + \prod_{i<j,i,j\in I} \nabla_i - \nabla_j\Big); \end{aligned} \tag{7.5.4}$$

see [12, Proposition A]. Note that the products in the exponential are the Alexander-Čech-Whitney cup products in the Čech complex $\check{C}^*(\mathfrak{U}, (\mathrm{End}(E) \otimes \Omega^\bullet_X))$.

The generalization of the above work to the negative cyclic homology is in [5]. After proving that Brown and Walker's map (7.3.6) can be extended as a chain map to a case when $d_A \neq 0$; (see [5, Theorem 5.2]), Kuerak Chung, Bumsig Kim, and the author prove the following.

Theorem 7.5.1 ([5, Theorem 1.2]) *Let $D_{\check{C}}(X, h)$ be the Čech model of the category of matrix factorizations over (X, h). Then the $k[\![u]\!]$-linear map*

$$\mathrm{tr}_\nabla : (C(D_{\check{C}}(X, h))[\![u]\!], b + uB) \to (\check{C}(\mathfrak{U}, \Omega^\bullet_{X/k})[\![u]\!], d_{\check{C}ech} - dh + ud)$$

is a quasi-isomorphism compatible with a HKR-type isomorphism.

As a corollary, for a global matrix factorization (E, δ_E) for (X, h) a negative cyclic homology valued Chern character, as a generalization of formula (7.5.4), is given by

$$\begin{aligned} \mathrm{ch}^{\mathsf{global}}_{HN}(E) &= \mathrm{str}\exp(-\mathrm{R}) \text{ where} \\ \mathrm{R} &:= \prod_{i\in I}(u\nabla_i^2 + [\nabla_i, \delta_E]) + \prod_{i<j,i,j\in I} (\nabla_i - \nabla_j); \end{aligned}$$

see [5, Theorem 1.1].

One can also find a similar result for the global equivariant case of a finite group; see [5, Section 6.3] and a formula for the localized Chern character of a global matrix factorization; see [5, Section 6.2].

7.5.2 Global HRR Theorem

In this section, ch is denoted by a global Chern character for simplicity's sake. For the Čech hyper-cohomology case an explicit formula is given in formula (7.5.4). We have all the pieces to make the HRR theorem global except an explicit expression of the canonical pairing. Assume that h is a function on a smooth variety X of dimension n such that the critical locus of h is in $h^{-1}(0)$ and proper over k. It implies that $\mathrm{MF}(X, h)$ is proper and smooth. From the global version of the Künneth formula and the HKR map

$$HH_*(\mathrm{MF}(X \times X, \widetilde{h})) \to \mathbb{H}^{-*}(\Omega^{\bullet}_{X\times X}, d\widetilde{h}),$$

one has a global Chern character $\mathrm{ch}(\mathcal{O}^{\widetilde{h}}_{\Delta_X}) \in \mathbb{H}^0(\Omega^{\bullet}_{X\times X}, d\widetilde{h})$ where $\Delta_X \subset X \times X$ denotes the diagonal, $\widetilde{h} := h \otimes 1 - 1 \otimes h$, and $(\Delta_X)_*\mathcal{O}_X := \mathcal{O}^{\widetilde{h}}_{\Delta_X}$ from a natural functor

$$(\Delta_X)_* : \mathrm{MF}(X, 0) \to \mathrm{MF}(X \times X, \widetilde{h}).$$

Letting the *wedge product* $-\wedge-$ of twisted Hodge cohomology classes be the composition of the Künneth map and the localization, we define a pairing $\langle -, -\rangle$ by the composition of maps

$$\begin{aligned} &\mathbb{H}^*(X, (\Omega^{\bullet}_X, -dh)) \times \mathbb{H}^*(X, (\Omega^{\bullet}_X, dh)) \\ &\qquad \xrightarrow{-\wedge-} \mathbb{H}^*_Z(X, (\Omega^{\bullet}_X, 0)) \xrightarrow{\wedge \mathrm{td}(X)} \mathbb{H}^*_Z(X, (\Omega^{\bullet}_X, 0)) \\ &\qquad \xrightarrow{proj} \mathbb{H}^0_Z(X, \Omega^n_X[n]) \to \mathrm{H}^0_c(X, \Omega^n_X[n]) \xrightarrow{(-1)^{\binom{n+1}{2}} \int_X} k \end{aligned} \tag{7.5.5}$$

where Z is the critical locus of h and H^*_c is a compactly supported hypercohomology. Note that the definition and properties of $\int_X$ are given in [8, Section 3.6] and the definition of the Todd class $\mathrm{td}(X)$ is given in [8, Section 3.2]. The global version of HRR theorem in terms of formula (7.2.6) can be stated as follow.

Theorem 7.5.2 ([8, Theorem 1.2]) *The canonical pairing* $\langle -, -\rangle_{\mathsf{can}}$ *under a HKR type isomorphism corresponds to the composition of maps in* (7.5.5),

$$\langle -, -\rangle := (-1)^{\binom{n+1}{2}} \int_X (- \wedge - \wedge \mathrm{td}(X)).$$

To prove the global HRR theorem, it suffices to show that $\langle -, -\rangle$ satisfies (7.2.5). For $\gamma \in \mathbb{H}^*(\Omega_X^\bullet, -dh)$ and $\gamma' \in \mathbb{H}^*(\Omega_X^\bullet, dh)$, we have

$$\sum_i \langle \gamma, T^i \rangle \langle T_i, \gamma' \rangle = \int_{X\times X} (\gamma \otimes \gamma') \wedge \mathrm{ch}(\mathcal{O}_{\Delta_X}^{\widetilde{h}}) \wedge (\mathrm{td}(X) \otimes \mathrm{td}(X)). \quad (7.5.6)$$

Note that $\int_X \otimes_k \int_X = \int_{X\times X}$ okun and

$$\mathrm{ch}(\mathcal{O}_{\Delta_X}^{\widetilde{h}}) = \sum_i T^i \otimes T_i \in \bigoplus_{q\in\mathbb{G}} \mathbb{H}^q(X, (\Omega_X^\bullet, dh)) \otimes \mathbb{H}^{-q}(X, (\Omega_X^\bullet, -dh)).$$

Thus, one needs to show that the following is true:

$$\int_{X\times X} (\gamma \otimes \gamma') \wedge \mathrm{ch}(\mathcal{O}_{\Delta_X}^{\widetilde{h}}) \wedge (\mathrm{td}(X) \otimes \mathrm{td}(X)) = \langle \gamma, \gamma' \rangle. \quad (7.5.7)$$

The main idea to prove formula (7.5.7) is to use the properties of $\int_X$ in Hodge cohomology and the deformation of $X \times X$ to the normal cone $N_{\Delta_X/X\times X}$ of Δ_X which is isomorphic to the total space of the tangent bundle T_X. Let $\overline{\pi}$ denote the projection $\mathbb{P}(T_X \oplus \mathcal{O}_X) \to X$. In [8, Theorem 1.2], it is proved that

$$\text{RHS of (7.5.6)} = \int_{\mathbb{P}(T_X\oplus\mathcal{O}_X)} \overline{\pi}^*(\gamma \wedge \gamma' \wedge \mathrm{td}(X)) \wedge \mathrm{ch}(\mathrm{Kos}(\overline{s})) \wedge \mathrm{td}(\mathcal{Q}) \quad (7.5.8)$$

where $\mathcal{Q}$ is the universal quotient bundle on $\mathbb{P}(T_X \oplus \mathcal{O}_X)$, $\overline{s}$ is a section of $\mathcal{Q}$ defined as the composition of tautological homomorphisms

$$\mathcal{O} \xrightarrow{(0,-\mathrm{id})} \overline{\pi}^* T_X \oplus \mathcal{O} \xrightarrow{quot} \mathcal{Q},$$

and $\mathrm{Kos}(\overline{s})$ is the Koszul complex $(\bigwedge^\bullet \mathcal{Q}^\vee, \iota_{\overline{s}})$ associated to $\overline{s}$. We note that $X = \{\overline{s} = 0\}$. One can observe that the restriction of $\mathrm{ch}(\mathrm{Kos}(\overline{s})) \wedge \mathrm{td}(\mathcal{Q})$ to the fiber of $\overline{\pi}$ is nothing but $c_n(Q)$ where Q is the tautological quotient bundle of $\mathbb{P}_k^n$ and $c_n(Q)$ is the n-th Chern class; see [7, Chapter 15] for details. Using the observation and some properties of pushforwards in Hodge cohomology which are provided with proofs in [8, Section 3.6], one can prove that

$$\text{RHS of (7.5.8)} = \langle \gamma, \gamma' \rangle.$$

Thus, one concludes that the pairing $\langle -, -\rangle$ in (7.5.5) is indeed a canonical pairing. Now the explicit realization of the global HRR theorem is obtained by showing that the right hand side of formula (7.2.6) is the Euler characteristic of $\mathrm{Hom}_{\mathrm{MF}}(E, F)$ where $E, F \in \mathrm{MF}(X, h)$; see [8, Theorem 1.1] for details.

Remark 7.5.3 Based on the arguments in [8], one can generalize the global HRR theorem, i.e., Theorem 7.5.2, to the case of the periodic cyclic homology, i.e., the twisted de Rham cohomology after application of the HKR type map. In [10] H. Kim and the author provide its proof. In addition, we note that the stacky version of global HRR theorem is obtained by D. Choa, B. Kim, and B. Sreedhar; see [4] for details.

7.5.3 Epilogue

The late Professor Bumsig Kim strove to apply and to generalize the techniques of algebraic geometry to the study of noncommutative geometry. To make noncommutativity commutative was one of main interests in his later mathematical life. Globalization of Chern characters and canonical pairings is no exception along this line. In the study, there is a possible interesting application back to algebraic geometry.

Grothendieck's Standard Conjecture D is an important open problem in algebraic geometry, which states that the numerical equivalence of algebraic cycles of a smooth projective variety X over a field k coincides with the homological equivalence. To date this conjecture remains essentially open besides cases under some conditions, for instance, when X is a complete intersection, $\dim_{\mathbb{C}} X \leq 4$, or X an abelian variety.

M. Marcolli and G. Tabuada proposed a noncommutative geometric generalization of Standard Conjecture D to smooth and proper dg-categories over a field, which states that for $E \in K_0(\mathrm{Perf}(X))$ the Euler pairing $\chi(F, E) = 0$ for all $F \in K_0(\mathrm{Perf}(X))$ if and only if $\mathrm{ch}_{\mathrm{HP}}(E) = 0$. Moreover, a theorem of G. Tabuada [19, Theorem 1.1] implies that, for a smooth projective variety X over a field of characteristic 0, the Standard Conjecture D holds for X if and only if its noncommutative version holds.

Instead of $\mathrm{Perf}(X)$, for the category $\mathrm{MF}(A, h)$ of matrix factorizations over $(A, -h)$ where A is a smooth k-algebra and $h \in A$ is a non-zero-divisor such that the singular locus of $\mathrm{Spec}(A/(h))$ is a finite set of points, M. Brown and M. Walker prove that the conjecture holds; see [2] for details. Subsequently, Bumsig Kim and the author prove that the conjecture holds for local stacky matrix factorizations; see [9] for details. Both proofs in [2, 9] use two ingredients. The first is the explicit formula of Chern character and the second is the HRR theorem. In a similar vein, the works by Bumsig Kim and others about the globalization of Chern character formula and the HRR theorem in [5, 8] may possibly provide tools to attack the conjecture in the global case. This application from noncommutative geometry to commutative geometry would be an agreeable outcome and Bumsig Kim's mathematical legacy may reside in it.

I cannot close this note without expressing my special and deepest gratitude to the late Professor Bumsig Kim for sharing his friendship and mathematical insights not easily found in papers and books. It is a true honor and privilege for me to be part of what he pursued in life and mathematics.

Acknowledgments The author thanks the anonymous referee for careful reading of the manuscript and the helpful comments and suggestions.

References

1. Brown, M., Walker, M.: A Chern-Weil formula for the Chern character of a perfect curved module. J. Noncommut. Geom. **14**(2), 709–772 (2020)
2. Brown, M., Walker, M.: Standard conjecture D for matrix factorizations. Adv. Math. **366**, 40 pp. (2020)
3. Căldăraru, A., Tu, J.: Curved A_∞ algebras and Landau-Ginzburg models. New York J. Math. **19**, 305–342 (2013)
4. Choa, D., Kim, B., Sreedhar, B.: Riemann-Roch for stacky matrix factorizations. Forum Math. Sigma **10** (2022), arXiv:2202.04418
5. Chung, K., Kim, B., Kim, T.: A Chain-level HKR-type map and a Chern character formula, arxiv:2109.14372
6. Chung, K., Kim, B., Kim, T.: Chern characters for curved dg-algebras, arXiv:2202.11403
7. Fulton, W.: Intersection Theory. Ergeb. Math. Grenzgeb. (3), vol. 2. Springer, Berlin (1998)
8. Kim, B.: Hirzebruch-Riemann-Roch for global matrix factorizations. Adv. Math., to appear, arXiv:2106.00435
9. Kim, B., Kim, T.: Standard conjecture D for local stacky matrix factorizations. Homol. Homotopy Appl. **26**(2), 193–207 (2024)
10. Kim, H., Kim, T.: Canonical pairing and Hirzebruch-Riemann-Roch formula for matrix factorizations, arXiv:2304.11675
11. Kim, H., Kim, T.: Higher residue pairing and canonical pairing on the twisted de Rham cohomology. Proc. Am. Math. Soc. **152**(10), 4187–4201 (2024)
12. Kim, B., Polishchuk, A.: Atiyah class and Chern character for global matrix factorizations. J. Inst. Math. Jussieu **21**, 1445–1470 (2022)
13. Loday, J.-L.: Cyclic Homology. Grundlehren der Mathematischen Wissenschaften, vol. 301. Springer, Berlin (1992)
14. Polishchuk, A., Positselski, L.: Hochschild (co)homology of the second kind I. Trans. Amer. Math. Soc. **364**(10), 5311–5368 (2012)
15. Polishchuk, A., Vaintrob, A.: Chern characters and Hirzebruch-Riemann-Roch formula for matrix factorizations. Duke Math. J. **161**(10), 1863–1926 (2012)
16. Segal, E.: The closed state space of affine Landau-Ginzburg B-models. J. Noncommut. Geom. **7**(3), 857–883 (2013)
17. Shklyarov, D.: Hirzebruch-Riemann-Roch-type formula for DG algebras. Proc. Lond. Math. Soc. (3) **106**(1), 1–32 (2013)
18. Shklyarov, D.: Matrix factorizations and higher residue pairings. Adv. Math. **292**, 181–209 (2016)
19. Tabuada, G.: A note on Grothendieck's standard conjectures of type C^+ and D. Proc. Am. Math. Soc. **146**(4), 1389–1399 (2018)

Open Access This chapter is licensed under the terms of the Creative Commons Attribution-NonCommercial-NoDerivatives 4.0 International License (http://creativecommons.org/licenses/by-nc-nd/4.0/), which permits any noncommercial use, sharing, distribution and reproduction in any medium or format, as long as you give appropriate credit to the original author(s) and the source, provide a link to the Creative Commons license and indicate if you modified the licensed material. You do not have permission under this license to share adapted material derived from this chapter or parts of it.

The images or other third party material in this chapter are included in the chapter's Creative Commons license, unless indicated otherwise in a credit line to the material. If material is not included in the chapter's Creative Commons license and your intended use is not permitted by statutory regulation or exceeds the permitted use, you will need to obtain permission directly from the copyright holder.

Chapter 8
Bordered Contact Instantons and Their Fredholm Theory and Generic Transversalities

Yong-Geun Oh

In memory of Bumsig Kim

Abstract In this article, we first establish the Fredholm theory for the bordered contact instantons defined on the punctured Riemann surfaces with prescribed asymptotic condition near the boundary punctures. We then prove the generic mapping transversality under the perturbation of Legendrian boundary condition. We also establish their generic (0-jet) evaluation transversality results of their moduli space under the perturbations of CR almost complex structures and of Legendrian boundary conditions. These are fundamental ingredients of the construction of the moduli space of bordered contact instantons and their applications.

Keywords Contact instantons · Fredholm theory · Mapping transversality · Perturbation of Legendrian boundary condition · Evaluation transversality

2010 Mathematics Subject Classification Primary 53D42, Secondary 58J32

8.1 Introduction

A contact manifold (M, ξ) is a $2n + 1$ dimensional manifold equipped with a completely non-integrable distribution of rank $2n$, called a contact structure. Complete non-integrability of ξ can be (locally) expressed by the non-vanishing property

---

Y.-G. Oh (✉)
Center for Geometry and Physics, Institute for Basic Science (IBS), 79 Jigok-ro 127beon-gil, nam-gu, Pohang-si, South Korea

POSTECH, Pohang, South Korea
e-mail: yongoh1@postech.ac.kr

© The Author(s) 2026

N.-G. Kang et al. (eds.), *Categorical and Enumerative Aspects of Mirror Symmetry*, KIAS Springer Series in Mathematics 5,
https://doi.org/10.1007/978-981-95-0385-8_8

$$\lambda \wedge (d\lambda)^n \neq 0$$

for a one-form λ. When ξ is coorientable, we can choose such a one-form globally so that $\ker \lambda = \xi$ is called a contact form associated to ξ. In the present article, we will always assume that (M, ξ) is coorientable equipped with a coorientation without mentioning further.

Each contact form λ of ξ canonically induces a splitting

$$TM = \mathbb{R}\langle R_\lambda \rangle \oplus \xi.$$

Here R_λ is the Reeb vector field of λ, which is uniquely determined by the equations

$$R_\lambda \rfloor \lambda \equiv 1, \quad R_\lambda \rfloor d\lambda \equiv 0.$$

We denote by $\Pi = \Pi_\lambda : TM \to TM$ the idempotent, i.e., an endomorphism satisfying $\Pi^2 = \Pi$ such that $\ker \Pi = \mathbb{R}\langle R_\lambda \rangle$ and $\operatorname{Im} \Pi = \xi$. Denote by $\pi = \pi_\lambda : TM \to \xi$ the associated projection.

Definition 8.1.1 Let (M, ξ) be a contact manifold and let λ be a contact form of ξ. Let $J \in \operatorname{End}(TM)$ be an endomorphism satisfying $J^2 = -\Pi$ such that $d\lambda(\cdot, J\cdot)$ is positive definite on ξ. We say that such J is *adapted to* λ. We define the set

$$\mathcal{J}_\lambda(M, \xi) = \{J : \xi \to \xi \mid J^2 = -\Pi, \ J \text{ adapted to } \lambda\} \tag{8.1.1}$$

Following [26], we call any such triple (M, λ, J) a contact triad of (M, ξ). For each given contact triad, we equip M with the triad metric

$$g = d\lambda(\cdot, J\cdot) + \lambda \otimes \lambda.$$

We denote by Σ be a compact Riemann surface (Σ, j) with or without boundary, and denote by $\dot\Sigma$ a punctured Riemann surface with a finite number of punctures which may be either from the interior or from the boundary of Σ.

For a given map $w : \dot\Sigma \to M$, we can decompose its derivative du, regarded as a w^*TM-valued one-form on $\dot\Sigma$, into

$$dw = d^\pi w + w^*\lambda \otimes R_\lambda \tag{8.1.2}$$

where $d^\pi w := \pi dw$. Furthermore $d^\pi w$ is decomposed into

$$d^\pi w = \overline{\partial}^\pi w + \partial^\pi w \tag{8.1.3}$$

where $\overline{\partial}^\pi w := (dw^\pi)^{(0,1)}_J$ (resp. $\partial^\pi w := (dw^\pi)^{(1,0)}_J$) is the anti-complex linear part (resp. the complex linear part) of $d^\pi w : (T\dot\Sigma, j) \to (\xi, J|_\xi)$. (For the simplicity of notation, we will abuse our notation by often denoting $J|_\xi$ by J. We also simply write $((\cdot)^\pi)_J^{(0,1)} = (\cdot)^{\pi(0,1)}$ and $((\cdot)^\pi)_J^{(1,0)} = (\cdot)^{\pi(1,0)}$ in general, unless there is a reason to specify J in notation.)

Definition 8.1.2 (Contact Instanton) Let Σ be as above. We call a pair of (j, w) of a complex structure on Σ and a map $w : \dot\Sigma \to M$ a a *contact Cauchy-Riemann map* if $\overline{\partial}^\pi w = 0$, and a *contact instanton* if it satisfies

$$\overline{\partial}^\pi w = 0, \quad d(w^*\lambda \circ j) = 0. \tag{8.1.4}$$

To avoid notional complexity and for the simplicity of exposition, we will assume that Σ is a compact surface of genus zero with or without boundary. We denote a marked Riemann surface by

$$(\Sigma, (z_1, \ldots, z_k))$$

where $(z_1, \ldots, z_k)$ are boundary marked points, unless said otherwise. They are ordered counterclockwise. We denote by $\overline{z_i z_{i+1}}$ the arc segment between z_i and z_{i+1}

In [19], the present author introduced the open-string version, the boundary value problem

$$\begin{cases} \overline{\partial}^\pi w = 0, & d(w^*\lambda \circ j) = 0 \\ w(\overline{z_i z_{i+1}}) \subset R_i, & i = 1, \ldots, k \end{cases} \tag{8.1.5}$$

for a map $w : (\dot\Sigma, \partial\dot\Sigma) \to (M, \vec R)$ for the Legendrian boundary condition

$$\vec R = \{R_1, \cdots, R_k\},$$

with a suitable asymptotic boundary condition at the punctures, and established its ellipticity by deriving relevant a priori estimates. (See Sect. 8.2.3 for the precise description of the asymptotic boundary condition.)

The Fredholm theory of contact instantons of closed-string version has been established by the present author in [23]. One of the main purposes of the present article is to extend the story to the open-string case and establish the Fredholm theory for the bordered contact instantons. We also establish the necessary generic transversality results of contact instantons under the perturbations of

(1) contact forms λ,
(2) the adapted CR almost complex structure J and
(3) the Legendrian boundary condition R_i's.

These are fundamental analytical ingredients needed for the applications to construct the moduli space of contact instantons with prescribed asymptotic limits and to establish the gluing theorem for the contact instanton Floer trajectories similarly as in the Floer theory of Lagrangian intersections [7] under the change of J's and [16] under the change of boundary condition.

Remark 8.1.3 We also need to make these studies for the Hamiltonian-perturbed contact instantons

$$\begin{cases} (du - X_H \otimes dt)^{\pi(0,1)} = 0, \quad d(e^{g_H(u)}(u^*\lambda + H\,dt)\circ j) = 0 \\ u(\tau, 0) \in R_0, \quad u(\tau, 1) \in R_1 \end{cases} \tag{8.1.6}$$

under the change of H too: Here the function $g_H : \mathbb{R} \times [0, 1] \to \mathbb{R}$ is some canonically defined function associated to H. (See [19].) This equation is the contact counterpart of the celebrated Floer's Hamiltonian-perturbed Cauchy-Riemann equation in symplectic geometry [8]. Similarly as in symplectic geometry, such an extension is an easy generalization of the arguments employed in the present article in that the presence of H does not play much role and so omitted for the clarity and simplicity of the exposition.

8.1.1 Fredholm Theory of Moduli Spaces of Bordered Contact Instantons

To develop a relevant Fredholm theory of the moduli spaces of bordered contact instantons with Legendrian boundary, we closely follow that of [23] by incorporating the boundary condition in the off-shell function spaces. For this purpose, we also need to establish all generic transversality results of the Reeb chords $\gamma^\pm$ and of the moduli space $\mathcal{M}(M, \lambda, J; R; \gamma^-, \gamma^+)$. Since such a transversality result under the perturbation of boundary Legendrian submanifolds are not considered in the general situation before, especially in relation to the present context of bordered contact instantons, we give their full details in Appendix A for the Reeb chords under the perturbation of contact forms (see [1, Appendix] for the proof of this generic nondegeneracy for the closed Reeb orbits), and in Part I for the moduli space of contact instantons under the perturbation of CR almost complex structures and under that of Legendrian boundary conditions, respectively.

One point we would like to highlight in the study of generic nondegeneracy of Reeb chords is that we consider the chords *in the sense of Moore paths* whose domains vary and so whose elements are represented by the pairs (γ, T) such that

$$T \in \mathbb{R}, \quad \gamma : [0, T] \to M, \quad T = \int \gamma^*\lambda \tag{8.1.7}$$

including $T = 0$. We emphasize here that we include the zero period $T = 0$ and consider the constant paths $(0, \gamma)$. Such a path exists only in the non-generic situation when $\psi(R)$ intersects R, e.g., when $\psi = id$. This transversality is important for the calculation of contact instanton homology constructed in [20, 22] and its applications [20, 21] to Sandon-Shelukhin type quantitative contact topology [32, 33].

Remark 8.1.4 It is an interesting open problem to equip Kuranishi structures on the compactified moduli spaces of contact instantons which are suitably compatible so that they give rise to the (Legendrian) contact DGA that appears in the case of trivial symplectic cobordism, i.e., the case of symplectization of contact manifolds. (See [4, 11, 30].) We believe that the general abstract framework of the Kuranishi structure from [9] or some variation of that of [4, 30] applies to the current case of contact instantons too. We hope to come back to this elsewhere.

We will closely follow the off-shell analytical framework from [23, 24] which handle the context of closed strings. In particular in terms of the decomposition $d\pi = d^\pi w + w^*\lambda\, R_\lambda$ and $Y = Y^\pi + \lambda(Y) R_\lambda$, an explicit tensorial formula of the linearized operator $D\Upsilon(w)$ is derived in [23, Theorem 10.1] which is the starting point of the Fredholm theory. (See also [24, Theorem 1.15] and Theorem 8.2.19 in the present paper.) We note that there are three kinds of perturbations we can think of as mentioned above, i.e., J, λ and the boundary Legendrian submanifolds R_i's. The study of perturbations of J is given in [23] which is of no change in the present open string case, and perturbation of contact forms is largely subsumed into that of J. *Therefore we will focus on the perturbation of the boundary in the present paper after establishment of the Fredholm theory for the bordered contact instantons.*

For this purpose, similarly as in symplectic geometry [16], we consider the universal moduli space

$$\mathcal{M}^{\text{univ}}(M, \lambda; \overline{\gamma}, \underline{\gamma})$$

consisting of the triples $(w, (J, \vec{R}))$ satisfying (8.1.5) with $\vec{R} = \{R_i\}_{i=1}^k$ *with fixed asymptotics* $(\overline{\gamma}, \underline{\gamma})$ *at the punctures of* $\dot{\Sigma}$. Denote by $\mathfrak{Leg}(M, \xi)$ the set of smooth Legendrian submanifolds the set of whose Reeb chords for this given fixed asymptotics.

We regard the assignment

$$\Upsilon^{\text{univ}} : w \mapsto \left(\overline{\partial}^\pi_J w, d(w^*\lambda \circ j)\right), \quad \Upsilon := (\Upsilon_1, \Upsilon_2)$$

as a section of the (infinite dimensional) vector bundle

$$\mathcal{CD}^{\text{univ}} \to \mathcal{F}^{\text{univ}}(M, \lambda; \overline{\gamma}, \underline{\gamma}) \tag{8.1.8}$$

where we put

$$\mathcal{F}^{\text{univ}}(M, \lambda; \overline{\gamma}, \underline{\gamma}) := \bigcup_{\vec{R} \in \mathfrak{Leg}(M,\xi)} \mathcal{F}\left(M, \lambda; \vec{R}; \overline{\gamma}, \underline{\gamma}\right) \times \mathcal{J}_\lambda(M, \xi).$$

We denote by Υ^{univ} the parameterized section of (8.1.8) defined by

$$\Upsilon^{\text{univ}}((w, J), \vec{R}) = \left(\overline{\partial}^{\pi}_J w, d(w^*\lambda \circ j)\right).$$

Here $\mathcal{F}(M, \lambda; \vec{R}; \overline{\gamma}, \underline{\gamma})$ is the off-shell function space associated to the moduli space $\mathcal{M}(M, \lambda; J; \vec{R}; \overline{\gamma}, \underline{\gamma})$. (See Definition 8.2.12.) We refer readers to Notation 8.2.13 for the definition of $\overline{\mathcal{C}\mathcal{D}}$. Then we summarize the main Fredholm results established in the present article into the following. This is the open-string counterpart of the Fredholm result established in [23] for the closed string case.

Theorem 8.1.5 *Let $\ell > 0$ be a given sufficiently large integer. Then*

(1) *Υ^{univ} is a smooth submersion on the open subset of $\mathcal{F}^{\text{univ}}$ consisting of somewhere injective map w.*
(2) *The parameterized moduli space $\mathcal{M}^{\mathcal{J}_\lambda}(M, \lambda, \vec{R}; \overline{\gamma}, \underline{\gamma})$ over $\mathcal{J}_\lambda(M, \xi)$ with $\vec{R}$ fixed is an infinite dimensional C^ℓ Banach manifold.*
(3) *The projection*

$$(\Upsilon^{\mathcal{J}_\lambda})^{-1}(0) \to \mathcal{J}_\lambda(M, \xi)$$

(with $\vec{R}$ fixed) is a Fredholm map and its index is the same as that of $D\Upsilon(w)$ for a (and so any) $w \in \mathcal{M}(M, \lambda, \vec{R}; J; \overline{\gamma}, \underline{\gamma})$.

Notation 8.1.6 We will denote by $\mathcal{M}^{(\cdot)}$ or $\mathcal{M}(\cdots; (\cdot))$ for various *parameterized* moduli spaces over the parameter space $(\cdot)$ such as $\mathcal{J}_\lambda = \mathcal{J}_\lambda(M, \xi)$, $\mathfrak{Leg}(M, \xi)$ and others. We apply similar notations for the associated off-shell function spaces. We also denote by $\Upsilon^{(\cdot)}$ the associated parameterized section of Υ.

8.1.2 Transversality Under the Perturbation of Boundary Condition

In [16], the present author established a transversality result of the open string version of the Gromov-Witten-Floer theory under the perturbation of Lagrangian boundary conditions. We also need to study the transversality result under the perturbation of Legendrian boundaries for the construction of Fukaya-type category on contact manifolds [15]. Such a study is carried out by Ekholm et al. [6] in the study of Chekanov-Eliashberg DGA of Legendrian submanifolds through symplectization adapting that of [16].

In the present section, we develop the Legendrian counterpart for the contact instantons imitating the arguments used in [16]. Although the overall scheme of the proof largely follows that of [16] its details are much more subtle and nontrivial. This is largely because the nature of contact instanton equation is more complicated than the pseudoholomorphic curves, especially because the study of adjoint problem of the linearized equation involves much more nontrivial systematic tensorial calculations than that of [16]. We attract readers' attention that this is another place where *the framework of contact instantons exhibits its naturality and compatibility with the existing contact geometry*, in that the contact distribution component and the Reeb component of the system (8.1.5) well interact with each other through the optimal covariant tensor calculus leading to the proof of the parametric transversality result under the perturbation of boundaries. (See Sect. 8.5 for such tensorial calculations.)

As in the proof of [16, Theorem I], we transforming the problem of perturbing boundaries by ambient contact isotopies, and study the following fibration

$$\Pi^{\mathcal{L}eg} : \mathcal{M}^{\mathcal{L}eg}(M, \lambda; \overline{\gamma}, \underline{\gamma}) \to \mathfrak{Leg}(M, \xi)$$

with J fixed. We then prove the following generic transversality result.

Theorem 8.1.7 (Theorem 8.4.3) *Let (M, ξ) be a contact manifold, and let λ a contact form be given. We consider* (8.1.5). *Fix J and k and consider transversal Legendrian link $\vec{R}$. Then there exists a residual subset of J's with $\vec{R}$ fixed (resp. of $\vec{R} = (R_1, \ldots, R_k)$ of Legendrian submanifolds with J fixed) such that the moduli space $\mathcal{M}(M, \lambda, \vec{R}; \overline{\gamma}, \underline{\gamma})$ is transversal.*

We refer to Appendix A for the definition of *transversal link*.

8.1.3 Generic Evaluation Transversality

Another crucial general analytical ingredient is the *evaluation map transversality*. Such an evaluation transversality will be important for the application to contact topology, for example, in the proof of Shelukhin's conjecture [21] and in the construction of Fukaya-type category of contact manifolds generated by Legendrian submanifolds in [15], similarly as in the study of pseudoholomorphic curves in symplectic geometry. A rigorous proof of the evaluation transversality is rather subtle even in the pseudoholomorphic curve theory as already mentioned in [17]. A conceptually canonical proof of the evaluation map transversality is given in [18, Section 10.5], which in turn followed the scheme provided by Le and Ono [13] and Zhu and the author [29] in their studies of one-jet evaluation transversality which is based on a standard structure theorem of the distributions with point support. (See Theorem 8.7.1 below.) Naturality of the proof in [17, 29] enables us to adapt it to the current context of contact instantons. However the proof of the evaluation transversality study for contact instantons is significantly more nontrivial in its details than the case of pseudoholomorphic curves thanks to the different nature of the equation which involves a system of partial differential equations of mixed degree.

We first recall the off-shell setting of the study of evaluation transversality. For given Legendrian link $\vec{R} = (R_1, \cdots, R_k)$, we consider the moduli space

$$\mathcal{M}((\dot\Sigma, \partial\dot\Sigma), (M, \vec{R}); J)$$

of finite energy maps $w : \dot\Sigma \to M$ satisfying the Eq. (8.1.5) as before. (We refer readers [20, 21] for the definition of relevant energies.)

We will treat the two cases, evaluation at an interior marked point and one at a boundary marked point, separately. We denote by the subindex (ℓ, k) the number of interior and boundary marked points respectively. Consider the parameterized moduli space

$$\begin{aligned}&\mathcal{M}_{(1,0)}((\dot\Sigma, \partial\dot\Sigma), (M, \vec{R}); \mathcal{J}_\lambda)\\ &= \{((j, w), J, z) \mid w : \Sigma \to M,\ \Upsilon(J, (j, w)) = 0,\ w(\partial\dot\Sigma) \subset \vec{R},\ z \in \operatorname{Int}\dot\Sigma\}.\end{aligned}$$

The evaluation map $\mathrm{ev}^+ : \mathcal{M}_{(1,0)}((\dot\Sigma, \partial\dot\Sigma, (M, \vec{R})) \to M$ is defined by

$$\mathrm{ev}^+((j, w), z) = w(z).$$

We then have the fibration

$$\widetilde{\mathcal{M}}_{(1,0)}((\dot\Sigma, \partial\dot\Sigma), (M, \vec{R}); \mathcal{J}_\lambda) = \bigcup_{J \in \mathcal{J}_\lambda} \widetilde{\mathcal{M}}_{(1,0)}((\dot\Sigma, \partial\dot\Sigma), (M, \vec{R}); J) \to \mathcal{J}_\lambda$$

and

$$\widetilde{\mathcal{M}}^{\mathrm{inj}}_{(1,0)}((\dot\Sigma, \partial\dot\Sigma), (M, \vec{R}))$$

to be the open subset of $\widetilde{\mathcal{M}}_{(1,0)}(M, \lambda, \vec{R})$ consisting of somewhere injective contact instantons. We have the universal (0-jet) evaluation map

$$\mathrm{Ev}^+ : \widetilde{\mathcal{M}}_{(1,0)}((\dot\Sigma, \partial\dot\Sigma), (M, \vec{R}); \mathcal{J}_\lambda) \to M.$$

We also consider the boundary evaluation map

$$\mathrm{Ev}_\partial : \widetilde{\mathcal{M}}_{(0,1)}((\dot\Sigma, \partial\dot\Sigma), (M, \vec{R}); \mathcal{J}_\lambda) \to \vec{R}.$$

The basic generic transversality is the following.

Theorem 8.1.8 (**0-Jet Evaluation Transversality, Theorem 8.6.1**) *Both evaluation maps* Ev^+ *and* Ev_∂ *are submersions on the open subset consisting of somewhere injective elements of* $\mathcal{M}^{\mathcal{J}_\lambda}((\dot\Sigma, \partial\dot\Sigma), (M, \vec{R}))$.

8.1.4 Review of the Contact Triad Connection

In this subsection, we give a brief review of the notion of canonical connection of contact triad (M, λ, J) that was introduced in [25]. This connection suits best for our tensorial calculations performed in the study of various component of the analyses of the moduli space of contact instanons which lead to various output equations that enable us to analyse their L^2-adjoint problem related to the application of Hahn-Banach theorem entering in the transversality analysis of the moduli spaces. (See Sect. 8.5 for the relevant tensor calculations.)

Theorem 8.1.9 (Contact Triad Connection [25]) *For every contact triad (M, λ, J), there exists a unique affine connection ∇, called the contact triad connection, satisfying the following properties:*

(1) *The connection ∇ is metric with respect to the contact triad metric, i.e., $\nabla g = 0$;*
(2) *The torsion tensor T of ∇ satisfies $T(R_\lambda, \cdot) = 0$;*
(3) *The covariant derivatives satisfy $\nabla_{R_\lambda} R_\lambda = 0$, and $\nabla_Y R_\lambda \in \xi$ for any $Y \in \xi$;*
(4) *The projection $\nabla^\pi := \pi \nabla|_\xi$ defines a Hermitian connection of the vector bundle $\xi \to M$ with Hermitian structure $(d\lambda|_\xi, J)$;*
(5) *The ξ-projection of the torsion T, denoted by $T^\pi := \pi T$ satisfies the following property:*

$$T^\pi(JY, Y) = 0 \tag{8.1.9}$$

for all Y tangent to ξ;

(6) *For $Y \in \xi$, we have the following*

$$\partial_Y^\nabla R_\lambda := \frac{1}{2}(\nabla_Y R_\lambda - J\nabla_{JY} R_\lambda) = 0.$$

From this theorem, we see that the contact triad connection ∇ canonically induces a Hermitian connection ∇^π for the Hermitian vector bundle (ξ, J, g_ξ), and we call it the *contact Hermitian connection.*

Moreover, the following fundamental properties of the contact triad connection was proved in [25], which will be useful to perform tensorial calculations later.

Corollary 8.1.10 *Let ∇ be the contact triad connection. Then*

(1) *For any vector field Y on M,*

$$\nabla_Y R_\lambda = \frac{1}{2}(\mathcal{L}_{R_\lambda} J)JY; \tag{8.1.10}$$

(2) $\lambda(T) = d\lambda$.

We refer readers to [25] for more discussion on the contact triad connection and its relation with other related canonical type connections.

Part 1: Generic Mapping Transversality Under the Perturbation of Boundaries

8.2 Fredholm Theory of (Relative) Contact Instantons

We start with setting-up the proper framework for the study of generic nondegeneracy results for the Reeb orbits and chords.

8.2.1 Set-Up for the Study of Generic Nondegeneracy of Reeb Orbits and Chords

We first introduce the following definition

Definition 8.2.1 Let $T \geq 0$ and consider a curve $\gamma : [0, 1] \to M$ be a smooth curve. We say (γ, T) an *iso-speed Reeb trajectory* if the pair satisfies

$$\dot{\gamma}(t) = T R_\lambda(\gamma(t)), \quad \int \gamma^*\lambda = T$$

for all $t \in [0, 1]$. If $\gamma(1) = \gamma(0)$, we call (γ, T) an iso-speed closed Reeb orbit and T the *action* of γ.

Remark 8.2.2 We remark that this representation of a Moore path is different from that of the one (8.1.7) given in the introduction of the present paper. The relationship is via the coordinate transformation

$$(\gamma, T) \mapsto (T, \gamma((\cdot)/T))$$

which transforms the pair (γ, T) in Definition 8.2.1 to the one in (8.1.7) as long as $T \neq 0$. This way of representing a Reeb chord as a Moore path is useful for the study of transversality of Reeb chords in that now the domains of the paths γ are fixed to $[0, 1]$. (See Appendix A.)

We start with the case of closed orbits.

Definition 8.2.3 Let (γ, T) be an iso-speed closed Reeb orbit in the sense as above. When $|T| > 0$ is minimal among such that $\gamma(1) = \gamma(0)$ with $\int \gamma^*\lambda \neq 0$, we call the pair (γ, T) a *simple* iso-speed closed Reed orbit.

We consider the relative version thereof.

Definition 8.2.4 (Iso-Speed Reeb Chords [20]) Let (R_0, R_1) be a pair of Legendrian submanifolds of (M, ξ) and $T \geq 0$. For given contact form λ, we say a pair (γ, T) with $\gamma : [0, 1] \to M$ is an iso-speed Reeb chord from R_0 to R_1 if

$$\dot{\gamma}(t) = T R_\lambda(\gamma(t)), \quad \gamma(0) \in R_0, \, \gamma(1) \in R_1.$$

We call such a pair (γ, T) *positive* (resp. *negative*) if $T \geq 0$ (resp. if $T < 0$).

We alert readers that the constant curve is not a Reeb trajectory in the standard sense in that it does not satisfy the Reeb trajectory equation $\dot{x} = R_\lambda(x)$, while it satisfies $\dot{x} = 0 = 0 \cdot R_\lambda(x)$ which shows that any constant curve valued at a point from $R_0 \cap R_1$ is a iso-speed Reeb chord with speed 0. When $T > 0$, the reparameterization $\gamma_T : [0, T] \to M$

$$\gamma_T(t) := \gamma(t/T)$$

satisfies $\dot{x} = R_\lambda(x)$ with the period $T > 0$, i.e., satisfies $\gamma_T(0) = \gamma_T(T)$.

Remark 8.2.5 Note that when $R_0 = R_1$, we have 'lots of iso-speed Reeb chords' arising from the constant chords. We will show that this component of constant chords is nondegenerate in the Bott-Morse sense. This is important in our study of contact instanton Legendrian Floer homology we introduce in [20, 22] and [28], especially in its calculation when R_1 is contact isotopic to R_0 and C^1-close thereto. This is the main reason why our generic nondegeneracy includes the constant trajectories defined over the fixed domain [0, 1] and emphasizes the *iso-speed* formulation of the Reeb chords given in Definition 8.2.4.

We now study the property of nondegeneracy of the pair (γ, T) by formulating the notion of nondegeneracy precisely including the case of constant trajectories, i.e., the case with $T = 0$.

Let (γ, T) be a closed Reeb orbit of action T. By definition, we can write $\gamma(T) = \phi^T_{R_\lambda}(\gamma(0))$ for the Reeb flow $\phi^T = \phi^T_{R_\lambda}$ of the Reeb vector field R_λ. In particular $p = \gamma(0)$ is a fixed point of the diffeomorphism ϕ^T. Since $\mathcal{L}_{R_\lambda}\lambda = 0$, ϕ^T is a contact diffeomorphism and so induces an isomorphism

$$\Psi_\gamma := d\phi^T(p)|_{\xi_p} : \xi_p \to \xi_p$$

which is the linearization restricted to ξ_p of the Poincaré return map.

Definition 8.2.6 Let $T > 0$. We say a T-closed Reeb orbit (T, λ) is *nondegenerate* if $\Psi_\gamma : \xi_p \to \xi_p$ with $p = \gamma(0)$ has not eigenvalue 1.

When $T = 0$, it is well-known that the constant loop is nondegenerate in the Bott-Morse sense.

For $T > 0$, the following generic nondegeneracy result is well-known to the experts, at least for the case of closed Reeb orbits. (See [1, Appendix A] for its proof.)

Theorem 8.2.7 (Albers-Braam-Wendl) *There exists a residual subset*

$$\mathcal{C}^{\text{reg}}(M, \xi) \subset \mathcal{C}(M, \xi)$$

such that for any $\lambda \in \mathcal{C}^{\text{reg}}(M, \xi)$ *all the closed Reeb orbits* λ *are nondegenerate if* $T > 0$.

The main purpose of Appendix A is to prove the following generic nondegeneracy result for Reeb chords which extends the above nondegeneracy results to the case of Reeb chords and to the Bott-Morse situation of constant chords.

Theorem 8.2.8 *Let* (M, ξ) *be a contact manifold. Let* (R_0, R_1) *be a pair of Legendrian submanifolds where either* $R_0 \cap R_1 = \emptyset$ *or* $R_0 = R_1$.

(1) *For a given pair* (R_0, R_1), *there exists a residual subset*

$$\mathcal{C}^{\text{reg}}(\xi; R_0, R_1) \subset \mathcal{C}(M, \xi)$$

such that for any $\lambda \in \mathcal{C}^{\text{reg}}(\xi; R_0, R_1)$ *all Reeb chords from* R_0 *to* R_1 *are nondegenerate for* $T > 0$ *when* $R_0 \cap R_1 = \emptyset$, *and Bott-Morse nondegenerate for* $R_0 = R_1$ *with* $T = 0$.

(2) *For a given contact form* λ, *there exists a residual subset of pairs* (R_0, R_1) *of Legendrian submanifolds such that all Reeb chords from* R_0 *to* R_1 *are nondegenerate for* $T > 0$ *and Bott-Morse nondegenerate when* $T = 0$.

8.2.2 Asymptotic Convergence and Vanishing of Asymptotic Charge

Next we recall from [26] (resp. from [27]) the asymptotic convergence result of contact instantons of finite energy $E(w) = E^\pi(w) + E^\perp(w) < \infty$ for the closed string case (resp. with Legendrian boundary condition of pair (R_0, R_1) for the open string case) near the punctures of a Riemann surface $\dot\Sigma$, respectively. (We refer to [20, 23, 28] for the precise definition of total energy.)

Let $\dot\Sigma$ be a punctured Riemann surface with punctures

$$\{p_i^+\}_{i=1,\cdots,l^+} \cup \{p_j^-\}_{j=1,\cdots,l^-}$$

equipped with a metric h with cylinder-like ends (resp. *strip-like ends* for the open string case) outside a compact subset K_Σ. Let $w : \dot\Sigma \to M$ be any such smooth map.

Under the hypotheses of nondegeneracy λ (resp. of the pair $(\lambda, (R_0, R_1)$ for the open string case) and of asymptotic convergence at the punctures, we can associate two natural asymptotic invariants at each puncture defined as

$$T := \lim_{r\to\infty} \int_{\{r\}\times S^1} (w|_{\{r\}\times S^1})^*\lambda \tag{8.2.1}$$

$$Q := \lim_{r\to\infty} \int_{\{r\}\times S^1} ((w|_{\{r\}\times S^1})^*\lambda \circ j) \tag{8.2.2}$$

at each puncture. (Here we only look at positive punctures. The case of negative punctures is similar.) As in [26], we call T the *asymptotic contact action* and Q the *asymptotic contact charge* of the contact instanton w at the given puncture.

The proof of the following subsequence convergence result is given in [26, Theorem 6.4].

Theorem 8.2.9 (Subsequence Convergence-Closed Strings, Theorem 6.4 [26]) *Let $w : [0, \infty) \times S^1 \to M$ satisfy the contact instanton equations* (8.1.5) *of finite energy. Then for any sequence $s_k \to \infty$, there exists a subsequence, still denoted by s_k, and a massless instanton $w_\infty(\tau, t)$ (i.e., $E^\pi(w_\infty) = 0$) on the cylinder $\mathbb{R} \times S^1$ that satisfies the following:*

(1) $\overline{\partial}^\pi w_\infty = 0$ *and*

$$\lim_{k\to\infty} w(s_k + \tau, t) = w_\infty(\tau, t)$$

in the $C^l(K \times S^1, Q)$ sense for any l, where $K \subset [0, \infty)$ is an arbitrary compact set.

(2) $w_\infty^*\lambda = -Q\, d\tau + T\, dt$

In general $Q = 0$ does not necessarily hold for the closed string case. When $Q \neq 0$ combined with $T = 0$ happens, we say w has the bad limit of *appearance of spiraling instantons along the Reeb core*. It is also proven in [23] that If $Q = 0 = T$, then the puncture is removable. When $Q = 0$, which is always the case when contact instanton is exact such as those arising from the symplectization case, w_τ converges to a Reeb orbit of period $|T|$ exponentially fast.

Now we make the corresponding statement for the open string case proved in [19, 27].

Theorem 8.2.10 (Subsequence Convergence; The Case of Open Strings) *Let $w : [0, \infty) \times [0, 1] \to M$ satisfy the contact instanton equations* (8.1.5). *Then for any sequence $s_k \to \infty$, there exists a subsequence, still denoted by s_k, and a massless instanton $w_\infty(\tau, t)$ (i.e., $E^\pi(w_\infty) = 0$) on the cylinder $\mathbb{R} \times [0, 1]$ such that*

$$\lim_{k\to\infty} w(s_k + \tau, t) = w_\infty(\tau, t)$$

in the $C^l(K \times [0,1], Q)$ *sense for any* l, *where* $K \subset [0, \infty)$ *is an arbitrary compact set. Furthermore,* w_∞ *has* $Q = 0$ *and the formula* $w_\infty(\tau, t) = \gamma(T\,t)$ *with asymptotic action* T, *where* γ *is some Reeb chord joining* R_0 *and* R_1 *of period* $|T|$.

Corollary 8.2.11 (Vanishing Charge) *Assume the pair* $(\lambda, \vec{R})$ *is nondegenerate. Let* w *be as above with finite energy. Suppose that* $w(\tau, \cdot)$ *converges as* $\tau \to \infty$ *in the strip-like coordinate at a puncture* $p \in \partial\Sigma$ *with associated Legendrian pair* (R, R'). *Then its asymptotic charge* Q *vanishes at* p.

8.2.3 Off-Shell Description of Moduli Spaces

For the exposition of this section, we adapt the one given in [23] to the current context of bordered contact instantons by incorporating the Legendrian boundary condition. In particular, we consider general bordered compact surfaces of arbitrary genus to be consistent with that of [23] (for the closed case), and order the marked points starting from $k = 1$, not from $k = 0$ as in (8.1.5).

We will be mainly interested in the two cases:

(1) A generic nondegenerate case of $R_1, \cdots, R_k$ which in particular are mutually disjoint,
(2) The case where $R_1, \cdots, R_k = R$.

We now choose a λ-adapted CR-almost complex structure J. Let (Σ, j) be a bordered compact Riemann surface, and let $\dot\Sigma$ be the punctured Riemann surface with $\{z_1, \ldots, z_k\} \subset \partial\Sigma$, we consider the moduli space

$$\mathcal{M}((\dot\Sigma, \partial\dot\Sigma), (M, \vec{R}); J), \quad \vec{R} = (R_1, \cdots, R_k)$$

of finite energy maps $w : \dot\Sigma \to M$ satisfying the Eq. (8.1.5).

The second case is transversal in the Bott-Morse sense both for the Reeb chords and for the moduli space of contact instantons, which is rather straightforward and easier to handle, and so omitted.

For the first case, all the asymptotic Reeb chords are nonconstant and have nonzero action $T \neq 0$. In particular, the relevant punctures z_i are not removable. Therefore we have the decomposition of the finite energy moduli space

$$\mathcal{M}((\dot\Sigma, \partial\dot\Sigma), (M, \vec{R}); J) = \bigcup_{\vec{\gamma} \in \prod_{i=0}^{k-1} \mathfrak{Reeb}(R_i, R_{i+1})} \mathcal{M}((\dot\Sigma, \partial\dot\Sigma), (M, \vec{R}); J; \vec{\gamma})$$

by the asymptotic convergence result from [19]. Depending on the choice of strip-like coordinates we divide the punctures

$$\{z_1, \cdots, z_k\} \subset \partial\Sigma$$

into two subclasses

$$p_1, \cdots, p_{s^+}, q_1, \cdots, q_{s^-} \in \partial\Sigma$$

as the positive and negative boundary punctures. We write $k = s^+ + s^-$.

Let γ_i^+ for $i = 1, \cdots, s^+$ and γ_j^- for $j = 1, \cdots, s^-$ be two given collections of Reeb chords at positive and negative punctures respectively. Following the notations from [3, 5] (but applied to the Reeb chords instead of closed Reeb orbits), we denote by $\underline{\gamma}$ and $\overline{\gamma}$ the corresponding collections

$$\underline{\gamma} = \{\gamma_1^+, \cdots, \gamma_{s^+}^+\}$$
$$\overline{\gamma} = \{\gamma_1^-, \cdots, \gamma_{s^-}^-\}.$$

For each p_i (resp. q_j), we associate the strip-like coordinates $(\tau, t) \in [0, \infty) \times [0, 1]$ (resp. $(\tau, t) \in (-\infty, 0] \times [0, 1]$) on the punctured disc $D_{e^{-2\pi K_0}}(p_i) \setminus \{p_i\}$ (resp. on $D_{e^{-2\pi K_0}}(q_i) \setminus \{q_i\}$) for some sufficiently large $K_0 > 0$.

Definition 8.2.12 We define

$$\mathcal{F}((\dot\Sigma, \partial\dot\Sigma), (M, \vec{R}); \underline{\gamma}, \overline{\gamma}) =: \mathcal{F}(M, \vec{R}; \underline{\gamma}, \overline{\gamma}) \tag{8.2.3}$$

to be the set of smooth maps satisfying the boundary condition

$$w(z) \in R_i \quad \text{for } z \in \overline{z_{i-1}z_i} \subset \partial\dot\Sigma \tag{8.2.4}$$

and the asymptotic condition

$$\lim_{\tau\to\infty} w((\tau, t)_i) = \gamma_i^+(T_i(t + t_i)), \qquad \lim_{\tau\to-\infty} w((\tau, t)_j) = \gamma_j^-(T_j(t - t_j)) \tag{8.2.5}$$

for some $t_i,\ t_j \in [0, 1]$, where

$$T_i = \int_0^1 (\gamma_i^+)^*\lambda, \quad T_j = \int_0^1 (\gamma_j^-)^*\lambda.$$

Here $t_i,\ t_j$ depends on the given analytic coordinate and the parameterizations of the Reeb chords.

We will fix the domain complex structure j and its associated Kähler metric h. We regard the assignment

$$\Upsilon : w \mapsto \left(\overline{\partial}^\pi w, d(w^*\lambda \circ j)\right), \quad \Upsilon := (\Upsilon_1, \Upsilon_2) \tag{8.2.6}$$

as a section of the (infinite dimensional) vector bundle: We first formally linearize and define a linear map

$$D\Upsilon(w) : \Omega^0(w^*TM, (\partial w)^*T\vec{R}) \to \Omega^{(0,1)}(w^*\xi) \oplus \Omega^2(\Sigma) \tag{8.2.7}$$

where we have the tangent space

$$T_w\mathcal{F} = \Omega^0(w^*TM, (\partial w)^*T\vec{R}).$$

For the simplicity of notation, we also introduce the following notation.

Notation 8.2.13 (Codomain of Υ) For given fixed $\vec{R}$, we define

$$\mathcal{CD}_{(J,\vec{R}),(j,w)} := \Omega_J^{(0,1)}(w^*\xi) \oplus \Omega^2(\Sigma) = \mathcal{H}_w^{\pi\,(0,1)} \oplus \Omega^2(\Sigma) \tag{8.2.8}$$

and

$$\begin{aligned} \mathcal{CD}_{(J,\vec{R})} &= \bigcup_{(j,w)\in\mathcal{F}} \{(j,w)\} \times \mathcal{CD}_{(J,\vec{R}),(j,w)} \\ \mathcal{CD}_{\vec{R}} &= \bigcup_{J,(j,w)\in\mathcal{F}} \{J\} \times \mathcal{CD}_{(J,\vec{R})}. \end{aligned}$$

Here $\mathcal{CD}$ stands for 'codomain'.

Since we will not vary $\vec{R}$ in the present discussion, we omit $\vec{R}$ from notation and simply write $\mathcal{CD} = \mathcal{CD}_{\vec{R}}$. We also denote by $\mathcal{CD}^{\text{univ}}$ as the further union

$$\mathcal{CD}^{\text{univ}} = \bigcup_{(J,\vec{R})} \{(J,\vec{R})\} \times \mathcal{CD}_{(J,\vec{R})}. \tag{8.2.9}$$

Let $k \geq 2$ and $p > 2$. We denote by

$$\mathcal{W}^{k,p} := \mathcal{W}^{k,p}((\dot{\Sigma}, \partial\dot{\Sigma}), (M, \vec{R}); \underline{\gamma}, \overline{\gamma}) \tag{8.2.10}$$

the completion of the off-shell function space (8.2.3). It has the structure of a Banach manifold modeled by the Banach space given by the following

Definition 8.2.14 (Tangent Space $T_w\mathcal{W}^{k,p}$) We define

$$W^{k,p}(w^*TM, (\partial w)^*T\vec{R}; \underline{\gamma}, \overline{\gamma})$$

to be the set of vector fields $Y = Y^\pi + \lambda(Y)R_\lambda$ along w that satisfy

$$\begin{cases} Y^\pi \in W^{k,p}\left((\dot{\Sigma}, \partial\dot{\Sigma}), (w^*\xi, (\partial w)^*T\vec{R})\right), \\ Y^\pi(z) \in T\vec{R} \quad \text{for } z \in \partial\dot{\Sigma} \end{cases} \tag{8.2.11}$$

and

$$\begin{cases} \lambda(Y) \in W^{k,p}((\dot\Sigma, \partial\dot\Sigma), (\mathbb{R}, \{0\})), \\ \lambda(Y)(z) = 0 \quad \text{for } z \in \partial\dot\Sigma \end{cases} \tag{8.2.12}$$

Here we use the splitting

$$TM = \xi \oplus \operatorname{span}_{\mathbb{R}}\{R_\lambda\}$$

where $\operatorname{span}_{\mathbb{R}}\{R_\lambda\} := \mathcal{L}$ is a trivial line bundle and so

$$\Gamma(w^*\mathcal{L}) \cong C^\infty\left((\dot\Sigma, \partial\dot\Sigma), (\mathbb{R}, \{0\})\right).$$

The above Banach space is decomposed into the direct sum

$$W^{k,p}((\dot\Sigma, \partial\dot\Sigma), (w^*\xi, (\partial w)^* T\vec{R})) \bigoplus W^{k,p}((\dot\Sigma, \partial\dot\Sigma), (\mathbb{R}, \{0\})) \otimes R_\lambda : \tag{8.2.13}$$

by writing $Y = (Y^\pi, gR_\lambda)$ with a real-valued function $g = \lambda(Y(w))$ on $\dot\Sigma$. Here we measure the various norms in terms of the triad metric of the triad (M, λ, J).

Now for each given J and $w \in \mathcal{W}^{k,p}((\dot\Sigma, \partial\dot\Sigma), (M, \vec{R}); \underline{\gamma}, \overline{\gamma})$, we consider the Banach space

$$\Omega^{(0,1)}_{k-1,p}(w^*\xi; J) := W^{k-1,p}(\Lambda^{(0,1)}_J(w^*\xi))$$

the $W^{k-1,p}$-completion of $\Omega^{(0,1)}(w^*\xi) = \Gamma(\Lambda^{(0,1)}(w^*\xi)$ and form the bundle

$$\mathcal{H}^{(0,1)}_{k-1,p}(M, J) := \bigcup_{w \in \mathcal{W}^{k,p}} \Omega^{(0,1)}_{k-1,p}(w^*\xi; J) \tag{8.2.14}$$

over $\mathcal{W}^{k,p}$.

Definition 8.2.15 We associate the Banach space

$$\mathcal{CD}^{(0,1)}_{k-1,p}(M, \lambda; J)|_w := \Omega^{(0,1)}_{k-1,p}(w^*\xi; J) \oplus \Omega^2_{k-2,p}(\dot\Sigma) \tag{8.2.15}$$

to each $w \in \mathcal{W}^{k,p}$ and form the bundle

$$\begin{aligned} \mathcal{CD}^{(0,1)}_{k-1,p}(M, \lambda; J) &:= \bigcup_{w \in \mathcal{W}^{k,p}} \mathcal{CD}^{(0,1)}_{k-1,p}(M, \lambda)|_w \\ &\cong \mathcal{H}^{(0,1)}_{k-1,p}(M, J) \bigoplus \left(\mathcal{W}^{k,p} \times \Omega^2_{k-2,p}(\dot\Sigma)\right) \end{aligned}$$

over the Banach manifold $\mathcal{W}^{k,p}$ given in (8.2.10).

Then we can regard the assignment

$$\Upsilon_1 : w \mapsto \overline{\partial}^\pi w$$

as a smooth section of the bundle $\mathcal{H}^{(0,1)}_{k-1,p}(M,\lambda) \to \mathcal{W}^{k,p}$. Furthermore the assignment

$$\Upsilon_2 : w \mapsto d(w^*\lambda \circ j)$$

defines a smooth section of the trivial bundle

$$\Omega^2_{k-2,p}(\Sigma) \times \mathcal{W}^{k,p} \to \mathcal{W}^{k,p}.$$

We summarize the above discussion into the following lemma.

Lemma 8.2.16 *Consider the vector bundle*

$$\mathcal{CD}_{k-1,p}(M,\vec{R};J) \to \mathcal{W}^{k,p}.$$

The map Υ continuously extends to a continuous section still denoted by

$$\Upsilon : \mathcal{W}^{k,p} \to \mathcal{CD}_{k-1,p}(M,\vec{R};J).$$

With these preparations, the following is a consequence of the exponential estimates established in [19]. (See [26] for the closed string case of vanishing charge.)

Proposition 8.2.17 *Assume λ is nondegenerate. Let $w : \dot\Sigma \to M$ be a contact instanton and let $w^*\lambda = a_1\, d\tau + a_2\, dt$. Suppose*

$$\begin{aligned} \lim_{\tau\to\infty} a_{1,i} &= 0, \quad \lim_{\tau\to\infty} a_{2,i} = T(p_i) \\ \lim_{\tau\to-\infty} a_{1,j} &= 0, \quad \lim_{\tau\to-\infty} a_{2,j} = T(q_j) \end{aligned} \tag{8.2.16}$$

at each puncture p_i and q_j. Then $w \in \mathcal{W}^{k,p}(M,\vec{R};\underline{\gamma},\overline{\gamma})$.

Now we are ready to define the moduli space of contact instantons with prescribed asymptotic condition.

Definition 8.2.18 Consider the zero set of the section Υ

$$\widetilde{\mathcal{M}}(M,\lambda,\vec{R};\underline{\gamma},\overline{\gamma};J) = \Upsilon^{-1}(0) \tag{8.2.17}$$

in the Banach manifold $\mathcal{W}^{k,p}(M, \lambda, \vec{R}; \underline{\gamma}, \overline{\gamma})$. We write $w \sim w'$ for two elements therefrom if there is a biholomorphism φ of the punctured bordered Riemann surfaces $(\dot\Sigma, \partial\dot\Sigma)$ such that $w' = w \circ \varphi$, and define the quotient space and

$$\mathcal{M}(M, \lambda, \vec{R}; \underline{\gamma}, \overline{\gamma}; J) = \widetilde{\mathcal{M}}(M, \lambda, \vec{R}; \underline{\gamma}, \overline{\gamma}; J)/\sim \tag{8.2.18}$$

to be the set of equivalence classes of contact instantons w under the equivalence relation $\sim$.

This definition does not depend on the choice of k or p as long as $k \geq 2,\ p > 2$. We call an equivalence class $[w]$ an isomorphism class of contact instantons and often just write it as w by an abuse of notation whose meaning should be clear from the give context.

8.2.4 Linearized Operator and its Ellipticity

Let $(\dot\Sigma, j)$ be a punctured Riemann surface, the set of whose punctures may be empty, i.e., $\dot\Sigma = \Sigma$ is either a closed or a punctured Riemann surface. In this subsection and the next, we lay out the precise relevant off-shell framework of functional analysis, and establish the Fredholm property of the linearization map. (We recall that we have already computed the linearization of Υ for the closed string case in [23].)

We recall that both for the elliptic regularity esimates in [19, 27] and for the optimal expression of the linearization map and its relevant calculations in [23], we have been using the contact triad connection ∇ of (M, λ, J) and the contact Hermitian connection ∇^π for (ξ, J) introduced in [25, 26]. Likewise we will utilize the contact triad connection in the following presentation of the linearization of contact instantons.

Then we have the following explicit formulae thereof.

Theorem 8.2.19 (Theorem 10.1 [23]; See Also Theorem 1.15 [24]) *In terms of the decomposition $dw = d^\pi w + w^*\lambda\, R_\lambda$ and $Y = Y^\pi + \lambda(Y) R_\lambda$, we have*

$$D\Upsilon_1(w)(Y) = \overline{\partial}^{\nabla^\pi} Y^\pi + B^{(0,1)}(Y^\pi) + T^{\pi,(0,1)}_{dw}(Y^\pi) + \frac{1}{2}\lambda(Y)(\mathcal{L}_{R_\lambda} J)J(\partial^\pi w) \tag{8.2.19}$$

$$D\Upsilon_2(w)(Y) = -\Delta(\lambda(Y))\, dA + d((Y^\pi \rfloor d\lambda) \circ j) \tag{8.2.20}$$

where $B^{(0,1)}$ and $T^{\pi,(0,1)}_{dw}$ are the $(0,1)$-components of B and T^π_{dw}, where $B,\ T^\pi_{dw}$: $\Omega^0(w^*TM) \to \Omega^1(w^*\xi)$ are zero-order differential operators given by

$$B(Y) = -\frac{1}{2} w^*\lambda \otimes \big((\mathcal{L}_{R_\lambda} J)JY\big) \tag{8.2.21}$$

and

$$T^{\pi}_{dw}(Y) = \pi T(Y, dw) \tag{8.2.22}$$

respectively.

From the above expression of the covariant linearization of of the section $\Upsilon = (\Upsilon_1, \Upsilon_2)$, the linearization continuously extends to a bounded linear map

$$D\Upsilon_{(\lambda,T)}(w) : T\mathcal{W}^{k,p} \to \mathcal{CD}_{k-1,p}(M, \lambda)$$

where we recall

$$T\mathcal{W}^{k,p} = \Omega^0_{k,p}\left(w^*TM, (\partial w)^*T\vec{R}\right)$$
$$\mathcal{CD}_{k-1,p}(M, \lambda) = \Omega^{(0,1)}_{k-1,p}(w^*\xi; J) \oplus \Omega^2_{k-2,p}(\Sigma)$$

for any choice of $k \geq 2,\ p > 2$. Using the decomposition

$$\Omega^0_{k,p}(w^*TM, (\partial w)^*T\vec{R}) \cong \Omega^0_{k,p}(w^*\xi, (\partial w)^*T\vec{R}) \oplus \Omega^0_{k,p}(\dot{\Sigma}, \partial\dot{\Sigma}) \otimes R_\lambda,$$

$D\Upsilon(w)$ can be written into the matrix form

$$\begin{pmatrix} \bar{\partial}^{\nabla^\pi} + T^{\pi,(0,1)}_{dw} + B^{(0,1)} & d\left((\cdot)\rfloor d\lambda\right) \circ j \\ \frac{1}{2}\lambda(\cdot)(\mathcal{L}_{R_\lambda} J)J\partial^\pi w & -\Delta(\lambda(\cdot))\, dA \end{pmatrix}. \tag{8.2.23}$$

It follows that the map $D\Upsilon(w)$ is a partial differential operator whose principal symbol map is given by $\sigma(D\Upsilon) = \sigma(D\Upsilon_1) \oplus \sigma(D\Upsilon_2)$ where

$$\sigma(D\Upsilon_1(w))(\eta) = J\Pi^*\eta$$
$$\sigma(D\Upsilon_2(w))(\eta) = \langle\lambda, \eta\rangle^2 = (\eta(R_\lambda))^2 \tag{8.2.24}$$

where η is a cotangent vector in $T^*M \setminus \{0\}$ and has decomposition

$$\eta = \eta^\pi + \eta(R_\lambda) \otimes w^*\lambda. \tag{8.2.25}$$

(See [12] for the discussion of general elliptic operators of mixed degree on noncompact manifolds with cylindrical ends.)

In particular we note that the restriction $D\Upsilon_1(w)|_{\Omega^0(w^*\xi)}$ has the same principal symbol as that of

$$\bar{\partial}^{\nabla^\pi} : \Omega^0(w^*\xi, (\partial w)^*\xi) \to \Omega^{(0,1)}(w^*\xi; J)$$

which is the first order elliptic operator of Cauchy-Riemann type, and that $D\Upsilon_2(w)$ has the symbol of the Hodge Laplacian acting on zero forms

$$*\,\Delta : \Omega^0(\dot\Sigma, \partial\dot\Sigma) \to \Omega^2(\dot\Sigma).$$

8.2.5 Fredholm Theory on Punctured Bordered Riemann Surfaces

By the (local) ellipticity shown in the previous subsection, it remains to examine the Fredholm property of the linearized operator $D\Upsilon(w)$. For this purpose, we need to examine the asymptotic behavior of the operator near punctures in strip-like coordinates.

We first decompose the section $Y \in w^*TM$ into

$$Y = Y^\pi + \lambda(Y) \otimes R_\lambda$$

as before. Then the matrix (8.2.23) has its entries given by

$$D\Upsilon_1^1(w)(Y^\pi) = \overline{\partial}^{\nabla^\pi} Y^\pi + B^{(0,1)}(Y^\pi) + T_{dw}^{\pi,(0,1)}(Y^\pi), \tag{8.2.26}$$

$$D\Upsilon_2^1(w)(Y^\pi) = d((Y^\pi \rfloor d\lambda) \circ j), \tag{8.2.27}$$

$$D\Upsilon_1^2(w)(\lambda(Y)R_\lambda) = \frac{1}{2}\lambda(Y)\mathcal{L}_{R_\lambda} J J \partial^\pi w, \tag{8.2.28}$$

$$D\Upsilon_2^2(w)(\lambda(Y)R_\lambda) = -\Delta(\lambda(Y))\, dA. \tag{8.2.29}$$

Noting that Y^π and $\lambda(Y)$ are independent of each other, we write

$$Y = Y^\pi + f R_\lambda, \quad f := \lambda(Y)$$

where $f : \dot\Sigma \to \mathbb{R}$ is an arbitrary function satisfying the boundary condition

$$Y^\pi(\partial\dot\Sigma) \subset T\vec{R}, \quad f|_{\partial\dot\Sigma} = 0$$

by the Legendrian boundary condition satisfied by Y. The following is obvious from the expression of the $D\Upsilon_i^j(w)$.

Lemma 8.2.20 *Suppose that w is a solution to* (8.1.5). *The operators $D\Upsilon_i^j(w)$ have the following continuous extensions:*

$$D\Upsilon_1^1(w)(Y^\pi) : \Omega^0_{k,p}(w^*\xi, (\partial w)^*T\vec{R}) \to \Omega^{(0,1)}_{k-1,p}(w^*\xi; J)$$

$$D\Upsilon_2^1(w)(Y^\pi) : \Omega^0_{k,p}(w^*\xi, (\partial w)^*T\vec{R}) \to \Omega^2_{k-1,p}(\dot\Sigma) \hookrightarrow \Omega^2_{k-2,p}(\dot\Sigma)$$

$$D\Upsilon_1^2(w)((\cdot)R_\lambda) : \Omega^0_{k,p}(\dot\Sigma, \partial\dot\Sigma) \to \Omega^{(0,1)}_{k,p}(w^*\xi; J) \hookrightarrow \Omega^{(0,1)}_{k-1,p}(w^*\xi; J)$$
$$D\Upsilon_2^2(w)((\cdot)R_\lambda) : \Omega^0_{k,p}(\dot\Sigma, \partial\dot\Sigma) \to \Omega^2_{k-2,p}(\Sigma).$$

We regard the domains of $D\Upsilon_i^2$ for $i = 1, 2$ as $C^\infty(\dot\Sigma, \partial\dot\Sigma)$ using the isomorphism

$$C^\infty(\dot\Sigma, \partial\dot\Sigma) \cong \Omega^0(\dot\Sigma, \partial\dot\Sigma) \otimes R_\lambda. \tag{8.2.30}$$

We now establish the following Fredholm property of the linearized operator.

Proposition 8.2.21 *Suppose that w is a solution to* (8.1.5). *Consider the completion of $D\Upsilon(w)$, which we still denote by $D\Upsilon(w)$, as a bounded linear map from $\Omega^0_{k,p}(w^*TM, (\partial w)^*T\vec R)$ to $\Omega^{(0,1)}(w^*\xi) \oplus \Omega^2(\Sigma)$ for $k \geq 2$ and $p \geq 2$. Then*

(1) *The off-diagonal terms of $D\Upsilon(w)$ are relatively compact operators against the diagonal operator.*
(2) *The operator $D\Upsilon(w)$ is homotopic to the operator*

$$\begin{pmatrix} \overline\partial^{\nabla^\pi} + T^{\pi,(0,1)}_{dw} + B^{(0,1)} & 0 \\ 0 & -\Delta(\lambda(\cdot))\,dA \end{pmatrix} \tag{8.2.31}$$

via the homotopy

$$s \in [0,1] \mapsto \begin{pmatrix} \overline\partial^{\nabla^\pi} + T^{\pi,(0,1)}_{dw} + B^{(0,1)} & s\,d\left((\cdot)\rfloor d\lambda\right)\circ j \\ \frac{s}{2}\lambda(\cdot)(\mathcal{L}_{R_\lambda}J)J(\pi dw)^{(1,0)} & -\Delta(\lambda(\cdot))\,dA \end{pmatrix} =: L_s \tag{8.2.32}$$

which is a continuous family of Fredholm operators.
(3) *And the principal symbol*

$$\sigma(z,\eta) : w^*TM|_z \to w^*\xi|_z \oplus \Lambda^2(T_z\Sigma), \quad 0 \neq \eta \in T_z^*\Sigma$$

of (8.2.31) *is given by the matrix*

$$\begin{pmatrix} \frac{\eta + i\eta\circ j}{2} Id & 0 \\ 0 & |\eta|^2 \end{pmatrix}.$$

Proof Statement (1) is a consequence of the exponential decay near the puncture [27], and the compactness of Sobolev embeddings

$$\Omega^2_{k-1,p}(\dot\Sigma) \hookrightarrow \Omega^2_{k-2,p}(\dot\Sigma), \quad \Omega^2_{k,p}(\dot\Sigma) \hookrightarrow \Omega^2_{k-2,p}(\dot\Sigma).$$

When $\partial\dot{\Sigma} = \emptyset$, the same kind of statement is proved in [23]. Essentially the same proof applies by incorporating the boundary condition. We leave some details and explanation on the requirement $d(w^*\lambda \circ j) = 0$ to [22]. □

Now we are ready to wrap-up the discussion of the Fredholm property of the linearization map

$$D\Upsilon_{(\lambda,T)}(w) : \Omega^0_{k,p}(w^*TM, (\partial w)^*T\vec{R}; \underline{\gamma}, \overline{\gamma}) \to \Omega^{(0,1)}_{k-1,p}(w^*\xi; J) \oplus \Omega^2_{k-2,p}(\dot{\Sigma})$$

by proving Statement (1) of Proposition 8.2.21.

The following proposition can be derived from the arguments used by Lockhart and McOwen [12] with the incorporation of Legendrian boundary condition which is an elliptic boundary valued problem as shown in [19].

Proposition 8.2.22 (Proposition 11.6 [23]) *Assume that* $\underline{\gamma}$, $\overline{\gamma}$ *are nondegenerate. Then the operator* (8.2.23) *is Fredholm.*

Then by the continuous invariance of the Fredholm index, we obtain

$$\text{Index}\, D\Upsilon_{(\lambda,T)}(w) = \text{Index}\left(\overline{\partial}^{\nabla^\pi} + T^{\pi,(0,1)}_{dw} + B^{(0,1)}\right) + \text{Index}(-\Delta). \tag{8.2.33}$$

The computation of index is given in [23] for the closed string case and is given for the current open string case in [27].

Remark 8.2.23 Suppose $\delta > 0$ satisfies the inequality

$$0 \leq \delta < \min\left\{\frac{\text{gap}(\vec{\gamma})}{p}, \frac{\pi}{p}\right\}$$

where $\text{gap}(\vec{\gamma})$ is the spectral gap,

$$\text{gap}(\overline{\gamma}, \underline{\gamma}) := \min_{\gamma_i, \gamma_j}\left\{d_{\text{H}}(\text{Spec} A_{(T_i,\gamma_i)}, 0), d_{\text{H}}(\text{Spec} A_{(T_j,\gamma_j)}, 0)\right\} \tag{8.2.34}$$

of the asymptotic operators $A_{(T_j,z_j)}$ or $A_{(T_i,z_i)}$ associated to the corresponding punctures. Then the above Fredholm property also holds in the weighted Sobolev space setting with exponential weight $e^{\delta|\tau|}$.

8.3 Generic Mapping Transversality Under the Perturbation of J's

In this section, we briefly recall the perturbation result under the perturbation of CR almost complex structures from [23] proved for the closed string case, and adapt it to the case of boundary perturbations.

Let a contact manifold (M, ξ) be given. We consider the contact forms λ of (M, ξ) such that all Reeb chords are nondegenerate. The set of such contact forms is residual in $\mathcal{C}(M, \xi)$. (See Appendix A for the proof.)

Then we involve the set $\mathcal{J}_\lambda(M, \xi)$ of adapted J's. We study the linearization of the map Υ^{univ} which is the map Υ augmented by the argument $J \in \mathcal{J}_\lambda(M, \xi)$. More precisely, we define the universal section

$$\Upsilon^{\text{univ}} : \mathcal{F} \times \mathcal{J}_\lambda(M, \xi) \to \mathcal{CD}^{\text{univ}}(M, \lambda)$$

given by

$$\Upsilon^{\text{univ}}((j, w), J) = \left(\overline{\partial}^\pi_J w, d(w^*\lambda \circ j)\right) \tag{8.3.1}$$

and study its linearization at each $(j, w, J) \in (\Upsilon^{\text{univ}})^{-1}(0)$. In the discussion below, we will fix the complex structure j on Σ, and so suppress j from the argument of Υ^{univ}.

The following universal linearization formula plays a crucial role in the generic transversality result as in the case of pseudoholomorphic curves in symplectic geometry.

Lemma 8.3.1 (Theorem 1.10 [23]) *Denote by $L := \delta J$ the first variation of J. We have the linearization*

$$D_{(w,J)}\Upsilon^{\text{univ}} : T_w\mathcal{F} \oplus T_J\mathcal{J}_\lambda(M, \xi) \to \Omega^{(0,1)}(w^*\xi) \bigoplus \Omega^2(\dot{\Sigma}) \otimes R_\lambda$$

whose explicit formula is given by

$$D_{(w,J)}\Upsilon^{\text{univ}}(Y, L) = D_1\Upsilon^{\text{univ}}(Y) + D_2\Upsilon^{\text{univ}}(L)$$

where we have partial derivatives

$$D_1\Upsilon^{\text{univ}}(Y) = D\Upsilon(Y), \quad D_2\Upsilon^{\text{univ}}(L) = \frac{1}{2}L(d^\pi u \circ j) \tag{8.3.2}$$

Proof This is straightforward from the definition

$$\overline{\partial}^\pi w = \frac{d^\pi w + J d^\pi w \circ j}{2}$$

and the fact that the projection π does not depend on the choice of J but depends only on λ. We omit its proof. □

Following the procedure of considering the set $\mathcal{J}^\ell(M, \lambda)$ of λ-adapted C^ℓ CR-almost complex structures J inductively as ℓ grows (see [14] and [18, Section 10.4] for the detailed explanation), we denote the zero set $(\Upsilon^{\text{univ}})^{-1}(0)$ by

$$\begin{aligned}\mathcal{M}(M, \lambda, \vec{R}; \overline{\gamma}, \underline{\gamma}; \mathcal{J}_\lambda) \\ = \left\{(w, J) \in \mathcal{W}^{k,p}(M, \vec{R}; \overline{\gamma}, \underline{\gamma}) \times \mathcal{J}^\ell_\lambda(M, \xi) \,\middle|\, \Upsilon^{\text{univ}}(w, J) = 0\right\}\end{aligned}$$

which we call the universal moduli space. Denote by

$$\Pi_2 : \mathcal{W}^{k,p}(M, \vec{R}; \underline{\gamma}, \overline{\gamma}) \times \mathcal{J}^\ell_\lambda(M, \xi) \to \mathcal{J}^\ell_\lambda(M, \xi)$$

the projection. Then we have

$$\mathcal{M}(J; \underline{\gamma}, \overline{\gamma}) = \mathcal{M}(M, \lambda, \vec{R}; \underline{\gamma}, \overline{\gamma}; J) = \Pi_2^{-1}(J) \cap \mathcal{M}(M, \lambda, \vec{R}; \underline{\gamma}, \overline{\gamma}). \tag{8.3.3}$$

We state the following standard statement that often occurs in this kind of generic transversality statement via the Sard-Smale theorem.

Theorem 8.3.2 *Let* $0 < \ell < k - \frac{2}{p}$. *Consider the moduli space* $\mathcal{M}(M, \lambda; \underline{\gamma}, \overline{\gamma})$. *Then*

(1) $\mathcal{M}(M, \lambda, \vec{R}; \underline{\gamma}, \overline{\gamma})$ *is an infinite dimensional* C^ℓ *Banach manifold.*
(2) *The projection*

$$\Pi_2|_{(\Upsilon^{\text{univ}})^{-1}(0)} : (\Upsilon^{\text{univ}})^{-1}(0) \to \mathcal{J}^\ell_\lambda(M, \xi)\mathcal{J}^\ell(M, \lambda)$$

is a Fredholm map and its index is the same as that of $D\Upsilon(w)$ *for a (and so any)* $w \in \mathcal{M}(M, \lambda, \vec{R}; J; \underline{\gamma}, \overline{\gamma})$.

An immediate corollary of Sard-Smale theorem is that for a generic choice of J

$$\Pi_2^{-1}(J) \cap (\Upsilon^{\text{univ}})^{-1}(0) = \mathcal{M}(J; \underline{\gamma}, \overline{\gamma})$$

is a smooth manifold: One essential ingredient for the generic transversality under the perturbation of $J \in \mathcal{J}_\lambda(M, \xi)$ is the usage of the following unique continuation result.

Proposition 8.3.3 (Unique Continuation Lemma; Proposition 12.3 [23]) *Any non-constant contact Cauchy-Riemann map does not have an accumulation point in the zero set of* dw.

Remark 8.3.4 The proof given in [23] utilizes the unique continuation result through the (local) symplectization of contact instantons which become a (local) pseudoholomorphic curves in the symplectization. We refer readers to the proof of Lemma 8.4.5 for the unique continuation result in the context of linearized problem whose proof is given purely in terms of the analysis of contact instantons without taking the symplectization. A similar proof can be also given to the nonlinear problem of contact instantons by adapting the proof of Lemma 8.4.5.

8.4 Generic Transversality Under the Perturbation of Boundaries

In this section, we study the problem of generic transversality under the boundary Legendrian submanifolds imitating the arguments used in [16] in the framework of perturbations of Lagrangian submanifolds in symplectic geometry.

We put the following generic configuration of the Legendrian link $\vec{R} = (R_1, \cdots, R_n)$ for the study of generic transversality problems for the moduli space of bordered contact instantons.

Definition 8.4.1 (General Position) We say that a Legendrian link $\vec{R}$ is in general position if the following hold:

(1) Each pair (R_i, R_j) is nondegenerate in the sense of Theorem 8.2.8.
(2) There is no triple (R_i, R_j, R_k) for which no triple of Reeb chords that simultaneously overlaps on an (relatively) open subsets of the images thereof.

This definition of general position being mentioned, we can perturb the link $\vec{R} = \{R_1, \cdots, R_n\}$ componentwise for the transversality studies under the perturbation of boundary conditions, and may restrict ourselves to the case where $\vec{R}$ is a one-component link.

Similarly as in [16, p. 511], we fix a Legendrian submanifold R_0 and represent each Legendrian submanifold C^∞-close to the given R_0 as the one-jet graph

$$\text{Image } j^1 f := \{(x, df(x), f(x)) \in J^1 R_0 \mid x \in R_0\}$$

of a smooth function $f : R_0 \to \mathbb{R}$, contained in a neighborhood of the zero section $U_{R_0} \subset J^1 R_0$ via the Darboux-Weinstein chart

$$\Phi_{R_0} : V_{R_0} \subset M \to U_{R_0} \subset J^1 R_0$$

where $V_{R_0} \subset M$ is a neighborhood of R_0. Then we can canonically represent each Legendrian submanifold R C^∞-close to R_0 as $R = \phi_R(R_0)$ where

$$\phi_R := \Phi_{R_0}^{-1} \circ j^1 f : R_0 \to M$$

where f is the unique function on R_0 with Image $j^1 f = \Phi_{R_0}(R)$. This map ϕ_R can be extended to an ambient contact isotopy $\psi_t : M \to M$ so that $\phi_R = \psi_1|_{R_0}$ so that $\psi_1 \in \text{Cont}_0(M, \xi)$ in particular.

We denote by $\mathcal{N}(R_0) = \mathcal{N}(R_0; \Phi_{R_0})$ the set of Legendrian submanifolds given by

$$\begin{aligned}\mathcal{N}(R_0) := \{R \in \mathfrak{Leg}(M,\xi) \mid R = \Phi_{R_0}^{-1}(\text{Image}\, j^1 f),\\ f : R_0 \to \mathbb{R},\ \text{Image}\, j^1 f \subset U_{R_0}\}.\end{aligned} \tag{8.4.1}$$

Then we consider the subset

$$\{(w, R) \mid R \subset U_{R_0},\ w(\partial\dot{\Sigma}) \subset R\} \subset \mathcal{F} \times \mathcal{N}(R_0)$$

and

$$\mathcal{M} = (\mathcal{F} \times \mathcal{N}(R_0)) \cap \Upsilon^{-1}(0). \tag{8.4.2}$$

Following [16], we introduce the following notion.

Definition 8.4.2 (Boundary Somewhere Injectivity) We say a map $w : \dot{\Sigma} \to M$ is boundary somewhere injective if there exists a (open) subset $A \subset \partial\dot{\Sigma}$ such that

$$w^{-1}(w(z)) \cap \partial\dot{\Sigma} = \{z\} \quad \text{for } z \in A.$$

We consider the set of pairs

$$(w, R) \in \mathcal{F}((\dot{\Sigma}, \partial\dot{\Sigma}), (M, R)) \times \mathfrak{Leg}(M, \xi)$$

and the universal section map $\Upsilon = (\Upsilon_1, \Upsilon_2)$ of the bundle $\mathcal{CD} \to \mathcal{F}$ given by

$$\Upsilon_1(w, R) = \overline{\partial}^{\pi} w, \quad \Upsilon_2(w, R) = d(w^*\lambda \circ j).$$

Under the above general position hypothesis, we prove the following

Theorem 8.4.3 *Let (M, ξ) be a contact manifold equipped with a a contact form λ. We consider the same equation considered in Theorem 8.3.2. Fix J and k and consider $\vec{R}$ in general position given above. Then the subset*

$$\mathcal{M} \subset \mathcal{F} \times \mathcal{N}(R_0)$$

is a smooth C^∞ submanifold (as a Frechet submanifold) near w satisfying the boundary somewhere injectivity.

Proof By the implicit function theorem, it is enough to prove the universal section map Υ^{univ} is a submersion at $((w, J), \vec{R})$ for any boundary somewhere injective w.

As usual in this kind of analysis, we consider the fiber bundle

$$\mathcal{W}^{k,p}(M, (\cdot); \underline{\gamma}, \overline{\gamma}) \to \mathcal{L}eg(M, \xi)$$

whose fiber at $(w, J, \vec{R})$ is given by

$$\mathcal{W}^{k,p}(M, \vec{R}; \underline{\gamma}, \overline{\gamma}).$$

For the simplicity of exposition, we fix all R_i's except R_0. We consider a perturbation of R_0 under the contact isotopy of the type

$$\psi(R_0), \quad \psi = \psi_H^1 \in \mathrm{Cont}_0(M, \xi)$$

described as above.

Now we can modify the argument used in the proof of the main theorem in [16] as follows. We denote by $\mathcal{L}(M)$ and $\mathcal{L}(R_0)$ the spaces of smooth Moore paths (γ, T) on M and R_0 respectively. (In [16], free loops are considered the set of which are denoted by $\Omega(M)$ and $\Omega(L_0)$ respectively.)

We consider the parameterized smooth map

$$\Upsilon^{\mathcal{L}eg} : \mathcal{F} \times \mathcal{N}(R_0) \to \mathcal{CD} \times \mathcal{L}(M)$$

by

$$\Upsilon^{\mathcal{L}eg}(w, R) := \left(\Upsilon(w), \phi_R^{-1} \circ w|_{\partial \dot\Sigma}\right).$$

Then by definition we have

$$\mathcal{M} = (\Upsilon^{\mathcal{L}eg})^{-1}(\{0\} \times \mathcal{L}(R_0)).$$

At this point, we can duplicate the rest of the proof of [16, pp. 514–516] with obvious modification which is now in order.

To prove surjectivity, we will prove the L^2-cokernel vanishes, i.e.,

$$\left(\mathrm{Image}\, D\Upsilon^{\mathcal{L}eg}(w, R) + \{0\} \oplus T_{\phi_R^{-1} \circ w|_{\partial\dot\Sigma}} \mathcal{L}(R_0)\right)^{\perp} = 0$$

as a subset of

$$\Omega^{(1,0)}(w^*\xi) \oplus C^\infty(\dot\Sigma) \bigoplus T_{\phi_{R_0}^{-1} \circ w} \mathcal{L}(M)$$

after a suitable Sobolev completion: For the space $\Omega^{(1,0)}(w^*\xi)$, we take the dual weighted Sobolev space $W_{-k,q;-\delta}$ for $W_{k,p;\delta}$ with $1/p + 1/q = 1$. (See [12], [16, p. 515] for the detailed explanation which however will not play much role in the calculation henceforth.)

Denote by NR_0 the normal bundle of R_0 (with respect to any given adapted metric g). If

$$((\eta, g), \alpha) \in \left(\text{Image}\, D\Upsilon^{\mathcal{L}eg}(w, R) + \{0\} \oplus T_{\phi_R^{-1} \circ w|_{\partial\dot\Sigma}} \mathcal{L}(R_0)\right)^{\perp},$$

it in particular implies that α satisfies

$$\alpha(z) \in N_{\phi_R \circ w(z)} R_0, \quad z \in \partial\dot\Sigma. \tag{8.4.3}$$

(See [16, p.516] for the relevant calculations.)

Furthermore we denote by $\eta^{(1,0)}$ the value of the $(1,0)$-form η evaluated against

$$\frac{\partial}{\partial z} \cong \frac{1}{2}\left(\frac{\partial}{\partial x} - j\frac{\partial}{\partial y}\right)$$

$$\eta^{(1,0)} := \eta_x + J\eta_y \tag{8.4.4}$$

where $\eta = \eta_x dx + \eta_y dy$ in an isothermal coordinate (x, y) of $\dot\Sigma$ adapted to $\partial\dot\Sigma$.

By somewhat more complicated calculations involving the integration by parts similarly as in [16, p. 516], we derive the following L^2-adjoint equation of associated linearized equation. (See the calculations given later in Sect. 8.8).) We postpone its proof till the next section, Sect. 8.5.

Proposition 8.4.4 *Consider the pair*

$$((\eta, g), \alpha) \in \mathcal{CD} \times T_{\phi_R^{-1} \circ w|_{\partial\dot\Sigma}} \mathcal{L}(M)$$

and recall the correspondence

$$(\eta, g) \longleftrightarrow \eta + g\, R_\lambda$$

given by the identification (8.2.30).

Assume that w is boundary somewhere injective and let $A \subset \partial\dot\Sigma$ be a nonempty (open) subset such that

$$w^{-1}(w(z)) \cap \partial\dot\Sigma = \{z\} \quad \text{for } z \in A.$$

Suppose that the pair $((\eta, g), \alpha)$ lies in the L^2-cokernel of $D\Upsilon^{\mathcal{L}eg}(w, R)$. Then the following hold and vice versa:

(1) *(Y, α) satisfies*

$$\begin{cases} (\delta^{\nabla^{\pi(0,1)}} + T^{\pi(1,0)} + B^{(1,0)})(\eta^{(1,0)}) - J\langle dg, dw\rangle = 0 & \text{on } \dot\Sigma \\ (J\eta^{(1,0)} + g dw)\left(\frac{\partial}{\partial \nu}\right) - \alpha = 0 & \text{on } \partial\dot\Sigma \end{cases} \tag{8.4.5}$$

and

$$\alpha^\perp = 0 \quad on\ A.$$

(2) *The function g satisfies*

$$\begin{cases} \Delta g - \frac{1}{2} g \left\langle (\mathcal{L}_{R_\lambda} J)(J d^\pi w), \eta \right\rangle = 0, & on\ \dot\Sigma \\ g = 0 = \frac{\partial g}{\partial \nu} & on\ \partial \dot\Sigma. \end{cases} \tag{8.4.6}$$

The Eqs. (8.4.5) and (8.4.6) are nothing but the analog of the one right above [16, Equation (3.13)]. In terms of the isothermal coordinates (x, y) adapted to $\partial \dot\Sigma$ and in the usual convention for the calculus of vector valued differential forms (see [35] or [26, Appendix B]), we can express

$$\langle dg, dw \rangle := dg \wedge *dw = \left(\frac{\partial g}{\partial x}\frac{\partial w}{\partial x} + \frac{\partial g}{\partial y}\frac{\partial w}{\partial y} \right)$$

$$J\eta^{(1,0)}\left(\frac{\partial}{\partial \nu}\right) = \eta_x - J\eta_y \tag{8.4.7}$$

for the one-form $\eta = \eta_x\, dx + \eta_y\, dy$.

Towards our goal of proving submersion property of the universal section map Υ^{univ}, i.e., surjectivity of the linearization map $\Upsilon^{\text{univ}}_{(w,J),\vec R}$ for boundary somewhere injective w, we first derive the following from the Eq. (8.4.6).

Lemma 8.4.5 $g = 0$.

Proof We note that g satisfies the following 2nd order linear elliptic equation of the Laplcian type

$$\Delta g + Ag = 0, \quad A = -\frac{1}{2} \left\langle (\mathcal{L}_{R_\lambda} J)(J d^\pi w), \eta \right\rangle \tag{8.4.8}$$

where the coefficient function A is smooth. We apply the unique continuation as follows.

Let $z_0 \in \partial \dot\Sigma$ be a point and $U \subset \dot\Sigma$ be its neighborhood. Using the isothermal coordinate (x, y) centered at z_0 we may identify U with a semi-disc $D_\delta = \{(x, y) \mid |x| < \delta,\ y \geq 0\}$. Then the boundary condition becomes

$$g(x, 0) = 0 = \frac{\partial g}{\partial y}(x, 0) \tag{8.4.9}$$

for all $(x, 0) \in \{y = 0\} \cap D_\delta$. Then by Aronszajin's unique continuation, more precisely [2, Remark 2] applied to (8.4.8) *with the Cauchy data* (8.4.9), we conclude $g \equiv 0$. (Note that the Cauchy data together with the Eq. (8.4.8) implies vanishing of infinity jets of g along the boundary $U \cap \partial \dot\Sigma$, in particular at z_0. Since z_0 lies in the boundary of the domain D_δ, we apply [2, Remark 2]), not the commonly used version that usually applies to an interior point.) □

Then (8.4.5) is reduced to

$$\begin{cases} \left(\delta^{\nabla^\pi(0,1)} + T^{\pi(1,0)} + B^{(1,0)}\right)(\eta^{(1,0)}) = 0 & \text{on } \dot\Sigma \\ -J\eta^{(1,0)} - \alpha = 0 & \text{on } \partial\dot\Sigma, \end{cases} \tag{8.4.10}$$

and

$$\alpha^\perp = 0 \quad \text{on } A.$$

By combining (8.4.3) and the last two boundary conditions, we derive

$$\eta^{(1,0)} = 0 \quad \text{on } A \subset \partial\dot\Sigma.$$

Then we derive $\eta = 0$ by the unique continuation from the first equation of (8.4.5)

$$-\partial^{\nabla^\pi}\eta + T^{\pi,(1,0)}_{dw}\eta + B^{(1,0)}\eta = 0$$

which is of Cauchy-Riemann type. Then by back substitution of $\eta = 0$, we derive $\alpha = 0$ again from $-J\eta^{(1,0)} - \alpha = 0$ on $\partial\dot\Sigma$. (See the proof of Proposition 3.4 in [16, p.516] for a similar argument used.)

Finally we examine (8.4.6). Recall $\dot\Sigma$ has non-empty boundary punctures each of which has strip-like coordinates $(\tau, t) \in \pm[0, \infty) \times [0, 1]$. By the asymptotic convergence result of finite energy contact instantons w and the given hypothesis, there exists a sufficiently large $R > 0$ such that $dw|_{\partial\dot\Sigma} \neq 0$ on $\dot\Sigma \cap \{(\tau, t) \mid t = 0, 1, |\tau| \geq R\}$ for a sufficiently large $R > 0$. It is immediate to check the vanishing of g by the finite energy condition that such a harmonic function satisfying $g = 0 = \frac{\partial g}{\partial \nu}$ must vanish which can be seen by expanding the finite energy harmonic function $g = g(\tau, t)$ by a trigonometric series on the strip-like region.

Combining all the above, we have finished the proof of Theorem 8.4.3. □

An immediate corollary of Sard-Smale theorem then is the following generic transversality result.

Theorem 8.4.6 *There exists a residual subset of $\vec{R} = (R_1, \ldots, R_k)$ of Legendrian submanifolds such that the moduli space $\mathcal{M}(M, \lambda, \vec{R}; \underline{\gamma}, \overline{\gamma}; J)$ is transversal so that it becomes a finite dimensional smooth manifold of given Fredholm index.*

It remains to prove that the L^2-cokernel element $((\eta, g), \alpha)$ is characterized by the Eqs. (8.4.5) and (8.4.6) which is now in order.

8.5 Derivation of Parametric L^2-Adjoint Equation

Our primary goal is to prove Proposition 8.4.4, i.e., to show that whenever $((\eta, g), \alpha)$ lies in the L^2-cokernel of $D\Upsilon^{\mathcal{L}eg}(w, \vec{R})$ satisfies

$$\langle D_w \Upsilon^{\mathcal{L}eg}(w, \vec{R})(Y, X_f), ((\eta, g), \alpha)\rangle = 0$$

for all Y, then it satisfies (8.4.5) and (8.4.6) in Proposition 8.4.4.

The entirety of the section will be occupied by the proof of this claim. We first recall the correspondence (8.2.30). Let $((\eta, g), \alpha)$ satisfy

$$\left\langle \left(D\Upsilon^{\mathcal{D}eg}(w, \vec{R})\right)^\dagger ((\eta, g), \alpha), (Y, X_f)\right\rangle = 0$$

for all $Y \in T_w\mathcal{F}$ and $f \in C^\infty(R_0)$. More explicitly, the equation is

$$\int_{\dot\Sigma} \langle D_w \Upsilon^{\mathcal{L}eg}(w, \vec{R})(Y, X_f), (\eta, g)\rangle + \int_{\partial\dot\Sigma} \langle \alpha, Y|_{\partial\dot\Sigma} - X_f(w)|_{\partial\dot\Sigma}\rangle = 0. \tag{8.5.1}$$

By definition, we have

$$D_w \Upsilon^{\mathcal{L}eg}(w, \vec{R})(Y, X_f) = D_\Upsilon(w)(Y)$$

where $D\Upsilon(w)$ is the linearization map (8.2.7) of the section Υ defined by (8.2.6) for the given Legendrian boundary condition $\vec{R}$.

Utilizing this, we decompose the first integral of (8.5.1) into

$$\int_{\dot\Sigma} \langle D_w \Upsilon^{\mathcal{L}eg}(w, \vec{R})(Y, X_f), (\eta, g)\rangle = \int_{\dot\Sigma} \langle D\Upsilon_1(w)(Y), \eta\rangle + \int_{\dot\Sigma} \langle D\Upsilon_2(w)(Y), g\, R_\lambda\rangle. \tag{8.5.2}$$

By the elliptic regularity, (η, g) is smooth and so we can do the integration by parts.

For the calculation of the first integral, we recall

$$D\Upsilon_1((w)(Y) = \left(\overline{\partial}^{\nabla^\pi} + B^{(0,1)} + T^{\pi,(0,1)}_{dw}\right)(Y^\pi) + \frac{1}{2}\lambda(Y)\left\langle (\mathcal{L}_{R_\lambda} J)(J\partial^\pi w), \eta\right\rangle.$$

Here we recall the property of the linear map $\mathcal{L}_{R_\lambda} J(\xi) \subset (\xi)$.

We take an isothermal coordinate (x, y) near the boundary so that $h = dx^2 + dy^2$, ∂_x is tangent to $\partial\dot\Sigma$ and ∂_y is inward normal thereto. In particular, for the area element dA on $\dot\Sigma$ and the arc-length element ds of $\dot\Sigma$ which are globally defined on $\dot\Sigma$, we have

$$dA = dx \wedge dy, \quad ds = dx \tag{8.5.3}$$

for any isothermal coordinate (x, y) along the boundary $\partial \dot{\Sigma}$ chosen as above.

Then we write $\eta = \eta_x\, dx + \eta_y\, dy$ and

$$\begin{aligned}\nabla^\pi Y^\pi &= \nabla^\pi_{\partial_x} Y^\pi\, dx + \nabla^\pi_{\partial_y} Y^\pi\, dy\\ (\nabla^\pi Y^\pi) \circ j &= -\nabla^\pi_{\partial_x} Y^\pi\, dy + \nabla^\pi_{\partial_y} Y^\pi\, dx.\end{aligned}$$

Therefore we obtain

$$\begin{aligned}\langle \overline{\partial}^{\nabla^\pi} Y^\pi, \eta\rangle\, dA &= \langle \overline{\partial}^{\nabla^\pi} Y^\pi, \eta\rangle dx \wedge dy\\ &= d\left(-\langle Y^\pi, \eta_y - J\eta_x\rangle\, dx + \langle Y^\pi, \eta_x + J\eta_y\rangle dy\right)\\ &\quad +\langle Y^\pi, (-\nabla^\pi_{\partial_x} + J\nabla^\pi_{\partial_y})(\eta_x + J\eta_y)\rangle\, dx \wedge dy.\end{aligned}$$

We write

$$(-\nabla^\pi_{\partial_x} + J\nabla^\pi_{\partial_y})(\eta_x + J\eta_y) =: \delta^{\nabla^{\pi(0,1)}}(\eta^{(1,0)}).$$

Substituting these into the first integral of (8.5.2), we obtain

$$\begin{aligned}\int_{\dot{\Sigma}} \langle D\Upsilon_1(w)(Y), (\eta, g)\rangle &= \int_{\dot{\Sigma}} \langle Y^\pi, (\delta^{\nabla^{\pi(0,1)}} + T^{\pi(1,0)} + B^{(1,0)})(\eta^{(1,0)})\rangle\\ &\quad + \int_{\partial\dot{\Sigma}} -\langle Y^\pi, \eta_y - J\eta_x\rangle\, dx\\ &\quad + \int_{\dot{\Sigma}} \frac{1}{2} g\, \lambda(Y) \left\langle (\mathcal{L}_{R_\lambda} J)(J\partial^\pi w), \eta\right\rangle.\end{aligned}$$

Rearranging the terms around, we get

$$\begin{aligned}&\int_{\dot{\Sigma}} \langle D\Upsilon_1((w, J), \vec{R})(Y), (\eta, g)\rangle\\ &= \int_{\dot{\Sigma}} \langle Y^\pi, (\delta^{\nabla^{\pi(0,1)}} + T^{\pi(1,0)} + B^{(1,0)})(\eta^{(1,0)})\rangle\\ &\quad + \int_{\dot{\Sigma}} \frac{1}{2} g\, \lambda(Y) \left\langle (\mathcal{L}_{R_\lambda} J)(J\partial^\pi w), \eta\right\rangle\, dA\\ &\quad + \int_{\partial\dot{\Sigma}} -\langle Y^\pi, \eta_y - J\eta_x\rangle\, dx\end{aligned} \tag{8.5.4}$$

Now we compute the second integral of (8.5.2)

$$\int_{\dot{\Sigma}} \langle D\Upsilon_2((w, J), \vec{R})(Y)\, g\rangle = \int_{\dot{\Sigma}} g(-\Delta(\lambda(Y)))\, dA + g d((Y^\pi \rfloor d\lambda) \circ j).$$

For the first, we apply Green's formula and the Legendrian boundary condition which implies $\lambda(Y) = 0$ on $\partial\dot{\Sigma}$, and get

$$\int_{\dot{\Sigma}} g(-\Delta(\lambda(Y)))\, dA = \int_{\dot{\Sigma}} -\Delta g\, (\lambda(Y)))\, dA + \int_{\partial\dot{\Sigma}} -g\frac{\partial\lambda(Y)}{\partial\nu} + \frac{\partial g}{\partial\nu}\lambda(Y). \tag{8.5.5}$$

For the second, we get

$$\begin{aligned}\int_{\dot{\Sigma}} gd((Y^\pi \rfloor d\lambda)\circ j) &= -\int_{\dot{\Sigma}} dg\wedge (Y^\pi\rfloor d\lambda)\circ j) + \int_{\partial\dot{\Sigma}} g(Y^\pi\rfloor d\lambda)\circ j)\\ &= \int_{\dot{\Sigma}} dg\circ j\wedge (Y^\pi\rfloor d\lambda) + \int_{\partial\dot{\Sigma}} gd\lambda\left(Y^\pi, \frac{\partial w}{\partial\nu}\right).\end{aligned}$$

By summing the above two, we get

$$\begin{aligned}&\int_{\dot{\Sigma}} g(-\Delta(\lambda(Y)))\, dA + gd((Y^\pi\rfloor d\lambda)\circ j)\\ &= \int_{\dot{\Sigma}} -\Delta g\, (\lambda(Y)))\, dA + \int_{\partial\dot{\Sigma}} -g\frac{\partial\lambda(Y)}{\partial\nu} + \frac{\partial g}{\partial\nu}\lambda(Y)\\ &\quad + \int_{\dot{\Sigma}} dg\circ j\wedge (Y^\pi\rfloor d\lambda) + \int_{\partial\dot{\Sigma}} gd\lambda\left(Y^\pi, \frac{\partial w}{\partial\nu}\right)\\ &= \int_{\dot{\Sigma}} -\Delta g\, (\lambda(Y)))\, dA + \int_{\dot{\Sigma}} d\lambda\left(Y^\pi, \frac{\partial g}{\partial x}\frac{\partial w}{\partial x} + \frac{\partial g}{\partial y}\frac{\partial w}{\partial y}\right)\\ &\quad + \int_{\partial\dot{\Sigma}} -g\frac{\partial\lambda(Y)}{\partial\nu} + \frac{\partial g}{\partial\nu}\lambda(Y) + \int_{\partial\dot{\Sigma}} gd\lambda\left(Y^\pi, \frac{\partial w}{\partial\nu}\right)\end{aligned} \tag{8.5.6}$$

By adding (8.5.4) and (8.5.6), we obtain

$$\begin{aligned}&\int_{\dot{\Sigma}} \langle D\Upsilon((w,J),\vec{R})(Y),(\eta,g)\rangle\\ &= \int_{\dot{\Sigma}} \langle Y^\pi, (\delta^{\nabla^{\pi(0,1)}} + T^{\pi(1,0)} + B^{(1,0)})(\eta^{(1,0)})\rangle\\ &\quad + \int_{\dot{\Sigma}} -\Delta g\, (\lambda(Y)))\, dA + \int_{\partial\dot{\Sigma}} -g\frac{\partial\lambda(Y)}{\partial\nu} + \frac{\partial g}{\partial\nu}\lambda(Y)\\ &\quad + \int_{\dot{\Sigma}} \frac{1}{2} g\,\lambda(Y)\left\langle(\mathcal{L}_{R_\lambda}J)(J\partial^\pi w),\eta\right\rangle dA\\ &\quad + \int_{\partial\dot{\Sigma}} -\langle Y^\pi, \eta_y - J\eta_x\rangle\, dx\\ &\quad + \int_{\dot{\Sigma}} d\lambda\left(Y^\pi, \frac{\partial g}{\partial x}\frac{\partial w}{\partial x} + \frac{\partial g}{\partial y}\frac{\partial w}{\partial y}\right) + \int_{\partial\dot{\Sigma}} gd\lambda\left(Y^\pi, \frac{\partial w}{\partial\nu}\right) ds.\end{aligned}$$

By rearranging terms, we get

$$\begin{aligned}
&\int_{\dot\Sigma} \langle D\Upsilon((w,J),\vec{R})(Y),(\eta,g)\rangle \\
&= \int_{\dot\Sigma} \langle Y^\pi, (\delta^{\nabla^{\pi(0,1)}} + T^{\pi(1,0)} + B^{(1,0)})(\eta^{(1,0)})\rangle \\
&\quad + \int_{\dot\Sigma} d\lambda\left(Y^\pi, \frac{\partial g}{\partial x}\frac{\partial w}{\partial x} + \frac{\partial g}{\partial y}\frac{\partial w}{\partial y}\right) dx\wedge dy \\
&\quad + \int_{\dot\Sigma} -\Delta g\, \lambda(Y)\, dA + \int_{\dot\Sigma} \frac{1}{2} g\, \lambda(Y) \left\langle (\mathcal{L}_{R_\lambda} J)(J\partial^\pi w), \eta\right\rangle dA \\
&\quad + \int_{\partial\dot\Sigma} \left(-g\frac{\partial\lambda(Y)}{\partial\nu} + \frac{\partial g}{\partial \nu}\lambda(Y)\right) ds \\
&\quad + \int_{\partial\dot\Sigma} -\langle Y^\pi, \eta_y - J\eta_x\rangle\, dx + \int_{\partial\dot\Sigma} g\, d\lambda\left(Y^\pi, \frac{\partial w}{\partial \nu}\right) ds. \qquad (8.5.7)
\end{aligned}$$

Substituting this into (8.5.1) followed by some rearrangement, we obtain

$$\begin{aligned}
0 &= \int_{\dot\Sigma} \left\langle Y^\pi, (\delta^{\nabla^{\pi(0,1)}} + T^{\pi(1,0)} + B^{(1,0)})(\eta^{(1,0)}) - J\left(\frac{\partial g}{\partial x}\frac{\partial w}{\partial x} + \frac{\partial g}{\partial y}\frac{\partial w}{\partial y}\right)\right\rangle dA \\
&\quad + \int_{\dot\Sigma} \left(-\Delta g + \frac{1}{2} g \left\langle (\mathcal{L}_{R_\lambda} J)(J\partial^\pi w), \eta\right\rangle\right) \lambda(Y)\, dA \\
&\quad + \int_{\partial\dot\Sigma} \left(-g\frac{\partial\lambda(Y)}{\partial\nu} + \frac{\partial g}{\partial \nu}\lambda(Y)\right) ds \\
&\quad + \int_{\partial\dot\Sigma} -\left\langle Y^\pi, \eta_y - J\eta_x + gJ\frac{\partial w}{\partial \nu} - \alpha\right\rangle ds \\
&\quad - \int_{\partial\dot\Sigma} \langle \alpha, X_f(w|_{\partial\dot\Sigma})\rangle\, ds
\end{aligned}$$

for all (Y, f). From this and the fact that the choices of η and of $\lambda(Y)$ are completely independent of each other, we have derived the following equation

$$\begin{cases} (\delta^{\nabla^{\pi(0,1)}} + T^{\pi(1,0)} + B^{(1,0)})(\eta^{(1,0)}) - J\left(\frac{\partial g}{\partial x}\frac{\partial w}{\partial x} + \frac{\partial g}{\partial y}\frac{\partial w}{\partial y}\right) = 0 & \text{on } \dot\Sigma \\ \eta_y - J\eta_x - gJ\frac{\partial w}{\partial\nu} - \alpha = 0 & \text{on } \partial\dot\Sigma \end{cases} \qquad (8.5.8)$$

and

$$\begin{cases} \Delta g - \frac{1}{2} g\left\langle (\mathcal{L}_{R_\lambda} J)(J\partial^\pi w), \eta\right\rangle = 0 & \text{on } \dot\Sigma \\ g = 0 = \frac{\partial g}{\partial \nu} & \text{on } \partial\dot\Sigma \end{cases} \qquad (8.5.9)$$

and

$$\alpha^{\perp} = 0 \quad \text{on } A.$$

Here the vanishing $g = 0 = \frac{\partial g}{\partial \nu}$ follows from the integral

$$\int_{\partial \dot{\Sigma}} \left(-g \frac{\partial \lambda(Y)}{\partial \nu} + \frac{\partial g}{\partial \nu} \lambda(Y) \right) ds$$

since we can freely choose the function $f := \lambda(Y)$ so that the value of $\frac{\partial f}{\partial \nu}$ can be made arbitrary with the value of f fixed and vice versa.

Recalling the coordinate expression given in (8.4.7), we have finished the proof of Proposition 8.4.4.

Part 2: Generic Evaluation Transversality

8.6 The Evaluation Transversality: Statement

In this section, we start with the discussion on another important general ingredient of the applications of contact instantons to contact topology, the evaluation map transversality, similarly as in the case of pseudoholomorphic curves.

We first recall the off-shell setting of the study of linearized operator in Theorem 8.2.19: Let (M, ξ) be a contact manifold and consider contact triads (M, λ, J) and let $\vec{R} = (R_1, R_2, \ldots, R_k)$ be a Legendrian link. We consider the associated contact instanton equation

$$\begin{cases} \overline{\partial}^{\pi} w = 0, & d(w^* \lambda \circ j) = 0 \\ w(\overline{z_i z_{i+1}}) \subset R_i, & i = 1, \ldots, k \end{cases} \tag{8.6.1}$$

for a map $w : (\dot{\Sigma}, \partial \dot{\Sigma}) \to (M, \vec{R})$ with the boundary condition given as above.

We consider the moduli space

$$\mathcal{M}(M, \lambda, \vec{R}; J) := \mathcal{M}((\dot{\Sigma}, \partial \dot{\Sigma}), (M, \vec{R}); J), \quad \vec{R} = (R_1, \cdots, R_k)$$

of finite energy maps $w : \dot{\Sigma} \to M$ satisfying the Eq. (8.6.1).

We also consider the space given in (8.2.3)

$$\mathcal{F}(M, \lambda, \vec{R}; \underline{\gamma}, \overline{\gamma})$$

consisting of smooth maps satisfying the boundary condition (8.2.4) and the asymptotic condition (8.2.5). Since we will not vary the pair $\underline{\gamma}, \overline{\gamma}$, we will often just write

$$\mathcal{F} = \mathcal{F}(M, \lambda, \vec{R}) = \mathcal{F}(M, \lambda, \vec{R}; \underline{\gamma}, \overline{\gamma})$$

and denote its marked version with ℓ interior and k boundary marked points by

$$\mathcal{F}_{(\ell,k)}(M, \lambda, \vec{R}). \tag{8.6.2}$$

We again consider the covariant linearized operator

$$D\Upsilon(w) : \Omega^0(w^*TM, (\partial w)^*T\vec{R}) \to \Omega^{(0,1)}(w^*\xi) \oplus \Omega^2(\Sigma)$$

of the section

$$\Upsilon : w \mapsto \left(\overline{\partial}^\pi w, d(w^*\lambda \circ j)\right), \quad \Upsilon := (\Upsilon_1, \Upsilon_2)$$

as before.

We will treat the two cases, evaluation at an interior marked point and one at a boundary marked point, separately. We denote by the subindex (ℓ, k) the number of interior and boundary marked points respectively.

Consider the parameterized marked moduli space

$$\begin{aligned} &\mathcal{M}_{(1,0)}(M, \lambda, \vec{R}; \mathcal{J}_\lambda) \\ &= \{((j, w), J, z) \mid w : \Sigma \to M,\ \Upsilon(J, (j, w)) = 0,\ w(\partial\dot{\Sigma}) \subset \vec{R},\ z \in \operatorname{Int}\dot{\Sigma}\}. \end{aligned}$$

The evaluation map $\mathrm{ev}^+ : \mathcal{M}_{(1,0)}(M, \lambda, \vec{R}; J) \to M$ is defined by

$$\mathrm{ev}^+((j, w), z) = w(z).$$

We then have the fibration

$$\widetilde{\mathcal{M}}_{(1,0)}(M, \lambda, \vec{R}; \mathcal{J}_\lambda) = \bigcup_{J \in \mathcal{J}_\lambda} \widetilde{\mathcal{M}}_{(1,0)}((\dot{\Sigma}, \partial\dot{\Sigma}), (M, \vec{R}); J) \to \mathcal{J}_\lambda$$

and

$$\widetilde{\mathcal{M}}^{\mathrm{inj}}_{(1,0)}(M, \lambda, \vec{R}; \mathcal{J}_\lambda)$$

to be the open subset of $\widetilde{\mathcal{M}}_{(1,0)}(M, \lambda, \vec{R}; \mathcal{J}_\lambda)$ consisting of somewhere injective contact instanton pairs $((j, w), J)$. We have the universal (0-jet) evaluation map

$$\mathrm{Ev}^+ : \widetilde{\mathcal{M}}_{(1,0)}(M, \lambda, \vec{R}; \mathcal{J}_\lambda) \to M.$$

The basic generic transversality is the following.

Theorem 8.6.1 (0-Jet Evaluation Transversality) *The evaluation map*

$$\mathrm{Ev}^+ : \widetilde{\mathcal{M}}_{(1,0)}(M, \lambda, \vec{R}; \mathcal{J}_\lambda) \to M$$

is a submersion. The same holds for the boundary evaluation map

$$\mathrm{Ev}_\partial : \widetilde{\mathcal{M}}_{(0,1)}(M, \lambda, \vec{R}; \mathcal{J}_\lambda) \to \vec{R}.$$

8.7 The Interior Evaluation Transversality: Proof

We closely follow the scheme exercised for the proof of evaluation transversality given in [18, Section 10.5] which in turn follows the scheme of the generic 1-jet transversality results proved in [17, 29] for the case of pseudoholomorphic curves in symplectic geometry.

An important ingredient in their proofs is the following structure theorem of the distributions with point support from [10, Section 4.5], [31, Theorem 6.25] whose proof we refer readers thereto.

Theorem 8.7.1 (Distribution with Point Support) *Suppose ψ is a distribution on open subset $\Omega \subset \mathbb{R}^n$ with* $\operatorname{supp}\psi \subset \{p\}$ *and of finite order $N < \infty$. Then ψ has the form*

$$\psi = \sum_{|\alpha| \leq N} D^\alpha \delta_p$$

where δ_p is the Dirac-delta function at p and $\alpha = (\alpha_1, \ldots, \alpha_n)$ is the multi-indices.

We start with the case of interior marked point and consider the map

$$\begin{aligned} \aleph_0 : \mathcal{J}_\lambda \times \mathcal{M}_{\dot\Sigma} \times \widetilde{\mathcal{F}}_{(1,0)}(\dot\Sigma, M) &\to \mathcal{CD} \times M \\ (J, (j, w), z_0) &\mapsto (\Upsilon^{\mathrm{univ}}(J, (j, w)), w(z_0)). \end{aligned} \tag{8.7.1}$$

Here the subindex 0 in $\aleph_0$ stands for the '0-jet', and the map Υ^{univ} is the map given in (8.3.1). (The higher-jet transversality can be also proved by adapting the proof of [17] and the present proof. We postpone its proof and an application elsewhere.) Since the boundary condition is irrelevant for the discussion of the present section, we omit $\vec{R}$ from the notation.

We denote by π_i the projection from $\mathcal{J}_\lambda \times \widetilde{\mathcal{F}}_{(1,0)}(\dot\Sigma, M)$ to the i-th factor with $i = 1, 2$. Then we introduce

$$\begin{aligned} \widetilde{\mathcal{M}}_{(1,0)}(\dot\Sigma, M; \{p\}; \mathcal{J}_\lambda) &= \aleph_0^{-1}(o_{\mathcal{CD}} \times \{p\}) \\ \widetilde{\mathcal{M}}_{(1,0)}(\dot\Sigma, M; \{p\}; J) &= \widetilde{\mathcal{M}}_{(1,0)}(\dot\Sigma, M; \{p\}; \mathcal{J}_\lambda) \bigcap \pi_1^{-1}(J). \end{aligned}$$

The following is a fundamental proposition for the proof of Theorem 8.1.8 as in the standard strategy exercised in the similar transversality result for the study of pseudoholomorphic curves in [18, Section 10.5] which in turn follows the scheme used in [29] for the 1-jet transversality proof for the case of pseudoholomorphic curves. We apply the same scheme with the replacement of pseudoholomorphic curves by contact instantons first for the 0-jet case in this part. Because the nature of equation is different, especially *because the contact instanton equation involves the second derivatives*, the proof involves additional complication beyond that of [29].

Proposition 8.7.2 *The map $\aleph_0$ is transverse to the submanifold*

$$o_{\mathcal{CD}} \times \{p\} \subset \mathcal{CD} \times M.$$

Proof Its linearization $D\aleph_0(J, (j, w), z)$ is given by the map

$$(L, (b, Y), v) \mapsto \Big(D_{J,(j,w)}\Upsilon^{\text{univ}}(L, (b, Y)), Y(w(z)) + dw(z)(v)\Big) \tag{8.7.2}$$

for

$$L \in T_J\mathcal{J}_\lambda,\; b \in T_j\mathcal{M}_{\dot\Sigma},\; v \in T_z\dot\Sigma,\; Y \in T_w\mathcal{F}(\Sigma, M).$$

This defines a linear map

$$T_J\mathcal{J}_\lambda \times T_j\mathcal{M}_{\dot\Sigma} \times T_w\mathcal{F}(\Sigma, M) \times T_z\dot\Sigma \to \mathcal{CD}_{(J,(j,w))} \times T_{w(z)}M$$

on $W^{1,p}$. But for the map $\aleph_0$ to be differentiable, we need to choose the completion $W^{k,p}(\dot\Sigma, M)$ of $\mathcal{F}(\dot\Sigma, M)$ with $k \geq 2$.

We take the Sobolev completion in the $W^{k,p}$-norm for at least $k \geq 2$. We take $k = 2$. We would like to prove that this linear map is a submersion at every element $(J, j, w, z_0) \in \widetilde{\mathcal{M}}_1(\dot\Sigma, M)$ i.e., at the pair (w, z_0) satisfying

$$\Upsilon^{\text{univ}}(J, (j, w)) = 0, \quad w(z_0) = p.$$

For this purpose, we need to study solvability of the system of equations

$$D_{J,(j,w)}\Upsilon^{\text{univ}}(L, (b, Y)) = (\gamma, \omega), \quad Y(w(z_0)) + dw(v) = X_0 \tag{8.7.3}$$

for any given $(\gamma, \omega) \in \mathcal{CD}_w$ and X_0, i.e.,

$$\gamma \in \Omega^{\pi(0,1)}_{(j,J)}(w^*TM),\; \omega \in \Omega^2(\dot\Sigma), \quad X_0 \in T_{w(z_0)}M.$$

For the current study of evaluation transversality, the domain complex structure j does not play much role in our study. Especially it does not play any role throughout our calculations except that it appears as a parameter. Therefore we will fix j

throughout the proof. Then it will be enough to consider the case $b = 0$. Then the above equation is reduced to

$$D_{J,w}\Upsilon^{\text{univ}}(L, Y) = (\gamma, \omega), \quad Y(w(z_0)) + dw(v) = X_0. \tag{8.7.4}$$

Firstly, we study (8.7.4) for $Y \in W^{2,p}$. We regard

$$\mathcal{CD}_{(J,(j,w))} \times T_{w(z_0)}M$$

as a Banach space with the norm $\|\cdot\|_{1,p} + \|\cdot\|_p + |\cdot|$, where $|\cdot|$ is any norm induced by an inner product on $T_x M$.

We will show that the image of the map (8.7.2) restricted to the elements of the form

$$(L, (0, Y), v)$$

is onto as a map

$$T_J \mathcal{J}_\lambda \times \Omega^0_{2,p}(w^*TM) \to \mathcal{CD}^{1,p}_{J,w} \times T_{w(z_0)}M$$

where (w, j, z_0, J) lies in $(\Upsilon^{\text{univ}}_1)^{-1}(o_{\mathcal{H}''} \times \xi)$, and we set

$$\mathcal{CD}^{1,p}_{J,w} := \Omega^{(0,1)}_{1,p}(w^*\xi) \times \Omega^2_p(\dot\Sigma). \tag{8.7.5}$$

For the clarification of notations, we denote the natural pairing

$$\mathcal{B} \times \mathcal{B}^* \to \mathbb{R}$$

by $\langle\cdot,\cdot\rangle$ for any Banach space $\mathcal{B}$ and the inner product on $T_x M$ by $(\cdot,\cdot)_x$.

We now prove the following which will then finish the proof by the ellipticity of the linearization map. The remaining part of proof will be occupied by the proof of this statement.

Proposition 8.7.3 *The subspace*

$$\text{Image}\,\aleph_0 \subset \mathcal{CD}_0 \oplus T_{w(z_0)}R$$

is dense.

Proof For the proof, we will use the Hahn-Banach lemma in an essential way. Let $((\eta, f), X_p) \in \mathcal{CD}^* \times T_p M$ satisfy

$$\left\langle D_w\Upsilon(Y) + \left(\frac{1}{2}L \cdot d^\pi w \circ j, 0\right), (\eta, f)\right\rangle + \langle Y, \delta_{z_0}X_p\rangle = 0 \tag{8.7.6}$$

for all $Y \in \Omega^0_{2,p}(w^*TM)$ and L where δ_{z_0} is the Dirac-delta function supported at z_0. By the Hahn-Banach lemma, it will be enough to prove

$$(\eta, f) = 0, \quad X_p = 0. \tag{8.7.7}$$

In the derivation of (8.7.6), we have used the formula

$$\Upsilon(J, (j, w)) = (\overline{\partial}^\pi_J w, d(w^* \circ \lambda \circ j))$$

to compute the linearization of $\aleph_0$. In particular, we have

$$D_1\aleph_0(L) = \left(\frac{1}{2}L \cdot d^\pi w \circ j, 0\right)$$

in the direction of J where the second factor of the value of Υ does not depend on J. Obviously, we have $D_2\aleph_0(Y) = D\Upsilon_{J,j}(Y)$.

Under this assumption, we would like to show (8.7.7). Without loss of any generality, we may assume that Y is smooth since $C^\infty(w^*TM) \hookrightarrow \Omega^0_{2,p}(w^*TM)$ is dense. Taking $L = 0$ in (8.7.6), we obtain

$$\langle D_w\Upsilon(Y), (\eta, f)\rangle + \langle Y, \delta_{z_0} X_p\rangle = 0 \quad \text{for all } Y \text{ of } C^\infty. \tag{8.7.8}$$

Therefore by definition of the distribution derivatives, η satisfies

$$(D_w\Upsilon)^\dagger(\eta, f) - \delta_{z_0} X_p = 0$$

as a distribution, i.e.,

$$(D_w\Upsilon)^\dagger(\eta, f) = \delta_{z_0} X_p$$

where

$$(D_w\Upsilon)^\dagger = (D_w\Upsilon(J, (j, w)))^\dagger$$

is the formal adjoint of $D_w\Upsilon(J, (j, w))$ whose symbol is of the same type as $D_w\Upsilon_{(j,J)}$ and so is an elliptic first order differential operator. (See (8.2.23) for the linearization formula and recall that $(\overline{\partial}^\pi_J)^\dagger = -\partial^\pi_J$ modulo zero order operators.) By the elliptic regularity, (η, f) is a classical solution on $\Sigma \setminus \{z_0\}$.

On the other hand, by setting $Y = 0$ in (8.7.6), we get

$$\langle L \cdot dw \circ j, \eta\rangle = 0 \tag{8.7.9}$$

for all $L \in T_J\mathcal{J}_\lambda$. From this identity, the argument used in the transversality proven in the previous section shows that $\eta = 0$ in a small neighborhood of any somewhere injective point in $\Sigma \setminus \{z_0\}$. Such a somewhere injective point exists by the hypothesis of w being somewhere injective and the fact that the set of somewhere injective points is open and dense in the domain under the given hypothesis. Then by the unique continuation theorem, we conclude that $\eta = 0$ on $\Sigma \setminus \{z_0\}$ and so the support of η as a distribution on Σ is contained at the one-point subset $\{z_0\}$ of Σ.

The following lemma will conclude the proof of Proposition 8.7.3. We postpone the proof of the lemma till the next section.

Lemma 8.7.4 *(η, f) is a distributional solution of $(D_w \Upsilon)^\dagger(\eta, f) = 0$ on Σ and so continuous. In particular, we have $(\eta, f) = 0$ in $(\mathcal{CD})^*$.*

Once we know $(\eta, f) = 0$, the Eq. (8.7.6) is reduced to the finite dimensional equation

$$(Y(z_0), X_p)_{z_0} = 0 \tag{8.7.10}$$

It remains to show that $X_p = 0$. For this, we have only to show that the image of the evaluation map

$$Y \mapsto Y(z_0)$$

is surjective onto $T_p M$, which is now obvious.

Now it remains to prove Lemma 8.7.4.

8.8 Proof of Lemma 8.7.4

Our primary goal is to prove

$$\langle D_w \Upsilon(Y), (\eta, f)\rangle = 0 \tag{8.8.1}$$

for all smooth $Y \in \Omega^0(w^* TM)$, i.e., (η, f) is a distributional solution of

$$(D_w \Upsilon(J, (j, w)))^\dagger(\eta, f) = 0$$

on the whole Σ, not just on $\Sigma \setminus \{z_0\}$. This will imply that (η, f) is a solution smooth everywhere by the elliptic regularity.

We start with (8.7.8)

$$\langle D_w \Upsilon(Y), (\eta, f)\rangle + \langle Y, \delta_{z_0} X_p\rangle = 0 \quad \text{for all } Y \in C^\infty. \tag{8.8.2}$$

We first simplify the expression of the pairing $\langle D_w \Upsilon(Y), (\eta, f)\rangle$ knowing that $\mathrm{supp}(\eta, f) \subset \{z_0\}$.

Let z be a complex coordinate centered at a fixed marked point z_0 and

$$(x_1, y_1, x_2, y_2, \cdots, x_n, y_n, \eta)$$

be a Darboux coordinates so that $\lambda = d\eta - \sum_{i=1} y_i dx_i$ on a neighborhood of $p \in M$.

Remark 8.8.1 In the proof of [18, Section 10.5], we chose a complex coordinates $(w_1, \cdots, w_n)$ identifying a neighborhood of p with an open subset of $\mathbb{C}^n$.

We consider the standard metric

$$h = \frac{\sqrt{-1}}{2} dz d\bar{z}$$

on a neighborhood $U \subset \dot{\Sigma}$ of z_0.

The following lemma will be crucial in our proof.

Lemma 8.8.2 *Let η be as above. For any smooth section Y of $w^*(TM)$ and η of $\left(\Omega^{(0,1)}_{1,p}(w^*\xi)\right)^*$*

$$\langle D\overline{\partial}^{\pi}_J(Y), \eta\rangle = \langle \overline{\partial} Y^{\pi}, \eta\rangle$$

where $\overline{\partial}$ is the standard Cauchy-Riemann operators on $\mathbb{R}^{2n} \cong \mathbb{C}^n$ in the above coordinate.

Proof We have already shown that η is a distribution with $\operatorname{supp}(\eta, f) \subset \{z_0\}$. By the structure theorem on the distribution supported at a point z_0 Theorem 8.7.1, we have

$$\eta = P\left(\frac{\partial}{\partial s}, \frac{\partial}{\partial t}\right)(\delta_{z_0})$$

where $z = s + it$ is the given complex coordinates at z_0 and $P\left(\frac{\partial}{\partial s}, \frac{\partial}{\partial t}\right)$ is a differential operator associated by the polynomial P of two variables with coefficients in

$$\Lambda^{(0,1)}_{(j_{z_0}, J_p)}(w^*\xi).$$

Furthermore since $\eta \in (W^{1,p})^* \cong W^{-1,q}$, the degree of P *must be zero* and so we obtain

$$\eta = \beta_{z_0} \cdot \delta_{z_0} \tag{8.8.3}$$

for some constant vector $\beta_{z_0} \in \Lambda^{(0,1)}(\xi_p)$.

We then have the expression

$$D\overline{\partial}^{\pi}_J Y = \overline{\partial} Y + E \cdot \partial Y + F \cdot Y$$

near z_0 in coordinates where E and F are zero-order matrix operators satisfying

$$E(z_0) = 0 = F(z_0).$$

(See [29, p. 331] and [34] for such a derivation.) Therefore by (8.8.3), we derive

$$\begin{aligned}\langle E \cdot \partial Y^\pi + F \cdot Y^\pi, \eta\rangle &= \langle E \cdot \partial Y^\pi + F \cdot Y^\pi, \beta_{z_0}\delta_{z_0}\rangle \\ &= (E(z_0)\partial Y^\pi(z_0) + F(z_0)Y^\pi(z_0), \beta_{z_0})_{z_0} = 0\end{aligned}$$

and we obtain

$$\langle D_w\overline{\partial}_J(Y), \eta\rangle = \langle \overline{\partial} Y^\pi + E \cdot \partial^\pi Y^\pi + F \cdot Y^\pi, \eta\rangle = \langle \overline{\partial}^\pi Y^\pi, \eta\rangle$$

since $\operatorname{supp} \eta \subset \{z_0\}$. This finishes the proof.

By this lemma, (8.8.2) becomes

$$\langle \overline{\partial} Y^\pi, \eta\rangle + \langle -\Delta(\lambda(Y))dA + d((Y^\pi \rfloor d\lambda) \circ j, f\rangle + \langle Y, \delta_{z_0} X_p\rangle = 0 \tag{8.8.4}$$

for all Y. We next rewrite the middle summand by integration by parts as in Part I but with the letter g replaced by f here.

Lemma 8.8.3 *We have*

$$\langle -\Delta(\lambda(Y))dA + d((Y^\pi \rfloor d\lambda) \circ j, f\rangle = -\int \lambda(Y)\, \Delta f\, dA + \int df \circ j \wedge (Y^\pi \rfloor d\lambda)$$

Recalling $Y = Y^\pi + \lambda(Y)\, R_\lambda$ and noting that $\lambda(Y)$ and Y^π are independent and arbitrary, we rearrange the summand of (8.8.4) into

$$\begin{aligned}0 = &-\int \lambda(Y)\, \Delta f\, dA + \langle \lambda(Y)\, R_\lambda, \delta_{z_0} X_p\rangle \\ &+ \int \langle \overline{\partial}^\pi Y^\pi, \eta\rangle + \int df \circ j \wedge (Y^\pi \rfloor d\lambda) \\ &+ \langle Y^\pi, \delta_{z_0} X_p\rangle\end{aligned}$$

Recall the decomposition $Y = Y^\pi + \lambda(Y)\, R_\lambda$. By considering Y with $Y^\pi = 0$ and with $\lambda(Y) = 0$ separately which are arbitrary, we have derived

$$0 = -\int \lambda(Y)\, \Delta f\, dA + \langle \lambda(Y)\, R_\lambda, \delta_{z_0} X_p\rangle \tag{8.8.5}$$

$$0 = \int \langle \overline{\partial} Y^\pi, \eta \rangle + \int df \circ j \wedge (Y^\pi \rfloor d\lambda) + \langle Y^\pi, \delta_{z_0} X_p \rangle. \tag{8.8.6}$$

We decompose Y as

$$Y(z) = (Y(z) - \chi(z)Y(z_0)) + \chi(z)Y(z_0)$$

on U where χ is a cut-off function with $\chi \equiv 1$ in a small neighborhood $V \subset U$ of z_0 and satisfies $\operatorname{supp} \chi \subset U$. It induces the corresponding decomposition of Y^π and $\lambda(Y)$.

We first examine the Eq. (8.8.6). Then the first summand $\widetilde{Y}^\pi$ defined by

$$\widetilde{Y}^\pi(z) := Y^\pi(z) - \chi(z)Y^\pi(z_0)$$

is a smooth section on Σ, and satisfies

$$\widetilde{Y}^\pi(z_0) = 0, \quad \overline{\partial}\widetilde{Y}^\pi = \overline{\partial} Y^\pi \quad \text{on } V$$

since $\chi(z)Y^\pi(z_0) \equiv Y^\pi(z_0)$ on V. Therefore applying (8.8.6) to $\widetilde{Y}^\pi$ instead of Y^π and recalling $\operatorname{supp}(\eta, f) \subset \{p\}$, we obtain

$$\langle \overline{\partial}\widetilde{Y}^\pi, \eta \rangle + \langle \widetilde{Y}^\pi, \delta_{z_0} X_p \rangle = 0.$$

Again using the support property $\operatorname{supp} \eta \subset \{z_0\}$ and (8.8.2), we derive

$$\langle \widetilde{Y}^\pi, \delta_{z_0} X_p \rangle = \langle \widetilde{Y}^\pi(z_0), X_p \rangle = 0 \tag{8.8.7}$$

and so $\langle \overline{\partial}\widetilde{Y}^\pi, \eta \rangle = 0$. But we also have

$$\langle \overline{\partial} Y^\pi, \eta \rangle = \langle \overline{\partial}\widetilde{Y}^\pi, \eta \rangle \tag{8.8.8}$$

since $\overline{\partial}^\pi \widetilde{Y}^\pi = \overline{\partial} Y^\pi$ on V and $\operatorname{supp} \eta \subset \{z_0\}$. Hence we obtain

$$\langle \overline{\partial}^\pi_J Y^\pi, \eta \rangle = 0$$

for all Y^π.

Applying similar reasoning to (8.8.5), we have derived

$$\int \lambda(Y) \Delta f \, dA = 0$$

for all $\lambda(Y)$. Combining the two, we have proved (η, f) is a weak solution of

$$(\overline{\partial}^\pi_J)^\dagger \eta = 0, \quad \Delta f = 0$$

on whole Σ. Therefore we have finished the proof of (8.8.1) by Lemma 8.8.2. By the elliptic regularity, (η, f) is a smooth solution. In particular it is continuous. Since we have already shown $(\eta, f) = 0$ on $\Sigma \setminus \{z_0\}$, continuity of η proves $(\eta, f) = 0$ on the whole Σ. This finishes the proof. □

This in turn finishes the proof of Proposition 8.7.3. □

8.9 The Case of the Boundary Evaluation Map

In this section, we explain how we can augment the arguments used in the proof of generic evaluation transversality to handle the case of boundary evaluation maps. We will also write R for $\vec{R}$ in the present section.

We now consider the map

$$\begin{aligned} \aleph_0^\partial &: \mathcal{J}_\lambda \times \mathcal{M}_{\dot\Sigma} \times \widetilde{\mathcal{F}}_{(0,1)}(\dot\Sigma, R) \to \mathcal{CD} \times R \\ &(J, (j, w), z_0) \mapsto (\Upsilon(J, (j, w)), w(z_0)). \end{aligned} \tag{8.9.1}$$

Then for a given point $p \in R$,

$$\begin{aligned} \widetilde{\mathcal{M}}_{(0,1)}(M, \lambda, R; \{p\}; \mathcal{J}_\lambda) &= \aleph_0^{-1}(o_{\mathcal{CD}} \times \{p\}) \\ \widetilde{\mathcal{M}}_{(1,0)}(M, \lambda, R; \{p\}; J) &= \widetilde{\mathcal{M}}_{(1,0)}(M, \lambda, R; \{p\}; \mathcal{J}_\lambda) \cap \pi_1^{-1}(J). \end{aligned}$$

We now establish the following boundary analog to Proposition 8.7.2.

Proposition 8.9.1 *The map $\aleph_0$ is transverse to the submanifold*

$$o_{\mathcal{CD}} \times \{p\} \subset \mathcal{CD} \times R.$$

Proof Its linearization $D\aleph_0(J, (j, w), z)$ is given by the map

$$(L, (b, Y), v) \mapsto \big(D_{J,(j,w)}\Upsilon(L, (b, Y)), Y(w(z)) + dw(z)(v)\big) \tag{8.9.2}$$

for

$$L \in T_J\mathcal{J}_\lambda,\ b \in T_j\mathcal{M}_{\dot\Sigma},\ v \in T_z\dot\Sigma,\ Y \in T_w\mathcal{F}(M, \lambda, R)).$$

But this time, $(L, (b, Y), v)$ satisfies the boundary condition

$$Y(\partial\dot\Sigma) \subset TR, \quad v \in T\partial\dot\Sigma. \tag{8.9.3}$$

This defines a linear map

$$T_J\mathcal{J}_\lambda \times T_j\mathcal{M}_{\dot\Sigma} \times T_w\mathcal{F}(M, \lambda, R) \times T_z\dot\Sigma \to \mathcal{CD}_{(J,(j,w))} \times T_{w(z)}R.$$

We take the Sobolev completion in the $W^{k,p}$-norm for with $k = 2$. We would like to prove that this linear map is a submersion. For this purpose, we again need to study solvability of the system of equations

$$D_{J,(j,w)}\Upsilon(L,(b,Y),v) = (\gamma,\omega), \quad Y(w(z_0)) + dw(v) = X_0 \tag{8.9.4}$$

for any given $(\gamma,\omega) \in \mathcal{CD}_w$ and X_0, i.e.,

$$\gamma \in \Omega^{\pi(0,1)}_{(j,J)}(w^*TM),\ \omega \in \Omega^2(\dot\Sigma), \quad X_0 \in T_{w(z_0)}R.$$

Again we put $b = 0 = v$ obtain the equation

$$D_{J,w}\Upsilon(L,Y) = (\gamma,\omega), \quad Y(w(z_0)) = X_0 \tag{8.9.5}$$

for $Y \in T_w\mathcal{F}(M,\lambda,R)$.

We will show that the image of the map (8.9.2) restricted to the elements of the form

$$(L,(0,Y),v)$$

is onto as a map

$$T_J\mathcal{J}_\lambda \times \Omega^0_{2,p}(w^*TM) \times T_z(\partial\dot\Sigma) \to \mathcal{CD}^{1,p}_{J,w} \times T_{w(z_0)}R$$

where (w,j,z_0,J) lies in $\Upsilon_1^{-1}(o_{\mathcal{H}''} \times \xi)$.

We now prove the following boundary analog to Proposition 8.7.3 which will then finish the proof.

Proposition 8.9.2 *The subspace*

$$\text{Image}\,\aleph^\partial_0 \subset \mathcal{CD} \oplus T_{w(z_0)}R$$

is dense.

Proof Let $((\eta,f),X_p) \in (\mathcal{CD})^* \times T_p\vec{R}$ satisfy

$$\left\langle D_w\Upsilon(Y) + \left(\frac{1}{2}L\cdot d^\pi w \circ j, 0\right), (\eta,f)\right\rangle + \langle Y, \delta_{z_0}X_p\rangle = 0 \tag{8.9.6}$$

for all $Y \in \Omega^0_{2,p}(w^*TM,(\partial w)^*TR)$ and L where δ_{z_0} is the Dirac-delta function supported at z_0. We again would like to show

$$(\eta,f) = 0, \quad X_p = 0. \tag{8.9.7}$$

Taking $L = 0$ in (8.7.6), we obtain

$$\langle D_w \Upsilon(Y), (\eta, f)\rangle + \langle Y, \delta_{z_0} X_p\rangle = 0 \tag{8.9.8}$$

for all Y of C^∞ satisfying the boundary condition

$$Y(\partial \dot{\Sigma}) \subset TR. \tag{8.9.9}$$

Therefore by definition of the distribution derivatives, η satisfies

$$(D_w \Upsilon(J, (j, w)))^\dagger(\eta, f) - \delta_{z_0} X_p = 0$$

as a distribution, i.e.,

$$(D_w \Upsilon(J, (j, w)))^\dagger(\eta, f) = \delta_{z_0} X_p$$

where $(D_w \Upsilon(J, (j, w)))^\dagger$ *is the formal* L^2*-adjoint of* $D_w \Upsilon(J, (j, w))$. The following lemma provides a description of the formal adjoint.

Lemma 8.9.3 *The* L^2*-adjoint* $(D_w \Upsilon(J, (j, w)))^\dagger$ *is an linear elliptic operator whose domain is given by the pairs* (η, f) *such that*

$$\eta \in \Omega^{(1,0)}_{-1,q}(M, \lambda; J), \quad f \in L^q$$

satisfying the elliptic boundary condition.

By the similar reasoning by considering the variations L with $Y = 0$, we again arrive at the following which will finish the proof by the same reason as for the interior case.

Lemma 8.9.4 η *is a distributional solution of* $(D_w \Upsilon(J, (j, w)))^\dagger(\eta, f) = 0$ *on* Σ *and so continuous. In particular, we have* $(\eta, f) = 0$ *in* $(\mathcal{CD})^*$.

Now it remains to prove Lemma 8.9.4. (In fact, we have only to establish just near $\{z_0\}$ since the support of (η, f) is concentrated at a point z_0.) Our primary goal is to prove (η, f) is a distributional solution of

$$(D_w \Upsilon(J, (j, w)))^\dagger(\eta, f) = 0$$

on an open set including z_0 (and so on the whole Σ) by the same reason. The rest of the section will be occupied by the proof of this goal.

We start with (8.9.8)

$$\langle D_w \Upsilon(Y), (\eta, f)\rangle + \langle Y, \delta_{z_0} X_p\rangle = 0$$

for all $Y \in C^\infty$ satisfying (8.2.4), $Y(\partial \dot\Sigma) \subset TR$. Again knowing that

$$\mathrm{supp}(\eta, f) \subset \{z_0\},$$

we can simplify the expression of the pairing $\langle D_w \Upsilon(Y), (\eta, f)\rangle$ to

$$\langle \overline{\partial}^\pi Y^\pi, \eta\rangle + \langle -\Delta(\lambda(Y))\, dA + d((Y^\pi \rfloor d\lambda) \circ j, f\rangle + \langle Y, \delta_{z_0} X_p\rangle = 0 \tag{8.9.10}$$

for all Y satisfying $Y(\partial\dot\Sigma) \subset TR$.

Now the boundary analog to Lemma 8.8.3 involves the boundary contribution which is again by integration by parts combined with $\lambda(Y) \equiv 0$ on $\partial\dot\Sigma$ by the Legendrian boundary condition of Y. (*This is different from* (8.5.5) *in that* $\lambda(Y) = 0$ *here while it was arbitrary therein.*)

Lemma 8.9.5 *We have*

$$\begin{aligned}
&\langle -\Delta(\lambda(Y)) dA + d((Y^\pi \rfloor d\lambda) \circ j, f\rangle \\
&= -\int \lambda(Y)\, \Delta f\, dA + \int df \circ j \wedge (Y^\pi \rfloor d\lambda) + \int_{\partial\dot\Sigma} f \frac{\partial}{\partial \nu}(\lambda(Y))\, d\theta - f\, Y^\pi \rfloor d\lambda.
\end{aligned}$$

Now we derive the boundary analogs to (8.8.5) and (8.8.6) respectively:

$$\begin{aligned}
0 = &-\int \lambda(Y)\, \Delta f\, dA + \int_{\partial\dot\Sigma} f \frac{\partial}{\partial \nu}(\lambda(Y))\, d\theta \\
&+ \langle \lambda(Y)\, R_\lambda, \delta_{z_0} X_p\rangle
\end{aligned} \tag{8.9.11}$$

$$\begin{aligned}
0 = &\int \langle \overline{\partial}^\pi Y^\pi, \eta\rangle + \int df \circ j \wedge (Y^\pi \rfloor d\lambda) - \int f\, Y^\pi \rfloor d\lambda \\
&+ \langle Y^\pi, \delta_{z_0} X_p\rangle.
\end{aligned} \tag{8.9.12}$$

Again by replacing Y by $\widetilde{Y}$ as before, we have now derived the following boundary analogs to (8.8.5) and (8.8.6)

$$\langle \overline{\partial}^\pi_J Y^\pi, \eta\rangle + \int df \circ j \wedge (Y^\pi \rfloor d\lambda) - \int f\, Y^\pi \rfloor d\lambda = 0$$

and

$$\int \lambda(Y) \Delta f\, dA + \int_{\partial\dot\Sigma} f \frac{\partial}{\partial \nu}(\lambda(Y))\, d\theta = 0$$

respectively for all Y satisfying $Y(\partial\dot\Sigma) \subset R$. Knowing that Y^π and $\lambda(Y)$ are independent functions, we have derived the equation

$$\Delta f = 0, \quad f|_{\partial\dot\Sigma} = 0.$$

By the elliptic regularity, f is continuous and hence $f \equiv 0$. (Recall $\operatorname{supp} f \subset \{p\}$. Substituting this into the first, we get the equation for η which satisfies

$$(\overline{\partial}^{\pi})^{\dagger}\eta = 0, \quad \eta|_{\partial\dot{\Sigma}} \perp TR$$

which is again an elliptic boundary value problem. Therefore η is also smooth at z_0 and hence continuous. Again we conclude $\eta \equiv 0$. Combining the two, we have finished the proof of Proposition 8.9.2. □

This in turn finishes the proof of Proposition 8.9.1. □

A Appendix: Generic Nondegeneracy of Reeb Chords

Let (M, ξ) be a contact manifold and (R_0, R_1) be a pair of Legendrian submanifolds.

We consider contact triads (M, λ, J) and consider the boundary value problem for (γ, T) with $\gamma : [0, 1] \to M$

$$\begin{cases} \dot{\gamma}(t) = T R_{\lambda}(\gamma(t)), \\ \gamma(0) \in R_0, \quad \gamma(1) \in R_1. \end{cases} \tag{A.1}$$

First we introduce the following nondegeneracy definition.

Definition A.1 We say a Reeb chord (γ, T) of (R_0, R_1) is nondegenerate if the linearization map $\Psi_{\gamma} = d\phi^T(p) : \xi_p \to \xi_p$ satisfies

$$\Psi_{\gamma}(T_{\gamma(0)}R_0) \pitchfork T_{\gamma(1)}R_1 \quad \text{in } \xi_{\gamma(1)}.$$

Remark A.2 In [20], the notion of *Reeb trace* denoted by Z_R of a Legendrian submanifold is introduced

$$Z_R := \bigcup_{t \in \mathbb{R}} \phi^t_{R_{\lambda}}(R)$$

which is an immersed submanifold of dimension $\dim R + 1$. Then the above nondegneracy is equivalent to the transversal intersection property

$$\Psi_{\gamma}(T_{\gamma(0)}R_0) \pitchfork T_{\gamma(1)}Z_{R_1} \quad \text{in} \quad T_{\gamma(1)}M.$$

Similarly as in the problem of closed Reeb chords, we first consider the following relative version of Reeb spectrum.

Definition A.3 Let λ be a contact form of contact manifold (M, ξ) and $R \subset M$ a connected Legendrian submanifold. Denote by $\mathfrak{Reeb}(M, \lambda)$ (resp. $\mathfrak{Reeb}(M, R; \lambda)$) the set of closed Reeb chords (resp. the set of self Reeb chords of R).

(1) We define $\mathrm{Spec}(M, \lambda)$ to be the set

$$\mathrm{Spec}(M, \lambda) = \left\{ \int_\gamma \lambda \mid \lambda \in \mathfrak{Reeb}(M, \lambda) \right\}$$

and call the *action spectrum* of (M, λ).

(2) We define the *period gap* to be the constant given by

$$T(M, \lambda) := \inf \left\{ \int_\gamma \lambda \mid \lambda \in \mathfrak{Reeb}(M, \lambda) \right\} > 0.$$

We define $\mathrm{Spec}(M, R; \lambda)$ and the associated $T(M, \lambda; R)$ similarly using the set $\mathfrak{Reeb}(M, R; \lambda)$ of Reeb chords of R.

We set $T(M, \lambda) = \infty$ (resp. $T(M, \lambda; R) = \infty$) if there is no closed Reeb orbit (resp. no (R_0, R_1)-Reeb chord). Then we define

$$T_\lambda(M; R) := \min\{T(M, \lambda), T(M, \lambda; R)\} \tag{A.2}$$

and call it the *(chord) period gap* of R in M.

We denote by

$$\mathcal{L}eg(M, \xi)$$

the set of Legendrian submanifold and by $\mathcal{L}eg(M, \xi; R)$ its connected component containing $R \in \mathcal{L}eg(M, \xi)$, i.e, the set of Legendrian submanifolds Legendrian isotopic to R. We denote by

$$\mathcal{P}(\mathcal{L}eg(M, \xi))$$

the monoid of Legendrian isotopies $[0, 1] \to \mathcal{L}eg(M, \xi)$. We have natural evaluation maps

$$\mathrm{ev}_0, \mathrm{ev}_1 : \mathcal{P}(\mathcal{L}eg(M, \xi)) \to \mathcal{L}eg(M, \xi)$$

and denote by

$$\mathcal{P}(\mathcal{L}eg(M, \xi), R) = \mathrm{ev}_0^{-1}(R) \subset \mathcal{P}(\mathcal{L}eg(M, \xi))$$

and

$$\mathcal{P}(\mathcal{L}eg(M,\xi),(R_0,R_1)) = (\mathrm{ev}_0 \times \mathrm{ev}_1)^{-1}(R_0,R_1) \subset \mathcal{P}(\mathcal{L}eg(M,\xi)).$$

Finally we vary λ $\mathfrak{Reeb}(M;\lambda)$ (resp. $\mathfrak{Reeb}(M,R;\lambda)$) for the given (M,ξ) (resp. $((M,R),\xi)$) and form the union

$$\mathfrak{Reeb}(M,\xi) = \bigcup_{\lambda \in \mathfrak{Cont}(M,\xi)} \mathfrak{Reeb}(M;\lambda) \tag{A.3}$$

and

$$\mathfrak{Reeb}(M,R,\xi) = \bigcup_{\lambda \in \mathfrak{Cont}(M,R,\xi)} \mathfrak{Reeb}(M,R,\lambda). \tag{A.4}$$

A.1 Under the Perturbation of Contact Forms

In this subsection, we prove the following relative version of Theorem 8.2.7.

Theorem A.4 *Let (M,ξ) be a contact manifold. Let (R_0,R_1) be a pair of Legendrian submanifolds allowing the case $R_0 = R_1$. There exists a residual subset $\mathrm{Cont}_1^{\mathrm{reg}}(M,\xi) \subset \mathcal{C}(M,\xi)$ such that for any $\lambda \in \mathrm{Cont}_1^{\mathrm{reg}}(M,\xi)$ all Reeb chords from R_0 to R_1 are nondegenerate for $T > 0$ and Bott-Morse nondegenerate when $T = 0$.*

The case $R_0 = R_1$ with $T = 0$ is easy to prove which we omit referring its details to [22]. This being mentioned, we will focus on the case $R_0 \cap R_1 = \emptyset$ in the following discussion.

Denote by $\mathcal{L}(M;R_0,R_1)$ the space of paths

$$\gamma : ([0,1],\{0,1\}) \to (M;R_0,R_1).$$

We consider the assignment

$$\Phi : (T,\gamma,\lambda) \mapsto \dot\gamma - T R_\lambda(\gamma) \tag{A.5}$$

as a section of the Banach vector bundle over

$$(0,\infty) \times \mathcal{L}^{1,2}(M;R_0,R_1) \times \mathrm{Cont}(M,\xi)$$

where $\mathcal{L}^{1,2}(M;R_0,R_1)$ is the $W^{1,2}$-completion of $\mathcal{L}(M;R_0,R_1)$. We have

$$\dot\gamma - T R_\lambda(\gamma) \in \Gamma(\gamma^*TM; T_{\gamma(0)}R_0, T_{\gamma(1)}R_1).$$

We define the vector bundle

$$\mathcal{L}^2(R_0, R_1) \to (0, \infty) \times \mathcal{L}^{1,2}(M; R_0, R_1) \times \mathrm{Cont}(M, \xi)$$

whose fiber at (T, γ, λ) is $L^2(\gamma^* TM)$. We denote by π_i, $i = 1, 2, 3$ the corresponding projections as before.

We denote $\mathfrak{Reeb}(M, \lambda; R_0, R_1) = \Phi_\lambda^{-1}(0)$, where

$$\Phi_\lambda := \Phi|_{(0,\infty)\times\mathcal{L}^{1,2}(M; R_0, R_1)\times\{\lambda\}}.$$

Then by definition (A.3), we have

$$\mathfrak{Reeb}(\lambda; R_0, R_1) = \Phi_\lambda^{-1}(0) = \mathfrak{Reeb}(M, \xi) \cap \pi_3^{-1}(\lambda).$$

Proposition A.5 *Suppose $R_0 \cap R_1 = \emptyset$. A Reeb chord (γ, T) of (R_0, R_1) is nondegenerate if and only if the linearization*

$$d_{(\gamma,T)}\Phi : \mathbb{R} \times W^{1,2}(\gamma^* TM; T_{\gamma(0)}R_0, T_{\gamma(1)}R_1) \to L^2(\gamma^* TM)$$

is surjective.

Proof We have the formula for $d_{(\gamma,T)}\Phi$ from (A.5)

$$d_{(\gamma,T)}\Phi(a, \xi) = \frac{D\xi}{dt} - T DR_\lambda(\gamma)(\xi) - aR_\lambda(\gamma)$$

acting on ξ satisfying the boundary condition

$$\xi(0) \in T_{\gamma(0)}R_0, \quad \xi(1) \in T_{\gamma(1)}R_1.$$

Then by the Fredholm alternative, we derive the L^2-cokernel of the operator $d_{(\gamma,T)}\Phi$ is given by

$$\ker(d_{(\gamma,T)}\Phi)^\dagger = \Big\{\eta \in \Gamma(\gamma^* TM) \mid$$

$$\begin{cases} \frac{D\eta}{dt} + T\, DR_\lambda(\gamma)^\dagger \eta = 0, \\ \eta(0) \in N_{\gamma(0)}R_0, \quad \eta(1) \in N_{\gamma(1)}R_1 \end{cases} \tag{A.6}$$

$$\int_0^1 \langle aR_\lambda(\gamma(t)), \eta(t)\rangle\, dt = 0 \forall a \in \mathbb{R}\Big\}. \tag{A.7}$$

We first derive the following lemma.

Lemma A.6 *We have*

$$DR_\lambda(\gamma)^\dagger = J(\gamma) DR_\lambda(\gamma) J(\gamma)$$

pointwise, where $J \in \mathrm{End}(TM)$ *is given by* $J = J_\xi \oplus id$ *with respect to the splitting* $TM = \xi \oplus \mathrm{span}\{R_\lambda\}$.

Proof Since ϕ^t preserves λ, we obtain

$$(\phi^t)^* d\lambda = d\lambda$$

i.e., we have

$$d\lambda(\phi^t(p))(d\phi^t(v_1), d\phi^t(v_2)) = d\lambda(p)(v_1, v_2).$$

Regard $t \mapsto d_x\phi^t$ as a section of $\mathrm{Hom}(T_xM, T_{\phi^{(\cdot)}(x)}M) \to \mathbb{R}$ of the vector bundle over $\mathbb{R}$. Then by taking the covariant derivative with respect to the connection ∇ preserving J and $d\lambda$ and utilizing the identity

$$d(\phi^t)^{-1}\frac{D}{dt}d\phi^t + \frac{D}{dt}d\phi^t d(\phi^t)^{-1} = 0,$$

we obtain

$$d\lambda\left(\frac{D}{dt}d\phi^t(v_1), d\phi^t(v_2)\right) + d\lambda\left(d\phi^t(v_1), \frac{D}{dt}d\phi^t(v_2)\right) = 0.$$

for all $v_1,\ v_2$. Therefore and so

$$d\lambda\left(\frac{D}{dt}d\phi^t \circ (d\phi^t)^{-1}(v_1), v_2\right) + d\lambda\left(v_1, \frac{D}{dt}d\phi^t \circ (d\phi^t)^{-1}(v_2)\right) = 0.$$

Since we have $DR_\lambda(p)(v) = \frac{D}{dt}d\phi^t(p) \circ (d\phi^t)^{-1}(v)$ by definition, we obtain

$$d\lambda(DR_\lambda(p)(v_1), v_2) + d\lambda(v_1, DR_\lambda(p)(v_2)) = 0.$$

In terms of the metric $g = d\lambda(\cdot, J_\xi \cdot)$, this can be rewritten as

$$-g(DR_\lambda(p)(v_1), J_\xi v_2) + g(Jv_1, DR_\lambda(p)(v_2)) = 0.$$

Hence we have

$$g(DR_\lambda(p)(v_1), J_\xi v_2) = -g(v_1, JDR_\lambda(p)(v_2))$$

By setting $v_2' = J_\xi v_2$, this is equivalent to

$$g(DR_\lambda(p)(v_1), v_2') = g(v_1, JDR_\lambda(p)J(v_2')).$$

This proves $DR_\lambda(p)^\dagger = J(p)DR_\lambda(p)J(p)$. □

Using this we derive

Lemma A.7 *For any* $\eta \in \ker(d_{(\gamma,T)}\Phi)^\dagger$, *we have* $\eta(t) \perp R_\lambda(\gamma(t))$ *for all* $t \in [0, 1]$.

Proof By the hypothesis $R_0 \cap R_1 = \emptyset$, γ cannot be a closed orbit and there exists an interval I open in $[0, 1]$ such that

$$\#\gamma^{-1}(t) \equiv 1 \mod 2$$

for all $t \in I$. Note that we can choose I so that it is either $I = [0, b)$ or $I = (a, 1]$. Then (A.7) implies $\eta(t) \perp R_\lambda(\gamma(t))$ for all $t \in I$.

On the other hand, using (A.6), we compute

$$\frac{d}{dt}\langle R_\lambda(\gamma(t)), \eta(t)\rangle = \langle R_\lambda(\gamma(t)), \frac{D\eta}{dt}\rangle = \langle R_\lambda(\gamma(t)), T DR_\lambda(\gamma)^\dagger \eta(t)\rangle$$

for all $t \in [0, 1]$, with respect to the contact triad connection ∇. We write

$$\eta(t) = \eta^\pi(t) + \lambda(\eta(t))R_\lambda(\gamma(t))$$

and recall $\nabla_{R_\lambda}(\xi) \subset (\xi)$ and $\nabla_{R_\lambda} R_\lambda = 0$. Then, substituting $DR_\lambda(\gamma)^\dagger = JDR_\lambda(\gamma)J$ on ξ and noting $DR_\lambda(\gamma)(\xi) \subset \xi$, we derive that the operator $DR_\lambda(\gamma)^\dagger$ preserves the splitting $TM = \xi \oplus \mathbb{R}\langle R_\lambda\rangle$. Then we rewrite

$$\begin{aligned}\langle R_\lambda(\gamma(t)), T DR_\lambda(\gamma)^\dagger \eta(t)\rangle &= \langle R_\lambda(\gamma(t)), T\lambda(\eta)R_\lambda(\gamma(t))\rangle \\ &= T\lambda(\eta) = T\langle R_\lambda, \eta\rangle.\end{aligned}$$

In conclusion, the function $g(t) := \langle R_\lambda(\gamma(t)), \eta(t)\rangle$ satisfies the linear 1-st order ODE

$$\dot g(t) - Tg(t) = 0. \tag{A.8}$$

On the other hand, on I, we have $\lambda(\eta(t)) = \langle R_\lambda(\gamma(t)), \eta(t)\rangle = 0$ for all $t \in I$. This implies $g(t) \equiv 0$ for all $t \in I$. Since g satisfies (A.8), this implies $g(t) = 0$ for all $t \in [0, 1]$, which finishes the proof of the lemma. □

This lemma implies $\eta(t) \in \xi_{\gamma(t)}$. Then using the identity

$$DR_\lambda(\gamma)^\dagger = J(\gamma)DR_\lambda(\gamma)J(\gamma)$$

on ξ, it follows

$$\frac{D\eta}{dt} + T\, JDR_\lambda(\gamma)J\eta = 0, \quad \eta(t) \in \xi_{\gamma(t)},\ \eta(0) \in N_{\gamma(0)}R_0 \cap \xi_{\gamma(0)}$$

i.e.,

$$\frac{DJ\eta}{dt} - TDR_\lambda(\gamma)J\eta = 0, \quad \eta(t) \in \xi_{\gamma(t)},\ \eta(0) \in N_{\gamma(0)}R_0 \cap \xi_{\gamma(0)}. \tag{A.9}$$

We consider the family

$$v(t) = (d\phi^{Tt})^{-1}(\gamma(t))J(\gamma)\eta(t)) \in \xi_p$$

and differentiate

$$\begin{aligned}\frac{dv}{dt} &= (d\phi^{Tt})^{-1}\frac{D(J\eta)}{dt} - T(d\phi^{Tt})^{-1}\frac{D(d\phi^{Tt})}{dt}(d\phi^{t})^{-1}(J\eta(t))\\ &= (d\phi^{Tt})^{-1}\left(\frac{D(J\eta)}{dt} - T\frac{D(d\phi^{Tt})}{dt}(d\phi^{Tt})^{-1}(J\eta(t))\right).\end{aligned}$$

But by definition, we have

$$\frac{D(d\phi^{Tt})}{dt}(d\phi^{Tt})^{-1} = DR_\lambda(\gamma(t))$$

and hence we obtain $\frac{dv}{dt} \equiv 0$. Therefore we have

$$v(1) = v(0), \text{ i.e., } (d\phi^{T})^{-1}(J\eta(1)) = J\eta(0). \tag{A.10}$$

Since $\eta(0) \in N_{\gamma(0)}R_0 \cap \xi_{\gamma(0)}$ and $\eta(1) \in N_{\gamma(1)}R_0 \cap \xi_{\gamma(1)}$, we have

$$J\eta(0) \in T_{\gamma(0)}R_0, \quad J\eta(1) \in T_{\gamma(1)}R_1.$$

Then (A.10) shows that

$$\ker DR_\lambda(\gamma)^\dagger = 0$$

is equivalent to

$$d\phi^T(T_{\gamma(0)}R_0) \cap T_{\gamma(1)}R_1 = \{0\}$$

which is equivalent to saying that γ is a nondegenerate Reeb chord from R_0 to R_1. The converse also holds by reading the above proof backwards. This finishes the proof of Proposition A.5.

Motivated by Proposition A.5, we now consider the full derivative $d\Phi$. It remains to compute $d_\lambda\Phi$. For this purpose, we recall the defining equation of R_λ:

$$X\rfloor\lambda = 1, \ X\rfloor d\lambda = 0$$

Consider the small perturbation $\lambda_\varepsilon = \lambda + \varepsilon\mu$ and write the corresponding Reeb vector field by $R_{\lambda_\varepsilon} = R_\lambda + \varepsilon Y(\mod o(\varepsilon))$. Then we have

$$(R_\lambda + \varepsilon Y)\rfloor(\lambda + \varepsilon\mu) = 1,\ (R_\lambda + \varepsilon Y)\rfloor d(\lambda + \varepsilon\mu) = 0 \qquad \mod o(\varepsilon).$$

By collecting the terms of order ε, we obtain

$$Y\rfloor\lambda + R_\lambda\rfloor\mu = 0,\ R_\lambda\rfloor d\mu + Y\rfloor d\lambda = 0.$$

Hence the variation $\delta_\lambda R_\lambda(\mu) =: Y_\mu$ is uniquely determined by the equation

$$Y_\mu\rfloor\lambda = -R_\lambda\rfloor\mu, \quad Y_\mu\rfloor d\lambda = -R_\lambda\rfloor d\mu. \tag{A.11}$$

Now we are ready to study $d_\lambda\Phi(\mu)$. We have

$$d_\lambda\Phi(\mu) = -T\, Y_\mu(\gamma).$$

Therefore if $\eta \in \operatorname{Coker}(d\Phi(\gamma))$, we must have

$$\int_0^1 \langle Y_\mu(\gamma(t)), \eta(t)\rangle\, dt = 0$$

for all μ. Since $\eta(t) \in \xi_{\gamma(t)}$, we have $\langle Y_\mu(\gamma(t)), \eta(t)\rangle = d\lambda(Y_\mu(t), J_\xi(\gamma(t))\eta(t))$. And (A.11) implies

$$d\lambda(Y_\mu(t), J_\xi(\gamma(t))\eta(t)) = -d\mu(R_\lambda(\gamma), J_\xi(\gamma)\eta).$$

Therefore we have

$$0 = \int_0^1 \langle Y_\mu(\gamma(t)), \eta(t)\rangle\, dt = -\int_0^1 d\mu(R_\lambda(\gamma(t)), J_\xi(\gamma(t))\eta(t))\, dt \tag{A.12}$$

for any one-form μ. Now the following lemma will finish the proof.

Lemma A.8 *Let $q \in M$ and consider $\xi_q \subset T_qM$. Denote by $\{R_\lambda(q)\}^\perp \subset T_q^*M$ the annihilator of $R_\lambda(q)$. Then we have*

$$\{d\mu(R_\lambda(q), \cdot) \in T_q^*M \mid d\mu \in \Gamma(S^2(T_q^*M))\} = \{R_\lambda(q)\}^\perp \cong \xi_q^*.$$

Proof Obviously we have

$$\{d\mu(R_\lambda(q), \cdot) \in T_q^*M \mid d\mu \in \Gamma(S^2(T_q^*M))\} \subset \{R_\lambda(q)\}^\perp.$$

For the opposite inclusion, it is enough to note in terms of local coordinates that for any nonzero vector $v \in \mathbb{R}^n$, the map

$$A \mapsto Av; \quad \Lambda^2(\mathbb{R}^n) \to \mathbb{R}^n$$

is surjective. Here $\Lambda^2(\mathbb{R}^n)$ is the set of skew-symmetric matrices and $n \geq 2$. This finishes the proof. □

Once we have this lemma, we can conclude (A.12) and the unique continuation for the Eq. (A.9) imply $\eta \equiv 0$. This finishes the proof of the theorem. □

For the later purpose, we also need the following theorem.

Theorem A.9 *Let* $\mathrm{Cont}^{\mathrm{reg}}(M, \xi; R_0, R_1; < N)$ *be the set of* λ*'s such that all* λ*-Reeb chords* γ *from* R_0 *to* R_1 *with* $\mathcal{A}_\lambda(\gamma) < N$. *Then it is open in* $\mathrm{Cont}(M, \xi)$ *for each given* $N > 0$.

Proof Consider the two projection

$$\begin{array}{ccc} & \Phi^{-1}(0) & \\ \Pi_1 \swarrow & & \searrow \Pi_3 \\ \mathbb{R}_+ & & \mathrm{Cont}(M, \xi; R_0, R_1) \end{array}$$

where $\Pi_i = \pi_i|_{\Phi^{-1}(0)}$. We denote

$$\mathfrak{Reeb}(M, \xi; R_0, R_1; < N) := \Pi_1^{(0,N)}.$$

Then $\mathrm{Cont}^{\mathrm{reg}}(M, \xi; R_0, R_1; N)$ is the set of regular values of the map

$$\Pi_3|_{\mathfrak{Reeb}(M,\xi;R_0,R_1;<N)} : \mathfrak{Reeb}(M, \xi; R_0, R_1; < N) \to \mathrm{Cont}(M, \xi; R_0, R_1).$$

Now let $\lambda \in \mathrm{Cont}^{\mathrm{reg}}(M, \xi; R_0, R_1; N)$. Then the set

$$\mathfrak{Reeb}(M, \xi; R_0, R_1; < N) \cap \Pi_3^{-1}(\lambda)$$

is compact. Therefore by the tube lemma, there exists an open neighborhood $\mathcal{V}$ of λ in $\mathrm{Cont}(M, \xi; R_0, R_1)$ such that all λ'-Reeb chords in $(\Pi_3|_{\mathfrak{Reeb}(M,\xi;R_0,R_1;<N)})^{-1}(\mathcal{V})$ are nondegenerate and hence $\mathcal{V} \subset \mathrm{Cont}^{\mathrm{reg}}(M, \xi; R_0, R_1; < N)$. This finishes the proof. □

Theorem A.10 *Let* (M, ξ) *be a contact manifold. Let* (R_0, R_1) *be a pair of Legendrian submanifolds allowing the case* $R_0 = R_1$. *For a given contact form* λ*, there exists a residual subset of pairs* (R_0, R_1) *of Legendrian submanifolds such that all Reeb chords from* R_0 *to* R_1 *are nondegenerate for* $T > 0$ *and Bott-Morse nondegenerate when* $T = 0$.

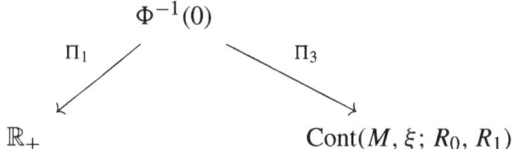

A.2 Under the Perturbation of Boundaries

In this section, we prove the following generic perturbation problem of the boundary Legendrian submanifolds by transforming the problem to that of perturbation of contact forms. Since we gave complete details of the proof of Theorem A.4, we will just indicate the differences in the proof of the following theorem therefrom.

Theorem A.11 *Let* (M, ξ) *be a contact manifold,* λ *a contact form and* $R_1 \in \mathcal{L}eg(M, \xi)$*. Then there exists a residual subset*

$$R_0 \in \mathcal{L}eg^{\text{reg}}(M, \xi) \subset \mathcal{L}eg(M, \xi)$$

of Legendrian submanifolds such that for all $R_0 \in \mathcal{L}eg(M, \xi)$ *all Reeb chords from* R_0 *to* R_1 *are nondegenerate for* $T > 0$ *and Bott-Morse nondegenerate when* $T = 0$.

Proof This time we consider the fiber bundle $\mathcal{L}^{1,2}(M; R_1)$ over

$$(0, \infty) \times \mathcal{L}eg(M, \xi)$$

whose fiber at (T, R_0) is given by

$$\mathcal{L}^{1,2}_{(T,R_0)}(M; R_1) = \mathcal{L}^{1,2}(M; R_0, R_1).$$

Then we consider the assignment

$$\Phi : (T, \gamma, R_0) \mapsto \dot{\gamma} - T R_\lambda(\gamma)$$

as a section of the Banach vector bundle

$$\mathcal{L}^2(M; R_1) \to \mathcal{L}^{1,2}(M; R_1)$$

whose fiber at (T, γ, R_0) is given by the vector space

$$L^2(\gamma^* TM, T_{\gamma(0)} R_0, T_{\gamma(1)} R_1).$$

Now we consider a perturbation of R_0 under the contact isotopy of the type

$$\psi(R_0), \quad \psi = \psi_H^1 \in \text{Cont}_0(M, \xi).$$

Then for given $\gamma \in \mathcal{L}^{1,2}(M; \psi(R_0), R_1)$, the composition path

$$\widetilde{\gamma}(t) := (\psi_H^t)^{-1}(\gamma(t))$$

satisfies the perturbed equation

$$\begin{cases} \dot{\widetilde{\gamma}}(t) = -X_H(\widetilde{\gamma}(t)) + (\psi_H^t)^* R(\widetilde{\gamma}(t) \\ \widetilde{\gamma}(0) \in R_0, \quad \widetilde{\gamma}(1) \in R_1 \end{cases}$$

with *fixed* boundary condition. Then we can duplicate the proof of Theorem A.4 by replacing perturbation of λ by that of Hamiltonian H above with almost same kind computation and so we omit the details. This then finishes the proof of Theorem A.11. □

Acknowledgments We would like to thank the referees for all her/his careful reading of manuscript and pointing out multitude of inconsistent notations, providing helpful suggestions to improve readability of the paper. Their suggestions and questions much improve the exposition of the paper and hence readability thereof.

This work is supported by the IBS project # IBS-R003-D1

References

1. Albers, P., Bramham, B., Wendl, C.: On nonseparating contact hypersurfaces in symplectic 4-manifolds. Algebr. Geom. Topol. **10**(2), 697–737 (2010)
2. Aronszajin, N.: A unique continuation theorem for solutions of elliptic partial differential equations of inequalities of second order. J. Math. Pures Appl. **36**, 235–249 (1957)
3. Bourgeois, F., Eliashberg, Y., Wysocki, K., Hofer, H., Zehnder, E.: Compactness results in symplectic field theory. Geom. Topol. **7**, 799–888 (2003)
4. Bao, E., Honda, K.: Semi-global Kuranishi charts and the definition of contact homology. Adv. Math. **414**, 108864, 148pp. (2023)
5. Bourgeois, F.: A Morse-Bott approach to contact homology. Ph.D. Dissertation, Stanford University (2002)
6. Ekholm, T., Etnyre, J., Sullivan, M.: The contact homology of Legendrian submanifolds in $\mathbb{R}^{2n+1}$. J. Differ. Geom. **71**(2), 177–305 (2005)
7. Floer, A.: The unregularized gradient flow of the symplectic action. Commun. Pure Appl. Math. **41**(6), 775–813 (1988)
8. Floer, A.:, Symplectic fixed points and holomorphic spheres. Commun. Math. Phys. **120**(4), 575–611 (1989)
9. Fukaya, K., Oh, Y.G., Ohta, H., Ono, K.: Kuranishi Structures and Virtual Fundamental Chains. Springer Monographs in Mathematics 638pp. Springer, Singapore (2020)
10. Gelfand, I.M., Shilov, G.E.: Generalized Functions, vol. 2. Academic Press, New York (1968)
11. Ishikawa, S.: Construction of general symplectic field theory (2018). arXiv:1807.09455
12. Lockhart, R., McOwen, R.: Elliptic differential operators on noncompact manifolds. Ann. Scuola Norm. Sup. Pisa Cl. Sci. **12**(3), 409–447 (1985)
13. Le, H.V., Ono, K.: Perturbation of pseudo-holomoprhic curves, addendum to "notes on symplectic 4-manifolds with $b_2^+ = 1$, II. Int. J. Math. **7**(6), 771–774 (1996)
14. McDuff, D., Salamon, D.: J-Holomorphic Curves and Symplectic Topology. American Mathematical Society Colloquium Publications, vol. 52. American Mathematical Society, Providence (2004)
15. Oh, Y.G.: Rational contact instantons and Legendrian Fukaya category (preprint, 2024). arXiv:2411.13830
16. Oh, Y.G.: Fredholm theory of holomorphic discs under the perturbation of boundary conditions. Math. Z. **222**(3), 505–520 (1996)

17. Oh, Y.G.: Higher jet evaluation transversality of J-holomorphic curves. J. Korean Math. Soc. **48**(2), 341–365 (2011)
18. Oh, Y.G.: Symplectic Topology and Floer Homology. Vol. 1.. New Mathematical Monographs, vol. 28. Cambridge University Press, Cambridge (2015)
19. Oh, Y.G.: Contact Hamiltonian dynamics and perturbed contact instantons with Legendrian boundary condition (2021, preprint). arXiv:2103.15390(v2)
20. Oh, Y.G.: Geometry and analysis of contact instantons and entangement of Legendrian links I (2021, preprint). arXiv:2111.02597
21. Oh, Y.G.: Contact instantons, anti-contact involution and proof of Shelukhin's conjecture (2022). arXiv:2212.03557
22. Oh, Y.G.: Gluing theories of contact instantons and of pseudoholomoprhic curves in SFT (2022, preprint). arXiv:2205.00370
23. Oh, Y.G.: Analysis of contact Cauchy-Riemann maps III: energy, bubbling and Fredholm theory. Bullet. Math. Sci. **13**(1), 2250011, 61pp. (2023)
24. Oh, Y.G., Savelyev, Y.: Pseudoholomoprhic Curves on the $\mathfrak{LCS}$-Fication of Contact Manifolds. Adv. Geom. **23**(2), 153–190 (2023)
25. Oh, Y.G., Wang, R.: Canonical connection on contact manifolds. In: Real and Complex Submanifolds, Springer Proceedings in Mathematics & Statistics, vol. 106, pp. 43–63 (2014). arXiv:1212.4817 in its full version
26. Oh, Y.G., Wang, R.: Analysis of contact Cauchy-Riemann maps I: a priori C^k estimates and asymptotic convergence. Osaka J. Math. **55**(4), 647–679 (2018)
27. Oh, Y.G., Yu, S.: Contact instantons with Legendrian boundary condition: a priori estimates, asymptotic convergence and index formula (2022). arXiv:2301.06023
28. Oh, Y.G., Yu, S.: Legendrian spectral invariants on the one-jet space via perturbed contact instantons (2023). arXiv:2301.06704
29. Oh, Y.G., Zhu, K.: Embedding property of J-holomorphic curves in Calabi-Yau manifolds for generic J. Asian J. Math. **13**(3), 323–340 (2009)
30. Pardon, J.: Contact homology and virtual fundamental cycles. J. Am. Math. Soc. **32**(3), 825–919 (2019)
31. Rudin, W.: Functional Analysis. McGraw-Hill, New York (1973)
32. Sandon, S.: On iterated translated points for contactomorphisms of $\mathbb{R}^{2n+1}$ and $\mathbb{R}^{2n} \times S^1$. Int. J. Math. **23**(2), 1250042, 14pp. (2012)
33. Shelukhin, E.: The Hofer norm of a contactomorphism. J. Symplect. Geom. **15**(4), 1173–1208 (2017)
34. Sikorav, J.C.: Some properties of holomorphic curves in almost complex manifolds. In: Audin, M., Lafontaine, J. (eds.) Holomorphic Curves in Symplectic Geometry, chap. V. Birkhäuser, Basel. 'Progr. Math. **117**, 165–189 (1994)
35. Wells, R.O.: Differential Analysis on Complex Manifolds. Graduate Texts in Mathematics, vol. 65, 3rd edn. Springer, New York (2008). With a new appendix by Oscar Garcia-Prada

Open Access This chapter is licensed under the terms of the Creative Commons Attribution-NonCommercial-NoDerivatives 4.0 International License (http://creativecommons.org/licenses/by-nc-nd/4.0/), which permits any noncommercial use, sharing, distribution and reproduction in any medium or format, as long as you give appropriate credit to the original author(s) and the source, provide a link to the Creative Commons license and indicate if you modified the licensed material. You do not have permission under this license to share adapted material derived from this chapter or parts of it.

The images or other third party material in this chapter are included in the chapter's Creative Commons license, unless indicated otherwise in a credit line to the material. If material is not included in the chapter's Creative Commons license and your intended use is not permitted by statutory regulation or exceeds the permitted use, you will need to obtain permission directly from the copyright holder.

Chapter 9
Duality for Landau-Ginzburg Models

Claude Sabbah

To the memory of Bumsig Kim

Abstract This article surveys various duality statements attached to a pair consisting of a smooth complex quasi-projective variety and a regular function on it.

Keywords Twisted de Rham complex · Kontsevich complex

9.1 Introduction

9.1.1 Comparing Cohomologies of Complexes of Differential Forms

The origin of this note is a question of Bumsig Kim: given a regular function f on a smooth connected complex quasi-projective variety U of dimension n, to compare two kinds of cohomologies attached to f together with their natural duality pairings:

- the hypercohomology of the twisted de Rham complex $(\Omega^\bullet_U, \mathrm{d} + \mathrm{d}f)$ together with a variant with compact support (to be defined since it is not a complex in $\mathsf{Mod}(\mathcal{O}_U)$) and the natural pairing between them,
- the cohomology and cohomology with compact support of the complex $(\Omega^\bullet_U, \mathrm{d}f)$, together with the Serre duality pairing between them (see e.g. [12]).

A first observation is that, although the first (hyper)cohomologies are finite dimensional, the second ones are not, unless some assumption on f is added, e.g. the

C. Sabbah (✉)
CMLS, CNRS, École polytechnique, Institut Polytechnique de Paris, Palaiseau Cedex, France
e-mail: Claude.Sabbah@polytechnique.edu

© The Author(s) 2026

N.-G. Kang et al. (eds.), *Categorical and Enumerative Aspects of Mirror Symmetry*,
KIAS Springer Series in Mathematics 5,
https://doi.org/10.1007/978-981-95-0385-8_9

e.g. the critical set, which is the support of the cohomology sheaves $\mathcal{H}^j(\Omega_U^\bullet, \mathrm{d}f)$, is compact anyway contained in a finite number of fibers of f). This leads us to choose a good projectivization of (U, f) as a projective morphism $f : Y \to \mathbb{A}^1$, so that Y is smooth quasi-projective and $H := Y \smallsetminus U$ is a divisor with normal crossings, and to consider suitable cohomologies on Y.

Furthermore, one way to compare them is to introduce a parameter u and to consider the twisted de Rham complex $(\Omega_U^\bullet[u], u\mathrm{d} + \mathrm{d}f)$, where, all along this note, the notation $[u]$ means the tensor product $\otimes_{\mathbb{C}} \mathbb{C}[u]$. The main question is then whether the hypercohomology $\boldsymbol{H}^k(U, (\Omega_U^\bullet[u], u\mathrm{d} + \mathrm{d}f))$ of the latter complex is a free $\mathbb{C}[u]$-module of finite rank.

(i) This property does not hold in general, as shown by the following simple example. Let $U = \mathbb{A}^1$ with coordinate t and set $f = 0$. Then

$$H^1(U, u\mathrm{d} + \mathrm{d}f) = \operatorname{coker}\Big[\mathbb{C}[t, u] \xrightarrow{u\partial_t} \mathbb{C}[t, u]\Big]$$

and this $\mathbb{C}[u]$-module is identified with the $\mathbb{C}$-vector space $\mathbb{C}[t]$ on which u acts by zero, hence is not of finite type.

(ii) This property holds if f is *proper*, but we are mainly interested in cases where f is not proper.

(iii) This property holds if we replace $\mathbb{C}[u]$ with the ring of Laurent polynomials $\mathbb{C}[u, u^{-1}]$, meaning that the problem of comparison only occurs the origin $u = 0$ (see (9.2)).

(iv) As indicated above, a way to prevent the infinite dimensionality at $u = 0$ is to replace the twisted de Rham complex with parameter u by a subcomplex that solves the comparison. This can be done by choosing a good projectivization (Y, f) as above and by considering the twisted logarithmic complex $(\Omega_Y^\bullet(\log H)[u], u\mathrm{d} + \mathrm{d}f)$ with parameter u (a meromorphic version of this complex, called the de Rham complex of the Brieskorn lattice, is considered in [24, §8]). This will be our starting point.

The abutment of the pairing we look for is obtained by means of a smooth projectivization X of Y such that f extends as a morphism $f : X \to \mathbb{P}^1$ and $D = X \smallsetminus U$ is a normal crossing divisor. The de Rham hypercohomologies $\boldsymbol{H}^k(X, (\Omega_X^\bullet, \mathrm{d}))$ and $\boldsymbol{H}^k(X, (\Omega_X^\bullet, 0))$ have the same dimension due to Hodge degeneration. It follows that the hypercohomology of the complex $(\Omega_X^\bullet[u], u\mathrm{d})$ is $\mathbb{C}[u]$-free of finite rank: its has finite type over $\mathbb{C}[u]$ and for any $u_o \in \mathbb{C}$, the dimension of the hypercohomology of $(\Omega_X^\bullet, u_o\mathrm{d})$ is independent of u_o, so that the assertion follows from Lemma 9.1.3 below. We can then identify $\boldsymbol{H}^{2n}(X, (\Omega_X^\bullet[u], u\mathrm{d}))$ with

$$H^{2n}_{\mathrm{dR}}(X, (\Omega_X^\bullet, \mathrm{d})) \otimes \mathbb{C}[u] \simeq H^n(X, \Omega_X^n) \otimes \mathbb{C}[u] \simeq \mathbb{C}[u].$$

Theorem A *The* $\mathbb{C}[u]$*-modules*

$$\boldsymbol{H}^k\big(Y, (\Omega_Y^\bullet(\log H)[u], u\mathrm{d}{+}\mathrm{d}f)\big) \text{ and } \boldsymbol{H}^k\big(Y, (\Omega_Y^\bullet(\log H)(-H)[u], u\mathrm{d}{-}\mathrm{d}f)\big),$$

are $\mathbb{C}[u]$*-free of finite rank, and equipped with a meromorphic connection having a pole of order at most 2 at* $u = 0$*, a regular singularity at infinity and no other pole. Furthermore, there is a natural perfect pairing*

$$\begin{aligned}&\boldsymbol{H}^{n+k}\big(Y, (\Omega_Y^\bullet(\log H)[u], u\mathrm{d} + \mathrm{d}f)\big)\\ &\qquad \otimes_{\mathbb{C}[u]} \boldsymbol{H}^{n-k}\big(Y, (\Omega_Y^\bullet(\log H)(-H)[u], u\mathrm{d} - \mathrm{d}f)\big) \longrightarrow \mathbb{C}[u], \qquad (9.1)\end{aligned}$$

which is compatible with the connections. All these objects are independent of the choice of the good projectivization (Y, f) *of* (U, f)*.*

Note that the freeness of $\boldsymbol{H}^k\big(Y, (\Omega_Y^\bullet(\log H)[u], u\mathrm{d} + \mathrm{d}f)\big)$ also follows from a variant of the Barannikov-Kontsevich theorem [20, §0.6]. The independence on the choice of the good projectivization follows from [30] and [6, Prop. 2.3].

Let us emphasize that, by restricting modulo $u\mathbb{C}[u]$ according to Lemma 9.1.3 below, we find that the perfect pairing

$$\boldsymbol{H}^{n+k}(Y, (\Omega_Y^\bullet(\log H), \mathrm{d}f)) \otimes_{\mathbb{C}} \boldsymbol{H}^{n-k}(Y, (\Omega_Y^\bullet(\log H)(-H), -\mathrm{d}f)) \longrightarrow \mathbb{C}$$

coincides with Serre's duality pairing as constructed in [12] since the complexes involved are compactly supported (they are supported on the union of the closures of the critical locus of the restriction of f to each stratum of (Y, H), hence in a finite number of fibers of f).

If $f : U \to \mathbb{A}^1$ is proper, the divisor H is empty and $U = Y$, so this theorem the assertion (ii). If on the other hand f satisfies a tameness condition (see Definition 9.4.3 below), then the twist $(-H)$ on the second term of (9.1) can be omitted (see Corollary 9.4.6).

(v) One can give a $\mathcal{D}$-module-theoretic interpretation of the previous results (see Sect. 9.4). Letting $\mathcal{D}_U$ denote the ring of algebraic differential operators on U with its filtration $F_\bullet$ by the order, we consider the Rees ring $R_F\mathcal{D}_U = \bigoplus_{k\geq 0} F_k\mathcal{D}_U \cdot u^k$. We regard $(\mathcal{O}_U[u], u\mathrm{d} + \mathrm{d}f)$ as a coherent $R_F\mathcal{D}_U$-module and $\boldsymbol{H}^k(U, (\Omega_U^\bullet[u], u\mathrm{d} + \mathrm{d}f))$ as isomorphic to the de Rham cohomology of this $R_F\mathcal{D}_U$-module. Although the ring $R_F\mathcal{D}_U$ has properties similar to those of $\mathcal{D}_U$, it does not satisfy Bernstein's inequality because of the possible occurrence of u-torsion, and this prevents us to apply Bernstein's results to deduce finiteness of the de Rham cohomology.

For (iii), we first notice that the second term in (9.1) plays the role of hypercohomology with compact support. In order to give a meaning to this remark, it is worthwhile working with modules over the ring of differential operators. More precisely, we consider the ring $\mathcal{D}_U[u, u^{-1}] = \mathcal{D}_U \otimes_{\mathbb{C}} \mathbb{C}[u, u^{-1}]$ of algebraic differential operator on U with coefficients in $\mathcal{O}_U[u, u^{-1}]$, so that the base ring is $\mathbb{C}[u, u^{-1}]$ instead of the field $\mathbb{C}$ (it would be equivalent to consider differential operators on $U \times \mathbb{G}_m$ relative to the projection to $\mathbb{G}_m$). We denote by $E_U^{f/u}$ the left $\mathcal{D}_U[u, u^{-1}]$-module $(\mathcal{O}_U[u, u^{-1}], \mathrm{d} + \mathrm{d}f/u)$.

We consider the two extensions in the sense of $\mathcal{D}$-modules, denoted $E_Y^{f/u}(*H)$ and $E_Y^{f/u}(!H)$ of $E_U^{f/u}$ by the open inclusion $U \hookrightarrow Y$, corresponding to the full extension and the extension with "proper support" (see Sect. 9.3 for details). Note that $E_Y^{f/u}$ also exists. We will show that, for any $k \in \mathbb{Z}$, both $\mathbb{C}[u, u^{-1}]$-modules $H^k_{\mathrm{dR}}(Y, E_Y^{f/u}(*H))$ and $H^k_{\mathrm{dR}}(Y, E_Y^{f/u}(!H))$ are free $\mathbb{C}[u, u^{-1}]$-modules of finite rank equipped with a connection having a regular singularity at $u = \infty$, that we also denote by $H^k_{\mathrm{dR}}(U, E_U^{f/u})$ and $H^k_{\mathrm{dR,c}}(U, E_U^{f/u})$ respectively. By definition of $E_Y^{f/u}(*H)$, we have

$$H^k_{\mathrm{dR}}(U, E_U^{f/u}) = \boldsymbol{H}^k(U, (\Omega^\bullet_U[u, u^{-1}], \mathrm{d} + \mathrm{d}f/u)).$$

On the other hand, we will see the isomorphism in (9.3.1 **):

$$\begin{aligned}\boldsymbol{H}^k\big(Y, (\Omega^\bullet_Y(\log H)[u, u^{-1}], \mathrm{d} + \mathrm{d}f/u)\\ \simeq \boldsymbol{H}^k\big(Y, (\Omega^\bullet_Y(\log H)[u], \mathrm{d} + \mathrm{d}f/u) \otimes_{\mathbb{C}[u]} \mathbb{C}[u, u^{-1}],\end{aligned}$$

compatible with the connections, and similarly after twisting the complexes by $(-H)$.

Theorem B *Restriction to U induces isomorphisms of free $\mathbb{C}[u, u^{-1}]$-modules of finite rank with connection*

$$\boldsymbol{H}^{n+k}\big(Y, (\Omega^\bullet_Y(\log H)[u, u^{-1}], \mathrm{d} + \mathrm{d}f/u)\big) \xrightarrow{\sim} H^k_{\mathrm{dR}}(U, E_U^{f/u}) \qquad \text{(B*)}$$

$$\boldsymbol{H}^{n-k}\big(Y, (\Omega^\bullet_Y(\log H)(-H)[u, u^{-1}], \mathrm{d} - \mathrm{d}f/u)\big) \xrightarrow{\sim} H^{-k}_{\mathrm{dR,c}}(U, E_U^{-f/u}), \qquad \text{(B!)}$$

giving rise to a perfect pairing compatible with the connections, by means of (9.1)*:*

$$H^k_{\mathrm{dR}}(U, E_U^{f/u}) \otimes_{\mathbb{C}[u,u^{-1}]} H^{-k}_{\mathrm{dR,c}}(U, E_U^{-f/u}) \longrightarrow \mathbb{C}[u, u^{-1}]. \qquad (9.2)$$

This theorem makes clear the independence of the choice of the good projectivization (Y, f) in Theorem A if we restrict to $\mathbb{G}_m = \operatorname{Spec} \mathbb{C}[u, u^{-1}]$. It is similar to [29, Th. 3], where mixed Hodge modules are considered (see also [8]). Note that a perfect pairing like that of Theorem B can be obtained by means of $\mathcal{D}$-module theory (see [18, App. 2]) but we will not compare these two ways of defining a perfect pairing. Theorem A provides each term in (B*) and (B!) of a canonical lattice, i.e., a free $\mathbb{C}[u]$-submodule which has the same rank as the corresponding $\mathbb{C}[u, u^{-1}]$-

module, on which the connection has a pole of order two. It is a generalization of the *Brieskorn lattice* of singularity theory [5, 19].

Remark 9.1.1 Such twisted de Rham complexes have been considered from many points of view and their cohomology sometimes takes the name of Dwork cohomology. Also, since $\mathrm{d}f$ is the main object used to define the twisted de Rham complex, one could consider a pair (U, ω), where ω is any closed regular 1-form, instead of a pair (U, f). In addition to the references used in this text, let us mention only a few articles representing other directions of research: [1, 3, 7, 10, 14, 17, 26] and [2], [22, §2.4].

9.1.2 Formalization with Respect to u

If the critical set of f in U is assumed to be compact, then the cohomology $\boldsymbol{H}^k(U, (\Omega_U^\bullet, \mathrm{d}f))$ is finite-dimensional since the complex of coherent sheaves is supported on this critical set. One would naturally expect that, consequently, $\boldsymbol{H}^k(U, (\Omega_U^\bullet[u], u\mathrm{d} + \mathrm{d}f))$ is of finite type over $\mathbb{C}[u]$. We do not know whether this property holds. However, as shown in Theorem $\hat{\mathrm{C}}$ below, it holds if we replace $\mathbb{C}[u]$ with the ring $\mathbb{C}[\![u]\!]$ of formal power series in a suitable way, under the following condition:[1]

(9.3) There exists a projectivization $f_Z : Z \to \mathbb{A}^1$ of $f : U \to \mathbb{A}^1$ with Z smooth (no other assumption on $Z \smallsetminus U$), such that the critical set of f_Z is contained in U (in particular, it is compact).

Some care has to be taken when working with formal power series. For a coherent $\mathcal{O}_U$-module $\mathcal{F}$, we have $\mathcal{F}[u] := \mathbb{C}[u] \otimes_{\mathbb{C}} \mathcal{F}$ and we set $\mathcal{F}[\![u]\!] = \varprojlim_\ell (\mathcal{F}[u]/u^\ell \mathcal{F}[u])$. This is in general not equal to $\mathbb{C}[\![u]\!] \otimes_{\mathbb{C}} \mathcal{F}$ and for $x \in U$ we have a strict inclusion $\mathcal{F}[\![u]\!]_x \subsetneq \mathcal{F}_x[\![u]\!]$: a germ of section of $\mathcal{F}[\![u]\!]$ at $x \in U$ consists of a formal power series $\sum_n f_n u^n$ where f_n are sections of $\mathcal{F}$ defined on a fixed neighbourhood of x, while for $\mathcal{F}_x[\![u]\!]$ we allow the neighbourhood to be shrunk when $n \to \infty$. In particular, there is a natural morphism $\mathbb{C}[\![u]\!] \otimes_{\mathbb{C}} \mathcal{F} \to \mathcal{F}[\![u]\!]$. We then set $\mathcal{F}(\!(u)\!) := \mathcal{F}[\![u]\!][u^{-1}] = \mathbb{C}(\!(u)\!) \otimes_{\mathbb{C}[\![u]\!]} \mathcal{F}[\![u]\!]$.

Theorem $\hat{\mathbf{A}}$ *The statement of Theorem A holds if we replace everywhere $[u]$ with $[\![u]\!]$, and the free $\mathbb{C}[\![u]\!]$-modules and pairings are obtained from those of Theorem A by tensoring with $\mathbb{C}[\![u]\!]$ over $\mathbb{C}[u]$.*

[1] Added on May 2023: T. Mochizuki has sent me a proof assuming only compactness of the critical set and which extends to the case where U is complex analytic and Kähler, and f holomorphic.

Theorem $\hat{\mathrm{C}}$ *Under Condition* (9.3), *the* $\mathbb{C}[\![u]\!]$*-modules*

$$\boldsymbol{H}^k\big(U, (\Omega_U^\bullet[\![u]\!], u\mathrm{d} + \mathrm{d}f)\big)$$

are $\mathbb{C}[\![u]\!]$*-free, coincide via the restriction morphism with the corresponding* $\mathbb{C}[\![u]\!]$*-modules in Theorem* $\hat{\mathrm{A}}$*, and the pairing* (9.1) *induces a perfect pairing*

$$\boldsymbol{H}^{n+k}\big(U, (\Omega_U^\bullet[\![u]\!], u\mathrm{d} + \mathrm{d}f)\big) \otimes_{\mathbb{C}[u]} \boldsymbol{H}^{n-k}\big(U, (\Omega_U^\bullet[\![u]\!], u\mathrm{d} - \mathrm{d}f)\big) \longrightarrow \mathbb{C}[\![u]\!] \tag{9.4}$$

which itself induces, by working modulo $u\mathbb{C}[\![u]\!]$*, the Serre duality pairing*

$$\boldsymbol{H}^{n+k}\big(U, (\Omega_U^\bullet, \mathrm{d}f)\big) \otimes_{\mathbb{C}} \boldsymbol{H}^{n-k}\big(U, (\Omega_U^\bullet, -\mathrm{d}f)\big) \longrightarrow \mathbb{C}.$$

See Proposition 9.5.2 for a more precise result. The way (9.1) induces (9.4) will be explained in detail in Sect. 9.5.

Remark 9.1.2 The pairing (9.4) can be regarded as a global version (with respect to U) of K. Saito's higher residue pairings [25] for a germ of holomorphic function with an isolated singularity.

9.1.3 Setting and Notation

Let U be a smooth connected quasi-projective variety of dimension n and let $f \in \mathcal{O}(U)$ be any regular function on U, that we regard as a morphism $f : U \to \mathbb{A}^1$. Let u be a new variable. We consider the *twisted de Rham complex* $(\Omega_U^\bullet[u], u\mathrm{d} + \mathrm{d}f)$, with $\Omega_U^\bullet[u] := \Omega_U^\bullet \otimes_{\mathbb{C}} \mathbb{C}[u]$, whose hypercohomology on U is $\boldsymbol{H}^k(U, (\Omega_U^\bullet[u], u\mathrm{d} + \mathrm{d}f))$. We sometimes make use of the isomorphic subcomplex $(u^{-\bullet}\Omega_U^\bullet[u], \mathrm{d}+\mathrm{d}f/u)$ of $(\Omega_U^\bullet[u, u^{-1}], \mathrm{d}+\mathrm{d}f/u)$, the isomorphism being obtained by multiplying the degree k term by u^{-k}. We will also define the cohomology with compact support $\boldsymbol{H}_{\mathrm{c}}^k(U, (\Omega_U^\bullet[u], u\mathrm{d}+\mathrm{d}f)) \simeq \boldsymbol{H}_{\mathrm{c}}^k(U, (u^{-\bullet}\Omega_U^\bullet(-D)[u], \mathrm{d}+\mathrm{d}f/u))$.

It is convenient to choose a *good projectivization* of (U, f), namely, a pair (X, f) consisting smooth projective variety X containing U as a Zariski open subset and such that

(a) $D := X \smallsetminus U$ is a normal crossing divisor in X,
(b) $f : U \to \mathbb{A}^1$ extends as a morphism $f : X \to \mathbb{P}^1$.

We then set $P = f^*(\infty)$ (the pole divisor of f), we denote by $|P|$ its support, which is contained in D, and we decompose $D = |P| \cup H$, where H is some normal crossing divisor in X having no irreducible component contained in $|P|$. We let $j : U := X \smallsetminus D \hookrightarrow X$ denote the open inclusion. We note that f induces a projective morphism $f : X \smallsetminus |P| =: Y \to \mathbb{A}^1$. We can also regard f as a global section of $\mathcal{O}_X(*D)$. We will keep the notation H for $H \cap Y$.

The following lemma will be of constant use.

Lemma 9.1.3 *Let $K_u^\bullet$ be a bounded complex of sheaves on X of $\mathbb{C}[u]$-modules such that $\boldsymbol{H}^j(X, K_u^\bullet)$ has finite type over $\mathbb{C}[u]$ for every j. Then the following properties are equivalent:*

1. *For every $u_o \in \mathbb{C}$ and every j, $\dim \boldsymbol{H}^j(X, K_u^\bullet/(u-u_o)K_u^\bullet)$ is independent of u_o.*
2. *For every j, $\boldsymbol{H}^j(X, K_u^\bullet)$ is a free $\mathbb{C}[u]$-module.*

In such a case, for every $u_o \in \mathbb{C}$ and every j, we have

$$\boldsymbol{H}^j(X, K_u^\bullet/(u-u_o)K_u^\bullet) = \boldsymbol{H}^j(X, K_u^\bullet)/(u-u_o)\boldsymbol{H}^j(X, K_u^\bullet). \tag{9.1.1 *}$$

Furthermore, for any morphism $\varphi : K_u^\bullet \to L_u^\bullet$ between two such complexes satisfying (1) *or* (2)*, if the induced morphism*

$$\boldsymbol{H}^j(X, K_u^\bullet/(u-u_o)K_u^\bullet) \longrightarrow \boldsymbol{H}^j(X, L_u^\bullet/(u-u_o)L_u^\bullet)$$

is an isomorphism for any $u_o \in \mathbb{C}$ and any $j \in \mathbb{Z}$, then $\boldsymbol{R}\Gamma(X, \varphi)$ is a quasi-isomorphism, that is,

$$\boldsymbol{H}^j(X, \varphi) : \boldsymbol{H}^j(X, K_u^\bullet) \longrightarrow \boldsymbol{H}^j(X, L_u^\bullet)$$

is an isomorphism of free $\mathbb{C}[u]$-modules for any $j \in \mathbb{Z}$.

Proof Let $\mathbb{C}[u]_{\mathrm{loc}}$ be a localization of $\mathbb{C}[u]$ such that $\mathbb{C}[u]_{\mathrm{loc}} \otimes_{\mathbb{C}[u]} \boldsymbol{H}^j(X, K_u^\bullet)$ is $\mathbb{C}[u]_{\mathrm{loc}}$-free for every j. If u_o is a closed point of $\operatorname{Spec} \mathbb{C}[u]_{\mathrm{loc}}$, the maps $(u-u_o)$ in the long exact sequence

$$\cdots \boldsymbol{H}^j(X, K_u^\bullet) \xrightarrow{u-u_o} \boldsymbol{H}^j(X, K_u^\bullet) \longrightarrow \boldsymbol{H}^j(X, K_u^\bullet/(u-u_o)K_u^\bullet) \longrightarrow \cdots$$

are all injective. By decreasing induction on j, one identifies $\boldsymbol{H}^j(X, K_u^\bullet/(u-u_o)K_u^\bullet)$ with $\boldsymbol{H}^j(X, K_u^\bullet)/(u-u_o)\boldsymbol{H}^j(X, K_u^\bullet)$ for all j, i.e., that (9.1.1 *) holds for such an u_o.

If (2) holds, then the above property holds for any $u_o \in \mathbb{C}$ and the dimension of $\boldsymbol{H}^j(X, K_u^\bullet/(u-u_o)K_u^\bullet)$ is constant and equal to the rank of $\boldsymbol{H}^j(X, K_u^\bullet)$, so that (1) also holds.

Assume now that (1) holds. We argue by decreasing induction on j. Assume that $u-u_o : \boldsymbol{H}^{j+1}(X, K_u^\bullet) \to \boldsymbol{H}^{j+1}(X, K_u^\bullet)$ is injective for any u_o. Then the exact sequence above implies that (9.1.1 *) holds in degree j for any u_o. Since the dimension is independent of u_o, the $\mathbb{C}[u]$-module $\boldsymbol{H}^j(X, K_u^\bullet)$ is $\mathbb{C}[u]$-free, so that $u-u_o$ is injective on it for any $u_o \in \mathbb{C}$ and we conclude by induction since $\boldsymbol{H}^k(X, K_u^\bullet) = 0$ for $k \gg 0$.

For the last assertion, the assumption and the first part imply that φ induces an isomorphism $\boldsymbol{H}^j(X, K_u^\bullet)/(u-u_o)\boldsymbol{H}^j(X, K_u^\bullet) \xrightarrow{\sim} \boldsymbol{H}^j(X, L_u^\bullet)/(u-u_o)\boldsymbol{H}^j(X, L_u^\bullet)$ for any $u_o \in \mathbb{C}$. We conclude by applying a variant of Nakayama's lemma: if a morphism between free $\mathbb{C}[u]$-modules of finite rank induces an isomorphism after restriction to any $u_o \in \mathbb{C}$, then it is an isomorphism. □

9.2 Freeness and Duality for the Kontsevich Complexes

Before considering the hypercohomology $\boldsymbol{H}^k(U, (\Omega^\bullet_U[u], u\mathrm{d} + \mathrm{d}f))$, it is useful to gather some properties of a variant of this de Rham cohomology where the computation is made on the projective variety X and the terms of the de Rham complexes are $\mathcal{O}_X$-coherent. If $f = 0$, this amounts to computing the hypercohomology of the logarithmic de Rham complex instead of that of the meromorphic de Rham complex on X (see Notation of Sect. 9.1.3).

9.2.1 Kontsevich Complexes [16]

For $k \geq 0$, we set

$$\Omega^k_f = \{\omega \in \Omega^k_X(\log D) \mid \mathrm{d}f \wedge \omega \in \Omega^{k+1}_X(\log D)\}.$$

Since d sends $\Omega^k_X(\log D)$ to $\Omega^{k+1}_X(\log D)$, we obtain the *Kontsevich complex*

$$(\Omega^\bullet_f, \mathrm{d} + \mathrm{d}f) \quad (\text{denoted } \Omega^\bullet_X(\log D, f) \text{ in } [16]).$$

We will also consider the twisted complex $(\Omega^\bullet_f(-|P|), \mathrm{d} + \mathrm{d}f)$, whose terms are $\Omega^k_f(-|P|)$.

One can equip these complexes with the decreasing filtration $\sigma^\bullet$ by the stupidly truncated subcomplexes. The inclusion of filtered complexes

$$(\Omega^\bullet_f(-|P|), \mathrm{d} + \mathrm{d}f, \sigma^\bullet) \hookrightarrow (\Omega^\bullet_f, \mathrm{d} + \mathrm{d}f, \sigma^\bullet)$$

and

$$(\Omega^\bullet_f(-D), \mathrm{d} + \mathrm{d}f, \sigma^\bullet) \hookrightarrow (\Omega^\bullet_f(-H), \mathrm{d} + \mathrm{d}f, \sigma^\bullet)$$

are filtered quasi-isomorphisms (see [9, Prop. 1.4.2] and [16, Proof of Lem. 2.12]).

The local computation of Ω^k_f (see [9, (1.3.1)]) shows that the wedge product

$$\Omega^k_f \otimes_{\mathcal{O}_X} \Omega^{n-k}_f(-D) \longrightarrow \Omega^n_X(\log D)(-D) = \Omega^n_X \tag{9.5}$$

is a perfect pairing. It also induces a pairing of complexes

$$(\Omega^\bullet_f, \mathrm{d} + \mathrm{d}f) \otimes_{\mathbb{C}} (\Omega^\bullet_f(-D), \mathrm{d} - \mathrm{d}f) \longrightarrow (\Omega^\bullet_X, \mathrm{d}), \tag{9.6}$$

where the termwise product is induced by

$$\Omega_X^k(\log D) \otimes \Omega_X^\ell(\log D)(-D) \longrightarrow \Omega_X^{k+\ell}(\log D)(-D) \hookrightarrow \Omega_X^{k+\ell}.$$

Proposition 9.2.1 (J.-D. Yu [30]) *The corresponding cohomological pairing*

$$\boldsymbol{H}^{n+k}\big(X, (\Omega_f^\bullet, \mathrm{d}+\mathrm{d}f)\big) \otimes_{\mathbb{C}} \boldsymbol{H}^{n-k}\big(X, (\Omega_f^\bullet(-D), \mathrm{d}-\mathrm{d}f)\big) \longrightarrow H^{2n}_{\mathrm{dR}}(X)$$

is perfect.

Corollary 9.2.2 *Through the quasi-isomorphism* $(\Omega_f^\bullet(-|P|), \mathrm{d}+\mathrm{d}f) \hookrightarrow (\Omega_f^\bullet, \mathrm{d}+\mathrm{d}f)$, *the pairing obtained from that of Proposition 9.2.1:*

$$\boldsymbol{H}^{n+k}\big(X, (\Omega_f^\bullet, \mathrm{d}+\mathrm{d}f)\big) \otimes_{\mathbb{C}} \boldsymbol{H}^{n-k}\big(X, (\Omega_f^\bullet(-H), \mathrm{d}-\mathrm{d}f)\big) \longrightarrow H^{2n}_{\mathrm{dR}}(X)$$

is perfect. □

Proof of Proposition 9.2.1 We will make use of the following lemma. □

Lemma 9.2.3 (J.-D. Yu) *Let* $A^\bullet, B^\bullet$ *be bounded complexes of* $\mathcal{O}_X$*-modules equipped with*

- *finite exhaustive decreasing filtrations* $F^\bullet$,
- *a pairing* $A^\bullet \otimes B^\bullet \to (\Omega_X^\bullet, \mathrm{d})$

satisfying the two conditions

1. *the pairing induces a well-defined pairing* $F^pA^\bullet \otimes (B^\bullet/F^{n+1-p}B^\bullet) \to (\Omega_X^\bullet, \mathrm{d})$ *for each* p,
2. *the induced pairing* $\boldsymbol{H}^{n+k}\big(X, \mathrm{gr}_F^p A^\bullet\big) \otimes_{\mathbb{C}} \boldsymbol{H}^{n-k}\big(X, \mathrm{gr}_F^{n-p} B^\bullet\big) \to H^{2n}_{\mathrm{dR}}(X)$ *is perfect for each* k, p.

Then the induced pairing $\boldsymbol{H}^{n+k}\big(X, A^\bullet\big) \otimes_{\mathbb{C}} \boldsymbol{H}^{n-k}\big(X, B^\bullet\big) \to H^{2n}_{\mathrm{dR}}(X)$ *is perfect for each* k.

Proof We argue by induction on p. We consider the commutative diagram (omitting X in the notation and setting $F^p := F^pA^\bullet$ and $G_{n+1-p} := B^\bullet/F^{n+1-p}B^\bullet$)

$$\begin{array}{ccccccccc}
\boldsymbol{H}^{n+k-1}(\mathrm{gr}_F^p) & \longrightarrow & \boldsymbol{H}^{n+k}(F^{p+1}) & \longrightarrow & \boldsymbol{H}^{n+k}(F^p) & \longrightarrow & \boldsymbol{H}^{n+k}(\mathrm{gr}_F^p) & \longrightarrow & \boldsymbol{H}^{n+k+1}(F^{p+1}) \\
\downarrow & & \downarrow & & \downarrow & & \downarrow & & \downarrow \\
\boldsymbol{H}^{n-k+1}(\mathrm{gr}_F^{n-p})^\vee & \to & \boldsymbol{H}^{n-k}(G_{n-p})^\vee & \to & \boldsymbol{H}^{n-k}(G_{n+1-p})^\vee & \to & \boldsymbol{H}^{n-k}(\mathrm{gr}_F^{n-p})^\vee & \to & \boldsymbol{H}^{n-k-1}(G_{n-p})^\vee
\end{array}$$

By decreasing induction on p and Condition (2), the vertical morphisms except maybe the middle one are isomorphisms. Hence the middle one is so. For $p \gg 0$, both terms $\boldsymbol{H}^{n+k}(F^p)$ and $\boldsymbol{H}^{n-k}(G_{n+1-p})^\vee$ are zero, and for $p \ll 0$,

$$\boldsymbol{H}^{n+k}(F^p) = \boldsymbol{H}^{n+k}(X, A^\bullet) \quad \text{and} \quad \boldsymbol{H}^{n-k}(G_{n+1-p})^\vee = \boldsymbol{H}^{n-k}(X, B^\bullet).$$

□

If we equip each complex in (9.6) with the filtration $\sigma^\bullet$, the pairing (9.6) clearly satisfies 9.2.3(1). Furthermore, 9.2.3(2) follows from Serre's duality applied to (9.5). Therefore, an application of Lemma 9.2.3 concludes the proof.

Remark 9.2.4

1. A similar argument with the complex $(\Omega_X^\bullet, 0)$ instead of $(\Omega_X^\bullet, \mathrm{d})$ yields a perfect pairing induced by Serre's duality:

$$\boldsymbol{H}^{n+k}\big(X, (\Omega_f^\bullet, \mathrm{d}f)\big) \otimes_{\mathbb{C}} \boldsymbol{H}^{n-k}\big(X, (\Omega_f^\bullet(-D), -\mathrm{d}f)\big) \longrightarrow \boldsymbol{H}^n(X, \Omega_X^n).$$

As for the Kontsevich complex, the inclusion $(\Omega_f^\bullet(-|P|), \mathrm{d}f) \hookrightarrow (\Omega_f^\bullet, -\mathrm{d}f)$ is a quasi-isomorphism, as well as $(\Omega_f^\bullet(-D), \mathrm{d}f) \hookrightarrow (\Omega_f^\bullet(-H), -\mathrm{d}f)$, so that we deduce a perfect pairing

$$\boldsymbol{H}^{n+k}\big(X, (\Omega_f^\bullet, \mathrm{d}f)\big) \otimes_{\mathbb{C}} \boldsymbol{H}^{n-k}\big(X, (\Omega_f^\bullet(-H), -\mathrm{d}f)\big) \longrightarrow \boldsymbol{H}^n(X, \Omega_X^n).$$

2. There exist natural perfect pairings

$$\boldsymbol{H}^{n+k}\big(Y, (\Omega^\bullet(\log H), \mathrm{d}+\mathrm{d}f)\big) \otimes_{\mathbb{C}} \boldsymbol{H}^{n-k}\big(Y, (\Omega^\bullet(\log H)(-H), \mathrm{d}-\mathrm{d}f)\big) \longrightarrow \mathbb{C}$$

and

$$\boldsymbol{H}^{n+k}\big(Y, (\Omega^\bullet(\log H), \mathrm{d}f)\big) \otimes_{\mathbb{C}} \boldsymbol{H}^{n-k}\big(Y, (\Omega^\bullet(\log H)(-H), -\mathrm{d}f)\big) \longrightarrow \mathbb{C}.$$

This will be shown with a parameter in Lemma 9.2.7 below, by identifying the source of these pairings respectively with the sources of the pairing of Proposition 9.2.1 and that of the previous remark.

9.2.2 Kontsevich Complexes with the Variable *u*

We now replace $(\Omega_f^\bullet, \mathrm{d}+\mathrm{d}f)$ with $(\Omega_f^\bullet[u], u\mathrm{d}+\mathrm{d}f)$. If we make $u^2\partial_u$ act on $\Omega_f^k[u]$ by

$$u^2\partial_u(\eta \otimes h(u)) = \eta \otimes (u^2\partial_u + ku)(h(u)) - f\eta \otimes h(u),$$

then this action commutes with the differential of the complex and induces a natural action on its hypercohomology, i.e., a meromorphic connection with a pole of order two at $u = 0$ and no other pole except at infinity.

Corollary 9.2.5 *The cohomologies*

$$\boldsymbol{H}^{n+k}\big(X, (\Omega_f^\bullet[u], u\mathrm{d} + \mathrm{d}f)\big) \quad \textit{and} \quad \boldsymbol{H}^{n-k}\big(X, (\Omega_f^\bullet(-D)[u], u\mathrm{d} + \mathrm{d}f)\big)$$

are $\mathbb{C}[u]$-free of finite rank, and the pairing

$$\boldsymbol{H}^{n+k}\big(X, (\Omega^{\bullet}_f[u], u\mathrm{d}+\mathrm{d}f)\big) \otimes_{\mathbb{C}[u]} \boldsymbol{H}^{n-k}\big(X, (\Omega^{\bullet}_f(-D)[u], u\mathrm{d}+\mathrm{d}f)\big) \longrightarrow H^{2n}_{\mathrm{dR}}(X)[u]$$

is perfect and compatible with the natural meromorphic action of ∂_u. Moreover, the natural morphism

$$\boldsymbol{H}^{n-k}\big(X, (\Omega^{\bullet}_f(-D)[u], u\mathrm{d}+\mathrm{d}f)\big) \longrightarrow \boldsymbol{H}^{n-k}\big(X, (\Omega^{\bullet}_f(-H)[u], u\mathrm{d}+\mathrm{d}f)\big)$$

is an isomorphism which induces, by means of the previous pairing, a perfect pairing

$$\boldsymbol{H}^{n+k}\big(X, (\Omega^{\bullet}_f[u], u\mathrm{d}+\mathrm{d}f)\big) \otimes_{\mathbb{C}[u]} \boldsymbol{H}^{n-k}\big(X, (\Omega^{\bullet}_f(-H)[u], u\mathrm{d}+\mathrm{d}f)\big) \longrightarrow H^{2n}_{\mathrm{dR}}(X)[u].$$

Proof of Corollary 9.2.5 Recall that $\dim \boldsymbol{H}^k\big(X, (\Omega^{\bullet}_f, u_o\mathrm{d}+\mathrm{d}f)\big)$ is independent of u_o (see [9, Th. 1.3.2]). Since the $\mathbb{C}[u]$-finiteness is clear by a spectral sequence argument owing to the fact that each term of the complex is $\mathcal{O}_X[u]$-coherent, the $\mathbb{C}[u]$-freeness of $\boldsymbol{H}^k\big(X, (\Omega^{\bullet}_f[u], u\mathrm{d}+\mathrm{d}f)\big)$ follows from Lemma 9.1.3. By duality (Proposition 9.2.1 and Remark 9.2.4), $\dim \boldsymbol{H}^k\big(X, (\Omega^{\bullet}_f(-D), u_o\mathrm{d}-\mathrm{d}f)\big)$ is independent of u_o, and since $\mathbb{C}[u]$-finiteness is also clear, $\mathbb{C}[u]$-freeness follows.

We deduce the perfectness of the pairing by tensoring with $\mathbb{C}[u]/(u-u_o)$ for any u_o, where it follows from loc. cit. The last assertion follows then from Remark 9.2.4. □

Remark 9.2.6 From the point of view developed in Theorem B and the other proof given in Sect. 9.4, it is convenient to consider the complexes $(u^{-\bullet}\Omega^{\bullet}_f[u], \mathrm{d}\pm\mathrm{d}f/u)$ with degree k term $u^{-k}\Omega^k_f[u] \subset \Omega^k_f[u,u^{-1}]$. Multiplication by u^k on the degree k term induces an isomorphism with $(\Omega^{\bullet}_f[u], u\mathrm{d}\pm\mathrm{d}f)$. We deduce isomorphisms

$$\boldsymbol{H}^k\big(X, (u^{-\bullet}\Omega^{\bullet}_f[u], \mathrm{d}\pm\mathrm{d}f/u)\big) \simeq \boldsymbol{H}^k\big(X, (\Omega^{\bullet}_f[u], u\mathrm{d}\pm\mathrm{d}f)\big).$$

Due to the perfect pairing $u^{-j}\Omega^j_f[u]\otimes u^{j-n}\Omega^{n-j}_f(-D)[u] \to u^{-n}\Omega^n_X[u]$ obtained from (9.5), we see that the perfect pairing between these free $\mathbb{C}[u]$-modules takes values in $u^{-n}H^{2n}_{\mathrm{dR}}(X)[u]$.

9.2.3 Proof of Theorem A

A first part of the theorem, namely $\mathbb{C}[u]$-freeness and finiteness, as well as perfectness of (9.1), follows from Corollary 9.2.5, according to the next lemma.

Lemma 9.2.7 *For each k, the natural morphisms*

$$\begin{aligned}\boldsymbol{H}^k\big(X, (\Omega^\bullet_f[u], u\mathrm{d}+\mathrm{d}f\big) &\longrightarrow \boldsymbol{H}^k\big(X, (\Omega^\bullet_f(*P)[u], u\mathrm{d}+\mathrm{d}f)\big)\\ &\longrightarrow \boldsymbol{H}^k\big(Y, (\Omega^\bullet_Y(\log H)[u], u\mathrm{d}+\mathrm{d}f)\big),\end{aligned}$$

and the similar ones after twisting the complexes by $(-H)$*, are isomorphisms.*

Proof We will treat the case without twist, the other case being treated similarly. The proof of [9, Cor. 1.4.3] shows that, for any $u_o \in \mathbb{C}$ and any $\ell \geq 1$, the inclusion of complexes

$$(\Omega^\bullet_f, u_o\mathrm{d}+\mathrm{d}f) \hookrightarrow (\Omega^\bullet_f(\ell P), u_o\mathrm{d}+\mathrm{d}f)$$

is a quasi-isomorphism. Since $\boldsymbol{H}^k\big(X, (\Omega^\bullet_f(\ell P)[u], u\mathrm{d}+\mathrm{d}f)\big)$ has finite type over $\mathbb{C}[u]$, we can apply Lemma 9.1.3 to deduce that the morphism

$$\boldsymbol{H}^k\big(X, (\Omega^\bullet_f[u], u\mathrm{d}+\mathrm{d}f)\big) \longrightarrow \boldsymbol{H}^k\big(X, (\Omega^\bullet_f(\ell P)[u], u\mathrm{d}+\mathrm{d}f)\big)$$

is an isomorphism for any k and ℓ. Since $\Omega^k_f(*P) = \varinjlim_\ell \Omega^k_f(\ell P)$, we only need to justify the commutation of direct limits and hypercohomology. □

Lemma 9.2.8 *Let* $(K^\bullet_\ell, \delta)_\ell$ *be an inductive system of complexes of fixed amplitude on an algebraic variety Z, whose terms are quasi-coherent* $\mathcal{O}_X$*-modules. Then, for each k, we have*

$$\varinjlim_\ell \boldsymbol{H}^k(Z, (K^\bullet_\ell, \delta)) \xrightarrow{\sim} \boldsymbol{H}^k(Z, \varinjlim_\ell (K^\bullet_\ell, \delta)).$$

Proof We filter the complexes by stupid truncation. We have such a morphism at each level of the corresponding spectral sequence and it is enough to prove the assertion for the first page of the spectral sequence. This amounts to show the isomorphism

$$\varinjlim_\ell H^k(Z, K^j_\ell) \xrightarrow{\sim} H^k(Z, \varinjlim_\ell K^j_\ell),$$

which follows from Noetherianity of Z, since K^j_ℓ are quasi-coherent. □

To show that the second morphism is an isomorphism, we argue with the same spectral sequence argument. We are thus reduced to showing that the restriction morphism

$$H^k(X, \Omega^j_f(*P)[u]) \longrightarrow H^k(Y, \Omega^j_Y(\log H)[u])$$

is an isomorphism. By the commutation with inductive limits, we are left with showing $H^k(X, \Omega_f^j(*P)) \xrightarrow{\sim} H^k(Y, \Omega_Y^j(\log H))$, which is clear since $\Omega_Y^k(\log H)$ is the restriction to Y of $\Omega_f^k(*P)$.

Proof of Theorem A, End The existence of a compatible action of $u^2\partial_u$ has been seen in Sect. 9.2.2. The regularity of the connection at $u = \infty$ will be seen in Remark 9.3.2.

It remains to show the independence of the good projectivization. It is enough to consider a morphism $\pi : (Y', H') \to (Y, H)$ of such kind which is the identity on U, and set $f' = f \circ \pi$. We have natural morphisms of complexes compatible with the actions of $u^2\partial_u$ and $u^2\partial_u$ respectively:

$$\begin{aligned}(\Omega_Y^\bullet(\log H)[u], u\mathrm{d} + \mathrm{d}f) &\longrightarrow \boldsymbol{R}\pi_*\pi^{-1}(\Omega_Y^\bullet(\log H)[u], u\mathrm{d} + \mathrm{d}f)\\ &\longrightarrow \boldsymbol{R}\pi_*(\pi^*\Omega_Y^\bullet(\log H)[u], u\mathrm{d} + \mathrm{d}f)\\ &\longrightarrow \boldsymbol{R}\pi_*(\Omega_{Y'}^\bullet(\log H')[u], u\mathrm{d} + \mathrm{d}f')\end{aligned}$$

inducing a natural morphism of free $\mathbb{C}[u]$-modules of finite rank with meromorphic connection:

$$\boldsymbol{H}^k(Y, (\Omega_Y^\bullet(\log H)[u], u\mathrm{d} + \mathrm{d}f)) \longrightarrow \boldsymbol{H}^k(Y', (\Omega_{Y'}^\bullet(\log H')[u], u\mathrm{d} + \mathrm{d}f')).$$

and a similar property after twisting by $(-H)$ and $(-H')$ respectively. The perfect pairings are also compatible with these morphisms. We are thus reduced to showing that such a morphism and its analogue after the twist is an isomorphism.

Its restriction to any $u_o \neq 0$ is an isomorphism, according to [9, Cor. 1.4.3]. The argument for the twisted case will be given in the proof of Theorem B. To conclude with Lemma 9.1.3 it remains to show that the natural morphism

$$\boldsymbol{H}^k(Y, (\Omega_Y^\bullet(\log H), \mathrm{d}f)) \longrightarrow \boldsymbol{H}^k(Y', (\Omega_{Y'}^\bullet(\log H'), \mathrm{d}f'))$$

is an isomorphism, and similarly after a twist by $(-H)$ and $(-H')$ respectively. The non twisted case is proved in [6, Prop. 2.3]. The twisted case can be obtain by duality, by showing that the perfect pairings considered in Remark 9.2.4 are compatible with the isomorphisms induced by π. This ends the proof of Theorem A.

□

Remark 9.2.9 (Computation of the Rank) For $f : U \to \mathbb{A}^1$ as above, let $j_Y : U \hookrightarrow Y$ denote the inclusion. We consider the complex $\boldsymbol{R}j_{Y*}\mathbb{C}_U$ on Y, and for any $c \in \mathbb{C}$, the vanishing cycle complex $\phi_{f-c}\boldsymbol{R}j_{Y*}\mathbb{C}_U$. It follows from [20, Th. 2] that

$$\operatorname{rk} \boldsymbol{H}^k\big(Y, (\Omega_Y^\bullet(\log H)[u], u\mathrm{d} + \mathrm{d}f)\big) = \sum_{c\in\mathbb{C}} \dim \boldsymbol{H}^{k-1}(f^{-1}(c), \phi_{f-c}\boldsymbol{R}j_{Y*}\mathbb{C}_U).$$

9.3 The Generic Pairing: Proof of Theorem B

In Theorem A, one can replace the logarithmic complex $\Omega_Y^\bullet(\log H)$ with the meromorphic complex $\Omega_Y^\bullet(*H)$ if one also replaces polynomials in u with Laurent polynomials in u, and we will instead consider the $\mathbb{C}[u, u^{-1}]$-module

$$\boldsymbol{H}^k(U, (\Omega_U^\bullet[u, u^{-1}], \mathrm{d} + \mathrm{d}f/u))$$

with its connection induced by the action of ∂_u coming from that of $\partial_u - f/u^2$ on each term of the complex.

Lemma 9.3.1 *The restriction morphisms*

$$\begin{aligned}\boldsymbol{H}^k(X, (\Omega_X^\bullet(*D)[u, u^{-1}], \mathrm{d} + \mathrm{d}f/u))&\\ \longrightarrow \boldsymbol{H}^k(Y, (\Omega_Y^\bullet(*H)[u, u^{-1}], \mathrm{d} + \mathrm{d}f/u))&\\ \longrightarrow \boldsymbol{H}^k(U, (\Omega_U^\bullet[u, u^{-1}], \mathrm{d} + \mathrm{d}f/u))& \qquad (9.3.1\,*)\end{aligned}$$

are isomorphisms of $\mathbb{C}[u, u^{-1}]$*-modules compatible with the action of* ∂_u *for each* k*, as well as the natural morphisms*

$$\begin{aligned}\boldsymbol{H}^k\big(Y, (u^{-\bullet}\Omega_Y^\bullet(\log H)[u], \mathrm{d} + \mathrm{d}f/u)\big) \otimes_{\mathbb{C}[u]} \mathbb{C}[u, u^{-1}]&\\ \longrightarrow \boldsymbol{H}^k\big(Y, (u^{-\bullet}\Omega_Y^\bullet(\log H)[u, u^{-1}], \mathrm{d} + \mathrm{d}f/u)\big)&\\ \longrightarrow \boldsymbol{H}^k(Y, (\Omega_Y^\bullet(*H)[u, u^{-1}], \mathrm{d} + \mathrm{d}f/u)).& \qquad (9.3.1\,{**})\end{aligned}$$

Proof Compatibility with the action of ∂_u is clear, as it already holds at the level of complexes. The isomorphism property follows Lemma 9.2.8, except for the last morphism of (9.3.1 **).

For the latter, its left-hand side is $\mathbb{C}[u, u^{-1}]$-free of finite rank, after Theorem A. Its rank is given in Remark 9.2.9. The right-hand side is interpreted as the localized Fourier transform of the pushforward by f (at a suitable degree) of the $\mathcal{D}_Y$-module $\mathcal{O}_Y(*H)$. It is well-known to be $\mathbb{C}[u, u^{-1}]$-free of the rank given by Remark 9.2.9. If we fix $u_o \neq 0$, then the corresponding morphism in cohomology is an isomorphism, as follows from [30, Cor. 1.4] and noticed in [9, §1.2]. The conclusion follows from Lemma 9.1.3 (over the ring $\mathbb{C}[u, u^{-1}]$ instead of $\mathbb{C}[u]$). □

Remark 9.3.2 The interpretation in terms of Fourier transform shows that the action of ∂_u has a regular singularity at $u = \infty$, since the Gauss-Manin systems of f have regular singularity at each of their singularities.

In order to interpret $\boldsymbol{H}^k\big(Y, (\Omega_Y^\bullet(\log H)(-H)[u,u^{-1}], \mathrm{d}+\mathrm{d}f/u)\big)$ (to which the first isomorphism of (9.3.1 **) applies in a similar way) in meromorphic terms, we work in the category of holonomic $\mathcal{D}$-modules on Y. More precisely, we consider the ring

$$\mathcal{D}_Y^u := \mathcal{D}_Y[u,u^{-1}] = \mathcal{D}_Y \otimes_{\mathbb{C}} \mathbb{C}[u,u^{-1}]$$

of algebraic differential operator on Y with coefficients in $\mathcal{O}_Y^u := \mathcal{O}_Y[u,u^{-1}]$, so that the base ring is $\mathbb{C}[u,u^{-1}]$ instead of the field $\mathbb{C}$ (it would be equivalent to consider differential operators on $Y \times \mathbb{G}_{\mathrm{m}}$ relative to the projection to $\mathbb{G}_{\mathrm{m}}$).

The $\mathcal{D}_Y^u$-module $E_Y^{f/u} = (\mathcal{O}_Y^u, \mathrm{d}+\mathrm{d}f/u)$ comes with two localizations along H, denoted by $E_Y^{f/u}(*H)$ and $E_Y^{f/u}(!H)$. They satisfy the following properties:

- $E_Y^{f/u}(*H)$ is generated, as a $\mathcal{D}_Y^u$-module, by $E_Y^{f/u}(H)$ which is a $\mathcal{D}_Y^u(\log H)$-module.[2] It satisfies moreover

$$E_Y^{f/u}(*H) \simeq \mathcal{D}_Y^u \otimes_{\mathcal{D}_Y^u(\log H)} (E_Y^{f/u}(H)).$$

- $E_Y^{f/u}(!H)$ is defined as

$$E_Y^{f/u}(!H) := \mathcal{D}_Y^u \otimes_{\mathcal{D}_Y^u(\log H)} (E_Y^{f/u}(-H)),$$

  where $E_Y^{f/u}(-H)$ is regarded as a $\mathcal{D}_Y^u(\log H)$-module.

Lemma 9.3.3 *The $\mathbb{C}[u,u^{-1}]$-modules $H^k_{\mathrm{dR}}(Y, E_Y^{f/u}(*H))$ and $H^k_{\mathrm{dR}}(Y, E_Y^{f/u}(!H))$ are free of finite rank for each k.*

Proof For a *regular holonomic* $\mathcal{D}$-module M on the affine line $\mathbb{A}^1$ with affine coordinate t, the de Rham cohomology $H^k_{\mathrm{dR}}(\mathbb{A}^1, M[u,u^{-1}] \otimes E^{t/u})$ is nonzero in degree $k = 1$ at most, and this cohomology is $\mathbb{C}[u,u^{-1}]$-locally free of finite rank (see e.g. [18]). On noting that, for $\star = *, !$, we have $E_Y^{f/u}(\star H) \simeq \mathcal{O}_Y(\star H) \otimes_{\mathcal{O}_Y} E_Y^{f/u}$ (where $\mathcal{O}_Y(\star H)$ is defined in a way similar to that of $E_Y^{f/u}(\star H)$), we obtain the statement by applying the previous result to the $\mathcal{D}$-module pushforward $M = \mathcal{H}^{k-1-n} f_+ \mathcal{O}_Y(\star H)$, which is known to be regular holonomic. □

Proof of Theorem B As already noticed, the right-hand side in (9.3.1 **) can be rewritten as $H^{k-n}_{\mathrm{dR}}(Y, E_Y^{f/u}(*H))$, and this yields the first line (B*) of Theorem B.

We now prove the isomorphism of the second line (B!). The proof can also be adapted to the first line, and thereby gives another way to obtain (9.3.1 **). For a $\mathcal{D}_Y^u$-module M, the de Rham complex is a realization (up to a shift $[n]$) of

[2] $\mathcal{D}_Y^u(\log H)$ is locally generated by vector fields which are logarithmic along H, and the notation $(-\log H)$ should be more adapted.

$\omega_Y^u \otimes^L_{\mathcal{D}_Y^u} M$ and, similarly, for a $\mathcal{D}_Y^u(\log H)$-module N, the logarithmic de Rham complex of N is a realization (up to a shift $[n]$) of $\omega_Y^u \otimes^L_{\mathcal{D}_Y^u(\log H)} N$, where the canonical sheaf $\omega_Y^u = \omega_Y \otimes_{\mathbb{C}} \mathbb{C}[u, u^{-1}]$ is equipped with its natural structure of right $\mathcal{D}_Y^u$-module. We interpret (up to a shift $[n]$) the left-hand side of (B!) as the hypercohomology on Y of $\omega_Y^u \otimes^L_{\mathcal{D}_Y^u(\log H)} E_Y^{f/u}(-H)$ and the right hand side as that of $\omega_Y^u \otimes^L_{\mathcal{D}_Y^u} E_Y^{f/u}(!H)$. Due to the isomorphism

$$\omega_Y^u \otimes^L_{\mathcal{D}_Y^u} (\mathcal{D}_Y^u \otimes^L_{\mathcal{D}_Y^u(\log H)} E_Y^{f/u}(-H)) \simeq \omega_Y^u \otimes^L_{\mathcal{D}_Y^u(\log H)} E_Y^{f/u}(-H),$$

the isomorphism (B!) would follow from the isomorphism

$$\mathcal{D}_Y^u \otimes_{\mathcal{D}_Y^u(\log H)} E_Y^{f/u}(-H) \simeq \mathcal{D}_Y^u \otimes^L_{\mathcal{D}_Y^u(\log H)} E_Y^{f/u}(-H).$$

Although $\mathcal{D}_Y^u$ is not $\mathcal{D}_Y^u(\log H)$-flat, one can use the criterion of [8, Prop. B.5]: this isomorphism holds if in any local coordinate system adapted to the divisor H where $H = \{x_1 \cdots x_\ell = 0\}$, any subsequence of the sequence $(x_1, \dots, x_\ell)$ is a regular sequence for $E_Y^{f/u}(-H)$. In the present setting, this criterion is easily checked, and this ends the proof of (B!). By replacing $E_Y^{f/u}(-H)$ with $E_Y^{f/u}(H)$, one would obtain another proof of (B*). □

9.4 Another Approach to Theorem A

9.4.1 Application of the Theory of Mixed Hodge Module

We will make use of the theory of mixed Hodge modules of M. Saito [27, 28] in order to prove $\mathbb{C}[u]$-freeness in Theorem A.

We consider the algebraic mixed Hodge modules on Y whose underlying filtered $\mathcal{D}_Y$-modules are respectively $(\mathcal{O}_Y(*H), F_\bullet\mathcal{O}_Y(*H))$ and $(\mathcal{O}_Y(!H), F_\bullet\mathcal{O}_Y(!H))$. There exists a natural morphism of mixed Hodge modules inducing the natural morphism of filtered $\mathcal{D}_Y$-modules:

$$(\mathcal{O}_Y(!H), F_\bullet\mathcal{O}_Y(!H)) \longrightarrow (\mathcal{O}_Y(*H), F_\bullet\mathcal{O}_Y(*H)).$$

It is understood that the Hodge filtrations $F_\bullet\mathcal{O}_Y(!H)$, $F_\bullet\mathcal{O}_Y(*H)$ are coherent filtrations with respect to the filtration $F_\bullet\mathcal{D}_Y$ of $\mathcal{D}_Y$ by the order of differential operators.

We consider the Rees module construction, by setting $R_F\mathcal{D}_Y = \bigoplus_k F_k\mathcal{D}_Y u^k$ and similarly for filtered $\mathcal{D}_Y$-modules. In particular, $R_F\mathcal{O}_Y = \mathcal{O}_Y[u]$.

We set $G_0[\star H] = R_F(\mathcal{O}_Y(\star H))$, with $\star = !, *$, that we consider as an $R_F\mathcal{D}_Y$-module. We obtain corresponding de Rham complexes of $\mathbb{C}[u]$-modules:

$$\mathrm{DR}_Y\, G_0[\star H] = \big\{0 \to R_F(\mathcal{O}_Y(\star H)) \xrightarrow{\mathrm{d}+\mathrm{d}f/u} \cdots \xrightarrow{\mathrm{d}+\mathrm{d}f/u} u^{-n}\Omega^n_Y \otimes R_F(\mathcal{O}_Y(\star H)) \to 0\big\}.$$

Proposition 9.4.1 *We have a commutative diagram for each k:*

$$\begin{array}{ccc} \boldsymbol{H}^k\big(Y, (u^{-\bullet}\Omega^\bullet_Y(\log H)(-H)[u], \mathrm{d}+\mathrm{d}f/u)\big) & \longrightarrow & \boldsymbol{H}^k\big(Y, (u^{-\bullet}\Omega^\bullet_Y(\log H)[u], \mathrm{d}+\mathrm{d}f/u)\big) \\ \wr\big\downarrow & & \big\downarrow\wr \\ \boldsymbol{H}^k\big(Y, \mathrm{DR}_Y\, G_0[!H]\big) & \longrightarrow & \boldsymbol{H}^k\big(Y, \mathrm{DR}_Y\, G_0[*H]\big) \end{array}$$

Proof The proof of the isomorphisms is similar to that of Theorem B by using the sheaves $R_F\mathcal{D}_Y$ and $R_F\mathcal{D}_Y(\log H)$ (denoted respectively $\widetilde{\mathcal{D}}_Y$ and $\widetilde{\mathcal{D}}_Y(\log H)$ in [8]) instead of the sheaves $\mathcal{D}^u_Y$ and $\mathcal{D}^u_Y(\log H)$. The criterion of [8, Prop. B.5] applies in a straightforward way. □

Remark 9.4.2 The notation $G_0(*H)$ (instead of $G_0[*H]$) would mean $(R_F\mathcal{O}_Y)(*H)$ (instead of $R_F(\mathcal{O}_Y(*H))$). Thus $G_0(*H) = \mathcal{O}_Y(*H)[u]$, and the associated twisted de Rham complex is that considered at the end of Sect. 9.1.3. Its hypercohomology may not be of finite type, as we have seen in Sect. 9.1.1(i). This means that, in order to obtain finite type, one needs to take into account the Hodge filtration at infinity on U (with f remaining finite). A similar comment applies to $G_0(!H)$. On the other hand, Theorem B shows that this distinction disappears if we invert u.

9.4.2 *The Case of Tame Functions*

For tame functions, we will show that there is no need to twist by $(-H)$ in the pairing (9.1) of Theorem A, similarly to what occurs in the proper case.

Definition 9.4.3 (Katz Tameness, See [15, Prop. 14.13.3]) We say that $f : U \to \mathbb{C}$ is Katz-tame if the cone of the natural morphism of complexes

$$\boldsymbol{R}f_!\mathbb{C}_U \longrightarrow \boldsymbol{R}f_*\mathbb{C}_U$$

has constant cohomology sheaves.

Proposition 9.4.4 *Assume that $f: U \to \mathbb{C}$ is Katz-tame. Then, for every k, the natural morphism*

$$\boldsymbol{H}^k\big(Y, \mathrm{DR}_Y\, G_0[!H]\big) \longrightarrow \boldsymbol{H}^k\big(Y, \mathrm{DR}_Y\, G_0[*H]\big) \tag{9.4.2 *}$$

is an isomorphism.

Proof If M is a left $\mathcal{D}_{\mathbb{A}^1}$-module on the affine line with coordinate t, we also consider it as an $\mathcal{O}_{\mathbb{A}^1}$-module with connection ∇. Let $(M, F_\bullet M)$ be the coherent filtered $\mathcal{D}$-module and let $R_F M$ denote the associated $R_F\mathcal{D}_{\mathbb{A}^1}$-module equipped with its u-connection $u\nabla$. The twisted de Rham complex

$$0 \longrightarrow R_F M \xrightarrow{\nabla + \mathrm{d}t/u} u^{-1}\Omega^1_{\mathbb{A}^1} \otimes R_F M \longrightarrow 0$$

has nonzero cohomology in degree 1 at most (i.e., $\nabla + \mathrm{d}t/u$ is injective). Furthermore, it also has nonzero hypercohomology in degree 1 at most: hypercohomology is computed by means of global sections on $\mathbb{A}^1$ of the latter complex, and injectivity of $\nabla + \mathrm{d}t/u$ is checked similarly.

On the other hand, for each j, the j-th pushforward by the map $f: Y \to \mathbb{A}^1$ of the Rees modules $R_F\mathcal{O}_Y(\star H)$ ($\star = !, *$) takes the form $R_F M$ for some filtered $\mathcal{D}_{\mathbb{A}^1}$-module M: this property is equivalent to the degeneration at E_1 of the spectral sequence attached to the proper pushforward of a coherent filtered $\mathcal{D}_Y$-module when the latter underlies a mixed Hodge module [28]. It follows that $\boldsymbol{H}^k\big(Y, \mathrm{DR}_Y\, G_0[\star H]\big)$ can be computed as the hypercohomology of $R_F M^k_\star$, where $R_F M^k_\star$ is the $R_F\mathcal{D}_{\mathbb{A}^1}$-module underlying the pushforward (of the suitable degree) of $R_F\mathcal{O}_Y(\star H)$.

The morphism of constructible complexes in the assumption of Katz-tameness comes from a morphism of the corresponding objects in the derived category $\mathsf{D}^{\mathrm{b}}(\mathsf{MHM}(\mathbb{A}^1))$ (see [28]), hence of the corresponding cohomology mixed Hodge modules $\mathcal{M}^j_!$ and $\mathcal{M}^j_*$. □

Lemma 9.4.5 *Under the assumption of Katz-tameness, for each k, the kernel and cokernel of the natural morphism $\mathcal{M}^j_! \to \mathcal{M}^j_*$ are constant mixed Hodge modules (i.e., whose associated perverse sheaf is the constant sheaf up to a shift).*

Proof It is a matter of proving that a constructible complex on $\mathbb{A}^{1\,\mathrm{an}}$ whose cohomology is constant has also constant perverse cohomology. This is standard (e.g. by using that a constructible complex on $\mathbb{A}^{1\,\mathrm{an}}$ has constant cohomology resp. perverse cohomology if and only if for any $c \in \mathbb{A}^{1\,\mathrm{an}}$ the associated complex of vanishing cycles at c is isomorphic to zero). □

A constant mixed Hodge module on $\mathbb{A}^1$ has a finite filtration (the weight filtration) whose pure graded Hodge modules are also constant, and the associated filtered $\mathcal{D}_{\mathbb{A}^1}$-modules are isomorphic to $\mathcal{O}_{\mathbb{A}^1}$ with its standard filtration possibly shifted, so that $R_F\mathcal{O}_{\mathbb{A}^1} \simeq \mathcal{O}_{\mathbb{A}^1}[u]$. With the notation above, for each k, the kernel and cokernel of the natural morphism $R_F M^k_! \to R_F M^k_*$ are of that form.

By the first part of the proof, that the natural morphism (9.4.2 *) is an isomorphism will thus be proved if we prove that the hypercohomology of the twisted de Rham complex associated to $R_F \mathcal{O}_{\mathbb{A}^1}$ is zero, that is, the kernel and cokernel of

$$\mathbb{C}[t,u] \xrightarrow{\partial_t + 1/u} u^{-1}\mathbb{C}[t,u]$$

are zero. This is a simple check (for the vanishing of the cokernel, one uses that a polynomial of degree d in $\mathbb{C}[t]$ is annihilated by ∂_t^{d+1}).

It follows immediately from Proposition 9.4.1 that, in the tame case, we can omit the twist by $(-H)$ in (9.1):

Corollary 9.4.6 *Assume that $f : U \to \mathbb{C}$ is Katz-tame. Then, for each k, the pairing* (9.1) *induces a nondegenerate pairing between free $\mathbb{C}[u]$-modules of finite rank:*

$$\begin{aligned}&\boldsymbol{H}^{n+k}\big(Y, (u^{-\bullet}\Omega_Y^\bullet(\log H)[u], \mathrm{d} + \mathrm{d}f/u)\big)\\ &\qquad\otimes_{\mathbb{C}[u]} \boldsymbol{H}^{n-k}\big(Y, (u^{-\bullet}\Omega_Y^\bullet(\log H)[u], \mathrm{d} - \mathrm{d}f/u)\big) \longrightarrow u^{-n}\mathbb{C}[u].\end{aligned}$$

According to (9.3.1 **), this implies:

Corollary 9.4.7 *Assume that $f : U \to \mathbb{C}$ is Katz-tame. Then, for each k, the pairing* (9.1) *induces a nondegenerate pairing between free $\mathbb{C}[u, u^{-1}]$-modules of finite rank:*

$$\begin{aligned}&\boldsymbol{H}^{n+k}(U, (\Omega_U^\bullet[u, u^{-1}], \mathrm{d} + \mathrm{d}f/u))\\ &\qquad\otimes_{\mathbb{C}[u,u^{-1}]} \boldsymbol{H}^{n-k}(U, (\Omega_U^\bullet[u, u^{-1}], \mathrm{d} + \mathrm{d}f/u)) \longrightarrow \mathbb{C}[u, u^{-1}].\end{aligned}$$

Example 9.4.8 Assume that H is smooth and $f_{|H}$ is smooth. It follows that for each $t \in \mathbb{A}^1$, $f^{-1}(t)$ is smooth near H and cuts H transversally. Let $j : U \hookrightarrow Y$ denote the open inclusion and $i : H \hookrightarrow Y$ the complementary closed inclusion. Then the cone of the natural morphism $j_!\mathbb{C}_U \to \boldsymbol{R}j_*\mathbb{C}_U$ is supported on H and its restriction to H is isomorphic to $i^{-1}\boldsymbol{R}j_*\mathbb{C}_U$, which has constant cohomology sheaves. It follows that the Katz-tameness condition is satisfied by f on U (use e.g. the argument with vanishing cycles and the good behaviour of the vanishing cycle functor by proper pushforward). Let us also notice that the assumptions in Theorem $\hat{\mathrm{C}}$ are also fulfilled in this example.

9.5 The Formal Pairing

9.5.1 Proof of Theorem $\hat{A}$

We first make formal the Kontsevich complex with the u-parameter. For that purpose, we set $\Omega^k_f[\![u]\!] := \varprojlim_\ell \Omega^k_f[u]/u^\ell\Omega^k_f[u]$. We refer to [21] for some properties of this construction.

Proposition 9.5.1 *For any k, we have*

$$\boldsymbol{H}^k\big(X, (\Omega^\bullet_f[\![u]\!], u\mathrm{d} + \mathrm{d}f)\big) = \mathbb{C}[\![u]\!] \otimes_{\mathbb{C}[u]} \boldsymbol{H}^k\big(X, (\Omega^\bullet_f[u], u\mathrm{d} + \mathrm{d}f)\big)$$

and a similar property for $\Omega^\bullet_f(-D)$, $\Omega^\bullet_f(-H)$, and the pairs $(Y, \Omega^\bullet_Y(\log H))$ and $(Y, \Omega^\bullet_Y(\log H)(-H))$.

Proof We prove the case of $\Omega^\bullet_f$, the other cases being proved similarly (by using Lemma 9.2.7 for the latter two pairs). By a straightforward induction, arguing as in Lemma 9.1.3 (due to the freeness property in Corollary 9.2.5), we find for each $\ell \geq 1$:

$$\begin{aligned}\boldsymbol{H}^k&\big(X, (\Omega^\bullet_f[u]/u^\ell\Omega^\bullet_f[u], u\mathrm{d} + \mathrm{d}f)\big)\\ &\simeq \boldsymbol{H}^k\big(X, (\Omega^\bullet_f[u], u\mathrm{d} + \mathrm{d}f)\big)/u^\ell\boldsymbol{H}^k\big(X, (\Omega^\bullet_f[u], u\mathrm{d} + \mathrm{d}f)\big),\end{aligned}$$

so that the conclusion follows. □

Proof of Theorem $\hat{A}$ We just apply the functor $\mathbb{C}[\![u]\!]\otimes_{\mathbb{C}[u]}$ to the statements of Theorem A, according to Proposition 9.5.1. □

9.5.2 Proof of Theorem $\hat{C}$

We consider the complex $(\Omega^\bullet_U[\![u]\!], u\mathrm{d} + \mathrm{d}f)$. The $\mathbb{C}$-constructible complex of vanishing cycles $\phi_{f-c}\mathbb{C}_{U^{\mathrm{an}}}$ (which is perverse up to a shift) will come into play.

Proposition 9.5.2 *Assume that the critical set of f is compact. Then the $\mathbb{C}[\![u]\!]$-module $\boldsymbol{H}^k\big(U, (\Omega^\bullet_U[\![u]\!], u\mathrm{d} + \mathrm{d}f)\big)$ is of finite type for each k. It is $\mathbb{C}[\![u]\!]$-free for every k if and only if*

$$\dim \boldsymbol{H}^k\big(U, (\Omega^\bullet_U, \mathrm{d}f)\big) = \sum_{c\in\mathbb{C}} \dim \boldsymbol{H}^{k-1}(f^{-1}(c)^{\mathrm{an}}, \phi_{f-c}\mathbb{C}_{U^{\mathrm{an}}}) \quad \forall k.$$

Note that we do not assert the existence of an isomorphism similar to that of Proposition 9.5.1 for $\boldsymbol{H}^k\big(U, (\Omega^\bullet_U[\![u]\!], u\mathrm{d} + \mathrm{d}f)\big)$, since we do not know whether the assumption of finite-dimensionality of $\boldsymbol{H}^k\big(U, (\Omega^\bullet_U, \mathrm{d}f)\big)$ is enough to ensure that $\boldsymbol{H}^k\big(U, (\Omega_U^\bullet[\![u]\!], u\mathrm{d} + \mathrm{d}f)\big)$ is of finite type over $\mathbb{C}[u]$. We first review some results of [4, 21, 23].

Lemma 9.5.3 *For each k, the natural restriction morphism*

$$\boldsymbol{H}^k\big(X, (\Omega_X^\bullet(*D)[\![u]\!], u\mathrm{d} + \mathrm{d}f)\big) \longrightarrow \boldsymbol{H}^k\big(U, (\Omega_U^\bullet[\![u]\!], u\mathrm{d} + \mathrm{d}f)\big)$$

is an isomorphism.

Proof By a spectral sequence argument, it is enough to prove that for any j, k, the restriction morphism

$$H^k\big(X, \Omega_X^j(*D)[\![u]\!]\big) \longrightarrow H^k\big(U, \Omega_U^j[\![u]\!]\big)$$

is an isomorphism. It follows from [13, §4], as noted in [21, §2.a], that in such cases, $[\![u]\!]$ commutes with taking cohomology. Then the assertion is clear. □

Lemma 9.5.4 (Algebraic/Analytic Comparison) *For each k, the natural morphism*

$$\boldsymbol{H}^k\big(X, (\Omega_X^\bullet(*D)[\![u]\!], u\mathrm{d} + \mathrm{d}f)\big) \longrightarrow \boldsymbol{H}^k\big(X^{\mathrm{an}}, (\Omega_{X^{\mathrm{an}}}^\bullet(*D)[\![u]\!], u\mathrm{d} + \mathrm{d}f)\big)$$

is an isomorphism.

Proof We consider the morphism between the spectral sequences associated to the filtration of the de Rham complexes by the stupid truncation. The morphism at the E_1 level is

$$\boldsymbol{H}^k\big(X, \Omega_X^j(*D)[\![u]\!]\big) \longrightarrow \boldsymbol{H}^k\big(X^{\mathrm{an}}, \Omega_{X^{\mathrm{an}}}^j(*D)[\![u]\!]\big).$$

By [21, (2.3)], this morphism is an isomorphism, which implies the lemma by a spectral sequence argument. □

Lemma 9.5.5 *The natural morphism of complexes*

$$(\Omega_{X^{\mathrm{an}}}^\bullet(*D)[\![u]\!], u\mathrm{d} + \mathrm{d}f) \longrightarrow \boldsymbol{R}j_*(\Omega_{U^{\mathrm{an}}}^\bullet[\![u]\!], u\mathrm{d} + \mathrm{d}f)$$

is a quasi-isomorphism.

Proof Although the result of [21, Prop. 4.1] is stated for $(\!(u)\!)$ instead of $[\![u]\!]$, one can follow its proof with $[\![u]\!]$ instead of $(\!(u)\!)$. □

Lemma 9.5.6 *The order of the u-torsion of each cohomology sheaf*

$$\mathcal{H}^j(\Omega_{U^{\mathrm{an}}}^\bullet[\![u]\!], u\mathrm{d}+\mathrm{d}f)$$

is locally bounded and, modulo its torsion, this sheaf is a constructible sheaf of $\mathbb{C}[\![u]\!]$*-modules of finite type.*

Proof This is essentially [4, Th. 1] if we work with the isomorphic complex $(u^{-\bullet}\Omega_U^\bullet[u], \mathrm{d} + \mathrm{d}f/u)$. Let us give details on the reduction to loc. cit. In [23, (1.5.2)], two complexes are considered, with the notation ∂_t^{-1} for our notation u (and the function $-f$ is considered, instead of f). One is $F_0 = (u^{-\bullet}\Omega_U^\bullet[u], \mathrm{d} + \mathrm{d}f/u)$ and for each $\ell \geq 0$, the subcomplex F_ℓ with terms $u^\ell(u^{-\bullet}\Omega_U^\bullet[u])$. The other one is G_0 and similarly G_ℓ, with inclusions $u^{\ell+1}F_0 \subset u^\ell G_0 \subset u^\ell F_0$ for each $\ell \geq 0$, so that F_0/G_0 has u-torsion of order one. The result of [4, Th. 1] together with the identification of [23, (1.5.5)] implies the assertion of the lemma for the complex $\varprojlim_\ell(G_0/G_\ell)$. On the other hand, the complex occurring in the lemma is $\varprojlim_\ell(F_0/F_\ell)$, so the inclusions above yield the statement of the lemma. □

Proof of Proposition 9.5.2 We start with the finiteness statement. According to Lemmas 9.5.3–9.5.5, we are reduced to proving the $\mathbb{C}[\![u]\!]$-finiteness of $\boldsymbol{H}^k\big(U^{\mathrm{an}}, (\Omega_{U^{\mathrm{an}}}^\bullet[\![u]\!], u\mathrm{d} + \mathrm{d}f)\big)$ for each k.

The complex $(\Omega_{U^{\mathrm{an}}}^\bullet[\![u]\!], u\mathrm{d}+\mathrm{d}f)$ is supported on the critical set of $f : U \to \mathbb{A}^1$: indeed, this follows from the fact that, on the product $\Delta \times V^{\mathrm{an}}$ of an open disc Δ with coordinate x by a complex manifold V^{an}, the morphism

$$\mathcal{O}(\Delta \times V^{\mathrm{an}})[\![u]\!] \xrightarrow{\;u\partial_x + 1\;} \mathcal{O}(\Delta \times V^{\mathrm{an}})[\![u]\!]$$

is an isomorphism, which is easily checked. Since the critical set of f is compact by assumption, we conclude from Lemma 9.5.6 that the order of the u-torsion of each cohomology sheaf is bounded, say by N.

We consider the spectral sequence with

$$E_2^{p,q} = H^p\big(U^{\mathrm{an}}, \mathcal{H}^q(\Omega_{U^{\mathrm{an}}}^\bullet[\![u]\!], u\mathrm{d} + \mathrm{d}f)\big).$$

Since it can be realized as the spectral sequence of a bounded double complex, it converges at a finite step. Let $\mathcal{T}^q$ be the $\mathbb{C}[\![u]\!]$-torsion of $\mathcal{H}^q := \mathcal{H}^q(\Omega_{U^{\mathrm{an}}}^\bullet[\![u]\!], u\mathrm{d} + \mathrm{d}f)$ and $\mathcal{I}^q$ the quotient $\mathcal{H}^q/\mathcal{T}^q$. Since $\mathcal{H}^q$ is supported on a compact set, Lemma 9.5.6 implies that $H^p(U^{\mathrm{an}}, \mathcal{I}^q)$ has finite type over $\mathbb{C}[\![u]\!]$ and $H^p(U^{\mathrm{an}}, \mathcal{T}^q)$ is of N-torsion. Therefore, $E_2^{p,q}$ has $\mathbb{C}[\![u]\!]$-torsion of order bounded by N and its quotient by torsion has finite type over $\mathbb{C}[\![u]\!]$. This property goes through the spectral sequence, and we conclude that it holds for $\boldsymbol{H}^k\big(U^{\mathrm{an}}, (\Omega_{U^{\mathrm{an}}}^\bullet[\![u]\!], u\mathrm{d} + \mathrm{d}f)\big)$.

On the other hand, the quotient complex $(\Omega_{U^{\mathrm{an}}}^\bullet[\![u]\!]/u^N\Omega_{U^{\mathrm{an}}}^\bullet[\![u]\!], u\mathrm{d} + \mathrm{d}f)$ has a finite filtration $F^\bullet$ whose graded terms are all isomorphic to the complex $(\Omega_{U^{\mathrm{an}}}^\bullet, \mathrm{d}f)$. The latter is a complex in $\mathsf{Mod}_{\mathrm{coh}}(\mathcal{O}_{U^{\mathrm{an}}})$ supported on the compact critical set of f. Therefore, $\boldsymbol{H}^k\big(U^{\mathrm{an}}, \mathrm{gr}_F(\Omega_{U^{\mathrm{an}}}^\bullet[\![u]\!]/u^N\Omega_{U^{\mathrm{an}}}^\bullet[\![u]\!], u\mathrm{d}+\mathrm{d}f)\big)$ is finite dimensional, and so is $\boldsymbol{H}^k\big(U^{\mathrm{an}}, (\Omega_{U^{\mathrm{an}}}^\bullet[\![u]\!]/u^N\Omega_{U^{\mathrm{an}}}^\bullet[\![u]\!], u\mathrm{d} + \mathrm{d}f)\big)$. We conclude the proof of the first statement by considering the hypercohomology exact sequence deduced from the exact sequence

$$0 \longrightarrow (\Omega^\bullet_{U^{\mathrm{an}}}[\![u]\!], u\mathrm{d}+\mathrm{d}f) \xrightarrow{u^N} (\Omega^\bullet_{U^{\mathrm{an}}}[\![u]\!], u\mathrm{d}+\mathrm{d}f) \\ \longrightarrow (\Omega^\bullet_{U^{\mathrm{an}}}[\![u]\!]/u^N\Omega^\bullet_{U^{\mathrm{an}}}[\![u]\!], u\mathrm{d}+\mathrm{d}f) \longrightarrow 0.$$

We now consider the second statement. Since $\mathbb{C}(\!(u)\!)$ is $\mathbb{C}[\![u]\!]$-flat, tensoring with $\mathbb{C}(\!(u)\!)$ commutes with taking cohomology, and we have

$$\begin{aligned} \boldsymbol{H}^k\big(U, (\Omega^\bullet_U(\!(u)\!), u\mathrm{d}+\mathrm{d}f)\big) &= \mathbb{C}(\!(u)\!) \otimes_{\mathbb{C}[\![u]\!]} \boldsymbol{H}^k\big(U, (\Omega^\bullet_U[\![u]\!], u\mathrm{d}+\mathrm{d}f)\big) \\ &= \boldsymbol{H}^k\big(U, (\Omega^\bullet_U[\![u]\!], u\mathrm{d}+\mathrm{d}f)\big)[u^{-1}]. \end{aligned}$$

Indeed, this is seen first for each $H^k(U, \Omega^j_U(\!(u)\!))$ by Noetherianity of U, and then deduced for $\boldsymbol{H}^k\big(U, (\Omega^\bullet_U(\!(u)\!), u\mathrm{d}+\mathrm{d}f)\big)$ by a spectral sequence argument already used.

We now apply the property that, for a $\mathbb{C}[\![u]\!]$-module of finite type M, M is $\mathbb{C}[\![u]\!]$-free if and only if

$$\dim_{\mathbb{C}} M/uM = \dim_{\mathbb{C}(\!(u)\!)} M[u^{-1}].$$

Set $M^k = \boldsymbol{H}^k\big(U, (\Omega^\bullet_U[\![u]\!], u\mathrm{d}+\mathrm{d}f)\big)$. We argue by induction on the length of the long exact sequence of hypercohomology associated with the short exact sequence

$$0 \longrightarrow (\Omega^\bullet_U[\![u]\!], u\mathrm{d}+\mathrm{d}f) \xrightarrow{u} (\Omega^\bullet_U[\![u]\!], u\mathrm{d}+\mathrm{d}f) \longrightarrow (\Omega^\bullet_U, \mathrm{d}f) \longrightarrow 0.$$

Let k be such that $M^j = 0$ for $j > k$. Then $M^k/uM^k = \boldsymbol{H}^k\big(U, (\Omega^\bullet_U, \mathrm{d}f)\big)$. On the other hand, $\dim_{\mathbb{C}(\!(u)\!)} M^k[u^{-1}] = \dim \sum_{c\in\mathbb{C}} \dim \boldsymbol{H}^{k-1}(f^{-1}(c)^{\mathrm{an}}, \phi_{f-c}\mathbb{C}_{U^{\mathrm{an}}})$ by the main theorem of [23] (see also [21, Th. 1.1]). We conclude that M^k is $\mathbb{C}[\![u]\!]$-free if and only if both dimensions are equal. In such a case, $u : M^k \to M^k$ is injective. We can thus truncate the long exact sequence mentioned above after $k-1$ and we conclude by decreasing induction on k. □

Proof of the First Part of Theorem $\hat{\mathbf{C}}$ In view of Proposition 9.5.2, we only need to show the equality of dimensions occurring in that proposition under Condition (9.3). This is precisely [20, Th. 2]. □

Proof of the Second Part of Theorem $\hat{\mathbf{C}}$ In view of Theorem Â, it is enough to prove that the natural morphisms

$$\boldsymbol{H}^k\big(X, (\Omega^\bullet_f(-D)[\![u]\!], u\mathrm{d}+\mathrm{d}f)\big) \longrightarrow \boldsymbol{H}^k\big(X, (\Omega^\bullet_f[\![u]\!], u\mathrm{d}+\mathrm{d}f)\big) \\ \longrightarrow \boldsymbol{H}^k\big(X, (\Omega^\bullet_X(*D)[\![u]\!], u\mathrm{d}+\mathrm{d}f)\big)$$

are isomorphisms. Together with Lemma 9.5.3, this shows the correspondence with the $\mathbb{C}[\![u]\!]$-modules of Theorem Â. These are free $\mathbb{C}[\![u]\!]$-modules of finite rank, according to the first part of Theorem $\hat{\mathrm{C}}$. It is thus enough to prove this modulo $u\mathbb{C}[\![u]\!]$. We are left with the morphisms

$$\boldsymbol{H}^k\big(X, (\Omega_f^\bullet(-D), \mathrm{d}f)\big) \longrightarrow \boldsymbol{H}^k\big(X, (\Omega_f^\bullet, \mathrm{d}f)\big) \longrightarrow \boldsymbol{H}^k\big(X, (\Omega_X^\bullet(*D), \mathrm{d}f)\big).$$

Since X is compact, we can replace the complexes which one takes hypercohomology of by their analytic counterpart on X^{an} by GAGA. These analytic complexes are supported on the critical set of f, which is contained in U, hence they coincide in a neighbourhood of this set. The assertion follows. □

9.6 Geometry

Let $\mathbb{A}^{n+1}$ be the affine chart with coordinates $(x_0, \ldots, x_n)$ in $\mathbb{P}^{n+1}$ with complement $\mathbb{P}^n_\infty$, and let $\varpi : \widetilde{\mathbb{P}}^{n+1} \to \mathbb{P}^{n+1}$ be the blow-up of $\mathbb{P}^{n+1}$ at the origin, with exceptional divisor $\varpi^{-1}(0) = \mathbb{P}^n$. We identify $\varpi^{-1}(\mathbb{A}^{n+1})$ with the total space $\mathrm{Tot}(\mathcal{O}_{\mathbb{P}^n}(-1))$ and $\widetilde{\mathbb{P}}^{n+1}$ to $\mathbb{P}(\mathcal{O}_{\mathbb{P}^n}(-1) \oplus \mathbf{1})$.

For $d \geq 1$, we consider the action of μ_d on $\mathbb{A}^{n+1}$ by $x \mapsto \zeta x$ ($\zeta \in \mu_d$). This action lifts to $\widetilde{\mathbb{P}}^{n+1}$. We note that μ_d acts trivially on $\varpi^{-1}(0)$ and $\mathbb{P}^n_\infty$. Moreover,

$$\mathrm{Tot}(\mathcal{O}_{\mathbb{P}^n}(-1))/\mu_d \simeq \mathrm{Tot}(\mathcal{O}_{\mathbb{P}^n}(-d)), \qquad \mathbb{P}(\mathcal{O}_{\mathbb{P}^n}(-1) \oplus \mathbf{1})/\mu_d \simeq \mathbb{P}(\mathcal{O}_{\mathbb{P}^n}(-d) \oplus \mathbf{1}).$$

The quotient space $\mathbb{P}^{n+1}/\mu_d$ is smooth away from the origin. We have a commutative diagram

$$\begin{array}{ccc}
\mathbb{P}(\mathcal{O}_{\mathbb{P}^n}(-1) \oplus \mathbf{1}) & \xrightarrow{\ \varpi\ } & \mathbb{P}^{n+1} \\
{\scriptstyle \widetilde{\rho}_d}\big\downarrow & & \big\downarrow{\scriptstyle \rho_d} \\
\mathbb{P}(\mathcal{O}_{\mathbb{P}^n}(-d) \oplus \mathbf{1}) & \xrightarrow{\ \varpi_d\ } & \mathbb{P}^{n+1}/\mu_d
\end{array}$$

Let $f \in \mathbb{C}[x_0, \ldots, x_n]$ be a homogeneous polynomial of degree d, such that $f^{-1}(0)$ has an isolated singularity at the origin. Let $V \subset \mathbb{P}^n$ denote the smooth hypersurface defined by f and let X denote the closure of the graph $\{t - f(x) = 0\} \subset \mathbb{C}^{n+1} \times \mathbb{A}^1_t$ in $\mathbb{P}^{n+1} \times \mathbb{A}^1_t$. Since f is invariant by μ_d, it descends as a regular function f_d on $\mathbb{C}^{n+1}/\mu_d$ whose graph in $(\mathbb{C}^{n+1}/\mu_d) \times \mathbb{A}^1_t$ has closure $X_d := X/\mu_d \subset (\mathbb{P}^{n+1}/\mu_d) \times \mathbb{A}^1_t$. Similarly $\widetilde{f} := f \circ \varpi$ descends as a function on $\mathrm{Tot}(\mathcal{O}_{\mathbb{P}^n}(-d))$ whose graph has closure $\widetilde{X}_d$ in $\mathbb{P}(\mathcal{O}_{\mathbb{P}^n}(-d) \oplus \mathbf{1}) \times \mathbb{A}^1_t$. There is a natural proper modification which is an isomorphism away from the origin:

$$\pi_d : \widetilde{X}_d \longrightarrow X_d.$$

Lemma 9.6.1 *The space X_d is smooth away from the origin, and the projection $p : X_d \to \mathbb{A}^1_t$ is smooth away from the origin.* □

As a consequence, the composition $\widetilde{f}_d : \widetilde{X}_d \to \mathbb{A}^1_t$ is smooth away from $\pi_d^{-1}(0) \simeq \mathbb{P}^n$.

Lemma 9.6.2 *The critical fiber $\widetilde{f}_d^{-1}(0)$ is a reduced divisor with two components, one being $\pi_d^{-1}(0) \simeq \mathbb{P}^n$, intersecting normally along the smooth projective variety $V \subset \pi_d^{-1}(0)$.* □

As a consequence, the vanishing cycle complex $\phi_{\widetilde{f}_d}\mathbb{C}_{\widetilde{X}_d}$ is a complex of sheaves supported on V and has cohomology in degree 1 only. Moreover, $\mathcal{H}^1\phi_{\widetilde{f}_d}\mathbb{C}_{\widetilde{X}_d}$ is a local system of rank one, which is constant if $n \geq 3$, since V is 1-connected.

We apply the results of Sect. 9.2 to $Y = \widetilde{X}_d$ and $\widetilde{f}_d : \widetilde{X}_d \to \mathbb{A}^1_t$. By Theorem A, $H^k(\widetilde{X}_d, (\Omega^\bullet_{\widetilde{X}_d}[u], ud + d\widetilde{f}_d))$ is $\mathbb{C}[u]$-free of finite rank. Its rank is given by the computation of Remark 9.2.9, that is,

$$\dim \boldsymbol{H}^{k-1}(V, \phi_{\widetilde{f}_d}\mathbb{C}_{\widetilde{X}_d}) = \dim \boldsymbol{H}^{k-1}(V, \mathcal{H}^1\phi_{\widetilde{f}_d}\mathbb{C}_{\widetilde{X}_d}[-1]) = \dim H^{k-2}(V, \mathbb{C}).$$

Proposition 9.6.3 *The function $\widetilde{f}_d : Y = \widetilde{X}_d \to \mathbb{A}^1_t$ together with $U :=$ $\mathrm{Tot}(\mathcal{O}_{\mathbb{P}^n}(-d))$ and $H = Y \smallsetminus U$ satisfies the tameness property of Example 9.4.8.*

Proof Since $\widetilde{X}_d$ and X_d are equal in the neighbourhood of H, it is enough to work with X_d and f_d. Recall that X denotes the closure of the graph of f in $\mathbb{P}^{n+1} \times \mathbb{A}^1_t$. Then $X \smallsetminus (\mathbb{A}^{n+1} \times \mathbb{A}^1_t) = X \cap (\mathbb{P}^n_\infty \times \mathbb{A}^1_t)$ is the product $V \times \mathbb{A}^1_t$ and the restriction of f to it is simply the projection. It is thus obviously smooth, and the same property remains true after taking the quotient by μ_d, since μ_d acts as the identity on $\mathbb{P}^n_\infty$. □

We conclude that Corollaries 9.4.6, 9.4.7, and Theorem $\hat{\mathrm{C}}$ apply to $\widetilde{f}_d : Y = \widetilde{X}_d \to \mathbb{A}^1_t$.

Acknowledgments This work grew out from discussions with Bumsig Kim during my visit at KIAS on February 2019. It was supposed to be a starting point to understanding some questions related to gauged linear sigma models like the one described in [11, Ex. 2.4]. It was Bumsig Kim who insisted to make precise the comparison between the various dualities occurring in this context. I would like to thank KIAS for the excellent working conditions during this visit. I thank Jeng-Daw Yu for explaining some parts of his article [30] and for his comments, and the referee for noticing a mistake in the original proof of Proposition 9.5.2 and providing a correction.

References

1. Adolphson, A., Sperber, S.: On twisted de Rham cohomology. Nagoya Math. J. **146**, 55–81 (1997)
2. Arapura, D.: Geometry of cohomology support loci for local systems. I. J. Algebraic Geom. **6**(3), 563–597 (1997)

3. Baldassarri, F., D'Agnolo, A.: On Dwork cohomology and algebraic $\mathcal{D}$-modules. In: Geometric Aspects of Dwork Theory, pp. 245–253. Walter de Gruyter, Berlin (2004)
4. Barlet, D., Saito, M.: Brieskorn modules and Gauss-Manin systems for non-isolated hypersurface singularities. J. Lond. Math. Soc. (2) **76**(1), 211–224 (2007)
5. Brieskorn, E.: Die Monodromie der isolierten Singularitäten von Hyperflächen. Manuscripta Math. **2**, 103–161 (1970)
6. Chen, K.C., Yu, J.D.: The Künneth formula for the twisted de Rham and Higgs cohomologies. SIGMA Symmetry Integrability Geom. Methods Appl. **14**, Article no. 055, 14 p. (2018)
7. Dimca, A., Maaref, F., Sabbah, C., Saito, M.: Dwork cohomology and algebraic $\mathcal{D}$-modules. Math. Ann. **318**(1), 107–125 (2000)
8. Esnault, H., Sabbah, C.: Good lattices of algebraic connections. Documents Math. **24**, 175–205 (2019)
9. Esnault, H., Sabbah, C., Yu, J.D.: E_1-degeneration of the irregular Hodge filtration (with an appendix by M. Saito). J. Reine Angew. Math. **729**, 171–227 (2017)
10. Fan, H.: Schrödinger equations, deformation theory and tt^*-geometry (2011). arXiv:1107.1290
11. Favero, D., Kim, B.: General GLSM invariants and their cohomological field theories (2020). arXiv:2006.12182
12. Hartshorne, R.: Cohomology with compact supports for coherent sheaves on an algebraic variety. Math. Ann. **195**, 199–207 (1972)
13. Hartshorne, R.: On the de Rham cohomology of algebraic varieties. Publ. Math. Inst. Hautes Études Sci. **45**, 5–99 (1975)
14. Hertling, C.: Formes bilinéaires et hermitiennes pour des singularités: un aperçu. In: Singularités, vol. 18, pp. 1–17. Institut Élie Cartan, Nancy (2005). English transl.: arXiv:2011.10099
15. Katz, N.: Exponential Sums and Differential Equations, Ann. of Math. studies, vol. 124. Princeton University Press, Princeton, NJ (1990)
16. Katzarkov, L., Kontsevich, M., Pantev, T.: Bogomolov-Tian-Todorov theorems for Landau-Ginzburg models. J. Differential Geom. **105**(1), 55–117 (2017)
17. Li, S., Wen, H.: On the L^2-Hodge theory of Landau-Ginzburg models. Adv. Math. **396**, Article no. 108165, 48 p. (2022)
18. Malgrange, B.: Équations différentielles à coefficients polynomiaux, Progress in Math., vol. 96. Birkhäuser, Basel (1991)
19. Pham, F.: La descente des cols par les onglets de Lefschetz avec vues sur Gauss-Manin. In: Galligo, A., Granger, J.M., Maisonobe, Ph. (eds.) Systèmes différentiels et singularités (Luminy, 1983), *Astérisque*, vol. 130, pp. 11–47. Société Mathématique de France (1985)
20. Sabbah, C.: On a twisted de Rham complex. Tohoku Math. J. **51**, 125–140 (1999)
21. Sabbah, C.: On a twisted de Rham complex, II (2010). arXiv:1012.3818
22. Sabbah, C.: Vanishing cycles and their algebraic computation (2013). Lecture notes, Notre Dame. `sabbah_notredame1305.pdf`
23. Sabbah, C., Saito, M.: Kontsevich's conjecture on an algebraic formula for vanishing cycles of local systems. Algebraic Geom. **1**(1), 107–130 (2014)
24. Sabbah, C., Yu, J.D.: On the irregular Hodge filtration of exponentially twisted mixed Hodge modules. Forum Math. Sigma **3**, Article no. e9, 7 p. (2015)
25. Saito, K.: The higher residue pairings $K_F^{(k)}$ for a family of hypersurfaces singular points. In: Singularities, Proc. of Symposia in Pure Math., vol. 40, pp. 441–463. American Mathematical Society, Providence (1983)
26. Saito, K., Takahashi, A.: From primitive forms to Frobenius manifolds. In: R. Donagi, K. Wendland (eds.) From Hodge theory to integrability and TQFT: tt*-geometry, Proc. Symposia in Pure Math., vol. 78, pp. 31–48. American Mathematical Society, Providence, RI (2008)
27. Saito, M.: Modules de Hodge polarisables. Publ. RIMS, Kyoto Univ. **24**, 849–995 (1988)
28. Saito, M.: Mixed Hodge Modules. Publ. RIMS, Kyoto Univ. **26**, 221–333 (1990)
29. Wei, C.: Logarithmic comparison with smooth boundary divisor in mixed Hodge modules. Michigan Math. J. **69**(1), 201–223 (2020)
30. Yu, J.D.: Irregular Hodge filtration on twisted de Rham cohomology. Manuscripta Math. **144**(1–2), 99–133 (2014)

Open Access This chapter is licensed under the terms of the Creative Commons Attribution-NonCommercial-NoDerivatives 4.0 International License (http://creativecommons.org/licenses/by-nc-nd/4.0/), which permits any noncommercial use, sharing, distribution and reproduction in any medium or format, as long as you give appropriate credit to the original author(s) and the source, provide a link to the Creative Commons license and indicate if you modified the licensed material. You do not have permission under this license to share adapted material derived from this chapter or parts of it.

The images or other third party material in this chapter are included in the chapter's Creative Commons license, unless indicated otherwise in a credit line to the material. If material is not included in the chapter's Creative Commons license and your intended use is not permitted by statutory regulation or exceeds the permitted use, you will need to obtain permission directly from the copyright holder.

Chapter 10
A Kleiman Criterion for GIT Stack Quotients

Mark Shoemaker

In memory of Bumsig Kim

Abstract Kleiman's criterion states that, for X a projective scheme, a divisor D is ample if and only if it pairs positively with every non-zero element of the closure of the cone of curves. In other words, the cone of ample divisors in $N^1(X)$ is the interior of the nef cone. In this paper we present an analogous statement for a variety X acted on by a reductive group G with a choice of G-linearization $L \to X$. In this new context, the ample cone of X is replaced by a cell in the variation of GIT decomposition of the G-ample cone, and curves in X are replaced by quasimaps to $[X/G]$.

10.1 Introduction

Let X be a projective scheme over an algebraically closed field. Let $N^1(X)_{\mathbb{R}}$ denote the group of $\mathbb{R}$-divisors modulo numerical equivalence and denote by $\mathrm{Amp}(X) \subset N^1(X)_{\mathbb{R}}$ the convex cone spanned by ample divisors. Let $N_1(X)_{\mathbb{R}}$ be the group of 1-cycles in X up to numerical equivalence. Define the cone of curves $\mathrm{NE}(X)$ to be the convex cone spanned by effective curve classes in $N_1(X)_{\mathbb{R}}$.

It is a remarkable fact that the ampleness of a divisor can be determined solely based on its intersection with 1-cycles.

Theorem 10.1.1 (Kleiman's Criterion [9]) *A divisor D on X is ample if and only if*

$$D \cdot \gamma > 0$$

M. Shoemaker (✉)
Department of Mathematics, Colorado State University, Fort Collins, CO, USA
e-mail: mark.shoemaker@colostate.edu

© The Author(s) 2026

N.-G. Kang et al. (eds.), *Categorical and Enumerative Aspects of Mirror Symmetry*, KIAS Springer Series in Mathematics 5,
https://doi.org/10.1007/978-981-95-0385-8_10

for all $\gamma \in \overline{\mathrm{NE}(X)} \setminus \mathbf{0}$. *In other words, the ample cone is the interior of the dual of the cone of curves:*

$$\mathrm{Amp}(X) = \mathrm{int}\left(\mathrm{NE}(X)^{\vee}\right). \tag{10.1.1}$$

The goal of this paper is to obtain an analogous statement in the setting of variation of GIT.

10.1.1 Variation of GIT

Let G be a reductive group acting on X. As is now well-understood [3, 10, 12], there is not a canonical algebraic variety representing the "quotient" of X by G; one must first fix a linearization, i.e., a G-equivariant line bundle $L \to X$. Given such a choice of L, one can construct the *GIT quotient* $X /\!\!/_L G$, which is a categorical quotient of the L-semi-stable locus $X^{ss}(L)$, an open G-invariant subset of X.

Fix an ample G-equivariant line bundle $L \to X$ such that the semi-stable locus $X^{ss}(L)$ is nonempty. We will think of the triple of data of (X, G, L) as a *G-linearized variety* X. We would like to understand if there is an analogue of Theorem 10.1.1 for (X, G, L).

If two G-equivariant ample line bundles $L, L' \to X$ have the same semi-stable locus we say L and L' are GIT equivalent. In this case, the GIT quotients $X /\!\!/_L G$ and $X /\!\!/_{L'} G$ are isomorphic. Denote by $\mathrm{NS}^G(X)$ the group of G-equivariant line bundles up to G-algebraic equivalence (see Sect. 10.2.2) and let $\mathrm{Amp}^G(X) \subset \mathrm{NS}^G(X)_{\mathbb{Q}}$ denote the cone generated by ample line bundles. Define $C^{\circ}(L)$ to be the cone of classes of G-ample line bundles which are GIT equivalent to L:

$$C^{\circ}(L) = \{[L'] \in \mathrm{Amp}^G(X) | X^{ss}(L) = X^{ss}(L')\}.$$

Whenever $[L'] \in C^{\circ}(L)$, the line bundle $L' \to X$ descends to an ample line bundle on the GIT quotient $X /\!\!/_L G$. In fact, under sufficiently nice conditions the cone $C^{\circ}(L)$ may be identified with $\mathrm{Amp}(X /\!\!/_L G)$. It is reasonable, therefore, to view $C^{\circ}(L)$ as the analogue of the ample cone for the triple (X, G, L).

On the other hand, in general the equivalence classes

$$\{C^{\circ}(L)\}_{L \in \mathrm{Amp}^G(X)}$$

carry more information than just the isomorphism class of the GIT quotient $X /\!\!/_L G$, as illustrated by the following example.

Example 10.1.2 For simplicity we give a toric example. Let $T = \mathbb{G}_m^2$ act on $V = \mathbb{A}^3$ with charge matrix

$$\begin{pmatrix} 2 & 1 & 0 \\ 0 & 1 & 1 \end{pmatrix}.$$

Consider the characters $\theta_+, \theta_- \in \chi(T)$ with the following weights

$$\theta_+ = \begin{pmatrix} 4 \\ 2 \end{pmatrix}, \qquad \theta_- = \begin{pmatrix} 2 \\ 4 \end{pmatrix}.$$

The characters θ_+ and θ_- define T-linearizations of the trivial line bundle $\mathcal{O}_V$, which we denote by $\mathcal{O}_V(\theta_+)$ and $\mathcal{O}_V(\theta_-)$. Although θ_+ and θ_- are not GIT equivalent, the respective GIT quotients $V /\!\!/_{\theta_+} T$ and $V /\!\!/_{\theta_-} T$ are both isomorphic to $\mathbb{P}^1$. Furthermore, the T-equivariant line bundles $\mathcal{O}_V(\theta_+)$ and $\mathcal{O}_V(\theta_-)$ both descend to $\mathcal{O}_{\mathbb{P}^1}(1)$ on the GIT quotient.

In fact, the GIT equivalence classes carry not just the information of the GIT quotient, they also distinguish the various *GIT stack quotients*. In this case, the GIT stack quotients $[V^{ss}(\theta_+)/T]$ and $[V^{ss}(\theta_-)/T]$ are not isomorphic. Over $\mathbb{C}$ the first is the "football" $[\mathbb{P}^1_{\mathbb{C}}/\mu_2]$ while the second is the "teardrop" $\mathbb{WP}_{\mathbb{C}}(2, 1)$.

10.1.2 Quasimaps

If the GIT equivalence class $C^\circ(L)$ plays the role of the ample cone for the triple (X, G, L), what is the correct analogue of the cone of curves? From Example 10.1.2, we see that it is not sufficient to simply consider the cone of curves of the GIT quotient, $\mathrm{NE}(X /\!\!/_L G)$, because 1-cycles in $\mathbb{P}^1 = V /\!\!/_{\theta_+} T = V /\!\!/_{\theta_-} T$ cannot distinguish between $\mathcal{O}_V(\theta_+)$ and $\mathcal{O}_V(\theta_-)$.

We propose that the correct replacement for curves in this context is given by *L-quasimaps*. An L-quasimap is a morphism

$$f : C \to [X/G]$$

from a smooth curve C to the stack quotient $[X/G]$ such that the preimage of the L-semi-stable locus $f^{-1}([X^{ss}(L)/G])$ is a dense open subset of C.

There is a notion of degree for a morphism $f : C \to [X/G]$, which naturally gives an element $\deg(f) \in \mathrm{Hom}(\mathrm{NS}^G(X), \mathbb{Z})$. Define

$$\mathrm{NE}(L) \subset \mathrm{Hom}(\mathrm{NS}^G(X), \mathbb{Q})$$

to be the cone generated by the degrees of L-quasimaps. This plays the role of the cone of curves for (X, G, L).

In Sect. 10.4.1, we show that a certain class of quasimaps is closely related to the Hilbert–Mumford criterion. This is the main tool of the paper.

10.1.3 Results

Let G be a reductive group acting on a normal projective variety X over an algebraically closed field k. Fix an ample G-equivariant line bundle $L \to X$ such that the semi-stable locus $X^{ss}(L)$ is nonempty. Our main result is the following analogue of Kleiman's criterion.

Theorem 10.1.3 (Theorem 10.4.4) *The following cones are equal*

$$C^{\circ}(L) = \operatorname{relint}\left(\mathrm{NE}(L)^{\vee}\right) \cap \mathrm{Amp}^{G}(X), \tag{10.1.2}$$

where relint *denotes the relative interior.*

In particular, one can test whether a G-ample line bundle $L' \to X$ is GIT equivalent to L by intersecting it with classes in $\overline{\mathrm{NE}(L)}$. The relative interior is necessary, as the cone $C^{\circ}(L)$ may be contained in a proper subspace of $\mathrm{NS}^{G}(X)_{\mathbb{Q}}$.

If we restrict our attention to linearizations of a fixed ample line bundle $P \to X$, the statement takes a form even closer to (10.1.1). Let $\mathrm{NS}^{G}_{P}(X)$ denote the group of all linearizations of all powers of P.

Theorem 10.1.4 (Theorem 10.4.10) *Let $L \in \mathrm{NS}^{G}_{P}(X)$ be a linearization of a positive power of P. Let $A_P(L) = C^{\circ}(L) \cap \mathrm{NS}^{G}_{P}(X)$ be the cone of all linearizations of powers of P which are GIT equivalent to L and let $\mathrm{NE}_P(L)$ be the cone of degrees of L-quasimaps on $\mathrm{NS}^{G}_{P}(X)$. If $X^{ss}(L)$ is nonempty, then*

$$A_P(L) = \operatorname{relint}\left(\mathrm{NE}_P(L)^{\vee}\right).$$

Corollary 10.1.5 *If $\mathrm{NS}(X)_{\mathbb{Q}} \cong \mathbb{Q}$, then*

$$C^{\circ}(L) = \operatorname{relint}\left(\mathrm{NE}(L)^{\vee}\right).$$

Next, we look at what can be said beyond the ample cone. If $\mathrm{Pic}(X)_0$ is torsion, then $\mathrm{NS}^{G}(X)_{\mathbb{Q}} = \mathrm{Pic}^{G}(X)_{\mathbb{Q}}$. In this case GIT equivalence classes are well defined on all of $\mathrm{Pic}^{G}(X)_{\mathbb{Q}}$ and not just in the ample cone $\mathrm{Amp}^{G}(X)$. Define

$$A(L) := \{L' \in \mathrm{Pic}^{G}(X)_{\mathbb{Q}} \mid X^{ss}(L') = X^{ss}(L)\}$$

to be the GIT equivalence class of L in $\mathrm{Pic}^{G}(X)_{\mathbb{Q}}$.

We obtain results in two different contexts. The first is for generalized flag varieties.

Theorem 10.1.6 (Proposition 10.4.7) *If $X = H/P$ is a generalized flag variety, then*

$$\overline{A(L)} = \mathrm{NE}(L)^\vee. \tag{10.1.3}$$

The closure is necessary here, as the equivalence class $A(L)$ may contain points of its boundary. We expect that (10.1.3) holds more generally.

Finally, we consider the slightly different setting of a quotient of a normal affine variety V. In this case, $\mathrm{Pic}^G(V) = \chi(G)$. We prove:

Theorem 10.1.7 (Theorem 10.5.3) *If $V^{ss}(\theta)$ is nonempty, then*

$$A(\theta) = \mathrm{relint}\left(\mathrm{NE}(\theta)^\vee\right).$$

Example 10.1.8 Let us illustrate the theorem in a simple example. Let $T = \mathbb{G}_m$ act on $V = \mathbb{A}^1$ by scaling, and let $\theta = \mathrm{id} \in \chi(T)$ be the identity character. The GIT quotient $V /\!\!/_\theta T$ is a point. In this case the cone $A(\theta)$ consists of positive rational multiples of θ. After identifying $\chi(T) \otimes \mathbb{Q}$ with $\mathbb{Q}$, $A(\theta)$ is the cone $\mathbb{Q}_{>0}$.

For any $d \in \mathbb{Z}_{\geq 0}$, consider the θ-quasimap $f : \mathbb{P}^1 \to [V/T]$ given in coordinates by

$$[s : t] \mapsto [s^d].$$

It follows that the cone $\mathrm{NE}(\theta)$ contains the ray

$$\mathbb{Q}_{\geq 0} \subset \mathbb{Q} \cong \mathrm{Hom}(\chi(T), \mathbb{Q}).$$

From here one can check that $\mathrm{NE}(\theta)$ in fact equals $\mathbb{Q}_{\geq 0}$. We conclude that

$$A(\theta) = \mathrm{relint}\left(\mathrm{NE}(\theta)^\vee\right),$$

in agreement with Theorem 10.1.7.

10.2 GIT Setup

Fix G a reductive group acting on a normal projective algebraic variety X, both defined over an algebraically closed field k. Let $\pi : L \to X$ be a G-linearized line bundle, i.e. a line bundle with a G-action for which π is G-equivariant. We recall the definition of semi-stable and stable points of X with respect to L.

Definition 10.2.1 A geometric point x in X is L-semi-stable if, for some $m \geq 0$, there exists a section $\sigma \in \Gamma(X, L^{\otimes m})^G$ such that $\sigma(x) \neq 0$. We denote the associated open subset of X by $X^{ss}(L)$. A geometric point x in X is L-stable if x is L-semi-stable, the orbit $G \cdot x$ is closed in $X^{ss}(L)$, and $|G_x| < \infty$. We denote the associated open subset by $X^{s}(L)$. The L-unstable locus is $X^{us}(L) := X \setminus X^{ss}(L)$. When L is fixed we will sometimes refer to these as simply the semi-stable, stable, and unstable loci respectively.

Definition 10.2.2 Define the *GIT quotient* of X (with respect to L) to be:

$$X /\!\!/_L G := \operatorname{Proj}\left(\bigoplus_{r\geq 0} H^0(X, L^{\otimes r})^G\right).$$

Let $[X/G]$ denote the stack quotient of X by G. Define the *GIT stack quotient* to be

$$[X /\!\!/_L G] := [X^{ss}(L)/G].$$

The map $X^{ss}(L) \to X /\!\!/_L G$ is a categorical quotient, and consequently $X /\!\!/_L G$ depends only on $X^{ss}(L)$ and not L itself. This motivates the following definition.

Definition 10.2.3 Two G-linearized line bundles L and L' on X are said to be *GIT equivalent* if

$$X^{ss}(L) = X^{ss}(L').$$

By the work of Dolgachev–Hu [3], there are only a finite number of GIT equivalence classes within the ample cone.

10.2.1 A Test for Stability

A *one parameter subgroup (1-PS) of* G is a non-trivial group homomorphism $\lambda : \mathbb{G}_m \hookrightarrow G$. Given a 1-PS λ and a point $x \in X$, by properness of X the map

$$\begin{aligned} \lambda_x : \mathbb{G}_m &\to X \\ t &\mapsto \lambda(t) \cdot x \end{aligned}$$

extends uniquely to a morphism $\bar{\lambda}_x : \mathbb{A}^1 \to X$. Let $x_0 = \bar{\lambda}_x(0)$. By construction, x_0 is fixed by the action of $\mathbb{G}_m$, so $\mathbb{G}_m$ acts on $L|_{x_0}$ with some weight:[1]

$$\lambda(t) \cdot v = t^{\rho^L_{(\lambda,x)}} v$$

for all $v \in L|_{x_0}$ and $t \in \mathbb{G}_m$.

[1] There are different sign conventions for $\rho^L_{(\lambda,x)}$ based on whether one defines the line bundle associated to an invertible sheaf $\mathcal{L}$ to be $\operatorname{Spec}(\operatorname{Sym}^\bullet \mathcal{L})$ or $\operatorname{Spec}(\operatorname{Sym}^\bullet \mathcal{L}^\vee)$. We take the latter convention.

The celebrated Hilbert–Mumford criterion asserts that one can test the (semi-)stability of a point $x \in X$ using the asymptotics of all 1-PS's.

Theorem 10.2.4 (Hilbert–Mumford Criterion [10]) *If* $L \to X$ *is an* ample *G-linearized line bundle, then*

$$x \in X^{ss}(L) \iff \rho^L_{(\lambda,x)} \geq 0 \textit{ for all 1-PS's } \lambda : \mathbb{G}_m \hookrightarrow G;$$

$$x \in X^{s}(L) \iff \rho^L_{(\lambda,x)} > 0 \textit{ for all 1-PS's } \lambda : \mathbb{G}_m \hookrightarrow G.$$

10.2.2 Variation of GIT

Notation 10.2.5 Let W be a finite-dimensional vector space over $\mathbb{Q}$, and let

$$W^\vee = \mathrm{Hom}(W, \mathbb{Q})$$

denote the dual vector space. For $C \subset W$ a convex cone, define the dual cone $C^\vee \subset W^\vee$ by

$$C^\vee = \{f \in W^\vee | f(c) \geq 0 \text{ for all } c \in C\}.$$

Denote by $\mathrm{relint}(C)$ the relative interior of C.

Let $\mathrm{Pic}^G(X)$ denote the Picard group of G-linearized isomorphism classes of line bundles on X. We say $L_1, L_2 \in \mathrm{Pic}^G(X)$ are *G-algebraically equivalent* [12] if there is a connected variety T, points $t_1, t_2 \in T$, and a G-linearized line bundle $L \to T \times X$ (where the action of G on $T \times X$ is induced by the action on X), such that

$$L|_{t_1 \times X} \cong L_1, \qquad L|_{t_2 \times X} \cong L_2.$$

Define $\mathrm{NS}^G(X)$ to be the set of G-algebraic equivalence classes in $\mathrm{Pic}^G(X)$. By [12, Proposition 2.1], after tensoring with $\mathbb{Q}$ we have an exact sequence

$$0 \to \chi(G)_\mathbb{Q} \to \mathrm{NS}^G(X)_\mathbb{Q} \to \mathrm{NS}(X)_\mathbb{Q} \to 0,$$

where $\chi(G)$ is the group of characters of G and the last map forgets the G-linearization. Consequently, $\mathrm{NS}^G(X)_\mathbb{Q}$ is a finitely generated abelian group. By the Hilbert–Mumford criterion, if two ample G-linearized line bundles L and L' define the same element in $\mathrm{NS}^G(X)$, then they are GIT equivalent.

Denote by $\mathrm{Amp}^G(X) \subset \mathrm{NS}^G(X)_\mathbb{Q}$ the convex cone spanned by the classes of G-linearized ample line bundles. Denote by $C^G(X) \subset \mathrm{Amp}^G(X)$ the cone of G-

effective ample G-linearized line bundles, i.e. those $[L] \in \mathrm{Amp}^G(X)$ for which $X^{ss}(L)$ is nonempty. Such classes are called *G-ample.*

Definition 10.2.6 For $[L]$ a class in $C^G(X)$, define

$$C(L) = \{[L'] \in C^G(X) | X^{ss}(L) \subset X^{ss}(L')\}.$$

Let $C^\circ(L)$ denote the cone of G-ample line bundles which are GIT equivalent to L:

$$C^\circ(L) = \{[L'] \in C^G(X) | X^{ss}(L) = X^{ss}(L')\}.$$

Following the work of Thaddeus [12] and Dolgachev–Hu [3], Ressayre proved the following:

Theorem 10.2.7 ([11]) *The sets $C(L)$ are closed convex rational polyhedral cones in* $\mathrm{Amp}^G(X)$ *which form a fan covering $C^G(X)$. The GIT equivalence class $C^\circ(L)$ is the relative interior* $\mathrm{relint}(C(L))$.

10.3 Quasimaps

Fix a G-ample line bundle $L \to X$ as above.

Definition 10.3.1 An L-quasimap $f : C \to [X/G]$ is a morphism from a smooth curve C to the stack $[X/G]$ for which $f^{-1}([X^{ss}(L)/G])$ is a dense open subset.

A G-linearized line bundle $N \to X$ induces a line bundle on the stack quotient $[X/G]$ which we will denote by $[N/G] \to [X/G]$. Given a map

$$f : C \to [X/G]$$

one can pull back $[N/G]$ to C and take degree to obtain an integer

$$\deg(f^*[N/G]).$$

Definition 10.3.2 We define the degree of f to be the element of $\mathrm{Hom}(\mathrm{NS}^G(X), \mathbb{Z})$ given by

$$N \mapsto \deg(f^*[N/G]).$$

Those classes $\beta \in \mathrm{Hom}(\mathrm{NS}^G(X), \mathbb{Z})$ which are realized as the degree of an L-quasimap will be called L-quasimap classes. Denote by

$$\mathrm{NE}(L) \subset \mathrm{Hom}(\mathrm{NS}^G(X), \mathbb{Q})$$

the cone generated by L-quasimap classes.

Proposition 10.3.3 *Given an L-quasimap $f : C \to [X/G]$, the degree $\deg(f^*[L/G])$ is non-negative.*

Proof This was proven in [2, Lemma 3.2.1] in the case that X is an affine variety over $\mathbb{C}$. We recall the argument here with the necessary modifications.

Assume without loss of generality that C is irreducible. Choose a point $c \in C$ such that $f(c)$ lies in $[X^{ss}(L)/G]$. Such a point exists because f is assumed to be an L-quasimap. Let $x \in X^{ss}(L)$ be a point mapping to $f(c)$ and let $\sigma \in \Gamma(X, L^{\otimes m})^G$ be a G-invariant section of a positive power of L such that $\sigma(x) \neq 0$. The section σ descends to a section

$$\bar{\sigma} \in \Gamma([X/G], [L^{\otimes m}/G])$$

which is nonzero at $f(c)$. The degree $\deg(f^*[L^{\otimes m}/G])$ is therefore non-negative because it has a non-vanishing section $f^*(\bar{\sigma})$. We conclude by noting that $\deg(f^*[L/G]) = \frac{1}{m}\deg(f^*[L^{\otimes m}/G])$. □

Remark 10.3.4 Proposition 10.3.3 holds with the same proof for X a quasiprojective variety with an action of G.

Corollary 10.3.5 *We have the following inclusion of cones:*

$$C(L) \subset \mathrm{NE}(L)^{\vee}.$$

Proof If $[N] \in C(L)$, then $X^{ss}(L) \subset X^{ss}(N)$, thus any L-quasimap $f : C \to [X/G]$ is also an N-quasimap. By the proposition, we see that $\deg(f^*[N/G])$ will then be non-negative. □

10.4 Hilbert–Mumford Criterion and Quasimaps

10.4.1 The Main Construction

In this section we relate the weights $\rho^L_{(\lambda,x)}$ to the degrees of certain quasimaps.

Construction 10.4.1 Fix $\lambda : \mathbb{G}_m \hookrightarrow G$ a 1-PS and $x \in X^{ss}(L)$ a semi-stable point. Following Sect. 10.2.1, the morphism $\lambda_x : \mathbb{G}_m \to X$ given by $t \mapsto \lambda(t) \cdot x$ extends to a morphism

$$\bar{\lambda}_x : \mathbb{A}^1 \to X.$$

Define the map $\hat{\phi}_{(\lambda,x)} : \mathbb{A}^2 \setminus \{0\} \to X$ by

$$\hat{\phi}_{(\lambda,x)}(s,t) \mapsto \bar{\lambda}_x(t).$$

Note that when $t \neq 0$, (s, t) maps to $X^{ss}(L)$. If we let $\mathbb{G}_m$ act on $\mathbb{A}^2 \setminus \{0\}$ by scaling and on X via λ, then $\hat{\phi}_{(\lambda,x)}$ is $\mathbb{G}_m$-equivariant. It therefore induces an L-quasimap

$$\phi_{(\lambda,x)} : \mathbb{P}^1 = [\mathbb{A}^2 \setminus \{0\}/\mathbb{G}_m] \to [X/G].$$

Lemma 10.4.2 *Given a G-linearized line bundle $N \to X$, the degree of $\phi^*_{(\lambda,x)}([N/G])$ is $\rho^N_{(\lambda,x)}$.*

Proof The morphism $\phi_{(\lambda,x)}$ factors through the projection onto the second factor:

$$\phi_{(\lambda,x)} : [\mathbb{A}^2 \setminus \{0\}/\mathbb{G}_m] \xrightarrow{\pi_2} [\mathbb{A}^1/\mathbb{G}_m] \xrightarrow{[\bar{\lambda}_x/\mathbb{G}_m]} [X/G]$$

The pullback of N to $\mathbb{A}^1$ via $\bar{\lambda}_x$ is trivial, so the isomorphism class of $[\bar{\lambda}_x/\mathbb{G}_m]^*([N/G])$ is determined by the weight of the action of $\mathbb{G}_m$ on $\bar{\lambda}^*_x(N)|_0$, which is $\rho^N_{(\lambda,x)}$. □

10.4.2 A Kleiman Criterion, and Some Questions

Construction 10.4.1 gives a partial converse to Corollary 10.3.5.

Proposition 10.4.3 *If L and N are ample G-linearized line bundles such that $[N]$ is not contained in $C(L)$ then there exists an L-quasimap $f : C \to [X/G]$ such that* $\deg(f^*[N/G]) < 0$. *In other words,*

$$\mathrm{Amp}^G(X) \setminus C(L) \subset (\mathrm{NE}(L)^\vee)^c.$$

Proof If $[N]$ is not contained in $C(L)$, then there exists a point

$$x \in X^{ss}(L) \setminus X^{ss}(N).$$

By the Hilbert–Mumford criterion, there must exist a 1-PS $\lambda : \mathbb{G}_m \hookrightarrow G$ for which $\rho^N_{(\lambda,x)} < 0$. The associated quasimap $\phi_{(\lambda,x)}$ then has the desired properties. □

Combining Theorem 10.2.7, Corollary 10.3.5, and Proposition 10.4.3, we obtain the following analogue of Kleiman's criterion.

Theorem 10.4.4 *Fix a G-linearized normal projective variety (X, G, L), with $L \to X$ a G-ample line bundle. Then*

$$C^\circ(L) = \mathrm{relint}\left(\mathrm{NE}(L)^\vee\right) \cap \mathrm{Amp}^G(X). \tag{10.4.1}$$

In other words, a G-linearized ample line bundle $L' \to X$ is GIT equivalent to L if and only if

$$L' \cdot \gamma \textit{ is } \begin{Bmatrix} > 0 \textit{ for all } \gamma \in \overline{\mathrm{NE}(L)} \textit{ such that } -\gamma \notin \overline{\mathrm{NE}(L)} \\ = 0 \textit{ for all } \gamma \in \overline{\mathrm{NE}(L)} \textit{ such that } -\gamma \in \overline{\mathrm{NE}(L)} \end{Bmatrix}.$$

Assume $\mathrm{Pic}(X)_0$ is torsion, so $\mathrm{NS}^G(X)_{\mathbb{Q}} = \mathrm{Pic}^G(X)_{\mathbb{Q}}$. This guarantees that GIT equivalence classes are well-defined on all of $\mathrm{Pic}^G(X)_{\mathbb{Q}}$ and not just on the ample cone $\mathrm{Amp}^G(X)$. Under this assumption, let

$$A(L) := \{L' \in \mathrm{Pic}^G(X)_{\mathbb{Q}} | X^{ss}(L') = X^{ss}(L)\}$$

denote the GIT equivalence class of L. If L is G-ample, then

$$A(L) \cap \mathrm{Amp}^G(X) = C^\circ(L).$$

In this case, Theorem 10.4.4 may be rewritten as:

$$A(L) \cap \mathrm{Amp}^G(X) = \mathrm{relint}\left(\mathrm{NE}(L)^\vee\right) \cap \mathrm{Amp}^G(X). \tag{10.4.2}$$

Theorem 10.4.4 and Eq. (10.4.2) may be viewed as an analog of Kleiman's criterion on the ample cone. Here $A(L)$ plays the role of the ample cone with respect to the linearization L, and $\mathrm{NE}(L)$ plays the role of the cone of curves. It would be nice to understand when this relationship between cones extends beyond $\mathrm{Amp}^G(X)$.

Question 10.4.5 Assume $\mathrm{Pic}(X)_0$ is torsion. For which triples (X, G, L) does

$$A(L) = \mathrm{relint}\left(\mathrm{NE}(L)^\vee\right)? \tag{10.4.3}$$

It is not hard to construct examples where $A(L)$ contains some points of its boundary. (Consider, for instance, the action of $G = \mathbb{G}_m$ on $\mathbb{P}^1 \times \mathbb{P}^1$ which scales one of the $\mathbb{P}^1$ factors.) We are hopeful, however, that the following weaker condition is more common:

Question 10.4.6 Find conditions on X, G, and L, such that

$$\overline{A(L)} = \mathrm{NE}(L)^\vee. \tag{10.4.4}$$

We remark in passing that if L is not ample, Question 10.4.6 is already interesting when the group G is trivial.

One case where (10.4.4) holds is when X is a generalized flag variety.

Proposition 10.4.7 *Suppose $X = H/P$ where P is a parabolic subgroup of a reductive group H. If $L \to X$ is a G-ample line bundle then*

$$\overline{A(L)} = \mathrm{NE}(L)^\vee.$$

Proof The proof of Corollary 10.3.5 shows that $\overline{A(L)} \subset \mathrm{NE}(L)^\vee$. Thus by (10.4.2), it suffices to show $\mathrm{NE}(L)^\vee \subset \overline{\mathrm{Amp}^G(X)}$. Suppose

$$N \in \mathrm{Pic}^G(X)_{\mathbb{Q}} \setminus \overline{\mathrm{Amp}^G(X)}.$$

We will construct an L-quasimap f such that $\deg(f^*[N/G]) < 0$.

By Kleiman's criterion (Theorem 10.1.1), there exists an irreducible curve C with $N \cdot [C] < 0$. We can move C using the action of H on X so that it intersects the L-semi-stable locus. In particular, there exists an element $h \in H$ and a curve C' which intersects $X^{ss}(L)$ such that $C' = h \cdot C$. Let $\hat{C} \to C'$ be the normalization, and define $f : \hat{C} \to [X/G]$ to be the composition

$$\hat{C} \to C' \hookrightarrow X \to [X/G].$$

Then $\deg(f^*[N/G])$ is a positive multiple of $N \cdot [C']$ which is negative. By construction of C', f is an L-quasimap. □

We conclude this section by mentioning another line of inquiry which might be of interest. The definition of the semi-stable locus (Definition 10.2.1) naturally extends to the setting of $\mathcal{X}$ an Artin stack with a fixed line bundle $\mathcal{L} \to \mathcal{X}$ [1]. On the other hand, the L-quasimaps $\phi_{(\lambda,x)}$ of Construction 10.4.1 are closely related to Θ-stability for Artin stacks as defined in [6] and $\mathcal{L}$-stability of [7]. Questions 10.4.5 and 10.4.6 above can therefore be extended to Artin stacks, and Theorem 10.4.4 may be viewed as concerning the special case that $\mathcal{X} = [X/G]$ is the stack quotient of a projective variety by a reductive group.

Question 10.4.8 Let $\mathcal{X}$ be an Artin stack and let $\mathcal{L} \to \mathcal{X}$ be a line bundle. Find conditions on $\mathcal{X}$ and $\mathcal{L}$ such that

$$\overline{A(\mathcal{L})} = \mathrm{NE}(\mathcal{L})^\vee.$$

10.4.3 The Case of a Fixed Ample Line Bundle

For general X, the result is cleaner if we restrict our attention to powers of a fixed ample line bundle $P \to X$ and allow the linearization to vary.

Let $P \to X$ be an ample line bundle *without* a choice of G-linearization.

Definition 10.4.9 Let $\mathrm{NS}^G_P(X)_{\mathbb{Q}}$ denote the subspace of $\mathrm{NS}^G(X)_{\mathbb{Q}}$ spanned by all G-linearizations of powers of P. Define

$$A_P(L) := C^\circ(L) \cap \mathrm{NS}^G_P(X)_{\mathbb{Q}};$$
$$C^G_P(X) := C^G(X) \cap \mathrm{NS}^G_P(X)_{\mathbb{Q}}.$$

Elements of $\mathrm{Hom}(\mathrm{NS}^G(X), \mathbb{Q})$ naturally restrict to $\mathrm{NS}^G_P(X)$. Let $\mathrm{NE}_P(L)$ denote the image of $\mathrm{NE}(L)$ in $\mathrm{Hom}(\mathrm{NS}^G_P(X), \mathbb{Q})$.

Theorem 10.4.10 *In* $\mathrm{NS}^G_P(X)_\mathbb{Q}$, *if* $L \in C^G_P(X)$, *then*

$$A_P(L) = \mathrm{relint}\left(\mathrm{NE}_P(L)^\vee\right).$$

Proof By intersecting (10.4.1) with $\mathrm{NS}^G_P(X)_\mathbb{Q}$, we have that

$$A_P(L) = \mathrm{relint}\left(\mathrm{NE}_P(L)^\vee\right) \cap \mathrm{Amp}^G(X).$$

Thus, it suffices to show that if $L \in C^G_P(X)$ and

$$0 \neq N \in \mathrm{NS}^G_P(X)_\mathbb{Q} \setminus \mathrm{Amp}^G(X),$$

then there exists an L-quasimap f for which $\deg(f^*[N/G]) < 0$.

If N is not ample then after forgetting the linearization it is a non-positive power of P and therefore has non-positive degree. If $\deg(N) < 0$, let $X \hookrightarrow \mathbb{P}^r$ be the projective embedding of X determined by a very ample power of P. Intersecting X with a general linear subspace of the correct dimension, we obtain a curve $C \subset X$ which intersects the semi-stable locus $X^{ss}(L)$. If C is reducible, choose an irreducible component C' which intersects the semi-stable locus $X^{ss}(L)$. If C' is not smooth, let $\hat{C} \to C'$ be a resolution. Let $f : \hat{C} \to [X/G]$ denote the composition

$$\hat{C} \longrightarrow C' \hookrightarrow X \longrightarrow [X/G].$$

As in the proof of Proposition 10.4.7, f is an L-quasimap and $\deg(f^*[N/G]) < 0$.

If N has degree zero then it is $\mathcal{O}_X(\theta)$ for some non-trivial character $\theta \in \chi(G)$. Choose a 1-parameter subgroup $\lambda : \mathbb{G}_m \hookrightarrow G$ such that $\theta \circ \lambda$ has weight $-d$ for some $d > 0$. Choose $x \in X^{ss}(L)$ and consider $\phi_{(\lambda,x)}$ as in Construction 10.4.1. Then $\phi_{(\lambda,x)}$ is an L-quasimap, and $\deg(\phi^*_{(\lambda,x)}[N/G]) = -d < 0$. □

This immediately implies the following special case.

Corollary 10.4.11 *If* $\mathrm{NS}(X)_\mathbb{Q} \cong \mathbb{Q}$, *and* $L \in C^G(X)$, *then*

$$C^\circ(L) = \mathrm{relint}\left(\mathrm{NE}(L)^\vee\right).$$

10.5 Quotients of Affine Varieties

In this section we extend Theorem 10.4.4 to the affine setting. Let G be a reductive group acting linearly on a normal affine variety V. In this case $\mathrm{Pic}^G(V) = \chi(G)$.

With this identification we will use θ to denote both a character of G as well as the associated G-linearized line bundle on V.

Given $\theta \in \chi(G)$, define a θ-quasimap

$$f : C \to [V/G]$$

exactly as in Sect. 10.3. The degree of a morphism $f : C \to [V/G]$ lies in $\mathrm{Hom}(\chi(G), \mathbb{Z})$. Let $\mathrm{NE}(\theta)$ denote the cone of degrees of θ-quasimaps to $[V/G]$.

Following Definition 10.2.6, for θ a G-effective character in $\chi(G)_{\mathbb{Q}}$, let

$$C_V(\theta) = \{\theta' \in \chi(G)_{\mathbb{Q}} | V^{ss}(\theta) \subset V^{ss}(\theta')\}.$$

Let $A(\theta)$ denote the GIT equivalence class for a character $\theta \in \chi(G)$.

To follow the arguments of the previous section, we require appropriate modifications of Ressayre's theorem (Theorem 10.2.7) and the Hilbert–Mumford criterion in the case of a group acting on an affine variety. The generalization of Theorem 10.2.7 to the affine setting was given by Halic [5].

Theorem 10.5.1 ([5]) *The sets $C_V(\theta)$ are closed convex rational polyhedral cones which form a fan in $\chi(G)_{\mathbb{Q}}$. The GIT equivalence class $A(\theta)$ is the relative interior of $C_V(\theta)$.*

Next we state the analogue of the Hilbert–Mumford criterion for affine varieties, which was obtained by King in [8]. As with the Hilbert–Mumford criterion for projective varieties (Theorem 10.2.4), this is a consequence of the fact that when G is a reductive group stability can be tested using 1-PS's. In this case, for a 1-PS $\lambda : \mathbb{G}_m \hookrightarrow G$, there is no guarantee that the map $\lambda_v : \mathbb{G}_m \to V$ given by $t \mapsto \lambda(t) \cdot v$ extends to a morphism $\bar{\lambda}_v : \mathbb{A}^1 \to V$. It turns out that one can effectively ignore those 1-PS's for which λ_v does not extend when testing for stability. When $\bar{\lambda}_v : \mathbb{A}^1 \to V$ exists, the weight $\rho^{\theta}_{(\lambda,v)}$ is defined as in Sect. 10.2.1.

Theorem 10.5.2 ([4, 8]) *A point $v \in V$ is θ-semi-stable if and only if $\rho^{\theta}_{(\lambda,v)} \geq 0$ for all 1-PS's $\lambda : \mathbb{G}_m \hookrightarrow G$ such that $\lambda_v : \mathbb{G}_m \to V$ extends to $\mathbb{A}^1$.*

We conclude with the following analogue of Theorem 10.4.4.

Theorem 10.5.3 *Let G be a reductive group acting linearly on a normal affine variety V. Then for each $\theta \in \chi(G)$ with $V^{ss}(\theta) \neq \emptyset$, we have*

$$A(\theta) = \mathrm{relint}\left(\mathrm{NE}(\theta)^{\vee}\right).$$

Proof The argument follows the same steps as the proof of Theorem 10.4.4. By [2, Lemma 3.2.1] (or Remark 10.3.4), if a morphism $f : C \to [V/G]$ is a θ-quasimap, then $\deg(f^*[\mathcal{O}_V(\theta)/G]) \geq 0$. Therefore,

$$C_V(\theta) \subset \mathrm{NE}(\theta)^{\vee}.$$

Next, assume $\kappa \notin C_V(\theta)$. Then there exists a point $v \in V^{ss}(\theta) \setminus V^{ss}(\kappa)$. By Theorem 10.5.2, there exists a 1-PS $\lambda : \mathbb{G}_m \hookrightarrow G$ such that $\bar{\lambda}_v : \mathbb{A}^1 \to V$ exists and $\rho^{\kappa}_{(\lambda,v)} < 0$. Using $\bar{\lambda}_v$, we can define $\phi_{(\lambda,v)} : \mathbb{P}^1 \to [V/G]$ as in Construction 10.4.1. Then $\phi_{(\lambda,v)}$ is a θ-quasimap such that $\deg(\phi^*_{(\lambda,v)}[\mathcal{O}_V(\kappa)/G]) < 0$.

Consequently,

$$C_V(\theta)^c \subset (\mathrm{NE}(\theta)^\vee)^c,$$

and therefore

$$C_V(\theta) = \mathrm{NE}(\theta)^\vee.$$

By Theorem 10.5.1, $A(\theta) = \mathrm{relint}(C_V(\theta))$. This proves the theorem. □

Acknowledgments I am grateful to Andres Fernandez Herrero and Victoria Hoskins for helpful correspondences and conversations, and to Jeff Achter and the anonymous referee for their valuable comments and suggestions on earlier drafts. This work was partially supported by Simons Foundation Travel Grant 958189.

This paper is dedicated to the memory of Prof. Bumsig Kim. Through the beauty of his mathematics and the generosity with which he shared both his time and ideas, he had a great influence on me and so many others. I am grateful for our time together.

References

1. Alper, J.: Good moduli spaces for Artin stacks. Ann. Inst. Fourier (Grenoble) **63**(6), 2349–2402 (2013)
2. Ciocan-Fontanine, I., Kim, B., Maulik, D.: Stable quasimaps to GIT quotients. J. Geom. Phys. **75**, 17–47 (2014)
3. Dolgachev, I.V., Hu, Y.: Variation of geometric invariant theory quotients. Inst. Hautes Études Sci. Publ. Math. **87**, 5–56 (1998). With an appendix by Nicolas Ressayre
4. Gulbrandsen, M.G., Halle, L.H., & Hulek, K.: A relative Hilbert-Mumford criterion. Manuscripta Math. **148**(3–4), 283–301 (2015)
5. Halic, M.: Quotients of affine spaces for actions of reductive groups. https://arxiv.org/abs/math/0412278 (2004)
6. Halpern-Leistner, D.: On the structure of instability in moduli theory. https://arxiv.org/abs/1411.0627 (2014)
7. Heinloth, J.: Hilbert-Mumford stability on algebraic stacks and applications to G-bundles on curves. Épijournal Géom. Algébrique **1**, Art. 11, 37 (2017)
8. King, A.D.: Moduli of representations of finite-dimensional algebras. Q. J. Math. Oxford Ser. (2) **45**(180), 515–530 (1994)
9. Kleiman, S.L.: Toward a numerical theory of ampleness. Ann. Math. (2) **84**, 293–344 (1966)
10. Mumford, D., Fogarty, J., Kirwan, F.: Geometric Invariant Theory, volume 34 of Ergebnisse der Mathematik und ihrer Grenzgebiete (2) [Results in Mathematics and Related Areas (2)] (3rd edn.). Springer, Berlin (1994)
11. Ressayre, N.: The GIT-equivalence for G-line bundles. Geom. Dedicata **81**(1–3), 295–324 (2000)
12. Thaddeus, M.: Geometric invariant theory and flips. J. Amer. Math. Soc. **9**(3), 691–723 (1996)

Open Access This chapter is licensed under the terms of the Creative Commons Attribution-NonCommercial-NoDerivatives 4.0 International License (http://creativecommons.org/licenses/by-nc-nd/4.0/), which permits any noncommercial use, sharing, distribution and reproduction in any medium or format, as long as you give appropriate credit to the original author(s) and the source, provide a link to the Creative Commons license and indicate if you modified the licensed material. You do not have permission under this license to share adapted material derived from this chapter or parts of it.

The images or other third party material in this chapter are included in the chapter's Creative Commons license, unless indicated otherwise in a credit line to the material. If material is not included in the chapter's Creative Commons license and your intended use is not permitted by statutory regulation or exceeds the permitted use, you will need to obtain permission directly from the copyright holder.

Chapter 11
Derived Categories of Quot Schemes of Zero-Dimensional Quotients on Curves

Yukinobu Toda

To the memory of Professor Bumsig Kim

Abstract We prove the existence of semiorthogonal decompositions of derived categories of Quot schemes of zero-dimensional quotients on curves in terms of derived categories of symmetric products of curves. The above result is a categorical analogue of a similar formula for the class of Quot schemes in the Grothendieck ring of varieties by Bagnarol-Fantechi-Perroni. It is a special case of a more general Quot formula of relative dimension one, which is regarded as a Bosonic counterpart of the Quot formula conjectured by Jiang and proved by the author. The proof involves categorical wall-crossing formula for framed one loop quiver, which itself is motivated and has applications to categorical wall-crossing formula of Donaldson-Thomas invariants.

11.1 Introduction

11.1.1 Derived Categories of Quot Schemes over Curves

For $(r, d) \in \mathbb{Z}^2_{\geq 0}$, let $\mathrm{Gr}(r, d)$ be the Grassmannian variety parameterizing quotients $\mathbb{C}^r \twoheadrightarrow Q$ with $\dim Q = d$. In [24], Kapranov proved the existence of a full strong exceptional collection

$$D^b(\mathrm{Gr}(r, d)) = \langle E_\alpha : \alpha \in \mathbb{B}(r, d) \rangle \tag{11.1.1}$$

where $\mathbb{B}(r, d)$ is the set of Young diagrams with width $\leq r - d$ and height $\leq d$.

Y. Toda (✉)
Kavli Institute for the Physics and Mathematics of the Universe (WPI), University of Tokyo, Kashiwa, Japan
e-mail: yukinobu.toda@ipmu.jp

© The Author(s) 2026

N.-G. Kang et al. (eds.), *Categorical and Enumerative Aspects of Mirror Symmetry*, KIAS Springer Series in Mathematics 5,
https://doi.org/10.1007/978-981-95-0385-8_11

The Grassmannian $\mathrm{Gr}(r, d)$ is regarded as a Grothendieck Quot scheme over the point. By replacing the point by a curve, we obtain the Quot scheme of points over a curve. Let C be a smooth projective curve over $\mathbb{C}$, and $\mathcal{E} \to C$ a vector bundle on it of rank r. We consider the Quot scheme parameterizing zero-dimensional quotients of $\mathcal{E}$ with length d

$$\mathrm{Quot}_C(\mathcal{E}, d) = \{\mathcal{E} \twoheadrightarrow \mathcal{Q} : \dim \mathrm{Supp}(\mathcal{Q}) = 0, \mathrm{length}(\mathcal{Q}) = d\}.$$

The Quot scheme $\mathrm{Quot}_C(\mathcal{E}, d)$ is a smooth projective variety of dimension rd. When $r = 1$, it is isomorphic to the symmetric product $\mathrm{Sym}^d(C)$. In this paper, we prove the following structure of the derived category of $\mathrm{Quot}_C(\mathcal{E}, d)$, which gives a one dimensional analogue of (11.1.1):

Theorem 11.1.1 *There is a semiorthogonal decomposition of the form*

$$D^b(\mathrm{Quot}_C(\mathcal{E}, d)) = \left\langle D^b(\mathrm{Sym}^{d_1}(C) \times \cdots \times \mathrm{Sym}^{d_r}(C)) : d_1 + \cdots + d_r = d \right\rangle. \tag{11.1.2}$$

Here the order of the above semiorthogonal summands is given by a lexicographic order of $(d_1, \ldots, d_r) \in \mathbb{Z}^r_{\geq 0}$.

In [3, Proposition 4.5], Bagnarol-Fantechi-Perroni proved the following identity of the class of $\mathrm{Quot}_C(\mathcal{E}, d)$ in the Grothendieck ring of varieties (also see [4, 39]):

$$[\mathrm{Quot}_C(\mathcal{E}, d)] = \sum_{d_1 + \cdots + d_r = d} [\mathrm{Sym}^{d_1}(C)] \times \cdots \times [\mathrm{Sym}^{d_r}(C)] \times [\mathbb{A}^1]^{l_r}$$

where $l_r := \sum_{i=1}^r (i-1)d_i$. The semiorthogonal decomposition in Theorem 11.1.1 gives a categorical analogue of the above identity. We also refer to [32] on enumerative geometry related to $\mathrm{Quot}_C(\mathcal{E}, d)$. We also note that each semiorthogonal summand (11.1.2) may be further decomposed (see Remark 11.4.10).

11.1.2 Quot Formula of Relative Dimension One

The result of Theorem 11.1.1 is a special case of the Quot formula of relative dimension one, which is described below. Let S be a smooth quasi-projective scheme and

$$\pi : \mathcal{C} \to S$$

be a smooth projective morphism of relative dimension one. For $\mathcal{E} \in \mathrm{Coh}(\mathcal{C})$ with rank r, we consider the S-relative Quot scheme

$$\mathrm{Quot}_{\mathcal{C}/S}(\mathcal{E}, d) \to S \tag{11.1.3}$$

whose fiber over $s \in S$ is the Quot scheme $\mathrm{Quot}_{\mathcal{C}_s}(\mathcal{E}_s, d)$ over the curve $\mathcal{C}_s$. As we do not require $\mathcal{E}$ to be locally free nor S-flat, several geometric properties of $\mathrm{Quot}_{\mathcal{C}_s}(\mathcal{E}_s, d)$ (e.g. dimension, smoothness) may depend on $s \in S$.

Let $\mathcal{H} := \mathcal{E}xt^1_{\mathcal{C}}(\mathcal{E}, \mathcal{O}_{\mathcal{C}})$. As a dual side of (11.1.3), we also consider the S-relative Quot scheme $\mathrm{Quot}_{\mathcal{C}/S}(\mathcal{H}, d) \to S$. If we furthermore assume that $\mathcal{E}$ has homological dimension less than or equal to one, there exist quasi-smooth derived schemes

$$\mathbf{Quot}_{\mathcal{C}/S}(\mathcal{E}, d) \to S \leftarrow \mathbf{Quot}_{\mathcal{C}/S}(\mathcal{H}, d)$$

whose classical truncations are $\mathrm{Quot}_{\mathcal{C}/S}(\mathcal{E}, d)$, $\mathrm{Quot}_{\mathcal{C}/S}(\mathcal{H}, d)$, with virtual dimensions $\dim S + rd$, $\dim S - rd$, respectively (if non-empty). We also denote by

$$\mathrm{Sym}^k(\mathcal{C}/S) := \overbrace{(\mathcal{C} \times_S \cdots \times_S \mathcal{C})}^{k} / \mathfrak{S}_k \to S$$

the relative symmetric product, whose fiber at $s \in S$ is $\mathrm{Sym}^k(\mathcal{C}_s)$. We prove the following:

Theorem 11.1.2 *There is a semiorthogonal decomposition of the form*

$$D^b(\mathbf{Quot}_{\mathcal{C}/S}(\mathcal{E}, d)) = \Big\langle D^b(\mathrm{Sym}^{d_1}(\mathcal{C}/S) \times_S \cdots \times_S \mathrm{Sym}^{d_r}(\mathcal{C}/S) \times_S \mathbf{Quot}_{\mathcal{C}/S}(\mathcal{H}, d - \sum_{i=1}^{r} d_i)) : (d_1, \ldots, d_r) \in \mathbb{Z}^r_{\geq 0} \Big\rangle.$$

When $\mathcal{C} = S$, the following Quot formula is conjectured in [22] and proved in [45]:

$$D^b(\mathbf{Quot}_S(\mathcal{E}, d)) = \left\langle \binom{r}{k}\text{-copies of } D^b(\mathbf{Quot}_S(\mathcal{H}, d - k)) : 0 \leq k \leq \min\{d, r\} \right\rangle. \tag{11.1.4}$$

The result of Theorem 11.1.2 gives a relative dimension one analogue of the above semiorthogonal decomposition. Note that Theorem 11.1.1 is obtained from Theorem 11.1.2 by taking S to be the point and $\mathcal{E}$ to be a vector bundle. In some sense, the semiorthogonal decomposition in Theorem 11.1.2 may be regarded as a bosonic counterpart of the semiorthogonal decomposition (11.1.4) (see Remark 11.4.8).

11.1.3 Categorical Wall-Crossing Formula for Framed One Loop Quiver

The proof of Theorem 11.1.2 is similar to the proof of the formula (11.1.4) in [45]. Namely we construct relevant functors using global categorified Hall products, and reduce the problem to a local situation. In [45], the required local statement is semiorthogonal decomposition under Grassmannian flip proved in [44]. In the case of Theorem 11.1.2, the required local statement is semiorthogonal decomposition under wall-crossing of framed one loop quivers.

We denote by Q the quiver with one vertex $\{1\}$ and one loop. For $a, b \in \mathbb{Z}_{\geq 0}$ with $r := a - b \geq 0$, we denote by $Q_{a,b}$ the quiver with vertices $\{\infty, 1\}$, where there are a-arrows from ∞ to 1, b-arrows from 1 to ∞:

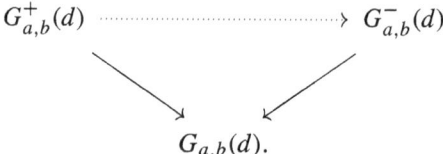

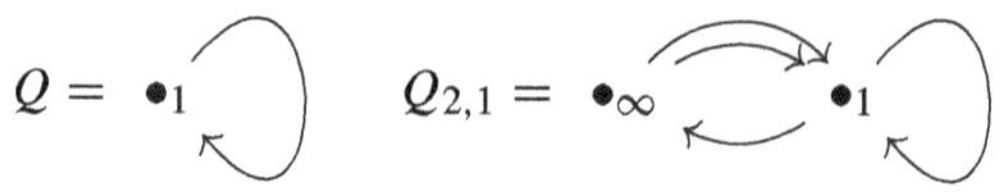

For $d \in \mathbb{Z}$, we denote by $\mathcal{M}_Q(d)$ the moduli stack of Q-representations with dimension vector d, $\mathcal{G}_{a,b}(d)$ the $\mathbb{C}^*$-rigidified moduli stack of $Q_{a,b}$-representations with dimension vector $(1, d)$. There are two GIT stable loci $G^{\pm}_{a,b}(d) \subset \mathcal{G}_{a,b}(d)$ with respect to the characters $\chi_0^{\pm}$, where $\chi_0\colon \mathrm{GL}(d) \to \mathbb{C}^*$ is the determinant character. They are related by a flip ($a > b$), flop ($a = b$)

$$\begin{array}{ccc} G^{+}_{a,b}(d) & \dashrightarrow & G^{-}_{a,b}(d) \\ & \searrow \quad \swarrow & \\ & G_{a,b}(d). & \end{array}$$

Here $\mathcal{G}_{a,b}(d) \to G_{a,b}(d)$ is the good moduli space. The D/K principle by Bondal-Orlov [5] and Kawamata [25] predicts a fully-faithful functor $D^b(G^{-}_{a,b}(d)) \hookrightarrow D^b(G^{+}_{a,b}(d))$. This is not difficult to prove, but we need more: we need to describe its semiorthogonal complements in terms of $D^b(G^{-}_{a,b}(d'))$ for $d' < d$. Similarly to the argument in [44], we prove it using categorified Hall products introduced in [35]. It is a functor

$$*\colon D^b(\mathcal{M}_Q(d_1)) \boxtimes D^b(\mathcal{G}_{a,b}(d_2)) \to D^b(\mathcal{G}_{a,b}(d_1 + d_2)). \tag{11.1.5}$$

The above functor is defined by the stack of short exact sequences of $Q_{a,b}$-representations, and categorifies cohomological Hall algebras in [29].

We will use the following two subcategories for $c \in \mathbb{Z}$:

$$\mathbb{W}(d) \subset D^b(\mathcal{M}_Q(d)), \ \mathbb{W}_c(d) \subset D^b(\mathcal{G}_{a,b}(d)). \tag{11.1.6}$$

The first subcategory first appeared in Špenko-Van den Bergh [41] in their construction of non-commutative resolutions, and was used in [36] to give an approximation of BPS sheaf in the context of cohomological DT theory in [13] (in [37], a similar category for the triple loop quiver with a super-potential is called a quasi-BPS category). For a general symmetric quiver it seems difficult to investigate it, but in our situation of the quiver Q the structure of $\mathbb{W}(d)$ is very simple: it is just generated by $\mathcal{O}_{\mathcal{M}_Q(d)}$ and equivalent to $D^b(M_Q(d))$ for the good moduli space

$$\mathcal{M}_Q(d) \to M_Q(d) = \mathrm{Sym}^d(\mathbb{A}^1) \cong \mathbb{A}^d.$$

The second one is a window subcategory introduced and studied in [2, 18, 19]. It has the property that the composition functors

$$\mathbb{W}_c(d) \hookrightarrow D^b(\mathcal{G}_{a,b}(d)) \twoheadrightarrow D^b(G^{\pm}_{a,b}(d))$$

are equivalences for $(c = a, +)$, $(c = b, -)$. The definitions of the subcategories (11.1.6) are similar, but because of the generic stabilizers of $\mathcal{M}_Q(d)$ they play different roles in this paper.

We show that, for each $(d_1, \ldots, d_r) \in \mathbb{Z}^r_{\geq 0}$, the categorified Hall product induces the fully-faithful functor

$$*: \mathbb{W}(d_1) \boxtimes (\mathbb{W}(d_2) \otimes \chi_0) \boxtimes \cdots \boxtimes (\mathbb{W}(d_l) \otimes \chi_0^{r-1}) \boxtimes (\mathbb{W}_b(d - \sum_{i=1}^{r} d_i) \otimes \chi_0^r) \to \mathbb{W}_a(d)$$

whose essential images form a semiorthogonal decomposition (see Theorem 11.3.12). In particular, we have the following:

Theorem 11.1.3 (Corollary 11.3.13) *There is a semiorthogonal decomposition of the form*

$$D^b(G^+_{a,b}(d)) = \Big\langle D^b(M_Q(d_1)) \boxtimes \cdots \boxtimes D^b(M_Q(d_r)) \boxtimes D^b(G^-_{a,b}(d - \sum_{i=1}^{r} d_i)) \\ : (d_1, \ldots, d_r) \in \mathbb{Z}^r_{\geq 0}\Big\rangle.$$

In [44], a categorical wall-crossing formula for framed zero-loop quivers is applied to give a categorical wall-crossing formula for the resolved conifold. In [37], a categorical wall-crossing formula for framed triple-loop quiver is obtained to give a categorical wall-crossing formula for $\mathbb{C}^3$. The result of Theorem 11.1.3 can be also used to give a categorical wall-crossing formula in other situations, e.g. a CY 3-fold which contracts a divisor to a curve.

11.1.4 Notation and Convention

In this paper, all the schemes and (derived) stacks are defined over $\mathbb{C}$. For a derived stack $\mathcal{X}$, we denote by $D^b(\mathcal{X})$ the homotopy category of ∞-category of quasi-coherent sheaves on $\mathcal{X}$ with coherent cohomologies. For a scheme S and derived stacks $\mathcal{X}_i \to S$ over S for $i = 1, 2$, we denote by $D^b(\mathcal{X}_1) \boxtimes_S D^b(\mathcal{X}_2) = D^b(\mathcal{X}_1 \times_S \mathcal{X}_2)$. For triangulated subcategories $\mathcal{C}_i \subset D^b(\mathcal{X}_i)$, we denote by $\mathcal{C}_1 \boxtimes_S \mathcal{C}_2$ the triangulated subcategory of $D^b(\mathcal{X}_1) \boxtimes_S D^b(\mathcal{X}_2)$ split generated by $C_1 \boxtimes C_2$ for $C_i \in \mathcal{C}_i$. For a morphism of schemes $T \to S$ and an Artin stack $\mathcal{X} \to S$, we write $\mathcal{X}_T := \mathcal{X} \times_S T$. For a perfect complex $\mathcal{E}$ on $\mathcal{X}$, we denote by $\mathcal{E}_T$ its pull-back to $\mathcal{X}_T$.

Let $S = \operatorname{Spec} R$ for a complete local $\mathbb{C}$-algebra R, and $T_i = \operatorname{Spec} A_i$ for complete local R-algebras A_i for $i = 1, 2$. We denote by $A_1 \widehat{\otimes}_R A_2$ the complete tensor product, and write $T_1 \widehat{\times}_R T_2 := \operatorname{Spec}(A_1 \widehat{\otimes}_R A_2)$. For derived stacks $\mathcal{X}_i \to T_i$ over T_i, we denote by $\mathcal{X}_1 \widehat{\times}_R \mathcal{X}_2 \to T_1 \widehat{\times}_R T_2$ the pull-back of $\mathcal{X}_1 \times_R \mathcal{X}_2 \to T_1 \times_R T_2$ via $T_1 \widehat{\times}_R T_2 \to T_1 \times_R T_2$. We denote by $D^b(\mathcal{X}_1) \widehat{\boxtimes}_R D^b(\mathcal{X}_2) = D^b(\mathcal{X}_1 \widehat{\times}_R \mathcal{X}_2)$. The triangulated subcategory $\mathcal{C}_1 \widehat{\boxtimes}_R \mathcal{C}_2$ in $D^b(\mathcal{X}_1) \widehat{\boxtimes}_R D^b(\mathcal{X}_2)$ is defined to be split generated by $(C_1 \boxtimes C_2)|_{\mathcal{X}_1 \widehat{\times}_R \mathcal{X}_2}$ for $C_i \in \mathcal{C}_i$. These notation also apply to categories of ($\mathbb{C}^*$-equivariant) factorizations in Sect. 11.2.5 in an obvious way.

Let G be a reductive algebraic group with maximal torus T. We denote by M the character lattice of T and N the cocharacter lattice of T. For a G-representation Y, we denote by $\mathrm{wt}_T(Y) \subset M$ the set of T-weights of Y. There is a perfect pairing

$$\langle -, - \rangle : N \times M \to \mathbb{Z}.$$

The Weyl group of G is denoted by W, and $M^W \subset M$ is defined to be the fixed part of W-action on M. We fix a Borel subgroup $B \subset G$ and set roots of B to be negative roots. We denote by $M^+ \subset M$ the dominant chamber, and for $\chi \in M^+$ we denote by $V(\chi)$ the irreducible G-representation with highest weight χ. We also define $\rho \in M_{\mathbb{Q}}$ to be the half sum of positive roots. For $\chi \in M$ and $w \in W$, define

$$w * \chi := w(\chi + \rho) - \rho.$$

If $\chi + \rho$ has a trivial stabilizer in W, there is a unique $w \in W$ such that $w * \chi \in M^+$, and in that case we set $\chi^+ := w * \chi$. Otherwise we set $V(\chi^+)$ to be zero.

11.2 Preliminary

11.2.1 Attracting Loci

Let Y be a smooth affine scheme with an action of a reductive algebraic group G. For a one parameter subgroup $\lambda\colon \mathbb{C}^* \to G$, let $Y^{\lambda\geq 0}$, $Y^{\lambda=0}$ be defined by

$$Y^{\lambda\geq 0} := \{y \in Y : \lim_{t\to 0} \lambda(t)(y) \text{ exists }\},$$

$$Y^{\lambda=0} := \{y \in Y : \lambda(t)(y) = y \text{ for all } t \in \mathbb{C}^*\}.$$

The Levi subgroup and the parabolic subgroup

$$G^{\lambda=0} \subset G^{\lambda\geq 0} \subset G$$

are also similarly defined by the conjugate G-action on G, i.e. $g\cdot(-) = g(-)g^{-1}$. The G-action on Y restricts to the $G^{\lambda\geq 0}$-action on $Y^{\lambda\geq 0}$, and the $G^{\lambda=0}$-action on $Y^{\lambda=0}$. We note that λ factors through $\lambda\colon \mathbb{C}^* \to G^{\lambda=0}$, and it acts on $Y^{\lambda=0}$ trivially. So we have the decomposition into fixed λ-weight subcategories

$$D^b([Y^{\lambda=0}/G^{\lambda=0}]) = \bigoplus_{j\in\mathbb{Z}} D^b([Y^{\lambda=0}/G^{\lambda=0}])_{\lambda\text{-wt}=j}.$$

We have the diagram of attracting loci

$$\begin{array}{ccc} [Y^{\lambda\geq 0}/G^{\lambda\geq 0}] & \xrightarrow{p_\lambda} & [Y/G] \\ \sigma_\lambda \uparrow \downarrow q_\lambda & & \\ [Y^{\lambda=0}/G^{\lambda=0}]. & & \end{array} \tag{11.2.1}$$

Here p_λ is induced by the inclusion $Y^{\lambda\geq 0} \subset Y$, and q_λ is given by taking the $t \to 0$ limit of the action of $\lambda(t)$ for $t \in \mathbb{C}^*$. The morphism σ_λ is a section of q_λ induced by inclusions $Y^{\lambda=0} \subset Y^{\lambda\geq 0}$ and $G^{\lambda=0} \subset G^{\lambda\geq 0}$. We will use the following lemma:

Lemma 11.2.1 ([18, Corollary 3.17, Amplification 3.18])

(i) For $\mathcal{E}_i \in D^b([Y^{\lambda\geq 0}/G^{\lambda\geq 0}])$ with $i = 1, 2$, suppose that

$$\sigma_\lambda^*\mathcal{E}_1 \in D^b([Y^{\lambda=0}/G^{\lambda=0}])_{\lambda\text{-wt}\geq j},\ \sigma_\lambda^*\mathcal{E}_2 \in D^b([Y^{\lambda=0}/G^{\lambda=0}])_{\lambda\text{-wt}<j}$$

for some j. Then $\mathrm{Hom}(\mathcal{E}_1, \mathcal{E}_2) = 0$.

(ii) For $j \in \mathbb{Z}$, the functor

$$q_\lambda^* \colon D^b([Y^{\lambda=0}/G^{\lambda=0}])_{\lambda\text{-wt}=j} \to D^b([Y^{\lambda\ge0}/G^{\lambda\ge0}])$$

is fully-faithful.

11.2.2 Kempf-Ness Stratification

Here review Kempf-Ness stratifications associated with GIT quotients of reductive algebraic groups following the convention of [18, Section 2.1]. Let Y and G be as in the previous subsection. For an element $l \in \operatorname{Pic}([Y/G])_{\mathbb{R}}$, we have the open subset of l-semistable points

$$Y^{l\text{-ss}} \subset Y$$

characterized by the set of points $y \in Y$ such that for any one parameter subgroup $\lambda \colon \mathbb{C}^* \to G$ such that the limit $z = \lim_{t\to0} \lambda(t)(y)$ exists in Y, we have $\operatorname{wt}(l|_z) \ge 0$. Let $|*|$ be the Weyl-invariant norm on $N_{\mathbb{R}}$. The above subset of l-semistable points fits into the *Kempf-Ness (KN) stratification*

$$Y = S_1 \sqcup S_2 \sqcup \cdots \sqcup S_N \sqcup Y^{l\text{-ss}}. \tag{11.2.2}$$

Here for each $1 \le i \le N$ there exists a one parameter subgroup $\lambda_i \colon \mathbb{C}^* \to T \subset G$, an open and closed subset Z_i of $(Y \setminus \cup_{i'<i} S_{i'})^{\lambda_i=0}$ (called *center* of S_i) such that

$$S_i = G \cdot Y_i,\ Y_i := \{y \in Y^{\lambda_i\ge0} : \lim_{t\to0} \lambda_i(t)(y) \in Z_i\}.$$

Moreover by setting the slope to be

$$\mu_i := -\frac{\operatorname{wt}(l|_{Z_i})}{|\lambda_i|} \in \mathbb{R} \tag{11.2.3}$$

we have the inequalities $\mu_1 > \mu_2 > \cdots > 0$. We have the following diagram (see [18, Definition 2.2])

$$\begin{array}{ccccc}
[Y_i/G^{\lambda_i\ge0}] & \xrightarrow{\cong} & [S_i/G] & \overset{q_i}{\hookrightarrow} & [(Y \setminus \cup_{i'<i} S_{i'})/G] \\
\downarrow & \swarrow_{p_i} & & \nearrow_{\tau_i} & \\
[Z_i/G^{\lambda_i=0}]. & & & &
\end{array} \tag{11.2.4}$$

Here the left vertical arrow is given by taking the $t \to 0$ limit of the action of $\lambda_i(t)$ for $t \in \mathbb{C}^*$, and τ_i, q_i are induced by the embedding $Z_i \hookrightarrow Y$, $S_i \hookrightarrow Y$ respectively.

Halpern-Leistner [17] extended the above notion of Kempf-Ness stratifications to Θ-stratifications for more general Artin stacks (see [17, Definition 2.2]). Let $\mathcal{N}$ be a classical Artin stack locally of finite type and with affine stabilizers. Suppose that it admits a good moduli space $\mathcal{N} \to N$ (see [1]). Then for any $l \in \mathrm{Pic}(\mathcal{N})_{\mathbb{Q}}$ and a positive definite $b \in H^4(\mathcal{N}, \mathbb{Q})$ (which corresponds to Weyl-invariant norm on $N_{\mathbb{R}}$ in the above setting), there is an associated Θ-stratification (see [16, Theorem 4.1.3])

$$\mathcal{N} = \mathcal{S}_1 \sqcup \mathcal{S}_2 \sqcup \cdots \sqcup \mathcal{S}_N \sqcup \mathcal{N}^{l\text{-ss}}.$$

Here b is called positive definite if for any non-degenerate $f \colon B\mathbb{C}^* \to \mathcal{N}$, we have $q^{-2} f^* b > 0$, where q is the generator of $H^*(B\mathbb{C}^*) = \mathbb{Q}[q]$. Similarly to KN stratifications, there are associated closed substacks $\mathcal{Z}_i \subset \mathcal{S}_i$ (called center of $\mathcal{S}_i$) with canonical $\mathbb{C}^*$-stabilizers at each point of $\mathcal{Z}_i$, and a diagram of attracting loci similar to (11.2.4)

$$\begin{array}{ccc} \mathcal{S}_i & \hookrightarrow & \mathcal{N} \setminus \bigcup_{i'<i} \mathcal{S}_{i'} \\ \downarrow & \nearrow_{\tau_i} & \\ \mathcal{Z}_i. & & \end{array} \tag{11.2.5}$$

11.2.3 Window Theorem

In the diagram (11.2.4), let $\eta_i \in \mathbb{Z}$ be defined by

$$\eta_i := \mathrm{wt}_{\lambda_i}(\det(N^{\vee}_{S_i/Y}|_{Z_i})). \tag{11.2.6}$$

In the case that Y is a G-representation, it is also written as

$$\eta_i = \langle \lambda_i, (Y^{\vee})^{\lambda_i>0} - (\mathfrak{g}^{\vee})^{\lambda_i>0} \rangle.$$

Here for a G-representation W and a one parameter subgroup $\lambda \colon \mathbb{C}^* \to T$, we denote by $W^{\lambda>0} \in K(BT)$ the subspace of W spanned by weights which pair positively with λ. We will use the following version of window theorem:

Theorem 11.2.2 ([2, 18]) *For each i, we take $m_i \in \mathbb{R}$. Let*

$$\mathbb{W}^l_{m_\bullet}([Y/G]) \subset D^b([Y/G]) \tag{11.2.7}$$

be the subcategory of objects $\mathcal{P}$ satisfying the condition

$$\tau_i^*(\mathcal{P}) \in \bigoplus_{j \in [m_i, m_i + \eta_i)} D^b([Z_i/G^{\lambda_i=0}])_{\lambda_i\text{-wt}=j} \tag{11.2.8}$$

for all $1 \le i \le N$. Then the composition functor

$$\mathbb{W}^l_{m_\bullet}([Y/G]) \hookrightarrow D^b([Y/G]) \twoheadrightarrow D^b([Y^{l\text{-ss}}/G])$$

is an equivalence.

Suppose that Y is a symmetric G-representation, i.e. $Y \cong Y^\vee$ as G-representations. In this case, there is another version of window theorem [19], called *magic window theorem*. We denote by $\Sigma \subset M_{\mathbb{R}}$ the subset

$$\Sigma = \sum_{\gamma \in \mathrm{wt}_T(Y)} [0, 1] \cdot \gamma \subset M_{\mathbb{R}}. \tag{11.2.9}$$

Namely Σ is the convex hull of T-weights of $\bigwedge^*(Y)$. For $\delta \in M_{\mathbb{R}}^W$, the magic window subcategory

$$\mathbb{W}_\delta \subset D^b([Y/G]) \tag{11.2.10}$$

is defined to be split generated by $V(\chi) \otimes \mathcal{O}_Y$ for $\chi \in M^+$ satisfying

$$\chi + \rho \in \frac{1}{2}\Sigma + \delta.$$

An element in $M_{\mathbb{R}}^W$ is called Σ-*generic* if it lies in the linear span of Σ but is not parallel to any face of Σ.

Theorem 11.2.3 ([19, Theorem 3.2]) *We take $\delta, l \in M_{\mathbb{R}}^W$ such that $\partial(\Sigma/2 + \delta) \cap M^+ = \emptyset$. Then the composition functor*

$$\mathbb{W}_\delta \hookrightarrow D^b([Y/G]) \twoheadrightarrow D^b([Y^{l\text{-ss}}/G])$$

is fully-faithful. If there is a Σ-generic element in $M_{\mathbb{R}}^W$, then the above functor is an equivalence whenever $Y^{l\text{-ss}} = Y^{l\text{-st}}$, where $Y^{l\text{-st}}$ is the l-stable part.

11.2.4 Categorified Hall Product

In the setting of the diagram (11.2.1), since p_λ is proper we have the functor

$$p_{\lambda *} q_\lambda^* \colon D^b([Y^{\lambda=0}/G^{\lambda=0}]) \to D^b([Y^{\lambda \ge 0}/G^{\lambda \ge 0}]) \to D^b([Y/G]). \tag{11.2.11}$$

In the case that Y is a moduli stack of representations of quivers, the above functor gives a categorified Hall product [35]. The image of $V(\chi) \otimes \mathcal{O}_{Y^{\lambda=0}}$ under the above functor is calculated using Borel-Weil-Bott theorem. We will use the following fact:

Proposition 11.2.4 ([19, Proposition 3.8]) *For $\chi \in M^+$, the object*

$$p_{\lambda*}q_\lambda^*(V(\chi) \otimes \mathcal{O}_{Y^{\lambda=0}}) \in D^b([Y/G])$$

is a successive extension of objects of the form $V((\chi - \sigma_I)^+) \otimes \mathcal{O}_Y[\sharp I - l(I)]$. Here I is a finite subset

$$I \subset \{\beta \in \mathrm{wt}_T(Y) : \langle \beta, \lambda \rangle < 0\},$$

σ_I is the sum of $\beta \in I$ and $l(I)$ is the length of $w \in W$ with $w * (\chi - \sigma_I) \in M^+$. Moreover if $(\chi - \sigma_I)^+ = \chi$ implies $\sharp I = 0$, then $V(\chi) \otimes \mathcal{O}_Y$ appears exactly once.

11.2.5 The Category of Factorizations

Let $\mathcal{Y}$ be a smooth noetherian algebraic stack over $\mathbb{C}$ and take $w \in \Gamma(\mathcal{O}_\mathcal{Y})$. A (coherent) factorization of w consists of

$$\mathcal{P}_0 \underset{\alpha_1}{\overset{\alpha_0}{\rightleftarrows}} \mathcal{P}_1, \quad \alpha_0 \circ \alpha_1 = \cdot w, \ \alpha_1 \circ \alpha_0 = \cdot w, \tag{11.2.12}$$

where each $\mathcal{P}_i$ is a coherent sheaf on $\mathcal{Y}$ and α_i is a morphism of coherent sheaves. The category of coherent factorizations naturally forms a dg-category, whose homotopy category is denoted by $\mathrm{HMF}(\mathcal{Y}, w)$. The subcategory of absolutely acyclic objects

$$\mathrm{Acy}^{\mathrm{abs}} \subset \mathrm{HMF}(\mathcal{Y}, w)$$

is defined to be the minimum thick triangulated subcategory which contains totalizations of short exact sequences of coherent factorizations of w. The triangulated category of factorizations of w is defined by (cf. [14, 34, 38])

$$\mathrm{MF}(\mathcal{Y}, w) := \mathrm{HMF}(\mathcal{Y}, w)/\mathrm{Acy}^{\mathrm{abs}}.$$

If $\mathcal{Y}$ is an affine scheme, then $\mathrm{MF}(\mathcal{Y}, w)$ is equivalent to Orlov's triangulated category of matrix factorizations of w [33].

Let $\mathcal{Y} = [Y/G]$ for a smooth quasi-projective scheme Y and G is an affine algebraic group acting on Y. Suppose that there is an auxiliary $\mathbb{C}^*$-action on Y which commutes with the G-action, and the regular function $w \in \Gamma(\mathcal{O}_\mathcal{Y})$ is of $\mathbb{C}^*$-weight one. A $\mathbb{C}^*$-equivariant (coherent) factorization of w consists of (11.2.12) such that α_0 is of $\mathbb{C}^*$-weight zero and α_1 is of $\mathbb{C}^*$-weight one. The triangulated category of $\mathbb{C}^*$-equivariant factorizations of w is also similarly defined, and denoted by

$$\mathrm{MF}^{\mathbb{C}^*}(\mathcal{Y}, w).$$

If $\mathbb{C}^*$ acts on Y trivially and $w = 0$, then $\mathrm{MF}^{\mathbb{C}^*}(\mathcal{Y}, 0)$ is equivalent to $D^b(\mathcal{Y})$.

11.2.6 Koszul Duality

For a derived Artin stack $\mathfrak{M}$, its (-1)-shifted cotangent is defined by

$$\Omega_{\mathfrak{M}}[-1] := \mathrm{Spec}\,\mathrm{Sym}(\mathbb{T}_{\mathfrak{M}}[1]).$$

Here $\mathbb{T}_{\mathfrak{M}}$ is the tangent complex of $\mathfrak{M}$. In the case that $\mathfrak{M}$ is a derived complete intersection, the classical truncation of $\Omega_{\mathfrak{M}}[-1]$ has the following critical locus description. Let $\mathcal{Y} = [Y/G]$ for a smooth quasi-projective scheme Y and G is an affine algebraic group acting on Y. Let $\mathcal{F} \to \mathcal{Y}$ be a vector bundle on it with a section s. Suppose that $\mathfrak{M}$ is a derived zero locus of s. Let w be the function

$$w \colon \mathcal{F}^{\vee} \to \mathbb{A}^1, \ w(y, v) = \langle s(y), v \rangle$$

for $y \in \mathcal{Y}$ and $v \in \mathcal{F}^{\vee}|_y$. Then $t_0(\Omega_{\mathfrak{M}}[-1])$ is isomorphic to the critical locus $\mathrm{Crit}(w)$ (see [23, Proposition 2.8], [45, Lemma 2.5]). The above construction is summarized in the following diagram

$$\begin{array}{ccc} \mathcal{Y} & \xhookrightarrow{0} & \mathcal{F} \\ \uparrow & \square & \downarrow \uparrow s \\ \mathfrak{M} & \hookrightarrow & \mathcal{Y}, \end{array} \qquad \begin{array}{ccccccc} t_0(\Omega_{\mathfrak{M}}[-1]) & \xrightarrow{\cong} & \mathrm{Crit}(w) & \hookrightarrow & \mathcal{F}^{\vee} & & \\ \downarrow & & \downarrow & & \downarrow & \searrow^{w} & \\ \mathfrak{M} & = & \mathfrak{M} & \hookrightarrow & \mathcal{Y} & & \mathbb{A}^1. \end{array} \tag{11.2.13}$$

Let $\mathbb{C}^*$ acts on fibers of $\mathcal{F}^{\vee} \to Y$ with weight one. In the above setting, the Koszul duality equivalence in [20, Proposition 4.8] (also see [21, 40, 43]) is the following:

Theorem 11.2.5 ([20, 21, 40, 43]) *There is an equivalence*

$$D^b(\mathfrak{M}) \xrightarrow{\sim} \mathrm{MF}^{\mathbb{C}^*}(\mathcal{F}^{\vee}, w).$$

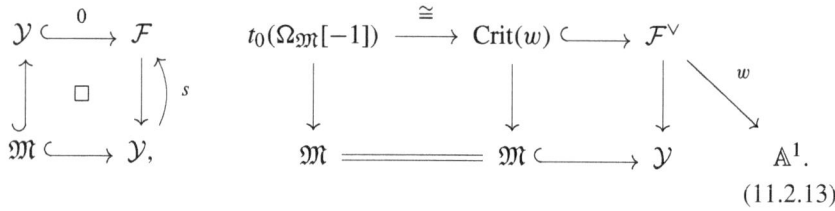

Let $\psi\colon F_0 \to F_1$ be a morphism of G-equivariant vector bundles on Y and set $\mathcal{F}_i = [F_i/G]$. We denote by

$$(\mathcal{F}_0 \xrightarrow{\psi} \mathcal{F}_1) \in D^b(\mathcal{Y})$$

the associated two term complex. We set $F^{-i} = F_i^{\vee}$, $\mathcal{F}^{-i} = \mathcal{F}_i^{\vee}$, and consider the following total space

$$\mathcal{F}_0 \times_{\mathcal{Y}} \mathcal{F}^{-1} = \left[(F_0 \times_Y F^{-1})/G\right].$$

There is a regular function w on it

$$w\colon \mathcal{F}_0 \times_{\mathcal{Y}} \mathcal{F}^{-1} \to \mathbb{A}^1,\ (y, u, v) \mapsto \langle \psi|_y(u), v\rangle = \langle u, \psi|_y^{\vee}(v)\rangle. \tag{11.2.14}$$

Here $y \in Y$, $u \in F_0|_y$, $v \in F^{-1}|_y$, and $\psi^{\vee}\colon F^{-1} \to F^0$ is the dual of ψ. We have two auxiliary $\mathbb{C}^*$-actions on $F_0 \times_Y F^{-1}$ which commute with the G-action: the one is acting on the fibers of $F^{-1} \to Y$ by weight one, and the another is acting on the fibers of $F_0 \to Y$ by weight one. The function w is of weight one with respect to both of the above actions, so we obtain two $\mathbb{C}^*$-equivariant categories of factorizations

$$\mathrm{MF}^{\mathbb{C}^*}(\mathcal{F}_0 \times_{\mathcal{Y}} \mathcal{F}^{-1}, w),\ \mathrm{MF}^{\mathbb{C}^*}(\mathcal{F}_0 \times_{\mathcal{Y}} \mathcal{F}^{-1}, w)'$$

where the left one is defined by the $\mathbb{C}^*$-action on F^{-1}, and the right one is defined by the $\mathbb{C}^*$-action on F_0.

Suppose that there is a one dimensional subtorus $\mathbb{C}^* \subset G$ which lies in the center of G and acts on Y trivially. Then $\mathbb{C}^* \subset G$ acts on fibers of $F_i \to Y$, and we assume that they are of weight one. In this situation, the following lemma is implicit in [30, Subsection 2.2].

Lemma 11.2.6 *In the above situation, there is an equivalence*

$$\mathrm{MF}^{\mathbb{C}^*}(\mathcal{F}_0 \times_{\mathcal{Y}} \mathcal{F}^{-1}, w) \simeq \mathrm{MF}^{\mathbb{C}^*}(\mathcal{F}_0 \times_{\mathcal{Y}} \mathcal{F}^{-1}, w)'. \tag{11.2.15}$$

Proof The isomorphism of algebraic groups

$$G \times \mathbb{C}^* \xrightarrow{\cong} G \times \mathbb{C}^*,\ (g, t) \mapsto (t^{-1}g, t) \tag{11.2.16}$$

gives an isomorphism of stacks

$$\left[(F_0 \times_Y F^{-1})/(G \times \mathbb{C}^*)\right] \xrightarrow{\cong} \left[(F_0 \times_Y F^{-1})/'(G \times \mathbb{C}^*)\right]. \tag{11.2.17}$$

Here in the left hand side the second $\mathbb{C}^*$ acts on the fibers of $F^{-1} \to Y$ by weight one, and in the right hand side it acts on the fibers of $F_0 \to Y$ by weight one. Since (11.2.16) commutes with the second projection, the isomorphism (11.2.17) induces the equivalence (11.2.15). □

Remark 11.2.7 Let $\mathfrak{U}$, $\mathfrak{U}^{\vee}$ be the derived zero loci

$$\mathfrak{U} = (\psi = 0) \subset \mathcal{F}_0, \ \mathfrak{U}^{\vee} = (\psi^{\vee} = 0) \subset \mathcal{F}^{-1}.$$

Then Theorem 11.2.5 together with Lemma 11.2.6 implies an equivalence $D^b(\mathfrak{U}) \simeq D^b(\mathfrak{U}^{\vee})$, which recovers linear Koszul duality in [31].

11.3 Categorical Wall-Crossing for Framed One Loop Quiver

11.3.1 One Loop Quiver

We denote by Q the one loop quiver, i.e. it consists of one vertex $\{1\}$ and one loop:

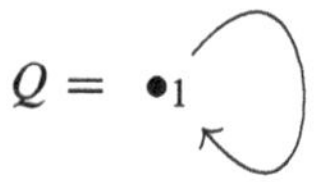

(11.3.1)

For $d \in \mathbb{Z}_{\geq 0}$, let V be a d-dimensional vector space. The quotient stack

$$\mathcal{M}_Q(d) := [\mathrm{End}(V)/\,\mathrm{GL}(V)]$$

is the moduli stack of Q-representations of dimension d. We have the good moduli space

$$\pi_Q \colon \mathcal{M}_Q(d) \to M_Q(d) := \mathrm{End}(V)/\!\!/\,\mathrm{GL}(V) \xrightarrow{\cong} \mathbb{A}^d. \tag{11.3.2}$$

Here the last isomorphism is given by assigning $A \in \mathrm{End}(V)$ to the coefficients of its characteristic polynomial. We refer to [1] for good moduli spaces.

Remark 11.3.1 The stack $\mathcal{M}_Q(d)$ is isomorphic to the stack of zero-dimensional sheaves Q on $\mathbb{A}^1$ with $\chi(Q) = d$. The morphism (11.3.2) is identified with the Hilbert-Chow morphism sending Q to the support of Q in $\mathrm{Sym}^d(\mathbb{A}^1) \cong \mathbb{A}^d$.

We fix a basis of V, and a maximal torus $T \subset \mathrm{GL}(V)$ to be consisting of diagonal matrices. We also set a Borel subgroup $B \subset \mathrm{GL}(V)$ to be consisting of upper triangular matrices, where roots of B are set to be negative roots. The character lattice M for T is given by $M = \mathbb{Z}^d$, and the dominant chamber $M^+ \subset M$ is given by

$$M_{\mathbb{R}}^+ = \{(x_1, x_2, \ldots, x_d) \in \mathbb{R}^d : x_1 \le x_2 \le \cdots \le x_d\}.$$

We also often denote the standard basis of M as $\{e_1, \ldots, e_d\}$ and write an element of $M_{\mathbb{R}}$ as $x_1e_1 + \cdots + x_de_d$. The half sum of positive roots ρ is given by

$$\rho = \frac{1}{2}\sum_{i>j}(e_i - e_j) = \left(-\frac{d-1}{2}, -\frac{d-3}{2}, \ldots, \frac{d-1}{2}\right).$$

The Weyl group of $\mathrm{GL}(V)$ is the symmetric group S_d, and the Weyl-invariant part M^W is generated by $\chi_0 = (1, \ldots, 1)$, where χ_0 corresponds to the character

$$\chi_0 \colon \mathrm{GL}(V) \to \mathbb{C}^*, \ g \mapsto \det(g). \tag{11.3.3}$$

Let $\Sigma(d) \subset M_{\mathbb{R}}$ be the subset in (11.2.9) for the symmetric $\mathrm{GL}(V)$-representation $\mathrm{End}(V)$. Explicitly it is

$$\Sigma(d) = \sum_{-1 \le c_{ij} \le 1, i>j} c_{ij} \cdot (e_i - e_j).$$

For $\delta = t\chi_0 \in M_{\mathbb{R}}^W$ with $t \in \mathbb{R}$, the subcategory

$$\mathbb{W}_\delta(d) \subset D^b(\mathcal{M}_Q(d)) \tag{11.3.4}$$

is defined as in (11.2.10), i.e. it is split generated by $V(\chi) \otimes \mathcal{O}_{\mathcal{M}_Q(d)}$, where $\chi \in M^+$ satisfies $\chi + \rho \in \Sigma(d)/2 + \delta$.

Lemma 11.3.2 *The triangulated subcategory (11.3.4) is non-zero if and only if $\delta = t\chi_0$ for $t \in \mathbb{Z}$. In this case, it is split generated by $\chi_0^{\otimes t}$. Here we have regarded χ_0 as a line bundle on $\mathcal{M}_Q(d)$.*

Proof Note that we have

$$\frac{1}{2}\Sigma(d) - \rho + t\chi_0 = \sum_{-1 \le c_{ij} \le 0, i>j} c_{ij} \cdot (e_i - e_j) + \sum_{i=1}^{d} t \cdot e_i.$$

Therefore any element χ in the LHS is written as

$$\chi = (-c_{21} - \cdots - c_{d1} + t)e_1 + (c_{21} - c_{32} - \cdots - c_{d2} + t)e_2 + \cdots$$
$$+ (c_{d1} + \cdots + c_{dd-1} + t)e_d.$$

If χ lies in the dominant chamber, we have $-c_{21}-\cdots-c_{d1}\leq c_{d1}+\cdots+c_{dd-1}$, hence $c_{21}=\cdots=c_{d1}=c_{d2}=\cdots=c_{dd-1}=0$ as $c_{ij}\leq 0$ for all i, j. By applying the same argument for other coefficients of χ, we conclude that $c_{ij}=0$ for all i, j. Therefore we have

$$\left(\frac{1}{2}\Sigma(d)-\rho+t\chi_0\right)\cap M_{\mathbb{R}}^{+}=t\chi_0,$$

and the lemma holds. □

Below we set

$$\mathbb{W}(d):=\mathbb{W}_{\delta=0}(d)\subset D^b(\mathcal{M}_Q(d)),$$

which is generated by $\mathcal{O}_{\mathcal{M}_Q(d)}$ by Lemma 11.3.2. Note that for $t\in\mathbb{Z}$, we have $\mathbb{W}_{\delta=t\chi_0}(d)=\mathbb{W}(d)\otimes\chi_0^{\otimes t}$.

Lemma 11.3.3 *The pull-back by the morphism (11.3.2) induces the equivalence*

$$\pi_Q^*\colon D^b(M_Q(d))\xrightarrow{\sim}\mathbb{W}(d). \tag{11.3.5}$$

Proof As π_Q is a good moduli space morphism, we have $\pi_{Q*}\mathcal{O}_{\mathcal{M}_Q(k)}=\mathcal{O}_{M_Q(k)}$, see [1, Definition 4.1]. it induces the isomorphism

$$\pi_Q^*\colon \mathrm{Hom}_{M_Q(k)}(\mathcal{O}_{M_Q(k)},\mathcal{O}_{M_Q(k)})\xrightarrow{\cong}\mathrm{Hom}_{\mathcal{M}_Q(k)}(\mathcal{O}_{\mathcal{M}_Q(k)},\mathcal{O}_{\mathcal{M}_Q(k)}).$$

Since the triangulated category $D^b(M_Q(d))$ is generated by $\mathcal{O}_{M_Q(d)}$, it follows that the functor (11.3.5) is fully-faithful. By Lemma 11.3.2, the functor (11.3.5) is also essentially surjective. □

For a one parameter subgroup $\lambda\colon\mathbb{C}^*\to\mathrm{GL}(V)$, we have the diagram of attracting loci

$$\begin{array}{ccc}\mathcal{M}_Q(d)^{\lambda\geq 0} & \xrightarrow{p_\lambda} & \mathcal{M}_Q(d)\\ \big\downarrow{\scriptstyle q_\lambda} & & \\ \mathcal{M}_Q(d)^{\lambda=0}. & & \end{array} \tag{11.3.6}$$

For $d=d_1+d_2$, let $\lambda\colon\mathbb{C}^*\to T\subset\mathrm{GL}(V)$ be given by

$$\lambda(t)=(\overbrace{t,\ldots,t}^{d_1},\overbrace{1,\ldots,1}^{d_2}). \tag{11.3.7}$$

Then we have

$$\mathcal{M}_Q(d)^{\lambda=0} = \mathcal{M}_Q(d_1) \times \mathcal{M}_Q(d_2)$$

and the functor (11.2.11) gives the associative categorified Hall product

$$* = p_{\lambda *} q_\lambda^* \colon D^b(\mathcal{M}_Q(d_1)) \boxtimes D^b(\mathcal{M}_Q(d_2)) \to D^b(\mathcal{M}_Q(d)). \tag{11.3.8}$$

Remark 11.3.4 Using Proposition 11.2.4, one can check that the functor (11.3.8) induces the functor

$$*\colon \overbrace{\mathbb{W}(1) \boxtimes \cdots \boxtimes \mathbb{W}(1)}^{d} \to \mathbb{W}(d)$$

such that $\mathbb{W}(d)$ is split generated by the image of the above functor. However the above functor is not fully-faithful.

11.3.2 Moduli Stacks of Representations of Framed One Loop Quiver

Let us take

$$(a, b) \in \mathbb{Z}^2_{\geq 0}, \ r := a - b \geq 0.$$

We denote by $Q_{a,b}$ the extended quiver of Q, that is the vertex $\{\infty\}$, the a-arrows from ∞ to 1 and the b-arrows from 1 to ∞ are added:

$Q_{2,1} =$ 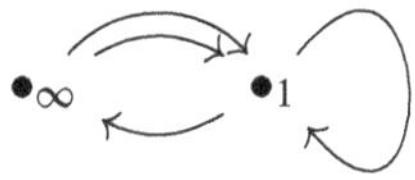

Let A, B, V be vector spaces such that

$$\dim A = a, \ \dim B = b, \ \dim V = d.$$

We set the following $\mathrm{GL}(V)$-representation

$$Y_{a,b}(d) := \mathrm{Hom}(A, V) \oplus \mathrm{Hom}(V, B) \oplus \mathrm{End}(V), \tag{11.3.9}$$

and form the following quotient stack

$$\mathcal{G}_{a,b}(d) := [(\mathrm{Hom}(A, V) \oplus \mathrm{Hom}(V, B) \oplus \mathrm{End}(V)) \,/\, \mathrm{GL}(V)]\,.$$

The above stack is the $\mathbb{C}^*$-rigidified moduli stack of $Q_{a,b}$-representations with dimension vector $(1, d)$. For a one parameter subgroup $\lambda \colon \mathbb{C}^* \to T$, we use the following notation for the diagram of attracting loci

$$\begin{array}{ccc} \mathcal{G}_{a,b}(d)^{\lambda \geq 0} & \xrightarrow{p_\lambda} & \mathcal{G}_{a,b}(d) \\ \downarrow {\scriptstyle q_\lambda} & & \\ \mathcal{G}_{a,b}(d)^{\lambda=0}. & & \end{array} \tag{11.3.10}$$

There exist two GIT quotients with respect to $\chi_0^{\pm 1}$ given by open substacks

$$G^{\pm}_{a,b}(d) \subset \mathcal{G}_{a,b}(d),$$

which are smooth quasi-projective varieties. Here χ_0-semistable locus $G^{+}_{a,b}(d)$ consists of

$$(\alpha, \beta, \gamma) \in \mathrm{Hom}(A, V) \oplus \mathrm{Hom}(V, B) \oplus \mathrm{End}(V)$$

such that, by setting V_γ to be the $\mathbb{C}[Q]$-module structure on V determined by γ, the image of $\alpha \colon A \to V$ generates V_γ as a $\mathbb{C}[Q]$-module. Similarly χ_0^{-1}-semistable locus $G^{-}_{a,b}(d)$ consists of (α, β, γ) such that the image of $\beta^\vee \colon B^\vee \to V^\vee$ generates $V^\vee_{\gamma^\vee}$ as a $\mathbb{C}[Q]$-module (see [43, Lemma 5.1.9]).

Let $G_{a,b}(d)$ be the good moduli space for $\mathcal{G}_{a,b}(d)$:

$$G_{a,b}(d) := (\mathrm{Hom}(A, V) \oplus \mathrm{Hom}(V, B) \oplus \mathrm{End}(V)) /\!\!/ \mathrm{GL}(V).$$

We have the diagram

$$\begin{array}{ccc} G^{+}_{a,b}(d) & \dashrightarrow & G^{-}_{a,b}(d) \\ & \searrow \quad \swarrow & \\ & G_{a,b}(d) & \end{array} \tag{11.3.11}$$

which is a flip if $a > b > 0$, flop if $a = b > 0$ (see [42, Lemma 7.11]).

Remark 11.3.5 When $b = 0$, then $G^{-}_{a,0}(d) = \emptyset$ and $G^{+}_{a,0}(d)$ is the Quot scheme on $\mathbb{A}^1$ which parametrizes quotients $\mathcal{O}_{\mathbb{A}^1}^{\oplus a} \twoheadrightarrow Q$ such that Q is zero-dimensional of length d.

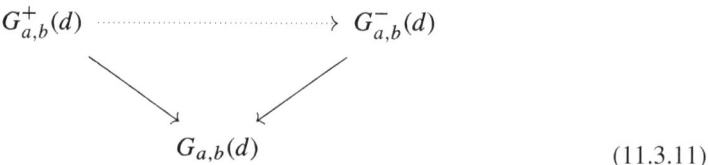

We take the Weyl-invariant norm on $N_{\mathbb{R}} = \mathbb{R}^d$ to be $|\lambda|^2 = \lambda_1^2 + \cdots + \lambda_d^2$. By [43, Lemma 5.1.9], we have the KN stratifications with respect to $(\chi_0^{\pm}, |*|)$

$$\mathcal{G}_{a,b}(d) = \mathcal{S}_0^{\pm} \sqcup \mathcal{S}_1^{\pm} \sqcup \cdots \sqcup \mathcal{S}_{d-1}^{\pm} \sqcup G_{a,b}^{\pm}(d) \tag{11.3.12}$$

where $\mathcal{S}_i^+$ consists of (α, β, γ) such that the image of $\alpha \colon A \to V$ generates i-dimensional $\mathbb{C}[Q]$-submodule of V_γ, $\mathcal{S}_i^-$ consists of (α, β, γ) such that the image of $\beta^\vee \colon B^\vee \to V^\vee$ generates i-dimensional $\mathbb{C}[Q]$-submodule of $V_{\gamma^\vee}^\vee$. The associated one parameter subgroups $\lambda_i^{\pm} \colon \mathbb{C}^* \to T$ are taken as

$$\lambda_i^+(t) = (\overbrace{1, \dots, 1}^{i}, \overbrace{t^{-1}, \dots, t^{-1}}^{d-i}),\ \lambda_i^-(t) = (\overbrace{t, \dots, t}^{d-i}, \overbrace{1, \dots, 1}^{i}) \tag{11.3.13}$$

with associated slopes (11.2.3) to be $\mu_i^{\pm} = \sqrt{d-i}$.

11.3.3 Window Subcategories

For $c \in \mathbb{Z}$, we set

$$\mathbb{B}_c(d) := \{(x_1, x_2, \dots, x_d) \in M^+ : 0 \le x_i \le c - 1\}. \tag{11.3.14}$$

We define the triangulated subcategory

$$\mathbb{W}_c(d) \subset D^b(\mathcal{G}_{a,b}(d)) \tag{11.3.15}$$

to be the smallest thick triangulated subcategory which contains $V(\chi) \otimes \mathcal{O}_{\mathcal{G}_{a,b}(d)}$ for $\chi \in \mathbb{B}_c(d)$. Note that $V(\chi)$ is the Schur power of V associated with the Young diagram corresponding to χ.

Proposition 11.3.6 *The following composition functors are equivalences*

$$\begin{aligned} \mathbb{W}_a(d) \subset D^b(\mathcal{G}_{a,b}(d)) &\twoheadrightarrow D^b(G_{a,b}^+(d)), \\ \mathbb{W}_b(d) \subset D^b(\mathcal{G}_{a,b}(d)) &\twoheadrightarrow D^b(G_{a,b}^-(d)). \end{aligned} \tag{11.3.16}$$

Proof Let λ_i^+ be the one parameter subgroup in (11.3.13). Then η_i^+ given in (11.2.6) is

$$\begin{aligned} \eta_i^+ &= \langle \lambda_i^+, (\mathrm{Hom}(A, V)^\vee \oplus \mathrm{Hom}(V, B)^\vee \oplus \mathrm{End}(V)^\vee)^{\lambda_i^+ > 0} - \mathrm{End}(V)^{\lambda_i^+ > 0} \rangle \\ &= a(d - i). \end{aligned}$$

Let $\chi' = (x'_1, \ldots, x'_d)$ be a T-weight of $V(\chi)$ for $\chi \in \mathbb{B}_a(d)$. Then we have $0 \leq x'_j \leq a-1$ for $1 \leq j \leq d$, so

$$-\eta_i^+ = -a(d-i) < \langle \chi', \lambda_i^+ \rangle = -\sum_{j=i+1}^{d} x'_j \leq 0.$$

Therefore by setting $m_i = -\eta_i^+ + \varepsilon$ for $0 < \varepsilon \ll 1$ and $l = \chi_0$ in (11.2.7), we have

$$\mathbb{W}_a(d) \subset \mathbb{W}_{m_\bullet}^{\chi_0}(\mathcal{G}_{a,b}(d)) \subset D^b(\mathcal{G}_{a,b}(d)). \tag{11.3.17}$$

It follows that the first composition functor in (11.3.16) is fully-faithful. A similar argument also shows that, by setting $m_i = 0$, the second composition functor in (11.3.16) is also fully-faithful.

It remains to show that (11.3.16) are essentially surjective. By setting $a = b$ in (11.3.9), we have the symmetric $\mathrm{GL}(V)$-representation $Y_{a,a}$. Note that $\mathcal{G}_{a,a}(d) = [Y_{a,a}/\mathrm{GL}(V)]$. The subset of weights (11.2.9) for $Y_{a,a}$ is given by

$$\Sigma_a(d) = \sum_{-a \leq c_i \leq a} c_i \cdot e_i + \sum_{-1 \leq c_{ij} \leq 1, i > j} c_{ij} \cdot (e_i - e_j).$$

Therefore we have

$$\frac{1}{2}\Sigma_a(d) - \rho + t\chi_0 = \sum_{t-a/2 \leq c_i \leq t+a/2} c_i \cdot e_i + \sum_{-1 \leq c_{ij} \leq 0, i > j} c_{ij} \cdot (e_i - e_j).$$

Therefore any element χ in the above set is written as

$$\begin{aligned}\chi = (-c_{21} - \cdots - c_{d1} + c_1)e_1 + (c_{21} - c_{32} - \cdots - c_{d2} + c_2)e_2 + \cdots \\ + (c_{d1} + \cdots + c_{dd-1} + c_d)e_d.\end{aligned}$$

We write it as $\chi = \alpha_1 e_1 + \cdots + \alpha_d e_d$. If it lies in $M_{\mathbb{R}}^+$, then as $c_{ij} \leq 0$ we have

$$t - \frac{a}{2} \leq c_1 \leq \alpha_1 \leq \cdots \leq \alpha_d \leq c_d \leq t + \frac{a}{2}.$$

It follows that we have

$$\left(\frac{1}{2}\Sigma_a(d) - \rho + t\chi_0\right) \cap M_{\mathbb{R}}^+ = \left(\sum_{t-a/2 \leq c_i \leq t+a/2} c_i \cdot e_i\right) \cap M_{\mathbb{R}}^+.$$

By setting $t = a/2 - \varepsilon$ for $0 < \varepsilon \ll 1$, we conclude that

$$\left(\frac{1}{2}\Sigma_a(d) - \rho + t\chi_0\right) \cap M^+ = \{(x_1, x_2, \dots, x_d) \in M : 0 \le x_1 \le \cdots \le x_d \le a-1\}. \tag{11.3.18}$$

Since χ_0 is $\Sigma_a(d)$-generic and $\Sigma_a(d)/2 - \rho + t\chi_0$ does not contain integer points on the boundary, the composition (11.3.16) is an equivalence by Theorem 11.2.3 when $a = b$.

When $a > b$, we fix a decomposition into the direct sum $A = B \oplus B'$. Then the projection $A \twoheadrightarrow B$ and the inclusion $B \hookrightarrow A$ define the projections p and the zero sections i

$$\mathcal{G}_{a,a}(d) \underset{p}{\overset{i}{\rightleftarrows}} \mathcal{G}_{a,b}(d) \underset{p}{\overset{i}{\rightleftarrows}} \mathcal{G}_{b,b}(d).$$

Since the χ_0-stability (resp. χ_0^{-1}-stability) on $\mathcal{G}_{a,b}(d)$ does not impose constraint on $\mathrm{Hom}(V, B)$-factor (resp. $\mathrm{Hom}(A, V)$-factor), we have Cartesian squares

$$\begin{array}{ccccccc} G^+_{a,a}(d) & \underset{p}{\overset{i}{\rightleftarrows}} & G^+_{a,b}(d) & \quad & G^-_{a,b}(d) & \underset{p}{\overset{i}{\rightleftarrows}} & G^-_{b,b}(d) \\ \downarrow & \square & \downarrow & & \downarrow & \square & \downarrow \\ \mathcal{G}_{a,a}(d) & \underset{p}{\overset{i}{\rightleftarrows}} & \mathcal{G}_{a,b}(d), & & \mathcal{G}_{a,b}(d) & \underset{p}{\overset{i}{\rightleftarrows}} & \mathcal{G}_{b,b}(d). \end{array}$$

Here the vertical arrows are open immersions. Note that each morphism p is an affine bundle. We have the functors

$$i^* \colon D^b(G^+_{a,a}(d)) \to D^b(G^+_{a,b}(d)), \ p^* \colon D^b(G^-_{b,b}(d)) \to D^b(G^-_{a,b}(d)).$$

Since the images of the above functors generate $D^b(G^+_{a,b}(d))$, $D^b(G^-_{a,b}(d))$ respectively, from the essentially surjectivity of the functors (11.3.16) for $a = b$, we also have the essentially surjectivity of (11.3.16) for $a > b$. $\square$

Remark 11.3.7 The proof of Proposition 11.3.6 implies that the first inclusion in (11.3.17) is an equal, i.e.

$$\mathbb{W}_a(d) = \mathbb{W}_{m_i = -\eta_i^+ + \varepsilon}(\mathcal{G}_{a,b}(d)), \ \mathbb{W}_b(d) = \mathbb{W}_{m_i = 0}(\mathcal{G}_{a,b}(d)).$$

This fact will be used in Lemma 11.4.5.

11.3.4 Computations of Categorified Hall Products

Let $\lambda\colon \mathbb{C}^* \to T \subset \mathrm{GL}(V)$ be the one parameter subgroup given by

$$\lambda(t) = (\overbrace{t, \dots, t}^{d_1}, 1, \dots, 1). \tag{11.3.19}$$

Then we have

$$\mathcal{G}_{a,b}^{\lambda=0}(d) = \mathcal{M}_Q(d_1) \times \mathcal{G}_{a,b}(d - d_1).$$

We have the diagram of attracting loci (11.3.10) and the associated categorified Hall product

$$*\colon D^b(\mathcal{M}_Q(d_1)) \boxtimes D^b(\mathcal{G}_{a,b}(d - d_1)) \to D^b(\mathcal{G}_{a,b}(d)). \tag{11.3.20}$$

By the iteration, we have the categorified Hall product

$$*\colon D^b(\mathcal{M}_Q(d_1)) \boxtimes \cdots \boxtimes D^b(\mathcal{M}_Q(d_l)) \boxtimes D^b\left(\mathcal{G}_{a,b}(d - d_1 - \cdots - d_l)\right) \to D^b(\mathcal{G}_{a,b}(d)). \tag{11.3.21}$$

We remark that, for $A_i \in D^b(\mathcal{M}_Q(d_i))$ and $B \in D^b(\mathcal{G}_{a,b}(d - d_1 - \cdots - d_l))$, the above functor satisfies that

$$(A_1 * \cdots * A_l * B) \otimes \chi_0 \cong (A_1 \otimes \chi_0) * \cdots (A_l \otimes \chi_0) * (B \otimes \chi_0). \tag{11.3.22}$$

Here χ_0 in the LHS is the determinant character for $\mathrm{GL}(d)$ and by abuse of notation χ_0 in the RHS are determinant characters for $\mathrm{GL}(d_i)$ and $\mathrm{GL}(d - d_1 - \cdots - d_l)$. The above isomorphism follows immediately from the definition of categorical Hall products.

We fix $c > b$. For $d' < d$, we fix the following embedding

$$\mathbb{B}_c(d') \hookrightarrow \mathbb{B}_c(d),\ (x_{d-d'+1}, \dots, x_d) \mapsto (x_1 = \cdots = x_{d-d'} = 0, x_{d-d'+1}, \dots, x_d). \tag{11.3.23}$$

We regard an element of $\mathbb{B}_c(d')$ as an element of $\mathbb{B}_c(d)$ by the above embedding. For $0 \le k \le d$, let $\mathbb{B}_{c,k}(d) \subset \mathbb{B}_c(d)$ be the subset defined by

$$\mathbb{B}_{c,k}(d) = \{(x_1, \dots, x_d) \in \mathbb{B}_c(d) : x_1 = \cdots = x_k = 0, x_{k+1} > 0\}.$$

Note that we have $\mathbb{B}_{c,d}(d) = \{0\}$ and

$$\mathbb{B}_{c,0}(d) = \mathbb{B}_{c-1}(d) + \chi_0$$

as $\chi_0 = (1, 1, \dots, 1)$. We have the decomposition into the disjoint union

$$\mathbb{B}_c(d) = \mathbb{B}_{c,0}(d) \sqcup \mathbb{B}_{c,1}(d) \sqcup \cdots \sqcup \mathbb{B}_{c,d}(d).$$

For $d' < d$ and $d - d' \leq k \leq d$, the embedding (11.3.23) induces the bijection

$$\mathbb{B}_{c,k-d+d'}(d') \xrightarrow{\cong} \mathbb{B}_{c,k}(d). \tag{11.3.24}$$

We define the subcategory

$$\mathbb{W}_{c,k}(d) \subset \mathbb{W}_c(d)$$

to be split generated by $V(\chi) \otimes \mathcal{O}_{\mathcal{G}_{a,b}(d)}$ for $\chi \in \mathbb{B}_{c,k}(d)$.

Proposition 11.3.8 *For $1 \leq k \leq d$ and $\chi \in \mathbb{B}_{c,0}(d-k)$, the object*

$$\mathcal{O}_{\mathcal{M}_Q(k)} * (V(\chi) \otimes \mathcal{O}_{\mathcal{G}_{a,b}(d-k)}) \in D^b(\mathcal{G}_{a,b}(d))$$

is generated by $V(\chi) \otimes \mathcal{O}_{\mathcal{G}_{a,b}(d)}$, where χ is regarded as an element of $\mathbb{B}_{c,k}(d)$ by (11.3.24), and $V(\chi') \otimes \mathcal{O}_{\mathcal{G}_{a,b}(d)}$ for $\chi' \in \mathbb{B}_{c,<k}(d)$. Moreover $V(\chi) \otimes \mathcal{O}_{\mathcal{G}_{a,b}(d)}$ appears exactly once.

Proof Let $Y_{a,b}(d)$ be the $\mathrm{GL}(V)$-representation (11.3.9), and λ the one parameter subgroup (11.3.19) for $d_1 = k$. Then we have

$$\{\beta \in \mathrm{wt}_T(Y_{a,b}(d)) : \langle \lambda, \beta \rangle < 0\} = \bigcup_{1 \leq i \leq k} \overbrace{\{-e_i, \dots, -e_i\}}^{b} \bigcup_{1 \leq j \leq k, k < i \leq d} \{(e_i - e_j)\}.$$

Let I be a subset of weights in the above set. Then in the notation of Proposition 11.2.4, for $\chi = (0, \dots, 0, x_{k+1}, \dots, x_d) \in \mathbb{B}_{c,k}(d)$, the element $\chi - \sigma_I + \rho$ is of the form

$$\chi - \sigma_I + \rho = \sum_{i=k+1}^{d} x_i e_i + \sum_{i=1}^{k} s_i e_i - \sum_{k < i \leq d, 1 \leq j \leq k} s_{ij}(e_i - e_j) + \frac{1}{2} \sum_{i > j} (e_i - e_j) \tag{11.3.25}$$

for some $s_i \in \mathbb{Z}$ with $0 \leq s_i \leq b$ and $s_{ij} \in \{0, 1\}$. Therefore we have

$$\chi - \sigma_I + \rho \in \left(\sum_{-c/2 \leq c_i \leq c/2} c_i e_i + \sum_{i > j, -1/2 \leq c_{ij} \leq 1/2} c_{ij}(e_i - e_j) \right) + t\chi_0 \tag{11.3.26}$$

for $t = c/2 - \varepsilon$ with $0 < \varepsilon \ll 1$. Suppose that $(\chi - \sigma_I)^+ \in M^+$ is defined. By its definition, there is unique $w \in S_d$ such that

$$w(\chi - \sigma_I + \rho) = (\chi - \sigma_I)^+ + \rho. \tag{11.3.27}$$

Since the right hand side of (11.3.26) is invariant under the Weyl group action, we have

$$\begin{aligned}(\chi - \sigma_I)^+ &\in \left(\sum_{t-c/2 \le c_i \le t+c/2} c_i e_i + \sum_{i>j, -1 \le c_{ij} \le 0} c_{ij}(e_i - e_j) \right) \cap M^+ \\ &= \{(x_1'', \dots, x_d'') \in M : 0 \le x_1'' \le \cdots \le x_d'' \le c - 1\}.\end{aligned}$$

Here the last identity follows as in (11.3.18). Therefore we have $(\chi - \sigma_I)^+ \in \mathbb{B}_c(d)$.

We show that $(\chi - \sigma_I)^+ \in \mathbb{B}_{c,\le k}(d)$, and $(\chi - \sigma_I)^+ \in \mathbb{B}_{c,k}(d)$ if and only if $I = \emptyset$. Then the proposition follows from Proposition 11.2.4. Let us write

$$(\chi - \sigma_I)^+ = \sum_{i=1}^{d} x_i' e_i,\ 0 \le x_1' \le \cdots \le x_d' \le c - 1.$$

Then by setting $y_i = x_i' - d/2 - 1/2 + i$, we have

$$(\chi - \sigma_I)^+ + \rho = \sum_{i=1}^{d} y_i e_i,\ -\frac{d-1}{2} \le y_1 < y_2 < \cdots < y_d < c + \frac{d-1}{2}.$$

Note that $(\chi - \sigma_I)^+ \in \mathbb{B}_{c,0}(d)$ if and only if $y_i > -(d-1)/2$ for all i.

Let $(\chi - \sigma_I + \rho)_{e_i}$ be the coefficient of $(\chi - \sigma_I + \rho)$ at e_i. From (11.3.25), for $1 \le j \le k$ we have

$$(\chi - \sigma_I + \rho)_{e_j} = -\frac{d}{2} - \frac{1}{2} + j + s_j + \sum_{i=k+1}^{d} s_{ij} \ge -\frac{d-1}{2} \tag{11.3.28}$$

and the equality holds if and only if $j = 1$, $s_1 = 0$ and $s_{i1} = 0$ for all $k < i \le d$. Also for $k < i \le d$, we have

$$(\chi - \sigma_I + \rho)_{e_i} = -\frac{d}{2} - \frac{1}{2} + i + x_i - \sum_{j=1}^{k} s_{ij} > -\frac{d-1}{2} + i - k - 1 \ge -\frac{d-1}{2}. \tag{11.3.29}$$

Here the first inequality is strict since $x_i > 0$. Therefore by the identity (11.3.27), we have either $(\chi - \sigma_I)^+ \in \mathbb{B}_{c,0}(d)$, or $s_1 = s_{i1} = 0$ for all $k < i \le d$.

Suppose that $s_1 = s_{i1} = 0$ for all $k < i \leq d$, so that $y_1 = -(d-1)/2$. For $2 \leq j \leq k$, from (11.3.28) we have

$$(\chi - \sigma_I + \rho)_{e_j} \geq -\frac{d-3}{2}$$

and the equality holds only if $j = 2$, $s_2 = 0$ and $s_{i2} = 0$ for $k < i \leq d$. Moreover for $k < i \leq d$, the inequality (11.3.29) is improved as

$$(\chi - \sigma_I + \rho)_{e_i} = -\frac{d}{2} - \frac{1}{2} + i + x_i - \sum_{j=2}^{k} s_{ij} > -\frac{d-3}{2} + i - k - 1 \geq -\frac{d-3}{2}.$$

It follows that we have either $y_2 > -(d-3)/2$, i.e. $(\chi - \sigma_I)^+ \in \mathbb{B}_{c,1}(d)$, or $s_2 = s_{i2} = 0$ for all $k < i \leq d$.

Repeating the above argument, we conclude that $(\chi - \sigma_I)^+ \in \mathbb{B}_{c,<k}(d)$, or $s_1 = \cdots = s_k = 0$ and $s_{ij} = 0$ for all $1 \leq j \leq k$ and $k < i \leq d$. In the latter case, we have $I = \emptyset$ and $(\chi - \sigma_I)^+ = \chi \in \mathbb{B}_{c,k}(d)$. □

Lemma 11.3.9 *The subcategory $\mathbb{W}_c(d) \subset D^b(\mathcal{G}_{a,b}(d))$ is generated by $\mathbb{W}(k) * (\mathbb{W}_{c-1}(d-k) \otimes \chi_0)$ for $0 \leq k \leq d$.*

Proof Since $\mathbb{W}_{c-1}(d-k) \otimes \chi_0 = \mathbb{W}_{c,0}(d-k)$ and $\mathbb{W}(k)$ is generated by $\mathcal{O}_{\mathcal{M}_Q(k)}$ by Lemma 11.3.2, we have

$$\mathbb{W}(k) * (\mathbb{W}_{c-1}(d-k) \otimes \chi_0) \subset \mathbb{W}_c(d),\ 0 \leq k \leq d \tag{11.3.30}$$

by Proposition 11.3.8. It is enough to show that for any $\chi \in \mathbb{B}_c(d)$ the object $V(\chi) \otimes \mathcal{O}_{\mathcal{G}_{a,b}}(d)$ is generated by the LHS in (11.3.30) for $0 \leq k \leq d$. If $\chi \in \mathbb{B}_{c,0}(d)$, then $V(\chi) \otimes \mathcal{O}_{\mathcal{G}_{a,b}}(d)$ is an object in the LHS in (11.3.30) for $k = 0$. For $\chi \in \mathbb{B}_{c,k}(d)$ with $k > 0$, by Proposition 11.3.8 $V(\chi) \otimes \mathcal{O}_{\mathcal{G}_{a,b}}(d)$ is generated by $\mathbb{W}(k) * \mathbb{W}_{c,0}(d-k) = \mathbb{W}(k) * (\mathbb{W}_{c-1}(d-k) \otimes \chi_0)$ and $V(\chi') \otimes \mathcal{O}_{\mathcal{G}_{a,b}(d)}$ for $\chi' \in \mathbb{B}_{c,<k}(d)$. Therefore by the induction of k, $V(\chi) \otimes \mathcal{O}_{\mathcal{G}_{a,b}}(d)$ is generated by the LHS in (11.3.30). □

Proposition 11.3.10 *The subcategory $\mathbb{W}_c(d) \subset D^b(\mathcal{G}_{a,b}(d))$ is generated by the subcategories*

$$\mathbb{W}(d_\bullet) := \mathbb{W}(d_1) * (\mathbb{W}(d_2) \otimes \chi_0) * \cdots * (\mathbb{W}(d_l) \otimes \chi_0^{l-1}) * (\mathbb{W}_b(d - d_1 - \cdots - d_l) \otimes \chi_0^l) \tag{11.3.31}$$

for $l = c - b > 0$ and $(d_1, \ldots, d_l) \in \mathbb{Z}^l_{\geq 0}$.

Proof Suppose that the proposition holds for $c - 1$. Then for any $d_1 \geq 0$, the category $\mathbb{W}_{c-1}(d - d_1)$ is generated by

$$\mathbb{W}(d_2) * (\mathbb{W}(d_3) \otimes \chi_0) * \cdots * (\mathbb{W}(d_l) \otimes \chi_0^{l-2}) * (\mathbb{W}_b(d - d_1 - \cdots - d_l) \otimes \chi_0^{l-1})$$

for $(d_2, \dots, d_l) \in \mathbb{Z}_{\geq 0}^{l-1}$. Then by Lemma 11.3.9, $\mathbb{W}_c(d)$ is generated by

$$\begin{aligned}
&\mathbb{W}(d_1) * \left\{ \left(\mathbb{W}(d_2) * (\mathbb{W}(d_3) \otimes \chi_0) * \cdots * (\mathbb{W}(d_l) \otimes \chi_0^{l-2}) \right. \right. \\
&\quad \left. \left. * (\mathbb{W}_b(d - d_1 - \cdots - d_l) \otimes \chi_0^{l-1}) \right) \otimes \chi_0 \right\} \\
&\quad = \mathbb{W}(d_1) * (\mathbb{W}(d_2) \otimes \chi_0) * \cdots * (\mathbb{W}(d_l) \otimes \chi_0^{l-1}) * (\mathbb{W}_b(d - d_1 - \cdots - d_l) \otimes \chi_0^{l})
\end{aligned}$$

for $(d_1, d_2, \dots, d_l) \in \mathbb{Z}_{\geq 0}^{l}$. Then the proposition holds by the induction of c. □

11.3.5 Semiorthogonal Decomposition

Proposition 11.3.11 *For each $0 \leq k \leq d$, the functor*

$$*\colon \mathbb{W}(k) \boxtimes (\mathbb{W}_{c-1}(d-k) \otimes \chi_0)) \to \mathbb{W}_c(d) \tag{11.3.32}$$

is fully-faithful, such that we have the semiorthogonal decomposition

$$\mathbb{W}_c(d) = \langle \mathbb{W}(d) * (\mathbb{W}_{c-1}(0) \otimes \chi_0),\ \mathbb{W}(d-1) * (\mathbb{W}_{c-1}(1) \otimes \chi_0), \cdots, \mathbb{W}_{c-1}(d) \otimes \chi_0 \rangle.$$

Proof The generation is proved in Lemma 11.3.9. It is enough to show that the functor (11.3.32) is fully-faithful, and the images of the above functors are semiorthogonal.

Let us take $k \leq k'$ and $\chi \in \mathbb{B}_{c-1,0}(d-k)$, $\chi' \in \mathbb{B}_{c-1,0}(d-k')$. Let λ, λ' be the one parameter subgroups $\mathbb{C}^* \to T \subset \mathrm{GL}(V)$ given by

$$\lambda(t) = (\overbrace{t, \dots, t}^{k}, 1, \dots, 1),\ \lambda'(t) = (\overbrace{t, \dots, t}^{k'}, 1, \dots, 1).$$

We have the diagrams of attracting loci

$$\begin{array}{ccc}
\mathcal{G}_{a,b}(d)^{\lambda \geq 0} & \xrightarrow{p_\lambda} & \mathcal{G}_{a,b}(d) \\
\downarrow q_\lambda & & \\
\mathcal{M}_Q(k) \times \mathcal{G}_{a,b}(d-k), & &
\end{array}
\qquad
\begin{array}{ccc}
\mathcal{G}_{a,b}(d)^{\lambda' \geq 0} & \xrightarrow{p_{\lambda'}} & \mathcal{G}_{a,b}(d) \\
\downarrow q_{\lambda'} & & \\
\mathcal{M}_Q(k') \times \mathcal{G}_{a,b}(d-k'). & &
\end{array}
\tag{11.3.33}$$

Then we have

$$\begin{aligned}&\mathrm{Hom}(\mathcal{O} * (V(\chi) \otimes \mathcal{O}), \mathcal{O} * (V(\chi') \otimes \mathcal{O}))\\&\cong \mathrm{Hom}(p_{\lambda *}q_{\lambda}^{*}(\mathcal{O} \boxtimes (V(\chi) \otimes \mathcal{O}), p_{\lambda' *}q_{\lambda'}^{*}(\mathcal{O} \boxtimes (V(\chi') \otimes \mathcal{O}))\\&\cong \mathrm{Hom}(p_{\lambda'}^{*}p_{\lambda *}q_{\lambda}^{*}(\mathcal{O} \boxtimes (V(\chi) \otimes \mathcal{O}), q_{\lambda'}^{*}(\mathcal{O} \boxtimes (V(\chi') \otimes \mathcal{O})).\end{aligned} \tag{11.3.34}$$

The object $p_{\lambda *}q_{\lambda}^{*}(\mathcal{O} \boxtimes (V(\chi) \otimes \mathcal{O})$ is generated by $V(\chi) \otimes \mathcal{O}$ and $V(\chi'') \otimes \mathcal{O}$ for $\chi'' \in \mathbb{B}_{c,<k}(d)$ by Proposition 11.3.8. If $k < k'$, then any T-weight of $V(\chi)$ and $V(\chi'')$ pairs positively with λ'. Since any T-weight of $V(\chi')$ pairs zero with λ', it follows that (11.3.34) is zero by Lemma 11.2.1 (i). Therefore the images of the functors (11.3.11) are semiorthogonal.

Suppose that $k = k'$. Since any T-weight of $V(\chi'')$ pairs positively with λ, we have

$$\begin{aligned}(11.3.34) &\cong \mathrm{Hom}(p_{\lambda'}^{*}(V(\chi) \otimes \mathcal{O}), q_{\lambda}^{*}(\mathcal{O} \boxtimes (V(\chi') \otimes \mathcal{O})))\\&\cong \mathrm{Hom}(q_{\lambda'}^{*}(\mathcal{O} \boxtimes (V(\chi) \otimes \mathcal{O})), q_{\lambda}^{*}(\mathcal{O} \boxtimes (V(\chi') \otimes \mathcal{O})))\\&\cong \mathrm{Hom}(\mathcal{O} \boxtimes (V(\chi) \otimes \mathcal{O}), \mathcal{O} \boxtimes (V(\chi') \otimes \mathcal{O})).\end{aligned}$$

Here the second isomorphism follows since $\mathcal{G}_{a,b}(d)^{\lambda \geq 0}$ is the stack of exact sequences

$$0 \to \mathbb{V}^{\lambda>0} \to \mathbb{V} \to \mathbb{V}^{\lambda=0} \to 0$$

of $Q_{a,b}$-representations, so the only Schur power from $\mathbb{V}^{\lambda=0}$ survives by Lemma 11.2.1 (i). The last isomorphism follows from Lemma 11.2.1 (ii). Therefore the functor (11.3.32) is fully-faithful. □

For $d_\bullet = (d_1, \ldots, d_l) \in \mathbb{Z}_{\geq 0}^l$ and $d'_\bullet = (d'_1, \ldots, d'_l) \in \mathbb{Z}_{\geq 0}^l$, we take the following lexicographic order: $d'_\bullet \succ d_\bullet$ if and only if there is some m so that $d_i = d'_i$ for $i < m$ and $d_m > d'_m$.

Theorem 11.3.12 *We take $c > b$ with $l := c - b > 0$. Then for each $d_\bullet = (d_1, \ldots, d_l) \in \mathbb{Z}_{\geq 0}^l$, the categorified Hall product (11.3.21) induces the fully-faithful functor*

$$*\colon \mathbb{W}(d_1) \boxtimes (\mathbb{W}(d_2) \otimes \chi_0) \boxtimes \cdots \boxtimes (\mathbb{W}(d_l) \otimes \chi_0^{l-1}) \boxtimes (\mathbb{W}_b(d - \sum_{i=1}^{l} d_i) \otimes \chi_0^l) \to \mathbb{W}_c(d)$$

such that, by setting $\mathcal{C}(d_\bullet) \subset D^b(\mathcal{G}_{a,b}(d))$ to be the essential image of the above functor, we have the semiorthogonal decomposition

$$\mathbb{W}_c(d) = \langle \mathcal{C}(d_\bullet) : d_\bullet \in \mathbb{Z}_{\geq 0}^l \rangle.$$

Here $\mathrm{Hom}(\mathcal{C}(d'_\bullet), \mathcal{C}(d_\bullet)) = 0$ *for* $d'_\bullet \succ d_\bullet$.

Proof The result is proved for $c = b + 1$ in Proposition 11.3.11. Then similarly to the proof of Proposition 11.3.10, the theorem easily follows by the induction of c. □

By applying Theorem 11.3.12 to $c = a$ and $r = a - b$, we obtain the following:

Corollary 11.3.13 *There is a semiorthogonal decomposition of the form*

$$D^b(G^+_{a,b}(d)) = \left\langle D^b(M_Q(d_1)) \boxtimes \cdots \boxtimes D^b(M_Q(d_r)) \boxtimes D^b(G^-_{a,b}(d - \sum_{i=1}^r d_i)) \right.$$
$$\left. : (d_1, \ldots, d_r) \in \mathbb{Z}^r_{\geq 0} \right\rangle.$$

Proof The corollary follows from Theorem 11.3.12, Proposition 11.3.6, and Lemma 11.3.3. □

11.3.6 The Case of Multiple Quivers

For $m \geq 1$, let $Q^{(m)}$ be the quiver with m-vertices $\{1, \ldots, m\}$ with one loop at each vertex. We also define $Q^{(m)}_{a,b}$ to be the quiver with vertices $\{\infty, 1, \ldots, m\}$ with one loop at each vertex $\{1, \ldots, m\}$, a-arrows from ∞ to each $i \in \{1, \ldots, m\}$, b-arrows from each $i \in \{1, \ldots, m\}$ to ∞. See the following picture for $Q^{(3)}_{2,1}$:

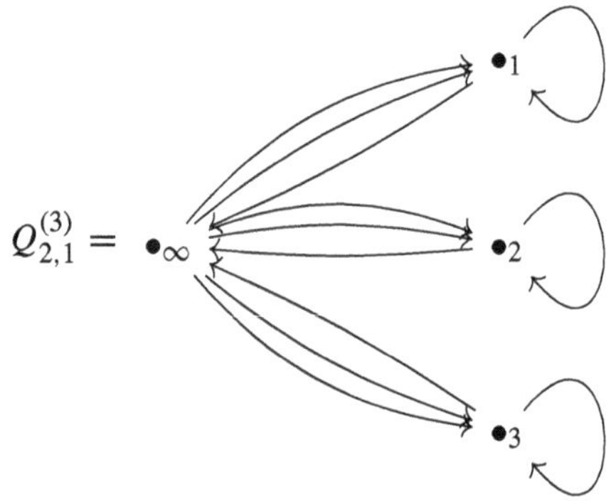

Let $d^{(*)} = (d^{(1)}, \ldots, d^{(m)}) \in \mathbb{Z}^m_{\geq 0}$ be a dimension vector of a $Q^{(m)}$-representation. We set

$$|d^{(*)}| := d^{(1)} + \cdots + d^{(m)}.$$

The moduli stack of $Q^{(m)}$-representations with dimension vector $d^{(*)}$ is given by

$$\mathcal{M}_{Q^{(m)}}(d^{(*)}) := \prod_{j=1}^{m} \mathcal{M}_Q(d^{(j)}) = \prod_{j=1}^{m} [\mathrm{End}(V^{(j)})/\,\mathrm{GL}(V^{(j)})]$$

where $\dim V^{(j)} = d^{(j)}$. The $\mathbb{C}^*$-rigidified moduli stack of $Q_{a,b}^{(m)}$-representations with dimension vector $(1, d^{(*)})$ is given by

$$\mathcal{G}_{a,b}^{(m)}(d^{(*)}) := \prod_{j=1}^{m} \mathcal{G}_{a,b}(d^{(j)}).$$

Let χ_0 be the determinant character for $\prod_{j=1}^{m} \mathrm{GL}(V^{(j)})$, i.e.

$$\chi_0 \colon \prod_{j=1}^{m} \mathrm{GL}(V^{(j)}) \to \mathbb{C}^*, \ \{g^{(j)}\}_{1 \le j \le m} \mapsto \prod_{j=1}^{m} \det(g^{(j)}).$$

Then the $\chi_0^{\pm}$-stable loci are

$$G_{a,b}^{(m)\pm}(d^{(*)}) := \prod_{j=1}^{m} G_{a,b}^{\pm}(d^{(j)}) \subset \mathcal{G}_{a,b}^{(m)}(d^{(*)}).$$

We take the Weyl-invariant norm $|*|$ for the cocharacter lattice of $\prod_{j=1}^{m} \mathrm{GL}(V^{(j)})$ to be

$$|(\lambda^{(1)}, \dots, \lambda^{(m)})|^2 = |\lambda^{(1)}|^2 + \cdots + |\lambda^{(m)}|^2.$$

We have the KN stratification of $\mathcal{G}_{a,b}^{(m)}(d^{(*)})$ with respect to $(\chi_0^{\pm 1}, |*|)$

$$\mathcal{G}_{a,b}^{(m)}(d^{(*)}) = \mathscr{S}_0^{\pm} \sqcup \cdots \sqcup \mathscr{S}_{|d^{(*)}|-1}^{\pm} \sqcup G_{a,b}^{(m)\pm}(d^{(*)}). \tag{11.3.35}$$

A strata $\mathscr{S}_i^{\pm}$ corresponds to $Q_{a,b}^{(m)}$-representations such that the images of arrows from the vertex ∞ (resp. duals of arrows going to ∞) generates i-dimensional $Q^{(m)}$-representations (see [43, Lemma 5.1.9]). Explicitly, by setting

$$\mathcal{G}_{a,b}(d^{(j)}) = \mathcal{S}_0^{(j)\pm} \sqcup \cdots \sqcup \mathcal{S}_{d^{(j)}-1}^{(j)\pm} \sqcup G_{a,b}^{\pm}(d^{(j)})$$

to be KN stratifications (11.3.12) for $d^{(j)}$, we have

$$\mathscr{S}_i^{\pm} = \coprod_{i_1+\cdots+i_m=i} \prod_{j=1}^{m} S_{i_j}^{(j)\pm}.$$

The corresponding one parameter subgroup is given by $\{\lambda_{i_j}^{\pm}\}_{1\le j\le m}$, where $\lambda_{i_j}^{\pm}$ is given in (11.3.13), with slope (11.2.3) given by $\mu_i^{\pm} = \sqrt{|d^{(*)}| - i}$.

We set

$$\begin{aligned} \mathbb{W}(d^{(*)}) &:= \boxtimes_{j=1}^{m} \mathbb{W}(d^{(j)}) \subset D^b(\mathcal{M}_{Q^{(m)}}(d^{(*)})), \\ \mathbb{W}_c(d^{(*)}) &:= \boxtimes_{j=1}^{m} \mathbb{W}_c(d^{(j)}) \subset D^b(\mathcal{G}_{a,b}^{(m)}(d^{(*)})). \end{aligned} \tag{11.3.36}$$

By Proposition 11.3.6, the following composition functors are equivalences:

$$\mathbb{W}_a(d^{(*)}) \subset D^b(\mathcal{G}_{a,b}^{(m)}(d^{(*)})) \twoheadrightarrow D^b(G_{a,b}^{(m)+}(d^{(*)})),$$
$$\mathbb{W}_b(d^{(*)}) \subset D^b(\mathcal{G}_{a,b}^{(m)}(d^{(*)})) \twoheadrightarrow D^b(G_{a,b}^{(m)-}(d^{(*)})).$$

For a decomposition $d^{(*)} = d_1^{(*)} + d_2^{(*)}$, let $\lambda \colon \mathbb{C}^* \to \prod_{j=1}^{m} \mathrm{GL}(V^{(j)})$ be the one parameter subgroup given by

$$\lambda = (\lambda^{(1)}, \ldots, \lambda^{(m)}),\ \lambda^{(j)}(t) = (\overbrace{t, \ldots, t}^{d_1^{(j)}}, \overbrace{1, \ldots, 1}^{d_2^{(j)}}).$$

We have the diagram of attracting loci

$$\begin{array}{ccccc} \prod_{j=1}^{m} \mathcal{G}_{a,b}(d^{(j)})^{\lambda^{(j)}\ge 0} & = & \mathcal{G}_{a,b}^{(m)}(d^{(*)})^{\lambda\ge 0} & \xrightarrow{q_\lambda} & \mathcal{G}_{a,b}^{(m)}(d^{(*)}) \\ & & \downarrow p_\lambda & & \\ \mathcal{M}_{Q^{(m)}}(d_1^{(*)}) \times \mathcal{G}_{a,b}^{(m)}(d_2^{(*)}) & = & \mathcal{G}_{a,b}^{(m)}(d^{(*)})^{\lambda=0}. & & \end{array} \tag{11.3.37}$$

which gives the categorified Hall product

$$* = q_{\lambda *} p_\lambda^* \colon D^b(\mathcal{M}_{Q^{(m)}}(d_1^{(*)})) \boxtimes D^b(\mathcal{G}_{a,b}^{(m)}(d_2^{(*)})) \to D^b(\mathcal{G}_{a,b}^{(m)}(d^{(*)})). \tag{11.3.38}$$

From Theorem 11.3.12, we have the following:

Theorem 11.3.14 *We take $c > b$ with $l := c - b > 0$. Then for each $d_\bullet = (d_1, \dots, d_l) \in \mathbb{Z}^l_{\geq 0}$, the categorified Hall product induces the fully-faithful functor*

$$*\colon \bigoplus_{\substack{(d_1^{(*)}, \dots, d_l^{(*)}) \\ |d_i^{(*)}| = d_i}} \mathbb{W}(d_1^{(*)}) \boxtimes (\mathbb{W}(d_2^{(*)}) \otimes \chi_0) \boxtimes \cdots \boxtimes (\mathbb{W}(d_l^{(*)}) \otimes \chi_0^{l-1})$$

$$\boxtimes\, (\mathbb{W}_b(d^{(*)} - \sum_{i=1}^{l} d_i^{(*)}) \otimes \chi_0^l) \to \mathbb{W}_c(d^{(*)})$$

such that, by setting $\mathcal{C}(d_\bullet) \subset D^b(\mathcal{G}^{(m)}_{a,b}(d^{(*)}))$ to be the essential image of the above functor, we have the semiorthogonal decomposition

$$\mathbb{W}_c(d^{(*)}) = \langle \mathcal{C}(d_\bullet) : d_\bullet \in \mathbb{Z}^l_{\geq 0} \rangle.$$

Here $\mathrm{Hom}(\mathcal{C}(d'_\bullet), \mathcal{C}(d_\bullet)) = 0$ *for $d'_\bullet \succ d_\bullet$.*

Proof The categorified Hall product (11.3.38) fits into the commutative diagram

$$\begin{array}{ccc}
D^b(\mathcal{M}_{Q^{(m)}}(d_1^{(*)})) \boxtimes D^b(\mathcal{G}^{(m)}_{a,b}(d_2^{(*)})) & \xrightarrow{\ *\ } & D^b(\mathcal{G}^{(m)}_{a,b}(d^{(*)})) \\
\downarrow \sim & & \downarrow \sim \\
\boxtimes_{j=1}^m \Big(D^b(\mathcal{M}_Q(d_1^{(j)})) \boxtimes D^b(\mathcal{G}_{a,b}(d_2^{(j)})) \Big) & \xrightarrow{\ \boxtimes *\ } & \boxtimes_{j=1}^m D^b(\mathcal{G}_{a,b}(d^{(j)})).
\end{array}$$

Here the left vertical arrow is the exchange of factors. Therefore from Theorem 11.3.12, for each $(d_1^{(*)}, \dots, d_l^{(*)})$ the functor

$$*\colon \mathbb{W}(d_1^{(*)}) \boxtimes (\mathbb{W}(d_2^{(*)}) \otimes \chi_0) \boxtimes \cdots \boxtimes (\mathbb{W}(d_l^{(*)}) \otimes \chi_0^{l-1})$$

$$\boxtimes\, (\mathbb{W}_b(d^{(*)} - \sum_{i=1}^{l} d_i^{(*)}) \otimes \chi_0^l) \to D^b(\mathcal{G}^{(m)}_{a,b}(d^{(*)}))$$

is fully-faithful such that, by setting $\mathcal{C}(d_\bullet^{(*)})$ to be essential image of the above functor, we have the semiorthogonal decomposition of the form

$$\mathbb{W}_c(d^{(*)}) = \langle \mathcal{C}(d_\bullet^{(*)}) : d_\bullet^{(*)} \in \mathbb{Z}^{lm}_{\geq 0} \rangle.$$

For $d_\bullet^{(*)} \in \mathbb{Z}^{lm}_{\geq 0}$, we set $|d_\bullet^{(*)}| := (|d_1^{(*)}|, \dots, |d_l^{(*)}|) \in \mathbb{Z}^l_{\geq 0}$. Then for $d_\bullet^{(*)} \neq d_\bullet^{(*)'}$ with $|d_\bullet^{(*)}| = |d_\bullet^{(*)'}|$, the subcategories $\mathcal{C}(d_\bullet^{(*)})$ and $\mathcal{C}(d_\bullet^{(*)'})$ are orthogonal. Indeed in this case there exist j, j' such that $d_\bullet^{(j)} \prec d_\bullet^{(j)'}$ and $d_\bullet^{(j)} \succ d_\bullet^{(j')'}$ so the orthogonality follows from the last statement of Theorem 11.3.12. Similarly if $|d_{\bullet}^{(*)'}| \succ |d_{\bullet}^{(*)}|$, then there is j such that $|d_{\bullet}^{(j)'}| \succ |d_{\bullet}^{(j)}|$ so we have $\mathrm{Hom}(\mathcal{C}(d_{\bullet}^{(*)'}), \mathcal{C}(d_{\bullet}^{(*)})) = 0$. Therefore the theorem holds. □

11.3.7 Some Versions for Categories of Factorizations

We will use some variants of Theorem 11.3.14 for categories of factorizations on some formal completions. Let

$$\mathcal{M}_{Q^{(m)}}(d^{(*)}) \to M_{Q^{(m)}}(d^{(*)})$$

be the good moduli space. Let R be a complete local $\mathbb{C}$-algebra with closed point $0 \in \operatorname{Spec} R$. We denote by $\widehat{M}_{Q^{(m)}}(d^{(*)})_R$ the following formal completion of $M_{Q^{(m)}}(d^{(*)}) \times \operatorname{Spec} R$

$$\widehat{M}_{Q^{(m)}}(d^{(*)})_R := \operatorname{Spec} \widehat{\mathcal{O}}_{M_{Q^{(m)}}(d^{(*)}) \times \operatorname{Spec} R, (0,0)},$$

and take the following formal fibers

$$\begin{array}{ccc}
\widehat{\mathcal{G}}_{a,b}^{(m)}(d^{(*)})_R & \longrightarrow & \mathcal{G}_{a,b}^{(m)}(d^{(*)}) \times \operatorname{Spec} R \\
\downarrow & \square & \downarrow \\
\widehat{\mathcal{M}}_{Q^{(m)}}(d^{(*)})_R & \longrightarrow & \mathcal{M}_{Q^{(m)}}(d^{(*)}) \times \operatorname{Spec} R \\
\downarrow & \square & \downarrow \\
\widehat{M}_{Q^{(m)}}(d^{(*)})_R & \longrightarrow & M_{Q^{(m)}}(d^{(*)}) \times \operatorname{Spec} R.
\end{array} \tag{11.3.39}$$

Here the upper right vertical arrow is the projection. We consider an auxiliary $\mathbb{C}^*$-action on $\mathcal{G}_{a,b}^{(m)}(d^{(*)})$ acting on maps from ∞ to each vertex $\{1, \ldots, m\}$ with weight one, and acts on $\operatorname{Spec} R$ trivially. Then it induces the $\mathbb{C}^*$-action on $\widehat{\mathcal{G}}_{a,b}^{(m)}(d^{(*)})_R$. We take a regular function

$$w \colon \widehat{\mathcal{G}}_{a,b}^{(m)}(d^{(*)})_R \to \mathbb{A}^1 \tag{11.3.40}$$

which is of weight one with respect to the above $\mathbb{C}^*$-action. We consider the triangulated category of $\mathbb{C}^*$-equivariant factorizations $\mathrm{MF}^{\mathbb{C}^*}(\widehat{\mathcal{G}}_{a,b}^{(m)}(d^{(*)})_R, w)$.

The diagram (11.3.37) extends to the diagram

$$\begin{array}{ccc}
\mathcal{G}^{(m)}_{a,b}(d^{(*)})^{\lambda\geq 0}\times \mathrm{Spec}\,R & \longrightarrow & \mathcal{G}^{(m)}_{a,b}(d^{(*)})\times \mathrm{Spec}\,R \\
\downarrow & & \Bigg\downarrow \\
\mathcal{M}_{Q^{(m)}}(d_1^{(*)})\times \mathcal{G}^{(m)}_{a,b}(d_2^{(*)})\times \mathrm{Spec}\,R & & \\
\downarrow & & \\
M_{Q^{(m)}}(d_1^{(*)})\times M_{Q^{(m)}}(d_2^{(*)})\times \mathrm{Spec}\,R & \xrightarrow{\oplus} & M_{Q^{(m)}}(d^{(*)})\times \mathrm{Spec}\,R.
\end{array} \tag{11.3.41}$$

Here the right vertical arrow and the left bottom vertical arrow are compositions in the right vertical arrow in (11.3.39). By taking pull-back via the bottom horizontal arrow in (11.3.39), and taking account of the function (11.3.40), we obtain the diagram

$$\begin{array}{ccccc}
\widehat{\mathcal{G}}^{(m)}_{a,b}(d^{(*)})^{\lambda\geq 0}_R & \longrightarrow & \widehat{\mathcal{G}}^{(m)}_{a,b}(d^{(*)})_R & & \\
\downarrow & & \Big\downarrow & \overset{w}{\searrow} & \\
\widehat{\mathcal{M}}_{Q^{(m)}}(d_1^{(*)})_R\widehat{\times}_R\widehat{\mathcal{G}}^{(m)}_{a,b}(d_2^{(*)})_R & & \xrightarrow{(0,w')} & & \mathbb{A}^1 \\
\downarrow & & \Big\downarrow & & \\
\widehat{M}_{Q^{(m)}}(d_1^{(*)})_R\widehat{\times}_R\widehat{M}_{Q^{(m)}}(d_2^{(*)})_R & \xrightarrow{\oplus} & \widehat{M}_{Q^{(m)}}(d^{(*)})_R. & &
\end{array} \tag{11.3.42}$$

Here in the above diagram, the function w descends to a function of the form $(0, w')$ in the middle horizontal arrow by the $\mathbb{C}^*$-weight one condition of w. By the abuse of notation, we also denote w' by w. From the above diagram, the Hall product (11.3.38) induces the one for formal completions

$$*\colon D^b(\widehat{\mathcal{M}}_{Q^{(m)}}(d_1^{(*)})_R)\widehat{\boxtimes}_R \mathrm{MF}^{\mathbb{C}^*}(\widehat{\mathcal{G}}^{(m)}_{a,b}(d_2^{(*)})_R, w)\to \mathrm{MF}^{\mathbb{C}^*}(\widehat{\mathcal{G}}^{(m)}_{a,b}(d^{(*)})_R, w). \tag{11.3.43}$$

The window subcategories

$$\widehat{\mathbb{W}}(d^{(*)})\subset D^b(\widehat{\mathcal{M}}_{Q^{(m)}}(d^{(*)})_R),\ \widehat{\mathbb{W}}_c(d^{(*)})\subset \mathrm{MF}^{\mathbb{C}^*}(\widehat{\mathcal{G}}^{(m)}_{a,b}(d^{(*)})_R, w) \tag{11.3.44}$$

are also defined similarly to (11.3.36): when $m = 1$ and $d^{(*)} = d$, they are the smallest thick triangulated subcategories which contain $\mathcal{O}_{\widehat{\mathcal{M}}_{Q}(d)_R}$, factorizations with entries direct sums of $V(\chi) \otimes \mathcal{O}_{B\mathbb{C}^*}(j) \otimes \mathcal{O}$ for $\chi \in \mathbb{B}_c(d)$ and $j \in \mathbb{Z}$, respectively. Here $\mathcal{O}_{B\mathbb{C}^*}(j)$ is the one dimensional $\mathbb{C}^*$-representation with weight j. For $m > 1$, they are defined to be the box-products of window subcategories of each factors as in (11.3.36). The result of Theorem 11.3.14 immediately implies the following variant of it (for example see the argument of [44, Corollary 4.22]):

Theorem 11.3.15 *We take $c > b$ with $l := c-b > 0$. For each $d_\bullet = (d_1, \ldots, d_l) \in \mathbb{Z}^l_{\geq 0}$, the functors (11.3.43) induce the fully-faithful functor*

$$*\colon \bigoplus_{\substack{(d_1^{(*)}, \ldots, d_l^{(*)}) \\ |d_i^{(*)}|=d_i}} \widehat{\mathbb{W}}(d_1^{(*)}) \widehat{\boxtimes}_R (\widehat{\mathbb{W}}(d_2^{(*)}) \otimes \chi_0) \widehat{\boxtimes}_R \cdots \widehat{\boxtimes}_R (\widehat{\mathbb{W}}(d_l^{(*)}) \otimes \chi_0^{l-1})$$

$$\widehat{\boxtimes}_R (\widehat{\mathbb{W}}_b(d - \sum_{i=1}^{l} d_i) \otimes \chi_0^l) \to \widehat{\mathbb{W}}_c(d^{(*)})$$

such that, by setting $\widehat{\mathcal{C}}(d_\bullet) \subset \mathrm{MF}^{\mathbb{C}^*}(\widehat{\mathcal{G}}^{(m)}_{a,b}(d^{(*)})_R, w)$ to be the essential image of the above functor, we have the semiorthogonal decomposition

$$\widehat{\mathbb{W}}_c(d^{(*)}) = \langle \widehat{\mathcal{C}}(d_\bullet) : d_\bullet \in \mathbb{Z}^l_{\geq 0} \rangle.$$

Here $\mathrm{Hom}(\widehat{\mathcal{C}}(d'_\bullet), \widehat{\mathcal{C}}(d_\bullet)) = 0$ *for $d'_\bullet \succ d_\bullet$.*

11.4 Quot Formula of Relative Dimension One

11.4.1 Relative Quot Schemes

Let S be a smooth quasi-projective scheme and

$$\pi : \mathcal{C} \to S$$

be a smooth projective morphism of relative dimension one. The stack of S-relative zero-dimensional sheaves of length d is given by the 2-functor

$$\mathcal{M}_{\mathcal{C}/S}(d) \colon (Sch/S) \to (Groupoid)$$

which sends $T \to S$ to the groupoid of T-flat sheaves $\mathcal{P} \in \mathrm{Coh}(\mathcal{C}_T)$ such that for any $t \in T$, the object $\mathcal{P}_t \in \mathrm{Coh}(\mathcal{C}_t)$ is zero-dimensional of length d. The stack $\mathcal{M}_{\mathcal{C}/S}(d)$ is a smooth Artin stack such that the structure morphism

$$\mathcal{M}_{\mathcal{C}/S}(d) \to S$$

is smooth whose fiber at $s \in S$ is the stack of zero-dimensional sheaves of length d on the smooth curve $\mathcal{C}_s$.

For $\mathcal{E} \in \mathrm{Coh}(\mathcal{C})$ with $r := \mathrm{rank}(\mathcal{E}) \geq 0$, let $\mathcal{M}_{\mathcal{C}/S}(\mathcal{E}, d)$ be the 2-functor

$$\mathcal{M}_{\mathcal{C}/S}(\mathcal{E}, d) \colon (Sch/S) \to (Groupoid) \tag{11.4.1}$$

by sending T to the groupoid of pairs $(\mathcal{P}, u)$ where $\mathcal{P}$ is a T-valued point of $\mathcal{M}_{\mathcal{C}/S}(d)$ and $u \colon \mathcal{E}_T \to \mathcal{P}$ is a morphism. The set of isomorphisms is given by commutative diagrams

$$\begin{array}{ccc} \mathcal{E}_T & \xrightarrow{u} & \mathcal{P} \\ \| & & \downarrow \cong \\ \mathcal{E}_T & \xrightarrow{u'} & \mathcal{P}'. \end{array}$$

The 2-functor (11.4.1) is an Artin stack with morphisms

$$\mathcal{M}_{\mathcal{C}/S}(\mathcal{E}, d) \to \mathcal{M}_{\mathcal{C}/S}(d) \to S \tag{11.4.2}$$

where the first arrow is forgetting u.

The stack $\mathcal{M}_{\mathcal{C}/S}(\mathcal{E}, d)$ contains an open substack

$$\mathrm{Quot}_{\mathcal{C}/S}(\mathcal{E}, d) \subset \mathcal{M}_{\mathcal{C}/S}(\mathcal{E}, d)$$

corresponding to $(\mathcal{P}, u)$ such that $u \colon \mathcal{E}_T \to \mathcal{P}$ is surjective. The substack $\mathrm{Quot}_{\mathcal{C}/S}(\mathcal{E}, d)$ is the S-relative Grothendieck Quot scheme of $\mathcal{E}$, and it is a projective scheme over S. The fiber at $s \in S$ is the Quot scheme which parametrizes surjections $\mathcal{E}_s \twoheadrightarrow \mathcal{Q}$ where $\mathcal{Q} \in \mathrm{Coh}(\mathcal{C}_s)$ is zero-dimensional of length d.

If $\mathcal{E}$ is locally free, then the first arrow in (11.4.2) is the total space of a vector bundle over $\mathcal{M}_{\mathcal{C}/S}(d)$. Indeed let

$$\mathcal{Q} \in \mathrm{Coh}(\mathcal{C} \times_S \mathcal{M}_{\mathcal{C}/S}(d)) \tag{11.4.3}$$

be the universal zero-dimensional sheaf, and p_1, p_2 be the projections from $\mathcal{C} \times_S \mathcal{M}_{\mathcal{C}/S}(d)$ onto the corresponding factors. Then $p_{2*}(\mathcal{H}om(p_1^*\mathcal{E}, \mathcal{Q}))$ is a locally free sheaf on $\mathcal{M}_{\mathcal{C}/S}(d)$, and we have

$$\mathcal{M}_{\mathcal{C}/S}(\mathcal{E}, d) = \mathrm{Tot}(p_{2*}(\mathcal{H}om(p_1^*\mathcal{E}, \mathcal{Q}))) = \mathrm{Spec}_{\mathcal{M}_{\mathcal{C}/S}(d)} \mathrm{Sym}(p_{2*}(\mathcal{E}^\vee \boxtimes \mathcal{Q})^\vee).$$

In particular in this case $\mathcal{M}_{\mathcal{C}/S}(\mathcal{E}, d)$ is smooth of relative dimension rd.

Let W be a d-dimensional vector space, and $\mathrm{GL}_S(W) := \mathrm{GL}(W) \times S \to S$ be the group scheme over S. Let

$$\mathrm{Quot}^{\circ}_{C/S}(W \otimes \mathcal{O}_C, d) \subset \mathrm{Quot}_{C/S}(W \otimes \mathcal{O}_C, d)$$

be the open subscheme whose T-valued points correspond to $u\colon W \otimes \mathcal{O}_{C_T} \twoheadrightarrow \mathcal{P}$ such that the induced morphism $W \otimes \mathcal{O}_T \to \pi_{T*}\mathcal{P}$ is an isomorphism. Since any zero-dimensional sheaf is globally generated, the first morphism in (11.4.2) induces the isomorphism over S

$$[\mathrm{Quot}^{\circ}_{C/S}(W \otimes \mathcal{O}_C, d)/\,\mathrm{GL}_S(W)] \stackrel{\cong}{\to} \mathcal{M}_{C/S}(d). \tag{11.4.4}$$

11.4.2 Derived Structures of Relative Quot Schemes

Suppose that $\mathcal{E}$ has homological dimension less than or equal to one. Then there is a locally free resolution

$$0 \to \mathcal{E}^{-1} \stackrel{\phi}{\to} \mathcal{E}^0 \to \mathcal{E} \to 0 \tag{11.4.5}$$

such that

$$\mathcal{H} := \mathcal{E}xt^1_{\mathcal{O}_C}(\mathcal{E}, \mathcal{O}_C) = \mathrm{Cok}(\phi^\vee : \mathcal{E}_0 \to \mathcal{E}_1). \tag{11.4.6}$$

Here we have set $\mathcal{E}_0 := (\mathcal{E}^0)^\vee$ and $\mathcal{E}_1 := (\mathcal{E}^{-1})^\vee$. The morphism ϕ induces the morphism of vector bundles on $\mathcal{M}_{C/S}(d)$

$$\phi\colon \mathcal{M}_{C/S}(\mathcal{E}^0, d) \to \mathcal{M}_{C/S}(\mathcal{E}^{-1}, d)$$

which, on T-valued points, is defined by

$$(u\colon \mathcal{E}^0_T \to \mathcal{P}) \mapsto (u \circ \phi\colon \mathcal{E}^{-1}_T \stackrel{\phi}{\to} \mathcal{E}^0_T \stackrel{u}{\to} \mathcal{P}). \tag{11.4.7}$$

We define the derived stack $\mathbf{M}_{C/S}(\mathcal{E}, d)$ by the derived Cartesian square

$$\begin{array}{ccc} \mathbf{M}_{C/S}(\mathcal{E}, d) & \longrightarrow & \mathcal{M}_{C/S}(d) \\ \downarrow & \square & \downarrow 0 \\ \mathcal{M}_{C/S}(\mathcal{E}^0, d) & \stackrel{\phi}{\longrightarrow} & \mathcal{M}_{C/S}(\mathcal{E}^{-1}, d). \end{array} \tag{11.4.8}$$

Here the right vertical arrow is the zero section of the vector bundle $\mathcal{M}_{\mathcal{C}/S}(\mathcal{E}^{-1}, d) \to \mathcal{M}_{\mathcal{C}/S}(d)$. The classical truncation of $\mathbf{M}_{\mathcal{C}/S}(\mathcal{E}, d)$ is isomorphic to $\mathcal{M}_{\mathcal{C}/S}(\mathcal{E}, d)$.

The surjection $\mathcal{E}^0 \twoheadrightarrow \mathcal{E}$ induces the closed immersion

$$\mathrm{Quot}_{\mathcal{C}/S,d}(\mathcal{E}) \hookrightarrow \mathrm{Quot}_{\mathcal{C}/S,d}(\mathcal{E}^0), \tag{11.4.9}$$

where the target is an open substack of $\mathcal{M}_{\mathcal{C}/S}(\mathcal{E}^0, d)$. We define the quasi-smooth derived scheme $\mathbf{Quot}_{\mathcal{C}/S}(\mathcal{E}, d)$ over S by the Cartesian square

$$\begin{array}{ccc} \mathbf{Quot}_{\mathcal{C}/S}(\mathcal{E}, d) & \hookrightarrow & \mathbf{M}_{\mathcal{C}/S}(\mathcal{E}, d) \\ \downarrow & \square & \downarrow \\ \mathrm{Quot}_{\mathcal{C}/S}(\mathcal{E}^0, d) & \hookrightarrow & \mathcal{M}_{\mathcal{C}/S}(\mathcal{E}^0, d). \end{array} \tag{11.4.10}$$

Here horizontal arrows are open immersions. Note that $\mathbf{Quot}_{\mathcal{C}/S}(\mathcal{E}, d)$ is a derived open substack of $\mathbf{M}_{\mathcal{C}/S}(\mathcal{E}, d)$, with virtual dimension $\dim S + rd$, and whose classical truncation is $\mathrm{Quot}_{\mathcal{C}/S}(\mathcal{E}, d)$.

By taking the dual of the sequence (11.4.5), we obtain the exact sequence

$$\mathcal{E}_0 \stackrel{\phi^\vee}{\to} \mathcal{E}_1 \to \mathcal{H} \to 0.$$

Similarly to (11.4.8) and (11.4.10), we define $\mathbf{M}_{\mathcal{C}/S}(\mathcal{H}, d)$ and $\mathbf{Quot}_{\mathcal{C}/S}(\mathcal{H}, d)$ by the derived Cartesian squares

$$\begin{array}{ccccc} \mathbf{Quot}_{\mathcal{C}/S}(\mathcal{H}, d) & \hookrightarrow & \mathbf{M}_{\mathcal{C}/S}(\mathcal{H}, d) & \longrightarrow & \mathcal{M}_{\mathcal{C}/S}(d) \\ \downarrow & \square & \downarrow & \square & \downarrow 0 \\ \mathrm{Quot}_{\mathcal{C}/S}(\mathcal{E}_1, d) & \hookrightarrow & \mathcal{M}_{\mathcal{C}/S}(\mathcal{E}_1, d) & \stackrel{\phi^\vee}{\longrightarrow} & \mathcal{M}_{\mathcal{C}/S}(\mathcal{E}_0, d). \end{array} \tag{11.4.11}$$

Here the bottom right arrow is defined similarly to (11.4.7) from $\phi^\vee : \mathcal{E}_0 \to \mathcal{E}_1$. The derived stack $\mathbf{Quot}_{\mathcal{C}/S}(\mathcal{H}, d)$ is a derived open substack of $\mathbf{M}_{\mathcal{C}/S}(\mathcal{H}, d)$ with virtual dimension $\dim S - rd$, and its classical truncation is $\mathrm{Quot}_{\mathcal{C}/S}(\mathcal{H}, d)$.

11.4.3 (−1)-Shifted Cotangent Construction

By applying the construction in (11.2.14) for the morphism of vector bundles (11.4.6), we obtain the stack $\mathcal{N}_{\mathcal{C}/S}(\mathcal{E}^\bullet, d)$ with a regular function w

$$\mathcal{N}_{\mathcal{C}/S}(\mathcal{E}^\bullet, d) := \mathcal{M}_{\mathcal{C}/S}(\mathcal{E}^0, d) \times_{\mathcal{M}_{\mathcal{C}/S}(d)} \mathcal{M}_{\mathcal{C}/S}(\mathcal{E}^{-1}, d)^\vee \stackrel{w}{\to} \mathbb{A}^1.$$

By the Grothendieck duality, the T-valued points of the stack $\mathcal{N}_{\mathcal{C}/S}(\mathcal{E}^\bullet, d)$ consist of

$$(\mathcal{E}_T^0 \overset{u}{\to} \mathcal{P} \overset{v}{\to} \mathcal{E}_T^{-1} \otimes \omega_{\mathcal{C}_T/T}[1]) \tag{11.4.12}$$

where $\mathcal{P}$ is a T-valued point of $\mathcal{M}_{\mathcal{C}/S}(d)$. For an affine T, the function w on the T-valued point (11.4.12) is given by

$$\mathrm{Tr}(\mathcal{E}_T^0 \overset{u}{\to} \mathcal{P} \overset{v}{\to} \mathcal{E}_T^{-1} \otimes \omega_{\mathcal{C}_T/T}[1] \overset{\phi}{\to} \mathcal{E}_T^0 \otimes \omega_{\mathcal{C}_T/T}[1]) \in H^1(\mathcal{C}_T, \omega_{\mathcal{C}_T/T}) = \mathcal{O}_T. \tag{11.4.13}$$

By the construction, we have the isomorphism (see Sect. 11.2.6)

$$t_0(\Omega_{\mathbf{M}_{\mathcal{C}/S}(\mathcal{E},d)}[-1]) \cong \mathrm{Crit}(w) \subset \mathcal{N}_{\mathcal{C}/S}(\mathcal{E}^\bullet, d).$$

By applying the similar construction for the bottom right arrow in (11.4.11), we obtain the stack $\mathcal{N}_{\mathcal{C}/S}(\mathcal{E}_\bullet, d)$ with a regular function $w^\vee$ on it

$$\mathcal{N}_{\mathcal{C}/S}(\mathcal{E}_\bullet, d) := \mathcal{M}_{\mathcal{C}/S}(\mathcal{E}_1, d) \times_{\mathcal{M}_{\mathcal{C}/S}(d)} \mathcal{M}_{\mathcal{C}/S}(\mathcal{E}_0, d)^\vee \overset{w^\vee}{\to} \mathbb{A}^1.$$

The T-valued points of the stack $\mathcal{N}_{\mathcal{C}/S}(\mathcal{E}_\bullet, d)$ consist of

$$((\mathcal{E}_1)_T \overset{u'}{\to} \mathcal{P} \overset{v'}{\to} (\mathcal{E}_0)_T \otimes \omega_{\mathcal{C}_T/T}[1]) \tag{11.4.14}$$

where $\mathcal{P}$ is a T-valued point of $\mathcal{M}_{\mathcal{C}/S}(d)$. For an affine T, the function $w^\vee$ on the T-valued point (11.4.14) is given by

$$\mathrm{Tr}((\mathcal{E}_1)_T \overset{u'}{\to} \mathcal{P} \overset{v'}{\to} (\mathcal{E}_0)_T \otimes \omega_{\mathcal{C}_T/T}[1] \overset{\phi^\vee}{\to} (\mathcal{E}_1)_T \otimes \omega_{\mathcal{C}_T/T}[1]) \in H^1(\mathcal{C}_T, \omega_{\mathcal{C}_T/T}) = \mathcal{O}_T. \tag{11.4.15}$$

We also have the isomorphism

$$t_0(\Omega_{\mathbf{M}_{\mathcal{C}/S}(\mathcal{H},d)}[-1]) \cong \mathrm{Crit}(w^\vee) \subset \mathcal{N}_{\mathcal{C}/S}(\mathcal{E}_\bullet, d).$$

For a T-valued point $\mathcal{P} \in \mathrm{Coh}(\mathcal{C}_T)$ of $\mathcal{M}_{\mathcal{C}/S}(d)$, we define

$$\mathcal{P}^\vee := \mathbf{R}\mathcal{H}om_{\mathcal{C}_T}(\mathcal{P}, \omega_{\mathcal{C}_T/T}[1]) = \mathcal{E}xt^1_{\mathcal{C}_T}(\mathcal{P}, \omega_{\mathcal{C}_T/T}) \in \mathrm{Coh}(\mathcal{C}_T).$$

The above object also gives a T-valued point of $\mathcal{M}_{\mathcal{C}/S}(d)$, so we obtain the involution isomorphism

$$\mathbb{D} \colon \mathcal{M}_{\mathcal{C}/S}(d) \overset{\cong}{\to} \mathcal{M}_{\mathcal{C}/S}(d), \quad \mathcal{P} \mapsto \mathcal{P}^\vee. \tag{11.4.16}$$

Lemma 11.4.1 *There is an isomorphism*

$$\mathbb{D}\colon \mathcal{N}_{\mathcal{C}/S}(\mathcal{E}^{\bullet}, d) \xrightarrow{\cong} \mathcal{N}_{\mathcal{C}/S}(\mathcal{E}_{\bullet}, d) \tag{11.4.17}$$

such that the following diagram commutes:

$$\begin{array}{ccc} & \mathbb{A}^1 & \\ {}^{w}\nearrow & & \nwarrow^{w^{\vee}} \\ \mathcal{N}_{\mathcal{C}/S}(\mathcal{E}^{\bullet}, d) & \xrightarrow[\mathbb{D}]{\cong} & \mathcal{N}_{\mathcal{C}/S}(\mathcal{E}_{\bullet}, d) \\ \downarrow & & \downarrow \\ \mathcal{M}_{\mathcal{C}/S}(d) & \xrightarrow[\mathbb{D}]{\cong} & \mathcal{M}_{\mathcal{C}/S}(d). \end{array} \tag{11.4.18}$$

Here the vertical arrows are projections.

Proof By applying $\mathbf{R}\mathcal{H}om_{\mathcal{C}_T/T}(-, \omega_{\mathcal{C}_T/T}[1])$ to (11.4.12), we obtain

$$((\mathcal{E}_T)_1 \xrightarrow{v^{\vee}} \mathcal{P}^{\vee} \xrightarrow{u^{\vee}} (\mathcal{E}_T)_0 \otimes \omega_{\mathcal{C}_T/T}[1]). \tag{11.4.19}$$

The isomorphism (11.4.17) is obtained by assigning (11.4.12) with (11.4.19). Then the diagram (11.4.18) obviously commutes. □

11.4.4 Θ-Stratification

The stack $\mathcal{M}_{\mathcal{C}/S}(d)$ is the S-relative moduli stack of zero-dimensional semistable sheaves on $\mathcal{C}$, so it admits a good moduli space (which is nothing but the symmetric product)

$$\pi_{\mathcal{M}}\colon \mathcal{M}_{\mathcal{C}/S}(d) \to M_{\mathcal{C}/S}(d) = \mathrm{Sym}^d(\mathcal{C}/S). \tag{11.4.20}$$

A point $p \in \mathrm{Sym}^d(\mathcal{C}/S)$ corresponds to a point $s \in S$ and an effective divisor on $\mathcal{C}_s$ of degree d

$$p = \sum_{j=1}^{m} d^{(j)} p^{(j)}, \tag{11.4.21}$$

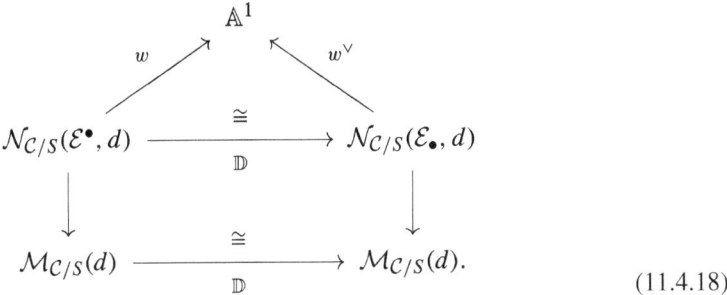

where $p^{(1)}, \ldots, p^{(m)}$ are distinct points in $\mathcal{C}_s$ and $d^{(j)} \geq 0$ with $d^{(1)} + \cdots + d^{(m)} = d$. Let R be the complete local $\mathbb{C}$-algebra $R = \widehat{\mathcal{O}}_{S,s}$. Formally locally at $p \in M_{C/S}(d)$, the stack $\mathcal{M}_{C/S}(d)$ is described by the following commutative diagram

$$\begin{array}{ccccc} & & \overset{\iota_p}{\frown\!\!\!\longrightarrow} & & \\ \widehat{\mathcal{M}}_{Q^{(m)}}(d^{(*)})_R & \xrightarrow{\cong} & \widehat{\mathcal{M}}_{C/S}(d)_p & \longrightarrow & \mathcal{M}_{C/S}(d) \\ \downarrow & & \downarrow \pi_p & \Box & \downarrow \pi_{\mathcal{M}} \\ \widehat{M}_{Q^{(m)}}(d^{(*)})_R & \xrightarrow{\cong} & \widehat{M}_{C/S}(d)_p & \longrightarrow & M_{C/S}(d). \end{array} \tag{11.4.22}$$

Here the middle vertical arrow is the formal fiber at p, and we have used the notation in Sect. 11.3.7 for the left vertical arrow.

The morphism $\mathcal{N}_{C/S}(\mathcal{E}^\bullet, d) \to \mathcal{M}_{C/S}(d)$ is an affine morphism, so it also admits a good moduli space $N_{C/S}(\mathcal{E}^\bullet, d)$ together with a commutative diagram

$$\begin{array}{ccc} \mathcal{N}_{C/S}(\mathcal{E}^\bullet, d) & \longrightarrow & \mathcal{M}_{C/S}(d) \\ \downarrow & \overset{g}{\searrow} & \downarrow \\ N_{C/S}(\mathcal{E}^\bullet, d) & \longrightarrow & M_{C/S}(d). \end{array} \tag{11.4.23}$$

Indeed $N_{C/S}(\mathcal{E}^\bullet, d) = \mathrm{Spec}\, g_* \mathcal{O}_{\mathcal{N}_{C/S}(\mathcal{E}^\bullet, d)}$, where g is the clockwise composition in the diagram (11.4.23).

For the universal sheaf $\mathcal{Q}$ in (11.4.3), we set

$$\mathcal{L} := \det(p_{2*}\mathcal{Q}) \in \mathrm{Pic}(\mathcal{M}_{C/S}(d)), \ b := \mathrm{ch}_2(p_{2*}\mathcal{Q}) \in H^4(\mathcal{M}_{C/S}(d), \mathbb{Q}).$$

We also regard them as elements of $\mathrm{Pic}(\mathcal{N}_{C/S}(\mathcal{E}^\bullet, d))$, $H^4(\mathcal{N}_{C/S}(\mathcal{E}^\bullet, d), \mathbb{Q})$ by pulling back them via $\mathcal{N}_{C/S}(\mathcal{E}^\bullet, d) \to \mathcal{M}_{C/S}(d)$.

Lemma 11.4.2 *The element $b \in H^4(\mathcal{N}_{C/S}(\mathcal{E}^\bullet, d), \mathbb{Q})$ is positive definite.*

Proof Let $f\colon B\mathbb{C}^* \to \mathcal{N}_{C/S}(\mathcal{E}^\bullet, d)$ be a non-degenerate morphism. It corresponds to a point $s \in S$ and a diagram

$$(\mathcal{E}_s^0 \to \mathcal{P}_0 \to \mathcal{E}_s^{-1} \otimes \omega_{C_s}[1]) \oplus \bigoplus_{j \neq 0} (0 \to \mathcal{P}_j \to 0)$$

where $\mathcal{P}_j$ are zero-dimensional sheaves on C_s and have $\mathbb{C}^*$-weight j. As f is non-degenerate, we have $\mathcal{P}_j \neq 0$ for some $j \neq 0$. Then we have

$$q^{-2} f^* b = \sum_j \frac{1}{2} j^2 \cdot \chi(\mathcal{P}_j) > 0.$$

$\square$

By the above lemma, there exist associated Θ-stratifications with respect to $(\mathcal{L}^{\pm 1}, b)$

$$\mathcal{N}_{C/S}(\mathcal{E}^{\bullet}, d) = \mathcal{S}_0^{\pm} \sqcup \mathcal{S}_1^{\pm} \sqcup \cdots \sqcup \mathcal{S}_{N^{\pm}}^{\pm} \sqcup \mathcal{N}_{C/S}(\mathcal{E}^{\bullet}, d)^{\mathcal{L}^{\pm 1}\text{-ss}}. \tag{11.4.24}$$

The stack $\mathcal{N}_{C/S}(\mathcal{E}^{\bullet}, d)$ and its Θ-stratifications are described formally locally on $M_{C/S}(d)$ in the following way. For $p \in M_{C/S}(d)$, we write it as (11.4.21), and set $R = \widehat{\mathcal{O}}_{S,s}$. We have the commutative diagram

$$\begin{array}{ccccc}
\multicolumn{5}{c}{\overset{\iota_p}{\xrightarrow{\hspace{14em}}}} \\
\widehat{\mathcal{G}}^{(m)}_{a,b}(d^{(*)})_R & \xrightarrow{\cong} & \widehat{\mathcal{N}}_{C/S}(\mathcal{E}^{\bullet}, d)_p & \longrightarrow & \mathcal{N}_{C/S}(\mathcal{E}^{\bullet}, d) \\
\downarrow & & \downarrow & \square & \downarrow g \\
\widehat{M}_{Q^{(m)}}(d^{(*)})_R & \xrightarrow{\cong} & \widehat{M}_{C/S}(d)_p & \longrightarrow & M_{C/S}(d).
\end{array} \tag{11.4.25}$$

Here $a := \operatorname{rank} \mathcal{E}^0$, $b := \operatorname{rank} \mathcal{E}^{-1}$, and we have used the notation in Sect. 11.3.7 for the left vertical arrow. Since $(\mathcal{L}, b)$ pulls back to $(\chi_0, |*|)$ under the top arrows in (11.4.25), the Θ-stratification pulls back to the KN stratifications in (11.3.35). In particular, we have $N^{\pm} = d - 1$.

11.4.5 Window Subcategories

For the good moduli space morphism (11.4.20), since $M_{C/S}(d)$ is smooth the functor

$$\pi_{\mathcal{M}}^{*} \colon D^b(M_{C/S}(d)) \to D^b(\mathcal{M}_{C/S}(d)) \tag{11.4.26}$$

is well-defined and it is fully-faithful since $\pi_{\mathcal{M}*}\mathcal{O}_{\mathcal{M}_{C/S}(d)} = \mathcal{O}_{M_{C/S}(d)}$ by the definition of good moduli space, see [1, Definition 4.1]. We define the triangulated subcategory

$$\mathbb{W}_{\mathrm{glob}}(d) \subset D^b(\mathcal{M}_{C/S}(d)) \tag{11.4.27}$$

to be the essential image of the functor (11.4.26).

Lemma 11.4.3 *An object $\mathcal{E} \in D^b(\mathcal{M}_{C/S}(d))$ lies in $\mathbb{W}_{\mathrm{glob}}(d)$ if and only if for any $p \in M_{C/S}(d)$ as in (11.4.21), we have $\iota_p^{*}\mathcal{E} \in \widehat{\mathbb{W}}(d^{(*)})$. Here ι_p is given in (11.4.22) and $\widehat{\mathbb{W}}(d^{(*)})$ is given in (11.3.44).*

Proof An object $\mathcal{E} \in D^b(\mathcal{M}_{C/S}(d))$ lies in $\mathbb{W}_{\mathrm{glob}}(d)$ if and only if the adjunction morphism

$$\pi_{\mathcal{M}}^* \pi_{\mathcal{M}*} \mathcal{E} \to \mathcal{E}$$

is an isomorphism. Let $\mathcal{F}$ be the cone of the above morphism. Then the last condition is equivalent to that $\mathcal{F} = 0$. This property is formally local on $M_{C/S}(d)$, so it is equivalent to that $\iota_p^* \mathcal{F} = 0$ for any $p \in M_{C/S}(d)$. By the base change, it is also equivalent to that

$$\pi_p^* \pi_{p*}(\iota_p^* \mathcal{E}) \to \iota_p^* \mathcal{E}$$

is an isomorphism, which is equivalent to that $\iota_p^* \mathcal{E} \in \widehat{\mathbb{W}}(d^{(*)})$. □

We take the $\mathbb{C}^*$-action on $\mathcal{N}_{C/S}(\mathcal{E}^\bullet, d)$ by the weight one action on the factor $\mathcal{M}_{C/S}(\mathcal{E}^0, d)$, i.e $t \in \mathbb{C}^*$ acts on (11.4.12) by

$$(\mathcal{E}_T^0 \stackrel{tu}{\to} \mathcal{P} \stackrel{v}{\to} \mathcal{E}_T^{-1} \otimes \omega_{C_T/T}[1]).$$

The function (11.4.13) is of weight one with respect to the above $\mathbb{C}^*$-action. So we have the triangulated category of $\mathbb{C}^*$-equivariant factorizations of w

$$\mathrm{MF}^{\mathbb{C}^*}(\mathcal{N}_{C/S}(\mathcal{E}^\bullet, d), w).$$

Let us consider Θ-stratifications (11.4.24). For each $1 \leq i \leq N^{\pm}$, let us take $m_i^{\pm} \in \mathbb{R}$. Similarly to (11.2.7), the window subcategory

$$\mathbb{W}_{m_\bullet^{\pm}}(d) \subset \mathrm{MF}^{\mathbb{C}^*}(\mathcal{N}_{C/S}(\mathcal{E}^\bullet, d), w)$$

is defined to be consisting of objects $\mathcal{P}$ so that for each $1 \leq i \leq N^{\pm}$ and center $\mathcal{Z}_i^{\pm} \subset \mathcal{S}_i^{\pm}$, we have the weight condition with respect to the canonical $\mathbb{C}^*$-stabilizer groups in $\mathcal{Z}_i^{\pm}$:

$$\mathrm{wt}(\mathcal{P}|_{\mathcal{Z}_i^{\pm}}) \subset [m_i^{\pm}, m_i^{\pm} + \eta_i^{\pm}) \qquad (11.4.28)$$

Here $\eta_i^{\pm} \in \mathbb{Z}$ are defined as in (11.2.6), i.e. the weight of the conormal bundle of $\mathcal{S}_i^{\pm}$ inside $\mathcal{N}_{C/S}(\mathcal{E}^\bullet, d)$ restricted to $\mathcal{Z}_i^{\pm}$.

Proposition 11.4.4 *There exist equivalences*

$$\mathbb{W}_{m_\bullet^+}(d) \simeq D^b(\mathbf{Quot}(\mathcal{E}, d)), \ \mathbb{W}_{m_\bullet^-}(d) \simeq D^b(\mathbf{Quot}(\mathcal{H}, d)). \qquad (11.4.29)$$

Proof A version of window theorem implies that the composition functors

$$\mathbb{W}_{m_\bullet^\pm}(d) \hookrightarrow \mathrm{MF}^{\mathbb{C}^*}(\mathcal{N}_{\mathcal{C}/S}(\mathcal{E}^\bullet, d), w) \twoheadrightarrow \mathrm{MF}^{\mathbb{C}^*}(\mathcal{N}_{\mathcal{C}/S}(\mathcal{E}^\bullet, d)^{\mathcal{L}^{\pm 1}\text{-ss}}, w) \tag{11.4.30}$$

are equivalences (see [16, Prop 2.4.2] for perfect complexes, and then use the argument of [18, Proposition 5.5] to deduce the result for factorization categories). From the formal local description of Θ-stratifications by the diagram (11.4.25), it follows that (11.4.12) is a T-valued point of $\mathcal{N}_{\mathcal{C}/S}(\mathcal{E}^\bullet, d)^{\mathcal{L}\text{-ss}}$ if and only if $u\colon \mathcal{E}_T^0 \to \mathcal{P}$ is surjective. Therefore from Theorem 11.2.5, the last category in (11.4.30) for $\mathcal{L}$-semistable part is equivalent to $D^b(\mathbf{Quot}(\mathcal{E}, d))$, so the first equivalence in (11.4.29) follows.

Under the isomorphism (11.4.16), the line bundle $\mathcal{L}$ pulls back to $\mathcal{L}^{-1}$. Therefore from the diagram (11.4.18), the isomorphism (11.4.17) restricts to the isomorphism

$$\mathbb{D}\colon \mathcal{N}_{\mathcal{C}/S}(\mathcal{E}^\bullet, d)^{\mathcal{L}^{-1}\text{-ss}} \stackrel{\cong}{\to} \mathcal{N}_{\mathcal{C}/S}(\mathcal{E}_\bullet, d)^{\mathcal{L}\text{-ss}}.$$

Similarly to above, the T-valued points of the right hand side consist of (11.4.14) such that $u'\colon (\mathcal{E}_1)_T \to \mathcal{P}$ is surjective. Also note that, under the isomorphism (11.4.4), the diagonal torus $\mathbb{C}^* \subset \mathrm{GL}(W)$ acts on $\mathrm{Quot}^\circ_{\mathcal{C}/S}(W \otimes \mathcal{O}_\mathcal{C}, d)$ trivially and acts on fibers of $\mathcal{M}_{\mathcal{C}/S}(\mathcal{E}^i, d) \to \mathcal{M}_{\mathcal{C}/S}(d)$ by weight one. Therefore by Theorem 11.2.5 and Lemma 11.2.6, the last category in (11.4.30) for $\mathcal{L}^{-1}$-semistable part is equivalent to $D^b(\mathbf{Quot}(\mathcal{H}, d))$, so the second equivalence in (11.4.29) follows. □

We take the special choices of $m_\bullet^\pm$ by $m_i^+ = -\eta_i + \varepsilon$ for $0 < \varepsilon \ll 1$ and $m_i^- = 0$, and define

$$\mathbb{W}^+_{\mathrm{glob}}(d) := \mathbb{W}_{m_i^+ = -\eta_i^+ + \varepsilon}(d), \ \mathbb{W}^-_{\mathrm{glob}}(d) := \mathbb{W}_{m_i^- = 0}(d). \tag{11.4.31}$$

Lemma 11.4.5 *An object $\mathcal{E} \in \mathrm{MF}^{\mathbb{C}^*}(\mathcal{N}_{\mathcal{C}/S}(\mathcal{E}^\bullet, d), w)$ lies in $\mathbb{W}^+_{\mathrm{glob}}(d)$ (resp. $\mathbb{W}^-_{\mathrm{glob}}(d)$) if and only if for any $p \in M_{\mathcal{C}/S}(d)$ as in (11.4.21), the object $\iota_p^*\mathcal{E} \in \mathrm{MF}^{\mathbb{C}^*}(\widehat{\mathcal{G}}^{(m)}_{a,b}(d^{(*)})_R, \iota_p^* w)$ lies in $\widehat{\mathbb{W}}_c(d^{(*)})$ for $c = a$ (resp. $c = b$). Here ι_p is given in (11.4.25) and $\widehat{\mathbb{W}}_c(d^{(*)})$ is given in (11.3.44).*

Proof The defining condition of $\mathbb{W}^\pm_{\mathrm{glob}}(d)$ is local on $M_{\mathcal{C}/S}(d)$, so $\mathcal{E}$ is an object in $\mathbb{W}^\pm_{\mathrm{glob}}(d)$ if and only if $\iota_p^*\mathcal{E}$ satisfies the same weight condition (11.4.28) with respect to the Θ-stratifications (11.4.24) pulled back via ι_p. As we already mentioned, they coincide with KN stratifications in (11.3.35). Therefore the argument of Proposition 11.3.6 implies the lemma (see Remark 11.3.7). □

11.4.6 Categorified Hall Product

For a decomposition $d = d_1 + d_2$, we define the stack over S

$$\mathcal{E}x_{\mathcal{C}/S}(d_1, d_2) \colon (Sch/S) \to (Groupoid)$$

whose T-valued points consist of exact sequences

$$0 \to \mathcal{P}_1 \to \mathcal{P}_3 \to \mathcal{P}_2 \to 0 \tag{11.4.32}$$

where $\mathcal{P}_i$ are T-valued points of $\mathcal{M}_{\mathcal{C}/S}(d_i)$ with $d_3 = d$. The stack $\mathcal{E}x_{\mathcal{C}/S}(d_1, d_2)$ is a smooth Artin stack of finite type over S. Indeed we have the obvious evaluation morphisms

$$\begin{array}{ccc} \mathcal{E}x_{\mathcal{C}/S}(d_1, d_2) & \xrightarrow{\mathrm{ev}_3} & \mathcal{M}_{\mathcal{C}/S}(d_3) \\ {\scriptstyle (\mathrm{ev}_1, \mathrm{ev}_2)} \downarrow & & \\ \mathcal{M}_{\mathcal{C}/S}(d_1) \times_S \mathcal{M}_{\mathcal{C}/S}(d_2) & & \end{array} \tag{11.4.33}$$

The left vertical arrow is smooth because of the vanishing of Ext^2 on curves, so $\mathcal{E}x_{\mathcal{C}/S}(d_1, d_2)$ is also smooth. Since every stacks in (11.4.33) are smooth over S and ev_3 is proper, we have the functor

$$* := \mathrm{ev}_{3*}(\mathrm{ev}_1, \mathrm{ev}_2)^* \colon D^b(\mathcal{M}_{\mathcal{C}/S}(d_1)) \boxtimes_S D^b(\mathcal{M}_{\mathcal{C}/S}(d_2)) \to D^b(\mathcal{M}_{\mathcal{C}/S}(d)) \tag{11.4.34}$$

giving categorified Hall algebra structure on

$$\bigoplus_{d \geq 0} D^b(\mathcal{M}_{\mathcal{C}/S}(d)). \tag{11.4.35}$$

We also define the stack over S

$$\mathcal{E}x_{\mathcal{C}/S}(\mathcal{E}^\bullet, d_1, d_2) \colon (Sch/S) \to (Groupoid)$$

whose T-valued points consist of diagrams

$$
\begin{array}{ccccc}
 & & 0 & & \\
 & & \downarrow & & \\
0 & \longrightarrow & \mathcal{P}_1 & \longrightarrow & 0 \\
\downarrow & & \downarrow & & \downarrow \\
\mathcal{E}_T^0 & \xrightarrow{u_3} & \mathcal{P}_3 & \xrightarrow{v_3} & \mathcal{E}_T^{-1} \otimes \omega_{\mathcal{C}_T/T}[1] \\
\| & & \downarrow & & \| \\
\mathcal{E}_T^0 & \xrightarrow{u_2} & \mathcal{P}_2 & \xrightarrow{v_2} & \mathcal{E}_T^{-1} \otimes \omega_{\mathcal{C}_T/T}[1] \\
 & & \downarrow & & \\
 & & 0 & &
\end{array}
\tag{11.4.36}
$$

where $\mathcal{P}_i$ are T-valued points of $\mathcal{M}_{\mathcal{C}/S}(d_i)$ with $d_3 = d$, the vertical arrows are exact sequences, the middle and the bottom arrows are T-valued points of $\mathcal{N}_{\mathcal{C}/S}(\mathcal{E}^\bullet, d_i)$ for $i = 3, 2$.

The stack $\mathcal{E}x_{\mathcal{C}/S}(\mathcal{E}^\bullet, d_1, d_2)$ is a smooth Artin stack of finite type. Indeed given an exact sequence (11.4.32), giving a diagram (11.4.36) is equivalent to giving morphisms $\mathcal{E}_T^0 \to \mathcal{P}_3$ and $\mathcal{P}_2 \to \mathcal{E}_T^{-1} \otimes \omega_{\mathcal{C}_T/T}[1]$, so $\mathcal{E}x_{\mathcal{C}/S}(\mathcal{E}^\bullet, d_1, d_2)$ is constructed as a Cartesian square

$$
\begin{array}{ccc}
\mathcal{E}x_{\mathcal{C}/S}(\mathcal{E}^\bullet, d_1, d_2) & \longrightarrow & \mathcal{E}x_{\mathcal{C}/S}(d_1, d_2) \\
\downarrow & \square & \downarrow {\scriptstyle (\mathrm{ev}_2, \mathrm{ev}_3)} \\
\mathcal{M}_{\mathcal{C}/S}(\mathcal{E}^{-1}, d_2)^\vee \times_S \mathcal{M}_{\mathcal{C}/S}(\mathcal{E}^0, d_3) & \longrightarrow & \mathcal{M}_{\mathcal{C}/S}(d_2) \times_S \mathcal{M}_{\mathcal{C}/S}(d_3).
\end{array}
$$

Since $\mathcal{E}x_{\mathcal{C}/S}(d_1, d_2)$ is smooth and the bottom horizontal arrow is a vector bundle, the above construction in particular implies that $\mathcal{E}x_{\mathcal{C}/S}(\mathcal{E}^\bullet, d_1, d_2)$ is smooth over S.

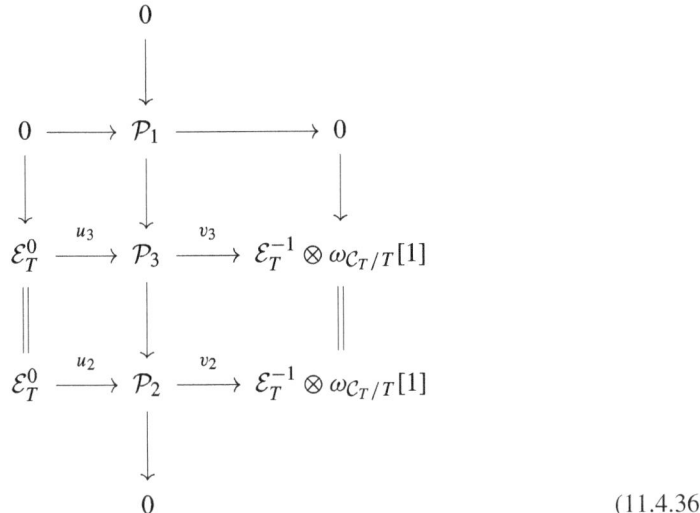

We have the obvious evaluation morphisms, commuting with super-potentials in (11.4.13)

$$
\begin{array}{ccccc}
\mathcal{E}x_{\mathcal{C}/S}(\mathcal{E}^\bullet, d_1, d_2) & \xrightarrow{\mathrm{ev}_3} & \mathcal{N}_{\mathcal{C}/S}(\mathcal{E}^\bullet, d_3) & & \\
\downarrow {\scriptstyle (\mathrm{ev}_1, \mathrm{ev}_2)} & & \downarrow & \searrow^{w} & \\
\mathcal{M}_{\mathcal{C}/S}(d_1) \times_S \mathcal{N}_{\mathcal{C}/S}(\mathcal{E}^\bullet, d_2) & & \xrightarrow{(0,w)} & & \mathbb{A}^1 \\
\downarrow & & \downarrow & & \\
M_{\mathcal{C}/S}(d_1) \times_S M_{\mathcal{C}/S}(d_2) & \xrightarrow{\oplus} & M_{\mathcal{C}/S}(d_3). & &
\end{array}
\tag{11.4.37}
$$

Here the right vertical arrow is the morphism g in (11.4.23), and the left bottom vertical arrow is the product of (11.4.20) with g. We have an auxiliary $\mathbb{C}^*$-action on $\mathcal{E}x_{\mathcal{C}/S}(\mathcal{E}^\bullet, d_1, d_2)$ acting on (11.4.36) by $(u_3, v_3, u_2, v_2) \mapsto (tu_3, v_3, tu_2, v_2)$. Then the diagram (11.4.37) is equivariant under the auxiliary $\mathbb{C}^*$-actions. Also note that every stacks in the upper left diagram in (11.4.37) are smooth and ev_3 is proper as any fiber is a closed subscheme of a Quot scheme. Therefore we have the categorified Hall product

$$
\begin{aligned}
* := \mathrm{ev}_{3*}(\mathrm{ev}_1, \mathrm{ev}_2)^* \colon & D^b(\mathcal{M}_{\mathcal{C}/S}(d_1)) \boxtimes_S \mathrm{MF}^{\mathbb{C}^*}(\mathcal{N}_{\mathcal{C}/S}(\mathcal{E}^\bullet, d_2), w) \\
& \to \mathrm{MF}^{\mathbb{C}^*}(\mathcal{N}_{\mathcal{C}/S}(\mathcal{E}^\bullet, d), w)
\end{aligned}
\tag{11.4.38}
$$

which gives a left module structure of (11.4.35) on

$$
\bigoplus_{d \geq 0} \mathrm{MF}^{\mathbb{C}^*}(\mathcal{N}_{\mathcal{C}/S}(\mathcal{E}^\bullet, d), w).
$$

11.4.7 Semiorthogonal Decomposition

Recall that $r = \mathrm{rank}(\mathcal{E})$, $a = \mathrm{rank}\, \mathcal{E}^0$, $b = \mathrm{rank}\, \mathcal{E}^{-1}$ so that $r = a - b$. We prove the following proposition:

Proposition 11.4.6 *For $(d_1, \ldots, d_r) \in \mathbb{Z}^r_{\geq 0}$, the categorified Hall product (11.4.38) induces the fully-faithful functor*

$$
\begin{aligned}
*\colon & \mathbb{W}_{\mathrm{glob}}(d_1) \boxtimes_S (\mathbb{W}_{\mathrm{glob}}(d_2) \otimes \mathcal{L}) \boxtimes_S \cdots \boxtimes_S (\mathbb{W}_{\mathrm{glob}}(d_r) \otimes \mathcal{L}^{r-1}) \\
& \boxtimes_S (\mathbb{W}^-_{\mathrm{glob}}(d - \sum_{i=1}^{r} d_i) \otimes \mathcal{L}^r) \to \mathbb{W}^+_{\mathrm{glob}}(d)
\end{aligned}
\tag{11.4.39}
$$

such that, by setting $\mathcal{C}^{-}_{\mathrm{glob}}(d_\bullet)$ *to be the essential image of the above functor, we have the semiorthogonal decomposition*

$$\mathbb{W}^{+}_{\mathrm{glob}}(d) = \langle \mathcal{C}^{-}_{\mathrm{glob}}(d_\bullet) : d_\bullet \in \mathbb{Z}^r_{\geq 0} \rangle.$$

Here $\mathrm{Hom}(\mathcal{C}_{\mathrm{glob}}(d'_\bullet), \mathcal{C}_{\mathrm{glob}}(d_\bullet)) = 0$ *for* $d'_\bullet \succ d_\bullet$.

Proof Following the argument of [44, Theorem 5.16], it is enough to show that the functor (11.4.39) is fully-faithful and forms a semiorthogonal decomposition formally locally on $M_{\mathcal{C}/S}(d)$. Let us take a closed point $p \in M_{\mathcal{C}/S}(d)$ written as in (11.4.21), and set $R = \widehat{\mathcal{O}}_{S,s}$. We write the formal completion $\widehat{M}_{\mathcal{C}/S}(d)_p$ as $\widehat{M}_{\mathcal{C}/S}(d)_{d^{(*)}p^{(*)}}$. Let $d = d_1 + d_2$ be a decomposition. Then we have the Cartesian square

$$\begin{array}{ccc}
\coprod_{d_1^{(*)}+d_2^{(*)}=d^{(*)}, |d_i^{(*)}|=d_i} \widehat{M}_{\mathcal{C}/S}(d_1)_{d_1^{(*)}p^{(*)}} \widehat{\times}_R \widehat{M}_{\mathcal{C}/S}(d_2)_{d_2^{(*)}p^{(*)}} & \longrightarrow & \widehat{M}_{\mathcal{C}/S}(d)_{d^{(*)}p^{(*)}} \\
\downarrow & & \downarrow \\
M_{\mathcal{C}/S}(d_1) \times_S M_{\mathcal{C}/S}(d_2) & \xrightarrow{\oplus} & M_{\mathcal{C}/S}(d).
\end{array}$$

Together with the diagrams (11.4.22), (11.4.25), the base change of the diagram (11.4.37) via $\widehat{M}_{\mathcal{C}/S}(d)_p \to M_{\mathcal{C}/S}(d)$ gives the diagram

$$\begin{array}{cccc}
\widehat{\mathcal{E}x}(\mathcal{E}^\bullet, d_1, d_2)_p & \longrightarrow & \widehat{\mathcal{G}}^{(m)}_{a,b}(d^{(*)})_R & \xrightarrow{\iota_p^* w} \ \mathbb{A}^1 \\
\downarrow & & \downarrow{\scriptstyle (0,\iota_p^* w)} & \\
\coprod_{d_1^{(*)}+d_2^{(*)}=d^{(*)}, |d_i^{(*)}|=d_i} \widehat{\mathcal{M}}_{Q^{(m)}}(d_1^{(*)})_R \widehat{\times}_R \widehat{\mathcal{G}}^{(m)}_{a,b}(d_2^{(*)})_R & \longrightarrow & & \mathbb{A}^1 \\
\downarrow & & \downarrow & \\
\coprod_{d_1^{(*)}+d_2^{(*)}=d^{(*)}, |d_i^{(*)}|=d_i} \widehat{M}_{Q^{(m)}}(d_1^{(*)})_R \widehat{\times}_R \widehat{M}_{Q^{(m)}}(d_2^{(*)})_R & \xrightarrow{\oplus} & \widehat{M}_{Q^{(m)}}(d^{(*)})_R. &
\end{array}$$

By comparing with the diagram (11.3.42), we see that the base change of the product functor (11.4.38) via $\widehat{M}_{\mathcal{C}/S}(d)_p \to M_{\mathcal{C}/S}(d)$ is identified with the direct sum of (11.3.43) for all the decompositions $d^{(*)} = d_1^{(*)} + d_2^{(*)}$ with $|d_i^{(*)}| = d_i$.

For $(d_1, \ldots, d_r) \in \mathbb{Z}^r_{\geq 0}$, the above argument shows that the base change of the product functor

$$\begin{aligned}
*\colon D^b(\mathcal{M}_{\mathcal{C}/S}(d_1)) \boxtimes_S \cdots \boxtimes_S D^b(\mathcal{M}_{\mathcal{C}/S}(d_r)) \boxtimes_S \mathrm{MF}^{\mathbb{C}^*}(\mathcal{N}_{\mathcal{C}/S}(\mathcal{E}^\bullet, d - \sum_{i=1}^r d_i), w) \\
\to \mathrm{MF}^{\mathbb{C}^*}(\mathcal{N}_{\mathcal{C}/S}(\mathcal{E}^\bullet, d), w)
\end{aligned}$$

via $\widehat{M}_{C/S}(d)_p \to M_{C/S}(d)$ gives the Hall product in Sect. 11.3.7

$$*: \bigoplus_{\substack{(d_1^{(*)},\ldots,d_r^{(*)})\\ |d_i^{(*)}|=d_i}} D^b(\widehat{\mathcal{M}}_{Q^{(m)}}(d_1^{(*)})_R)\widehat{\boxtimes}_R \cdots \widehat{\boxtimes}_R D^b(\widehat{\mathcal{M}}_{Q^{(m)}}(d_r^{(*)})_R)$$

$$\widehat{\boxtimes}_R \operatorname{MF}^{\mathbb{C}^*}(\widehat{\mathcal{G}}_{a,b}^{(m)}(d^{(*)} - \sum_{i=1}^{r} d_i^{(*)})_R, \iota_p^* w) \to \operatorname{MF}^{\mathbb{C}^*}(\widehat{\mathcal{G}}_{a,b}^{(m)}(d^{(*)})_R, \iota_p^* w).$$

Then from Lemmas 11.4.3 and 11.4.5, the required formal local statement is given in Theorem 11.3.15. □

The following is the main result in this paper:

Theorem 11.4.7 *There is a semiorthogonal decomposition of the form*

$$\begin{aligned} &D^b(\mathbf{Quot}_{C/S}(\mathcal{E}, d)) \\ &= \Big\langle D^b(\operatorname{Sym}^{d_1}(C/S) \times_S \cdots \times_S \operatorname{Sym}^{d_r}(C/S) \times_S \mathbf{Quot}_{C/S}(\mathcal{H}, d - \sum_{i=1}^{r} d_i)) \\ &\quad : (d_1, \ldots, d_r) \in \mathbb{Z}_{\geq 0}^r \Big\rangle. \end{aligned}$$

Proof The theorem follows from Proposition 11.4.4, Proposition 11.4.6, and the fact that $\mathbb{W}_{\mathrm{glob}}(d)$ is equivalent to $D^b(\operatorname{Sym}^d(C/S))$ by its definition. □

Remark 11.4.8 The number of semiorthogonal summands in Theorem 11.4.7 involving $\mathbf{Quot}_{C/S}(\mathcal{H}, d-k)$ is $\dim \operatorname{Sym}^k(\mathbb{C}^r)$, while that involving $\mathbf{Quot}_S(\mathcal{H}, d-k)$ in (11.1.4) is $\dim \operatorname{Sym}^k(\mathbb{C}^r[1])$. So in some sense the semiorthogonal decompositions in Theorem 11.4.7 and (11.1.4) are related by boson-fermion correspondence (see the related phenomena for cohomological DT theory in [12]).

For a smooth projective curve C, the Quot scheme of zero-dimensional quotients $\operatorname{Quot}_C(\mathcal{E}, d)$ is smooth of expected dimension rd. Therefore Theorem 11.4.7 implies the following:

Corollary 11.4.9 *There is a semiorthogonal decomposition of the form*

$$D^b(\operatorname{Quot}_C(\mathcal{E}, d)) = \left\langle D^b(\operatorname{Sym}^{d_1}(C) \times \cdots \times \operatorname{Sym}^{d_r}(C)) : d_1 + \cdots + d_r = d \right\rangle. \tag{11.4.40}$$

Remark 11.4.10 The semiorthogonal decomposition (11.4.40) may be further decomposed. Indeed for each $\delta > 0$, there is a semiorthogonal decomposition proved in [46, Corollary 5.11]

$$D^b(\mathrm{Sym}^{g-1+\delta}(C)) = \langle D^b(\mathrm{Sym}^{g-1-\delta}(C)), \overbrace{J(C), \ldots, J(C)}^{\delta}\rangle$$

where g is the genus of C and $J(C)$ is the Jacobian of C So if $d_i > g - 1$ for some i in (11.1.2), it can be further decomposed. In particular if $C = \mathbb{P}^1$, then (11.4.40) implies the existence of a full exceptional collection whose cardinality is

$$\sum_{d_1+\cdots+d_r=d} \prod_{i=1}^{r} (d_i + 1) = \binom{2r+d-1}{d}.$$

Acknowledgments This paper is dedicated to the memory of Professor Bumsig Kim. He kindly invited me to KIAS in Korea several times, and I thank him for his hospitality and discussions related to this paper. This paper in particular contains subjects of Quot schemes, matrix factorizations, GIT quotients and wall-crossing, and I learned many of them from the discussions with him or from his papers [6–11, 15, 26–28].

The author is also grateful to Tudor Păduraru for a joint work [37], whose subject is related to this paper. The author is supported by World Premier International Research Center Initiative (WPI initiative), MEXT, Japan, and Grant-in Aid for Scientific Research grant (No. 19H01779) from MEXT, Japan.

References

1. Alper, J.: Good moduli spaces for Artin stacks. Ann. Inst. Fourier (Grenoble) **63**(6), 2349–2402 (2013)
2. Ballard, M., Favero, D., Katzarkov, L.: Variation of geometric invariant theory quotients and derived categories. J. Reine Angew. Math. **746**, 235–303 (2019)
3. Bagnarol, M., Fantechi, B., Perroni, F.: On the motive of Quot schemes of zero-dimensional quotients on a curve. New York J. Math. **26**, 138–148 (2020)
4. Bifet, E.: Sur les points fixes du schéma $\mathrm{Quot}_{\mathcal{O}_X^r/X/k}$ sous l'action du tore $G^r_{m,k}$. C. R. Acad. Sci. Paris Sér. I Math. **309**(9), 609–612 (1989)
5. Bondal, A., Orlov, D.: Semiorthogonal decomposition for algebraic varieties, arXiv:9506012
6. Ciocan-Fontanine, I., Kim, B.: Quasimap wall-crossings and mirror symmetry. Publ. Math. Inst. Hautes Études Sci. **131**, 201–260 (2020)
7. Ciocan-Fontanine, I., Kim, B., Maulik, D.: Stable quasimaps to GIT quotients. J. Geom. Phys. **75**, 17–47 (2014)
8. Ciocan-Fontanine, I., Favero, D., Guéré, J., Kim, B., Shoemaker, M.: Fundamental factorization of a GLSM Part I: Construction. Mem. Amer. Math. Soc. **289**(1435), iv+96 (2023)
9. Chung, K., Kim, B., Kim, T.: A chain level HKR-type map and a Chern character formula, arXiv:2109.14372
10. Chung, K., Kim, B., Kim, T.: Chern characters for curved dg-algebras, arXiv:2202.11403
11. Choa, D., Kim, B., Sreedhar, B.: Riemann-Roch for stacky matrix factorizations. Forum Math. Sigma **10**, Paper No. e108, 29 (2022). MR 4519062

12. Davison, B.: A boson-fermion correspondence in cohomological Donaldson-Thomas theory, arXiv:2109.09788
13. Davison, B., Meinhardt, S.: Cohomological Donaldson-Thomas theory of a quiver with potential and quantum enveloping algebras. Invent. Math. **221**(3), 777–871 (2020)
14. Efimov, A.I., Positselski, L.: Coherent analogues of matrix factorizations and relative singularity categories. Algebra Number Theory **9**(5), 1159–1292 (2015)
15. Favero, D., & Kim, B.: General GLSM invariants and their Cohomological Field Theories, arXiv:2006.12182
16. Halpern-Leistner, D.: The D-equivalence conjecture for moduli spaces of sheaves on a K3 surface. Available in http://www.math.columbia.edu/~danhl/
17. Halpern-Leistner, D.: On the structure of instability in moduli theory, arXiv:1411.0627
18. Halpern-Leistner, D.: The derived category of a GIT quotient. J. Amer. Math. Soc. **28**(3), 871–912 (2015)
19. Halpern-Leistner, D., Sam, S.V.: Combinatorial constructions of derived equivalences. J. Amer. Math. Soc. **33**(3), 735–773 (2020)
20. Hirano, Y.: Derived Knörrer periodicity and Orlov's theorem for gauged Landau-Ginzburg models. Compos. Math. **153**(5), 973–1007 (2017)
21. Isik, M.U.: Equivalence of the derived category of a variety with a singularity category. Int. Math. Res. Not. IMRN **2013**(12), 2787–2808 (2013)
22. Jiang, Q.: Derived categories of Quot schemes of locally free quotients I, arXiv:2107.09193
23. Jiang, Y., & Thomas, R.: Virtual signed Euler characteristics. J. Algebraic Geom. **26**(2), 379–397 (2017)
24. Kapranov, M.: Derived category of coherent sheaves on Grasssmann manifolds (Russian). Izv. Akad. Nauk SSSR Ser. Mat. **48**, 192–202 (1984)
25. Kawamata, Y.: D-equivalence and K-equivalence. J. Differential Geom. **61**, 147–171 (2002)
26. Kim, B.: Hirbebruch-riemann-Roch for global matrix factorizations, arXiv:2106.00435
27. Kim, B., Kim, T.: Standard conjecture D for local stacky matrix factorizations, arXiv:2204.09849
28. Kim, B., & Polishchuk, A.: Atiyah class and Chern character for global matrix factorisations. J. Inst. Math. Jussieu **21**(4), 1445–1470 (2022)
29. Kontsevich, M., Soibelman, Y.: Cohomological Hall algebra, exponential Hodge structures and motivic Donaldson-Thomas invariants. Commun. Number Theory Phys. **5**(2), 231–352 (2011)
30. Koseki, N., Toda, Y.: Derived categories of Thaddeus pair moduli spaces via d-critical flips. Adv. Math. **391**, Paper No. 107965, 55 (2021)
31. Mirković, I., Riche, S.: Linear Koszul duality. Compos. Math. **146**(1), 233–258 (2010)
32. Oprea, D., Pandharipande, R.: Quot schemes of curves and surfaces: virtual classes, integrals, Euler characteristics, arXiv:1903.08787
33. Orlov, D.: Derived categories of coherent sheaves and triangulated categories of singularities. In Algebra, Arithmetic, and Geometry: In Honor of Yu. I. Manin. Vol. II, Progr. Math., vol. 270, pp. 503–531. Birkhäuser Boston, Boston, MA (2009)
34. Orlov, D.: Matrix factorizations for nonaffine LG-models. Math. Ann. **353**(1), 95–108 (2012)
35. Pădurariu, T.: Categorical and K-theoretic Hall algebras for quivers with potential. J. Inst. Math. Jussieu. **22**(6), 2717–2747 (2023)
36. Pădurariu, T.: Generators for K-theoretic Hall algebras of quivers with potential, arXiv:2108.07919
37. Padurariu, T., Toda, Y.: Categorical and K-theoretic Donaldson-Thomas theory of $\mathbb{C}^3$ (part I). Duke Math. J., to appear
38. Polishchuk, A., Vaintrob, A.: Matrix factorizations and singularity categories for stacks. Ann. Inst. Fourier (Grenoble) **61**(7), 2609–2642 (2011)
39. Ricolfi, A.: On the motive of the Quot scheme of finite quotients of a locally free sheaf. J. Math. Pures Appl. (9) **144**, 50–68 (2020)
40. Shipman, I.: A geometric approach to Orlov's theorem. Compos. Math. **148**(5), 1365–1389 (2012)

41. Špenko, Š., Van den Bergh, M.: Non-commutative resolutions of quotient singularities for reductive groups. Invent. Math. **210**(1), 3–67 (2017)
42. Toda, Y.: Birational geometry for d-critical loci and wall-crossing in Calabi-Yau 3-folds, arXiv:1805.00182
43. Toda, Y.: Categorical Donaldson-Thomas theory for local surfaces, arXiv:1907.09076
44. Toda, Y.: Categorical wall-crossing formula for Donaldson-Thomas theory on the resolved conifold, arXiv:2109.07064
45. Toda, Y.: Derived categories of Quot schemes of locally free quotients via categorified Hall products, arXiv:2110.02469
46. Toda, Y.: Semiorthogonal decompositions of stable pair moduli spaces via d-critical flips. J. Eur. Math. Soc. (JEMS) **23**(5), 1675–1725 (2021)

Open Access This chapter is licensed under the terms of the Creative Commons Attribution-NonCommercial-NoDerivatives 4.0 International License (http://creativecommons.org/licenses/by-nc-nd/4.0/), which permits any noncommercial use, sharing, distribution and reproduction in any medium or format, as long as you give appropriate credit to the original author(s) and the source, provide a link to the Creative Commons license and indicate if you modified the licensed material. You do not have permission under this license to share adapted material derived from this chapter or parts of it.

The images or other third party material in this chapter are included in the chapter's Creative Commons license, unless indicated otherwise in a credit line to the material. If material is not included in the chapter's Creative Commons license and your intended use is not permitted by statutory regulation or exceeds the permitted use, you will need to obtain permission directly from the copyright holder.

Appendix Some Research Papers of Professor Bumsig Kim

© The Author(s) 2026

N.-G. Kang et al. (eds.), *Categorical and Enumerative Aspects of Mirror Symmetry*,
KIAS Springer Series in Mathematics 5,
https://doi.org/10.1007/978-981-95-0385-8

A.1. Quantum Hyperplane Section Theorem for Homogeneous Spaces

Kim, B. Quantum hyperplane section theorem for homogeneous spaces. Acta Math **183**, 71–99 (1999). https://doi.org/10.1007/BF02392947

© 1999, Institut Mittag-Leffler. All rights reserved. Reprinted with permission.

© The Author(s) 2026

N.-G. Kang et al. (eds.), *Categorical and Enumerative Aspects of Mirror Symmetry*,
KIAS Springer Series in Mathematics 5,
https://doi.org/10.1007/978-981-95-0385-8

Acta Math., 183 (1999), 71–99
© 1999 by Institut Mittag-Leffler. All rights reserved

Quantum hyperplane section theorem for homogeneous spaces

by

BUMSIG KIM

Pohang University of Science and Technology
Pohang, Republic of Korea

1. Introduction

Quantum cohomology of a symplectic manifold is a certain deformed ring of the ordinary cohomology ring with parameter space given by the second cohomology group. It encodes enumerative geometry of rational curves on the manifold. In general it is difficult to compute the quantum cohomology structure. On the other hand, mirror symmetry predicts an answer to a question of counting the virtual numbers of rational curves of given degrees on a three-dimensional Calabi–Yau manifold, which amounts to knowing the structure of the quantum cohomology of the manifold. In the large class of Calabi–Yau manifolds, the complete intersections in toric manifolds or homogeneous spaces, this mirror symmetry prediction [7], [3], [1], [2], can be interpreted as a quantum cohomology counterpart of the weak Lefschetz hyperplane section theorem relating cohomology algebras of the ambient manifolds and their hyperplane sections. As it is mentioned in [13], the "quantum hyperplane section conjecture" can be formulated in intrinsic terms of Gromov–Witten theory on the ambient manifold and does not require a reference to its mirror partner. In this paper we formulate and prove the conjecture for homogeneous spaces. It would be one of the highly nontrivial functorial properties enjoyed by quantum cohomology algebras. One can compute the virtual numbers of rational curves on a Calabi–Yau 3-fold complete intersection, provided one knows the quantum cohomology algebra of the ambient space. In fact, one needs to know the quantum differential equations of the space, which are certain linear differential equations arising from the flat connection in the quantum cohomology algebra. The mirror symmetry prediction is that the quantum differential equations of a Calabi–Yau manifold are equivalent in a sense to the Picard–Fuchs differential equations of another Calabi–Yau manifold. In contrast, the proposed conjecture is that there is a certain relation between quantum differential equations of a manifold and those of a nonnegative smooth zero locus of a spanned decomposable vector bundle over the manifold. The formulation of the conjecture is given in [2]. When the ambient space is a symplectic toric manifold, the conjecture is a corollary of the Givental mirror theorem [14].

Let X be a compact homogeneous space of a semi-simple complex Lie group and let V be a vector bundle over X. Let $\beta \in H_2(X,\mathbf{Z})$ which is in the Mori cone Λ of X. Suppose that $V'_\beta := \pi_* e_1^* V$ becomes a vector orbi-bundle over the Kontsevich moduli space $\overline{M}_{0,0}(X,\beta)$, where e_1 is the evaluation map at the (first) marked point from $\overline{M}_{0,1}(X,\beta)$ to X, and π is the map from $\overline{M}_{0,1}(X,\beta)$ to $\overline{M}_{0,0}(X,\beta)$ associated with "forgetting the marked point" [17]. Then one might want to compute

$$\int_{\overline{M}_{0,0}(X,\beta)} \mathrm{Euler}(V'_\beta).$$

Introduce a formal parameter $\hbar$. Then it turns out that the classes

$$G^V_\beta := (e_1)_* \frac{\mathrm{Euler}(V_\beta)}{\hbar(\hbar - c)}$$

would be better considered [12], where $V_\beta = \pi^*(V'_\beta)$ and c (depending on β) are the first Chern classes of the universal cotangent line bundles. The classes are in $H^*(X)[\hbar^{-1}]$. They recover the original integrals which we want:

$$\int_X G^V_\beta = \frac{-2}{\hbar^3} \int_{\overline{M}_{0,0}(X,\beta)} \mathrm{Euler}(V'_\beta) + o(\hbar^{-3}).$$

Consider the classes

$$G^X_\beta := (e_1)_* \frac{1}{\hbar(\hbar - c)}$$

corresponding to X itself (without V). When V is a convex decomposable vector bundle $\bigoplus L_j$ of line bundles L_j, the main result of this paper proves some explicit relationship between $A := \{G^V_\beta \mid \beta \in \Lambda\}$ and $B := \{H^V_\beta \cup G^X_\beta \mid \beta \in \Lambda\}$, where

$$H^V_\beta = \prod_j \prod_{m=0}^{\langle c_1(L_j),\beta\rangle} (c_1(L_j) + m\hbar),$$

which is the key object introduced in this sequel.

We now formulate the precise result of this paper. Let $\{p_i\}_{i=1}^k$ denote the $\mathbf{Z}_+$-basis of the closed integral Kähler/ample cone of X. Let us introduce formal parameters q_i, $i=1,\dots,k$, and the ring $H^*(X)[\hbar^{-1}][[q_1,\dots,q_k]]$ of formal power series of q_i. Denote by q^β

$$\prod_{i=1}^k q_i^{\langle p_i,\beta\rangle}.$$

When $\beta=0$, let $G_0^X=1$ and $G_0^V=\text{Euler}(V)$. We want to compare generating functions J^V and I^V from A and B, respectively:

$$S^V := \sum_{\beta\in\Lambda} q^\beta G_\beta^V,$$
$$\Phi^V := \sum_{\beta\in\Lambda} q^\beta H_\beta^V \cup G_\beta^X.$$

We prove that one can be transformed to another by a unique "mirror" transformation. To describe the transformation, let

$$q_i = e^{t_i}, \quad \text{for } i=1,\ldots,k,$$

and introduce another formal variable t_0. Define the degree of q_i by

$$c_1(TX)-c_1(V)=\sum(\deg q_i)p_i.$$

Let

$$J^V := e^{(t_0+\sum_i p_i t_i)/\hbar} S^V$$

and

$$I^V := e^{(t_0+\sum_i p_i t_i)/\hbar} \Phi^V,$$

which are formal power series of $t_1,\ldots,t_k$, $e^{t_0/\hbar}, e^{t_1},\ldots,e^{t_k}$ over $H^*(X)[\hbar^{-1}]$.

THEOREM 1. *Assume that* $\deg q_i \geqslant 0$ *for all* i. *Then* J^V *and* I^V *coincide up to a unique weighted homogeneous change of variables:* $t_0 \mapsto t_0+f_0\hbar+f_{-1}$ *and* $t_i\mapsto t_i+f_i$, *where* $f_{-1},\ldots,f_k$ *are power series of* $q_1,\ldots,q_k$ *over* $\mathbf{Q}$ *without constant terms,* $\deg f_i=0$, $i=0,\ldots,k$, *and* $\deg f_{-1}=1$.

Remarks. (0) J^V will be a cohomological expression of solutions to quantum differential equations associated to (X,V), which is closely related to the quantum differential equations of the smooth zero locus of V. The similar theorem can be extended to the case of decomposable concavex vector bundles V (see [19], [15]).

(1) The change of variables must be understood as replacements of $e^{(t_0+\sum_i p_i t_i)/\hbar}$ by $e^{(t_0+\sum p_i t_i)/\hbar} e^{f_0} e^{f_{-1}/\hbar} e^{\sum p_i f_i/\hbar}$ and of q_i with $q_i e^{f_i}$. So,

$$S^V \quad \text{and} \quad e^{f_0} e^{f_{-1}/\hbar} e^{\sum p_i f_i/\hbar} \Phi^V(q_1 e^{f_1},\ldots,q_k e^{f_k})$$

are equal, where $f_{-1},\ldots,f_k$ are some power series of $q_1,\ldots,q_k$ over $\mathbf{Q}$ without constant terms, $\deg f_i=0$, $i=0,\ldots,k$, and $\deg f_{-1}=1$. In fact, f_i are uniquely well-determined by the coefficients of $1=(1/\hbar)^0$ and $1/\hbar$ in the expansions of S^V and Φ^V as power series of $1/\hbar$.

(2) In the case of a symplectic toric manifold X, the similar statement is a corollary of a mirror theorem in [14], where Φ^X is explicitly found.

(3) For the proof of Theorem 1 we follow the scheme of Givental's proof [12], [14] of the mirror theorem for nonnegative complete intersections in symplectic toric manifolds.

(4) The theorem verifies the prediction [1] of virtual numbers of rational curves in Calabi–Yau 3-fold complete intersections in Grassmannians.

(5) A mirror construction is established for complete intersections in partial flag manifolds [1], [2]. Because of the known quantum cohomology structure [8], in principle there is no essential difficulty in finding G^X_β for each partial flag manifold X, even though a general formula of it is unknown.

(6) In [21] the quantum hyperplane section principle is applied to a nonconvex, nontoric manifold.

Notation. X will always be a generalized flag manifold G/P, where G is a complex semi-simple Lie group and P is a parabolic subgroup. Let T be a maximal torus of G in P, and let T act on X from the left. Let a complex torus T' act on X trivially, and let V be a $T\times T'$-equivariant convex vector bundle over X. Consider E a multiplicative class and suppose that $E(V)\in H^*_{T\times T'}(X)$ is invertible in $H^*_{(T\times T')}(X):=H^*_{T\times T'}(X)\otimes H^*_{T\times T'}$, where $H^*_{(T\times T')}$ is the quotient field of $H^*_{T\times T'}(pt)$. In §2, we will not consider the T-action on X. In §6, additionally we will assume that V is decomposable. The convexity of V is by definition that $H^1(\mathbf{P}^1, f^*V)=0$ for any morphism $f\colon \mathbf{P}^1\to X$. Let the $T\times T'$-equivariant line bundles U_i, $i=1,\dots,k$, form an ample basis of the ordinary Picard group. We denote $\int_X ABE(V)$ by $\langle A,B\rangle_0^V$, for $A,B\in H^*_{(T\times T')}(X)$, and also we use $\int_V A:=\int_X AE(V)$ (equivariant push forwards). The Mori cone of X will be denoted by Λ, which can be identified with $\mathbf{Z}^k_+$ with respect to the coordinates $p_i:=c_1(U_i)$. On the additive group $\mathbf{Z}^k$ we will give the standard partial ordering, so that $d:=(d_1,\dots,d_k)\geqslant 0$ means $d_i\geqslant 0$ for all i. Let v be a fixed point of X with respect to the T-action, and let ϕ_v denote the equivariant pushforward of class 1 under the embedding i_v of the fixed point v to (X,V,E); this (X,V,E), by definition, has a Frobenius structure by the pairing $\langle\cdot,\cdot\rangle_0^V$, so that $A_v:=\langle A,\phi_v\rangle_0^V=i_v^*(A)$ for $A\in H^*_{(T\times T')}(X)$. For a G-manifold M, let M^G denote the set of G-fixed points of M. We will say simply degree and dimension for complex degree and complex dimension, respectively. Let $\sum_a T_a\otimes T^a$ be the equivariant diagonal class of (X,V,E) in $X\times X$. That is, $\langle T_a,T^b\rangle_0^V=\delta_{a,b}$. In the paper we will consider various rings: the $H^*_{T\times T'}[[\hbar^{-1}]][[q]]$-formal power series ring of $\hbar^{-1},q$ over $H^*_{T\times T'}$, the $H^*_{(T\times T')}[[\hbar^{-1}]][[q]]$-formal power series ring of $\hbar^{-1},q$ over $H^*_{(T\times T')}$, and the $H^*_{(T\times T')}(\hbar)[[q]]$-formal power series ring of q over the quotient field of $H^*_{T\times T'}[\hbar]$.

Structure of the paper. In §2, we recall a general theory of Gromov–Witten invariants and quantum cohomology. We introduce the Givental correlators S^V. In §3, we show that the equivariant correlators satisfy a certain recursion relation. In §4, we introduce the double construction and show that the correlators satisfy a polynomiality in a "double construction". In §5, we introduce a class $\mathcal{P}(X, V, \text{Euler})$ of series of $q=(q_1, ..., q_k), \hbar^{-1}$ over $H^*_{T\times T'}(X)$, where a "mirror" group acts freely and transitively. In §6, we introduce a modified correlator of S^X. It will also belong to the class $\mathcal{P}(X, V, \text{Euler})$. The modification is given by the hypergeometric correcting Euler classes H^V_β according to the decomposition type of V. In §§7 and 8, we analyze the torus T-action on a generalized flag manifold, its one-dimensional orbits and the representations of the global section spaces of equivariant line bundles restricted to the orbits. The analysis would be useful to find the explicit expression of Φ^X.

Acknowledgments. I am grateful to A. Givental and Y.-P. Lee for helping me to understand the paper [12], and to V. Batyrev, I. Ciocan-Fontaine, W. Fulton, B. Kreußler, E. Tjøtta, K. Wirthmüller for useful discussions on the papers [12], [14]. Also, I would like to thank the Institut Mittag-Leffler for the financial support during the yearlong program "Enumerative Geometry and its Interactions with Theoretical Physics" in 1996/97. My special thanks go to D. van Straten for numerous comments which improve the clarity of the paper.

2. Mirror symmetry

2.1. *The moduli space of stable maps.* To fix notation we recall the definition of stable maps and some elementary properties of the moduli spaces of stable maps to X [17], [10], [6]. The notion of stable maps is due to M. Kontsevich. We recommend the (survey) paper of W. Fulton and R. Pandharipande [10].

A prestable rational curve C is a connected arithmetic genus-0 projective curve with possibly nodes. The curve is not necessary irreducible. A prestable map $(f, C; x_1, ..., x_n)$ is a morphism f from C to X with fixed ordered n-many marked distinct smooth points $x_i \in C$. We will identify $(f, C; \{x_i\})$ with $(f', C', \{x'_i\})$ if there is an isomorphism h from C to C' preserving the configuration of marked points such that $f = f' \circ h$. A stable map $(f, C; \{x_i\})$ is a prestable map with only finitely many automorphisms.

Let $\overline{M}_{0,n}(X, \beta)$ be the (coarse moduli) space of all stable maps $(f, C; \{x_i\}_{i=1}^n)$ with the fixed homology type $\beta = f_*([C]) \in H_2(X, \mathbf{Z})$. Whenever it is nonempty, the moduli space is a connected(1) compact complex *orbifold* with complex dimension $\dim X +$

(1) For a proof of the connectedness see [20].

$\langle c_1(TX), \beta\rangle + n - 3$.

More precisely, locally near a stable map the moduli space has data of a quotient of a holomorphic domain by the (finite) group action of all automorphisms of the stable map. In the paper [10] are constructed smooth open complex domains V with finite groups Γ which act on V such that V/Γ are naturally glued together in the moduli space of stable maps. Let $X \subset_i \mathbf{P}^N$, $\beta \neq 0$, and $(X, i_*(\beta)) \neq (\mathbf{P}^1, [\text{line}])$. Here [line] denotes the line class of $H_2(\mathbf{P}^N)$. Given a stable map (f, C) (without marked points for simplicity), choose hyperplanes H_j in $\mathbf{P}^N$ satisfying that $\{H_j\}$ gives rise to a basis of $H^0(\mathbf{P}^N, \mathcal{O}(1))$, f is transversal to the hyperplanes, and their inverse images $\{x_{i,j}\}_i = f^{-1}(H_j)$ contain no nodes of C. Then the data $(C; \{x_{i,j}\})$ determines a point in the moduli space of marked stable curves. Conversely, a point in a suitable closed subvariety of an open smooth domain of the moduli space of marked stable curves naturally determines a stable map f with the extra choices of elements in $(\mathbf{C}^\times)^N$. If G is the product of the symmetric group of the elements of each group $\{x_{i,j}\}_i$, then this G has an action sending the data $(f, C; \{x_{i,j}\})$ to another by permutations of the new marked points. A $(\mathbf{C}^\times)^N$-bundle of the smooth closed subvariety is an algebraic local chart of the moduli space of stable maps at f with the induced G-action.

Example. Let $X = \mathbf{P}^2$ and f be a stable map without marked points such that f is transversal to the hyperplanes $x=0$, $y=0$ and $z=0$. Assume that no singular points of C are mapped into the hyperplanes, and $f_*[C] = 2[\text{line}]$. Consider their inverse images (Cartier divisors), a_1, a_2, b_1, b_2, c_1, c_2 in C. This information $(C; a_1, \ldots, c_2)$ as a stable curve will determine f uniquely with $(\mathbf{C}^\times)^2$-ambiguity. This $(\mathbf{C}^\times)^2$-bundle over some open subset of the smooth space $\overline{M}_{0,6}$ is the local smooth chart. Notice that for instance, $(C; a_2, a_1, b_1, b_2, c_2, c_1)$ gives rise to the same f up to isomorphism. Thus we have to take account of the quotient by the finite group permuting the elements of sets $\{a_1, a_2\}$, $\{b_1, b_2\}$ and $\{c_1, c_2\}$.

CLAIM. *The stabilizer subgroup $G_{(C;\{x_{i,j}\})}$ of G is exactly the automorphism group* $\mathrm{Aut}(f, C)$ *of* (f, C).

Proof. We shall construct a correspondence between $G_{(C;\{x_{i,j}\})}$ and $\mathrm{Aut}(f, C)$. Let $g \in G_{(C;\{x_{i,j}\})}$ which is given by one of the suitable permutations of $x_{i,j}$. So, $g(C; \{x_{i,j}\}) = (C; \{g(x_{i,j})\})$. Since the permutation does not change the stable curve $(C; \{x_{i,j}\})$, there is an isomorphism h from $(C; \{x_{i,j}\})$ to $(C; \{g(x_{i,j})\})$. The isomorphism h is unique since there is no nontrivial automorphism in the stable curve of genus 0. Of course this h gives rise to an automorphism of (f, C).

Conversely, if h is an automorphism of (f, C), then it induces an isomorphism from $(C; \{x_{i,j}\})$ to $(C; \{g(x_{i,j})\})$ for a unique permutation g which we allow. Thus we established a 1-1 correspondence, which can be easily seen to be a group homomorphism.

Remark. The action of $\mathrm{Aut}(f, C)$ may not be effective in general. For instance, see $\overline{M}_{0,0}(\mathbf{P}^1, 2[\text{line}])$.

2.2. *Gromov–Witten invariants and* $QH^*_{(T')}(V)$. There are natural morphisms on the moduli spaces, namely, evaluation maps e_i at the ith marked points and forgetting-marked-point maps π:

$$\begin{array}{ccc} \overline{M}_{0,n+1}(X,\beta) & \xrightarrow{e_{n+1}} & X \\ \downarrow{\scriptstyle \pi} & & \\ \overline{M}_{0,n}(X,\beta). & & \end{array}$$

If s_i are the universal sections for the marked points, then $e_i = e_{n+1} \circ s_i$ (here we assume that π is the forgetful map of the last marked point). In the orbifold charts, π gives the universal family of stable maps as a fine moduli space.

Consider, for a second homology class $\beta \neq 0$ and an integer $n \geqslant 0$, the vector orbi-bundle $V_\beta = \pi_*(e^*_{n+1}(V))$. Here π is a flat morphism in the level of orbifold charts. Thus indeed, V_β is vector orbi-bundle with the fiber $H^0(C, f^*(V))$ at $(f, C; \{x_i\})$. Notice that $V_\beta = \pi^*(V_\beta)$ (it has nothing to do with marked points).

Notation. For $A_i \in H^*_{(T')}(X)$,

$$V_0 := V,$$
$$\overline{M}_{0,i}(X,0) := X \quad \text{for } i = 0, 1, 2,$$
$$\langle A_1, \ldots, A_N \rangle^V_\beta := \int_{\overline{M}_{0,N}(X,\beta)} e_1^*(A_1) \cup \ldots \cup e_N^*(A_N) \cup E(V_\beta).$$

Then one can show that for all β

$$\sum_{\beta_1+\beta_2=\beta} \sum_a \langle A_1, A_2, T_a \rangle^V \langle T^a, A_3, A_4 \rangle^V$$

are totally symmetric in A_i (see §4 in [12]). This property will be equivalent to the associativity of the quantum cohomology of $QH^*_{(T')}(V)$ which we define below.

Let us choose a basis $\{p_i\}_{i=1}^k$ of $H^2(X)$ by classes in the closed Kähler cone.

Notation.

$$q^\beta := \prod_i q_i^{\langle p_i, \beta \rangle},$$
$$\langle A_1, \ldots, A_N \rangle^V := \sum q^\beta \langle A_1, \ldots, A_N \rangle^V_\beta.$$

The quantum multiplication $\circ$ is defined by the following simple requirement: for $A, B, C \in H^*_{(T')}(X)$,

$$\langle A \circ B, C\rangle_0^V = \langle A, B, C\rangle^V$$

which is a formal power series of the parameters q_i. Thus our quantum cohomology $QH^*_{(T')}(V)$ is defined as $H^*_{(T')}(X) \otimes_{\mathbf{Q}} \mathbf{Q}[[q_1, ..., q_k]]$ with a product structure.

2.3. *Givental's correlators.* We review the topic after [9], [12].

2.3.1. *The flat connections and the fundamental solutions.* Now let $q_i = e^{t_i}$ with the formal parameters t_i. We have a one-parameter family of the formal $\mathcal{D}$-module structures on $QH^*_{(T')}(V)$ by giving a flat connection $\nabla_i = \hbar \partial/\partial t_i - p_i \circ$ for any nonzero $\hbar$, $i=1, ..., k$. For the fundamental solutions we introduce $c_i \in H^*_{T'}(\overline{M}_{0,N}(X, \beta))$, so-called gravitational descendents. These c_i are the first Chern classes of the universal cotangent line bundles at the ith marked points. The line bundles are, by definition, the dual of the normal bundle of $s_i(\overline{M}_{0,N}(X, \beta))$ in $\overline{M}_{0,N+1}(X, \beta)$.

Notation. Let $f_i(x) \in H^*_{(T')}[y][[x]]$ for indeterminants x, y. Throughout the paper (see [12] and [22]),

$$\langle A_1 f_1(c), ..., A_N f_N(c); B\rangle_\beta^X := \int_{\overline{M}_{0,N}(X,\beta)} e_1^*(A_1) f_1(c_1) \dots e_N^*(A_N) f_N(c_N) B,$$

$$\langle A_1 f_1(c), ..., A_N f_N(c); B\rangle_\beta^V := \langle A_1 f_1(c), ..., A_N f_N(c); B E(V_\beta)\rangle_\beta^X,$$

$$\langle A_1 f_1(c), ..., A_N f_N(c); B\rangle^X := \sum_\beta q^\beta \langle A_1 f_1(c), ..., A_N f_N(c); B\rangle_\beta^X,$$

$$\langle A_1 f_1(c), ..., A_N f_N(c); B\rangle^V := \sum_\beta q^\beta \langle A_1 f_1(c), ..., A_N f_N(c); B\rangle_\beta^V,$$

where $A_i \in H^*_{(T')}(X)$, $B \in H^*_{(T')}(\overline{M}_{0,N}(X, \beta))$.

The system of the first-order equations $\nabla_i \vec{s} = 0$, $i=1, ..., k$, has the following complete set of $(\dim H^*(X))$-many solutions [12]:

$$\vec{s}_a := \sum_b \left\langle \frac{e^{pt/\hbar} T_a}{\hbar - c}, T_b \right\rangle^V T^b,$$

where pt denotes $\sum_{i=1}^k p_i t_i$ and $\hbar$ is a formal variable (but when $\beta = 0$, set $\hbar = 1$).

The following two formulas show that $\vec{s}_a$ are indeed solutions to the quantum differential system $\nabla_i \vec{s} = 0$.

Using that $c_i - \pi^*(c_i)$ is the fundamental class Δ_i represented by the section

$$s_i : \overline{M}_{0,n}(X, \beta) \to \overline{M}_{0,n+1}(X, \beta)$$

and $c_i \cup \Delta_i = 0$ (the image of s_i is isomorphic to $\overline{M}_{0,3}(X,0) \times_X \overline{M}_{0,n}(X,\beta)$), it is easy to derive the so-called fundamental class axiom and the divisor axiom [22], [12]. Let $f_i(x)$ be a polynomial with coefficients in $\pi^*(H^*_{(T')}(\overline{M}_{0,n}(X,d)))$. Let D be a divisor class in $H^*_{(T')}(X)$. Then (for $n>0$)

$$\langle f_1(c), \dots, f_n(c), 1\rangle_\beta^V = \sum_i \left\langle f_1(c), \dots, \frac{f_i(c)-f_i(0)}{c}, \dots, f_n(c)\right\rangle_\beta^V$$

(where we abuse the notation "$f_i(\pi^*(c)) = f_i(c)$") and

$$\begin{aligned}\langle f_1(c), \dots, f_n(c), D\rangle_\beta^V &= \langle D, \beta\rangle \langle f_1(c), \dots, f_n(c)\rangle_\beta^V \\ &\quad + \sum_i \left\langle f_1(c), \dots, f_{i-1}(c), D \cup \frac{f_i(c)-f_i(0)}{c}, f_{i+1}(c), \dots, f_n(c)\right\rangle_\beta^V.\end{aligned}$$

Consider

$$e^{pt/\hbar} S^V := \sum_a \langle \vec{s}_a, 1\rangle_0^V T^a = e^{pt/\hbar} \sum_a \left\langle \frac{T_a}{\hbar(\hbar - c)}\right\rangle^V T^a = e^{pt/\hbar}(1 + o(1/\hbar)),$$

which is the main object in this paper. This S^V will be called Givental's correlator for (X, V, E). It is an element in $H^*_{(T')}(X)[\hbar^{-1}][[q_1, \dots, q_k]]$. Notice that S^V for (X, V, Euler) is homogeneous of degree 0 if we let $\sum(\deg q_i)p_i = c_1(TX) - c_1(V)$, $\deg \hbar = 1$ and $\deg A = b$ if $A \in H^{2b}_{T'}(X)$.

The quantum $\mathcal{D}$-module of $QH^*_{(T')}(V)$ is defined by the $\mathcal{D}$-module generated by $\langle \vec{s}, 1\rangle_0^V$ for all flat sections $\vec{s}$. When there is no V considered, we denote by $QH^*(X)$ the quantum cohomology. That is, using $\langle \dots \rangle^X$, we define $QH^*(X)$.

Remark. Suppose that a differential operator $P(\hbar\partial/\partial t_i, e^{t_i}, \hbar)$ with coefficients in $H^*_{(T')}$ annihilates $\langle \vec{s}, 1\rangle_0^V$ for all flat sections $\vec{s}$. Then $P(p_1, \dots, p_k, q_1, \dots, q_k, 0)$ holds in $QH^*_{(T')}(V)$ [12].

2.3.2. *Examples.* The projective space $\mathbf{P}^n$: It is well known that in the quantum cohomology ring $QH^*(\mathbf{P}^n)$, $(p\circ)^{n+1} = q$, where $p = c_1(\mathcal{O}(1))$ and q is given with respect to the line class dual to p. The corresponding operator is $(\hbar d/dt)^{n+1} - e^t$. The solutions are explicitly known in [11]. S^X is

$$1 + \sum_{d>0} e^{dt} \frac{1}{((p+\hbar)(p+2\hbar)\dots(p+d\hbar))^{n+1}}.$$

The complete flag manifolds $F(n)$: Let $F(n)$ be the set of all complete flags $(\mathbf{C}^1 \subset \dots \subset \mathbf{C}^n)$ in $\mathbf{C}^n$. The usual cohomology ring is $\mathbf{Q}[x_1, x_2, \dots, x_n]/(I_1, \dots, I_n)$ where x_i are the Chern classes of $(S_i/S_{i-1})^*$, S_i are the universal subbundles with fibers $\mathbf{C}^i$, and I_i are the ith elementary symmetric polynomials of $x_1, ..., x_n$. Let us use as a basis of $H_2(F(n), \mathbf{Z})$ duals of the first Chern classes of $(S_i)^*$, $i=1, ..., n-1$. They are in the edges of the closed Kähler cone.

Let $A(x_i)$ be a matrix

$$\begin{pmatrix} x_1 & q_1 & 0 & 0 & \dots & 0 \\ -1 & x_2 & q_2 & 0 & \dots & \\ & \ddots & & & \ddots & \\ 0 & \dots & & -1 & x_{n-1} & q_{n-1} \\ 0 & \dots & & 0 & -1 & x_n \end{pmatrix}.$$

Then the quantum relations are generated by the coefficients of the characteristic polynomial of the matrix $A(x_i)$.

The corresponding differential operators turn out to be obtained by the same method using $A(x_i)$ with arguments $\hbar\partial/\partial t_1$ instead of x_1, $\hbar\partial/\partial t_i - \hbar\partial/\partial t_{i-1}$ instead of x_i, and $-\hbar\partial/\partial t_{n-1}$ instead of x_n [13], [16]. These differential operators are the integrals of the quantized Toda lattices. The quadratic differential operator of them can be easily derived. In fact, given a quantum relation of $F(n)$ between the divisors x_i, there is a unique operator satisfying that its symbol becomes the relation, and that it annihilates $\langle \vec{s}, 1\rangle_0$ for all flat sections $\vec{s}$. In general, the explicit cohomological expression S^X of solutions to the quantum differential operators are not known.

2.3.3. *The general quintic hypersurface in* $\mathbf{P}^4$. Let Y be a smooth degree-5 hypersurface in $\mathbf{P}^4$. Y is not a homogeneous space. However, using the virtual fundamental class $[\overline{M}_{0,n}(Y, \beta)]^{\mathrm{virt}}$ [4], [5], [18], one can define also the quantum cohomology $QH^*(Y)$ of Y. It is expected that

$$\langle A_1 f_1(c), ..., A_N f_N(c)\rangle^Y = \langle A_1 f_1(c), ..., A_N f_N(c)\rangle^{\mathcal{O}(5)}.$$

Let p be the induced class of the hyperplane divisor in $\mathbf{P}^4$. The quantum relation is $(p\circ)^4=0$. The corresponding operator is, however, *not* $(\hbar d/dt)^4$, but

$$\left(\hbar\frac{d}{dt}\right)^2\left(\frac{(\hbar d/dt)^2}{5+f(q)}\right),$$

where $\langle p\circ p, p\rangle_0^Y = 5+f(q)$. Notice that in this *Calabi–Yau* 3-fold case, we loose the whole information of quantum cohomology when one considers only the quantum relation, $(p\circ)^4=0$. The unknown $f(q)$ was conjectured by physicists [7]. The general idea of the prediction is the following. Roughly speaking, in theoretical physics, there are quantum field theories associated to Calabi–Yau 3-folds by the A-model and the B-model. What we have constructed so far are A-model objects for Calabi–Yau 3-folds. On the other hand, using a family of the so-called mirror manifolds which are also Calabi–Yau 3-folds, conjecturally one may construct the equivalent quantum field theory by the B-model. The corresponding mirror partner of a quantum differential equation/quantum $\mathcal{D}$-module is the Picard–Fuchs differential equation/Gauss–Manin connection of the mirror family. It was predicted that they are equivalent by a certain transformation. In [7] are obtained the conjectural mirror family of quintics, the Picard–Fuchs differential equation and the transformation. That is how the prediction is made. The prediction is now proven to be correct by Givental [12].

2.4. *The idea of the proof of Theorem* 1. To describe the idea, let us notice that Givental's proof [14] of the mirror conjecture for the nonnegative toric complete intersections can be divided into three parts. (He shows in the paper that the mirror phenomenon occurs also in non-Calabi–Yau manifolds.) Let X be a symplectic toric manifold with a big torus T, and V be a $T\times T'$-equivariant decomposable convex vector bundle over X, where T' acts on X trivially. Consider the $T\times T'$-equivariant correlator corresponding to S^V, which will be denoted also by S^V.

(1) In the A-part, it is proven that

(a) the $T\times T'$-equivariant solution vector $S^V \in H^*_{(T\times T')}(X)[[q,\hbar^{-1}]]$ has an "almost recursion relation",

(b) it satisfies the polynomiality in the so-called "double construction", and

(c) it is uniquely determined by the above two properties with the asymptotical behavior $S^V = 1 + o(1/\hbar)$.

(2) In the B-part, another ($T\times T'$-equivariant hypergeometric) vector Φ^V, presumably given by the mirror symmetry conjecture, is constructed. It is verified that it also satisfies (a) and (b) using a toric (naive) compactification of holomorphic maps from $\mathbf{P}^1$ to X.

(3) When $c_1(X) - c_1(V)$ is nonnegative and E=Euler, there is a suitable equivalence transformation between Φ^V and S^V.

In this paper, for a $T\times T'$-equivariant decomposable convex vector bundle V over any compact homogeneous X of a semi-simple complex Lie group G, we will show that S^V satisfies property (1) above. In this case, T is a maximal torus of G.

We define Φ^V which corresponds to Φ^V of the toric case in property (2): Let $\Phi^X = S^X = \sum_d \Phi^X_d q^d$. For Φ^V, we will find a modification $H'_d \in H^*_{T\times T'}(X)[\hbar]$ (depending on V and d) such that if $\Phi^V := \sum_d \Phi^X_d H'_d q^d$, then

(A) Φ^V (after the restriction to the fixed points) has the almost recursion relation exactly like S^V, and

(B) Φ^V has the polynomial property in the double construction.

In fact, we design H'_d to satisfy (A) and (B).

Finally, when E=Euler and $c_1(TX)-c_1(V)$ is nonnegative, we will prove that a certain operation will transform S^V to Φ^V, since they satisfy the same almost recursion relation and the polynomiality of the double construction.

3. The almost recursion relations

As in §2 let X be a homogeneous manifold G/P where G is a complex semi-simple Lie group and P is a parabolic subgroup. Let T be a maximal torus. The T-action has only isolated fixed points $\{v, w, ...\}$. The one-dimensional invariant orbit of T is analyzed in detail in §§7 and 8. For a moment we need the fact that the closures of orbits form a finite number of projective lines $\mathbf{P}^1$ connecting a fixed point v to another fixed point w. For a given equivariant vector bundle W over a $T\times T'$-space M, we use $[W]$ which denote the element in the K-group $K^0_{T\times T'}(M)$ corresponding to the $T\times T'$-vector bundle W.

The torus action on X induces the natural action on the moduli space of stable maps by the functorial property. Since the evaluation maps are $T\times T'$-equivariant, the pullbacks of $T\times T'$-bundles have natural actions in the orbifold sense. In turn, V_β has the induced $T\times T'$-action. *All ingredients in* §2 *are from now on the equivariant ones.* We would like to evaluate S^V as a specialization of the equivariant one corresponding to S^V. We use the same notation $S^V \in H^*_{(T\times T')}(X)[[\hbar^{-1}]][[q_1, ..., q_k]]$ for the equivariant one. Notice that S^V might have power series of $\hbar^{-1}$ in each coefficient of q^d, since c are not anymore nilpotent. Using the localization theorem, we shall find an "almost recursion relation" on the equivariant Givental correlator. To begin with, we summarize the fixed points of the induced action on the moduli space of stable maps. If a stable map represents a fixed point in the moduli space, the image of the map should lie in the closure of the one-dimensional orbits. The special points are mapped to isolated fixed points. Let us denote by $\alpha_{v,w}$ the character of the tangent space of a one-dimensional orbit connecting an isolated fixed point v to another w. Then, $-\alpha_{v,w}$ is the character of the tangent line of the one-dimensional orbit $o(v,w)$ at w. We use $\beta_{v,w}$ to stand for the second homology class represented by the ray. Denote by $o(v)$ the set of all fixed points $w \neq v$ which can be connected by a one-dimensional orbit with v.

LEMMA 1 (Recursion Lemma [12]). *Denote by ϕ_v the equivariant classes $i_*(1)$ at v, where i_v denotes the $T\times T'$-equivariant inclusion of the point v into (X,V). Then $S^V \in H^*_{T\times T'}(X)[[\hbar^{-1}]][[q_1, ..., q_k]]$ has an "almost" recursion relation, namely, for any $v \in X^T$:*

(0) $S^V_v(q,\hbar) := \langle S^V, \phi_v \rangle^V_0 \in H^*_{(T\times T')}(\hbar)[[q]]$ and the substitution $S^V_w(q, -\alpha_{v,w}/m)$ of $\hbar$ with $-\alpha_{v,w}$ in $S_w(q,\hbar)$ is well-defined;

(1) *the difference* R_v *of* $S_v^V(q,\hbar)$ *and the "recursion part" is a power series of* q *over the* polynomial *ring of* $1/\hbar$, *that is,*

$$R_v := S_v^V(q,\hbar) - \sum_{\substack{w\in o(v)\\ m>0}} q^{m\beta_{v,w}} \frac{(-\alpha_{v,w})/m}{\hbar(\alpha_{v,w}+m\hbar)} \cdot \frac{E(V_{v,w,m}) i_v^*(\phi_v)}{\mathrm{Euler}(N_{v,w,m})} S_w^V(q, -\alpha_{v,w}/m)$$

is in $H^*_{(T\times T')}[\hbar^{-1}][[q]]$, *where* $V_{v,w,m}$ *is the* $T\times T'$*-representation space* $H^0(\mathbf{P}^1, f^*V)$ *with* f *the totally ramified* m*-fold map onto* $o(v,w)$ *over* v *and* w, *and* $N_{v,w,m}$ *is the* $T\times T'$*-representation space* $[H^0(\mathbf{P}^1, f^*TX)] - [0]$; *and*

(2) *furthermore, for* S^X *itself, the first term* R_v *is* 1.

We will say that the statement (1) reveals the almost recursion relation of S^V. The statement (2) shows that S_v^X have recursion relations in the ordinary sense.

Proof. First of all, using the short exact sequence

$$0 \to \mathrm{Ker} \to V_d \to e_1^*(V) \to 0$$

over $\overline{M}_{0,1}(X,d)$, we see that S^V is indeed in $H^*_{T\times T'}[[\hbar^{-1}]][[q_1,\ldots,q_k]]$. (The last map in the sequence is given by the evaluation of global sections at the marked point.)

A connected component of the T-fixed loci of the moduli space $X_d := \overline{M}_{0,1}(X,d)$ is isomorphic to a product of Deline–Mumford spaces with marked points from the special points of the inverse image $f^{-1}(v)$ of the generic f in the component for all $v\in X^T$. Now fix a v and consider S_v^V. It is enough to count the fixed locus $F^{d,v}$ where the marked point x should be mapped to the fixed point v since ϕ_v can be supported only near the point. For a stable map $(f, C; x)$ denote by C_1 the irreducible component of C containing the marked point x. Then $F^{d,v}$ is the disjoint union of

$$F_1^{d,v} := \{(f,C;x)\in F^{d,v} \mid f(C_1) = v\}$$

and

$$F_2^{d,v} := \bigcup_{\substack{w\in o(v)\\ m=1,\ldots,m\beta_{v,w}\leqslant d}} F^{d,v,w,m},$$

where $F^{d,v,w,m}$ is

$$\{(f,C;x)\in F^{d,v} \mid w\in f(C_1), \deg f|_{C_1} = m\in\mathbf{N}\}.$$

S_v^V is an integral over $F^{d,v}$'s by a localization theorem for orbifolds. We claim that the integral of

$$\frac{E(V_d)e_1^*(\phi_v)}{\hbar(\hbar - c)}$$

over $F_1^{d,v}$ is in $H^*_{(T\times T')}[\hbar^{-1}]$. The reason is that the universal cotangent line bundle over $F_1^{d,v}$ in the moduli space has the trivial action. It implies that the equivariant class c restricted to $F_1^{d,v}$ is nilpotent.

Now we shall obtain the "almost recursion relation" from the contribution of the fixed loci $F_2^{d,v}$. Denote $d-m\beta_{v,w}$ by d'. Since C_1 is always one end of C for any $(f,C;x)\in F_2^{d,v}$, we can have a natural isomorphism from $F^{d',w}$ to $F^{d,v,w,m}$, where $F^{d',w}$ are fixed loci in $X_{d'}:=\overline{M}_{0,1}(X,d')$, consisting of the stable maps sending the marked points to w. We obtain the morphism, joining the m-covering of $o(v,w)$ to stable maps in $F^{d',w}$. By the m-covering of $o(v,w)$ we mean a totally m-ramified map from $\mathbf{P}^1\cong C_1$ to $o(v,w)$ over v and w. Let $x'=f^{-1}(w)\cap C_1$.

We claim that the normal bundles as in a K-group $K^0(F^{d,v,w,m}\cong F^{d',w})$ satisfy the equality

$$[N_{X_d/F^{d,v,w,m}}]-[N_{X_{d'}/F^{d',w}}]=[N_{v,w,m}]-[T_wX]+[T_{x'}C_1\otimes L|_{F^{d',w}}], \tag{1}$$

where L is the universal tangent line bundle over $X_{d'}$. The reason of the claim is as follows: Recall that each fixed component is isomorphic to the product of moduli spaces of stable curves (see §3 in [17] for detail). Hence, we conclude that $[N_{X_d/F^{d,v,w,m}}]-[N_{X_{d'}/F^{d',w}}]$ (over each fixed component) is equal to a trivial bundle with nontrivial actions. The twister by action can be computed by study of action on normal spaces at $(f,C_1\cup C_2;x)\in F^{d,v,w,m}$. Let N_1 be the normal space of $F^{d,v,w,m}$ at $(f,C_1\cup C_2;x)$ and N_2 be the normal space of $F^{d',w}$ at $(f|_{C_2},C_2;x':=C_1\cap C_2)$. Then as representation spaces,

$$\begin{aligned}[N_1]=[N_2]&+([H^0(C_1,f|^*_{C_1}TX)]-[H^0(C_1,TC_1)])-[T_wX]\\&+[T_{x'}C_1\otimes T_{x'}C_2]+[T_{x'}C_1]+[T_xC_1].\end{aligned}$$

Hence we conclude the claim (1) after canceling of $[H^0(C_1,TC_1)]=[0]+[T_{x'}C_1]+[T_xC_1]$.

On the other hand, the direct sum of the fiber of V_d at $(f,C_1\cup C_2;x)\in F^{d,v,w,m}$ and $V|_w$ is equal to the direct sum of the fiber of $V_{d'}$ at $(f|_{C_2},C_2;x')$ and $H^0(C_1,(f|_{C_1})^*V)$. Thus, applying the localization theorem we obtain

$$\int_{X_d}\frac{E(V_d)e_1^*(\phi_v)}{\hbar(\hbar-c)}=I+\sum_{\substack{w\in o(m)\\ 0<m\\ m\beta_{v,w}\leqslant d}}\frac{E(V_{v,w,m})i_v^*(\phi_v)(-\alpha_{v,w}/m)}{m\hbar(\alpha_{v,w}/m+\hbar)\,\mathrm{Euler}(N_{v,w,m})}\times\int_{X_{d-m\beta_{v,w}}}\frac{E(V_{d-m\beta_{v,w}})e_1^*(\phi_w)}{(-\alpha_{v,w}/m)(-\alpha_{v,w}/m-c)},$$

where I is the integral over $F_1^{d,v}$. The factor m in $m\hbar(\alpha_{v,w}/m+\hbar)$ comes from the nature of the orbifold localization theorem. (There are m automorphisms of $f|_{C_1}$.)

Using induction on $|d|=\sum d_i$, we may assume that the integral factors in the second term are well-defined and belong to $H^*_{(T\times T')}$. (The localization theorem itself also explains them.) So, statements (0) and (1) in the lemma are proven.

Now let us prove statement (2). Since $\langle c_1(TX),\beta\rangle\geqslant 2$ for all β, by degree counting we see that there are no contributions from the integral over $F_1^{d,v}$. The reason is that $\dim \overline{M}_{0,\sum d_i+1}=(\sum d_i)-2$ is less than $2(\sum d_i)-2$ if $(d_1,\ldots,d_k)\neq 0$ and $\dim \overline{M}_{0,1}(X,d)\geqslant 2\sum d_i+\dim X-2$. So, in the case of S^X, $R_v=1$.

4. The double construction

LEMMA 2 (Double Construction Lemma). *The double construction*

$$W(S^V):=\int_V S^V(qe^{\hbar z},\hbar)e^{\sum p_i z_i}S^V(q,-\hbar)$$

is a power series of $q_1,\ldots,q_k$ and $z_1,\ldots,z_k$ with coefficients in $H^*_{T\times T'}[\hbar]$.

A priori, $W(S^V)$ has coefficients in the Laurent power series ring of $\hbar^{-1}$ over $H^*_{T\times T'}$. For the proof we will make use of graph spaces and universal classes defined below.

4.1. *The main lemma.* Let L_d be the projective space of the collection of all $(f_0,\ldots,f_N)$ such that $f_i(z_0,z_1)$ are homogeneous polynomials of degree d. L_d is isomorphic to $\mathbf{P}^{(d+1)(N+1)-1}$. Given a stable map of degree $(d,1)$ from a prestable curve C to $\mathbf{P}^N\times\mathbf{P}^1$, there is a special irreducible component C_0 of C such that C_0 has degree $(d_0,1)$ under the stable map. This special component C_0 is parametrized by $\mathbf{P}^1$ in the target space. Thus we can identify C_0 with $\mathbf{P}^1$ and keep track where the other components intersect. Suppose that the other connected components $C_1,\ldots,C_l$ of $C-C_0$ intersect with $C_0=\mathbf{P}^1$ at $[x_1:y_1],\ldots,[x_l:y_l]$. If the degrees of C_i are d_i under the stable map, we now associate the stable map to

$$\prod_{i=1}^{l}(y_iz_0-x_iz_1)^{d_i}(f_0^0,\ldots,f_N^0),$$

where $(f_0^0,\ldots,f_N^0)$ are the polynomials coming from the data of the restriction of f to C_0.

MAIN LEMMA (Givental [12]). *The above "polynomial" mapping from $G_d(\mathbf{P}^N):=\overline{M}_{0,0}(\mathbf{P}^N\times\mathbf{P}^1,(d,1))$ to L_d is a $(\mathbf{C}^\times)^N\times\mathbf{C}^\times$-equivariant morphism, where $\mathbf{P}^N$ has the diagonal $(\mathbf{C}^\times)^N$-action and $\mathbf{P}^1$ has the $\mathbf{C}^\times$-action by $[z_0:z_1]\mapsto[tz_0:z_1]$ for $t\in\mathbf{C}^\times$.*

Notice that the $\mathbf{C}^\times$-action on L_d is given by

$$[f_0(z_0,z_1):\ldots:f_N(z_0,z_1)]\mapsto[f_0(t^{-1}z_0,z_1):\ldots:f_N(t^{-1}z_0,z_1)]$$

for $t\in\mathbf{C}^\times$.

4.2. *The universal class.* The $T\times T'$-equivariant spanned line bundle U_i over X gives rise to the $T\times T'$-equivariant morphism $\mu_0^i\colon X\to\mathbf{P}^N$, and so we obtain

$$\begin{array}{ccccc} (\mu_d^i)^*(\mathcal{O}(1)) & & & & \mathcal{O}(1) \\ \downarrow & & & & \downarrow \\ G_d(X) & \longrightarrow & G_{d_i}(\mathbf{P}^N) & \longrightarrow & L_{d_i} \end{array}$$

where $G_d(X)$ is the graph space $\overline{M}_{0,0}(X\times\mathbf{P}^1,(d,1))$, and μ_d^i is the $T\times T'\times\mathbf{C}^\times$-equivariant map from $G_d(X)$ to L_{d_i}. On $\mathcal{O}(1)$ we choose the lifted $\mathbf{C}^\times$-action coming from an action on the vector space of all $(N+1)$-multiples $(f_0,\dots,f_N)$ of degree-d_i homogeneous polynomials f_i, where the action is given by

$$(f_0(z_0,z_1),\dots,f_N(z_0,z_1))\mapsto(f_0(z_0,tz_1),\dots,f_N(z_0,tz_1))$$

for $t\in\mathbf{C}^\times$.

Denote by $P_i=c_1((\mu_d^i)^*\mathcal{O}(1))$ the $T\times T'\times\mathbf{C}^\times$-equivariant Chern class. It is said to be a universal class in the paper [14].

Denote by W_d the vector orbi-bundle over $G_d(X)$ with the fiber $H^0(C,\psi^*\pi_1^*V)$ at (C,ψ): Consider

$$\begin{array}{ccc} G_{d,1}(X) & \xrightarrow{e_1} & X\times\mathbf{P}^1 \\ {\scriptstyle\pi}\downarrow & & \uparrow \\ G_d(X) & & \pi_1^*V \end{array}$$

where $G_{d,1}(X)$ denotes the graph space with one marked point, and π_1 is the projection of $X\times\mathbf{P}^1$ to the first factor X. Then $W_d:=\pi_*e_1^*\pi_1^*V$.

4.3. *Proof of Lemma* 2. It is enough to show the equality

$$\sum_d q^d\int_{G_d(X)} e^{P_z}E(W_d)=\int_V S^V(q,\hbar)e^{pz}S^V(qe^{-\hbar z},-\hbar).$$

The left integral is a $T\times T'\times\mathbf{C}^\times$-equivariant push forward with $\hbar$ as $c_1(\mathcal{O}(1))$ over $\mathbf{P}^\infty$, and the right one is a $T\times T'$-equivariant push forward with a formal variable $\hbar$.

We will apply the localization theorem. Let us analyze the $\mathbf{C}^\times$-action fixed loci $G_d(X)^{\mathbf{C}^\times}$ of $G_d(X)$. $G_d(X)^{\mathbf{C}^\times}$ is isomorphic to

$$\sum_{d^{(1)}+d^{(2)}=d}\overline{M}_{0,1}(X,d^{(1)})\times_X\overline{M}_{0,1}(X,d^{(2)}).$$

Suppose $|d^{(1)}|+|d^{(2)}|\neq0$. The normal bundle is as follows:

When $|d^{(1)}|\cdot|d^{(2)}|=0$: The codimension is 2 (one from the nodal condition and the other from the condition of the image of the nodal point). Then the Euler class of the normal bundle is $\hbar(\hbar-c_0)$, or $-\hbar(-\hbar-c_\infty)$, where c_0 and c_∞ are the Chern classes of universal cotangent line bundles of the first marked point over $\overline{M}_{0,1}(X,d^{(1)})$ and $\overline{M}_{0,1}(X,d^{(2)})$, respectively. Here we assume the following convention: $0=[0:1]$, $\infty=[1:0]$, the associated equivariant line bundle to the character 1 of the group $\mathbf{C}^\times$ has $\hbar$ as its equivariant Chern class.

When $|d^{(1)}|\cdot|d^{(2)}|\neq 0$: The codimension is 4 and the Euler class is

$$\hbar(\hbar-c_0)(-\hbar)(-\hbar-c_\infty).$$

Here, for instance, $c_0\in H^2(\overline{M}_{0,1}(X,d^{(1)})\times_X\overline{M}_{0,1}(X,d^{(2)}))$ is the pullback of the Chern class of the universal cotangent line bundle of the first factor of $\overline{M}_{0,1}(X,d^{(1)})\times_X \overline{M}_{0,1}(X,d^{(2)})$.

Let us analyze P_i restricted to $G_d(X)^{\mathbf{C}^\times}$. Consider the commutative diagram

$$\begin{array}{ccc}
G_{d^{(1)},d^{(2)}}(X):=\overline{M}_{0,1}(X,d^{(1)})\times_X\overline{M}_{0,1}(X,d^{(2)}) & \xrightarrow{\mu^i_d} & L_{d_i}\ni z_0^{d_i^{(1)}}z_1^{d_i^{(2)}}[x_0:\ldots:x_N] \\
\downarrow{\scriptstyle \pi_2} & & \uparrow \qquad \uparrow \\
\overline{M}_{0,1}(X,d^{(2)}) & \xrightarrow{\mu^i_0\circ e_1} & \mathbf{P}^N\ni [x_0:\ldots:x_N],
\end{array}$$

where the first vertical map π_2 is the projection and under the second vertical map $\mathbf{P}^N$ is embedded into L_{d_i} as the $\mathbf{C}^\times$-action fixed locus of the part $\{z_0^{d^{(1)}}z_1^{d^{(2)}}[x_0:\ldots:x_N]\,|\,[x_0:\ldots:x_N]\in\mathbf{P}^N\}$. One concludes that $e_1^*\circ(\mu_0^i)^*(c_1(\mathcal{O}(1)|_{\mathbf{P}^N}))=e_1^*(p_i)-d_i^{(2)}\hbar$, and so

$$\sum P_i z_i|_{G_{d^{(1)},d^{(2)}}}=\sum(\pi_2^*e_1^*(p_i)-d_i^{(2)}\hbar)\,z_i.$$

Since

$$S^V(q,\hbar)\,e^{pz}S^V(qe^{-\hbar z},-\hbar)$$
$$=\sum_{a,b,d^{(1)},d^{(2)}}\left\langle\frac{T_a}{\hbar(\hbar-c)}\right\rangle^V_{d^{(1)}}T^a q^{d^{(1)}}e^{pz}\left\langle\frac{T^b}{-\hbar(-\hbar-c)}\right\rangle^V_{d^{(2)}}T_b q^{d^{(2)}}e^{-d^{(2)}\hbar z},$$

we see that

$$\int_V S^V(q,\hbar)\,e^{pz}S^V(qe^{-\hbar z},-\hbar)=\sum_{a,d^{(1)},d^{(2)}}\left\langle\frac{T_a}{\hbar(\hbar-c)}\right\rangle^V_{d^{(1)}}\left\langle\frac{T^a e^{pz-d^{(2)}\hbar z}}{-\hbar(-\hbar-c)}\right\rangle^V_{d^{(2)}}q^{d^{(1)}+d^{(2)}}$$
$$=\sum_d q^d\int_{G_{d^{(1)},d^{(2)}}(X)}\frac{e^{(\pi_2^*e_1^*p-d^{(2)}\hbar)z}E(W_d)}{[N_{G_d(X)/G_{d^{(1)},d^{(2)}}(X)}]}$$
$$=\sum_d q^d\int_{G_d(X)}e^{Pz}E(W_d),$$

after applying the localization theorem only for the $\mathbf{C}^\times$-action on $G_d(X)$.

5. The class $\mathcal{P}(\mathcal{C})$ and mirror transformations

5.1. *The class* $\mathcal{P}(\mathcal{C})$. Let $\mathcal{C}$ be the collection of given data of $C_{v,w,m} \in H^*_{(T\times T')}$, $\alpha_{v,w} \in H^*_{T\times T'}$ and $\beta_{v,w} \in \Lambda - 0$, for all $(v,w,m) \in X^T \times X^T \times \mathbf{N}$ with $v \in o(w)$. Here $\mathbf{N}$ is the set of positive integers. Assume that $(p_i)_w - (p_i)_v = -\langle p_i, \beta_{v,w}\rangle \alpha_{v,w}$ for all $i=1,\ldots,k$. Define the degree of $\hbar$ as 1. Let $q_1, \ldots, q_k$ be formal parameters with some given nonnegative degrees. Define the degree of a homogeneous class of $H^b_{T\times T'}(X)$ as $\frac{1}{2}b$. Let $\mathcal{P}(\mathcal{C})$ be the class of all $Z(q,\hbar) \in H^*_{T\times T'}(X)[[\hbar^{-1}, q]]$ of homogeneous degree 0 such that

(a) $Z(0,\hbar)=1$, $Z_v(q,\hbar) := \langle Z, \phi_v\rangle_0^V$ is in $H^*_{(T\times T')}(\hbar)[[q]]$ for any fixed point v, and $Z_w(q, -\alpha_{v,w}/m)$ are well-defined for all $v \in o(w)$, $m>0$ (m are positive integers);

(b) the almost recursion relation for each fixed point v holds, that is, by definition,

$$R_v := Z_v(q,\hbar) - \sum_{\substack{m>0\\ w\in o(v)}} q^{m\beta_{v,w}} \frac{C_{v,w,m}}{\hbar(\alpha_{v,w}+m\hbar)} Z_w(q, -\alpha_{v,w}/m)$$

is in $H^*_{(T\times T')}[\hbar^{-1}][[q]]$, where

$$q^{m\beta_{v,w}} := \prod_i q_i^{m\langle p_i, \beta_{v,w}\rangle};$$

and

(c) in the double construction,

$$W(Z)(q,z) := \int_V Z(qe^{\hbar z}, \hbar) e^{\sum p_i z_i} Z(q, -\hbar)$$

is in $H^*_{T\times T'}[\hbar][[z,q]]$. (Here we use the multi-index notation for $z=(z_1,\ldots,z_k)$ and $q=(q_1,\ldots,q_k)$.)

Whenever the data $\mathcal{C}$ come from (X,V,E) as in Lemma 1, we denote the class by $\mathcal{P}(X,V,E)$. So, in the case

$$C_{v,w,m} := C^V_{v,w,m} := \frac{(-\alpha_{v,w})/mE(V_{v,w,m}) i_v^*(\phi_v)}{\mathrm{Euler}(N_{v,w,m})},$$

$\alpha_{v,w}$ is the character of $T_v o(v,w)$, $\beta_{v,w} = [o(v,w)] \in H_2(X,\mathbf{Z})$, and

$$c_1(TX) - c_1(V) = \sum_{i=1,\ldots,k} (\deg q_i) p_i.$$

So far, we have proved that S^V for $E=$Euler is in class $\mathcal{P}(X,V,\mathrm{Euler})$.

Below we introduce on $\mathcal{P}(\mathcal{C})$ a transformation group generated by the following three types of operations.

(1) Multiplication by $f(q)$: Let $f(q) = \sum_{d\geqslant 0} f_d q^d$, where $f_d \in \mathbf{Q}$, $f(q)$ is homogeneous of degree 0, and $f(0)=1$. Then $f(q)Z \in \mathcal{P}(\mathcal{C})$.

(2) Multiplication by $\exp(f(q)/\hbar)$: Let $f(q)=\sum_{d>0} f_d q^d$, where f_d are in $H^*_{T\times T'}$. Suppose $\deg(f(q))=1$. Then $Z^{\rm new}:=\exp(f(q)/\hbar)Z$ is still in $\mathcal{P}(C)$.

(3) Coordinate changes: Consider a transformation

$$Z \to Z^{\rm new} := \exp\Big(\sum_i f_i(q)p_i/\hbar\Big) Z(q\exp(f(q)),\hbar),$$

where $f_i(q)=\sum_{d>0} f_i^{(d)} q^d$ of homogeneous degree 0, $f_i^{(d)} \in \mathbf{Q}$, and

$$q\exp(f(q)) = (q_1\exp(f_1(q)),\dots,q_k\exp(f_k(q))).$$

Then $Z^{\rm new}$ is still in $\mathcal{P}(C)$.

Let us call the transformation group the mirror group.

THEOREM 2 ([14]). *Suppose that* $\deg q_i$ *are nonnegative and that there is at least one element of the form* $1+o(\hbar^{-1})$ *in the class* $\mathcal{P}(C)$. *Then the mirror group action on* $\mathcal{P}(C)$ *is free and transitive.*

First, we will check (1), (2) and (3); and prove the so-called uniqueness lemma and then the theorem above.

Proof of (1). First, $Z^{\rm new}:=fZ$ is homogeneous of degree 0, $f(0)Z(0,\hbar)=1$, fZ_v are in $H^*_{(T\times T')}(\hbar)[[q]]$, and of course $Z^{\rm new}_w(q,-\alpha_{v,w}/m)$ are well-defined. Second,

$$Z^{\rm new}_v = f(q)R_v + \sum q^{m\beta_{v,w}} \frac{C_{v,w,m}}{\hbar(\alpha_{v,w}+m\hbar)} Z^{\rm new}_w(q,-\alpha_{v,w}/m).$$

Thus fZ has the almost recursion relation.

Finally,

$$W^{\rm new} := \int_V Z^{\rm new}(qe^{\hbar z},\hbar)e^{pz}Z^{\rm new}(q,-\hbar) = f(qe^{\hbar z})f(q)W,$$

which still has the polynomial coefficients in $H^*_{T\times T'}[\hbar]$.

Proof of (2). The new $Z^{\rm new}$ is homogeneous of degree 0, $Z^{\rm new}(0,\hbar)=1$, $Z^{\rm new}_v$ are in $H^*_{(T\times T')}(\hbar)[[q]]$, and $Z^{\rm new}_w(q,-\alpha_{v,w}/m)$ are well-defined. Since

$$\exp\left(\frac{f(q)}{\hbar}+\frac{mf(q)}{\alpha_{v,w}}\right) = 1+(\alpha_{v,w}+m\hbar)g_{\alpha_{v,w},m}$$

and $g_{\alpha_{v,w},m}$ is a q-series with polynomial coefficients in $H^*_{(T\times T')}[\hbar^{-1}]$, $Z^{\rm new}$ has the almost recursion relation.

Once again,

$$W^{\mathrm{new}}=\exp\left(\frac{1}{\hbar}(f(qe^{\hbar z})-f(q))\right)W.$$

But $f(qe^{\hbar z})-f(q)=\sum_{d>0} f_d((e^{\hbar z})^d-1)q^d$ is a (z,q)-series with polynomial coefficients in $\hbar H^*_{T\times T'}[\hbar]$.

Proof of (3). The Z^{new} is homogeneous of degree 0, $Z^{\mathrm{new}}(0,\hbar)=1$, Z^{new}_v are in $H^*_{(T\times T')}(\hbar)[[q]]$, and $Z^{\mathrm{new}}_w(q,-\alpha_{v,w}/m)$ make sense.

Since $(p_i)_w-(p_i)_v=-\langle p_i,\beta_{v,w}\rangle\alpha_{v,w}$,

$$\sum_i f_i(q)(p_i)_v/\hbar=\sum_i f_i(q)(p_i)_w/(-\alpha_{v,w}/m)$$
$$-m\sum_i\langle p_i,\beta_{v,w}\rangle f_i(q)+\sum_i\frac{f_i(q)(p_i)_v}{\alpha_{v,w}\hbar}(m\hbar+\alpha_{v,w}).$$

The exponential of the last term on the right can be denoted by $1+(\alpha_{v,w}+m\hbar)g_{\alpha_{v,w},m}$ where $g_{\alpha_{v,w},m}$ is a q-series with coefficients which are in $H^*_{(T\times T')}[\hbar^{-1}]$. Z^{new} satisfies the almost recursion relation.

Consider the double construction

$$W^{\mathrm{new}}(q,z)=\int_V e^{f(qe^{\hbar z})p/\hbar}Z(qe^{\hbar z}e^{f(qe^{\hbar z})},\hbar)e^{pz}e^{-f(q)p/\hbar}Z(qe^{f(q)},-\hbar)$$
$$=W\left(qe^{f(q)},z+\frac{f(qe^{\hbar z})-f(q)}{\hbar}\right).$$

But since $f(qe^{\hbar z})-f(q)$ is divisible by $\hbar$, W^{new} is a polynomial (q,z)-series.

By the way, the inverse transformation can be given by unique $g_i(q)\in\mathbf{Q}[[q]]$, $i=1,\dots,k$, satisfying $f_i(q_1e^{g_1(q)},\dots,q_ke^{g_k(q)})=-g_i(q)$ and $g_i(0)=0$ for all i.

LEMMA 3 (Uniqueness Lemma). *Let $Z=\sum_{d\geqslant 0}Z_dq^d$ and $Z'=\sum_{d\geqslant 0}Z'_dq^d$ be series in $\mathcal{P}(C)$. Suppose $Z\equiv Z'$ modulo $(1/\hbar)^2$. Then $Z'=Z$.*

Proof. We may suppose that $Z'_d=Z_d$ for all $0\leqslant d<d_0$ for some $d_0\neq(0,\dots,0)$. Let

$$D(\hbar):=Z'_{d_0}-Z_{d_0}=A\hbar^{-2r-1}+B\hbar^{-2r}+\dots=\hbar^{-2r}(A/\hbar+B+O(\hbar)),$$

where $A,B\in H^*_{T\times T'}(X)$. ($A$ might be 0.) This is possible since $\langle D,\phi_v\rangle_0$ for all v are polynomials of $1/\hbar$ over $H^*_{(T\times T')}$, and so D is a polynomial of $1/\hbar$ over $H^*_{T\times T'}(X)$. Consider the coefficient of q^{d_0} in $W(Z')-W(Z)$, which can be set $\delta(D)=\int_V e^{(p+d_0\hbar)z}D(\hbar)+e^{pz}D(-\hbar)$. If $r=0$, then $D=0$ since $D\equiv 0$ modulo $(1/\hbar)^2$. Assume $r\geqslant 1$. We shall show that $A=0=B$, which implies by induction that $D=0$. Notice that, since $\delta(D)$ is a polynomial of $\hbar$,

$$O(\hbar^2)=\hbar^{2r}\delta(D)=\int_V e^{(p+d_0\hbar)z}(A/\hbar+B+O(\hbar))+e^{pz}(-A/\hbar+B+O(\hbar))$$

$$=\int_V e^{pz}Ad_0z+2Be^{pz}+O(\hbar).$$

So,

$$0=d_0z\int_V e^{pz}A+2\int_V Be^{pz}=\sum_{v\in X^T}(d_0ze^{p_vz}A_v+2e^{p_vz}B_v)\frac{1}{i_v^*(\phi_v)},$$

where A_v and B_v are the restrictions of A and B to the fixed point v, respectively. Since p_vz are different as v are different (this can be seen in §§ 7 and 8), e^{p_vz} and $d_0ze^{p_vz}$ are linearly independent over $H^*_{(T\times T')}$. So we conclude that $A_v=0=B_v$ for all v, and hence $A=0=B$.

5.2. *Proof of Theorem* 2. It suffices to show the transitivity of the action. Let Z_1 and Z_2 be in class $\mathcal{P}(C)$ and let $Z_1=1+o(1/\hbar)$.

Since $\deg q\geqslant 0$, we may let

$$Z_2=Z_2^{(0)}+Z_2^{(1)}\frac{1}{\hbar}+o(1/\hbar),$$

where $Z_2^{(0)}\in H^*_{T\times T'}(X)[[q]]$ is homogeneous of degree 0 and $Z_2^{(1)}$ is homogeneous of degree 1. Furthermore, $Z_2^{(0)}(q)\in H^*_{T\times T'}(X)[[q]]$ is a q-series with coefficients in $\mathbf{Q}$, $Z_2^{(0)}(0)=1$, and $Z_2^{(1)}(q)$ is a q-series with coefficients in $H^*_{T\times T'}[p]$ by degree counting.

We may let

$$\frac{Z_2^{(1)}(q)}{Z_2^{(0)}(q)}=\sum_i(f_i(q)\cdot p_i)+g(q),$$

where $f_i(q)$ are pure q-series over $\mathbf{Q}$ of degree 0 and $g(q)$ are of degree 1 in $H^*_{T\times T'}[[q]]$. In addition, $f_i(0)=0=g(0)$.

Now, consider operations on Z_1: first, coordinate changes,

$$Z_1'=\exp(f(q)p/\hbar)Z_1(q\exp(f(q)),\hbar)=1+\frac{1}{\hbar}f(q)p+o(1/\hbar),$$

second, multiplication by $\exp(g(q)/\hbar)$,

$$Z_1''=\exp(g(q)/\hbar)Z_1'=1+\frac{1}{\hbar}(f(q)p+g(q))+o(1/\hbar),$$

and finally, multiplication by $Z_2^{(0)}(q)$,

$$Z_1'''=Z_2^{(0)}(q)Z_1''=Z_2^{(0)}+\frac{1}{\hbar}Z_2^{(1)}+o(1/\hbar).$$

According to the uniqueness lemma, the last one, Z_1''', must be equal to Z_2 since $Z_1'''\cong Z_2$ modulo $(1/\hbar)^2$.

5.3. *Transformation from J^V to I^V.* We explain the transformation introduced in the introduction. Let $\widetilde{Z}$ be the nonequivariant specialization of Z. Let Z_1 and Z_2 be in class $\mathcal{P}(C)$ and let $Z_1=1+o(1/\hbar)$. Now let us specialize the equivariant setting to the nonequivariant one. Let $J^V=e^{(t_0+pt)/\hbar}\widetilde{Z}_1(q)$ and $I^V=e^{(t_0+pt)/\hbar}\widetilde{Z}_2(q)$. Then, they are equivalent up to the unique coordinate change $t_0\mapsto t_0+f_0(q)\hbar+f_{-1}(q)$ and $t\mapsto t_i+f_i(q)$, $i=1,...,k$, where $f_j\in\mathbf{Q}[[q]]$ for all j, f_0 and f_i $(i=1,...,k)$ have degree 0, f_{-1} has degree 1, and $f_j(0)=0$ for all j.

6. The modified B-series

This second part departs in perspective from Givental's papers [12], [14].

Let X be a homogeneous manifold with the torus $T\times T'$-action. From now on let $V=L_1\oplus...\oplus L_l$ be an equivariant decomposable convex vector bundle over X, where L_i are line bundles. In this section, $\{T_a\}$ and $\{T^b\}$ denote bases of $H^*(X)$ dual to each other with respect to the usual Poincaré paring $\langle\cdot,\cdot\rangle_0^X$.

6.1. *The correcting Euler classes.* Let $x=(x_1,...,x_l)$ be indeterminants.

Define a polynomial of x over $\mathbf{Z}[\hbar]$ for $\beta\in\Lambda$:

$$H_\beta(x,\hbar):=\prod_{i=1}^{l}\prod_{m=0}^{\langle c_1(L_i),\beta\rangle}(x_i+m\hbar).$$

Set

$$H'_\beta(x,\hbar):=\frac{H_\beta(x,\hbar)}{\prod x_i}.$$

We treat each linear factor $(x_i+m\hbar)$ of H_β as a Chern character. Define

$$\Phi^V(q,\hbar):=\sum_{d\in\Lambda}\sum_a q^d\left\langle\frac{T_a}{\hbar(\hbar-c)}\right\rangle_d^X T^a E(H'_d(x,\hbar))(c_1(L),\hbar),$$

where $c_1(L)=(c_1(L_1),...,c_1(L_k))$.

CLAIM. (1) $(p_i)_w=(p_i)_v-\langle p_i,\beta_{v,w}\rangle\alpha_{v,w}$,
(2) $c_1(L)_w=c_1(L)_v-\langle c_1(L),\beta_{v,w}\rangle\alpha_{v,w}$,
(3) $E(V_{v,w,m})=E(H_{m\beta_{v,w}})(c_1(L)_v,-\alpha_{v,w}/m)$.

Proof. Let U be any equivariant convex line bundle. On the ray $o(v,w)$ $(\cong\mathbf{P}^1)$, we have a homogeneous coordinate $[z_0:z_1]$ such that the induced action on the ray is linear (because of the equivariant embedding theorem). We have also global sections $z_0^n,z_0^{n-1}z_1,...,z_1^n$ of the restriction $U|_{\mathbf{P}^1}$ of U to the ray, where $[1:0]=w$, $[0:1]=v$ and

$n=\langle c_1(U),\beta_{v,w}\rangle$. We know that z_0^n, z_1^n and z_0/z_1 have the characters $c_1(U)_v$, $c_1(U)_w$ and $\alpha_{v,w}$, respectively. This concludes the proof.

The first claim shows that we have a well-defined class $\mathcal{P}(X,V,E)$. (Otherwise, the mirror group transformation may not preserve the class $\mathcal{P}(X,V,E)$.)

THEOREM 3. *Suppose that* $c_1(TX)-c_1(V)$ *is in the closed ample cone, and let* $E=$ Euler. *Then* Φ^V *is in the class* $\mathcal{P}(X,V,\text{Euler})$.

Notice that for $\beta'\leqslant\beta$

$$H_\beta(x-\langle c_1(L),\beta'\rangle\hbar,\hbar)=H_{\beta'}(x,-h)\,H'_{\beta-\beta'}(x,\hbar), \tag{2}$$

$$H'_\beta(x,\hbar)=H'_{\beta'}(x,h)\,H'_{\beta-\beta'}(x+\langle c_1(L),\beta'\rangle\hbar,\hbar), \tag{3}$$

which will show the polynomiality of the double construction and the almost recursion relation for Φ^V, respectively.

6.2. *The proof of Theorem* 3. The homogeneity of Φ^V is clear when $E=$Euler, and the rest of the properties will be proven for general E.

First of all, it is a simple check to see

$$\Phi^V\in H^*_{T\times T'}[[\hbar^{-1}]][[q]].$$

For the polynomiality, consider

$$\int_V \Phi^V(q,\hbar)e^{pz}\Phi^V(qe^{-\hbar z},-\hbar)=\sum_d\sum_{d^{(1)}+d^{(2)}=d,a} q^{d^{(1)}}\left\langle\frac{T_aE(H_{d^{(1)}})(c_1(L),\hbar)}{E(V)\hbar(\hbar-c)}\right\rangle^X_{d^{(1)}}$$
$$\times q^{d^{(2)}}\left\langle\frac{T^ae^{(p-d^{(2)}\hbar)z}E(H_{d^{(2)}})(c_1(L),-\hbar)}{-\hbar(-\hbar-c)}\right\rangle^X_{d^{(2)}}, \tag{4}$$

where $\langle T_a,T^b\rangle_0^X=\delta_{a,b}$. Let us use the notation and facts in §4.3. Since

$$E(H_{d^{(1)}})(c_1(L),\hbar)\,E(H_{d^{(2)}})(c_1(L),-\hbar)=E(H_d(x-\langle c_1(L),d^{(2)}\rangle\hbar,\hbar))(c_1(L),\hbar)\,E(V)$$

from (2), the universal class $U(c_1(L))$ corresponding to $c_1(L)$ restricted to $G_{d_1,d_2}(X)$ is

$$\pi_2^*e_1^*(c_1(L))-\langle c_1(L),d^{(2)}\rangle\hbar,$$

and $e_1\circ\pi_1=e_1\circ\pi_2$, (4) is equal to

$$\sum_d q^d\int_{G_{d^{(1)},d^{(2)}}(X)}\frac{e^{(\pi_2^*e_1^*p-d^{(2)}\hbar)z}E(H_d)(U(c_1(L)),\hbar)}{[N_{G_d(X)/G_{d^{(1)},d^{(2)}}(X)}]}$$
$$=\sum_d q^d\int_{G_d(X)}e^{Pz}E(H_d)(U(c_1(L)),\hbar),$$

which shows the polynomiality.

Now let us check the almost recursion relation. Let

$$S_v^X(q,\hbar) := \langle S^X, \phi_v^X \rangle_0^X = \sum_d S_{v,d}^X(\hbar) q^d.$$

Since (if $d \neq 0$)

$$S_{v,d}^X(\hbar) = \sum_{\substack{w \in o(v) \\ 0<m \\ m\beta_{v,w} \leqslant d}} \frac{C_{v,w,m}^X}{\hbar(\alpha_{v,w}+m\hbar)} S_{w,d-m\beta_{v,w}}^X\left(-\frac{\alpha_{v,w}}{m}\right)$$

and

$$E(H'_\beta)\left(c_1(L)_v, -\frac{\alpha_{v,w}}{m}\right) = \frac{E(V_{v,w,m})}{E(V)_v} E(H'_{\beta-m\beta_{v,w}})\left(c_1(L)_w, -\frac{\alpha_{v,w}}{m}\right)$$

from (3) and the claim, we obtain that

$$\Phi_{v,d}^V(\hbar) := \langle \Phi_d^V(\hbar), \phi_v \rangle_0^V = R_{v,d} + \sum_{\substack{w \in o(v) \\ 0<m \\ m\beta_{v,w} \leqslant d}} \frac{C_{v,w,m}^X E(V_{v,w,m})}{E(V)_v \hbar(\alpha_{v,w}+m\hbar)}$$

$$\times \Phi_{w,d-m\beta_{v,w}}^X\left(-\frac{\alpha_{v,w}}{m}\right) E(H'_{d-m\beta_{v,w}})\left(c_1(L)_w, -\frac{\alpha_{v,w}}{m}\right),$$

where $R_{v,d}$ is indeed a polynomial of $1/\hbar$ over $H^*_{(T\times T')}$ since

$$\Phi_d^V = (e_1)_* \frac{c^{\langle c_1(X),d\rangle - 2}}{\hbar^{\langle c_1(X),d\rangle -1}(\hbar - c)} \in \frac{1}{\hbar^{\langle c_1(X),d\rangle}} H^*_{T\times T'}[[\hbar^{-1}]].$$

However, since

$$C_{v,w,m}^V = \frac{C_{v,w,m}^X E(V_{v,w,m})}{E(V)_v},$$

$\Phi_v^V(q,\hbar)$ has the same almost recursion coefficients $C_{v,w,m}^V$ with S^V.

6.3. *A proof of Main Theorem* 1. Recursion Lemma 1 and Double Construction Lemma 2 show that S^V is in class $\mathcal{P}(X,V,\text{Euler})$. Certainly S^V is of the form $1+o(\hbar^{-1})$. According to Theorem 3, Φ^V also belongs to $\mathcal{P}(X,V,\text{Euler})$. Then Theorem 2 results in a proof of the main theorem. (We use the condition that E=Euler, in order to make sure that S^V and Φ^V are homogeneous of degree 0.)

7. Grassmannians

7.1. *Notation.* Let $e_1, \ldots, e_n$ form the standard basis of $\mathbf{C}^n$, $T{=}(\mathbf{C}^\times)^n$ be the complex torus, and $X{:=}\mathrm{Gr}(k,n)$ be the Grassmannian, the set of all k-subspaces in $\mathbf{C}^n$. As usual, let T act on $\mathrm{Gr}(k,n)$ by the diagonal action. The fixed points $v{=}(i_1, \ldots, i_k)$ are then the k-planes generated by the vectors $e_{i_1}, \ldots, e_{i_k}$. Denote by $\mathbf{C}^n \times X$ the trivial vector bundle with the standard action. Then we may consider L, the determinant of the bundle dual to the T-equivariant universal k-subbundle of $\mathbf{C}^n \times X$. Define $V{=}L^{\otimes l}$, $l{>}0$. Denote by p the equivariant class $c_1(L)$. We may identify $H^*(BT)$ with $\mathbf{Q}[\varepsilon_1, \ldots, \varepsilon_n]$ by the correspondence that ε_i is also denoted the equivariant Chern class of the line bundle over a point equipped with T-action as the representation of the character ε_i. With respect to the Chern class of L, we shall write $d{\in}\mathbf{Z}{=}H_2(X,\mathbf{Z})$.

7.2. *A-series.*

7.2.1. *Fixed points.* Let v be, say, $(1,2,\ldots,k)$. Then around the point, a local chart can be described by

$$\begin{pmatrix} 1 & 0 & \ldots & 0 \\ 0 & 1 & & 0 \\ \vdots & \vdots & & \vdots \\ 0 & 0 & \ldots & 1 \\ x_{1,1} & x_{1,2} & \ldots & x_{n-k,k} \\ \vdots & \vdots & & \vdots \\ x_{n-k,1} & x_{n-k,2} & \ldots & x_{n-k,k} \end{pmatrix}.$$

For each complex value $(x_{i,j})$ the column vectors in the matrix span a k-plane which stands for a point in $\mathrm{Gr}(k,n)$. Then in the chart the action by $(t_1, \ldots, t_n){\in}T$ is described as

$$\begin{pmatrix} x_{1,1} & x_{1,2} & \ldots & x_{n-k,k} \\ \vdots & \vdots & \ddots & \vdots \\ x_{n-k,1} & x_{n-k,2} & \ldots & x_{n-k,k} \end{pmatrix} \mapsto \begin{pmatrix} t_1^{-1}t_{k+1}x_{1,1} & t_2^{-1}t_{k+1}x_{1,2} & \ldots & t_k^{-1}t_{k+1}x_{1,k} \\ \vdots & \vdots & \ddots & \vdots \\ t_1^{-1}t_n x_{n-k,1} & t_2^{-1}t_n x_{n-k,2} & \ldots & t_k^{-1}t_n x_{n-k,k} \end{pmatrix}.$$

In each isolated fixed point of the Grassmannian there are $\dim \mathrm{Gr}(k,n)$-many one-dimensional orbits (rays) passing through the point. For instance, if $v{=}(1,2,\ldots,k)$, then there is only one ray (v,w) from v to $w{=}(\ldots,\hat{\imath},\ldots,j)$ for any $i{\leqslant}k{<}j$. These rays have degree $1{\in}H_2(X,\mathbf{Z})$.

7.2.2. *The Euler classes.* Notice that the tangent space at $v{=}(1,2,\ldots,k)$ of the ray connecting v to $w{=}(\ldots,\hat{\imath},\ldots,j)$ has the character $\alpha(v,w){=}\varepsilon_j{-}\varepsilon_i$, where $j{>}k{\geqslant}i$. Similarly one can find out the characters for the other cases.

Let $f\colon \mathbf{P}^1 \to \mathrm{Gr}(k,n)$ be an m-fold morphism totally ramifying the ray over v and w. The T-representation space $H^0(f^*L^{\otimes l})$ has the orbi-characters

$$\frac{ap_v+bp_w}{m} = lp_v - \frac{\varepsilon_j - \varepsilon_i}{m} b \quad \text{for } a+b=lm,\ a \geqslant 0,\ b \geqslant 0,$$

where $p_v = -(\varepsilon_1 + \ldots + \varepsilon_k)$ and $p_w = -(\varepsilon_1 + \ldots + \hat{\varepsilon}_i + \ldots + \varepsilon_k + \varepsilon_j)$ are $p = c_1(L)$ restricted to the fixed points v and w, respectively.

8. The flag manifolds

We analyze fixed points of the maximal torus actions and the invariant curves connecting two fixed points. This explicit description would be useful also to find S^X explicitly.

8.1. *The complete flag manifolds.* Let X be the set of all Borel subgroups of a simply-connected semi-simple Lie group G. It is a homogeneous space with the G-action by conjugation. Then the maximal torus T-action (fix one) has isolated fixed points. They are exactly Borel subgroups containing T since the normalizer of a Borel subgroup is so itself. The fixed points are naturally one-to-one corresponding to the set of Weyl chambers. Each Borel subgroup containing T gives rise to negative roots (our convention) of B and so a chamber associated to the positive roots. Let C be the set of chambers. The tangent line subspace associated to the positive root α has the character α.

There is, if one fixes a fixed point v, a natural correspondence between the $H^2(X,\mathbf{Z})$ and the characters of T. Then the Kähler cone is exactly the positive Weyl chamber v. Notice that the fundamental roots span the Kähler cone. Consider co-roots $\alpha^\vee$. They span the Mori cone. We can identify the Mori cone Λ with the nonnegative integer span of co-roots.

8.2. *The generalized flag manifolds.* Let X be the set of all parabolic subgroups with a given conjugate type. Let T be a maximal torus of G. Then the fixed loci are isolated fixed points consisting of parabolic subgroups containing T.

8.2.1. *Rays.* Let us choose a fixed point $P \supset T$. Then the rays at the fixed point are described by the following way. (The rays are by definition the one-dimensional orbits of T passing through P.) Fix B a Borel subgroup in P containing T. First consider the T-equivariant fibration $G/B \to G/P$ and the rational map to G/B by $\exp(zX_\alpha) \in G$, $z \in \mathbf{C}$, where X_α is an eigenvector of the positive root α. Since $\exp H \exp(zX_\alpha) \exp -H = \exp(z \exp \mathrm{ad}\, H(X_\alpha)) = \exp(z \exp(\alpha(H)) X_\alpha)$ for $H \in \mathrm{Lie}\, T$, we conclude that it is a T-invariant stable map. By the composition of the fibration, we obtain all the rays. They are effectively labeled by the positive roots which are not roots of P. So there are exactly $\dim X$-many rays at each fixed point. The tangent line at the ray has the character α.

8.2.2. *The Kähler cone.* Here we need the Levi decomposition of P, and then consider simple roots $\{\alpha_i\}_{i\in P(\Delta)}$ which are not roots of the semi-simple part of P. Then the fundamental roots with respect to P is, by definition, $\{\lambda_i\}_{i\in P(\Delta)}$, where λ_i are dual to $\alpha_i^\vee$.

Choose a fixed point P. We may identify $H^2(X,\mathbf{Z})$ with the set of integral weights according to the Borel–Weil theorem. Then the Kähler cone is the set of all dominant integral weights *with respect to P*.

8.2.3. *Homogeneous line bundles.* One can produce all the very ample line bundles by homogeneous line bundles associated to irreducible representations of P with highest weights λ. The weights corresponding to the very ample line bundles are exactly the positive integral combinations of the fundamental weights with respect to P. We shall denote by $\mathcal{O}(\lambda)$ the homogeneous line bundle associated to the (one-dimensional) highest weight $\lambda=\sum_{i\in P(\Delta)} a_i\lambda_i$ representation of P. It is a very ample bundle if and only if $a_i>0$ for all i.

This also shows that the ray $\mathbf{P}^1$ associated to α has the homology class "$\alpha^\vee$", in the sense that $\langle \mathbf{P}^1, c_1(\mathcal{O}(\lambda))\rangle=(\alpha^\vee,\lambda)$. *We shall use* $\alpha^\vee$ *to denote the homology class.*

8.2.4. $\sum(p_i)_v z_i$ *are distinct for distinct fixed points* v. Consider a line bundle L associated to $\lambda=\sum_{i\in P(\Delta)} a_i\lambda_i$. (Here in advance, we have to fix $P\supset T$.) Let S_α denote the Weyl group element of the reflection associated to a positive root α which is not a root of P. Then the line bundle is $L=\mathcal{O}(S_\alpha(\lambda))$ if one looks at it with respect to another "origin" $P'=\exp\big(\frac{1}{2}\pi(X_\alpha-Y_\alpha)\big)P\exp\big(-\frac{1}{2}\pi(X_\alpha-Y_\alpha)\big)$, where $[X_\alpha,Y_\alpha]=H_\alpha$, $[H_\alpha,X_\alpha]=2X_\alpha$ and $[H_\alpha,Y_\alpha]=-2Y_\alpha$. This P' is the other T-fixed point which lies in the ray associated to α which passes through P. (Because of the $SL(2,\mathbf{C})$-equivariant map from $\mathbf{P}^1$ to the ray, it is enough to check it when $G=SL(2,\mathbf{C})$, which is obvious.) Now it is clear that $\sum(p_i)_v z_i$ are distinct for distinct fixed points v.

8.2.5. $V_{v,w,m}$ *and* $N_{v,w,m}$. Let $V=\mathcal{O}(\lambda)$. Let $\psi:\mathbf{P}^1\to X$ be a stable map totally ramifying one of the rays, passing through $P\supset T$. Suppose that the ray is associated to a positive root α with respect to P, and that f is an m-multiple branched cover representing an isolated T-fixed point of $\overline{M}_{0,0}(X,m\alpha^\vee)$. Then the T-representation space $H^0(\mathbf{P}^1,f^*(\mathcal{O}(\lambda))$ has the characters

$$\lambda-a\frac{\alpha}{m},\quad a=0,\dots,m(\lambda,\alpha^\vee).$$

To see it, use the coordinate $z\in\mathbf{C}$ around the fixed point, and the equalities

$$\exp H\exp(zE_\alpha)\exp -H=\exp(z\exp\operatorname{ad}H(E_\alpha))=\exp(z\exp(\alpha(H))E_\alpha).$$

Similarly, $N_{v,w,m}$ has the characters

$$\delta - a\frac{\alpha}{m} \quad \text{for } \alpha \neq \delta > 0,\ a = 0, ..., m(\delta, \alpha^\vee),$$

$$\alpha - a\frac{\alpha}{m} \quad \text{for } a = 0, ..., \widehat{m}, ..., 2m,$$

where $\delta > 0$ means that δ is a positive root with respect to P.

References

[1] BATYREV, V., CIOCAN-FONTAINE, I., KIM, B. & STRATEN, D. VAN, Conifold transitions and mirror symmetry for complete intersections in Grassmannians. *Nuclear Phys. B*, 514 (1998), 640–666.

[2] — Mirror symmetry and toric degenerations of partial flag manifolds. Preprint, 1998.

[3] BATYREV, V. & STRATEN, D. VAN, Generalized hypergeometric functions and rational curves on Calabi–Yau complete intersections in toric varieties. *Comm. Math. Phys.*, 168 (1995), 493–533.

[4] BEHREND, K., Gromov–Witten invariants in algebraic geometry. *Invent. Math.*, 127 (1997), 601–617.

[5] BEHREND, K. & FANTECHI, B., The intrinsic normal cone. *Invent. Math.*, 128 (1997), 45–88.

[6] BEHREND, K. & MANIN, YU., Stacks of stable maps and Gromov–Witten invariants. *Duke Math. J.*, 85 (1996), 1–60.

[7] CANDELAS, P., OSSA, X. DE LA, GREEN, P. & PARKES, L., A pair of Calabi–Yau manifolds as an exactly soluble superconformal theory. *Nuclear Phys. B*, 359 (1991), 21–74.

[8] CIOCAN-FONTAINE, I., On quantum cohomology rings of partial flag varieties. Institut Mittag-Leffler Report No. 12, 1996/1997.

[9] DUBROVIN, B., Geometry of 2D topological field theories, in *Integrable Systems and Quantum Groups* (Montecatini Terme, 1993), pp. 120–348. Lecture Notes in Math., 1620. Springer-Verlag, Berlin, 1996.

[10] FULTON, W. & PANDHARIPANDE, R., Notes on stable maps and quantum cohomology, in *Algebraic geometry* (Santa Cruz, 1995), pp. 45–96. Proc. Sympos. Pure Math., 62:2. Amer. Math. Soc., Providence, RI, 1997.

[11] GIVENTAL, A., Homological geometry, I: Projective hypersurfaces. *Selecta Math.* (*N.S.*), 1 (1995), 325–345.

[12] — Equivariant Gromov–Witten invariants. *Iternat. Math. Res. Notices*, 13 (1996), 613–663.

[13] — Stationary phase integrals, quantum Toda lattices, flag manifolds and the mirror conjecture, in *Topics in Singularity Theory*: *V. I. Arnold's* 60*th Anniversary Collection*, pp. 103–115. Amer. Math. Soc. Transl. Ser. 2, 180. Amer. Math. Soc., Providence, RI, 1997.

[14] — A mirror theorem for toric complete intersections, in *Topological Field Theory, Primitive Forms and Related Topics* (Kyoto, 1996), pp. 141–175. Progr. Math., 160. Birkhäuser Boston, Boston, MA, 1998.

[15] — Elliptic Gromov–Witten invariants and the generalized mirror conjecture. Preprint, 1998.

[16] KIM, B., Quantum cohomology of flag manifolds G/B and quantum Toda lattices. *Ann. of Math.* (*2*), 149 (1999), 129–148.

[17] KONTSEVICH, M., Enumeration of rational curves via torus actions, in *The Moduli Space of Curves* (Texel Island, 1994), pp. 335–368. Progr. Math., 129. Birkhäuser Boston, Boston, 1995.
[18] LI, J. & TIAN, G., Virtual moduli cycle and Gromov–Witten invariants of algebraic varieties. *J. Amer. Math. Soc.*, 11 (1998), 119–174.
[19] LIAN, B., LIU, K. & YAU, S.-T., Mirror principle, I. *Asian J. Math.*, 1 (1997), 729–763.
[20] THOMSEN, J., Irreducibility of $\overline{M}_{0,n}(G/P,\beta)$. *Internat. J. Math.*, 9 (1998), 367–376.
[21] TJØTTA, E., Rational curves on the space of determinantal nets of conics. Thesis, University of Bergen, 1997.
[22] WITTEN, E., Two-dimensional gravity and intersection theory on moduli space, in *Surveys in Differential Geometry* (Cambridge, MA, 1990), pp. 243–310. Lehigh Univ., Bethlehem, PA, 1991.

BUMSIG KIM
Department of Mathematics
Pohang University of Science and Technology
Pohang, 790-784
Republic of Korea
bumsig@euclid.postech.ac.kr

Received February 4, 1998

A.2. Mirror Symmetry and Toric Degenerations of Partial Flag Manifolds

Batyrev, V.V., Ciocan-Fontanine, I., Kim, B. et al. Mirror symmetry and toric degenerations of partial flag manifolds. Acta Math. **184**, 1–39 (2000). https://doi.org/10.1007/BF02392780

© 2000, Institut Mittag-Leffler. All rights reserved. Reprinted with permission.

© The Author(s) 2026

N.-G. Kang et al. (eds.), *Categorical and Enumerative Aspects of Mirror Symmetry*,
KIAS Springer Series in Mathematics 5,
https://doi.org/10.1007/978-981-95-0385-8

Acta Math., 184 (2000), 1–39
© 2000 by Institut Mittag-Leffler. All rights reserved

Mirror symmetry and toric degenerations of partial flag manifolds

by

VICTOR V. BATYREV

Eberhard-Karls-Universität Tübingen
Tübingen, Germany

IONUŢ CIOCAN-FONTANINE

Northwestern University
Evanston, IL, U.S.A.

BUMSIG KIM

Pohang University of Science and Technology
Pohang, The Republic of Korea

and

DUCO VAN STRATEN

Johannes Gutenberg-Universität Mainz
Mainz, Germany

Contents

1. Introduction

Using combinatorial dualities for reflexive polyhedra and Gorenstein cones together with the theory of generalized GKZ-hypergeometric functions, one can extend the calculation of the number n_d of rational curves of degree d on the generic quintic threefold in $\mathbf{P}^4$ by Candelas, de la Ossa, Green and Parkes [10] to the case of Calabi–Yau complete intersections in toric varieties [3], [4], [7], [9].

Another class of examples which includes Calabi–Yau quintic 3-folds are Calabi–Yau complete intersections in homogeneous Fano varieties G/P where G is a semisimple Lie group and P is its parabolic subgroup. It is a priori not clear how to find an appropriate mirror family for these varieties, because G/P is not a toric variety in general. In [6], we described a mirror construction (compatible with [12]) for complete intersections in the Grassmannian $G(k,n)$, which turned out to involve a degeneration of $G(k,n)$ to a certain singular toric Fano variety $P(k,n)$ introduced by Sturmfels in [28].

In this paper we consider the extension of our methods to the case of complete intersections in *arbitrary partial flag manifolds* and give complete proofs of statements from [6].

It turns out that the Plücker embedding of any such flag manifold $F{:=}F(n_1,...,n_l,n)$ admits a flat degeneration to a Gorenstein toric Fano variety $P(n_1,...,n_l,n)$. This deformation has been studied recently by Gonciulea and Lakshmibai in [24], [18], [19]. The "mirror-dual" toric variety $\mathbf{P}_{\Delta(n_1,...,n_l,n)}$ associated with a reflexive polyhedron $\Delta(n_1,...,n_l,n)$ has a nice combinatorial description in terms of a certain graph $\Gamma{:=}\Gamma(n_1,...,n_l,n)$ that was introduced by Givental for the case of the complete flag manifolds [16]. The idea of toric degenerations has been discussed in a more general framework in [4].

Using the residue formula, we compute explicitly a series $\Phi_F{:=}\Phi_F(q_1,...,q_l)$ associated with the graph Γ and conjecture that Φ_F gives a solution to the quantum $\mathcal{D}$-module associated with Gromov–Witten classes and quantum cohomology of the partial flag manifold F. We note that there is no essential difficulty in checking the conjecture in each particular case at hand, because it involves only calculations in the small quantum cohomology ring of F, for which explicit formulas are known [11], see also Remark 5.1.12 (ii). Applying the "trick with factorials" (see [6], or §4.2 below) to a Calabi–Yau complete intersection in F, we obtain Φ_F as a specialization of the toric GKZ-hypergeometric series, from which the instanton numbers (i.e., the virtual numbers of rational curves on the Calabi–Yau) can be computed via the standard procedure (see e.g. [7]). As the validity of this trick was shown recently for general homogeneous spaces [23], this implies that any instanton numbers computed via the usual "mirror symmetry method" are automatically proven to be correct in all cases for which our conjecture on Φ_F holds. The series Φ_F of complete flag manifolds has also been investigated by Schechtman [27].

The paper is organized as follows. In §2 we introduce main combinatorial notions used in the definition of a Gorenstein toric Fano variety $P(n_1,...,n_l,n)$ associated with a given partial flag manifold $F(n_1,...,n_l,n)$. In §3 we investigate singularities of $P(n_1,...,n_l,n)$ and show that these singularities can be smoothed by a flat deformation to the partial flag manifold $F(n_1,...,n_l,n)$. As a consequence of our results, we prove a generalized version of a conjecture of Gonciulea and Lakshmibai about the singular locus of $P(n_1, ..., n_l, n)$ [19]. In §4 we discuss quantum differential systems following ideas of Givental [15], [16], [17]. Finally, in §5 we explain the mirror construction for Calabi–Yau complete intersections in partial flag varieties F and the computations of the corresponding hypergeometric series Φ_F.

Acknowledgement. We would like to thank A. Givental, S. Katz, S.-A. Strømme, and E. Rødland for helpful discussions, and the Mittag-Leffler Institute for hospitality. The second and third named authors have been supported by Mittag-Leffler Institute postdoctoral fellowships.

2. Toric varieties associated with partial flag manifolds

In this section we explain how to associate to an arbitrary partial flag manifold $F(n_1, ..., n_l, n)$ certain combinatorial objects: a graph $\Gamma(n_1, ..., n_l, n)$, a reflexive polytope $\Delta(n_1, ..., n_l, n)$ and a Gorenstein toric Fano variety $P(n_1, ..., n_l, n)$.

2.1. The graph $\Gamma(n_1, ..., n_l, n)$

Let $k_1, k_2, ..., k_{l+1}$ be a fixed sequence of positive integers. We set $n_0=0$, $n_i:=k_1+...+k_i$ $(i=1, ..., l+1)$ and $n:=n_{l+1}$. Denote by $F(n_1, ..., n_l, n)$ the partial flag manifold parametrizing sequences of subspaces

$$0 \subset V_1 \subset V_2 \subset ... \subset V_l \subset \mathbf{C}^n,$$

with $\dim V_i = n_i$ $(i=1, ..., l)$. Then

$$\dim F(n_1, ..., n_l, n) = \sum_{i=1}^{l} (n_i - n_{i-1})(n - n_i).$$

To simplify notations, we shall often write F instead of $F(n_1, ..., n_l, n)$, if there is no confusion about the numbers $n_1, ..., n_l, n$. By a classical result of Ehresmann ([13]), a natural basis for the integral cohomology of F is given by the Schubert classes. These are Poincaré dual to the fundamental classes of the closed Schubert cells $C_w \subset F$, parametrized by permutations $w \in S_n$ modulo the subgroup

$$W(k_1, ..., k_{l+1}) := S_{k_1} \times ... \times S_{k_{l+1}} \subset S_n.$$

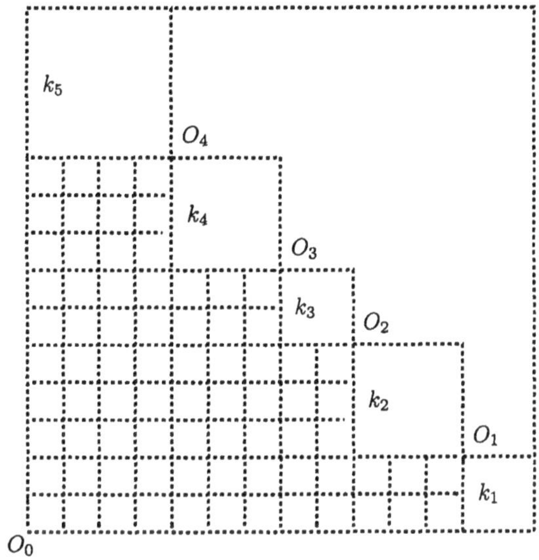

Fig. 1

In particular, the Picard group of F, which is isomorphic to $H^2(F, \mathbf{Z})$, is generated by l divisors $C_1, \ldots, C_l$, corresponding to the simple transpositions $\tau_i \in S_n$ exchanging n_i and n_i+1.

Definition 2.1.1. Denote by $\Lambda := \Lambda(n_1, \ldots, n_l, n)$ the standard *ladder diagram* consisting of unit squares (the number of unit squares in Λ is equal to the dimension of F) corresponding to the Schubert cell of maximal dimension in the flag manifold F. We place the ladder diagram Λ in the lower left corner of an $(n \times n)$-square Q. The lower left corner of Λ (or of Q) will be denoted by O_0. We denote by O_i ($i \in \{1, \ldots, l\}$) the common vertex of the diagonal squares Q_i of size $k_i \times k_i$, and Q_{i+1} of size $k_{i+1} \times k_{i+1}$ (Figure 1 illustrates the case $l=4$).

Definition 2.1.2. Let $\Lambda = \Lambda(n_1, \ldots, n_l, n)$ be the above ladder diagram. We *associate* with Λ the following:

(i) $D = D(n_1, \ldots, n_l, n)$, the set of centers of unit squares in Λ: we place a dot at the center of each unit square and call elements of D *dots*.

(ii) $S = S(n_1, \ldots, n_l, n)$, the set consisting of $l+1$ *stars*: an element of S is obtained by placing a star at the $(\frac{1}{2}, \frac{1}{2})$-shift of the lower left corner of each of the diagonal squares Q_i ($i \in \{1, \ldots, l+1\}$).

(iii) $E = E(n_1, \ldots, n_l, n)$, the set of oriented horizontal and vertical segments connect-

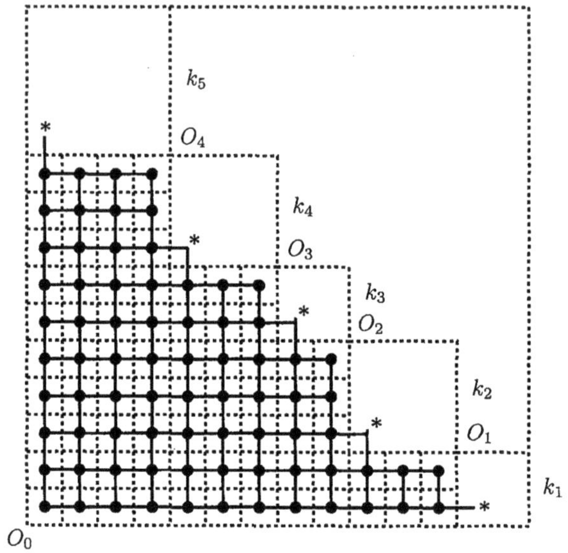

Fig. 2

ing adjacent elements of $D \cup S$: the vertical segments are oriented downwards, and the horizontal segments are oriented to the right.

Definition 2.1.3. $\Gamma := \Gamma(n_1, ..., n_l, n)$ is the oriented graph whose set of vertices is $D \cup S$, and whose set of oriented edges is E.

Such a graph Γ (without the orientation!) is shown in Figure 2. The edges of Γ are drawn with solid lines.

Definition 2.1.4. We denote by $L(D) \cong \mathbf{Z}^{|D|}$, $L(S) \cong \mathbf{Z}^{|S|}$ and $L(E) \cong \mathbf{Z}^{|E|}$ the free abelian groups (or lattices) generated by the sets D, S and E.

We remark that the lattices $L(D) \oplus L(S)$ and $L(E)$ can be viewed as the groups of 0-chains and 1-chains of the graph Γ. Then the boundary map in the chain complex is

$$\partial : L(E) \to L(D) \oplus L(S), \quad e \mapsto h(e) - t(e),$$

where $h, t : E \to D \cup S$ are the maps that associate to an oriented edge $e \in E$ its *head* and its *tail* respectively. See Figure 3.

Definition 2.1.5. A *box* b in Γ is a subset of 4 edges $\{e, f, g, h\} \subset E$ which form together with their endpoints a connected subgraph $\Gamma_b \subset \Gamma$ such that the topological

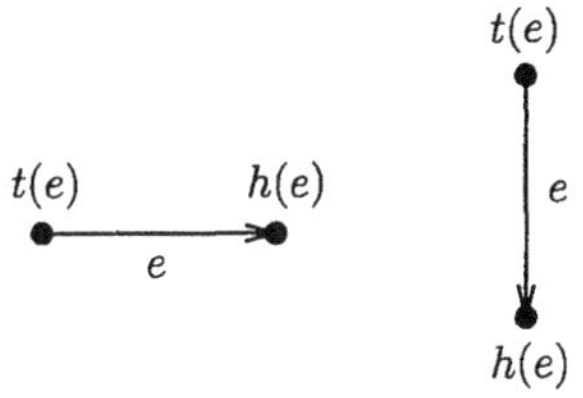

Fig. 3

space associated to Γ_b is homeomorphic to a circle. The set of boxes in Γ will be denoted by B.

It is easy to see that

$$H_0(\Gamma) = \mathrm{Coker}(\partial) \cong \mathbf{Z}, \quad H_1(\Gamma) = \mathrm{Ker}(\partial) \cong \mathbf{Z}^{|B|}.$$

We also consider the projection $\varrho: L(D) \oplus L(S) \to L(D)$ and the composed map

$$\delta := \varrho \circ \partial : L(E) \to L(D).$$

Since one can regard the groups $L(E)$ and $L(D)$ together with the homomorphism δ as the relative chain complex of the topological pair (Γ, S), we have

$$H_0(\Gamma, S) = \mathrm{Coker}(\delta) = 0, \quad H_1(\Gamma, S) = \mathrm{Ker}(\delta) \cong \mathbf{Z}^{|B|+l}.$$

Definition 2.1.6. A *roof* $\mathcal{R}_i$, $i \in \{1, 2, ..., l\}$, is the set of $k_i + k_{i+1}$ edges of Γ forming the oriented path that runs along the upper right "boundary" of Γ between the ith and the $(i+1)$st stars in S.

Definition 2.1.7. The *corner* C_b of a box $b = \{e, f, g, h\} \in B$ is the pair of edges $\{e, f\} \subset b$ meeting at the lower left vertex of Γ_b. So a corner C_b contains one vertical edge e and one horizontal edge f such that $h(e) = t(f)$.

The roofs and corners give a decomposition of the set E of edges of the graph Γ into a disjoint union of subsets:

$$E = \mathcal{R}_1 \cup ... \cup \mathcal{R}_l \cup \bigcup_{b \in B} C_b.$$

This decomposition is shown in Figure 4.

Definition 2.1.8. The *opposite corner* C_b^- of a box $b = \{e, f, g, h\} \in B$ is the pair of edges $\{g, h\} \subset b$ meeting at the upper right vertex of Γ_b. An opposite corner C_b^- contains one vertical edge h and one horizontal edge g such that $h(g) = t(h)$.

By elementary arguments one obtains:

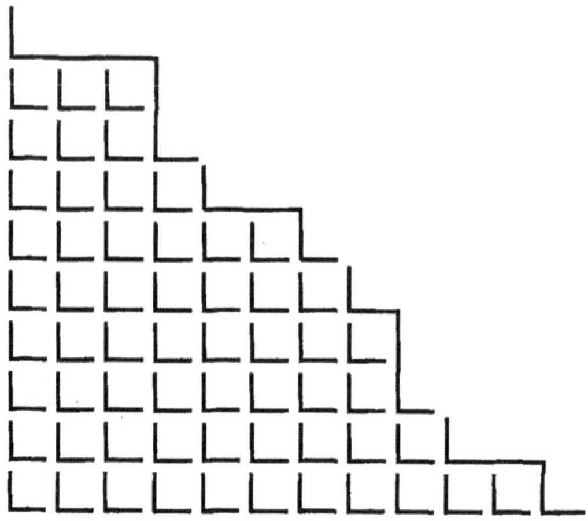

Fig. 4

PROPOSITION 2.1.9. *The elements*

$$\varrho_b = \sum_{e \in C_b} e - \sum_{e \in C_b^-} e,$$

where b runs over the set B, form a natural $\mathbf{Z}$*-basis of* $\mathrm{Ker}(\partial) \subset L(E)$. *Moreover, the elements*

$$\varrho_i = \sum_{e \in \mathcal{R}_i} e, \quad i \in \{1, ..., l\},$$

and

$$\varrho_b = \sum_{e \in C_b} e - \sum_{e \in C_b^-} e, \quad b \in B,$$

form a natural $\mathbf{Z}$*-basis of* $\mathrm{Ker}(\delta) \subset L(E)$. □

2.2. The toric variety $P(n_1, ..., n_l, n)$

We denote again by δ the $\mathbf{R}$-scalar extension $L(E) \otimes \mathbf{R} \to L(D) \otimes \mathbf{R}$ of the homomorphism $\delta: L(E) \to L(D)$.

Definition 2.2.1. The polyhedron $\Delta := \Delta(n_1, ..., n_l, n)$ associated to F is the convex hull of the set

$$\delta(E) \subset L(D) \otimes \mathbf{R},$$

where the set E is identified with the standard basis of $L(E) \otimes \mathbf{R} \cong \mathbf{R}^{|E|}$.

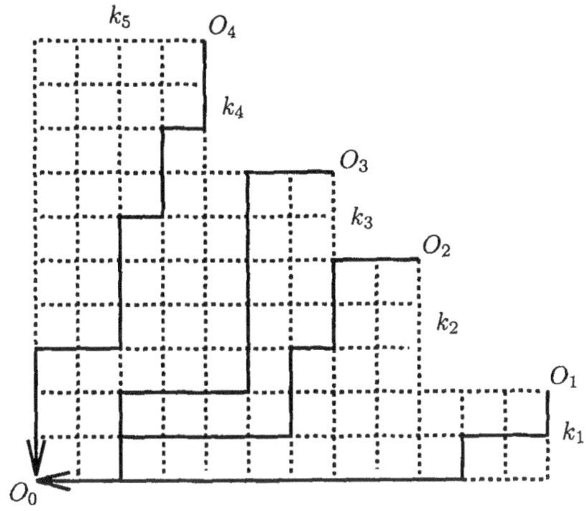

Fig. 5

In order to describe the faces of the polyhedron Δ we introduce some further combinatorial objects associated to the ladder diagram Λ.

Definition 2.2.2. (i) A *positive path* π in the diagram Λ is a path obtained by starting at one of the points O_i $(i=1,...,l)$ and moving either downwards, or to the left along some n edges of Λ, until the lower left corner O_0 is reached (see Figure 5). We denote by Π the set of positive paths, and by Π_i the set of positive paths connecting O_i and O_0, so that

$$\Pi = \Pi_1 \cup ... \cup \Pi_l.$$

Note that the number of elements in Π_i is

$$N_i = \binom{n}{n_i}.$$

(ii) A *meander* is a collection of positive paths $\{\pi_1,...,\pi_l\}$ $(\pi_i \in \Pi_i)$, with the property that the union

$$\pi_1 \cup ... \cup \pi_l$$

is a *tree* with endpoints $O_0, O_1, ..., O_l$.

The set of all meanders is denoted by $\mathcal{M}$.

THEOREM 2.2.3. *There is a natural bijection between the codimension-*1 *faces of* Δ *and the set* $\mathcal{M}$ *of meanders.*

Proof. Since every face Θ of Δ is given by its supporting hyperplane, it follows from the exact sequence

$$0 \to \mathrm{Ker}(\delta) \to L(E) \otimes \mathbf{R} \xrightarrow{\delta} L(D) \otimes \mathbf{R} \to 0$$

that this hyperplane can be described by a linear function

$$\lambda: L(E)\otimes\mathbf{R}\to\mathbf{R}$$

which vanishes on $\mathrm{Ker}(\delta)$ and satisfies the conditions

$$\lambda(v)\leqslant 1, \quad \text{for all } v\in L(E)\otimes\mathbf{R} \text{ with } \delta(v)\in\Delta,$$

and

$$\delta(v)\in\Theta \quad \text{if and only if} \quad \lambda(v)=1 \text{ and } \delta(v)\in\Delta.$$

Let us show that every meander $m=\{\pi_1,\dots,\pi_l\}\in\mathcal{M}$ defines such a linear function λ_m. We define the value of λ_m on $e\in E$ by the formula

$$\lambda_m(e):=1-\sum_{\{i:\pi_i\cap e\neq\varnothing\}}|\mathcal{R}_i|. \tag{1}$$

It follows that $\lambda_m(e)=1$ if the meander m does not intersect e, and $\lambda_m(e)$ is negative if m intersects e. Now we show that the linear function λ_m satisfies the requirement $\lambda_m|_{\mathrm{Ker}(\delta)}=0$. By Proposition 2.1.9, it suffices to prove that

$$\sum_{e\in\mathcal{R}_i}\lambda_m(e)=0, \tag{2}$$

for all $i\in\{1,2,\dots,l\}$, and

$$\sum_{e\in C_b}\lambda_m(e)=\sum_{e\in C_b^-}\lambda_m(e), \tag{3}$$

for all $b\in B$.

We remark first that every roof $\mathcal{R}_i$, $i\in\{1,2,\dots,l\}$, contains exactly one edge $e_i\in E$ intersecting the positive path $\pi_i\in m$, for which

$$\lambda_m(e_i):=1-|\mathcal{R}_i|<0.$$

On the other hand, $\lambda(e)=1$ for each $e\in\mathcal{R}_i$, $e\neq e_i$. It follows that

$$\sum_{e\in\mathcal{R}_i}\lambda_m(e)=0 \quad \text{for all } i\in\{1,2,\dots,l\}.$$

Now let $b\in B$ be an arbitrary box. Since the positive paths of the meander m form a tree, only the following three cases can occur:

Case 1. The meander m does not intersect edges in b. Then $\lambda_m(e)=1$ for all 4 edges of b, and hence (3) holds.

Case 2. The meander m intersects exactly two edges in b. Then m intersects exactly one edge $e' \in b$ belonging to $\mathcal{C}_b$ and exactly one edge $e'' \in b$ belonging to $\mathcal{C}_b^-$. By the formula (1) for λ_m, we have $\lambda_m(e') = \lambda_m(e'')$. So again the relation (3) holds.

Case 3. The meander m intersects exactly three edges in b. Then m intersects both edges $e', e'' \in b$ belonging to $\mathcal{C}_b^-$ and exactly one edge $e''' \in b$ belonging to $\mathcal{C}_b$. By (1),

$$\lambda_m(e''') = \lambda_m(e') + \lambda_m(e'') - 1.$$

Again the relation (3) holds.

Therefore, by Proposition 2.1.9, $\lambda_m|_{\mathrm{Ker}(\delta)} = 0$.

Let Θ_m be the face of Δ defined by the supporting affine hyperplane $\lambda_m(\cdot) = 1$. We claim that Θ_m has codimension 1. Since Θ_m is the convex hull of the lattice points in Δ corresponding to the edges $e \in E$ on which λ_m takes the value 1, it is sufficient to show that any linear function λ' satisfying $\lambda'|_{\mathrm{Ker}(\delta)} = 0$ and $\lambda'(v) = 1$ for all $v \in L(E) \otimes \mathbf{R}$ with $\delta(v) \in \Theta_m$ must coincide with λ_m. Indeed, by Proposition 2.1.9, the value of such a linear function λ' is uniquely determined on each edge e of each roof $\mathcal{R}_i$ $(1 \leqslant i \leqslant l)$:

$$\lambda'(e) = \begin{cases} 1 - |\mathcal{R}_i| & \text{if } \pi_i \cap e \neq \varnothing, \\ 1 & \text{otherwise.} \end{cases}$$

Next we remark that if for some box $b \in B$ we have shown that

$$\lambda'(e) = \lambda_m(e)$$

holds for all $e \in \mathcal{C}_b^-$, then, by Proposition 2.1.9 and (3), we obtain

$$\sum_{e \in \mathcal{C}_b} \lambda'(e) = \sum_{e \in \mathcal{C}_b} \lambda_m(e)$$

and therefore

$$\lambda'(e) = \lambda_m(e) \quad \text{for all } e \in \mathcal{C}_b,$$

since only one edge $e \in \mathcal{C}_b$ can be intersected by m (see Cases 1–3). Since we have established the equality $\lambda'(e) = \lambda_m(e)$ for all $e \in \mathcal{R}_1 \cup \ldots \cup \mathcal{R}_l$, the above arguments imply the equality $\lambda'(e) = \lambda_m(e)$ for all $e \in E$.

Now we prove that any codimension-1 face Θ of Δ can be obtained from some meander $m \in \mathcal{M}$. For this purpose, it suffices to show that if a supporting linear function λ defines a face $\Theta \subset \Delta$, then there exists a meander $m \in \mathcal{M}$ with $\Theta \subset \Theta_m$. The latter is equivalent to the condition $\lambda(e) < 1$ for all edges $e \in E$ such that $e \cap m \neq \varnothing$.

First we remark that the linear function λ cannot attain the value 1 on all edges of the roof $\mathcal{R}_1$, because λ vanishes on the element $\varrho_1 \in \mathrm{Ker}(\delta)$ (see Proposition 2.1.9).

Now start a positive path π_1 at O_1 whose first nonempty intersection with edges of the opposite corner C_b^- of some box $b \in B$ occurs on an edge $e_1 \in \mathcal{R}_1$ with $\lambda(e_1)<1$. Since

$$\sum_{e\in C_b} \lambda(e) = \sum_{e\in C_b^-} \lambda(e),$$

the value of λ on at least one of the two edges of C_b has to be strictly less than 1. We prolong our path through that edge and enter a next box, where the same reasoning applies. Continuing this, we complete a positive path π_1 from O_1 to O_0 crossing only edges where λ is strictly less than 1. Now we repeat this construction for each of the O_i in subsequent order, starting at O_2, etc. If in the process of constructing a positive path π_i we collide with some already constructed positive path π_j $(j<i)$, we just follow from this point the path π_j. In the end, we produce a meander with the required property.

We conclude that Θ_m $(m\in\mathcal{M})$ are all the codimension-1 faces of Δ. □

COROLLARY 2.2.4. *$\Delta(n_1, ..., n_l, n)$ is a reflexive polyhedron.*

Proof. The statement follows immediately from Theorem 2.2.3 and from the integrality of the supporting linear function λ_m (see Definition 4.1.5 in [3]). □

Definition 2.2.5. The complete rational polyhedral fan $\Sigma=\Sigma(n_1, ..., n_l, n)$ is the fan defined as the collection of cones over all faces of Δ. The toric variety $\mathbf{P}_\Sigma$ associated to the fan Σ will be denoted by $P=P(n_1, ..., n_l, n)$.

Using one of the equivalent characterizations of reflexive polyhedra (see Theorem 4.1.9 in [3]), we obtain from Corollary 2.2.4:

PROPOSITION 2.2.6. *$P(n_1, ..., n_l, n)$ is a Gorenstein toric Fano variety.* □

3. Further properties of $P(n_1, ..., n_l, n)$

3.1. Singular locus

Definition 3.1.1. Define $\widehat{P}=\widehat{P}(n_1, ..., n_l, n)$ to be the toric variety $\mathbf{P}_{\widehat{\Sigma}}$ associated to the fan $\widehat{\Sigma}$, obtained by refining the fan Σ to a simplicial one, whose one-dimensional cones are the same as the ones of Σ (i.e., they are generated by the lattice vectors $\{\delta(e), e\in E\}\subset L(D)$) and whose combinatorial structure is given by the following $|B|+l$ primitive collections:

$$\mathcal{R}_1,\ \mathcal{R}_2,\ ...,\ \mathcal{R}_l \text{ and } C_b,\quad b\in B.$$

In other words, the cones of maximal dimension of the fan $\widehat{\Sigma}$ are defined by taking all edges $e\in E$ except one from each roof and from each corner.

PROPOSITION 3.1.2. *The variety $\widehat{P}$ is a small toric desingularization of P.*

Proof. We have to show that each cone of $\widehat{\Sigma}$ is contained in a cone of Σ, and each cone of $\widehat{\Sigma}$ is generated by a part of a basis. It suffices to prove the above properties for cones of $\widehat{\Sigma}$ of maximal dimension.

Choose an edge e_i in each roof $\mathcal{R}_i$ $(i{=}1,...,l)$ and an edge f_b in each corner C_b, $b{\in}B$. This choice determines a $|D|$-dimensional cone σ in $\widehat{\Sigma}$. For each $i{=}1,...,l$ there exists a unique positive path from O_i to O_0 with the following two properties:

(i) π_i crosses the edge e_i;

(ii) if π_i enters a box b, then it crosses the edge f_b.

It is easy to see that the union $\pi_1\cup...\cup\pi_l$ of these paths is a meander. Indeed, if a union of positive paths as above is not a tree, then there must exist a box $b{\in}B$ with both edges of the corner C_b intersecting the union of positive paths. This contradicts the second of the above conditions. Therefore the set of edges $\{e_i\}\cup\{f_b\}$ defines uniquely a meander $m{\in}\mathcal{M}$, and the cone σ is contained in the cone over the face $\Theta_m{\subset}\Delta$. On the other hand, the elements $\{\varrho_i\}_{i=1,...,l}$ and $\{\varrho_b\}_{b\in B}$ together with the set

$$G_\sigma := E\backslash(\{e_i\}_{i=1,...,l}\cup\{f_b\}_{b\in B})$$

form a $\mathbf{Z}$-basis of $L(E)$. By Proposition 2.1.9, the set of generators of σ (i.e., the δ-image of G_σ) is a $\mathbf{Z}$-basis of $L(D)$.

The desingularization morphism $\widehat{P}{\to}P$ induced by the refinement $\widehat{\Sigma}$ of Σ is small (i.e., contracts no divisor), because the sets of 1-dimensional cones in $\widehat{\Sigma}$ and Σ are the same. □

There is another way to describe $\widehat{P}$, namely as an iterated toric fibration over $\mathbf{P}^1$: One starts with the product of projective spaces

$$\mathbf{P}^{|\mathcal{R}_1|-1}\times...\times\mathbf{P}^{|\mathcal{R}_l|-1}$$

corresponding to the roofs. Then one chooses a corner C_b of a box $b{\in}B$ whose opposite corner C_b^- belongs to a roof. This choice allows us to define a toric bundle over $\mathbf{P}^1$ with the fibre $\mathbf{P}^{|\mathcal{R}_1|-1}\times...\times\mathbf{P}^{|\mathcal{R}_l|-1}$. Then one adds a new corner $C_{b'}$ of a box $b'{\in}B$ whose opposite corner $C_{b'}^-$ is contained in the union of roofs and C_b, etc. At each stage of this process one gets a toric fibre bundle over $\mathbf{P}^1$, with fibre the space constructed in the previous step. Using this description of $\widehat{P}$, one obtains an alternative proof of the fact that the anticanonical divisor on P is Cartier and ample, i.e., that the polyhedron Δ is reflexive.

Definition 3.1.3. Let $b{\in}B$ be an arbitrary box. Define $W_b{\subset}P$ to be the closure of the torus orbit in P corresponding to the 3-dimensional cone σ_b generated by the δ-image of the 4-element set b.

THEOREM 3.1.4. *The singular locus of P consists of codimension-3 strata W_b, $b \in B$. These are conifold strata, i.e., transverse to a generic point of W_b the variety P has an ordinary double point.*

Proof. Since the desingularization morphism $\varphi: \widehat{P} \to P$ is small, P is smooth in codimension 2. Moreover, the singular locus of P is precisely the union of toric strata in P over which the morphism φ is not bijective. According to the main result of [26], the exceptional locus $Ex(\varphi) \subset \widehat{P}$ (i.e., $\varphi^{-1}(\mathrm{Sing}(P))$) is the union of toric strata covered by rational curves contracted by φ. On the other hand, since $\widehat{P}$ is an iterated toric bundle, the Mori cone $\overline{NE}(\widehat{P})$ is a simplicial cone generated by the classes of the primitive relations

$$\sum_{e \in \mathcal{R}_i} \delta(e) = 0, \quad i = 1, \dots, l,$$

and

$$\sum_{e \in C_b} \delta(e) = \sum_{e \in C_b^-} \delta(e), \quad b \in B,$$

(see §§2 and 4 in [2]). Since the morphism φ is defined by the semiample anticanonical class of $\widehat{P}$, it contracts exactly the extremal rays in $\overline{NE}(\widehat{P})$ defined by the primitive relations corresponding to the boxes $b \in B$. The rational curves representing each such class cover the codimension-2 strata $\widehat{W}_b$, $b \in B$, corresponding to the 2-dimensional cones in $\widehat{\Sigma}$ spanned by the δ-images of the edges forming the opposite corner C_b^-. These strata are contracted, with $\mathbf{P}^1$-fibres, to the codimension-3 strata W_b in P corresponding to the 3-dimensional cones $\sigma_b \in \Sigma$ over the quadrilateral faces Θ_b of Δ whose vertices are δ-images of the edges in b ($b \in B$). It follows that $\bigcup_{b \in B} W_b$ is exactly the singular locus of P. □

3.2. Canonical flat smoothing

Let $F = F(n_1, \dots, n_l, n)$ be a partial flag manifold. The semiample line bundles

$$\mathcal{O}(C_1), \ \dots, \ \mathcal{O}(C_l)$$

associated to the Schubert divisors $C_1, \dots, C_l$ define the Plücker embedding of F into a product of projective spaces:

$$\phi: F \hookrightarrow \mathbf{P}^{N_1 - 1} \times \dots \times \mathbf{P}^{N_l - 1}, \quad \text{where } N_i = \binom{n}{n_i}.$$

We will always consider F as a smooth projective variety together with this embedding.

We describe now an embedding of P in the same product of projective spaces.

Definition 3.2.1. For each $e \in E$, let H_e be the toric Weil divisor on P determined by the 1-dimensional cone of Σ spanned by the vector $\delta(e)$.

For every edge $e \in \bigcup_{i=1}^{l} \mathcal{R}_i$ which is part of a roof, denote by $U(e)$ the subset of E consisting of the edge e, together with all edges $f \in E$ which are either directly below e in the graph Γ, if e is horizontal, or directly to the left of e, if e is vertical.

Fix $1 \leqslant i \leqslant l$. For $e \in \mathcal{R}_i$ consider the Weil divisor $\sum_{f \in U(e)} H_f$.

LEMMA 3.2.2. *For each $e \in \mathcal{R}_i$, the Weil divisor $\sum_{f \in U(e)} H_f$ is Cartier. Moreover, if $e' \in \mathcal{R}_i$ is another edge in the same roof, then the associated divisor $\sum_{f' \in U(e')} H_{f'}$ is linearly equivalent to $\sum_{f \in U(e)} H_f$.*

Proof. To each edge $e \in \mathcal{R}_i$, and each positive path $\pi \in \Pi_i$ joining O_i with O_0, we associate a linear function $\pi[e]: L(E) \to \mathbf{Z}$ defined by

$$\pi[e](g) = \begin{cases} 0 & \text{if } \pi \cap g = \varnothing \text{ and } g \notin U(e), \\ 0 & \text{if } \pi \cap g \neq \varnothing \text{ and } g \in U(e), \\ -1 & \text{if } \pi \cap g = \varnothing \text{ and } g \in U(e), \\ 1 & \text{if } \pi \cap g \neq \varnothing \text{ and } g \notin U(e). \end{cases}$$

It is an elementary exercise to check that $\pi[e]$ vanishes on the elements

$$\varrho_j = \sum_{g \in \mathcal{R}_j} g, \quad j \in \{1, \dots, l\},$$

and

$$\varrho_b = \sum_{g \in C_b} g - \sum_{g \in C_b^-} g, \quad b \in B.$$

It follows from Proposition 2.1.9 that $\pi[e]$ descends to a linear function on $L(D)$.

To show that $\sum_{f \in U(e)} H_f$ is Cartier, it suffices to construct for each maximal-dimensional cone σ in Σ an integral linear function

$$\lambda_\sigma: L(E) \to \mathbf{Z}$$

which vanishes on $\ker(\delta)$ and satisfies

(i) $\lambda_\sigma(g) = 0$, for all $g \in E$ such that $\delta(g) \in \sigma$ and $g \notin U(e)$;

(ii) $\lambda_\sigma(g) = -1$, for all $g \in E$ such that $\delta(g) \in \sigma$ and $g \in U(e)$.

By Theorem 2.2.3, every maximal cone σ is determined by a meander $m = (\pi_1, \pi_2, \dots, \pi_l)$, and $\delta(g) \in \sigma$ if and only if the meander does not intersect the edge g (cf. the proof of Theorem 2.2.3). It follows that

$$\lambda_\sigma := \pi_i[e]$$

satisfies the above conditions, where π_i is the positive path in m which joins O_i with O_0. Hence $\sum_{f\in U(e)} H_f$ is Cartier. Note that the functional λ_σ defined above does not depend on the positive paths π_j $(j\neq i)$ in m that do not intersect the roof $\mathcal{R}_i$.

To prove the second part of the lemma, define an integral linear function $\mu\colon L(E)\to\mathbf{Z}$ by

$$\mu(g)=\begin{cases}-1 & \text{if } g\in U(e),\\ 1 & \text{if } g\in U(e'),\\ 0 & \text{otherwise.}\end{cases}$$

As above, one can easily check that μ vanishes on $\ker(\delta)$, and hence it descends to a linear function on $L(D)$. The descended linear function defines a rational function on P, whose divisor is $\sum_{f\in U(e)} H_f-\sum_{f'\in U(e')} H_{f'}$. This finishes the proof of the lemma. □

Definition 3.2.3. For each $i=1,2,\dots,l$, the line bundle associated to the roof $\mathcal{R}_i$ is

$$\mathcal{L}_i:=\mathcal{O}\Bigl(\sum_{f\in U(e)} H_f\Bigr),$$

for some edge $e\in\mathcal{R}_i$.

It follows from Lemma 3.2.2 that $\mathcal{L}_i$ does not depend on the choice of the edge $e\in\mathcal{R}_i$.

We note that for each maximal-dimensional cone σ the linear function λ_σ defined in the proof of Lemma 3.2.2 satisfies $\lambda_\sigma(g)\geqslant 0$ for all $g\in E$ such that $g\notin\sigma$. This implies that the line bundle $\mathcal{O}(\sum_{f\in U(e)} H_f)$ is generated by global sections (cf. [14, p. 68]). We will now identify the space of global sections.

The Cartier divisor $\sum_{f\in U(e)} H_f$ determines a rational convex polyhedron $\Delta[e]$ in the dual vector space $L(D)^*\otimes\mathbf{R}$, given by

$$\Delta[e]=\{\lambda\in L(D)^*\otimes\mathbf{R}:\lambda(\delta(g))\geqslant -1\ \forall g\in U(e),\ \lambda(\delta(g))\geqslant 0\ \forall g\in E\setminus U(e)\}.$$

The space of global sections of the line bundle $\mathcal{O}(\sum_{f\in U(e)} H_f)$ has a natural basis, indexed by the lattice points in $\Delta[e]$. By its very definition, for each positive path $\pi\in\Pi_i$, the linear function $\pi[e]$ introduced in the proof of Lemma 3.2.2 gives such a lattice point.

PROPOSITION 3.2.4. *For each $i=1,2,\dots,l$, the space of global sections of $\mathcal{L}_i$ has a natural basis parametrized by the set Π_i of positive paths connecting O_i and O_0.*

Proof. Choose an edge $e\in\mathcal{R}_i$. We have to show that the only lattice points in $\Delta[e]$ are the ones given by $\pi[e]$, $\pi\in\Pi_i$. Let $\lambda\colon L(E)\to\mathbf{Z}$ be any linear function vanishing on $\ker(\delta)$, and such that the descended linear function is in $\Delta[e]$.

Since on the one hand λ vanishes on every

$$\varrho_j=\sum_{g\in\mathcal{R}_j} g,\quad j\in\{1,\dots,l\},$$

and on the other hand λ can be negative only on edges in $U(e)$, there are exactly two possibilities:

(I) $\lambda(g)=0$ for all $g\in\bigcup_{j=1}^{l}\mathcal{R}_j$;

(II) $\lambda(e)=-1$, there exists an edge h in the ith roof $\mathcal{R}_i$ with $\lambda(h)=1$, and $\lambda(g)=0$ for all $g\in\bigcup_{j\neq i}\mathcal{R}_j\setminus\{e,f\}$.

If (I) holds, then we start a positive path π at O_i that intersects the roof $\mathcal{R}_i$ at the edge e. Let b be the box containing e in its opposite corner and let f be the other edge in $U(e)$ contained in this box. If $\lambda(f)=0$, we prolong the path through the edge f, and enter a next box b', where we have the same situation as before (i.e., there is another edge $f'\in U(e)$, and if $\lambda(f')=0$, then we prolong the path through f', etc.). So we may assume that $\lambda(f)=-1$. The edge f is part of the corner C_b of b. Let f'' be the other edge in C_b. Since λ vanishes on all elements

$$\varrho_b=\sum_{g\in C_b} g-\sum_{g\in C_b^-} g, \quad b\in B,$$

$\lambda(f'')$ must be strictly positive (hence at least 1). We prolong the path π through the edge f'', and enter a next box b'', for which f'' is part of the opposite corner. Now λ is nonnegative on all four edges of b'', and $\lambda(f'')\geqslant 1$. It follows that there must be an edge f''' in the corner $C_{b''}$, with $\lambda(f''')\geqslant 1$. We prolong the path through this edge, and enter a next box, where the same reasoning applies. Continuing this, we complete eventually a positive path π. Consider the linear function $\nu:=\lambda-\pi[e]$ on $L(E)$. By construction, and the definition of $\pi[e]$, the functional ν is nonnegative on all edges $g\in E$. On the other hand, ν vanishes on the generators of $\ker(\delta)$ described in Proposition 2.1.9, since both λ and $\pi[e]$ do. We claim that ν is identically zero on $L(E)\otimes\mathbf{R}$. Indeed, since ν is nonnegative and $\nu(\sum_{g\in\mathcal{R}_j} g)=0$ for $j=1,2,\ldots,l$, it follows that ν takes the value zero on each edge in the union of all roofs. Similarly, if ν vanishes on each of the edges of the opposite corner C_b^- of some box, then it must vanish on each of the edges of the corner C_b as well. From these two facts, one obtains inductively that ν takes the value zero on every $g\in E$. Hence $\lambda=\pi[e]$.

Assume now that (II) holds. In this case, we start a positive path π at O_i that intersects the roof $\mathcal{R}_i$ at the edge h. A reasoning entirely similar to that in case (I) shows that the path π can be completed such that the functional $\lambda-\pi[e]$ is nonnegative on every edge. Hence we obtain again $\lambda=\pi[e]$. □

Definition 3.2.5. The $(|D|+l)$-dimensional cone $C=C(n_1,\ldots,n_l,n)$ associated to the flag manifold F is the convex polyhedral cone in the space $\mathrm{Im}(\partial)\otimes\mathbf{R}$ spanned by the vectors

$$\partial(e)\in\mathrm{Im}(\partial)\otimes\mathbf{R}\cong\mathbf{R}^{|D|+|S|-1}$$

with $e \in E$. We denote by C^* the dual cone in the dual space $\mathrm{Im}(\partial)^* \otimes \mathbf{R}$.

Definition 3.2.6. Let $\pi \in \Pi$ be any positive path $\pi \in \Pi$. We associate to π a linear function

$$\lambda_\pi : L(E) \to \mathbf{Z}$$

by setting $\lambda_\pi(e)=1$ if the path π crosses the edge e, and $\lambda_\pi(e)=0$ if it does not.

Remark 3.2.7. If the path π enters a box $b \in B$, then it does so by crossing an edge which is part of the opposite corner C_b^-, and it has to leave b by crossing an edge which is part of the corner C_b^-. It follows that the corresponding functional λ_π is zero on $\mathrm{Ker}(\partial)=H_1(\Gamma)$, and hence it descends to a functional on $L(E)/\mathrm{Ker}(\partial)=\mathrm{Im}(\partial)$, still denoted by λ_π. By definition, λ_π is a lattice point in the dual cone $C^* \subset \mathrm{Im}(\partial)^* \otimes \mathbf{R}$.

THEOREM 3.2.8. *The semigroup of lattice points in C^* is minimally generated by the set of all λ_π, where π runs over the set Π of positive paths.*

Proof. Let $\lambda: L(E) \to \mathbf{Z}$ with $\lambda|_{\mathrm{Ker}(\partial)}=0$ and $\lambda(e) \geqslant 0$ for all $e \in E$. We define the *weight* of λ to be

$$w(\lambda) = \sum_{e \in E} \lambda(e).$$

It is clear that $w(\lambda) \geqslant 0$, and that $w(\lambda)=0$ if and only if $\lambda=0$. Note also that $w(\lambda_\pi)=n$ for all $\pi \in \Pi$.

The statement of the theorem will be proved if we show that $w(\lambda) \geqslant n$ for all nonzero integral linear functions $\lambda: L(E) \to \mathbf{Z}$ with $\lambda|_{\mathrm{Ker}(\partial)}=0$ and $\lambda(e) \geqslant 0$ for all $e \in E$, and, moreover, any such λ is a nonnegative integral linear combination of λ_π $(\pi \in \Pi)$.

By Proposition 2.1.9, the requirement $\lambda|_{\mathrm{Ker}(\partial)}=0$ is equivalent to $\lambda(\varrho_b)=0$ for all $b \in B$, or

$$\sum_{e \in C_b} \lambda(e) = \sum_{e \in C_b^-} \lambda(e) \quad \text{for all } b \in B.$$

As in the proof of Proposition 3.2.4, the above condition implies that if $\lambda \neq 0$, then there exists a roof $\mathcal{R}_i$ containing an edge e on which λ is nonzero (hence $\lambda(e) \geqslant 1$). We start to construct a positive path π_i from O_i by choosing its edges in such a way that e is the first edge of the graph Γ intersected by π_i. Let $b \in B$ be a box containing e in its opposite corner (i.e., $e \in C_b^-$). Since $\lambda(\varrho_b)=0$, there must be an edge $f \in C_b$ such that $\lambda(f)>0$ (hence $\lambda(f) \geqslant 1$). We prolong the path π_i through f and enter a next box b', for which $f \in C_{b'}^-$. Again there must exist an edge $g \in C_{b'}$ such that $\lambda(g) \geqslant 1$, etc. Continuing this process, we eventually obtain a positive path π_i, which only crosses edges e of Γ having the property $\lambda(e) \geqslant 1$. This shows that $\lambda' := \lambda - \lambda_{\pi_i}$ is again an integral nonnegative linear functional on C. On the other hand,

$$w(\lambda') = w(\lambda) - w(\lambda_{\pi_i}) = w(\lambda) - n.$$

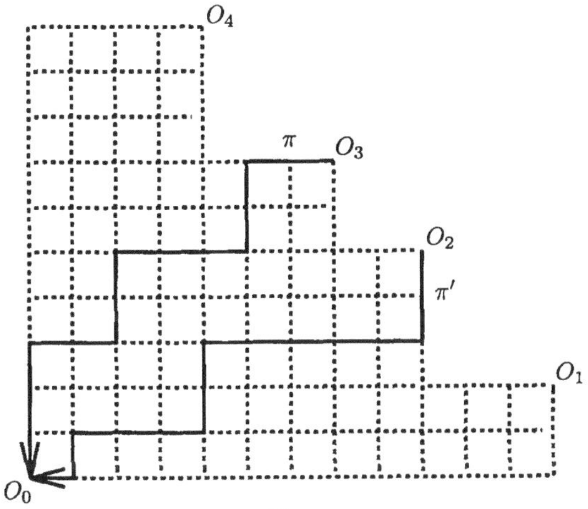

Fig. 6

Since $w(\lambda')\geqslant 0$, this shows that $w(\lambda)\geqslant n$. By induction on $w(\lambda)$, we can assume that λ' is already a nonnegative integral linear combination of λ_π, and hence so is $\lambda=\lambda'+\lambda_{\pi_i}$. □

Definition 3.2.9. We define a partial ordering on the set Π of positive paths by declaring that $\pi\geqslant\pi'$ if the path π runs above the path π'. See Figure 6.

Remark 3.2.10. It is easy to see that the set Π of positive paths together with the above partial ordering is a distributive lattice. The maximum $\max(\pi,\pi')$ for any two paths π and π' is the path bounding the union of the regions under π and π'; similarly, $\min(\pi,\pi')$ bounds the intersection of these regions.

Definition 3.2.11. Consider the partition of the set of independent variables $\{z_\pi\}_{\pi\in\Pi}$ into l disjoint subsets

$$\{z_\pi\}_{\pi\in\Pi_i}, \quad i=1,\dots,l,$$

and define $X=X(n_1,\dots,n_l,n)$ to be the subvariety of

$$\mathbf{P}^{N_1-1}\times\mathbf{P}^{N_2-1}\times\dots\times\mathbf{P}^{N_l-1}$$

given by the l-homogeneous quadratic equations

$$z_\pi z_{\pi'}-z_{\min(\pi,\pi')}z_{\max(\pi,\pi')}=0, \tag{4}$$

for all pairs of noncomparable elements $\pi,\pi'\in\Pi$.

The variety X has been investigated by N. Gonciulea and V. Lakshmibai in the papers [18], [19], where the following result has been proved:

THEOREM 3.2.12. (i) *$X(n_1,...,n_l,n)$ is a $|D|$-dimensional, irreducible, normal, toric variety.*

(ii) *There exists a flat deformation*

$$\varrho\colon \mathcal{X} \to \operatorname{Spec}(\mathbf{C}[t])$$

such that $\varrho^{-1}(0)=X(n_1,...,n_l,n)$ and $\varrho^{-1}(t)=F(n_1,...,n_l,n)$ for all $t\neq 0$.

The next theorem describes an isomorphism

$$X(n_1,...,n_l,n)\cong P(n_1,...,n_l,n).$$

THEOREM 3.2.13. *Let $P=P(n_1,...,n_l,n)$ be the toric variety associated with a partial flag manifold $F=F(n_1,...,n_l,n)$. The line bundles $\mathcal{L}_i$ $(i=1,...,l)$ define an embedding*

$$\psi\colon P\hookrightarrow \mathbf{P}^{N_1-1}\times\mathbf{P}^{N_2-1}\times...\times\mathbf{P}^{N_l-1},$$

whose image coincides with the toric variety $X(n_1,...,n_l,n)$.

Proof. We have $X=\operatorname{Proj}(\mathbf{C}[z_\pi;\pi\in\Pi]/\mathcal{I})$, with $\mathcal{I}$ the ideal generated by the quadratic polynomials in (4), and Proj is taken with respect to the $\mathbf{Z}^l$-grading given by

$$\deg(z_\pi)=(0,...,0,\overset{i}{1},0,...,0),\quad \text{if } \pi\in\Pi_i.$$

If we identify Π with the set $\{\lambda_\pi,\pi\in\Pi\}\subset\operatorname{Im}(\partial)^*$, then $\mathcal{I}$ is the *toric ideal* (see the definition in [28, p. 31]) associated to this set (this is a standard fact about the ideals associated to distributive lattices; see for example Theorem 4.3 in [18] for a proof). Let Y be the affine toric variety $\operatorname{Spec}(\mathbf{C}[z_\pi;\pi\in\Pi]/\mathcal{I})$. By Theorem 3.2.8 and Proposition 13.5 in [28], Y coincides with the affine toric variety defined by the cone $C\subset\operatorname{Im}(\partial)\otimes\mathbf{R}$, i.e., $\mathbf{C}[z_\pi;\pi\in\Pi]/\mathcal{I}$ can be identified with the ring $\mathbf{C}[S_C]$ determined by the semigroup S_C of lattice points in the dual cone C^*.

Pick an edge $e_i\in\mathcal{R}_i$ for each $1\leqslant i\leqslant l$, and identify the line bundle $\mathcal{L}_i$ with $\mathcal{O}(\sum_{f\in U(e_i)}H_f)$. For each $1\leqslant i\leqslant l$, let $\Delta[e_i]\subset L(D)^*\otimes\mathbf{R}$ be the supporting polyhedron for the global sections of the line bundle $\mathcal{L}_i$ (cf. Proposition 3.2.4); recall that the lattice points in $\Delta[e_i]$ are given by the linear functions $\pi[e_i]$ $(\pi\in\Pi_i)$ defined in the proof of Lemma 3.2.2. Define now for each i a linear function $v[e_i]\colon L(E)\to\mathbf{Z}$ by

$$v[e_i](f)=\begin{cases}1 & \text{if } f\in U(e),\\ 0 & \text{otherwise.}\end{cases}$$

It is clear that $v[e_i]$ descends to a functional on $\operatorname{Im}(\partial)$, and that for every path $\pi\in\Pi_i$ the functional λ_π in Definition 3.2.6 coincides with $\pi[e_i]+v[e_i]$. For each $1\leqslant i\leqslant l$, let

$\sigma_i \subset \mathrm{Im}(\partial)^* \otimes \mathbf{R}$ be the cone over the translated polyhedron $v[e_i]+\Delta[e_i]$. Then the Minkowski sum $\sigma := \sigma_1 + \ldots + \sigma_l$ of these cones coincides with the cone C^*, since both σ and C^* are generated by the vectors $\{\lambda_\pi, \pi \in \Pi\}$. It follows that $P \cong \mathrm{Proj}(\mathbf{C}[S_C])$, where Proj is taken with respect to the natural $\mathbf{Z}^l$-grading induced by the decomposition of C^* into the Minkowski sum of the σ_i.

For each $1 \leqslant i \leqslant l$, choose an ordering $\{\pi_{i,1}, \pi_{i,2}, \ldots, \pi_{i,N_i}\}$ of the set Π_i. Let $s_{\pi_{i,j}} \in H^0(P, \mathcal{L}_i)$ denote the section determined by $\pi_{i,j}$ $(i=1, \ldots, l,\ j=1, \ldots, N_i)$. The line bundles $\mathcal{L}_1, \ldots, \mathcal{L}_l$ define a morphism

$$\psi: P \to \mathbf{P}^{N_1-1} \times \mathbf{P}^{N_2-1} \times \ldots \times \mathbf{P}^{N_l-1},$$
$$x \mapsto ([s_{\pi_{1,1}}(x) : \ldots : s_{\pi_{1,N_1}}(x)], \ldots, [s_{\pi_{l,1}}(x) : \ldots : s_{\pi_{l,N_l}}(x)]).$$

By the above arguments, ψ is the isomorphism

$$\mathrm{Proj}(\mathbf{C}[S_C]) \to \mathrm{Proj}(\mathbf{C}[z_\pi; \pi \in \Pi]/\mathcal{I}),$$

and the theorem is proved. □

From Theorems 3.2.12 and 3.2.13 we obtain

COROLLARY 3.2.14. *There exists a flat deformation*

$$\varrho: \mathcal{X} \to \mathrm{Spec}(\mathbf{C}[t])$$

such that $\varrho^{-1}(0) = P(n_1, \ldots, n_l, n)$ *and* $\varrho^{-1}(t) = F(n_1, \ldots, n_l, n)$ *for all* $t \neq 0$.

Remark 3.2.15. A description of the singular locus of P was conjectured by N. Gonciulea and V. Lakshmibai in the case when F is a Grassmannian (see [19]). Our Theorem 3.1.4 proves this conjecture and its generalization for arbitrary partial flag manifolds F.

4. Quantum differential systems

4.1. Quantum $\mathcal{D}$-module

In order to explain our mirror construction, we give a short overview of the quantum cohomology $\mathcal{D}$-module. The reader is referred to [15], [23] for details.

Let V be a smooth projective variety. Denote by $\{T_a\}_a$ and $\{T^a\}_a$ two homogeneous bases of $H^*(V, \mathbf{Q})$, dual with respect to the Poincaré pairing, i.e., such that

$$\langle T_a, T^b \rangle = \delta_{a,b}.$$

We will consider only the even-degree part of $H^*(V, \mathbf{Q})$ and will assume that $H^2(V, \mathbf{Z})$ and $H_2(V, \mathbf{Z})$ are torsion-free. We denote by 1 the fundamental class of V.

To simplify the exposition, suppose that there is a basis $\{p_i, i=1,2,...,l\}$ of $H^2(V, \mathbf{Z})$ consisting of nef-divisors. Let $NE(V)$ be the Mori cone of V.

Introduce formal parameters q_i, $i=1,...,l$, and let $\mathbf{Q}[[q_1,...,q_l]]$ be the ring of formal power series. The small quantum cohomology ring of V will be denoted by $QH^*(V)$. This is the free $\mathbf{Q}[[q_1,...,q_l]]$-module $H^*(V, \mathbf{Q})\otimes_{\mathbf{Q}}\mathbf{Q}[[q_1,...,q_l]]$, together with a new multiplication given by

$$T_a \circ T_b = \sum_{\beta\in NE(V)} \prod_{i=1}^{l} q_i^{\langle p_i,\beta\rangle} \Big(\sum_c I^V_{3,\beta}(T_aT_bT_c)T^c\Big),$$

with $I^V_{3,\beta}(T_aT_bT_c)$ the 3*-point, genus-*0*, Gromov–Witten invariants* of V.

Remark 4.1.1. For the case of a partial flag manifold, the small quantum cohomology ring is well understood. A presentation of this ring is known ([1], [20], [21]), as well as explicit formulas for quantum multiplication ([11]).

The operators of quantum multiplication with the generators p_i give the *quantum differential system*, a consistent first-order partial differential system (see e.g. [15]):

$$\hbar\frac{\partial}{\partial t_i}\vec{S} = p_i\circ\vec{S}, \quad i=1,...,l,$$
$$\hbar\frac{\partial}{\partial t_0}\vec{S} = 1\circ\vec{S},$$

where $\vec{S}$ is an $H^*(V, \mathbf{Q})$-valued function in formal variables t_0 and $t_i = \log q_i$, $i=1,...,l$. Here $\hbar$ is an additional parameter.

Remarkably, a complete set of solutions to this system can be written down explicitly in terms of the so-called *gravitational descendants* [15]:

$$\vec{S}^V_a := e^{t_0/\hbar}\Big(e^{pt/\hbar}T_a + \sum_{\beta\in NE(V)-0} q^{\langle p,\beta\rangle}\sum_b T^b\int_{[\overline{M}_{0,2}(V,\beta)]}\frac{e_1^*(e^{pt/\hbar}T_a)}{\hbar - c}\cup e_2^*(T_b)\Big).$$

Here $\overline{M}_{0,2}(V,\beta)$ is Kontsevich's space of stable maps, with evaluation morphisms e_1, e_2: $\overline{M}_{0,2}(V,\beta)\to V$ at the two marked points, $[\overline{M}_{0,2}(V,\beta)]$ is the *virtual fundamental class* ([8], [25]), and c is the first Chern class of the line bundle over $\overline{M}_{0,2}(V,\beta)$ given by the cotangent line at the first marked point. Finally, pt and $q^{\langle p,\beta\rangle}$ are shorthand notations for $\sum_i p_it_i$ and $\prod_i q_i^{\langle p_i,\beta\rangle}$ respectively.

The *quantum $\mathcal{D}$-module* of V is the $\mathcal{D}$-module generated by the functions $\langle\vec{S}, 1\rangle$ for all solutions $\vec{S}$ to the above differential system.

A general conjecture about the structure of quantum $\mathcal{D}$-modules is Givental's version of the mirror conjecture [16]:

CONJECTURE 4.1.2. *There exists a family $(M_q, \mathcal{F}_q, \omega_q)$ of (possibly noncompact) complex manifolds M_q, having the same dimension as V, together with holomorphic functions $\mathcal{F}_q$ and holomorphic volume forms ω_q such that the $\mathcal{D}$-module generated by integrals*

$$\int_{\gamma\subset M_q} e^{(\mathcal{F}_q+t_0)/\hbar}\omega_q,$$

where γ are suitable Morse-theoretic middle-dimension cycles of the function $\mathrm{Re}(\mathcal{F}_q)$, *is equivalent to the quantum $\mathcal{D}$-module of V.*

4.2. Complete intersections

Now assume that V is Fano. Let X be the zero-locus of a generic section of a decomposable rank-r vector bundle

$$\mathcal{E} = \bigoplus_{j=1}^{r} L_j,$$

such that each L_j is generated by global sections. In such a situation one can also define a quantum ring $QH^*(\mathcal{E})$ over the coefficient ring $\mathbf{Q}[[q_1,...,q_l]]$ which encodes some of the enumerative geometry of rational curves on the complete intersection X. This leads to a quantum differential system for $(V,\mathcal{E})$ (see [17], [23]). We define degrees of q_i's by requiring that

$$c_1(TV)-c_1(\mathcal{E})=\sum(\deg q_i)p_i.$$

Furthermore, we suppose that all degrees of q_i are nonnegative (this is equivalent to the condition that $-K_X$ is nef). One can write down a similar complete set of solutions to the quantum differential system for $(V,\mathcal{E})$ [17]:

$$\vec{S}_a^{\mathcal{E}} := e^{t_0/\hbar}\left(e^{pt/\hbar}T_a + \sum_{\beta\in NE(V)-0} q^{\langle p,\beta\rangle}\sum_b T^b\int_{[\overline{M}_{0,2}(V,\beta)]}\frac{e_1^*(e^{pt/\hbar}T_a)}{\hbar-c}\cup e_2^*(T_b)\cup E_\beta\right),$$

where E_β is the Euler class of the vector bundle on $\overline{M}_{0,2}(V,\beta)$ whose fibre over a point $(C,\mu;x_1,x_2)$ is the subspace of $H^0(\mu^*\mathcal{E})$ consisting of sections vanishing at x_2, and the rest of the notations are as above.

Consider the cohomology-valued functions

$$S_V := \sum_a \langle \vec{S}_a^V, 1\rangle T^a$$

and

$$S_{\mathcal{E}} := \sum_a \langle \vec{S}_a^{\mathcal{E}}, c_r(\mathcal{E})\rangle T^a.$$

These functions are given explicitly by the expressions

$$S_V = e^{(t_0+pt)/\hbar}\left(1+\sum_{\beta\in NE(V)-0} q^{\langle p,\beta\rangle}(e_1)_*\left(\frac{1}{\hbar-c}\right)\right)$$

and

$$S_{\mathcal{E}} = e^{(t_0+pt)/\hbar}\left(c_r(\mathcal{E})+\sum_{\beta\in NE(V)-0} q^{\langle p,\beta\rangle}(e_1)_*\left(\frac{e_1^*(c_r(\mathcal{E}))\cdot E'_\beta}{\hbar-c}\right)\right),$$

where now E'_β is the Euler class of the vector bundle on $\overline{M}_{0,2}(V,\beta)$ whose fibre over a point $(C,\mu;x_1,x_2)$ is the subspace of $H^0(\mu^*\mathcal{E})$ consisting of sections vanishing at x_1.

Remark 4.2.1. If we view X as an abstract variety, the general theory in §4.1 gives an $H^*(X,\mathbf{Q})$-valued function S_X. The functions S_X and $S_{\mathcal{E}}$ are closely related. For example, if $i^*\colon H^2(V,\mathbf{Z})\tilde{\to} H^2(X,\mathbf{Z})$, where $i\colon X\hookrightarrow V$ is the inclusion, then $i_*(S_X)=S_{\mathcal{E}}$.

Now consider a new cohomology-valued function

$$I_{\mathcal{E}} = e^{(t_0+pt)/\hbar}\left(c_r(\mathcal{E})+\sum_{\beta\in NE(V)-0} q^{\langle p,\beta\rangle}\prod_j\prod_{m=0}^{\langle c_1(L_j),\beta\rangle}(c_1(L_j)+m\hbar)(e_1)_*\left(\frac{1}{\hbar-c}\right)\right).$$

In general it is very hard to compute S_X or $S_{\mathcal{E}}$ explicitly. However, note that $I_{\mathcal{E}}$ can be computed directly from the function S_V associated to the ambient manifold, which in many cases turns out to be more tractable. It is therefore extremely useful to have a result relating $S_{\mathcal{E}}$ and $I_{\mathcal{E}}$. Extending ideas of Givental, B. Kim [23] has recently proved the following theorem, which applies to the cases considered in this paper:

THEOREM 4.2.2. *If V is a homogeneous space and $X\subset V$ is the zero-locus of a generic section of a nonnegative decomposable vector bundle $\mathcal{E}$, then $S_{\mathcal{E}}$ and $I_{\mathcal{E}}$ coincide up to a weighted homogeneous triangular change of variables:*

$$t_0\to t_0+f_0(q)\hbar+f_{-1}(q),\quad \log q_i\to\log q_i+f_i(q),\quad i=1,\dots,l,$$

where $f_{-1},f_0,f_1,\dots,f_l$ are weighted homogeneous formal power series supported in $NE(V)-0$, with $\deg f_{-1}=1$ *and* $\deg f_i=0$, $i=0,1,\dots,l$.

In particular, this implies that the coefficient Φ_V of the cohomology class $1\in H^*(V,\mathbf{Q})$ in S_V, and the coefficient Φ_X of $c_r(\mathcal{E})$ in $I_{\mathcal{E}}$ (specialized to $\hbar=1$, $t_0=0$) are related in a very simple way. Namely, if

$$\Phi_V=\sum_{\beta\in NE(V)-0} a_\beta q^{\langle p,\beta\rangle},\quad \Phi_X=\sum_{\beta\in NE(V)-0} b_\beta q^{\langle p,\beta\rangle},$$

then

$$b_\beta = a_\beta \prod_{i=1}^{r} (\langle c_1(L_i), \beta \rangle !). \tag{5}$$

We will refer to Theorem 4.2.2 as the *quantum hyperplane section theorem.* The relation (5) above was called the "trick with factorials" in [6].

5. The mirror construction

In this section we give a partially conjectural mirror construction for partial flag manifolds, and use it to obtain an explicit hypergeometric series as the power-series expansion of the integral representation. The case of Calabi–Yau complete intersections is then discussed in some detail.

5.1. Hypergeometric solutions for partial flag manifolds

Let $F=F(n_1,...,n_l,n)$ be a partial flag manifold. In the notations of §2, we introduce l independent variables q_i, $i=1,2,...,l$ (each q_i corresponds to the roof $\mathcal{R}_i$), $|B|$ independent variables $\tilde{q}_b$, $b\in B$, and $|E|$ independent variables y_e, $e\in E$. Consider the following set of algebraically independent polynomial equations:

(i) Roof equations: for $i=1,2,...,l$,

$$\mathcal{F}_i := \prod_{e\in\mathcal{R}_i} y_e - q_i = 0. \tag{6}$$

(ii) Box equations: for $b=\{e,f,g,h\}\in B$,

$$\mathcal{G}_b := y_e y_f - \tilde{q}_b y_g y_h = 0, \tag{7}$$

where $\{e,f\}=\mathcal{C}_b$.

This set of equations was discussed by Givental [16], and was used to give an integral representation for the solutions to the quantum cohomology differential equations for the special case of complete flag manifolds. The results in that paper were the starting point for our investigations. We describe below Givental's result and our (conjectural) generalization to a general partial flag manifold.

Let $\mathbf{A}^{|E|}$ be the complex affine space with the coordinates y_e ($e\in E$). For fixed parameter values of

$$(q,\tilde{q}) := (q_1,...,q_l,...,\tilde{q}_b,...)$$

we obtain an affine variety

$$M_{q,\tilde{q}} := \{u\in\mathbf{A}^{|E|} : \mathcal{F}_i = 0, i=1,...,l, \text{ and } \mathcal{G}_b = 0, b\in B\}.$$

If all components of $(q, \tilde{q})$ are nonzero, $M_{q,\tilde{q}}$ is isomorphic to the torus $(\mathbf{C}^*)^{|D|}$.

One can define on $M_{q,\tilde{q}}$ a holomorphic volume form

$$\omega_{q,\tilde{q}} := \operatorname{Res}_{M_{q,\tilde{q}}}\left(\frac{\Omega}{\prod_{i=1}^{l} \mathcal{F}_i \prod_{b\in B} \mathcal{G}_b}\right),$$

where

$$\Omega := \bigwedge_{e\in E} dy_e.$$

Let $\mathcal{F} = \sum_{e\in E} y_e$. Consider the integral

$$I_\gamma(q, \tilde{q}) := \int_\gamma e^{\mathcal{F}} \omega_{q,\tilde{q}},$$

where $\gamma \in H_{|D|}(M_{q,\tilde{q}}, \operatorname{Re}(\mathcal{F}) = -\infty)$. We put

$$\Phi_\gamma(q_1, ..., q_l) := I_\gamma(q_1, ..., q_l, 1, 1, ..., 1).$$

We can now formulate a precise version of Conjecture 4.1.2:

CONJECTURE 5.1.1. *Let $\vec{S}$ be any solution to the quantum differential system for F. Then the component $\langle \vec{S}, 1\rangle$ can be expressed as $\Phi_\gamma(q)$ for some $\gamma \subset M_{q,1}$.*

Remark 5.1.2. This conjecture generalizes Givental's mirror theorem for complete flag manifolds [16].

Definition 5.1.3. Let W denote the set of edges in the diagram Λ that intersect Γ. We orient the vertical edges in W upwards and the horizontal edges to the right. Let $V := B \cup \{0, 1, 2, ..., l\}$. For $w \in W$, the *tail* $t(w)$ of w is defined to be the box $b_1 \in B$ where w starts. Similarly, the *head* $h(w)$ of w is the box $b_2 \in B$ where w ends. If w crosses the roof $\mathcal{R}_i$, so that its "head" is outside the graph Γ, we put $h(w) := i$, and if the "tail" of w is outside Γ, we put $t(w) = 0$. In the sense of duality of planar graphs, the graph with vertices V, edges W and incidence given by $h, t: W \to V$ is *dual* to the graph Γ with all stars collapsed to one point.

Definition 5.1.4. For each cone $\sigma \in \widehat{\Sigma}$ of maximal dimension we define a cycle $\gamma = \gamma_{q,\tilde{q}}(\sigma)$ in $M_{q,\tilde{q}}$ by

$$\gamma := \{y \in M_{q,\tilde{q}} : |y_e| = 1 \text{ for all } e \in E \text{ with } \delta(e) \in \sigma\}.$$

Note that the y_f with $\delta(f) \notin \sigma$ are determined uniquely by the y_e with $\delta(e) \in \sigma$ and the roof and box equations (6), (7).

The cycle γ is a real torus, of dimension equal to $\dim_{\mathbf{C}}(M_{q,\tilde{q}}) = \dim_{\mathbf{C}}(F)$. Since it is defined over the entire family of the $M_{q,\tilde{q}}$, it is invariant under monodromy. The integral over this special cycle will be denoted by $I(q, \tilde{q})$.

Definition 5.1.5. The specialization $\Phi_F(q) := I(q_1, ..., q_l, 1, ..., 1)$ is called the *hypergeometric series of the partial flag manifold F*.

It turns out that $I(q, \tilde{q})$ has a nice power-series expansion.

THEOREM 5.1.6.

$$I(q,\tilde{q}) = \sum_{m_1,\dots,m_l,\dots,m_b,\dots} A_{m_1,\dots,m_l,\dots,m_b,\dots}\, q_1^{m_1}\dots q_l^{m_l} \prod_{b\in B} \tilde{q}_b^{m_b},$$

with

$$A_{m_1,\dots,m_l,\dots,m_b,\dots} := \frac{1}{(m_1!)^{k_1+k_2}} \cdot \frac{1}{(m_2!)^{k_2+k_3}} \cdot \dots \cdot \frac{1}{(m_l!)^{k_l+k_{l+1}}} B_{m_1,\dots,m_l,\dots,m_b,\dots},$$

$$B_{m_1,\dots,m_l,\dots,m_b,\dots} := \prod_{w\in W} \binom{m_{h(w)}}{m_{t(w)}}.$$

Proof. By Leray's theorem, the integral is equal to

$$\int_{T(\gamma_{q,\tilde{q}}(\sigma))} e^{\mathcal{F}} \frac{\Omega}{\prod_{i=1}^{l} \mathcal{F}_i \prod_{b\in B} \mathcal{G}_b},$$

where T is the tube map. For $|q|<1$, $|\tilde{q}|<1$, the cycle $T(\gamma_{q,\tilde{q}}(\sigma))$ is homologous to the cycle

$$T := \{y \in \mathbf{A}^{|E|} : |y_e| = 1 \text{ for all } e \in E\}$$

in the complement of the hypersurfaces $y_e = 0$. We now expand all the terms in the integrand:

$$e^{\mathcal{F}} = \sum_{d=0}^{\infty} \frac{1}{d!} \mathcal{F}^d = \sum_{d_e \geqslant 0} \frac{\prod_{e\in E} y_e^{d_e}}{\prod_{e\in E} d_e!},$$

$$\frac{1}{\mathcal{F}_i} = \frac{1}{\prod_{e\in\mathcal{R}_i} y_e} \sum_{m_i\geqslant 0} \left(\frac{q_i}{\prod_{e\in\mathcal{R}_i} y_e}\right)^{m_i},$$

$$\frac{1}{\mathcal{G}_b} = \frac{1}{y_e y_f} \sum_{m_b\geqslant 0} \left(\frac{\tilde{q}_b y_g y_h}{y_e y_f}\right)^{m_b},$$

where $\{e, f\}$ makes up the corner and $\{g, h\}$ the opposite corner of the box $b = \{e, f, g, h\}$. The integral picks up precisely the constant coefficient of the following power series in the y_e's, with parameters the q's and $\tilde{q}$'s:

$$\sum_{d_e, m_i, m_b \geqslant 0} \frac{\prod_{e\in E} y_e^{d_e}}{\prod_{e\in E} d_e!} \prod_{i=1}^{l} \left(\frac{q_i}{\prod_{e\in\mathcal{R}_i} y_e}\right)^{m_i} \prod_{b\in B} \left(\frac{\tilde{q}_b y_g y_h}{y_e y_f}\right)^{m_b}.$$

Now there are three types of edges.

Type I. $e \in \mathcal{R}_i$ for some $i=1,\dots,l$. Then e is also edge of the opposite corner of a unique box b. Only the terms with

$$d_e = m_i - m_b$$

will give a contribution.

Type II. $e \in b \cap b'$ for two boxes b and b'. We can then assume that e is part of the corner of b, and the opposite corner of b'. Only the terms with

$$d_e = m_b - m_{b'}$$

will give a contribution.

Type III. e is contained in a unique $b \in B$. In this case e is part of the corner of b. Only the terms with

$$d_e = m_b$$

will give a contribution.

Hence we see that the integral is given by the series

$$\sum_{m_i, m_b \geqslant 0} \frac{1}{\prod_{e\in E} d_e!} \prod_{i=1}^{l} q_i^{m_i} \prod_{b \in B} \tilde{q}_b^{m_b},$$

where for each edge the number d_e is determined by the m_i and m_b by the above equations. We can rewrite this coefficient nicely in terms of binomial coefficients as follows. Each edge $w \in W$ of the diagram Λ intersects precisely one edge $e \in E$ of Type I or Type II. The corresponding coefficient d_e is then given by

$$d_e = m_{h(w)} - m_{t(w)}.$$

Trivially,

$$\frac{1}{\prod_{e\in E} d_e!} = \frac{\prod_{w\in W} m_{h(w)}!}{\prod_{w\in W} m_{h(w)}! \prod_{e\in E} d_e!}.$$

The heads of arrows $w \in W$ which are *not* tails are the heads of arrows intersecting the edges of Type I. The tails of arrows $w \in W$ which are not heads are in bijection to the edges of Type III. Hence, when we pull out a factor $\prod_{i=1}^{l} \prod_{e\in\mathcal{R}_i} m_e!$ from the denominator of the left-hand side of the above equality, the other terms in the numerator and the denominator can precisely be combined into the product

$$\prod_{w\in W} \binom{m_{h(w)}}{m_{t(w)}}.$$

This proves the result. □

Remark 5.1.7. Note that $I(q,\tilde{q})$ is the generalized hypergeometric series for the smooth toric variety $\widehat{P}$ defined in §3. The parameters $q_1, ..., q_l$ correspond to the generators of $\mathrm{Pic}(\widehat{P})$ coming from the singular variety P (the pullbacks of the line bundles $\mathcal{L}_1, ..., \mathcal{L}_l$), while $\tilde{q}_b$ correspond to the additional generators of $\mathrm{Pic}(\widehat{P})$.

Theorem 5.1.6 shows that it is very easy to write down the power-series expansion for $I(q,\tilde{q})$ directly from the diagram.

Example 5.1.8. $F(2,5)$ (the Grassmannian of 2-planes in $\mathbf{C}^5$):

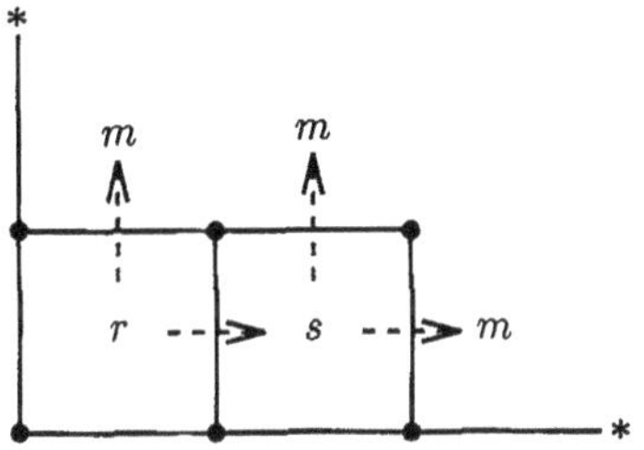

Hence we read off:

$$I(q,\tilde{q}) = \sum_{m,r,s\geqslant 0} \frac{1}{(m!)^5}\binom{s}{r}\binom{m}{r}\binom{m}{s}^2 q^m \tilde{q}_1^r \tilde{q}_2^s.$$

Example 5.1.9. $F(3,6)$ (the Grassmannian of 3-planes in $\mathbf{C}^6$):

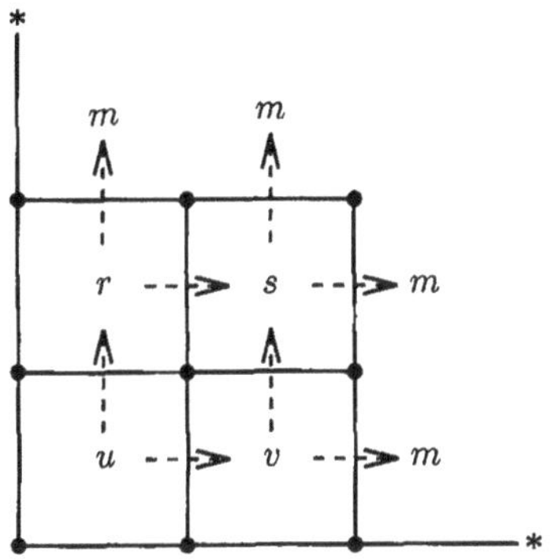

Hence we read off:

$$I(q,\tilde{q}) = \sum_{m,r,s,u,v} \frac{1}{(m!)^6}\binom{r}{u}\binom{v}{u}\binom{s}{r}\binom{s}{v}\binom{m}{r}\binom{m}{s}^2\binom{m}{v} q^m \tilde{q}_1^r \tilde{q}_2^s \tilde{q}_3^u \tilde{q}_4^v.$$

Example 5.1.10. $F(1,2,3,4)$ (the variety of complete flags in $\mathbf{C}^4$):

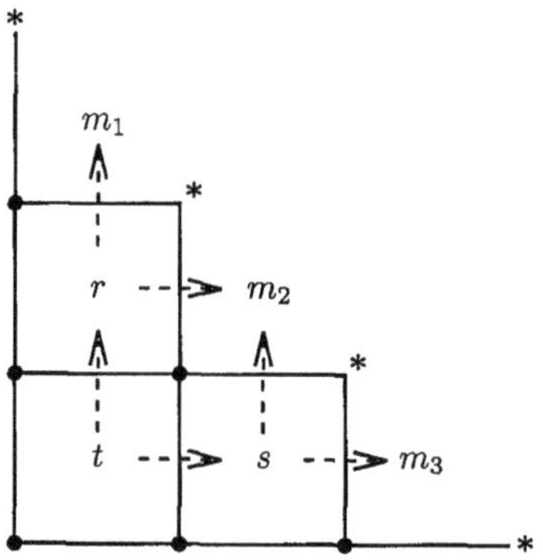

Hence we read off:

$$I(q,\tilde{q}) = \sum_{m_1,m_2,m_3,r,s,t} A_{m_1,m_2,m_3,r,s,t}\, q_1^{m_1} q_2^{m_2} q_3^{m_3} \tilde{q}_1^r \tilde{q}_2^s \tilde{q}_3^t,$$

with

$$A_{m_1,m_2,m_3,r,s,t} = \frac{1}{(m_1!)^2(m_2!)^2(m_3!)^2} \binom{r}{t}\binom{s}{t}\binom{m_1}{r}\binom{m_2}{r}\binom{m_2}{s}\binom{m_3}{s}.$$

A weaker version of Conjecture 5.1.1 is

CONJECTURE 5.1.11. *The series $\Phi_F := I(q,1)$ is the coefficient of the cohomology class 1 in the $H^*(F,\mathbf{Q})$-valued function S_F describing the quantum $\mathcal{D}$-module of F, i.e.,*

$$\Phi_F = 1 + \sum_{\overline{m}:=(m_1,\dots,m_l)\neq 0} \left(\int_{\overline{M}_{0,2}(F,\overline{m})} \frac{e_1^*(e^{Ct}\Omega_F)}{1-c} \cup e_2^*(1) \right) q_1^{m_1} \dots q_l^{m_l},$$

where Ct stands for $C_1t_1 + \dots + C_lt_l$, with $\{C_1,\dots,C_l\}$ the Schubert basis of $H^2(F,\mathbf{Q})$, and Ω_F is the cohomology class of a point.

Remark 5.1.12. (i) Besides the case of complete flag manifolds (cf. Remark 5.1.2), there is another case for which the above conjecture agrees with previously known results. Consider the partial flag manifold $F := F(1,n-1,n)$ of flags $V^1 \subset V^{n-1} \subset \mathbf{C}^n$. The Plücker embedding identifies F with a $(1,1)$-hypersurface in $\mathbf{P}^{n-1} \times \mathbf{P}^{n-1}$. The hypergeometric series for $\mathbf{P}^{n-1} \times \mathbf{P}^{n-1}$ is

$$\sum_{m_1,m_2 \geqslant 0} \frac{1}{(m_1!)^n (m_2!)^n} q_1^{m_1} q_2^{m_2}$$

(cf. [17]), and by the quantum hyperplane section theorem ([17], [23]) we obtain that the hypergeometric series for F is

$$\sum_{m_1,m_2\geqslant 0} \frac{(m_1+m_2)!}{(m_1!)^n(m_2!)^n} q_1^{m_1} q_2^{m_2}. \tag{8}$$

On the other hand, the recipe of Theorem 5.1.6 gives the formula

$$\sum_{m_1,m_2\geqslant 0} \frac{\sum_s \binom{m_1}{s}\binom{m_2}{s}}{(m_1!)^{n-1}(m_2!)^{n-1}} q_1^{m_1} q_2^{m_2} \tag{9}$$

for the hypergeometric series of F. The identity

$$\sum_s \binom{m_1}{s}\binom{m_2}{s} = \binom{m_1+m_2}{m_1}$$

implies that the series (8) and (9) coincide.

(ii) The quantum Pieri formula [11] gives explicitly the quantum product of a special Schubert class with a general one, and in particular the quantum product of a Schubert divisor with any other Schubert class. Using this, one can write down in reasonably low-dimensional cases the quantum differential system for F, and reduce this first-order system to higher-order differential equations satisfied by the components. In particular, one can write down the differential operators annihilating the component $\langle \vec{S}, 1\rangle$ of any solution $\vec{S}$, and check by direct computation that the hypergeometric series $\Phi_F(q)$ of Theorem 5.1.6 is annihilated by these operators. In [6] this is done for the Grassmannians containing complete-intersection Calabi–Yau 3-folds. For the complete flag manifolds, the operators are known to be the operators for the quantum Toda lattice (see [22]).

5.2. Calabi–Yau complete intersections in $F(n_1, ..., n_l, n)$

Recall that Pic(F) is generated by the line bundles $\mathcal{O}(C_i)$, $i=1,...,l$, which also generate the (closed) Kähler cone. Hence any line bundle $\mathcal{H}$ on F which is globally generated is of the form $\mathcal{O}(\bar{d}):=\mathcal{O}(\sum_{i=1}^{l} d^{(i)}C_i)$, with $d^{(i)}$ nonnegative. The common zero-locus of r general sections of the line bundles $\mathcal{O}(\bar{d}_1), ..., \mathcal{O}(\bar{d}_r)$ will be denoted by $X:=X_{\bar{d}_1,...,\bar{d}_r}$.

Assuming Conjecture 5.1.11, it follows from the quantum hyperplane section theorem that the hypergeometric series Φ_X has the expression

$$\Phi_X = \sum_m \prod_{j=1}^{r} \Big(\sum_{i=1}^{l} d_j^{(i)} m_i\Big)!\, A_{m_1,...,m_l}\, q_1^{m_1} \cdots q_l^{m_l}, \tag{10}$$

where $A_{m_1,\dots,m_l}$ are the coefficients of Φ_F in Theorem 5.1.6.

From now on the complete intersection $X_{\bar{d}_1,\dots,\bar{d}_r}$ is assumed to be a Calabi–Yau manifold. The construction of mirrors described in [6] for the case when F is a Grassmannian can be extended to the case of a general F as follows:

X can be regarded as the intersection of $F \subset \mathbf{P}^{N_1-1} \times \dots \times \mathbf{P}^{N_l-1}$ with r general hypersurfaces Z_j $(j=1,\dots,r)$ in $\mathbf{P}^{N_1-1} \times \dots \times \mathbf{P}^{N_l-1}$, with Z_j of multidegree $(d_j^{(1)},\dots,d_j^{(l)})$. Let Y be the Calabi–Yau complete intersection of the same hypersurfaces with the toric degeneration P of F.

For each edge $e \in \bigcup_{i=1}^{l} \mathcal{R}_i$ which is part of a roof, define polynomials

$$\varphi_e(y) := \sum_{f \in U(e)} c_f y_f,$$

where c_f are generically chosen complex numbers. (Recall that we have defined $U(e)$ as the set consisting of e, together with all edges in the graph Γ which are either directly below e, if e is horizontal, or directly to the left of e, if e is vertical.)

Partition each of the roofs $\mathcal{R}_i$, $i=1,\dots,l$, into r disjoint subsets

$$\mathcal{R}_i = \mathcal{R}_{i,1} \cup \dots \cup \mathcal{R}_{i,r}$$

such that $|\mathcal{R}_{i,j}| = d_i^{(j)}$. It follows from Definition 3.2.3 and Theorem 3.2.13 that the toric Weil divisor

$$\sum_{i=1}^{l} \sum_{e \in \mathcal{R}_{i,j}} H_e$$

is Cartier, and

$$\mathcal{O}\Big(\sum_{i=1}^{l} \sum_{e \in \mathcal{R}_{i,j}} H_e\Big) \cong \mathcal{L}_1^{\otimes d_j^{(1)}} \otimes \dots \otimes \mathcal{L}_l^{\otimes d_j^{(l)}}.$$

Consider the torus T in the affine space $\cong \mathbf{A}^{|E|}$ given by the following set of equations:

(i) Roof equations: for $i=1,2,\dots,l$,

$$\prod_{e \in \mathcal{R}_i} y_e = 1.$$

(ii) Box equations: for $b=\{e,f,g,h\} \in B$,

$$y_e y_f - y_g y_h = 0,$$

where $\{e,f\}$ form the corner C_b of b.

Introduce additional independent variables x_d, $d \in D$, one for each generator of the lattice $L(D)$. For every edge $e \in E$, set

$$x^{\delta(e)} := x_{h(e)} (x_{t(e)})^{-1},$$

where, as before, $h(e)$ (resp. $t(e)$) is the head (resp. tail) of e. The torus T can be identified with $\mathrm{Spec}(\mathbf{C}[x_d, x_d^{-1}; d \in D])$, with the embedding $T \hookrightarrow \mathbf{A}^{|E|}$ induced by the ring homomorphism

$$\mathbf{C}[y_e; e \in E] \to \mathbf{C}[x_d, x_d^{-1}; d \in D], \quad y_e \mapsto x^{\delta(e)}.$$

With this identification, we obtain Laurent polynomials

$$\varphi_e(x) := \sum_{f \in U(e)} c_f x^{\delta(f)}.$$

For $j = 1, \ldots, r$, let ∇_j be the Newton polyhedron of the Laurent polynomial

$$\mathcal{P}_j := 1 - \sum_{i=1}^{l} \sum_{e \in \mathcal{R}_{i,j}} \varphi_e(x).$$

The polyhedra ∇_j, $j = 1, \ldots, r$, define a nef-partition of the anticanonical class of P (see definitions in [9], [4]), and according to [7], [9], the mirror family Y^* of the Calabi–Yau complete intersection $Y \subset P$ consists of Calabi–Yau compactifications of the general complete intersections in T defined by the equations

$$1 - \sum_{i=1}^{l} \sum_{e \in \mathcal{R}_{i,j}} \varphi_e(x) = 0, \quad j = 1, \ldots, r. \tag{11}$$

CONJECTURE 5.2.1. *Let Y_0^* be a Calabi–Yau compactification of a general complete intersection in T defined by the equations* (11), *with the additional requirement that the coefficients satisfy the relation*

$$c_{f_1} c_{f_2} = c_{f_3} c_{f_4}$$

whenever $\{f_1, f_2, f_3, f_4\}$ make up a box $b \in B$, with $\{f_1, f_2\}$ forming the corner C_b of b. Then a minimal desingularization of Y_0^* is a mirror of a generic complete-intersection Calabi–Yau $X \subset F$.

The main period of the mirror Y^* of Y is given by

$$\Phi_Y = \int_\gamma \mathrm{Res}_{M_q, \bar{q}} \left(\frac{\Omega}{\prod_{j=1}^{r} \mathcal{E}_j \prod_{i=1}^{l} \mathcal{F}_i \prod_{b \in B} \mathcal{G}_b} \right),$$

where the extra factors $\mathcal{E}_j$ come from the nef-partition of the anticanonical class of P described above. Specifically,

$$\mathcal{E}_j := 1 - \sum_{i=1}^{l} \sum_{e \in \mathcal{R}_{i,j}} \sum_{f \in U(e)} y_f, \quad j = 1, ..., r.$$

By direct expansion of the integral defining Φ_Y (as in Theorem 5.1.6), followed by the specialization $\tilde{q}_b = 1$, $b \in B$, one gets exactly the hypergeometric series Φ_X.

Finally, we discuss some applications to the case when $X \subset F$ is a Calabi–Yau 3-fold.

First, as discussed in [6], our construction can be interpreted via *conifold transitions.* Indeed, by Theorem 3.1.4, if X is generic, then its degeneration $Y \subset P$ is a singular Calabi–Yau 3-fold, whose singular locus consists of finitely many nodes. The resolution of singularities $\widehat{P} \to P$ induces a small resolution $\widehat{Y} \to Y$. In other words the (nonsingular) Calabi–Yau's X and $\widehat{Y}$ are related by a conifold transition, and Conjecture 5.2.1 essentially states that their mirrors are related in a similar fashion.

Second, it is well understood (see e.g. [7]) that the knowledge of the hypergeometric series Φ_X for a Calabi–Yau 3-fold gives the virtual numbers of rational curves on X via a formal calculation. In [6] we have used the hypergeometric series (10) to compute these numbers for complete intersections in Grassmannians.

5.3. List of Calabi–Yau complete-intersection 3-folds

Recall that if $F := F(n_1, ..., n_l, n)$ is a partial flag manifold, then

$$\dim(F) = \sum_{i=1}^{l} (n_i - n_{i-1})(n - n_i). \tag{12}$$

In the Schubert basis of the Picard group, the anticanonical bundle of F is given by

$$\omega_F^{-1} = \mathcal{O}\left(\sum_{i=1}^{l} (n_{i+1} - n_{i-1}) C_i \right). \tag{13}$$

A (general) complete-intersection Calabi–Yau 3-fold in F is the common zero-locus of $r := \dim(F) - 3$ general sections $s_j \in H^0(F, \mathcal{O}(\bar{d}_j))$, where $\mathcal{O}(\bar{d}_j)$, $j = 1, 2, ..., r$, are line bundles with $\bigotimes_{j=1}^{r} \mathcal{O}(\bar{d}_j) = \omega_F^{-1}$. Hence, if F contains a complete-intersection Calabi–Yau 3-fold, then necessarily

$$\dim(F) \leqslant 3 + \sum_{i=1}^{l} (n_{i+1} - n_{i-1}) = n + n_l - n_1 + 3. \tag{14}$$

PROPOSITION 5.3.1. *If $F:=F(n_1,...,n_l,n)$ is a partial flag manifold containing a complete-intersection Calabi–Yau 3-fold, and F is not a projective space or one of the manifolds $F(1,n-1,n)$, then $n\leqslant 7$.*

Proof. Using (12), after some manipulation, one can rewrite the inequality (14) as

$$(n_1-1)(n-n_1-1)+(n_2-n_1)(n-n_2-1)+...+(n_l-n_{l-1})(n-n_l-1)\leqslant 4. \qquad (15)$$

There are two cases.

(1) $n_1>1$. Then it is easy to see that $(n_1-1)(n-n_1-1)>4$ for $n\geqslant 8$, unless $n_1=n-1$, in which case F is a projective space.

(2) $n_1=1$. If $l=1$, then F is a projective space, so we may assume $l\geqslant 2$. As above, $(n_2-1)(n-n_2-1)>4$ for $n\geqslant 8$, unless $n_2=n-1$, in which case $F=F(1,n-1,n)$. □

Remark 5.3.2. The flag manifold $F(1,n-1,n)$ sits as a $(1,1)$-hypersurface in $\mathbf{P}^{n-1}\times\mathbf{P}^{n-1}$. Hence these cases (as well as the case when F is projective space) can be viewed as particular instances of complete-intersection Calabi–Yau's in toric varieties.

We list below all the partial flag manifolds (not excluded by Proposition 5.3.1) for which the inequality (14) is satisfied. The anticanonical class of F, denoted by $-K_F$, is expressed in terms of the natural Schubert basis of the Picard group. The last column of the table below contains the possible splittings of the anticanonical class into $\dim(F)-3$ nonnegative divisors.

In general, there is a natural duality isomorphism

$$F(n_1,...,n_l,n)\cong F(n-n_l,...,n-n_1,n). \qquad (16)$$

This is taken into account by listing only one of the two isomorphic flag manifolds. It may also be that the flag manifold is self-dual, i.e., (16) is an automorphism, and two families of complete-intersection Calabi–Yau 3-folds corresponding to different splittings of the anticanonical class are interchanged by the duality automorphism. Whenever this happens (e.g., when F parametrizes complete flags), only one of the two splittings of $-K_F$ is listed.

n	F	$\dim(F)$	$-K_F$	splitting of $-K_F$
7	$F(2,7)$	10	7	7(1)
7	$F(1,2,7)$	11	(2, 6)	2(1, 0)+6(0, 1)
7	$F(1,5,7)$	14	(5, 6)	5(1, 0)+6(0, 1)
7	$F(1,2,6,7)$	15	(2, 5, 5)	2(1, 0, 0)+5(0, 1, 0)+5(0, 0, 1)
6	$F(2,6)$	8	6	(2)+4(1)
6	$F(3,6)$	9	6	6(1)
6	$F(1,2,6)$	9	(2, 5)	(2, 0)+5(0, 1)
				(1, 0)+(1, 1)+4(0, 1)
				2(1, 0)+(0, 2)+3(0, 1)
6	$F(1,3,6)$	11	(3, 5)	3(1, 0)+5(0, 1)
6	$F(1,4,6)$	11	(4, 5)	(2, 0)+2(1, 0)+5(0, 1)
				3(1, 0)+(1, 1)+4(0, 1)
				4(1, 0)+(0, 2)+3(0, 1)
6	$F(1,2,5,6)$	12	(2, 4, 4)	(2, 0, 0)+4(0, 1, 0)+4(0, 0, 1)
				(1, 0, 0)+(1, 1, 0)+3(0, 1, 0)+4(0, 0, 1)
				(1, 0, 0)+(1, 0, 1)+4(0, 1, 0)+3(0, 0, 1)
				2(1, 0, 0)+(0, 2, 0)+2(0, 1, 0)+4(0, 0, 1)
				2(1, 0, 0)+(0, 1, 1)+3(0, 1, 0)+3(0, 0, 1)
				2(1, 0, 0)+4(0, 1, 0)+(0, 0, 2)+2(0, 0, 1)
6	$F(1,3,5,6)$	13	(3, 4, 3)	3(1, 0, 0)+4(0, 1, 0)+3(0, 0, 1)
5	$F(2,5)$	6	5	(3)+2(1)
				2(2)+(1)
5	$F(1,2,5)$	7	(2, 4)	(2, 0)+(0, 2)+2(0, 1)
				(1, 0)+(1, 1)+(0, 2)+(0, 1)
				2(1, 1)+2(0, 1)
				2(1, 0)+2(0, 2)
				(1, 0)+(1, 2)+2(0, 1)
				(2, 1)+3(0, 1)
5	$F(2,3,5)$	8	(3, 3)	(1, 0)+(2, 0)+3(0, 1)
				2(1, 0)+(1, 1)+2(0, 1)
5	$F(1,3,5)$	8	(3, 4)	(3, 0)+4(0, 1)
				(1, 0)+(2, 1)+3(0, 1)
				(1, 1)+(2, 0)+3(0, 1)
				(1, 0)+2(1, 1)+2(0, 1)
				(1, 0)+(2, 0)+(0, 2)+2(0, 1)
				3(1, 0)+2(0, 2)
				3(1, 0)+(0, 1)+(0, 3)
				2(1, 0)+(1, 2)+2(0, 1)

n	F	$\dim(F)$	$-K_F$	splitting of $-K_F$
5	$F(1,2,4,5)$	9	$(2,3,3)$	$2(1,0,0)+(0,3,0)+3(0,0,1)$
				$2(1,0,0)+3(0,1,0)+(0,0,3)$
				$(2,1,0)+2(0,1,0)+3(0,0,1)$
				$(2,0,1)+3(0,1,0)+2(0,0,1)$
				$(1,2,0)+(1,0,0)+(0,1,0)+3(0,0,1)$
				$2(1,0,0)+(0,2,1)+(0,1,0)+2(0,0,1)$
				$2(1,0,0)+(0,1,2)+2(0,1,0)+(0,0,1)$
				$(1,0,0)+(1,0,2)+3(0,1,0)+(0,0,1)$
				$(1,1,1)+(1,0,0)+2(0,1,0)+2(0,0,1)$
				$(2,0,0)+(0,2,0)+(0,1,0)+3(0,0,1)$
				$(2,0,0)+(0,0,2)+3(0,1,0)+(0,0,1)$
				$2(1,0,0)+(0,2,0)+(0,1,0)+(0,0,2)+(0,0,1)$
				$2(1,1,0)+(0,1,0)+3(0,0,1)$
				$(1,1,0)+(1,0,1)+2(0,1,0)+2(0,0,1)$
				$(1,1,0)+(1,0,0)+(0,1,1)+(0,1,0)+2(0,0,1)$
				$2(1,0,0)+2(0,1,1)+(0,1,0)+(0,0,1)$
				$(1,0,0)+(1,0,1)+(0,1,1)+2(0,1,0)+(0,0,1)$
				$2(1,0,1)+3(0,1,0)+(0,0,1)$
				$(2,0,0)+(0,1,1)+2(0,1,0)+2(0,0,1)$
				$(1,1,0)+(0,2,0)+(1,0,0)+3(0,0,1)$
				$(1,0,1)+(0,2,0)+(1,0,0)+(0,1,0)+2(0,0,1)$
				$2(1,0,0)+(0,2,0)+(0,1,1)+2(0,0,1)$
				$(1,1,0)+(1,0,0)+2(0,1,0)+(0,0,2)+(0,0,1)$
				$(1,0,1)+(1,0,0)+3(0,1,0)+(0,0,2)$
				$2(1,0,0)+(0,1,1)+2(0,1,0)+(0,0,2)$
5	$F(1,2,3,5)$	9	$(2,2,3)$	$(2,0,0)+2(0,1,0)+3(0,0,1)$
				$(0,2,0)+2(1,0,0)+3(0,0,1)$
				$(0,0,2)+2(1,0,0)+2(0,1,0)+(0,0,1)$
				$(1,1,0)+(1,0,0)+(0,1,0)+3(0,0,1)$
				$(1,0,1)+(1,0,0)+2(0,1,0)+2(0,0,1)$
				$(0,1,1)+2(1,0,0)+(0,1,0)+2(0,0,1)$
5	$F(1,2,3,4,5)$	10	$(2,2,2,2)$	$(2,0,0,0)+2(0,1,0,0)+2(0,0,1,0)+2(0,0,0,1)$
				$2(1,0,0,0)+(0,2,0,0)+2(0,0,1,0)+2(0,0,0,1)$
				$(1,1,0,0)+(1,0,0,0)+(0,1,0,0)+2(0,0,1,0)+2(0,0,0,1)$
				$(1,0,0,1)+(1,0,0,0)+2(0,1,0,0)+2(0,0,1,0)+(0,0,0,1)$
				$(1,0,1,0)+(1,0,0,0)+2(0,1,0,0)+(0,0,1,0)+2(0,0,0,1)$
				$(0,1,1,0)+2(1,0,0,0)+(0,1,0,0)+(0,0,1,0)+2(0,0,0,1)$

n	F	$\dim(F)$	$-K_F$	splitting of $-K_F$
4	$F(2,4)$	4	4	(4)
4	$F(1,2,4)$	5	$(2,3)$	$(1,0)+(1,3)$
				$(1,1)+(1,2)$
				$(2,1)+(0,2)$
				$(2,2)+(0,1)$
4	$F(1,2,3,4)$	6	$(2,2,2)$	$(2,0,0)+(0,2,0)+(0,0,2)$
				$(1,1,0)+(1,0,1)+(0,1,1)$
				$(1,2,0)+(1,0,0)+(0,0,2)$
				$(1,2,0)+(1,0,1)+(0,0,1)$
				$(2,1,0)+(0,1,0)+(0,0,2)$
				$(2,1,0)+(0,1,1)+(0,0,1)$
				$(2,0,1)+(0,2,0)+(0,0,1)$
				$(2,0,1)+(0,1,1)+(0,1,0)$
				$2(1,1,0)+(0,0,2)$
				$2(1,0,1)+(0,2,0)$
				$(2,2,0)+2(0,0,1)$
				$(2,0,2)+2(0,1,0)$

References

[1] ASTASHKEVICH, A. & SADOV, V., Quantum cohomology of partial flag manifolds $F_{n_1,\dots,n_k}$. *Comm. Math. Phys.*, 170 (1995), 503–528.

[2] BATYREV, V. V., On classifications of smooth projective toric varieties. *Tôhoku Math. J.*, 43 (1991), 569–585.

[3] — Dual polyhedra and mirror symmetry for Calabi–Yau hypersurfaces in toric varieties. *J. Algebraic Geom.*, 3 (1994), 493–535.

[4] — Toric degenerations of Fano varieties and constructing mirror manifolds. *alg-geom*/9712034.

[5] BATYREV, V. V. & BORISOV, L. A., Dual cones and mirror symmetry for generalized Calabi–Yau manifolds, in *Mirror Symmetry*, Vol. II, pp. 71–86. AMS/IP Stud. Adv. Math., 1. Amer. Math. Soc., Providence, RI, 1997.

[6] BATYREV, V. V., CIOCAN-FONTANINE, I., KIM, B. & STRATEN, D. VAN, Conifold transitions and mirror symmetry for Calabi–Yau complete intersections in Grassmannians. *Nuclear Phys. B*, 514 (1998), 640–666 (*alg-geom*/9710022).

[7] BATYREV, V. V. & STRATEN, D. VAN, Generalized hypergeometric functions and rational curves on Calabi–Yau complete intersections in toric varieties. *Comm. Math. Phys.*, 168 (1995), 493–533 (*alg-geom*/9307010).

[8] BEHREND, K., Gromov–Witten invariants in algebraic geometry. *Invent. Math.*, 127 (1997), 601–617.

[9] BORISOV, L. A., Towards mirror symmetry of Calabi–Yau complete intersections in Gorenstein toric Fano varieties. *alg-geom*/9310001.

[10] CANDELAS, P., OSSA, X. DE LA, GREEN, P. & PARKES, L., A pair of Calabi–Yau manifolds as an exactly soluble superconformal theory. *Nuclear Phys. B*, 359 (1991), 21–74.

[11] CIOCAN-FONTANINE, I., On quantum cohomology rings of partial flag varieties. *Duke Math. J.*, 98 (1999), 485–524.

[12] EGUCHI, T., HORI, K. & XIONG, C.-S., Gravitational quantum cohomology. *Internat. J. Modern Phys. A*, 12 (1997), 1743–1782 (*hep-th*/9605225).

[13] EHRESMANN, C., Sur la topologie des certaines espaces homogènes. *Ann. of Math.*, 35 (1934), 396–443.

[14] FULTON, W., *Introduction to Toric Varieties*. Ann. of Math. Stud., 131. Princeton Univ. Press, Princeton, NJ, 1993.

[15] GIVENTAL, A., Equivariant Gromov–Witten invariants. *Iternat. Math. Res. Notices*, 1996, 613–663 (*alg-geom*/9603021).

[16] — Stationary phase integrals, quantum Toda lattices, flag manifolds and the mirror conjecture, in *Topics in Singularity Theory: V. I. Arnold's 60th Anniversary Collection*, pp. 103–115. Amer. Math. Soc. Transl. Ser. 2, 180. Amer. Math. Soc., Providence, RI, 1997 (*alg-geom*/9612001).

[17] — A mirror theorem for toric complete intersections, in *Topological Field Theory, Primitive Forms and Related Topics* (Kyoto, 1996), pp. 141–175. Progr. Math., 160. Birkhäuser Boston, Boston, MA, 1998 (*alg-geom*/9701016).

[18] GONCIULEA, N. & LAKSHMIBAI, V., Degenerations of flag and Schubert varieties to toric varieties. *Transform. Groups*, 1 (1996), 215–248.

[19] — Schubert varieties, toric varieties, and ladder determinantal varieties. *Ann. Inst. Fourier (Grenoble)*, 47 (1997), 1013–1064.

[20] KIM, B., Quantum cohomology of partial flag manifolds and a residue formula for their intersection pairings. *Internat. Math. Res. Notices*, 1995, 1–16.

[21] — On equivariant quantum cohomology. *Internat. Math. Res. Notices*, 1996, 841–851.

[22] — Quantum cohomology of flag manifolds G/B and quantum Toda lattices. *Ann. of Math.*, 149 (1999), 129–148.
[23] — Quantum hyperplane section theorem for homogeneous spaces. *Acta Math.*, 183 (1999), 71–99 (*alg-geom*/9712008).
[24] LAKSHMIBAI, V., Degenerations of flag varieties to toric varieties. *C. R. Acad. Sci. Paris Sér. I Math.*, 321 (1995), 1229–1234.
[25] LI, J. & TIAN, G., Virtual moduli cycles and Gromov–Witten invariants of algebraic varieties. *J. Amer. Math. Soc.*, 11 (1998), 119–174.
[26] REID, M., Decomposition of toric morphisms, in *Arithmetic and Geometry*, Vol. II, pp. 395–418. Progr. Math., 36. Birkhäuser Boston, Boston, MA, 1983.
[27] SCHECHTMAN, V., On hypergeometric functions connected with quantum cohomology of flag spaces. *q-alg*/9712049.
[28] STURMFELS, B., *Gröbner Bases and Convex Polytopes*. Univ. Lecture Ser., 8. Amer. Math. Soc., Providence, RI, 1996.

VICTOR V. BATYREV
Mathematisches Institut
Eberhard-Karls-Universität Tübingen
DE-72076 Tübingen
Germany
batyrev@bastau.mathematik.uni-tuebingen.de

IONUŢ CIOCAN-FONTANINE
Department of Mathematics
Northwestern University
Evanston, IL 60208
U.S.A.
ciocan@math.nwu.edu

BUMSIG KIM
Department of Mathematics
Pohang University of Science and Technology (Postech)
Pohang, 790-784
The Republic of Korea
bumsig@euclid.postech.ac.kr

DUCO VAN STRATEN
FB 17, Mathematik
Johannes Gutenberg-Universität Mainz
DE-55099 Mainz
Germany
straten@mathematik.uni-mainz.de

Received May 18, 1998

A.3. The Abelian/nonabelian Correspondence and Frobenius Manifolds

Ciocan-Fontanine, I., Kim, B. & Sabbah, C. The Abelian/Nonabelian correspondence and Frobenius manifolds. Invent. math. **171**, 301–343 (2008). https://doi.org/10.1007/s00222-007-0082-x

© 2007, Springer-Verlag All rights reserved. Reprinted with permission.

© The Author(s) 2026

N.-G. Kang et al. (eds.), *Categorical and Enumerative Aspects of Mirror Symmetry*, KIAS Springer Series in Mathematics 5,
https://doi.org/10.1007/978-981-95-0385-8

Invent. math. 171, 301–343 (2008)
DOI: 10.1007/s00222-007-0082-x

Inventiones mathematicae

The abelian/nonabelian correspondence and Frobenius manifolds

Ionuţ Ciocan-Fontanine[1], **Bumsig Kim**[2], **Claude Sabbah**[3]

[1] School of Mathematics, University of Minnesota, Minneapolis, MN 55455, USA (e-mail: ciocan@math.umn.edu)

[2] School of Mathematics, Korea Institute for Advanced Study, 207-43 Cheongnyangni 2-dong, Dongdaemun-gu, Seoul, 130-722, Korea (e-mail: bumsig@kias.re.kr)

[3] UMR 7640 du C.N.R.S., Centre de mathématiques Laurent Schwartz, École polytechnique, F-91128 Palaiseau cedex, France (e-mail: sabbah@math.polytechnique.fr)

Oblatum 9-X-2006 & 16-IX-2007
Published online: 12 October 2007 – © Springer-Verlag 2007

Abstract. We propose an approach via Frobenius manifolds to the study (began in [BCK2]) of the relation between rational Gromov–Witten invariants of nonabelian quotients $X//\mathbf{G}$ and those of the corresponding "abelianized" quotients $X//\mathbf{T}$, for $\mathbf{T}$ a maximal torus in $\mathbf{G}$. The ensuing conjecture expresses the Gromov–Witten potential of $X//\mathbf{G}$ in terms of the potential of $X//\mathbf{T}$. We prove this conjecture when the nonabelian quotients are partial flag manifolds.

1. Introduction

1.1. The paper [BCK2] conjectures a correspondence between the genus zero Gromov–Witten invariants of nonsingular projective GIT quotients $X//\mathbf{G}$ and $X//\mathbf{T}$, for $\mathbf{G}$ a complex reductive Lie group with a linearized action on a projective manifold X and $\mathbf{T}$ a maximal torus in $\mathbf{G}$. The correspondence expresses (descendant) Gromov–Witten invariants of $X//\mathbf{G}$ in terms of Gromov–Witten invariants of $X//\mathbf{T}$ *twisted* by (the top Chern class of) a certain decomposable vector bundle on $X//\mathbf{T}$.

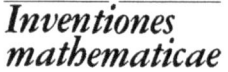

Our main goal in this paper is to give a natural reformulation of the correspondence in terms of the Frobenius structures describing the (big) quantum cohomology rings $QH^*(X//\mathbf{G}, \mathbb{C})$ and $QH^*(X//\mathbf{T}, \mathbb{C})$. This is accomplished in Sect. 3. To explain it, recall that a given cohomology class $\sigma \in H^*(X//\mathbf{G})$ can be lifted to a class $\tilde{\sigma}$ (of the same degree) in the Weyl group invariant subspace $H^*(X//\mathbf{T})^{\mathbf{W}}$. Such a lift is not unique, however, if ω is the fundamental $\mathbf{W}$-anti-invariant class, then $\tilde{\sigma} \cup \omega$ is uniquely determined by σ. Moreover, by results of Ellingsrud and Strømme when $X = \mathbb{P}^N$, and later Martin in full generality, this identification respects cup products:

$$(\widetilde{\sigma \cup_{X//\mathbf{G}} \sigma'}) \cup \omega = \tilde{\sigma} \cup (\tilde{\sigma}' \cup \omega) \in H^*(X//\mathbf{T}).$$

A naive guess might be that the identification respects quantum products as well, that is,

$$(\widetilde{\sigma \star_{X//\mathbf{G}} \sigma'}) \cup \omega = \tilde{\sigma} \star_{X//\mathbf{T}} (\tilde{\sigma}' \cup \omega),$$

after an appropriate specialization of quantum parameters. Indeed, as shown in [BCK1], Theorem 2.5, this is the case for *small* quantum products when $X//\mathbf{G}$ is a Grassmannian. At the level of Gromov–Witten invariants, this would translate into an appealing identity of the form

(1.1.1)
$$\langle \sigma_1, \sigma_2, \dots, \sigma_{n-1}, \sigma_n \rangle_{0,n,\beta}^{X//\mathbf{G}} = \pm \sum_{\tilde{\beta} \mapsto \beta} \langle \tilde{\sigma}_1, \dots, \tilde{\sigma}_{n-2}, \tilde{\sigma}_{n-1} \cup \omega, \tilde{\sigma}_n \cup \omega \rangle_{0,n,\tilde{\beta}}^{X//\mathbf{T}}.$$

It is not hard to convince oneself, however, that this fails for big quantum cohomology (already for the Grassmannian $Grass(2, 4)$), and that it has no reason to be true in general even for small quantum cohomology. Instead, we conjecture a generalization to quantum cohomology as follows:

Fix a lifting $\tilde{\bullet}$ of $H^*(X//\mathbf{G})$ to a subspace $U \subset H^*(X//\mathbf{T})^{\mathbf{W}}$. Let $\{t_i\}$ be the coordinates on $H^*(X//\mathbf{G})$, corresponding to a choice of basis, and let $\{\tilde{t}_i\}$ be the coordinates on U corresponding to the lifted basis. Let $N(X//\mathbf{G})$ and $N(X//\mathbf{T})$ be the Novikov rings for the two quotients.

The quantum product in $QH^*(X//\mathbf{G}, \mathbb{C})$ is a $N(X//\mathbf{G})[[t]]$-linear product on $H^*(X//\mathbf{G}, \mathbb{C}) \otimes_{\mathbb{C}} N(X//\mathbf{G})[[t]]$, while the quantum product in $QH^*(X//\mathbf{T}, \mathbb{C})$ is a $N(X//\mathbf{T})[[\tilde{t}, y]]$-linear product on $H^*(X//\mathbf{T}, \mathbb{C}) \otimes_{\mathbb{C}} N(X//\mathbf{T})[[\tilde{t}, y]]$, where $(\tilde{t}, y)$ is an extension of $\tilde{t}$ to coordinates on the entire $H^*(X//\mathbf{T}, \mathbb{C})$.

There is a natural specialization of Novikov variables $p : N(X//\mathbf{T}) \to N(X//\mathbf{G})$ which takes into account that there are more curve classes on $X//\mathbf{T}$. We denote by "$\star$" the quantum product on $X//\mathbf{T}$ with the Novikov variables specialized via p. Given $\sigma, \sigma' \in H^*(X//\mathbf{G})$, there are classes $\xi, \xi' \in U \otimes_{\mathbb{C}} N(X//\mathbf{G})[[\,\tilde{t}\,]]$, uniquely determined by $\xi \star \omega = \tilde{\sigma} \cup \omega$ and $\xi' \star \omega = \tilde{\sigma}' \cup \omega$ respectively.

Conjecture. There is an equality

$$((\widetilde{\sigma \star_{X//\mathbf{G}} \sigma'}) \cup \omega)(t) = \xi \star \xi' \star \omega\ (\tilde{t}, 0) = (\xi \star (\tilde{\sigma}' \cup \omega))(\tilde{t}, 0),$$

after an explicit change of variable $\tilde{t} = \tilde{t}(t)$.

At the level of Gromov–Witten invariants, the Conjecture says that the right-hand side of the naive formula (1.1.1) receives a correction term which is a sum of products of invariants of $X//\mathbf{T}$ of the same type (see the appendix for a discussion and some examples).

We should warn the reader that the above formulation is a translation of the actual Conjecture 3.7.1 in the body of the paper, which is stated in the conceptual framework of Frobenius structures. It is in this framework that one is naturally lead to the conjecture. Indeed, if N is the formal germ of the affine space over $N(X//\mathbf{G})$ associated to the subspace U, the general machinery of the infinitesimal period mapping in the theory of Frobenius–Saito structures (see e.g., [Sab]) gives a canonical Frobenius manifold structure on N. It is induced by the primitive homogeneous section ω of the (trivial) bundle with fiber the anti-invariant subspace $H^*(X//\mathbf{T})^a$ over N, together with the restriction to this bundle (in an appropriate sense) of the Frobenius structure on $H^*(X//\mathbf{T})$. Our conjecture says that this new Frobenius manifold is identified with the Frobenius manifold given by the Gromov–Witten theory of $X//\mathbf{G}$. The new flat metric ${}^{\omega}g$ on the sheaf Θ_N of vector fields satisfies

$$^{\omega}g(\tilde{\sigma}, \tilde{\sigma}') = g(\tilde{\sigma} \star \omega, \tilde{\sigma}' \star \omega).$$

It follows that the coordinates $\{\tilde{t}_i\}$ on N provided by lifting are *not* flat for the new Frobenius structure, or, equivalently, the liftings $\tilde{\sigma}$ are not horizontal vector fields. The vector fields ξ, ξ' appearing in the statement of the conjecture are precisely the horizontal vector fields corresponding to σ, σ' under the identification of flat coordinates of Frobenius structures. This identification of coordinates is the change variable $\tilde{t} = \tilde{t}(t)$.

In fact, we treat a more general situation in Sect. 3, by considering the *equivariant* Gromov–Witten theories in the presence of compatible actions of an additional torus $\mathbf{S}$ on $X//\mathbf{G}$ and $X//\mathbf{T}$. The corresponding Frobenius structures are more general than the ones considered in [Sab], as they lack an Euler vector field. However, a suitable modification of the notion of Euler vector field allows the application of the theory of infinitesimal period mappings in this case as well. We give an exposition of the relevant facts in Sects. 2.2–2.3.

This generalization is needed in Sect. 4, where we prove, by using reconstruction theorems for Gromov–Witten invariants (extended to the equivariant setting), that the conjecture above can be reduced in many cases to the abelian/nonabelian correspondence for *small* J-functions from [BCK2]. In particular, the following result is obtained:

Theorem. *Let $Fl = Fl(k_1, \ldots, k_r, n)$ be the flag manifold parameterizing flags of subspaces $\{\mathbb{C}^{k_1} \subset \cdots \subset \mathbb{C}^{k_r} \subset \mathbb{C}^n\}$, viewed as a GIT quotient $\mathbb{P}^l//\mathbf{G}$ for appropriate $l, \mathbf{G}$. Denote by Y the toric variety which is the corresponding abelian quotient $\mathbb{P}^l//\mathbf{T}$ (cf.* [BCK2]*). Then the conjecture is true for the pair (Fl, Y).*

The theorem implies that the genus zero Gromov–Witten invariants of a flag manifold (with any number of insertions) can be expressed in terms of Gromov–Witten invariants of the associated toric variety Y. In an appendix we write down explicit formulae in the simplest case of the Grassmannian $Grass(k, n)$, for which the abelian quotient is the product of k copies of $\mathbb{P}^{n-1}$.

In Sect. 5, we obtain an equivalent formulation (5.3.4) of the conjecture in terms of (big) J-functions of $X//\mathbf{T}$ and $X//\mathbf{G}$. It generalizes Conjecture 4.3 of [BCK2] and, by the above theorem, it holds for type A flag manifolds.

Finally, in Sect. 6 we extend the abelian/nonabelian correspondence to include Gromov–Witten invariants with an additional twist by homogeneous vector bundles. As an application, we describe the J-function of a generalized flag manifold for a simple complex Lie group of type B, C, or D as the twisted J-function of the abelianization of the corresponding flag manifold of type A.

Acknowledgements. Ciocan–Fontanine's work was partially supported by the NSF grant DMS-0303614. Part of the final writing of the paper was done during a visit by Ciocan-Fontanine to KIAS, whose support and hospitality are gratefully acknowledged. Kim's work is supported by KOSEF grant R01-2004-000-10870-0. Kim thanks staffs at École Polytechnique for their warm hospitality during his visit. Sabbah thanks KIAS for providing him with excellent working conditions during his visit.

2. Preliminaries on Frobenius structures

2.1. Formal Frobenius manifolds from Gromov–Witten theory. Let R be a $\mathbb{C}$-algebra and let K be a free R-module of rank m. We think of K as the affine m-space over R (precisely, the spectrum of the symmetric algebra of the dual module). Let $M := Spf(R[[K^\vee]])$ be the formal completion of K at the origin. M is a formal manifold over R. We denote by Θ_M its formal relative tangent sheaf over R. Note that it is canonically identified with $K \otimes_R \mathcal{O}_M$.

Definition 2.1.1. *The data $(M, \star, g, e, \mathfrak{E})$ is called a (conformal, even) formal Frobenius manifold over R if the following properties hold:*

- *g is an $\mathcal{O}_M$-linear, nondegenerate pairing such that its metric connection ∇ is flat*
- *$\star$ is an $\mathcal{O}_M$-linear, associative, commutative product on Θ_M*
- *e is a formal vector field on M over R which is the identity for the product $\star$, and such that $\nabla e = 0$*
- *∇c is symmetric, where the tensor c is defined by $c(u, v, w) = g(u \star v, w)$*
- *$\mathfrak{E}$ is a formal vector field on M over R satisfying*

$$\mathcal{L}_{\mathfrak{E}}(g) = Dg, \quad \mathcal{L}_{\mathfrak{E}}(\star) = \star, \quad \mathcal{L}_{\mathfrak{E}}(e) = -e,$$

where $\mathcal{L}_{\mathfrak{E}}$ denotes the Lie derivative and $D \in \mathbb{C}$ is a constant.

The fourth condition implies that there is a formal function F on M (the *potential* of the Frobenius manifold) such that the tensor c is given by the third derivatives of F in flat coordinates, and then associativity of $\star$ translates into the WDVV equations for F. The vector field $\mathfrak{E}$ is called an *Euler vector field.*

We recall here the formal Frobenius manifold structures determined by the genus zero GW-theories (ordinary and equivariant) of a projective manifold endowed with an action of an algebraic complex torus $\mathbf{S} \cong (\mathbb{C}^*)^\ell$. Detailed expositions can be found in [LP2] or [Man], to which we refer the reader.

Let Y be a smooth projective variety over $\mathbb{C}$. We assume for simplicity that $H_2(Y, \mathbb{Z})$ is torsion-free and that the odd cohomology $H^{2*+1}(Y, \mathbb{C})$ vanishes. We denote by $N(Y)$ the *Novikov ring* of Y. It can be described as the $\mathbb{C}$-algebra of "power series" $\{\sum_{\beta \in NE_1} c_\beta Q^\beta | c_\beta \in \mathbb{C}\}$, where $NE_1 \subset H_2(Y, \mathbb{Z})$ is the semigroup of effective curve classes.

The genus zero Gromov–Witten theory of Y determines a formal Frobenius manifold over $R = N(Y)$. We take

$$K = N(Y) \otimes_{\mathbb{C}} H^*(Y, \mathbb{C}),$$

so that $M = Spf(N(Y)[[K^\vee]])$. The metric g is given by the intersection pairing:

$$g(\gamma, \gamma') = \int_Y \gamma \cup \gamma'.$$

Let $\{1 = \gamma_0, \gamma_1, \dots, \gamma_r, \gamma_{r+1}, \dots, \gamma_{m-1}\}$ be a basis of $H^*(Y, \mathbb{C})$ consisting of integral homogeneous classes, such that $\gamma_1, \dots, \gamma_r$ form a basis of H^2. We write $\sigma = \sum t_i \gamma_i$ for a general cohomology class on Y. The functions t_i give flat coordinates on M. A potential function is defined using the genus zero Gromov–Witten invariants of Y

$$F(Q, t) := \sum_{\beta \in NE_1} \sum_{n \geq 0} Q^\beta \frac{1}{n!} \langle \underbrace{\sigma, \dots, \sigma}_{n} \rangle_{0,n,\beta},$$

where the unstable terms with $\beta = 0,\ n \leq 2$ are omitted in the sum. The tensor c is given in flat coordinates by

$$c_{ijk} = \partial_{t_i} \partial_{t_j} \partial_{t_k} F$$

and the product $\star$ is called the *big quantum product.* The unit vector field e is given by the class $\gamma_0 = 1$.

The following notation is customary:

$$\langle\langle \sigma_1, \dots, \sigma_r \rangle\rangle = \sum_{\beta \in NE_1} \sum_{n} Q^\beta \frac{1}{n!} \langle \sigma_1, \dots, \sigma_r, \underbrace{\sigma, \dots, \sigma}_{n} \rangle_{0,n+r,\beta},$$

where $\sigma_j \in H^*(Y, \mathbb{C})$ are given cohomology classes and $\sigma = \sum t_i \gamma_i$ is the general element in $H^*(Y, \mathbb{C})$ (so that $\langle\langle\ \rangle\rangle = F$). We extend this double bracket $\mathcal{O}_M$-linearly to general vector fields $\sigma_1, \ldots, \sigma_r$. It is easy to see that for any vector field ξ we have

$$\nabla_\xi(F) = \langle\langle \xi \rangle\rangle.$$

In particular, since $\nabla_{\partial_{t_i}} \partial_{t_j} = 0$, the quantum product can be written in our chosen basis

$$\gamma_i \star \gamma_j = \sum_k \langle\langle \gamma_i, \gamma_j, \gamma_k \rangle\rangle \gamma_k^\vee$$

where $\gamma_k^\vee = \sum_l g^{kl} \gamma_l$ with (g^{kl}) the inverse matrix of the metric g.

The divisor axiom for Gromov–Witten invariants implies that the Gromov–Witten potential has the special form

$$F = F_{cl} + \sum_{\beta \in NE_1, \beta \neq 0} Q^\beta e^{\beta \cdot t_{\text{small}}} F_\beta, \tag{2.1.1}$$

with F_{cl} a cubic polynomial in the t_i's and $F_\beta \in \mathbb{C}[[t_{r+1}, \ldots, t_{m-1}]]$ formal power series in the *non-divisorial* coordinates. Here we use the notation $\beta \cdot t_{\text{small}}$ for the intersection index of β with the general H^2-class,

$$\beta \cdot t_{\text{small}} := \int_\beta \sum_{i=1}^r t_i \gamma_i.$$

We will also use the notation F_q for $F - F_{cl}$.

Assume now that Y is acted upon by the torus $\mathbf{S} \cong (\mathbb{C}^*)^\ell$. The equivariant cohomology $H^*_{\mathbf{S}}(Y, \mathbb{C})$ is a module over the polynomial ring

$$H^*_{\mathbf{S}}(pt) = H^*(B\mathbf{S}) \cong \mathbb{C}[\lambda_1, \ldots, \lambda_\ell],$$

and it is in fact a free module by [Gin]. Taking

$$R = N(Y)[\lambda] := N(Y) \otimes_{\mathbb{C}} \mathbb{C}[\lambda_1, \ldots, \lambda_\ell]$$

and

$$K_{\mathbf{S}} = R \otimes_{\mathbb{C}[\lambda_1, \ldots, \lambda_\ell]} H^*_{\mathbf{S}}(Y, \mathbb{C})$$

we get similarly a formal Frobenius manifold over R. The metric g is now given by the ($\mathbb{C}[\lambda_1, \ldots, \lambda_\ell]$-valued) equivariant intersection pairing, while in F the GW-invariants are replaced by their $\mathbf{S}$-equivariant counterparts. The unit vector field and equivariant big quantum product are obtained analogously.

Localization with respect to $\mathbf{S}$ determines yet another Frobenius structure. Consider the localization of $H^*(B\mathbf{S})$, i.e., the field of fractions $\mathbb{C}(\lambda_1, \ldots, \lambda_\ell)$, and set

$$N(Y)[\lambda]_{(\lambda)} = N(Y) \otimes_{\mathbb{C}} \mathbb{C}(\lambda_1, \ldots, \lambda_\ell)$$
$$K^*_{\mathbf{S}} = N(Y)[\lambda]_{(\lambda)} \otimes_{N(Y)[\lambda]} K_{\mathbf{S}}.$$

Taking $M = Spf(N(Y)[\lambda]_{(\lambda)}[[K_{\mathbf{S}}^{*\vee}]])$ with the localized equivariant metric, potential function, and unit vector field determines a formal Frobenius manifold over $N(Y)[\lambda]_{(\lambda)}$(in other words, we simply consider the Frobenius structure induced by base change via $N(Y) \to N(Y)[\lambda]_{(\lambda)}$).

In both the equivariant and localized equivariant cases the potential function in flat coordinates t has the special form (2.1.1), with $F_\beta \in \mathbb{C}[\lambda][[t_{r+1}, \dots, t_{m-1}]]$.

Finally, we discuss the Euler vector fields. The Frobenius manifold defined by the (nonequivariant) Gromov–Witten theory of Y is conformal: the Euler vector field (with $D = 2 - \dim(Y)$) is explicitly

$$\mathfrak{E} = \sum_{i=0}^{m-1} \left(1 - \frac{\mathrm{cdeg}\gamma_i}{2}\right) t_i \partial_{t_i} + c_1(TY).$$

Here "cdeg" is the cohomological degree.

Consider the $\mathbf{S}$-equivariant version of this vector field

$$\mathfrak{E} = \sum_{i=0}^{m-1} \left(1 - \frac{\mathrm{cdeg}\gamma_i}{2}\right) t_i \partial_{t_i} + c_1^{\mathbf{S}}(TY)$$

with γ_i's now forming a basis of $H_{\mathbf{S}}^*(Y, \mathbb{C})$ over $H^*(B\mathbf{S})$. $\mathfrak{E}$ does not give a conformal structure (because equivariant Gromov–Witten invariants do not satisfy a dimension constraint). Nevertheless, we consider a variant of the Euler vector field in this context as well, by relaxing the requirement of linearity over $N(Y) \otimes \mathbb{C}[\lambda]$ and will define below an Euler vector field $\mathfrak{E}_{\mathbf{S}}$ as an $N(Y)$-derivation on $\mathcal{O}_M$ (that is, an $N(Y)$-derivation of $K_{\mathbf{S}}^\vee$ into itself). The flat coordinates $\{t_i\}$, together with $\{\lambda_1, \dots, \lambda_\ell\}$ form a coordinate system on M over $Spec(N(Y))$. Therefore

$$\mathcal{E}_{\mathbf{S}} := \sum_{i=1}^{\ell} \lambda_i \partial_{\lambda_i}$$

is a well-defined "absolute" vector field (i.e., $N(Y)$-linear derivation) and acts by Lie bracket on the relative vector fields Θ_M. Put

$$\mathfrak{E}_{\mathbf{S}} := \mathfrak{E} + \mathcal{E}_{\mathbf{S}}.$$

If $\eta \in \Theta_M$ is any relative vector field, then the commutator $[\mathfrak{E}_{\mathbf{S}}, \eta]$ is also in Θ_M. Hence Lie derivatives of tensors on Θ_M are well defined. The vector field $\mathfrak{E}_{\mathbf{S}}$ will still satisfy the conditions in Definition 2.1.1, again with $D = 2 - \dim(Y)$. The same $\mathfrak{E}_{\mathbf{S}}$ will be used for the localized structure as well.

We have

$$\mathcal{L}_{\mathfrak{E}_{\mathbf{S}}}(\partial_{t_i}) = \left(-1 + \frac{\mathrm{cdeg}\gamma_i}{2}\right)\partial_{t_i}, \ \mathcal{L}_{\mathfrak{E}_{\mathbf{S}}}(\lambda_i) = \lambda_i, \ \mathcal{L}_{\mathfrak{E}_{\mathbf{S}}}(F) = (3 - \dim(Y))F.$$

2.2. S-equivariant Frobenius manifolds over R. We extend the construction of Frobenius manifold through an infinitesimal period mapping to the previous setting. Let M be as above, with $\mathcal{O}_M = R[[K_{\mathbf{S}}^{*\vee}]]$ and $R = N(Y)[\lambda]$ or $R = N(Y)[\lambda]_{(\lambda)}$. Let E be a free $\mathcal{O}_M$-module of finite rank. An $\mathbf{S}$-*conformal connection* on E consists of a pair $\widetilde{\nabla} = (\nabla, \widetilde{\nabla}_{\mathfrak{E}_{\mathbf{S}}})$, where ∇ is an R-connection on E and $\widetilde{\nabla}_{\mathfrak{E}_{\mathbf{S}}}$ is a $N(Y)$-linear derivation satisfying $\widetilde{\nabla}_{\mathfrak{E}_{\mathbf{S}}}(\varphi e) = \mathcal{L}_{\mathfrak{E}_{\mathbf{S}}}(\varphi)e + \varphi\widetilde{\nabla}_{\mathfrak{E}_{\mathbf{S}}}e$ for any $e \in E$ and $\varphi \in \mathcal{O}_M$. We say that $\widetilde{\nabla}$ is *flat* if ∇ is flat and for any vector field $\xi \in \Theta_M$, $[\widetilde{\nabla}_{\mathfrak{E}_{\mathbf{S}}}, \nabla_\xi] = \nabla_{[\mathfrak{E}_{\mathbf{S}},\xi]}$. In coordinates (t_i) defined from an $N(Y)$-basis of K, the previous condition is equivalent to the pairwise commutation of the operators $\nabla_{\partial_{t_i}}$ and $\widetilde{\nabla}_{\mathfrak{E}_{\mathbf{S}}}$. Such a connection $\widetilde{\nabla}$ extends in a natural way to a similar object on $\hom_{\mathcal{O}_M}(E, E)$.

An $\mathbf{S}$-equivariant *pre-Saito structure* $(M, E, \widetilde{\nabla}, \Phi, R_0, g)$ of weight w over M consists of

- a free $\mathcal{O}_M$-module E of finite rank with a flat $\mathbf{S}$-conformal connection $\widetilde{\nabla}$,
- $\mathcal{O}_M$-linear morphisms $\Phi : \Theta_M \otimes_{\mathcal{O}_M} E \to E$ and $R_0 : E \to E$,
- an $\mathcal{O}_M$-bilinear form g on E,

satisfying, when expressed in coordinates (t_i), the following relations for all i, j:

$$\nabla_{\partial_{t_i}}(\Phi_{\partial_{t_j}}) = \nabla_{\partial_{t_j}}(\Phi_{\partial_{t_i}}), \quad [\Phi_{\partial_{t_i}}, \Phi_{\partial_{t_j}}] = 0, \quad [R_0, \Phi_{\partial_{t_i}}] = 0,$$
$$\Phi_{\partial_{t_i}} - \widetilde{\nabla}_{\mathfrak{E}_{\mathbf{S}}}(\Phi_{\partial_{t_i}}) + \nabla_{\partial_{t_i}}(R_0) = 0,$$
$$\nabla(g) = 0, \quad \widetilde{\nabla}_{\mathfrak{E}_{\mathbf{S}}}(g) = -wg, \quad \Phi^*_{\partial_{t_i}} = \Phi_{\partial_{t_i}}, \quad R_0^* = R_0,$$

where * means taking the g-adjoint and $\widetilde{\nabla}_{\mathfrak{E}_{\mathbf{S}}}(g)$ is defined as usual by the formula $\widetilde{\nabla}_{\mathfrak{E}_{\mathbf{S}}}(g)(\xi, \eta) = \mathcal{L}_{\mathfrak{E}_{\mathbf{S}}}(g(\xi, \eta)) - g(\widetilde{\nabla}_{\mathfrak{E}_{\mathbf{S}}}\xi, \eta) - g(\xi, \widetilde{\nabla}_{\mathfrak{E}_{\mathbf{S}}}\eta)$.

The pull-back of an $\mathbf{S}$-equivariant pre-Saito structure by a morphism $f : N \to M$ is well-defined only for morphisms f^* which commute with $\mathcal{L}_{\mathfrak{E}_{\mathbf{S}}}$.

The definition of an $\mathbf{S}$-*equivariant Frobenius manifold over* R is a variant of Definition 2.1.1: With the same data $(M, \star, g, e, \mathfrak{E})$, we set $\mathfrak{E}_{\mathbf{S}} = \mathfrak{E} + \mathfrak{E}_{\mathbf{S}}$, and we replace the homogeneity conditions by the following ones:

$$\mathcal{L}_{\mathfrak{E}_{\mathbf{S}}}(g) = Dg, \quad \mathcal{L}_{\mathfrak{E}_{\mathbf{S}}}(\star) = \star, \quad \mathcal{L}_{\mathfrak{E}_{\mathbf{S}}}(e) = -e.$$

Let $(M, E, \widetilde{\nabla}, \Phi, R_0, g)$ be an $\mathbf{S}$-equivariant *pre-Saito structure* of weight w and let ω be a ∇-horizontal section of E. It defines an $\mathcal{O}_M$-linear morphism $\varphi_\omega : \Theta_M \to E$ by $\xi \mapsto -\Phi_\xi(\omega)$. We say that such a section ω of E is

(1) *primitive* if the associated period mapping $\varphi_\omega : \Theta_M \to E$ is an isomorphism,
(2) *homogeneous* of degree $q \in \mathbb{C}$ if $\widetilde{\nabla}_{\mathfrak{E}_{\mathbf{S}}}\omega = q\omega$.

The data of an $\mathbf{S}$-equivariant pre-Saito structure and of a homogeneous primitive section ω is called an $\mathbf{S}$-equivariant Saito structure. As in [Sab, §4.3] and following K. Saito, we obtain the following results.

If ω is primitive and homogeneous, φ_ω induces a flat, torsion-free, R-connection ${}^\omega\nabla := \varphi_\omega^{-1}\nabla\varphi_\omega$ on Θ_M, and an associative and commutative $\mathcal{O}_M$-bilinear product $\star$ by $\xi \star \eta = -\Phi_\xi(\varphi_\omega(\eta))$, with $e = \varphi_\omega^{-1}(\omega)$ as unit, and ${}^\omega\nabla e = 0$. Moreover, ${}^\omega\nabla$ is the metric connection attached to the metric ${}^\omega g$ on Θ_M obtained from g through φ_ω, and ${}^\omega\nabla$ is $\mathbf{S}$-conformal and flat as such, setting ${}^\omega\widetilde{\nabla}_{\mathcal{E}_\mathbf{S}} = \varphi_\omega^{-1} \circ \widetilde{\nabla}_{\mathcal{E}_\mathbf{S}} \circ \varphi_\omega - \mathrm{Id}$.

The Euler field is $\mathfrak{E} = \varphi_\omega^{-1}(R_0(\omega))$. It is therefore a section of Θ_M. We have ${}^\omega\nabla\mathfrak{E} = \mathcal{L}_{\mathcal{E}_\mathbf{S}} - {}^\omega\widetilde{\nabla}_{\mathcal{E}_\mathbf{S}} + q\mathrm{Id}$. In particular, ${}^\omega\nabla({}^\omega\nabla\mathfrak{E}) = 0$.

If we put $D = 2q + 2 - w$, and if we set as above $\mathfrak{E}_\mathbf{S} = \mathfrak{E} + \mathcal{E}_\mathbf{S}$, we get

$$\mathcal{L}_{\mathfrak{E}_\mathbf{S}}(e) = -e, \quad \mathcal{L}_{\mathfrak{E}_\mathbf{S}}(\star) = \star, \quad \mathcal{L}_{\mathfrak{E}_\mathbf{S}}({}^\omega g) = D \cdot {}^\omega g.$$

Given an $\mathbf{S}$-equivariant pre-Saito structure $(M, E, \widetilde{\nabla}, \Phi, R_0, g)$ of weight w, the datum of a homogeneous primitive section ω of E having weight q induces on M, through φ_ω, the structure of a $\mathbf{S}$-equivariant Frobenius manifold of weight $D = 2q + 2 - w$.

Conversely, any $\mathbf{S}$-equivariant Frobenius manifold $(M, \star, g, e, \mathfrak{E})$ defines an $\mathbf{S}$-equivariant pre-Saito structure $(M, \Theta_M, \widetilde{\nabla}, \Phi, R_0, g)$ having e as homogeneous primitive form.

For instance, to give the correspondence $(M, \star, g, e, \mathfrak{E}) \mapsto (M, \Theta_M, \widetilde{\nabla}, \Phi, R_0, g)$ we take ∇ to be the Levi–Civita connection of g, and

$$\widetilde{\nabla}_{\mathcal{E}_\mathbf{S}} = \mathrm{Id} + \mathcal{L}_{\mathcal{E}_\mathbf{S}} - \nabla\mathfrak{E}, \quad \Phi_\xi(\eta) = -(\xi \star \eta),$$
$$R_0 = \mathfrak{E}\star = -\Phi_\mathfrak{E}, \quad q = 0, \quad w = 2 - D.$$

2.3. S-Equivariant Frobenius manifolds with finite group action. Let us consider an $\mathbf{S}$-equivariant Frobenius manifold $(M, \star, g, e, \mathfrak{E})$ of weight D over R. Let W be a finite group which acts by relative automorphisms on M, hence on Θ_M, in a compatible way with the $\mathbf{S}$-equivariant Frobenius structure. (To be precise, we assume that W acts trivially on R, and on M by automorphisms preserving the map to $Spec(R)$.) In particular, the action of W on Θ_M commutes with $\mathcal{L}_{\mathcal{E}_\mathbf{S}}$.

Let M^W be the fixed set of W on M. Then W acts by $\mathcal{O}_{M^W}$-linear isomorphisms on $\Theta_M|_{M^W}$. Moreover, the fixed set M^W is a smooth subscheme of M over R and the fixed bundle $(\Theta_M|_{M^W})^W$ is equal to Θ_{M^W}.

Let us moreover assume that W is equipped with a *non trivial* character $\mathrm{sgn} : W \to \{\pm 1\}$. We denote by $a : \Theta_M|_{M^W} \to \Theta_M|_{M^W}$ the antisymmetrization morphism and by E its image. Then E is a locally free $\mathcal{O}_{M^W}$-submodule of $\Theta_M|_{M^W}$ and we have a decomposition $\Theta_M|_{M^W} = E \oplus \ker a$. This decomposition is g-orthogonal, as $g(a\xi, a\eta) = g(\xi, \eta)$ for any ξ, η and g restricted to E is nondegenerate.

As the inclusion $M^W \hookrightarrow M$ commutes with $\mathcal{L}_{\mathcal{E}_\mathbf{S}}$, one can restrict to M^W the $\mathbf{S}$-equivariant pre-Saito structure associated to $(M, \star, g, e, \mathfrak{E})$ to get such an object with corresponding bundle $\Theta_M|_{M^W}$. One can moreover induce this structure on the $\mathcal{O}_{M^W}$-module E, as the following operators leave E invariant:

- the connection ∇ (i.e., $\nabla_\xi \eta$ is a section of E whenever ξ is a section of Θ_{M^W} and η a section of E),
- the Higgs field Φ, (i.e., $\xi \star \eta = -\Phi_\xi \eta$ is a section of E whenever ξ is a section of Θ_{M^W} and η a section of E),
- the operator $R_0 = -\Phi_{\mathfrak{E}} = \mathfrak{E}\,\star$,
- the operator $\widetilde{\nabla}_{g_S} = \mathrm{Id} + \mathcal{L}_{g_S} - \nabla\mathfrak{E}$ (i.e., $\nabla_\eta \mathfrak{E}$ is a section of E whenever η is a section of E).

The following is then clear:

Lemma 2.3.1. *The tuple* $(M^W, E, \widetilde{\nabla}, \Phi, R_0, g)$ *defines an* **S**-*equivariant pre-Saito structure of weight* $w = 2 - D$ *on* M^W.

Proposition 2.3.2. *Let us assume that there exists a section* ω *of* $E \subset \Theta_M|_{M^W}$ *which is* ∇-*horizontal and an eigenvector of* $\widetilde{\nabla}_{g_S}$ *(acting on* E *or on* $\Theta_M|_{M^W}$*) and such that the morphism*

$$\begin{aligned} \Theta_{M^W} &\longrightarrow E \\ \xi &\longmapsto \xi \star \omega \end{aligned}$$

is onto. Then, any smooth formal subscheme $N \subset M^W$ *over* R *defined by an ideal invariant under* $\mathcal{L}_{g_S}$ *and such that the induced morphism* $\Theta_N \to E|_N$ *is an isomorphism comes equipped with a natural structure of an* **S**-*equivariant Frobenius manifold of weight* D.

Proof. We restrict the **S**-equivariant pre-Saito structure $(M^W, E, \widetilde{\nabla}, \Phi, R_0, g)$ to N to get an object $(N, E|_N, \widetilde{\nabla}, \Phi, R_0, g)$ of the same kind. Then, as $\omega|_N$ is a ∇-horizontal section of $E|_N$ and as the morphism $\Theta_N \to E|_N$ given by $\xi \mapsto \xi \star \omega|_N = \varphi_\omega(\xi)$ is an isomorphism, ω is primitive. Moreover, ω is homogeneous in E hence $\omega|_N$ is so in $E|_N$. One can then apply the correspondence of Sect. 2.2. □

Some properties of the **S**-*equivariant Frobenius manifold structure on* N. Abusing notation, we denote by $- \star \omega^{-1}$ the inverse map of the isomorphism $\star\,\omega : \Theta_N \to E|_N$, i.e., we will write it also as operating on the right.

We denote by ${}^\omega g$, ${}^\omega\nabla$ the metric and connection on Θ_N coming from that on $E|_N$, and by $\circ$ the product on Θ_N induced by the Higgs field on $E|_N$.

(1) Let ξ, η be sections of Θ_N. The product $\xi \star \eta$ in $\Theta_M|_N$ may not be a section of Θ_N (it is only a section of $\Theta_{M^W}|_N$). We have $[\xi \star \eta - \xi \circ \eta] \star \omega = 0$. In fact, the composition

$$\Theta_{M^W}\big|_N \xrightarrow{\star\omega} E|_N \xrightarrow{\star\omega^{-1}} \Theta_N$$

induces a projection $\Theta_{M^W}|_N \to \Theta_N$, and $\xi \circ \eta$ is nothing else but the projection of $\xi \star \eta$ on Θ_N, so that we have the formula

$$\xi \circ \eta = (\xi \star \eta \star \omega) \star \omega^{-1}.$$

(2) Let us assume that we can find N such that the unit field e is *tangent* to N. This condition does not lead to a contradiction, as $e \star \omega = \omega \neq 0$. Then $e|_N$ is the unit field for the **S**-equivariant Frobenius manifold structure on N. Indeed, clearly, $e|_N \circ \eta = \eta$ for any section η of Θ_N. On the other hand, we have to check that e is ${}^{\omega}\nabla$-horizontal:

$$ {}^{\omega}\nabla e|_N := \nabla(e|_N \star \omega) \star \omega^{-1} = \nabla(\omega) \star \omega^{-1} = 0, \quad \text{as } \nabla(\omega) = 0. $$

(3) Let us assume that N is chosen so that the Euler vector field $\mathfrak{E}$ is tangent to N. Then $\mathfrak{E}|_N$ is the Euler vector field for the Frobenius manifold structure on N, as $R_0 = \mathfrak{E}\star$ leaves E invariant.

(4) We have ${}^{\omega}g(\xi, \eta) = g(\xi \star \omega, \eta \star \omega)$ for any $\xi, \eta \in \Theta_N$.

Remark 2.3.3. Given an R-basis $\boldsymbol{e}^o$ of $E/(t_0, \dots, t_{m-1})E$, there exists a unique system of flat coordinates (t_i) on N for which $\partial_{t_i} \star \omega \equiv e_i^o$ mod $(t_0, \dots, t_{m-1})E$. Given any other formal smooth subscheme N' over R satisfying the properties in Proposition 2.3.2, with corresponding coordinates (t_i'), we do not know whether the natural isomorphism $\mathcal{O}_N \to \mathcal{O}_{N'}$, $t_i \mapsto t_i'$, is compatible with the **S**-equivariant Frobenius structures. In other words, there is a priori no uniqueness in the construction resulting from Proposition 2.3.2. However, when this construction is applied to the setting of Sect. 3.1, Conjecture 3.7.1 also gives uniqueness.

3. The abelian/nonabelian correspondence for Frobenius structures

A precise relation between the genus zero Gromov–Witten theory (with descendants) of a quotient by a nonabelian group and a twist of the theory for the quotient by a maximal torus in the group was conjectured in [BCK2]. Here we formulate a version of this correspondence for the associated Frobenius structures.

3.1. Setting. Let X be a smooth projective variety over $\mathbb{C}$ with the (linearized) action of a complex reductive group $\mathbf{G}$, and let $\mathbf{T} \subset \mathbf{G}$ be a maximal torus. In this setting, there are two geometric invariant theory (GIT) quotients, $X/\!/\mathbf{T}$ and $X/\!/\mathbf{G}$. We assume (for both actions) that all semistable points are stable and that all isotropy groups of stable points are trivial, so that $X/\!/\mathbf{T}$ and $X/\!/\mathbf{G}$ are smooth projective varieties. Further, we assume that the $\mathbf{G}$-unstable locus $X \setminus X^s(\mathbf{G})$ has codimension at least 2 in X. (Note that this last condition is automatic when X is a projective space.)

There is a diagram

$$ \begin{array}{ccc} X/\!/\mathbf{T} = X^s(\mathbf{T})/\mathbf{T} & \xleftarrow[j]{\quad} & X^s(\mathbf{G})/\mathbf{T} \\ & & \big\downarrow{\scriptstyle \pi} \\ & & X/\!/\mathbf{G} = X^s(\mathbf{G})/\mathbf{G} \end{array} $$

with j an open immersion and π a $\mathbf{G}/\mathbf{T}$-fibration.

The above diagram leads to a comparison of the cohomology of the non-abelian quotient $X//\mathbf{G}$ to that of the abelian quotient $X//\mathbf{T}$ [ES,Mar,Kir]. We describe an equivariant version of it. Let another (possibly trivial) complex torus $\mathbf{S}$ act on X. Assume that the action commutes with the action of $\mathbf{G}$ and preserves $X^s(\mathbf{T})$ and $X^s(\mathbf{G})$. There is an induced action of $\mathbf{S}$ on the smooth projective varieties $X//\mathbf{T}$ and $X//\mathbf{G}$. The morphisms in the diagram are $\mathbf{S}$-equivariant. To the pair $(\mathbf{G}, \mathbf{T})$ we associate the usual Lie-theoretic data:

- the Weyl group $\mathbf{W} = N(\mathbf{T})/\mathbf{T}$ ($N(\mathbf{T})$ is the normalizer of $\mathbf{T}$ in $\mathbf{G}$).
- the root system Φ with decomposition $\Phi = \Phi_+ \cup \Phi_-$ into positive and negative roots.
- for each root $\alpha \in \Phi$ the 1-dimensional $\mathbf{T}$-representation $\mathbb{C}_\alpha$ with weight α.

The Weyl group acts on $X//\mathbf{T}$, hence also on the equivariant cohomology ring $H^*_{\mathbf{S}}(X//\mathbf{T}, \mathbb{C})$. The representations $\mathbb{C}_\alpha$ define $\mathbf{S}$-equivariant line bundles

$$L_\alpha := X^s(\mathbf{T}) \times_{\mathbf{T}} \mathbb{C}_\alpha$$

on $X//\mathbf{T}$ with equivariant first Chern classes $c_1^{\mathbf{S}}(L_\alpha) \in H^*_{\mathbf{S}}(X//\mathbf{T}, \mathbb{C})$. The $\mathbf{S}$-action on L_α is induced by the $\mathbf{S}$-action on $X^s(\mathbf{T})$ (and the trivial $\mathbf{S}$-action on $\mathbb{C}_\alpha$). Note that $L_{-\alpha} \cong L_\alpha^\vee$ for any pair $(\alpha, -\alpha)$ of opposite roots. The equivariant cohomology class

$$\omega := \sqrt{\frac{1}{|\mathbf{W}|} \prod_{\alpha\in\Phi} c_1^{\mathbf{S}}(L_\alpha)} = \sqrt{\frac{(-1)^{|\Phi_+|}}{|\mathbf{W}|}} \prod_{\alpha\in\Phi_+} c_1^{\mathbf{S}}(L_\alpha)$$

will play an important role in this paper. It is the fundamental $\mathbf{W}$-anti-invariant class in the equivariant cohomology of $X//\mathbf{T}$; any other $\mathbf{W}$-anti-invariant class ϕ can be written (non-uniquely) as $\gamma \cup \omega$, with $\gamma \in H^*_{\mathbf{S}}(X//\mathbf{T}, \mathbb{C})^{\mathbf{W}}$. (The reason for considering ω rather than the customary $\Delta = \prod_{\alpha\in\Phi_+} c_1^{\mathbf{S}}(L_\alpha)$ is one of convenience: we simply want to avoid having to insert the factor $(-1)^{|\Phi_+|}/|\mathbf{W}|$ in all formulae comparing Gromov–Witten invariants of $X//\mathbf{G}$ and $X//\mathbf{T}$.)

The following facts are known:

(3.1.1) π^* induces an isomorphism $H^*_{\mathbf{S}}(X//\mathbf{G}) \cong H^*_{\mathbf{S}}(X^s(\mathbf{G})/\mathbf{T})^{\mathbf{W}}$

(3.1.2) There is an exact sequence

$$0 \longrightarrow \ker(\cup\omega) \xrightarrow{\subset} H^*_{\mathbf{S}}(X//\mathbf{T})^{\mathbf{W}} \xrightarrow[(\pi^*)^{-1}\circ j^*]{} H^*_{\mathbf{S}}(X//\mathbf{G}) \longrightarrow 0$$

where $\ker(\cup\omega)$ is $\{\gamma \in H^*_{\mathbf{S}}(X//\mathbf{T})^{\mathbf{W}} \mid \gamma \cup \omega = 0\}$.

(3.1.3) The equivariant push-forwards satisfy the equality

$$\int_{X//\mathbf{T}} \omega^2 \tilde{\sigma} = \int_{X//\mathbf{G}} \sigma$$

for all $\sigma \in H^*_{\mathbf{S}}(X//\mathbf{G})$, $\tilde{\sigma} \in H^*_{\mathbf{S}}(X//\mathbf{T})$ with $j^*\tilde{\sigma} = \pi^*(\sigma)$. (Such $\tilde{\sigma}$ are called *lifts* of σ.)

(3.1.4) There is an identification of the **S**-equivariant relative tangent bundle T_π of $\pi : X^s(\mathbf{G})/\mathbf{T} \to X^s(\mathbf{G})/\mathbf{G}$ with $\oplus_{\alpha\in\Phi} L_\alpha|_{X^s(\mathbf{G})/\mathbf{T}}$.

In the nonequivariant case (that is, for **S**=1), (3.1.1) is a classical result, (3.1.2) is proved in [ES] for $X = \mathbb{P}^N$ and in [Kir] in general, (3.1.3) is proved in [Mar] and (3.1.4) can be seen by a direct computation. The extensions to the equivariant context are straightforward and left to the reader.

3.2. The W-induced Frobenius manifold. Applying the results in Sect. 2.3 to the Weyl group action on the **S**-equivariant Frobenius manifold given by the equivariant Gromov–Witten theory of $X//\mathbf{T}$, a new **S**-equivariant Frobenius manifold (of dimension over the base ring equal to the rank of $H^*_{\mathbf{S}}(X//\mathbf{G}, \mathbb{C})$) is obtained. In this subsection we spell out for concreteness the details of the construction and the main properties of the new Frobenius structure in this special case.

As mentioned in the introduction, a specialization of Novikov variables will be needed before comparing the new Frobenius structure with the one given by the equivariant Gromov–Witten theory of $X//\mathbf{G}$ and we start with this specialization.

Recall from (3.1.3) the notion of lift of cohomology classes from $X//\mathbf{G}$ to $X//\mathbf{T}$. By (3.1.2), one can always choose **W**-invariant lifts. These are not generally unique; however, the assumption that the **G**-unstable locus in X has codimension ≥ 2 implies that for divisor classes the **W**-invariant lifts are unique. This allows us to lift curve classes as well (cf. [BCK2]): the inclusion

$$\mathrm{Pic}(X//\mathbf{G}) \cong \mathrm{Pic}(X//\mathbf{T})^{\mathbf{W}} \subset \mathrm{Pic}(X//\mathbf{T})$$

induces by duality a projection

$$\varrho : NE_1(X//\mathbf{T}) \longrightarrow NE_1(X//\mathbf{G}).$$

We say that $\tilde{\beta}$ lifts $\beta \in NE_1(X//\mathbf{G})$ (and write $\tilde{\beta} \mapsto \beta$) if $\varrho(\tilde{\beta}) = \beta$. Note that any effective β has finitely many lifts. Define a projection on Novikov rings

(3.2.1)

$$p : N(X//\mathbf{T}) \to N(X//\mathbf{G}), \quad p\Big(\sum_{\tilde{\beta}} c_{\tilde{\beta}}\, Q^{\tilde{\beta}}\Big) = \sum_{\beta} (-1)^{\epsilon(\beta)} \Big(\sum_{\tilde{\beta}\mapsto\beta} c_{\tilde{\beta}}\Big) Q^{\beta},$$

where

$$\epsilon : NE_1(X//\mathbf{G}) \longrightarrow \mathbb{Z}_2$$

is defined by

$$\epsilon(\beta) = \Big(\int_{\tilde{\beta}} \sum_{\alpha\in\Phi_+} c_1^{\mathbf{S}}(L_\alpha)\Big) \pmod 2$$

with $\tilde{\beta}$ any lift of β. This makes sense, since the right-hand side does not depend on the choice of lift. Indeed, if α' is any simple root and $v_{\alpha'} \in \mathbf{W}$ is the corresponding reflection, then by standard properties of root systems

$$v_{\alpha'}\Big(\sum_{\alpha\in\Phi_+} c_1^{\mathbf{S}}(L_\alpha)\Big) = \sum_{\alpha\in\Phi_+} c_1^{\mathbf{S}}(L_\alpha) - 2c_1^{\mathbf{S}}(L_{\alpha'}),$$

so $\sum_{\alpha\in\Phi_+} c_1^{\mathbf{S}}(L_\alpha)$ is $\mathbf{W}$-invariant as a cohomology class with $\mathbb{Z}_2$-coefficients.

The sign in (3.2.1), which may seem rather mysterious, has its origin in the twisting bundle appearing in the abelian/nonabelian correspondence, as formulated in [BCK2, Conjecture 4.2].

Let Z be the formal Frobenius manifold defined by the $\mathbf{S}$-equivariant Gromov–Witten theory of $X//\mathbf{T}$, with potential function $F^{X//\mathbf{T},\mathbf{S}}$. Let M be the formal scheme over $N(X//\mathbf{G}) \otimes \mathbb{C}[\lambda]$ obtained by base change from Z by the morphism of Novikov rings (3.2.1). Let $\theta : M \longrightarrow Z$ be the base change map. We obtain a formal Frobenius structure over $N(X//\mathbf{G}) \otimes \mathbb{C}[\lambda]$ on (M, Θ_M) by pulling-back via θ the Frobenius structure on Z. Note that only the potential (and therefore the quantum product) changes under the pull-back, since the coefficients of the metric, the horizontal sections and the Euler vector field do not depend on the Novikov variables. Explicitly, the potential of the Frobenius structure on M is

(3.2.2)

$$F := \theta^*(F^{X//\mathbf{T},\mathbf{S}}) = \sum_{\beta\in NE_1(X//\mathbf{G})} (-1)^{\epsilon(\beta)} Q^\beta \sum_{n\geq 0} \frac{1}{n!}\Big(\sum_{\tilde{\beta}\mapsto\beta} \langle\underbrace{\gamma,\ldots,\gamma}_{n}\rangle_{0,n,\tilde{\beta}}^{X//\mathbf{T},\mathbf{S}}\Big).$$

Choose a homogeneous basis $\{\sigma_0 = 1, \sigma_1, \ldots, \sigma_r, \sigma_{r+1}, \ldots, \sigma_{m-1}\}$ of $H^*_{\mathbf{S}}(X//\mathbf{G})$ over $\mathbb{C}[\lambda] := \mathbb{C}[\lambda_1, \ldots, \lambda_\ell] = H^*(B\mathbf{S})$, such that $\{\sigma_1, \ldots, \sigma_r\}$ forms a basis of $H^2_{\mathbf{S}}(X//\mathbf{G})$ and *fix* homogeneous lifts $\gamma_i \in H^*_{\mathbf{S}}(X//\mathbf{T})^{\mathbf{W}}$ of σ_i. The fixed lifts give rise to a $\mathbb{C}$-linear embedding

(3.2.3) $$H^*_{\mathbf{S}}(X//\mathbf{G}, \mathbb{C}) \subset H^*_{\mathbf{S}}(X//\mathbf{T}, \mathbb{C})$$

(which may not in general be a homomorphism of equivariant cohomology rings).

The image of the embedding (3.2.3) determines a formal submanifold N of M over $N(X//\mathbf{T}) \otimes \mathbb{C}[\lambda]$.

Let

$$V := H^*_{\mathbf{S}}(X//\mathbf{T}, \mathbb{C})^a$$

be the subspace of $\mathbf{W}$-anti-invariant classes. The composition of (3.2.3) with the map

$$\cup\omega : H^*_{\mathbf{S}}(X//\mathbf{T}, \mathbb{C})^{\mathbf{W}} \to H^*_{\mathbf{S}}(X//\mathbf{T}, \mathbb{C})^a$$

is an isomorphism from $H^*_{\mathbf{S}}(X//\mathbf{G}, \mathbb{C})$ to V. Let $\mathcal{V} = V \otimes \mathcal{O}_N$ be the subsheaf of $\Theta_M|_N$ induced by V. Let $\star$ be the quantum product on Θ_M (that is, the pull-back by θ of the quantum product on $H^*_{\mathbf{S}}(X//\mathbf{T}, \mathbb{C})$). Consider the map

$$\star\omega : (\Theta_M|_N)^{\mathbf{W}} \longrightarrow \mathcal{V}, \quad \xi \mapsto (\hat{\xi} \star \omega)|_N,$$

with $\hat{\xi} \in \Theta_M^{\mathbf{W}}$ any extension of ξ to M. (It is well defined, since the quantum product of two vector fields at a point depends only on their values at the point.) This map reduces to $\cup\omega$ modulo the ideal generated by $\{Q^\beta | \beta \neq 0\}$. By Nakayama's lemma, $\star\omega$ induces an isomorphism $\Theta_N \to \mathcal{V}$. Let $\phi : \mathcal{V} \to \Theta_N$ be the inverse isomorphism. Abusing notation, when $\eta \in \mathcal{V}$ we write $\eta \star \omega^{-1}$ for $\phi(\eta)$. Hence we have for $\xi \in \Theta_N$

$$(\xi \star \omega) \star \omega^{-1} = \xi.$$

We now induce a structure of formal Frobenius manifold on N (over $N(X//\mathbf{G}) \otimes \mathbb{C}[\lambda]$) using the maps $\star\omega$ and $\star\omega^{-1}$. Explicitly:

(3.2.4) The metric ${}^\omega g$ on Θ_N is given by the composition

$$\Theta_N \otimes \Theta_N \hookrightarrow \Theta_M|_N \otimes \Theta_M|_N \xrightarrow{\star\omega \otimes \star\omega} \mathcal{V} \otimes \mathcal{V} \xrightarrow{g|_{\mathcal{V}}} \mathcal{O}_N,$$

that is,

$${}^\omega g(\xi, \eta) = g|_{\mathcal{V}}(\xi \star \omega, \eta \star \omega).$$

Note that $g|_{\mathcal{V}}$ is nondegenerate on $\mathcal{V}$ by Martin's formula (3.1.3).

(3.2.5) The Levi–Civita connection ${}^\omega\nabla$ of ${}^\omega g$ satisfies

$${}^\omega\nabla_\xi \eta = (\nabla_{\hat{\xi}}(\hat{\eta} \star \omega))|_N \star \omega^{-1}.$$

(3.2.6) The product of $\xi, \eta \in \Theta_N$ is defined by

$$\xi \circ \eta = (\xi \star \eta \star \omega) \star \omega^{-1}.$$

In other words, $\xi \circ \eta$ is the projection of $\xi \star \eta$ along $\ker(\star\omega)$.

(3.2.7) The unit is the vector field 1 restricted to N.

The symmetry of ${}^\omega\nabla({}^\omega g(\cdot \circ \cdot, \cdot))$ and the corresponding potential function are discussed in Sect. 3.5 below.

3.3. Flat coordinates. On N there are coordinates $\tilde{t}_0, \dots, \tilde{t}_{m-1}$ determined by the basis $\{\gamma_0 = 1, \gamma_1, \dots, \gamma_r, \dots, \gamma_{m-1}\}$ of lifts introduced above. These are just restrictions to N of coordinates on M which are flat for the connection ∇. Let

(3.3.1) $$\xi_i(\tilde{t}) := (\gamma_i \cup \omega) \star \omega^{-1}, \quad i = 0, \dots, m-1.$$

Equivalently, ξ_i is defined by the equality

(3.3.2) $$\xi_i \star \omega = \gamma_i \cup \omega.$$

The ξ_i's form a basis of Θ_N consisting of ${}^\omega\nabla$-horizontal vector fields. Denote by $s := (s_0, s_1, \dots, s_r, \dots, s_{m-1})$ the corresponding ${}^\omega\nabla$-flat coordinates on N (so that $\partial_{s_i} = \xi_i$). Note that

(3.3.3) $$s \equiv \tilde{t}, \text{ modulo the ideal generated by } \{Q^\beta | \beta \neq 0\}.$$

3.4. Euler vector field. Since

$$
\begin{aligned}
j^* c_1^{\mathbf{S}}(T(X//\mathbf{T})) &= c_1^{\mathbf{S}}(T(X^s(\mathbf{G})/\mathbf{T})) \\
&= \pi^*\big(c_1^{\mathbf{S}}(T(X//\mathbf{G}))\big) + \sum_{\alpha\in\Phi} c_1^{\mathbf{S}}(L_\alpha) \\
&= \pi^*\big(c_1^{\mathbf{S}}(T(X//\mathbf{G}))\big)
\end{aligned}
$$

and $\mathrm{Pic}_{\mathbf{S}}(X//\mathbf{T}) \cong \mathrm{Pic}_{\mathbf{S}}(X^s(\mathbf{G})/\mathbf{T})$ via j^*, we conclude that $c_1^{\mathbf{S}}(T(X//\mathbf{T}))$ is $\mathbf{W}$-invariant. Viewing $c_1^{\mathbf{S}}(T(X//\mathbf{T}))$ as a vector field on M, its restriction to N is therefore a section of $(\Theta_M|_N)^{\mathbf{W}}$. Moreover, this restriction is in fact tangent to N, since (by uniqueness of lifts of divisors) N contains the germ of linear subspace $H^2_{\mathbf{S}}(X//\mathbf{T})^{\mathbf{W}}$.

Define the Euler vector field by

$$
\mathfrak{E}_{\mathbf{S}} = \mathfrak{E} + \sum_{i=1}^{\ell} \lambda_i \partial_{\lambda_i}
$$

with

$$
\mathfrak{E} = \sum_{i=0}^{m-1}\left(1 - \frac{\deg \sigma_i}{2}\right)\tilde{t}_i \partial_{\tilde{t}_i} + c_1^{\mathbf{S}}(T(X//\mathbf{T}))|_N.
$$

Note that $\mathfrak{E}_{\mathbf{S}}$ is simply the restriction to N of the corresponding Euler vector field for $X//\mathbf{T}$ (see Sect. 2.2): Extend $\{\gamma_0, \ldots, \gamma_{m-1}\}$ to a basis of $H^*_{\mathbf{S}}(X//\mathbf{T})$ for the Euler vector field for $X//\mathbf{T}$.

Applying $\mathcal{L}_{\mathfrak{E}_{\mathbf{S}}}$ to the equality $(\xi_i \star \omega) = \partial_{\tilde{t}_i} \cup \omega$, we see that

$$
\mathcal{L}_{\mathfrak{E}_{\mathbf{S}}} \xi_i = \left(1 - \frac{\deg\, \sigma_i}{2}\right)\xi_i.
$$

Easy calculations show then that

$$
\mathcal{L}_{\mathfrak{E}_{\mathbf{S}}}({}^{\omega}g) = (2 - \dim(X//\mathbf{G}))\, {}^{\omega}g, \quad \mathcal{L}_{\mathfrak{E}_{\mathbf{S}}}(\circ) = \circ
$$

(and obviously $\mathcal{L}_{\mathfrak{E}_{\mathbf{S}}}(1) = -1$), hence $\mathfrak{E}_{\mathbf{S}}$ is indeed an Euler vector field. Also,

$$
\mathcal{L}_{\mathfrak{E}_{\mathbf{S}}}(s_i) = \deg(\tilde{t}_i) s_i,
$$

that is, $\deg s_i = \deg \tilde{t}_i$. In particular, $\deg s_1 = \cdots = \deg s_r = 0$.

3.5. Potential. Recall that we identify the cohomology classes on $X//\mathbf{T}$ with $\mathcal{O}_M$-linear vector fields on M. Denote by $\partial_{\tilde{t}_i \cup \omega}$ the vector field corresponding to $\gamma_i \cup \omega$. The components of the tensor "$\circ$" in the basis of ${}^{\omega}\nabla$-horizontal fields are

(3.5.1)
$$ {}^{\omega}g(\xi_i \circ \xi_j, \xi_k) = g|_V(\xi_i \star \xi_j \star \omega, \xi_k \star \omega) = g(\hat{\xi}_i \star (\gamma_j \cup \omega), \gamma_k \cup \omega)|_N $$

where $\hat{\xi}_i$ is any extension of ξ to a **W**-invariant vector field on M. Since

$$ \begin{aligned} g(\hat{\xi}_i \star (\gamma_j \cup \omega), \gamma_k \cup \omega)|_N &= g(\gamma_j \cup \omega, \hat{\xi}_i \star (\gamma_k \cup \omega))|_N \\ &= g|_V(\gamma_j \star \omega, \xi_i \star \xi_k \star \omega) = {}^{\omega}g(\xi_j, \xi_i \circ \xi_k), \end{aligned} $$

we see that the Frobenius algebra property

$$ {}^{\omega}g(\xi_i \circ \xi_j, \xi_k) = {}^{\omega}g(\xi_j, \xi_i \circ \xi_k) \tag{3.5.2} $$

holds. Recall the potential F (see (3.2.2)) of the formal Frobenius manifold M. We get from (3.5.1)

$$ \begin{aligned} {}^{\omega}g(\xi_i \circ \xi_j, \xi_k) &= g(\hat{\xi}_i \star (\gamma_j \cup \omega), \gamma_k \cup \omega)|_N \\ &= (\hat{\xi}_i(\partial_{\tilde{t}_j \cup \omega}\partial_{\tilde{t}_k \cup \omega}F))|_N = \xi_i((\partial_{\tilde{t}_j \cup \omega}\partial_{\tilde{t}_k \cup \omega}F)|_N). \end{aligned} \tag{3.5.3} $$

Note that

$$ \xi_l(\xi_i((\partial_{\tilde{t}_j \cup \omega}\partial_{\tilde{t}_k \cup \omega}F)|_N)) = \xi_i(\xi_l((\partial_{\tilde{t}_j \cup \omega}\partial_{\tilde{t}_k \cup \omega}F)|_N)), $$

since $[\xi_l, \xi_i] = 0$. Hence

$$ \xi_l\big({}^{\omega}g(\xi_i \circ \xi_j, \xi_k)\big) = \xi_i\big({}^{\omega}g(\xi_l \circ \xi_j, \xi_k)\big). $$

Combined with (3.5.2), this implies that the tensor $\xi_l({}^{\omega}g(\xi_i \circ \xi_j, \xi_k))$ is symmetric in the indices l, i, j, k, hence there is a (formal) function F' on N such that

$$ \partial_{s_i}\partial_{s_j}\partial_{s_k}F' = {}^{\omega}g(\xi_i \circ \xi_j, \xi_k). $$

Finally, a direct computation shows that $L_{\mathfrak{E}_{\mathbf{S}}}F' = (3 - \dim X//\mathbf{G})F'$ up to quadratic terms.

This finishes the construction of the induced structure of formal **S**-equivariant Frobenius manifold over $N(X//\mathbf{G}) \otimes \mathbb{C}[\lambda]$ on N.

3.6. More on the flat coordinates. For later use we record here some details about the change of coordinates $s_i(\tilde{t})$ on N. From the defining equation (3.3.1) for the horizontal vector fields ξ_i it follows that the jacobian matrix $A := (\partial s_i/\partial \tilde{t}_j)_{i,j}$ is given explicitly by

$$\frac{\partial s_i}{\partial \tilde{t}_j} = \left(\partial_{\tilde{t}_j}\partial^{\vee}_{\tilde{t}_i\cup\omega}\partial_{\tilde{t}_0\cup\omega}F\right)\Big|_N$$

where

$$\partial^{\vee}_{\tilde{t}_i\cup\omega} := \sum_k g^{ik}\partial_{\tilde{t}_k\cup\omega}$$

with $(g^{ik}) \in GL_m(\mathbb{C}[\lambda])$ the inverse matrix of the metric $g|_{\mathcal{V}}$. Using the divisor axiom for Gromov–Witten invariants of $X//\mathbf{T}$ in the formula (3.2.2) for the potential F, we see that the entries of the jacobian matrix have the form

$$\begin{aligned}(3.6.1)\quad \frac{\partial s_i}{\partial \tilde{t}_0} &= \delta_{i0}\\ \frac{\partial s_i}{\partial \tilde{t}_j} &= \delta_{ij} + \sum_{\beta\neq 0} Q^{\beta}e^{\beta\cdot\tilde{t}_{\text{small}}}c_{\beta,ij}(\tilde{t}_{r+1},\dots,\tilde{t}_{m-1}), \quad j\neq 0,\end{aligned}$$

where $c_{\beta,ij} \in \mathbb{C}[\lambda][[\tilde{t}_{r+1},\dots,\tilde{t}_{m-1}]]$ and

$$\beta\cdot\tilde{t}_{\text{small}} = \sum_{i=1}^{r}\tilde{t}_i\int_{\beta}\sigma_i.$$

By integrating (3.6.1) (with the initial condition $s(0) = 0$), we obtain a refined version of (3.3.3)

$$(3.6.2)\qquad s_i = \tilde{t}_i + \sum_{\beta\neq 0} Q^{\beta}e^{\beta\cdot\tilde{t}_{\text{small}}}b_{\beta,i}(\tilde{t}_{r+1},\dots,\tilde{t}_{m-1}),$$

with $b_{\beta,i} \in \mathbb{C}[\lambda][[\tilde{t}_{r+1},\dots,\tilde{t}_{m-1}]]$.

By considering the inverse jacobian matrix (which gives the map $\star\,\omega^{-1}$), it follows that the inverse coordinate change $\tilde{t}(s)$ is also of the type (3.6.2), hence the potential function F' in flat coordinates s_i has the special form (2.1.1) (up to quadratic terms)

$$(3.6.3)\qquad F' = F'_{cl} + \sum_{\beta\neq 0} Q^{\beta}e^{\beta\cdot s_{\text{small}}}F'_{\beta}(s_{r+1},\dots,s_{m-1})$$

where $\beta\cdot s_{\text{small}} = \sum_{i=1}^{r} s_i(\int_{\beta}\sigma_i)$.

Finally, we record what happens with the "small" parameter spaces under the change of coordinates.

Lemma 3.6.1. (*i*) *If $X//\mathbf{G}$ is Fano of index ≥ 2, then the subspaces of N given by the equations $\{s_0 = s_{r+1} = \cdots = s_{m-1} = 0\}$ and $\{\tilde{t}_0 = \tilde{t}_{r+1} = \cdots = \tilde{t}_{m-1} = 0\}$ coincide. Moreover, on this subspace we have $s_i = \tilde{t}_i$ for $i = 1,\dots,r$.*

(ii) If $c_1(T(X//\mathbf{G}))$ is nef, then the subspaces $\{s_{r+1} = \cdots = s_{m-1} = 0\}$ and $\{\tilde{t}_{r+1} = \cdots = \tilde{t}_{m-1} = 0\}$ coincide.

Proof. (*i*) Let $1 \leq i \leq r$. After restriction to $\tilde{t}_0 = \tilde{t}_{r+1} = \cdots = \tilde{t}_{m-1} = 0$ we obtain

$$\xi_i = \gamma_i + \sum_j \Big(\sum_\beta c_{\beta,ij} Q^\beta e^{\beta \cdot \tilde{t}_{\mathrm{small}}} \Big) \gamma_j .$$

Since $\deg \xi_i = 1$ and $\deg e^{\beta\cdot\tilde{t}_{\mathrm{small}}} = \int_\beta c_1(T(X//\mathbf{G})) \geq 2$, we must have $\xi = \gamma_i$ and the statement follows. The proof of (*ii*) is similar. □

3.7. Main conjecture. Let P be the formal $\mathbf{S}$-equivariant Frobenius manifold over $N(X//\mathbf{G}) \otimes \mathbb{C}[\lambda]$ defined by the genus zero $\mathbf{S}$-equivariant Gromov–Witten theory of $X//\mathbf{G}$, with flat coordinates $t_0, t_1, \ldots, t_r, \ldots, t_{m-1}$ corresponding to the $\mathbb{C}[\lambda]$-basis $\{\sigma_0 = 1, \sigma_1, \ldots, \sigma_r, \ldots, \sigma_{m-1}\}$ of $H^*_{\mathbf{S}}(X//\mathbf{G})$ and potential function $F^{X//\mathbf{G},\mathbf{S}}$. We are now ready to formulate the abelian/nonabelian correspondence:

Conjecture 3.7.1. Let $\varphi : P \longrightarrow N$ be the isomorphism of formal schemes over $N(X//\mathbf{G}) \otimes \mathbb{C}[\lambda]$ defined by $\varphi^*(s_i) = t_i$. Then φ induces an isomorphism of formal $\mathbf{S}$-equivariant Frobenius structures such that $\varphi^*(\xi_i) = \sigma_i$ and $\varphi^* F' = F^{X//\mathbf{G},\mathbf{S}}$ up to quadratic terms.

Note that $\varphi^*(\xi_i) = \sigma_i$ follows easily from (3.3.1). The main point of the conjecture is the identification of potentials. We also remark that the conjecture implies in particular that the new $\mathbf{W}$-induced Frobenius structure constructed in this section does not depend on the choice of the $\mathbf{W}$-invariant lift of $H^*_{\mathbf{S}}(X//\mathbf{G}, \mathbb{C})$.

4. Proof of Conjecture 3.7.1 for flag manifolds

4.1. Preliminaries. Let $0 < k_1 < \cdots < k_r < n = k_{r+1}$ be integers. Consider the vector space

$$\Omega := \bigoplus_{i=1}^{r} \mathrm{Mat}_{k_i \times k_{i+1}}(\mathbb{C})$$

where $\mathrm{Mat}_{k_i \times k_{i+1}}(\mathbb{C})$ is the space of matrices of size $k_i \times k_{i+1}$ with complex entries. Let $\mathbf{G} := \prod_{i=1}^r GL_{k_i}(\mathbb{C})$, with maximal torus $\mathbf{T}$ equal to the product of the subgroups of diagonal matrices. $\mathbf{G}$ acts on Ω by

$$(g_1, \ldots, g_r)(A_1, \ldots, A_r) = \big(g_1 A_1 g_2^{-1}, g_2 A_2 g_3^{-1}, \ldots, g_{r-1} A_{r-1} g_r^{-1}, g_r A_r\big).$$

This action descends to an action on $X := \mathbb{P}(\Omega)$, with a canonical linearization on $\mathcal{O}(1)$, and the GIT quotient $X//\mathbf{G}$ is the partial flag manifold $Fl(k_1, \ldots, k_r, n)$ parameterizing flags of subspaces $\{\mathbb{C}^{k_1} \subset \cdots \subset \mathbb{C}^{k_r} \subset \mathbb{C}^n\}$.

The corresponding abelian quotient $X//\mathbf{T}$ is a toric variety which can be realized as a tower of fibered products of projective bundles.

Let $\mathbf{S} \cong (\mathbb{C}^*)^n$ be the subgroup of diagonal matrices in $GL_n(\mathbb{C})$, acting on Ω by right-multiplication of A_r. There are induced $\mathbf{S}$-actions on $X//\mathbf{G}$ (which is just the usual action of the maximal torus in GL_n on the flag manifold) and on $X//\mathbf{T}$. See [BCK2, §5.1] for more details on $X//\mathbf{G}$, $X//\mathbf{T}$, and the $\mathbf{S}$-actions on them. As before, we let $\mathbb{C}[\lambda] = \mathbb{C}[\lambda_1, \dots, \lambda_n] = H^*(B\mathbf{S}, \mathbb{C})$, with quotient field $\mathbb{C}(\lambda)$. Our goal in this section is to prove

Theorem 4.1.1. *Conjecture 3.7.1 holds for*
(a) the usual Gromov–Witten theory of $Fl(k_1, \dots, k_r, n)$.
(b) the $\mathbf{S}$-equivariant Gromov–Witten theory of $Fl(k_1, \dots, k_r, n)$.

Remark 4.1.2. Note that part (a) follows from (b) by taking the non-equivariant limit $\lambda_1 = \cdots = \lambda_n = 0$ of the potential functions.

Our strategy for proving Theorem 4.1.1 is to use reconstruction theorems to reduce the statement to a comparison for 1-point invariants which was established in [BCK1,BCK2]. Typically, reconstruction results for Gromov–Witten invariants work under the assumption that the cohomology ring is generated by divisors. Our observation here is that in the torus-equivariant setting, this assumption needs only to hold after localization. This enlarges the class of varieties for which reconstruction is applicable. We begin with a simple lemma.

Lemma 4.1.3. *Let $\mathbb{P}^N$ be acted by a torus $\mathbf{S}$ and let Y be an $\mathbf{S}$-invariant smooth subvariety. Suppose that the natural map $H^*((\mathbb{P}^N)^{\mathbf{S}}) \to H^*(Y^{\mathbf{S}})$ is surjective (for example, this is true when the $\mathbf{S}$-fixed locus $(\mathbb{P}^N)^{\mathbf{S}}$ is isolated). Then the localized equivariant cohomology ring $H^*_{\mathbf{S}}(Y, \mathbb{C}) \otimes_{\mathbb{C}[\lambda]} \mathbb{C}(\lambda)$ is generated (as a $\mathbb{C}(\lambda)$-algebra) by divisors, i.e., by $\{c_1(L) \otimes 1 | L \in \mathrm{Pic}^{\mathbf{S}}(Y)\}$.*

Proof. There is a commutative diagram

$$\begin{array}{ccc} H^*_{\mathbf{S}}(\mathbb{P}^N) & \longleftarrow & H^*((\mathbb{P}^N)^{\mathbf{S}}) \otimes \mathbb{C}[\lambda] \\ \downarrow & & \downarrow \\ H^*_{\mathbf{S}}(Y) & \longleftarrow & H^*(Y^{\mathbf{S}}) \otimes \mathbb{C}[\lambda]. \end{array}$$

After tensor product with $\mathbb{C}(\lambda)$, the horizontal maps are isomorphism by the localization theorem. □

It is well-known that $X//\mathbf{G} = Fl(k_1, \dots, k_r, n)$ admits an $\mathbf{S}$-equivariant embedding into a product of projective spaces on which $\mathbf{S}$ acts with isolated fixed points. By Lemma 4.1.3, the localized equivariant cohomology

$$H^*_{\mathbf{S}}(X//\mathbf{G}, \mathbb{C}) \otimes_{\mathbb{C}[\lambda]} \mathbb{C}(\lambda)$$

is generated by divisor classes. Note that this is false in general without localization. For example the equivariant cohomology of Grassmannians is *not* generated by divisors. (On the other hand, since $X//\mathbf{T}$ is a toric variety, both the usual and $\mathbf{S}$-equivariant cohomology rings are already generated by divisors.)

Before going into the details of the proof, it is useful to discuss the base-change of Novikov rings (3.2.1) in the particular case of flag manifolds. By choosing the usual Schubert basis in $H_2(Fl(k_1, \dots, k_r, n), \mathbb{Z})$, the semigroup of effective curve classes is identified with $(\mathbb{Z}_+)^r$. We write $d = (d_1, \dots, d_r)$ for the typical element in this semigroup. The Novikov ring is simply the power series ring $\mathbb{C}[[Q_1, \dots, Q_r]]$. Similarly, effective curve classes on the toric variety $X//\mathbf{T}$ are described by tuples of non-negative integers

$$\tilde{d} = (d_{11}, \dots, d_{1k_1}, \dots, d_{r1}, \dots, d_{rk_r})$$

and the Novikov ring is identified with $\mathbb{C}[[Q_{ij} | 1 \le i \le r, 1 \le j \le k_i]]$. A class $\tilde{d}$ is a lift of d if and only if

$$d_i = \sum_{j=1}^{k_i} d_{ij}, \quad i = 1, \dots, r.$$

Finally, $\epsilon(d) = \sum_{i=1}^{r} (k_i - 1) d_i \pmod 2$. Hence the projection (3.2.1) of Novikov rings is

$$\text{(4.1.1)} \quad p : \mathbb{C}[[Q_{ij} | 1 \le i \le r,\ 1 \le j \le k_i]] \longrightarrow \mathbb{C}[[Q_1, \dots, Q_r]], \qquad Q_{ij} \longmapsto (-1)^{(k_i - 1)} Q_i.$$

4.2. Kontsevich–Manin reconstruction and reduction to 2-point invariants. This step involves an equivariant version of the Kontsevich–Manin reconstruction theorem. In its original formulation [KM], the reconstruction theorem states that if the cohomology ring $H^*(Y, \mathbb{C})$ is generated by divisors, then all Gromov–Witten invariants of Y can be reconstructed from 3-point invariants for which at least one insertion is a divisor class. These in turn are expressed in terms of 2-point invariants by using the divisor equation in Gromov–Witten theory. We give here an extension of reconstruction to the $\mathbf{S}$-equivariant setting.

Lemma 4.2.1. *Let Y be a smooth complex projective variety with $\mathbf{S}$-action. Let*

$$P := Spf\big((N(Y) \otimes \mathbb{C}[\lambda])\big[\big[H^*_{\mathbf{S}}(Y, \mathbb{C})^\vee\big]\big]\big),$$

endowed with the formal $\mathbf{S}$-equivariant Frobenius structure $(P, \star, g, 1, \mathfrak{E}_{\mathbf{S}})$ defined by the equivariant Gromov–Witten potential F^Y. Let $t = (t_0, t_1, \dots, t_r, \dots, t_{m-1})$ be the flat coordinates defined by a basis of $H^*_{\mathbf{S}}(Y, \mathbb{C})$, such *that* $t_{\text{small}} = (t_1, \dots, t_r)$ *are the coordinates on the small parameter space* $H^2_{\mathbf{S}}(Y, \mathbb{C})$. *Let* $G \in \mathcal{O}_P$ *be another formal function satisfying the WDVV equations. Assume that:*

(*i*) *In flat coordinates G has the form* (2.1.1)

$$G = G_{cl} + \sum_{\beta \in E, \beta \neq 0} Q^{\beta} e^{\beta \cdot t_{\text{small}}} G_{\beta},$$

with $G_\beta \in \mathbb{C}[\lambda][[t_{r+1}, \dots, t_{m-1}]]$ *and* G_{cl} *a cubic polynomial in the* t_i*'s (with coefficients in* $\mathbb{C}[\lambda]$*).*

(*ii*) $\mathcal{L}_{\mathfrak{E}_{\mathbf{S}}}(G) = (3 - \dim(Y))G$.

(*iii*) $G_{cl} = F^Y_{cl}$.

(*iv*) $\partial_{t_i} \partial_{t_j} G|_{t_{\text{small}}} = \partial_{t_i} \partial_{t_j} F^Y|_{t_{\text{small}}}$, *for all* i, j.

(*v*) *The localized equivariant cohomology ring* $H^*_{\mathbf{S}}(Y, \mathbb{C}) \otimes_{\mathbb{C}[\lambda]} \mathbb{C}(\lambda)$ *is generated by* $H^2_{\mathbf{S}}(Y, \mathbb{C})$ *as a* $\mathbb{C}(\lambda)$*-algebra.*

Then $G = F^Y$.

Proof. Let $P_{(\lambda)}$ be the **S**-equivariant Frobenius manifold defined by the localized Gromov–Witten theory of Y (see Sect. 2.1). The function G defines a formal **S**-equivariant Frobenius structure $(P, \circ, g, 1, \mathfrak{E}_{\mathbf{S}})$ over $N(Y) \otimes \mathbb{C}[\lambda]$, and a localized Frobenius structure over $N(Y) \otimes \mathbb{C}(\lambda)$ as well, by viewing it as a formal function on $P_{(\lambda)}$ via the natural (injective!) localization map $\iota : \mathcal{O}_P \to \mathcal{O}_{P_{(\lambda)}}$. It suffices to check that the localized potentials $F^Y_{(\lambda)} = \iota(F^Y)$ and $G_{(\lambda)} = \iota(G)$ are equal. The assumptions (*i*)–(*iii*) hold for the localized potentials as well (where in (*i*) we replace $\mathbb{C}[\lambda]$ by $\mathbb{C}(\lambda)$).

In the conformal case, a formal Frobenius structure satisfying (*i*) and (*ii*) is said to be of *qc-type* in [Man]. Such structure has "cup product", defined by

$$G_{cl} = \frac{1}{6} g\Big(\Big(\sum t_i \partial_{t_i}\Big) \cup \Big(\sum t_i \partial_{t_i}\Big), \sum t_i \partial_{t_i}\Big)$$

and "correlators"

$$I_{0,n,\beta}(\partial_{t_{i_1}}, \dots, \partial_{t_{i_n}}) = \partial_{t_{i_1}} \dots \partial_{t_{i_n}} G_\beta|_{t=0}$$

which satisfy the analogue of the divisor axiom in Gromov–Witten theory. See [Man, § 5.4]. The same will hold for the Frobenius structure defined by our potential G, or for its localized version. We may call them Frobenius structures of equivariant qc-type.

Assumption (v) and the usual Kontsevich–Manin reconstruction imply that the localized GW-potential $F^Y_{(\lambda)}$ is determined recursively by $\partial_{t_i} \partial_{t_j} F^Y_{(\lambda)}|_{t_{\text{small}}}$. The proof only uses properties of Gromov–Witten invariants which are shared by the correlators of any Frobenius structure of equivariant qc-type, hence it will work in the abstract case as well.

Assumption (*iii*) says that the abstract cup product coincides with the usual one on cohomology. By the above discussion reconstruction applies and we find that $G_{(\lambda)}$ is determined recursively by $\partial_{t_i}\partial_{t_j}G_{(\lambda)}|_{t_{\text{small}}}$, *with the same recursion coefficients* in $\mathbb{C}(\lambda)$ as those for $F^Y_{(\lambda)}$. By assumption (*iv*), we are done. □

We go back now to the proof of Theorem 4.1. We intend to apply Lemma 4.2.1 to $Y = Fl$, $G = \varphi^* F'$. Note that assumption (v) holds by Lemma 4.1.3, assumption (i) holds by (3.6.3), while (ii) and (iii) are immediate from the construction of F' in Sects. 3.2–3.5. Hence the theorem will be proved if we can show that (iv) holds as well, i.e.,

$$\partial_{t_i}\partial_{t_j}F^{Fl,\mathbf{S}}|_{t_0=t_{r+1}=\cdots=t_{m-1}=0} = \varphi^*(\partial_{s_i}\partial_{s_j}F'|_{s_0=s_{r+1}=\cdots=s_{m-1}=0}). \tag{4.2.1}$$

Recall that (in the notation of Sect. 2.2)

$$\partial_{t_i}\partial_{t_j}F^{Fl,\mathbf{S}}(t) = \langle\langle\sigma_i, \sigma_j\rangle\rangle.$$

Setting $t_0 = t_{r+1} = \cdots = t_{m-1} = 0$ and using the divisor axiom we get

$$\partial_{t_i}\partial_{t_j}F^{Fl,\mathbf{S}}|_{t_0=t_{r+1}=\cdots=t_{m-1}=0} = \sum_{d=(d_1,\ldots,d_r)}\prod_{l=1}^{r}\left(Q_l e^{t_l}\right)^{d_l}\langle\sigma_i, \sigma_j\rangle^{Fl,\mathbf{S}}_{0,2,d}. \tag{4.2.2}$$

On the other hand, since

$$\partial_{s_k}\partial_{s_i}\partial_{s_j}F' = \partial_{s_k}((\partial_{\tilde{t}_i\cup\omega}\partial_{\tilde{t}_j\cup\omega}F)|_N)$$

it follows that

$$\partial_{s_i}\partial_{s_j}F' = (\partial_{\tilde{t}_i\cup\omega}\partial_{\tilde{t}_j\cup\omega}F)|_N \tag{4.2.3}$$

up to a constant (in the base ring). By adding appropriate quadratic terms to F', we may assume that (4.2.3) holds exactly. (Recall that F is the $\mathbf{S}$-equivariant Gromov–Witten potential of $X//\mathbf{T}$ with the Novikov variables specialized as in (4.1.1).) Moreover, the first Chern class of the flag manifold satisfies

$$\int_d c_1(T(Fl)) = \sum_{l=1}^{r} d_l(k_{l+1}-k_{l-1}) \geq \min_l\{k_{l+1}-k_{l-1}\} \geq 2.$$

Therefore the specialization of the left-hand side of (4.2.3) to $s_0 = s_{r+1} = \cdots = s_{m-1} = 0$ is equal to its specialization at $\tilde{t}_0 = \tilde{t}_{r+1} = \cdots = \tilde{t}_{m-1} = 0$ by Lemma 3.6.1 (i). Using the divisor axiom as above in the right hand side of (4.2.3), we conclude that

(4.2.4)

$$\partial_{s_i}\partial_{s_j}F'|_{s_{r+1}=\cdots=s_{m-1}=0} = \sum_{d=(d_1,\ldots,d_r)}\prod_{l=1}^{r}\left(Q_l e^{\tilde{t}_l}\right)^{d_l}\Big(\sum_{\tilde{d}\mapsto d}(-1)^{\sum(k_l-1)d_l}\langle\gamma_i\cup\omega, \gamma_j\cup\omega\rangle^{X//\mathbf{T},\mathbf{S}}_{0,2,\tilde{d}}\Big).$$

Using (4.2.2) and (4.2.4), the proof of (4.2.1), and therefore of Theorem 4.1.1, is reduced to checking the following identity among 2-point invariants:

$$\langle \sigma_i, \sigma_j \rangle_{0,2,d}^{Fl,\mathbf{S}} = \sum_{\tilde{d} \mapsto d} (-1)^{\sum (k_l - 1) d_l} \langle \gamma_i \cup \omega, \gamma_j \cup \omega \rangle_{0,2,\tilde{d}}^{X//\mathbf{T},\mathbf{S}}. \tag{4.2.5}$$

4.3. Lee–Pandharipande reconstruction and reduction to 1-point invariants. There is another reconstruction theorem, due to Lee and Pandharipande [LP1], and independently to Bertram and Kley, which reduces in certain cases computations of (descendant) GW-invariants with any number of insertions to 1-point descendants. In fact, Lee and Pandharipande deduce the reconstruction result from universal relations they found among divisors in the Picard group of the moduli space $\overline{M}_{0,2}(\mathbb{P}^N, d[line])$ of 2-pointed stable maps to $\mathbb{P}^N$. We establish first a straightforward equivariant extension of their divisor relation.

Let Y be a projective variety with $\mathbf{S}$-action. Let $\overline{M}_{0,2}(Y, \beta)$ be the moduli space of 2-pointed genus zero stable maps with evaluation maps

$$ev_1, ev_2 : \overline{M}_{0,2}(Y, \beta) \longrightarrow Y.$$

The moduli space inherits an $\mathbf{S}$-action and the evaluation maps are equivariant. Let $\psi = \psi_1$ be the $\mathbf{S}$-equivariant first Chern class of the line bundle on $\overline{M}_{0,2}(Y, \beta)$ with fiber $T^*_{x_1} C$ over the point $[f : (C, x_1, x_2) \to Y]$.

There is a "boundary divisor" $D_{1,\beta_1|2,\beta_2}$ in $\overline{M}_{0,2}(Y, \beta)$ corresponding to maps with reducible domains and splitting type $\beta_1 + \beta_2 = \beta$. It is obtained as the image of the ($\mathbf{S}$-equivariant) gluing morphism

$$j_{\beta_1,\beta_2} : \overline{M}_{0,\{x_1,\bullet\}}(Y, \beta_1) \times_Y \overline{M}_{0,\{x_2,\bullet\}}(Y, \beta_2) \longrightarrow \overline{M}_{0,2}(Y, \beta)$$

and one defines its virtual fundamental class in the equivariant Chow group $A_*^{\mathbf{S}}(\overline{M}_{0,2}(Y, \beta), \mathbb{Q})$ as the push-forward of

$$[\overline{M}_{0,\{x_1,\bullet\}}(Y, \beta_1)]^{\mathrm{vir}} \boxtimes [\overline{M}_{0,\{x_2,\bullet\}}(Y, \beta_2)]^{\mathrm{vir}}.$$

Lemma 4.3.1. *For all $L \in \mathrm{Pic}^{\mathbf{S}}(Y)$, the relation*

$$ev_2^*(L) \cap [\overline{M}_{0,2}(Y, \beta)]^{\mathrm{vir}} - \Big(ev_1^*(L) + \Big(\int_\beta L\Big)\psi\Big) \cap [\overline{M}_{0,2}(Y, \beta)]^{\mathrm{vir}} + \sum_{\beta_1+\beta_2=\beta} \Big(\int_{\beta_2} L\Big) \cap [D_{1,\beta_1|2,\beta_2}]^{\mathrm{vir}} = 0$$

holds in $A_*^{\mathbf{S}}(\overline{M}_{0,2}(Y, \beta), \mathbb{Q})$.

Proof. As in [LP1], since the relation is linear in L and the equivariant Picard group is spanned over $\mathbb{Q}$ by $\mathbf{S}$-equivariant very ample line bundles, the Lemma will follow from the case $Y = \mathbb{P}^N$, $\beta = d[line]$ and the stronger statement

$$(4.3.1)\quad ev_2^*(L) - ev_1^*(L) - \Big(\int_\beta L\Big)\psi + \sum_{\beta_1+\beta_2=\beta} \Big(\int_{\beta_2} L\Big) D_{1,\beta_1|2,\beta_2} = 0$$

in $\mathrm{Pic}^{\mathbf{S}}(\mathbb{P}^N)$.

The relation (4.3.1) holds after passing to the non-equivariant limit $\lambda_i = 0$ by [LP1, Theorem 1]. Therefore the left-hand side is a linear polynomial in the λ_i's and the corresponding equivariant line bundle is just a trivial bundle twisted by a character of $\mathbf{S}$. To check that this character is trivial, it suffices to restrict to *any* $\mathbf{S}$-fixed point of $\overline{M}_{0,2}(\mathbb{P}^N, d[line])$. There are many possible choices of fixed points that will work. One particular such for which the computation is very easy is the point corresponding to a stable map with domain $C \cup D$ (the union of two irreducible components) such that $x_1, x_2 \in C$ and $f : C \cup D \to \mathbb{P}^N$ collapses C to a fixed point $p \in \mathbb{P}^N$ and maps D with degree d onto an $\mathbf{S}$-invariant line in $\mathbb{P}^N$ joining p to another fixed point q, such that the map is totally ramified at q. The classes ψ and $D_{1,\beta_1|2,\beta_2}$ vanish when restricted to this point, while $ev_1^* L$ and $ev_2^* L$ have the same restriction. Relation (4.3.1) and hence the lemma are proved. □

We will use Lemma 4.3.1 to obtain a reconstruction result in the context of the abelian/nonabelian correspondence. Recall that descendant (genus 0) Gromov–Witten invariants of a smooth projective Y are defined by

$$\langle \tau_{a_1}(\gamma_1), \ldots, \tau_{a_n}(\gamma_n)\rangle^Y_{0,n,\beta} := \int_{[M_{0,n}(Y,\beta)]^{\mathrm{vir}}} \prod_i \psi_i^{a_i} ev_i^*(\gamma_i),$$

where $\gamma_i \in H^*(Y)$ and ψ_i are the first Chern classes of the cotangent line bundles at the marked points. The definition extends to torus-equivariant descendants (which will be $\mathbb{C}[\lambda]$-valued). We establish first an auxiliary vanishing result for certain descendant invariants of $X//\mathbf{T}$.

Let $X//\mathbf{G}$, $X//\mathbf{T}$, $\mathbf{S}$ etc. be as in the setting Sect. 3.1. Let $\beta \in H_2(X//\mathbf{G}, \mathbb{Z})$ be fixed. Consider the moduli space

$$\mathcal{M}_\beta := \coprod_{\tilde\beta \mapsto \beta} \overline{M}_{0,n}(X//\mathbf{T}, \tilde\beta)$$

with the obvious evaluation maps $ev_i : \mathcal{M}_\beta \longrightarrow X//\mathbf{T}$, $i = 1, \ldots, n$ and virtual class $[\mathcal{M}_\beta]^{\mathrm{vir}}$. Note that

$$H^*_{\mathbf{S}}(\mathcal{M}_\beta, \mathbb{C}) \cong \bigoplus_{\tilde\beta \mapsto \beta} H^*_{\mathbf{S}}(\overline{M}_{0,n}(X//\mathbf{T}, \tilde\beta), \mathbb{C}).$$

Introduce "psi-classes" on $\mathcal{M}_\beta$ by

$$\psi_i := \sum_{\tilde{\beta}\mapsto\beta} \psi_{i,\tilde{\beta}}$$

and define for cohomology classes $\gamma_1, \ldots, \gamma_n \in H^*_{\mathbf{S}}(X//\mathbf{T})$.

$$\begin{aligned} I_{n,\beta}(\tau_{a_1}(\gamma_1), \ldots, \tau_{a_n}(\gamma_n)) &:= (-1)^{\epsilon(\beta)} \int_{[\mathcal{M}_\beta]^{\mathrm{vir}}} \prod_i \psi_i^{a_i} ev_i^*(\gamma_i) \\ &= (-1)^{\epsilon(\beta)} \sum_{\tilde{\beta}\mapsto\beta} \langle \tau_{a_1}(\gamma_1), \ldots, \tau_{a_n}(\gamma_n)\rangle_{0,n,\tilde{\beta}}^{X//\mathbf{T},\mathbf{S}}. \end{aligned}$$

Recall that the intersection form is non-degenerate on the $\mathbf{W}$-anti-invariant subspace $H^*_{\mathbf{S}}(X//\mathbf{T})^a$. We denote the orthogonal complement by $(H^*_{\mathbf{S}}(X//\mathbf{T})^a)^\perp$.

Lemma 4.3.2. *If $\tilde{\sigma}_1, \ldots, \tilde{\sigma}_{n-1}$ are $\mathbf{W}$-invariant lifts of classes σ_i in $H^*_{\mathbf{S}}(X//\mathbf{G})$ and $\gamma \in (H^*_{\mathbf{S}}(X//\mathbf{T})^a)^\perp$, then*

$$I_{n,\beta}(\tau_{a_1}(\tilde{\sigma}_1 \cup \omega), \tau_{a_2}(\tilde{\sigma}_2), \ldots, \tau_{a_{n-1}}(\tilde{\sigma}_{n-1}), \tau_{a_n}(\gamma)) = 0.$$

Proof. The $\mathbf{W}$-action on $X//\mathbf{T}$ induces a $\mathbf{W}$-action on $\mathcal{M}_\beta$, by composing stable maps with the automorphisms in $\mathbf{W}$. The evaluation maps are easily seen to be $\mathbf{W}$-equivariant. Note also that the psi-classes are $\mathbf{W}$-invariant. Hence the class

$$(ev_n)_*\Big(ev_1^*(\omega)\prod_{i=1}^{n}\psi_i^{a_i}\prod_{i=1}^{n-1} ev_i^*(\tilde{\sigma}_i) \cap [\mathcal{M}_\beta]^{\mathrm{vir}}\Big)$$

is $\mathbf{W}$-anti-invariant. The lemma now follows from the projection formula. □

Proposition 4.3.3. *Let $X//\mathbf{G}$, $X//\mathbf{T}$, $\mathbf{S}$ be as in the setting Sect. 3.1. Assume that the localized equivariant cohomology $H^*_{\mathbf{S}}(X//\mathbf{G}, \mathbb{C}) \otimes_{\mathbb{C}[\lambda]} \mathbb{C}(\lambda)$ is generated as a $\mathbb{C}(\lambda)$-algebra by divisors (that is, by $c_1(L) \otimes 1$ for $L \in \mathrm{Pic}^{\mathbf{S}}(X//\mathbf{G})$). Let σ_i, σ_j be any equivariant cohomology classes on $X//\mathbf{G}$, with $\mathbf{W}$-invariant lifts $\tilde{\sigma}_i, \tilde{\sigma}_j$ to $X//\mathbf{T}$. If the identity*

$$\langle \tau_a(\sigma_i), \sigma_j\rangle_{0,2,\beta}^{X//\mathbf{G},\mathbf{S}} = \sum_{\tilde{\beta}\mapsto\beta} (-1)^{\epsilon(\beta)} \langle \tau_a(\tilde{\sigma}_i \cup \omega), \tilde{\sigma}_j \cup \omega\rangle_{0,2,\tilde{\beta}}^{X//\mathbf{T},\mathbf{S}} \tag{4.3.2}$$

holds for $\sigma_j = 1$, then it holds in general.

Proof. It is enough to prove the Proposition for a *fixed* choice of lifts of cohomology classes on $X//\mathbf{G}$ to $X//\mathbf{T}$.

It follows immediately from Martin's integration formula (3.1.3) that

$$\widetilde{\sigma' \cup \sigma''} \cup \omega = \tilde{\sigma}' \cup \tilde{\sigma}'' \cup \omega \tag{4.3.3}$$

for any $\sigma', \sigma'' \in H^*_{\mathbf{S}}(X//\mathbf{G}, \mathbb{C})$ (see e.g. [BCK1, Cor. 2.3] for an argument).

Assume first that the equivariant cohomology ring of $X//\mathbf{G}$ is generated by divisors without localization (this happens for example when $X//\mathbf{G}$ is the complete flag manifold $Fl(1, 2, \ldots, n-1, n)$). Using Lemma 4.3.1 and the splitting axiom for GW-invariants we find that $\langle \tau_a(\sigma_i), \sigma_j \rangle^{X//\mathbf{G},\mathbf{S}}_{0,2,\beta}$ is expressed recursively (with $\mathbb{C}[\lambda]$-coefficients) in terms of invariants $\langle \tau_{a'}(\sigma'), 1 \rangle^{X//\mathbf{G},\mathbf{S}}_{0,2,\beta'}$ (these can be further reduced to 1-point descendants by the fundamental class axiom for GW-invariants). This is just the reconstruction of Lee-Pandharipande.

Recall the notation $I_{2,\beta}(\tau_a(\tilde{\sigma}_i \cup \omega), \tilde{\sigma}_j \cup \omega)$ introduced above for the right-hand side of the identity (4.3.2). The divisor relation in Lemma 4.3.1 can be extended in an obvious manner to the moduli space $\mathcal{M}_\beta$ for $\mathbf{W}$-invariant lifts $\tilde{L}$ of line bundles $L \in \mathrm{Pic}^{\mathbf{S}}(X//\mathbf{G})$. By Lemma 4.3.2, the reconstruction procedure applies to the invariants $I_{2,\beta}(\tau_a(\tilde{\sigma}_i \cup \omega), \tilde{\sigma}_j \cup \omega)$ and (by (4.3.3) and the equality $\epsilon(\beta_1 + \beta_2) = \epsilon(\beta_1) + \epsilon(\beta_2)$) it expresses them in terms of $I_{2,\beta'}(\tau_{a'}(\tilde{\sigma}' \cup \omega), \omega)$ with the <u>same</u> recursion coefficients. The proposition is proved in this case.

In the general case the same argument will work word for word, except that the recursion coefficients will now be rational functions rather than polynomials in the λ_i's. □

Remark 4.3.4. In view of Lemma 4.3.2, one might be tempted to try to extend the version of Lee–Pandharipande reconstruction above to descendants with any number of insertions and ψ-classes at all points. However, this is not possible, because an analogue of the fundamental class axiom does not hold for the invariants $I_{n,\beta}$ (indeed, in general $I_{n,\beta}(\tilde{\sigma}_1 \cup \omega, \tilde{\sigma}_2, \ldots, \tilde{\sigma}_{n-1}, \omega) \neq 0$). This is the reason for which the "twisting" by $\star\,\omega$ is necessary.

Corollary 4.3.5. *The following identity holds between Gromov–Witten invariants of $X//\mathbf{G} = Fl(k_1, \ldots, k_r, n)$ and those of the abelian quotient $X//\mathbf{T}$: for any $d = (d_1, \ldots, d_r) \in H_2(Fl, \mathbb{Z})$, any $a \geq 0$ and any equivariant cohomology classes σ_i, σ_j on Fl, with lifts γ_i, γ_j respectively,*

$$\langle \tau_a(\sigma_i), \sigma_j \rangle^{Fl,\mathbf{S}}_{0,2,d} = \sum_{\tilde{d} \mapsto d} (-1)^{\sum (k_a - 1) d_a} \langle \tau_a(\gamma_i \cup \omega), \gamma_j \cup \omega \rangle^{X//\mathbf{T},\mathbf{S}}_{0,2,\tilde{d}}.$$

Proof. By Lemmas 4.1.3 and 4.3.1, it suffices to check that

$$\langle \tau_a(\sigma_i), 1 \rangle^{Fl,\mathbf{S}}_{0,2,d} = \sum_{\tilde{d} \mapsto d} (-1)^{\sum (k_a - 1) d_a} \langle \tau_a(\gamma_i \cup \omega), \omega \rangle^{X//\mathbf{T},\mathbf{S}}_{0,2,\tilde{d}}. \tag{4.3.4}$$

This is (essentially) proved in [BCK1,BCK2]. However, since the actual statement is explicitly written (see formula (5) on p. 124 and Remark on p. 125 in [BCK1]) only for Grassmannians and non-equivariant invariants, we should say a few words here.

For the general flag manifold, a correspondence between the equivariant "small" J-functions of Fl and $X//\mathbf{T}$ is given by [BCK2, Theorem 1] (see the next section below for more about J-functions). Reading the argument in [BCK1, pp. 124–125] backwards[1], the equality (4.3.4) follows from the J-functions correspondence, provided that for any factorization

$$\omega = \left(\sqrt{\frac{(-1)^{|\Phi_+|}}{|\mathbf{W}|}}\right)\Big(\prod_{\alpha\in A} c_1^{\mathbf{S}}(L_\alpha)\Big) \cup \Big(\prod_{\alpha\in\Phi_+\setminus A} c_1^{\mathbf{S}}(L_\alpha)\Big)$$

we have

$$\left(\sqrt{\frac{(-1)^{|\Phi_+|}}{|\mathbf{W}|}}\right)\Big(\prod_{\alpha\in A} c_1^{\mathbf{S}}(L_\alpha)\Big) \star_{\text{small}} \Big(\prod_{\alpha\in\Phi_+\setminus A} c_1^{\mathbf{S}}(L_\alpha)\Big) = \omega, \tag{4.3.5}$$

where $\star_{\text{small}}$ is the small equivariant quantum product on $X//\mathbf{T}$, restricted to $H^2_{\mathbf{S}}(X//\mathbf{T}, \mathbb{C})^{\mathbf{W}}$ and with the Novikov variables specialized as in (4.1.1). By a simple degree counting, this last equality is always true when $X//\mathbf{G}$ (and hence $X//\mathbf{T}$, cf. Sect. 3.4) is a Fano variety. Indeed, the left-hand side of (4.3.5) is $\mathbf{W}$-anti-invariant, homogeneous, and of the form

$$\omega + \text{quantum corrections}.$$

However, ω is the unique class of lowest degree in $H^*_{\mathbf{S}}(X//\mathbf{T}, \mathbb{C})^a$, and in the Fano case the quantum parameters have positive degree. Hence the quantum corrections must vanish. □

It remains to observe that relation (4.2.5) is a special case of the corollary to conclude the proof of Theorem 4.1.1. □

Note that the only instance in this section where we have used that $X//\mathbf{G}$ is a flag manifold was in quoting the small J-function correspondence from [BCK2]. In other words, we have proved

Theorem 4.3.6. *Let $X, \mathbf{G}, \mathbf{T}, \mathbf{S}$ etc. be as in the setting Sect. 3.1. Assume that $X//\mathbf{G}$ is Fano of index ≥ 2 and that its equivariant cohomology is generated by divisors after localization. Then Conjecture 3.7.1 holds if and only if* (4.3.4) *holds, if and only if the abelian/nonabelian correspondence for small J-functions holds.*

[1] The specialization of the t_i-variables there corresponds exactly to our specialization (4.1.1) of the Novikov variables Q_i here.

A similar statement holds if we only assume that $c_1(T(X//\mathbf{G}))$ is nef, by using Lemma 3.6.1 *(ii)* in the argument just above (4.2.4). However, the change of coordinates $s(\tilde{t})$ will be nontrivial even for the restriction to subspace $\{s_{r+1} = \cdots = s_{m-1} = 0\}$, and coincides with the change of coordinates in the abelian/nonabelian correspondence for small J-functions (see [BCK2, Conjecture 4.3]). This is precisely analogous to the mirror theorem [Giv1] for hypersurfaces in projective space. We leave the precise formulation for the interested reader.

5. The abelian/nonabelian correspondence for J-functions

Our goal in this section is to explain why Conjecture 3.7.1 is equivalent to (an extension to the big parameter space of) the correspondence between the J-functions of $X//\mathbf{G}$ and $X//\mathbf{T}$ proposed in [BCK2, Conjecture 4.3]. In particular, by Theorem 4.1.1 and Corollary 5.3.4 below, the correspondence holds for the flag manifolds $Fl(k_1, \ldots, k_r, n)$.

5.1. Deformed flat coordinates. First we recall the definition of deformed flat coordinates following Dubrovin [Du1,Du2,Du3]. Let M be a Frobenius manifold (say, analytic, for simplicity), with Euler vector field. There is a deformed flat connection ∇^z on Θ_M given by

$$\nabla^z_\xi \eta := \nabla_\xi \eta - z^{-1} \xi \star \eta$$

(see [Du1, p. 189 and p. 323] and also [Du3]; however, we follow Givental for the convention on z). By identifying the cotangent sheaf Ω^1_M and the tangent sheaf Θ_M via the flat metric, a deformed flat connection is induced on Ω^1_M. A coordinate system J_i of M is called a *deformed flat coordinate system* if dJ_i are horizontal sections. In other words, J_i form a complete solution space to the second order linear PDE system

$$z\partial_{t_i}\partial_{t_j} J = \sum_{\gamma} c_{ij}^k \partial_{t_k} J \tag{5.1.1}$$

where t_i are flat coordinates and c_{ij}^k are structure constants of multiplications, i.e., $\partial_{t_i} \star \partial_{t_j} = \sum_k c_{ij}^k \partial_{t_k}$.

Suppose that the potential function F (defined up to quadratic terms) for the Frobenius structure is of the form $F = F_c + F_q$, with F_c a cubic form of the flat coordinates t_i and $F_q \in \mathbb{C}[[q_1, \ldots, q_r, t_{r+1}, \ldots, t_R]]$ such that $q_i = e^{t_i}$ and $F_q \equiv 0$ modulo the ideal $(q_1, \ldots, q_r)$ (cf. 2.1.1).

Consider the normalization condition

$$\sum J_i \partial_{t_i} \equiv z e^{t/z} = z\partial_{t_0} + t + O(z^{-1}) \quad (\text{mod } (q_1, \ldots, q_r)),$$

where $t = \sum t_i \partial_{t_i}$, the products of vector fields in the exponential are the "cup" products (determined by $F_{cl} = (1/6)g(t \cup t, t)$) and $1 = \partial_{t_0}$. The normalization uniquely determines deformed flat coordinates once the flat coordinates are chosen (see [Du2, Lemma 2.2]). We will call $\sum J_i \partial_i$ the *J-function* if it is normalized as above.

5.2. *J*-functions in quantum cohomology. The J-function plays an important role in Gromov–Witten theory. Let Y be a projective algebraic manifold. Then the J-function J_Y for the (formal) Frobenius structure defined by the quantum cohomology of Y can be constructed using descendant Gromov–Witten invariants. Let $\{\phi_i\}$ be a homogeneous basis of $H^*(Y)$, with Poincaré dual basis $\{\phi^i\}$. Let $t := \sum_i t_i \phi_i$. J_Y coincides with the assignment

$$H^*(Y) \ni t \mapsto z + t + \sum_i \phi^i \left\langle\!\!\left\langle \frac{\phi_i}{z-\psi} \right\rangle\!\!\right\rangle \quad \in z + t + \mathcal{H}_- \tag{5.2.1}$$

cf. [CG,Giv3], where $\mathcal{H}_- = \frac{1}{z} H^*(Y) \otimes_{\mathbb{C}} N[Y][[\frac{1}{z}]]$.

Here we use the double-bracket notation introduced in Sect. 2.2, so that

$$\left\langle\!\!\left\langle \frac{\phi_i}{z-\psi} \right\rangle\!\!\right\rangle = \sum_{\beta \in NE_1} Q^\beta \sum_{n \geq 0} \frac{1}{n!} \int_{[\overline{M}_{0,n+1}(Y,\beta)]^{\text{vir}}} \frac{ev_1^*(\phi_i)}{z-\psi} ev_2^*(t) \ldots ev_{n+1}^*(t)$$

where $\psi = \psi_1$ and $1/(z-\psi)$ is formally expanded as a geometric series.

The normalization condition

$$J_Y(t,z) \equiv z e^{t/z}$$

modulo quantum corrections follows from the well-known result

$$\int_{\overline{M}_{0,n}} \psi_1^{l_1} \ldots \psi_n^{l_n} = (n-3)!/l_1! \ldots l_n! \quad \text{if } \sum l_i = n-3.$$

(Note that in the paper [BCK2] $J_Y(t,z)/z$ is used for J-function, i.e., a different normalization.)

5.3. The abelian/nonabelian correspondence. Let $X, \mathbf{G}, \mathbf{T}$ be as in the setting Sect. 3.1. (For simplicity, we do not consider the equivariant theory here; the interested reader can readily make the necessary modifications to cover this case as well.) We have the $\mathbf{W}$-induced Frobenius structure over the Novikov ring $N(X//\mathbf{G})$ constructed in Sects. 3.2–3.6. We will keep the notations, and make liberal use of all its properties proved there. Moreover, *from now on, we assume that Conjecture 3.7.1 holds for* $X//\mathbf{G}$ *and* $X//\mathbf{T}$.

If $J_{X//\mathbf{G}} = \sum_{i=0}^{m-1} J_{i,X//\mathbf{G}}(t_0, \ldots, t_{m-1}, z)\sigma_i$ is the J function of $X//\mathbf{G}$, as given by (5.2.1), put

$$\tilde{J}_{X//\mathbf{G}}(t,z) := \sum_{i=0}^{m-1} J_{i,X//\mathbf{G}}(t_0, \ldots, t_{m-1}, z)\gamma_i.$$

(Recall that γ_i's are chosen $\mathbf{W}$-invariant lifts of the σ_i's.)

Lemma 5.3.1. $\tilde{J}_{X//G}(t,z) \cup \omega = (z\partial_\omega J_{X//T})\big|_{Q^{\tilde{\beta}}=(-1)^{\epsilon(\beta)}Q^\beta,N}(\varphi(t),z).$

Proof. Both sides satisfy the normalization condition $J \equiv ze^{t/z} \cup \omega$ modulo quantum corrections. Therefore it suffices to check that $\{\partial_\omega J_i\}_i$ forms a deformed flat coordinate system for $(N, \circ, {}^\omega g, e, \mathfrak{E})$ if J_δ is a deformed flat coordinate for $(M, \star, g, e, \mathfrak{E})$ such that $\{J_i|_N\}_i$ form a coordinate system of N. Indeed, by Conjecture 3.7.1, which we're assuming, the Frobenius manifolds P and N are isomorphic via φ.

First, we rewrite the PDE (5.1.1) as

$$z\partial_i\partial_j J = (\partial_i \star \partial_j)J. \tag{5.3.1}$$

This is useful in computations.

Next, if ξ and η are ${}^\omega\nabla$-horizontal vector fields, then

$$\begin{aligned} z\partial_{\xi\circ\eta}\partial_\omega J_i &= \partial_{(\xi\circ\eta)\star\omega} J_i \\ &= \partial_{\xi\star(\eta\star\omega)} J_i \\ &= z\partial_\xi \partial_{\eta\star\omega} J_i \\ &= z^2 \partial_\xi \partial_\eta \partial_\omega J_i \end{aligned}$$

since ω and $\eta \star \omega$ are ∇-horizontal. □

Remark 5.3.2. Lemma 5.3.1 reveals the relation between $\tilde{t}$ and the $s = \varphi(t)$:

$$\tilde{t} = s + \sum_{n=0}^{\infty} \frac{(-1)^{\epsilon(\beta)}Q^\beta}{n!} \sum_{i,\ 0\neq\tilde{\beta}\mapsto\beta} \gamma_i < \gamma^i \cup \omega, \omega, \underbrace{s, \dots, s}_{n} >_{0,n+2,\tilde{\beta}},$$

where $\{\gamma^j \cup \omega\}$ is the basis of $H^*(X//\mathbf{T})^a$ dual to $\{\gamma_i \cup \omega\}$, that is,

$$\int_{X//\mathbf{T}} \gamma_i \cup \omega \cup \gamma^j \cup \omega = \delta_i^j.$$

Define, for $\tau \in N$,

(5.3.2)

$$\begin{aligned} I(\tau,z) :&= \Big(\Big(\prod_{\alpha\in\Phi_+} z\partial_\alpha\Big) J_{X//T}\Big)\Big|_{Q^{\tilde{\beta}}=(-1)^{\epsilon(\beta)}Q^\beta,N}(\tau,z) \\ &= \sum_\beta (-1)^{\epsilon(\beta)}Q^\beta \sum_{\tilde{\beta}\mapsto\beta} \prod_{\alpha\in\Phi_+} \Big(c_1(L_\alpha) + z\int_{\tilde{\beta}} c_1(L_\alpha)\Big) J^{\tilde{\beta}}_{X//\mathbf{T}}\Big|_N(\tau,z) \end{aligned}$$

where ∂_α is the (∇-flat) vector field associated to $c_1(L_\alpha)$, the derivative of J is taken component-wise and $J^{\tilde{\beta}}_{X//\mathbf{T}}$ is the coefficient of $Q^{\tilde{\beta}}$ in $J_{X//\mathbf{T}}$ before specializing the Novikov variables. The latter equality follows from the divisor axiom.

Theorem 5.3.3. *There are unique $C^i(t,z) \in N(X//\mathbf{G})[z][[t]]$ such that*

$$I(\varphi(t), z) = \sum_i C^i(t,z) z\partial_{t_i} \tilde{J}_{X//\mathbf{G}}(t,z) \cup \omega.$$

Proof. For the proof we use Givental's description [Giv3] of the rational Gromov–Witten theory for a projective manifold Y by means of a certain Lagrangian cone $\mathcal{L}_Y$ with special properties (see [Giv3, Theorem 1]).

Let $s \in N$. By the very definition

$$I(s,-z) := \pm\Big(\Big(\prod_{\alpha\in\Phi_+} z\partial_\alpha\Big) J_{X//\mathbf{T}}\Big)\Big|_{Q^{\tilde{\beta}}=(-1)^{\epsilon(\beta)}Q^\beta}(s,-z) \in zL,$$

where $zL := zT_p\mathcal{L}_{X//\mathbf{T}}$ is the tangent space to the Lagrangian cone at the point $p = J_{X//\mathbf{T}}(s)$.

Let $\{\phi_\mu\}$ be a basis of $H^*(X//\mathbf{T})$ obtained by adjoining to the basis $\{\gamma_i \cup \omega\}$ of the $\mathbf{W}$-anti-invariant subspace $H^*(X//\mathbf{T})^a$ a basis of $(H^*(X//\mathbf{T})^a)^\perp$. Since $\{z\partial_\mu J_{X//\mathbf{T}}(s,-z)\}$ form a basis of zL/z^2L over $N(X//\mathbf{T})$,

$$I(s,z) = \sum C^\mu(s,z) z\partial_\mu J_{X//\mathbf{T}}|_{Q^{\tilde{\beta}}=(-1)^{\epsilon(\beta)}Q^\beta}(s,z)$$

for some unique $C^\mu(s,z) \in N(X//\mathbf{T})[z][[s]]$.

Since I is $\mathbf{W}$-anti-invariant by construction, the terms corresponding to the basis of $(H^*(X//\mathbf{T})^a)^\perp$ must vanish and we obtain

$$I(s,z) = \sum_i C^{\gamma_i\cup\omega}(s,z)(z\partial_{\gamma_i\cup\omega} J_{X//\mathbf{T}})|_{Q^{\tilde{\beta}}=(-1)^{\epsilon(\beta)}Q^\beta}(s,z). \tag{5.3.3}$$

Now $\partial_{s_i} \star \omega = \partial_{\gamma_i\cup\omega}$, therefore by (5.3.1)

$$\begin{aligned}\sum_i C^{\gamma_i\cup\omega}(s,z)&(z\partial_{\gamma_i\cup\omega} J_{X//\mathbf{T}})|_{Q^{\tilde{\beta}}=(-1)^{\epsilon(\beta)}Q^\beta}(s,z)\\ &= \sum_i C^{\gamma_i\cup\omega}(s,z)(z\partial_{s_i} z\partial_\omega J_{X//\mathbf{T}})|_{Q^{\tilde{\beta}}=(-1)^{\epsilon(\beta)}Q^\beta}(s,z).\end{aligned} \tag{5.3.4}$$

Finally, Lemma 5.3.1 gives

$$\begin{aligned}\sum_i C^{\gamma_i\cup\omega}(s,z)&(z\partial_{s_i} z\partial_\omega J_{X//\mathbf{T}})|_{Q^{\tilde{\beta}}=(-1)^{\epsilon(\beta)}Q^\beta}(s,z)\\ &= \sum_i C^{\gamma_i\cup\omega}(t,z) z\partial_{t_i} \tilde{J}_{X//\mathbf{G}}(t,z) \cup \omega,\end{aligned} \tag{5.3.5}$$

where $\varphi(t) = s$. The theorem follows from (5.3.3), (5.3.4) and (5.3.5). □

Corollary 5.3.4.

$$\tilde{J}_{X//\mathbf{G}}(t,z)\cup\omega = I(\tilde{t},z)+\sum_i C^i(\tilde{t},z)z\partial_{\tilde{t}_i}I(\tilde{t},z)$$

for some unique $C^i(\tilde{t},z)\in N(X//G)[[z,\tilde{t}]]$, *where* $\tilde{t}=\sum\tilde{t}_i\gamma_i$. *The expression of* t *in terms of* $\tilde{t}$ *is uniquely determined by the expansion of the right-hand side as* $z+t(\tilde{t})+O(z^{-1})$ *(and coincides with the formula* (3.6.2)*).*

Proof. The theorem above shows that, with the identification of cohomology spaces $H^*(X//\mathbf{T})^a$ with $H^*(X//\mathbf{G})$ by the map $\tilde{\sigma}\cup\omega\mapsto\sigma$, the I-function generates the Lagrangian cone $\mathcal{L}_{X//\mathbf{G}}$ describing the rational Gromov–Witten theory of $X//\mathbf{G}$ [Giv3]. Since $\{z\partial_{\tilde{t}_i}I(\tilde{t},-z)\}$ also form a basis of L/zL, where L is the tangent space of $\mathcal{L}_{X//\mathbf{G}}$ at the point $I(\tilde{t},-z)$, the corollary follows.

A constructive argument may also be given, using the "Birkhoff factorization" method. See [CG], Corollary 5 and the paragraph before it for details. □

Corollary 5.3.4 is a generalization of [BCK2, Conjecture 4.3] to the "big" parameter space. The arguments in this section can be reversed to show that the corollary implies Conjecture 3.7.1

6. Flag manifolds for other classical types

In this section, we extend the abelian/nonabelian correspondence in the presence of additional twists by homogeneous vector bundles and apply it to the case of generalized flag manifolds of Lie groups of types B, C, D.

6.1. Twisting by bundles. Let $\mathbf{S}\times\mathbf{G}$ act on X as in Sect. 3.1. Let $\mathcal{V}$ be a $\mathbf{G}$-representation space (as in [BCK2]). There are $\mathbf{S}\times\mathbf{G}\times\mathbb{C}^*$-actions on X and $\mathcal{V}$ (where $\mathbb{C}^*$ acts trivially on X and homothetically on $\mathcal{V}$), inducing $\mathbf{S}\times\mathbb{C}^*$-equivariant vector bundles

$$\mathcal{V}_{\mathbf{T}}:=X^s(\mathbf{T})\times_{\mathbf{T}}\mathcal{V},\quad \mathcal{V}_{\mathbf{G}}:=X^s(\mathbf{G})\times_{\mathbf{G}}\mathcal{V}$$

over nonsingular quotients $X//\mathbf{T}$ and $X//\mathbf{G}$, respectively. Put $\mathbb{C}[\lambda']:=H^*(B\mathbb{C}^*)$.

There is an $\mathbf{S}\times\mathbb{C}^*$-equivariant Frobenius structure on

$$Z':=Spf\Bigg(N(X//\mathbf{T})[\lambda] \otimes_{\mathbb{C}}\mathbb{C}\left(\left(\frac{1}{\lambda'}\right)\right)\left[\left[\left(H^*_{\mathbf{S}}(X//\mathbf{T})\otimes\left(N(X//\mathbf{T})[\lambda]\otimes_{\mathbb{C}}\mathbb{C}\left(\left(\frac{1}{\lambda'}\right)\right)\right)\right)^{\vee}\right]\right]\Bigg)$$

defined by the $\mathbf{S} \times \mathbb{C}^*$-equivariant genus zero Gromov–Witten invariants of $X//\mathbf{T}$ *twisted by (the equivariant Euler class of)* $\mathcal{V}_\mathbf{T}$. Here we introduce the extra coefficient ring $\mathbb{C}((\frac{1}{\lambda'}))$ to invert

$$c_{\text{top}}^{\mathbf{S}\times\mathbb{C}^*}(\mathcal{V}_\mathbf{T}) = \sum_{i=0}^{\text{rk}\mathcal{V}_\mathbf{T}} (\lambda')^{\text{rk}\mathcal{V}_\mathbf{T}-i} c_i^S(\mathcal{V}_\mathbf{T}).$$

We list some comments on this Frobenius structure for clarification, and refer the reader to [CG] for details.

- The twisted metric $g_{\mathcal{V}_\mathbf{T}}$ is given by

$$g_{\mathcal{V}_\mathbf{T}}(a, b) := \int_{X//\mathbf{T}} a \cup b \cup c_{\text{top}}^{\mathbf{S}\times\mathbb{C}^*}(\mathcal{V}_\mathbf{T}), \quad \text{for } a, b \in H_\mathbf{S}^*(X//\mathbf{T}).$$

- The twisted product is given by the requirement that

$$\begin{aligned} g_{\mathcal{V}_\mathbf{T}}(a *_{\mathcal{V}_\mathbf{T}} b, c) &= \langle\langle a, b, c\rangle\rangle_{\mathcal{V}_\mathbf{T}} \\ &:= \sum_{\tilde\beta \in NE_1(X//\mathbf{T})} \sum_n \frac{Q^{\tilde\beta}}{n!} \int_{[\overline{M}_{0,n+3}(X//\mathbf{T},\tilde\beta)]^{\text{vir}}} ev_1^*(a) ev_2^*(b) ev_3^*(c) \\ &\qquad \times ev_4^*(t) \ldots ev_{n+3}^*(t)\, c_{\text{vir.top}}^{\mathbf{S}\times\mathbb{C}^*}(R^\bullet\pi_* ev_{n+4}^* \mathcal{V}_\mathbf{T}), \end{aligned}$$

  where π denotes the projection $\overline{M}_{0,n+4}(X//\mathbf{T},\tilde\beta) \to \overline{M}_{0,n+3}(X//\mathbf{T},\tilde\beta)$ of moduli stacks of stable maps which forgets the last marked point.
- The Euler vector field is $\mathfrak{E}_{\mathcal{V}_\mathbf{T}} = \mathfrak{E} + \mathcal{E}_\mathbf{S} + \mathcal{E}_{\mathbb{C}^*} - c_1^\mathbf{S}(\mathcal{V}_\mathbf{T})$.
- The normalized ($\mathbf{S} \times \mathbb{C}^*$-equivariant) J-function is

$$J_{\mathcal{V}_\mathbf{T}}^{\mathbf{S}\times\mathbb{C}^*} : t \mapsto z + t + \sum_i \phi^i \left\langle\left\langle \frac{\phi_i}{z-\psi} \right\rangle\right\rangle_{\mathcal{V}_\mathbf{T}},$$

  where $\{\phi_i\}$ and $\{\phi^i\}$ are dual bases with respect to the twisted metric $g_{\mathcal{V}_\mathbf{T}}$.

Similarly, we construct an $\mathbf{S} \times \mathbb{C}^*$-equivariant Frobenius structure on the formal scheme P' associated to $H_\mathbf{S}^*(X//\mathbf{G}) \otimes (N(X//\mathbf{G})[\lambda] \otimes \mathbb{C}((\frac{1}{\lambda'})))$ using genus zero $\mathbf{S} \times \mathbb{C}^*$-equivariant Gromov–Witten invariants on $X//\mathbf{G}$ twisted by $\mathcal{V}_\mathbf{G}$.

Now, as in Sect. 3.2, we can further twist the Frobenius structure on Z' by $\omega := \sqrt{\frac{1}{|\mathbf{W}|}\prod_{\alpha\in\Phi} c_1^\mathbf{S}(L_\alpha)}$ in order to induce an $\mathbf{S} \times \mathbb{C}^*$-equivariant Frobenius structure on the formal scheme N' over $N(X//\mathbf{G})[\lambda] \otimes_\mathbb{C} \mathbb{C}((\frac{1}{\lambda'}))$ obtained as in loc. cit. by fixing a lift of $H_\mathbf{S}^*(X//\mathbf{G})$ to $H_\mathbf{S}^*(X//\mathbf{T})^\mathbf{W}$.

Conjecture 6.1.1. Let $\varphi : P' \to N'$ be the isomorphisms of formal schemes over $N(X//G)[\lambda] \otimes_\mathbb{C} \mathbb{C}((\frac{1}{\lambda'}))$ defined by $\varphi^*(s_i) = t_i$. Then φ induces an isomorphism of formal $\mathbf{S} \times \mathbb{C}^*$-equivariant Frobenius structures.

Theorem 6.1.2. *Conjecture 3.7.1 implies Conjecture 6.1.1, and furthermore,*

$$\tilde{J}^{\mathbf{S}\times\mathbb{C}^*}_{\mathcal{V}_\mathbf{G}}(t,z)\cup\omega = z\partial_\omega J^{\mathbf{S}\times\mathbb{C}^*}_{\mathcal{V}_\mathbf{T}}\Big|_{Q^{\tilde\beta}=(-1)^{\epsilon(\beta)}Q^\beta,N'}(\varphi(t),z).$$

Proof. It is enough to show the equality of J-functions above, since it implies that φ preserves the product structures.

Abusing notation, for $\gamma\in H^*_{\mathbf{S}\times\mathbb{C}^*}(X//\mathbf{T})^a$, $\sigma\in H^*_{\mathbf{S}\times\mathbb{C}^*}(X//\mathbf{G})$, denote σ by γ/ω if $\tilde\sigma\cup\omega=\gamma$. We also denote by $\mathcal{L}^{\mathbf{S}}_{X//\mathbf{G}}$, $\mathcal{L}^{\mathbf{S}\times\mathbb{C}^*}_{X//\mathbf{G}}$, and $\mathcal{L}^{\mathbf{S}\times\mathbb{C}^*}_{\mathcal{V}_\mathbf{G}}$ the Lagrangian cones given respectively by the $\mathbf{S}$-equivariant, $\mathbf{S}\times\mathbb{C}^*$-equivariant, and $\mathcal{V}_\mathbf{G}$-twisted, $\mathbf{S}\times\mathbb{C}^*$-equivariant rational GW-invariants of $X//\mathbf{G}$.

By (the $\mathbf{S}$-equivariant version of) Lemma 5.3.1

$$\frac{z\partial_\omega J^{\mathbf{S}}_{X//\mathbf{T}}\big|_{Q^{\tilde\beta}=(-1)^{\epsilon(\tilde\beta)}Q^\beta,N'}(-z)}{\omega}\in\mathcal{L}^{\mathbf{S}}_{X//\mathbf{G}}.$$

Hence

$$\Delta_{\mathcal{V}_\mathbf{G}}\frac{z\partial_\omega J^{\mathbf{S}}_{X//\mathbf{T}}\big|_{Q^{\tilde\beta}=(-1)^{\epsilon(\tilde\beta)}Q^\beta,N'}(-z)}{\omega}\in\Delta_{\mathcal{V}_\mathbf{G}}\mathcal{L}^{\mathbf{S}\times\mathbb{C}^*}_{X//G}=\mathcal{L}^{\mathbf{S}\times\mathbb{C}^*}_{\mathcal{V}_\mathbf{G}}$$

by [CG, Corollary 4], where

$$\Delta_{\mathcal{V}_\mathbf{G}}=\prod_{\rho_i:\ \text{Chern roots of }\mathcal{V}_G} b_{\rho_i}(\lambda',z),$$

$$b_\rho(\lambda',z)=\exp\left(\frac{(\lambda'+\rho)\ln(\lambda'+\rho)-(\lambda'+\rho)}{z}+\sum_{m>0}\frac{B_{2m}}{2m(2m-1)}\left(\frac{z}{\lambda'+\rho}\right)^{2m-1}\right)$$

(and B_{2m} are the Bernoulli numbers). Since

$$\Delta_{\mathcal{V}_\mathbf{G}}\frac{z\partial_\omega J^{\mathbf{S}}_{X//\mathbf{T}}\big|_{Q^{\tilde\beta}=(-1)^{\epsilon(\beta)}Q^\beta,N'}}{\omega}=\frac{z\partial_\omega\widetilde{\Delta}_{\mathcal{V}_\mathbf{G}}J^{\mathbf{S}}_{X//T}\big|_{Q^{\tilde\beta}=(-1)^{\epsilon(\beta)}Q^\beta,N'}}{\omega}$$

and

$$\widetilde{\Delta}_{\mathcal{V}_\mathbf{G}}=\Delta_{\mathcal{V}_\mathbf{T}}\pmod{\ker(\cup\omega)}$$

we conclude that

$$\text{(6.1.1)}\qquad\frac{z\partial_\omega J^{\mathbf{S}\times\mathbb{C}^*}_{\mathcal{V}_\mathbf{T}}\big|_{Q^{\tilde\beta}=(-1)^{\epsilon(\beta)}Q^\beta,N'}(-z)}{\omega}\in\mathcal{L}^{\mathbf{S}\times\mathbb{C}^*}_{\mathcal{V}_\mathbf{G}}.$$

Since the J-function $J_{\mathcal{V}_{\mathbf{G}}}(-z)$ is uniquely characterized by the intersection of the Lagrangian cone $\mathcal{L}_{\mathcal{V}_{\mathbf{G}}}$ with the subspace $-z+z\mathcal{H}_-$ as in [Giv3], it follows that (6.1.1) is the J-function for P'. That is,

$$J^{\mathbf{S}\times\mathbb{C}^*}_{\mathcal{V}_{\mathbf{G}}}(t,z)=\frac{z\partial_\omega J^{\mathbf{S}\times\mathbb{C}^*}_{\mathcal{V}_{\mathbf{T}}}|_{Q^{\tilde{\beta}}=(-1)^{\epsilon(\beta)}Q^{\beta},N'}}{\omega}(\tau(t),z)$$

for some unique $\tau(t)$. As in Corollary 5.3.4, the relation between τ and t is given by the expansion of the right-hand side with respect to z.

We have

$$\begin{aligned}g_{\mathcal{V}_{\mathbf{G}}}(\partial_{t_i},\partial_{t_j})+o(z)&=g_{\mathcal{V}_{\mathbf{G}}}(\partial_{t_i}J_{\mathcal{V}_{\mathbf{G}}},\partial_{t_j}J_{\mathcal{V}_{\mathbf{G}}})\\&=g_{\mathcal{V}_{\mathbf{T}}}(z\partial_{t_i}\partial_\omega J_{\mathcal{V}_{\mathbf{T}}},z\partial_{t_i}\partial_\omega J_{\mathcal{V}_{\mathbf{T}}})\\&=g_{\mathcal{V}_{\mathbf{T}}}(\partial_{\eta_i\star_{\mathcal{V}_{\mathbf{T}}}\omega},\partial_{\eta_j\star_{\mathcal{V}_{\mathbf{T}}}\omega})+o(z)\end{aligned}$$

where $\eta_i:=\partial_{t_i}(\tau)$. We conclude that $\eta_i\star_{\mathcal{V}_{\mathbf{T}}}\omega=\gamma_i\cup\omega$, hence $\tau(t)$ coincides with the map φ. □

Remark 6.1.3. If $\mathcal{V}_{\mathbf{G}}$ and $\mathcal{V}_{\mathbf{T}}$ are generated by $\mathbf{S}$-equivariant global sections, then $J^{\mathbf{S}}_{\mathcal{V}_{\mathbf{G}}}$ and $J^{\mathbf{S}}_{\mathcal{V}_{\mathbf{T}}}$ are well-defined without the auxiliary variable λ' (see [CG]) and hence the equality of J-functions in Theorem 6.1.2 also holds without λ'.

6.2. A simple lemma. Let X be a nonsingular projective variety with an $\mathbf{S}$-action whose fixed points are isolated, and let Y be a connected component of the nonsingular zero locus of a regular $\mathbf{S}$-equivariant section of a $\mathbf{S}$-equivariant bundle E. Suppose that E is generated by $\mathbf{S}$-equivariant global sections. Let i denote the inclusion of Y in X.

Lemma 6.2.1. *If $i^*(\tilde{t})=t$, then $J^{\mathbf{S}}_Y(t,z)|_{Q^{\mathbf{d}}=Q^{i_*\mathbf{d}}}=i^*J^{\mathbf{S}}_E(\tilde{t},z)$ where $|_{Q^{\mathbf{d}}=Q^{i_*\mathbf{d}}}$ denotes the Novikov ring base change given by the pushforward $i_*:NE_1(Y)\to NE_1(X)$.*

Proof. For each fixed point p_i of X under the $\mathbf{S}$-action, choose a nonzero class δ_i in $H^*_{\mathbf{S}}(X)\otimes\mathbb{C}(\lambda)$ supported near p_i, and let $\{\delta^j\}$ be the dual basis, that is, $\int_{c^{\mathbf{S}}_{\mathrm{top}}(E)\cap[X]}\delta^i\cup\delta_j=\delta_{ij}$. Note that for nonzero $\beta\in NE_1(X)$,

$$\begin{aligned}i^*J^{\mathbf{S},\beta}_E(\tilde{t},z)&=\sum_{k\,:\,p_k\in X^{\mathbf{S}}}\frac{i^*\delta_k}{n!}\int_{c^{\mathbf{S}}_{\mathrm{top}}(\pi_*ev^*_{n+2}E)\cap[\overline{M}_{0,n+1}(X,\beta)]^{\mathrm{vir}}}\frac{\delta^k}{z-\psi}\prod_{i=1}^n ev^*_{1+i}(\tilde{t})\\&=\sum_{k\,:\,p_k\in Y^{\mathbf{S}}}\frac{i^*\delta_k}{n!}\sum_{\mathbf{d}\in NE_1(Y)\,:\,i_*\mathbf{d}=\beta}\int_{[\overline{M}_{0,n+1}(Y,\mathbf{d})]^{\mathrm{vir}}}\frac{i^*\delta^k}{z-\psi}\prod_{i=1}^n ev^*_{1+i}(t),\end{aligned}$$

where the latter equality follows from [KKP]. Note that $i^*J^{\mathbf{S},\beta}_E(\tilde{t})=0$ if there is no $\mathbf{d}\in NE_1(Y)$ such that $i_*\mathbf{d}=\beta$. Since $\{i^*\delta_k\}$ and $\{i^*\delta^k\}$ form a dual pair of bases in $H^*_{\mathbf{S}}(Y)\otimes\mathbb{C}(\lambda)$ with respect to the equivariant Poincaré pairing, we are done. □

Remark 6.2.2. The above Lemma is true for the nonequivariant J-functions as well, since both sides of the identity can be specialized to $\lambda = 0$.

6.3. J-functions of flag manifolds of classical type. Let Y be a generalized flag manifold K/P, with K a simple complex Lie group of type B, C, or D and P a parabolic subgroup. It can be viewed as a connected component of the zero locus of a canonical section of a homogeneous bundle $\mathcal{V}_{\mathbf{G}}$ over an appropriate type A partial flag manifold $X//\mathbf{G} = Fl(k_1, \dots, k_r, n)$. Here

$$\mathcal{V} = \begin{cases} S^2(V^*) & \text{for types } B, D \\ \wedge^2(V^*) & \text{for type } C \end{cases},$$

where V is the fundamental representation space of $GL_{k_r}(\mathbb{C})$. Note that $\mathcal{V}_{\mathbf{T}}$ is decomposable into a direct sum of line bundles (since $\mathbf{T}$-representations are completely reducible).

Let $i : Y \subset X//\mathbf{G}$ be the natural inclusion and put

$$I_{\mathcal{V}_{\mathbf{G}}} := \frac{1}{\omega}\Big(\Big(\prod_{\alpha\in\Phi_+} z\partial_\alpha\Big) I_{\mathcal{V}_{\mathbf{T}}}\Big)\Big|_{Q^{\tilde{\beta}}=(-1)^{\epsilon(\beta)}Q^{\beta},N'},$$

$$I_{\mathcal{V}_{\mathbf{T}}} := \sum_{\tilde{\beta}\in NE_1(X//\mathbf{T})} \prod_{k=1}^{\int_{\tilde{\beta}}\rho_i} \prod_{\rho_i:\,\text{Chern roots of } \mathcal{V}_{\mathbf{T}}} (\rho_i + kz) J^{\tilde{\beta}}_{X//\mathbf{T}}.$$

Note that $I_{\mathcal{V}_{\mathbf{T}}}$ is a $H^*(X//\mathbf{T})$-valued series and $I_{\mathcal{V}_{\mathbf{G}}}$ is a $H^*(X//\mathbf{G})$-valued series.

Let $\mathbf{S}$ be a maximal abelian subgroup of the simple complex Lie group K. It acts on the flag manifold $Fl(k_1, \dots, k_r, n)$ with isolated fixed points and Y is an $\mathbf{S}$-invariant submanifold. Since bundles $\mathcal{V}_{\mathbf{G}}$ and $\mathcal{V}_{\mathbf{T}}$ are generated by $\mathbf{S}$-equivariant global sections and $i^* : H^*_{\mathbf{S}}(X//\mathbf{G}) \to H^*_{\mathbf{S}}(Y)$ (as well as $i^* : H^*(X//\mathbf{G}) \to H^*(Y)$) is surjective, we obtain the following

Corollary 6.3.1. *Fix a subspace N_Y of $H^*(X//\mathbf{T})^{\mathbf{W}}$ which is a lift of $H^*(Y)$ under the composite surjection $i^* \circ (\pi^*)^{-1} \circ j^*$. The J-function of Y can be expressed as*

$$J_Y(t, z)\big|_{Q^{\mathbf{d}}=Q^{i_*\mathbf{d}}} = I_{\mathcal{V}_{\mathbf{G}}}(\tau, z) + \sum_k C^k(\tau, z) i^*\left(\frac{z\partial_{\tilde{t}_k} I_{\mathcal{V}_{\mathbf{G}}}(\tau, z)}{\omega}\right)$$

for some unique $C^k(\tau, z) \in N(X//\mathbf{G})[[z, \tau]]$, where $\tilde{t}_k$ are coordinates of N_Y.

Proof. Due to Remark 6.2.2, $J_Y = i^* J_{\mathcal{V}_{\mathbf{G}}}$. Moreover, by Remark 6.1.3,

$$J_{\mathcal{V}_{\mathbf{G}}} = \frac{z\partial_\omega}{\omega} J_{\mathcal{V}_T}\Big|_{Q^{\tilde{\beta}}=(-1)^{\epsilon(\beta)}Q^{\beta},N'}.$$

Now apply the quantum Lefschetz theorem of Coates and Givental [CG] and use a similar argument to the one in the proof of Theorem 5.3.3 to conclude that $i^*\left(\frac{I_{V_{\mathbf{G}}}(-z)}{\omega}\right)$ generates the Lagrangian cone $\mathcal{L}_Y$. □

Remark 6.3.2. This in particular reproves the result on small J-function of flag manifolds of types B, C, D in [BCK2]. No coordinate change is necessary for the explicit description of this small $J|_{t_{\text{small}}}$.

7. Appendix: Multi-point GW-invariants of Grassmannians

Recall from Sect. 4.3 the notation

$$I_{n,\beta}(\gamma_1, \ldots, \gamma_n) = (-1)^{\epsilon(\beta)} \sum_{\tilde{\beta} \mapsto \beta} \langle \gamma_1, \ldots, \gamma_n \rangle_{0,n,\tilde{\beta}}^{X//\mathbf{T}}.$$

Theorem 4.1.1, together with (3.5.3) (or, better, (4.2.3)), imply that Gromov–Witten invariants of a flag manifold can be written in terms of invariants of the corresponding toric variety $X//\mathbf{T}$ by a formula of the form

$$\langle \sigma_{i_1}, \ldots, \sigma_{i_n} \rangle_{0,n,\beta}^{X//\mathbf{G}} = I_{n,\beta}(\tilde{\sigma}_{i_1}, \ldots, \tilde{\sigma}_{i_{n-2}}, \tilde{\sigma}_{i_{n-1}} \cup \omega, \tilde{\sigma}_{i_n} \cup \omega) + \text{correction}$$

where "correction" is an expression involving invariants $I_{n',\beta'}(\ldots, \tilde{\sigma}_a \cup \omega, \tilde{\sigma}_b \cup \omega)$ with $n' \leq n$ and $\beta' \leq \beta$. Without going into too many details, this can be seen as follows. Using the double bracket notation for derivatives of Gromov–Witten potentials mentioned in Sect. 2.1, one writes (4.2.3) as

$$\langle\langle \sigma_i, \sigma_j \rangle\rangle_{X//\mathbf{G}}(s) = \langle\langle \tilde{\sigma}_i, \tilde{\sigma}_j \rangle\rangle_{X//\mathbf{T}}(\tilde{t}(s)),$$

with $\tilde{t}(s)$ the inverse of the change of variables (3.6.2). This is an equality of power series in s-variables, and the formula for GW-invariants is obtained by identifying the coefficients of monomials in the s_j's. The coefficient of an s-monomial in the power series $\tilde{t}_k(s)$ can be explicitly expressed using the Lagrange Inversion Formula (see [GJ, Theorem 1.2.9]) in terms of the coefficients of $\tilde{t}$-monomials of *lower* total degree in the power series $s(\tilde{t})$ from (3.6.2).

The above discussion shows that the correction term will in general be quite complicated. Moreover, while it is possible in principle to give an exact expression, this will require the use of Lagrange inversion for computing the inverse $\tilde{t}(s)$ of the coordinate change (3.6.2), or, equivalently, the inverse (expressed in s-variables) of the matrix of quantum multiplication with ω on a lift of $H^*(X//\mathbf{G})$.

However, since flag manifolds are Fano of index ≥ 2, a different approach that uses Lemma 3.6.1(i) will allow us to reduce to computing only the inverse of the matrix of *small* quantum multiplication with ω. In the case of Grassmannians, when the associated abelian quotient is a product of projective spaces, it is an easy observation that the small quantum product with ω is trivial ([BCK1, Lemma 2.4]), hence no matrix inversion is necessary. We present the derivation of closed formulae for Grassmannians in this appendix.

Let $Gr := Grass(k, n)$ be the Grassmannian of k-planes in n-space, thought of as the GIT quotient $\mathrm{Hom}(\mathbb{C}^k, \mathbb{C}^n)//GL_k(\mathbb{C})$. The abelian quotient is $\mathbb{P} := (\mathbb{P}^{n-1})^k$. We consider the usual Schubert basis $\{\sigma_\lambda\}$ of $H^*(Gr, \mathbb{C})$, indexed by partitions λ whose Young diagrams fit in a $k \times (n-k)$ rectangle. We denote by $\mathcal{P}(k, n)$ the set of all such partitions. The intersection form in this basis is given by

$$\int_{Gr} \sigma_\lambda \cup \sigma_\mu = \delta_{\mu\lambda^\vee},$$

where $\lambda^\vee$ the complementary partition to λ in the $k \times (n-k)$ rectangle. The Grassmannian has Picard number 1, so the Novikov ring is $\mathbb{C}[[Q]]$. On the other hand, the Picard group of $\mathbb{P}$ is isomorphic to $\mathbb{Z}^k$ and is generated by $H_1, \ldots, H_k$, with H_j the pull-back of the hyperplane class on the jth factor. The Novikov ring of $\mathbb{P}$ is $\mathbb{C}[[Q_1, \ldots, Q_k]]$, and the specialization of Novikov variables is $Q_i = (-1)^{k-1}Q$. In this case we also have a "canonical" lifting of a class on Gr to a $\mathbf{W}$-invariant class on $\mathbb{P}$ by taking

$$\tilde{\sigma}_\lambda = S_\lambda(H_1, \ldots, H_k),$$

with S_λ the Schur polynomial of the partition λ. A curve class $\tilde{d} = (d_1, \ldots, d_k)$ on $\mathbb{P}$ is a lift of the curve class d on Gr if and only if $\sum_{i=1}^k d_i = d$. Finally, we have

$$\omega = \sqrt{\frac{(-1)^{\binom{k}{2}}}{k!}} \prod_{i<j}(H_i - H_j).$$

Let $\lambda^1, \ldots, \lambda^l$ be (not necessarily distinct) partitions. The generating function for the l-point invariants of Gr with σ_{λ^i}'s as insertions is

$$\langle\langle \sigma_{\lambda^1}, \ldots, \sigma_{\lambda^l} \rangle\rangle_{Gr}|_{t_{\text{small}}} = \sum_{d \geq 0} q^d \langle \sigma_{\lambda^1}, \ldots, \sigma_{\lambda^l} \rangle_{0,l,d}^{Gr},$$

where $q^d = (Qe^{t_{\text{small}}})^d$. We start with three-point invariants. Let ξ_λ be the horizontal vector field (for the connection ${}^\omega\nabla$) in Θ_N corresponding to σ_λ via the isomorphism φ of Frobenius manifolds in Theorem 4.1.1. We have (cf. (3.5.3))

$$\begin{aligned}\langle\langle \sigma_\lambda, \sigma_\mu, \sigma_\nu \rangle\rangle_{Gr}(t) &= \xi_\lambda(\xi_\mu(\xi_\nu(F')))(\varphi(t)) \\ &= \langle\langle \hat{\xi}_\lambda, \tilde{\sigma}_\mu \cup \omega, \tilde{\sigma}_\nu \cup \omega \rangle\rangle_{\mathbb{P}}|_{Q_i=(-1)^{k-1}Q,N}(\varphi(t))\end{aligned}$$

where $\hat{\xi}_\lambda$ is an extension of ξ_λ to a vector field on M. To unburden the notation, this extension of vector fields will be understood when necessary, and the same letter will be used for a vector field in Θ_N, or its extension to Θ_M. Moreover, the specialization of Novikov variables and the restriction to N will be denoted by $\langle\langle \dots \rangle\rangle^{\bar{}}$. Hence we rewrite the last equation as

$$\text{(7.0.1)} \qquad \langle\langle \sigma_\lambda, \sigma_\mu, \sigma_\nu \rangle\rangle_{Gr}(t) = \langle\langle \xi_\lambda, \tilde{\sigma}_\mu \cup \omega, \tilde{\sigma}_\nu \cup \omega \rangle\rangle^{\bar{}}_{\mathbb{P}}(\varphi(t)).$$

By Lemma 3.6.1 (i) we get

$$\langle\langle \sigma_\lambda, \sigma_\mu, \sigma_\nu \rangle\rangle_{Gr}(t)\Big|_{t_{\rm small}} = \langle\langle \xi_\lambda, \tilde{\sigma}_\mu \cup \omega, \tilde{\sigma}_\nu \cup \omega \rangle\rangle^{\bar{}}_{\mathbb{P}}\Big|_{\tilde{t}_{\rm small}}.$$

From the relation $\xi_\lambda \star \omega = \tilde{\sigma}_\lambda \cup \omega$, and the fact that $\tilde{\sigma}_\lambda \star \omega|_{\tilde{t}_{\rm small}} = \tilde{\sigma}_\lambda \cup \omega$, we obtain

$$\text{(7.0.2)} \qquad \xi_\lambda|_{\tilde{t}_{\rm small}} = \tilde{\sigma}_\lambda.$$

It follows that

$$\begin{aligned} \langle \sigma_\lambda, \sigma_\mu, \sigma_\nu \rangle^{Gr}_{0,3,d} &= I^{\mathbb{P}}_{3,d}(\tilde{\sigma}_\lambda, \tilde{\sigma}_\mu \cup \omega, \tilde{\sigma}_\nu \cup \omega) \\ &= (-1)^{(k-1)d} \sum_{d_1+\cdots+d_k=d} \langle \tilde{\sigma}_\lambda, \tilde{\sigma}_\mu \cup \omega, \tilde{\sigma}_\nu \cup \omega \rangle^{\mathbb{P}}_{0,3,(d_1,\dots,d_k)}, \end{aligned}$$

an equation which was proved in [BCK1].

To obtain 4-point invariants we take the derivative of the relation (7.0.1) and get

$$\begin{aligned} \text{(7.0.3)} \quad \langle\langle \sigma_\pi, \sigma_\lambda, \sigma_\mu, \sigma_\nu \rangle\rangle_{Gr}(t) &= \xi_\pi(\langle\langle \xi_\lambda, \tilde{\sigma}_\mu \cup \omega, \tilde{\sigma}_\nu \cup \omega \rangle\rangle^{\bar{}}_{\mathbb{P}}(\varphi(t))) \\ &= \langle\langle \xi_\pi, \xi_\lambda, \tilde{\sigma}_\mu \cup \omega, \tilde{\sigma}_\nu \cup \omega \rangle\rangle^{\bar{}}_{\mathbb{P}}(\varphi(t)) \\ &\quad + \langle\langle \nabla_{\xi_\pi}\xi_\lambda, \tilde{\sigma}_\mu \cup \omega, \tilde{\sigma}_\nu \cup \omega \rangle\rangle^{\bar{}}_{\mathbb{P}}(\varphi(t)), \end{aligned}$$

where $\nabla = \nabla^{\mathbb{P}}$ is the connection on M. Since

$$\begin{aligned} 0 = {}^{\omega}\nabla_{\xi_\pi}\xi_\lambda \star \omega &= \nabla_{\xi_\pi}(\xi_\lambda \star \omega) \\ &= (\nabla_{\xi_\pi}\xi_\lambda) \star \omega + \sum_{a \in \mathcal{P}(k,n)} \langle\langle \xi_\pi, \xi_\lambda, \omega, \tilde{\sigma}_a \cup \omega \rangle\rangle^{\bar{}}_{\mathbb{P}}(\tilde{\sigma}_{a^\vee} \cup \omega), \end{aligned}$$

it follows that

$$\text{(7.0.4)} \qquad \nabla_{\xi_\pi}\xi_\lambda = -\sum_{a \in \mathcal{P}(k,n)} \langle\langle \xi_\pi, \xi_\lambda, \omega, \tilde{\sigma}_a \cup \omega \rangle\rangle^{\bar{}}_{\mathbb{P}}\,\xi_{a^\vee}.$$

Combining with (7.0.3) we find

$$\begin{aligned} \text{(7.0.5)}\ & \langle\langle \sigma_\pi, \sigma_\lambda, \sigma_\mu, \sigma_\nu \rangle\rangle_{Gr}(t) \\ &= \langle\langle \xi_\pi, \xi_\lambda, \tilde{\sigma}_\mu \cup \omega, \tilde{\sigma}_\nu \cup \omega \rangle\rangle^{\bar{}}_{\mathbb{P}}(\varphi(t)) \\ &\quad - \sum_{a \in \mathcal{P}(k,n)} \langle\langle \xi_\pi, \xi_\lambda, \omega, \tilde{\sigma}_a \cup \omega \rangle\rangle^{\bar{}}_{\mathbb{P}} \langle\langle \xi_{a^\vee}, \tilde{\sigma}_\mu \cup \omega, \tilde{\sigma}_\nu \cup \omega \rangle\rangle^{\bar{}}_{\mathbb{P}}(\varphi(t)). \end{aligned}$$

Now we restrict to t_{small}, using (7.0.2), to get

$$\begin{aligned}
&\langle \sigma_\pi, \sigma_\lambda, \sigma_\mu, \sigma_\nu \rangle^{Gr}_{0,4,d} \\
&\quad = I^{\mathbb{P}}_{4,d}(\tilde{\sigma}_\pi, \tilde{\sigma}_\lambda, \tilde{\sigma}_\mu \cup \omega, \tilde{\sigma}_\nu \cup \omega) \\
&\qquad - \sum_{a \in \mathcal{P}(k,n)} \sum_{e+f=d} I^{\mathbb{P}}_{4,e}(\tilde{\sigma}_\pi, \tilde{\sigma}_\lambda, \omega, \tilde{\sigma}_a \cup \omega) I^{\mathbb{P}}_{3,f}(\tilde{\sigma}_{a^\vee}, \tilde{\sigma}_\mu \cup \omega, \tilde{\sigma}_\nu \cup \omega).
\end{aligned}$$

The following remark is in order: while the left-hand side of the last formula is manifestly invariant under permutations of the indices π, λ, μ, and ν, it is not at all obvious that the right-hand side has this property. The invariance can, however, be checked directly using the splitting axiom for Gromov–Witten invariants, the vanishing result in Lemma 4.3.2, and the triviality of the small quantum product with ω.

Taking another derivative in (7.0.5) we get

$$\begin{aligned}
&\langle\langle \sigma_\rho, \sigma_\pi, \sigma_\lambda, \sigma_\mu, \sigma_\nu \rangle\rangle_{Gr}(t) \\
&\quad = \langle\langle \xi_\rho, \xi_\pi, \xi_\lambda, \tilde{\sigma}_\mu \cup \omega, \tilde{\sigma}_\nu \cup \omega \rangle\rangle^{-}_{\mathbb{P}}(\varphi(t)) \\
&\qquad + \langle\langle \nabla_{\xi_\rho}\xi_\pi, \xi_\lambda, \tilde{\sigma}_\mu \cup \omega, \tilde{\sigma}_\nu \cup \omega \rangle\rangle^{-}_{\mathbb{P}}(\varphi(t)) \\
&\qquad + \langle\langle \xi_\pi, \nabla_{\xi_\rho}\xi_\lambda, \tilde{\sigma}_\mu \cup \omega, \tilde{\sigma}_\nu \cup \omega \rangle\rangle^{-}_{\mathbb{P}}(\varphi(t)) \\
&\qquad - \sum_a \Big(\langle\langle \xi_\rho, \xi_\pi, \xi_\lambda, \omega, \tilde{\sigma}_a \cup \omega \rangle\rangle^{-}_{\mathbb{P}} \langle\langle \xi_{a^\vee}, \tilde{\sigma}_\mu \cup \omega, \tilde{\sigma}_\nu \cup \omega \rangle\rangle^{-}_{\mathbb{P}} \\
&\qquad\qquad + \langle\langle \nabla_{\xi_\rho}\xi_\pi, \xi_\lambda, \omega, \tilde{\sigma}_a \cup \omega \rangle\rangle^{-}_{\mathbb{P}} \langle\langle \xi_{a^\vee}, \tilde{\sigma}_\mu \cup \omega, \tilde{\sigma}_\nu \cup \omega \rangle\rangle^{-}_{\mathbb{P}} \\
&\qquad\qquad + \langle\langle \xi_\pi, \nabla_{\xi_\rho}\xi_\lambda, \omega, \tilde{\sigma}_a \cup \omega \rangle\rangle^{-}_{\mathbb{P}} \langle\langle \xi_{a^\vee}, \tilde{\sigma}_\mu \cup \omega, \tilde{\sigma}_\nu \cup \omega \rangle\rangle^{-}_{\mathbb{P}} \\
&\qquad\qquad + \langle\langle \xi_\pi, \xi_\lambda, \omega, \tilde{\sigma}_a \cup \omega \rangle\rangle^{-}_{\mathbb{P}} \langle\langle \xi_\rho, \xi_{a^\vee}, \tilde{\sigma}_\mu \cup \omega, \tilde{\sigma}_\nu \cup \omega \rangle\rangle^{-}_{\mathbb{P}} \\
&\qquad\qquad + \langle\langle \xi_\pi, \xi_\lambda, \omega, \tilde{\sigma}_a \cup \omega \rangle\rangle^{-}_{\mathbb{P}} \langle\langle \nabla_{\xi_\rho}\xi_{a^\vee}, \tilde{\sigma}_\mu \cup \omega, \tilde{\sigma}_\nu \cup \omega \rangle\rangle^{-}_{\mathbb{P}} \Big)(\varphi(t)).
\end{aligned}$$

As above, we use (7.0.4) to replace the $\nabla_{\xi_\bullet}\xi_\bullet$ insertions, then restrict to t_{small} and obtain the following formula for 5-point invariants:

$$\begin{aligned}
&\langle \sigma_\rho, \sigma_\pi, \sigma_\lambda, \sigma_\mu, \sigma_\nu \rangle^{Gr}_{0,5,d} \\
&\quad = I^{\mathbb{P}}_{5,d}(\tilde{\sigma}_\rho, \tilde{\sigma}_\pi, \tilde{\sigma}_\lambda, \tilde{\sigma}_\mu \cup \omega, \tilde{\sigma}_\nu \cup \omega) \\
&\qquad - \sum_a \sum_{e+f=d} \Big(I^{\mathbb{P}}_{5,e}(\tilde{\sigma}_\rho, \tilde{\sigma}_\pi, \tilde{\sigma}_\lambda, \omega, \tilde{\sigma}_a \cup \omega) I^{\mathbb{P}}_{3,f}(\tilde{\sigma}_{a^\vee}, \tilde{\sigma}_\mu \cup \omega, \tilde{\sigma}_\nu \cup \omega) \\
&\qquad + I^{\mathbb{P}}_{4,e}(\tilde{\sigma}_\rho, \tilde{\sigma}_\pi, \omega, \tilde{\sigma}_a \cup \omega) I^{\mathbb{P}}_{4,f}(\tilde{\sigma}_{a^\vee}, \tilde{\sigma}_\lambda, \tilde{\sigma}_\mu \cup \omega, \tilde{\sigma}_\nu \cup \omega) \\
&\qquad + I^{\mathbb{P}}_{4,e}(\tilde{\sigma}_\rho, \tilde{\sigma}_\lambda, \omega, \tilde{\sigma}_a \cup \omega) I^{\mathbb{P}}_{4,f}(\tilde{\sigma}_{a^\vee}, \tilde{\sigma}_\pi, \tilde{\sigma}_\mu \cup \omega, \tilde{\sigma}_\nu \cup \omega) \\
&\qquad + I^{\mathbb{P}}_{4,e}(\tilde{\sigma}_\pi, \tilde{\sigma}_\lambda, \omega, \tilde{\sigma}_a \cup \omega) I^{\mathbb{P}}_{4,f}(\tilde{\sigma}_{a^\vee}, \tilde{\sigma}_\rho, \tilde{\sigma}_\mu \cup \omega, \tilde{\sigma}_\nu \cup \omega) \Big) \\
&\qquad + \sum_{a,b} \sum_{e+f+h=d} \Big(I^{\mathbb{P}}_{4,e}(\tilde{\sigma}_\rho, \tilde{\sigma}_\pi, \omega, \tilde{\sigma}_b \cup \omega) I^{\mathbb{P}}_{4,f}(\tilde{\sigma}_{b^\vee}, \tilde{\sigma}_\lambda, \omega, \tilde{\sigma}_a \cup \omega) \\
&\qquad\qquad\qquad \times I^{\mathbb{P}}_{3,h}(\tilde{\sigma}_{a^\vee}, \tilde{\sigma}_\mu \cup \omega, \tilde{\sigma}_\nu \cup \omega)
\end{aligned}$$

$$
\begin{aligned}
&+ I^{\mathbb{P}}_{4,e}(\tilde{\sigma}_\rho, \tilde{\sigma}_\lambda, \omega, \tilde{\sigma}_b \cup \omega) I^{\mathbb{P}}_{4,f}(\tilde{\sigma}_{b^\vee}, \tilde{\sigma}_\pi, \omega, \tilde{\sigma}_a \cup \omega) I^{\mathbb{P}}_{3,h} \\
&\quad \times (\tilde{\sigma}_{a^\vee}, \tilde{\sigma}_\mu \cup \omega, \tilde{\sigma}_\nu \cup \omega) \\
&+ I^{\mathbb{P}}_{4,e}(\tilde{\sigma}_\pi, \tilde{\sigma}_\lambda, \omega, \tilde{\sigma}_b \cup \omega) I^{\mathbb{P}}_{4,f}(\tilde{\sigma}_{b^\vee}, \tilde{\sigma}_\rho, \omega, \tilde{\sigma}_a \cup \omega) I^{\mathbb{P}}_{3,h} \\
&\quad \times (\tilde{\sigma}_{a^\vee}, \tilde{\sigma}_\mu \cup \omega, \tilde{\sigma}_\nu \cup \omega)\big).
\end{aligned}
$$

It is now clear how to proceed to obtain and prove by induction a formula for Gromov–Witten invariants with an arbitrary number of insertions. We leave this to the reader.

References

[BCK1] Bertram, A., Ciocan-Fontanine, I., Kim, B.: Two proofs of a conjecture of Hori and Vafa. Duke Math. J. **126**(1), 101–136 (2005)

[BCK2] Bertram, A., Ciocan-Fontanine, I., Kim, B.: Gromov–Witten invariants for non-abelian and abelian quotients. J. Algebr. Geom. to appear, math.AG/0407254

[CG] Coates, T., Givental, A.: Quantum Riemann-Roch, Lefschetz, and Serre. Ann. Math. (2) **165**(1), 15–53 (2007)

[Du1] Dubrovin, B.: Geometry of 2D topological field theories. In: Integrable Systems and Quantum Groups (Montecatini Terme, 1993). Lect. Notes Math., vol. 1620, pp. 120–348. Springer, Berlin (1996)

[Du2] Dubrovin, B.: Painlevé transcendents in two-dimensional topological field theory. In: The Painlevé Property. CRM Ser. Math. Phys., pp. 287–412. Springer, New York (1999)

[Du3] Dubrovin, B.: Geometry and analytic theory of Frobenius manifolds. In: Proceedings of the International Congress of Mathematicians, vol. II, Berlin, 1998. Doc. Math., Extra vol. II, pp. 315–326 (1998)

[ES] Ellingsrud, G., Strømme, S.A.: On the Chow ring of a geometric quotient. Ann. Math. **130**, 159–187 (1989)

[Gin] Ginzburg, V.A.: Equivariant cohomology and Kähler geometry (Russian). Funkts. Anal. Prilozh. **21**(4), 19–34, 96 (1987)

[Giv1] Givental, A.: Equivariant Gromov–Witten invariants. Int. Math. Res. Not. **1996**(13), 613–663 (1996)

[Giv2] Givental, A.: Elliptic Gromov–Witten invariants and the generalized mirror conjecture. In: Integrable Systems and Algebraic Geometry (Kobe/Kyoto, 1997), pp. 107–155. World Sci. Publ., River Edge, NJ (1998)

[Giv3] Givental, A.: Symplectic geometry of Frobenius structures. In: Frobenius Manifolds. Aspects Math., E36, pp. 91–112. Vieweg, Wiesbaden (2004)

[GJ] Goulden, I.P., Jackson, D.M.: Combinatorial Enumeration. In: Wiley-Interscience Series in Discrete Mathematics. John Wiley & Sons, Inc., New York (1983)

[KKP] Kim, B., Kresch, A., Pantev, T.: Functoriality in intersection theory and a conjecture of Cox, Katz, and Lee. J. Pure Appl. Algebra **179**, 127–136 (2003)

[Kir] Kirwan, F.: Refinements of the Morse stratification of the normsquare of the moment map. In: The Breadth of Symplectic and Poisson Geometry. Prog. Math., vol. 232, pp. 327–362. Birkhäuser Boston, Boston, MA (2005)

[KM] Kontsevich, M., Manin, Y.I.: Gromov–Witten classes, quantum cohomology, and enumerative geometry. Commun. Math. Phys. **164**(3), 525–562 (1994)

[LP1] Lee, Y.P., Pandharipande, R.: A reconstruction theorem in quantum cohomology and quantum K-theory. Am. J. Math. **126**(6), 1367–1379 (2004)

[LP2] Lee, Y.P., Pandharipande, R.: Frobenius manifolds, Gromov–Witten theory, and Virasoro constraints, Part I. available at www.math.utah.edu/~yplee and/or www.math.princeton.edu/~rahulp

[Man] Manin, Y.I.: Frobenius Manifolds, Quantum Cohomology, and Moduli Spaces. Am. Math. Soc. Colloq. Publ., vol. 47. Am. Math. Soc., Providence, RI (1999)
[Mar] Martin, S.: Symplectic quotients by a nonabelian group and by its maximal torus. math.SG/0001002
[Sab] Sabbah, C.: Frobenius manifolds: Isomonodromic deformations and infinitesimal period mappings. Expo. Math. **16**, 1–58 (1998)

A.4. Quasimap Wall-crossings and Mirror Symmetry

Ciocan-Fontanine, I., Kim, B. Quasimap wall-crossings and mirror symmetry. Publ. math. IHES **131**, 201–260 (2020). https://doi.org/10.1007/s10240-020-00114-0

© IHES and Springer-Verlag GmbH Germany, part of Springer Nature 2020. Reprinted with permission.

© The Author(s) 2026

N.-G. Kang et al. (eds.), *Categorical and Enumerative Aspects of Mirror Symmetry*, KIAS Springer Series in Mathematics 5,
https://doi.org/10.1007/978-981-95-0385-8

QUASIMAP WALL-CROSSINGS AND MIRROR SYMMETRY

by IONUŢ CIOCAN-FONTANINE *and* BUMSIG KIM

ABSTRACT

We state a wall-crossing formula for the virtual classes of ε-stable quasimaps to GIT quotients and prove it for complete intersections in projective space, with no positivity restrictions on their first Chern class. As a consequence, the wall-crossing formula relating the genus g descendant Gromov-Witten potential and the genus g ε-quasimap descendant potential is established. For the quintic threefold, our results may be interpreted as giving a rigorous and geometric interpretation of the holomorphic limit of the BCOV B-model partition function of the mirror family.

CONTENTS

1. Introduction

1.1. *Overview.* — Let W be a complex affine variety acted upon by a reductive algebraic group **G**. Fix a character θ of **G** for which the induced **G**-action on the θ-semistable locus W^{ss} is free. For the quasiprojective target $W/\!/_{\theta}\mathbf{G}$ and a rational number $\varepsilon > 0$, or for $\varepsilon = 0+$, the notion of ε-stable quasimaps to $W/\!/_{\theta}\mathbf{G}$ was introduced in [12], inspired by [6, 25, 26]. They are in fact suitable maps from curves to the stack quotient $[W/\mathbf{G}]$. The Deligne-Mumford moduli stack $Q^{\varepsilon}_{g,k}(W/\!/_{\theta}\mathbf{G}, \beta)$ of ε-stable quasimaps of type (g, k, β) is proper over $\mathbf{C}$ if $W/\!/_{\theta}\mathbf{G}$ is projective. Here g, k, and β are respectively the genus of the domain curve, the number of markings, and the numerical class $\beta \in \mathrm{Hom}_{\mathbf{Z}}(\mathrm{Pic}([W/\mathbf{G}], \mathbf{Z}))$ of the quasimaps. If W has at worst lci singularities and W^{ss} is

© IHES and Springer-Verlag GmbH Germany, part of Springer Nature 2020
https://doi.org/10.1007/s10240-020-00114-0

smooth (as always assumed in this paper), the moduli stacks carry canonical virtual fundamental classes. There are evaluation maps ev_j to $W/\!/_\theta \mathbf{G}$, as well as cotangent psi-classes ψ_j at the j-th marking. Hence, we may define descendant ε-quasimap invariants

$$(\mathbf{1.1.1}) \qquad \langle \gamma_1\psi_1^{a_1}, \dots, \gamma_k\psi_k^{a_k}\rangle^{\varepsilon}_{g,k,\beta} = \int_{[Q^{\varepsilon}_{g,k}(W/\!/_\theta \mathbf{G},\beta)]^{\mathrm{vir}}} \prod_{j=1}^{k} \psi_j^{a_j} ev_j^* \gamma_j$$

for $\gamma_i \in A^*(W/\!/_\theta \mathbf{G})_{\mathbf{Q}}$ and $a_i \in \mathbf{Z}_{\geq 0}$. Here and for the rest of the paper, the Chow cohomology $A^*(Y)_{\mathbf{Q}}$ of a Deligne-Mumford stack Y is the algebra $A^*(Y \stackrel{\mathrm{id}}{\to} Y)_{\mathbf{Q}}$ of bivariant classes, see [15, §17.3] and [29, §5].

There is a wall-and-chamber structure on the space $\mathbf{Q}_{>0}$ of stability parameters. Assuming for simplicity $(g, k) \neq (0, 0)$, the walls are at $\varepsilon = 1/n$ with $n \in \mathbf{N}$ and the moduli spaces stay constant in each chamber $(\frac{1}{n+1}, \frac{1}{n}]$. For $\varepsilon \in (1, \infty)$, they parametrize exactly stable maps to $W/\!/_\theta \mathbf{G}$. A conjectural wall-crossing formula for the invariants of *semi-positive* targets was stated in the paper [9], and was proved for semi-positive (quasiprojective) toric quotients by localization techniques. In this paper we propose a geometric wall-crossing formula *at the level of virtual classes* and *without* any positivity restrictions (which, as we show, implies the above mentioned semi-positive numerical wall-crossing, see Corollary 1.5). The main result of the paper is a proof of the virtual class wall-crossing formula for complete intersections in projective spaces.

The wall-crossing formula has important applications to Mirror Symmetry for Calabi-Yau threefolds at higher genus. This is explained in Section 1.5, the main point being that, assuming the Mirror Conjecture, the genus g partition function of quasimap theory for the $\varepsilon = 0+$ stability of a Calabi-Yau threefold is precisely *equal* to (the holomorphic limit of) the B-model partition function of the mirror Calabi-Yau family, introduced in string theory by Bershadsky, Cecotti, Ooguri, and Vafa.

1.2. *Geometric wall-crossing.* — To state the wall-crossing formula, we recall some facts from quasimap theory and fix some notation.

The monoid $\mathrm{Eff}(W, \mathbf{G}, \theta)$ of θ-effective numerical classes is the submonoid of the additive group $\mathrm{Hom}_{\mathbf{Z}}(\mathrm{Pic}([W/\mathbf{G}], \mathbf{Z})$ consisting of classes represented by θ quasimaps (possibly with disconnected domain curves). The Novikov ring of the theory is

$$\mathbf{Q}[[q]] := \left\{ \sum_{\mathrm{Eff}(W,\mathbf{G},\theta)} a_\beta q^\beta \;\middle|\; a_\beta \in \mathbf{Q} \right\},$$

the q-adic completion of the semigroup ring $\mathbf{Q}[\mathrm{Eff}(W, \mathbf{G}, \theta)]$.

The GIT set-up gives (see [7, §3.1] for details) a natural morphism $i : [W/\mathbf{G}] \to [\mathbf{C}^{m+1}/\mathbf{C}^*]$ for some $m \in \mathbf{Z}_+$, inducing a closed immersion $i : W/\!/_\theta \mathbf{G} \hookrightarrow \mathbf{P}^m$ and also a morphism (denoted by the same letter)

$$i : Q^{\varepsilon}_{g,k}(W/\!/_\theta \mathbf{G}, \beta) \to Q^{\varepsilon}_{g,k}\big(\mathbf{P}^m, d(\beta)\big),$$

where $d(\beta) := i_*(\beta) \in \mathrm{Hom}(\mathrm{Pic}([\mathbf{C}^{m+1}/\mathbf{C}^*]), \mathbf{Z}) \cong \mathbf{Z}$.

Fix a positive rational number ε_0 such that $1/\varepsilon_0$ is an integer and let $\varepsilon_+ > \varepsilon_0 \geq \varepsilon_-$ be rational numbers in the two adjacent stability chambers separated by the wall ε_0. There is a morphism

$$c : Q^{\varepsilon_+}_{g,k}(\mathbf{P}^m, d(\beta)) \to Q^{\varepsilon_-}_{g,k}(\mathbf{P}^m, d(\beta))$$

which contracts rational tails of degree $1/\varepsilon_0$, see [28].

Let A denote a finite index set of cardinality $1, 2, 3, \ldots$ Consider splittings $\beta = \beta_0 + \sum_{a\in A} \beta_a$ into θ-effective numerical classes such that $d(\beta_a) = 1/\varepsilon_0$ for all $a \in A$. There is a natural morphism

$$b_A : Q^{\varepsilon_-}_{g,k+A}(\mathbf{P}^m, d(\beta_0)) \to Q^{\varepsilon_-}_{g,k}(\mathbf{P}^m, d(\beta))$$

which trades the markings in A for base points of length $1/\varepsilon_0$ ([7, §3.2]).

Finally, recall from [12, §7] and [7, §5] that for every triple (W, G, θ), with associated quotient $X = W/\!\!/_\theta \mathbf{G}$, there is a corresponding *small* I*-function*, denoted $I_{sm}(q, z)$. The precise definition we will use in this paper is Definition 5.1.1 in [7], specialized at $\varepsilon = 0+$ and $\mathbf{t} = 0$.

The small I-function lies in a certain completion $A^*(X)_{\mathbf{Q}}[[q]]\{\{1/z, z\}\}$ of Laurent series in $1/z$. (Here z may be viewed as a formal variable of degree one, though it is more natural to interpret z as the generator of the $\mathbf{C}^*$-equivariant cohomology $A^*_{\mathbf{C}^*}(\mathrm{Spec}(\mathbf{C}))$.) It can be explicitly calculated for many targets. For abelian quotients, that is, for toric varieties and for complete intersections in them, the small I-function is precisely the cohomology-valued hypergeometric series introduced by Givental [18] (up to exponential factors). Closed formulas for I_{sm} in many examples with nonabelian $\mathbf{G}$ (e.g., complete intersections in flag varieties, but many others as well) can also be written down using the so-called *abelian/nonabelian correspondence*, see [4, 5, 10, 11].

Consider the expansion

$$I_{sm}(q, z) = O(1/z^2) + \frac{I_1(q)}{z} + I_0(q) + I_{-1}(q)z + I_{-2}(q)z^2 + \cdots$$

and set

$$\left[zI_{sm}(q, z) - z\right]_+ := I_1(q) + \left(I_0(q) - 1\right)z + I_{-1}(q)z^2 + \cdots$$

In general $[zI_{sm}(q, z) - z]_+$ is a power series in (q, z), but each q-coefficient is a polynomial in z. For each $0 \neq \beta \in \mathrm{Eff}(W, \mathbf{G}, \theta)$, let

$$\mu_\beta(z) \in A^*(X)_{\mathbf{Q}}[z]$$

denote the coefficient of q^β in $[zI_{sm}(q, z) - z]_+$. By easy dimension counting, $\mu_\beta(z)$ is homogeneous of degree $1 + \beta(K_{[W/\mathbf{G}]})$. Here z has degree one, the Chow cohomology classes are given their usual degrees, and $K_{[W/\mathbf{G}]} = -\det(T_W) \in \mathrm{Pic}^{\mathbf{G}}(W) = \mathrm{Pic}([W/\mathbf{G}])$ is the canonical line bundle of the quotient stack.

We are now ready to state the wall-crossing for virtual classes.

Conjecture **1.1.** — *There is an equality*

$$(\mathbf{1.2.1}) \qquad i_*\big[Q^{\varepsilon-}_{g,k}(X,\beta)\big]^{\mathrm{vir}} - c_* i_*\big[Q^{\varepsilon+}_{g,k}(X,\beta)\big]^{\mathrm{vir}}$$
$$= \sum_{|A|} \sum_{\beta=\beta_0+\sum_{a\in A}\beta_a} \frac{1}{|A|!} b_{A*}(c_A)_* i_* \Big(\prod_{a\in A} ev_a^* \mu_{\beta_a}(z)|_{z=-\psi_a} \cap \big[Q^{\varepsilon+}_{g,k+A}(X,\beta_0)\big]^{\mathrm{vir}} \Big)$$

in the Chow group $A_*(Q^{\varepsilon-}_{g,k}(\mathbf{P}^m, d(\beta)))_{\mathbf{Q}}$.

More generally, let $\delta_1,\ldots,\delta_k \in A^*(X)_{\mathbf{Q}}$ *be arbitrary homogeneous cohomology classes. Then*

$$(\mathbf{1.2.2}) \qquad i_*\Big(\prod_{j=1}^k ev_j^*\delta_j \cap \big[Q^{\varepsilon-}_{g,k}(X,\beta)\big]^{\mathrm{vir}}\Big) - c_* i_*\Big(\prod_{j=1}^k ev_j^*\delta_j \cap \big[Q^{\varepsilon+}_{g,k}(X,\beta)\big]^{\mathrm{vir}}\Big)$$
$$= \sum_{|A|} \sum_{\beta=\beta_0+\sum_{a\in A}\beta_a} \frac{1}{|A|!} b_{A*}(c_A)_* i_* \Big(\prod_{j=1}^k ev_j^*\delta_j \prod_{a\in A} ev_a^* \mu_{\beta_a}(z)|_{z=-\psi_a}$$
$$\cap \big[Q^{\varepsilon+}_{g,k+A}(X,\beta_0)\big]^{\mathrm{vir}} \Big)$$

in $A_*(Q^{\varepsilon-}_{g,k}(\mathbf{P}^m, d(\beta)))_{\mathbf{Q}}$.

In the above statement, $c_A : Q^{\varepsilon+}_{g,k+A}(\mathbf{P}^m, d(\beta_0)) \to Q^{\varepsilon-}_{g,k+A}(\mathbf{P}^m, d(\beta_0))$ is the contraction of rational tails of degree $d(\beta_a) = 1/\varepsilon_0$.

Remark **1.2.** — For X a *semi-positive* quasi-projective toric manifold, Conjecture 1.1 coincides with Theorem 4.2.1 in [9], and the result is valid for any GIT presentation of X, see [9, §5.9.2]. In fact, the localization argument of [9] extends with little effort to prove (1.2.2) for *all* toric manifolds (i.e., no positivity restriction), offering the first evidence for the validity of Conjecture 1.1. We will treat this extension elsewhere.

1.3. *Numerical consequences.* — In this subsection we assume that $(W, \mathbf{G}, \theta)$ is a triple for which Conjecture 1.1 holds. We work with arbitrary stability parameters $\varepsilon \in \mathbf{Q}_{>0} \cup \{0+\}$ and will write $\varepsilon = \infty$ for all parameters in the Gromov-Witten chamber $(1,\infty)$.

Consider a formal power series in one variable ψ,

$$\mathbf{t}(\psi) := t_0 + t_1\psi + t_2\psi^2 + t_3\psi^3 + \cdots,$$

with coefficients $t_j \in A^*(X)_{\mathbf{Q}}$ general Chow cohomology classes.

The *genus g, ε-descendent potential* of X is

$$F^{\varepsilon}_g\big(q, \mathbf{t}(\psi)\big) := \sum_{(\beta,k)} \frac{q^\beta}{k!} \big\langle \mathbf{t}(\psi_1), \mathbf{t}(\psi_2), \ldots \mathbf{t}(\psi_k)\big\rangle^{\varepsilon}_{g,k,\beta},$$

the sum over all pairs (β, k) for which the corresponding moduli spaces exist. If we choose a basis $\{\gamma_j\}$ in $A^*(X)_{\mathbf{Q}}$ and write $t_i = \sum_j t_{ij}\gamma_j$, $i = 0, 1, 2, \ldots$, then $F_g^\varepsilon(q, \mathbf{t}(\psi))$ is a formal power series in the infinitely many variables t_{ij}, whose Taylor coefficients are the ε-quasimap invariants (1.1.1). In particular, F_g^∞ is the generating series for all descendent genus g Gromov-Witten invariants of X.

1.3.1. *Wall-crossing from Gromov-Witten invariants to ε-quasimap invariants.* — Let $J_{sm}^\varepsilon(q, z)$ be the small J-function of X ([7, Definition 5.1.1], specialized at $\mathbf{t} = 0$). With this notation, $I_{sm} = J_{sm}^{0+}$. Let

$$[zJ_{sm}^\varepsilon - z]_+ := J_1^\varepsilon(q) + \big(J_0^\varepsilon(q) - 1\big)z + J_{-1}^\varepsilon(q)z^2 + \cdots$$

This is explicit for all ε, since it is a q-truncation of the corresponding expression for the small I-function:

$$[zJ_{sm}^\varepsilon(q, z) - z]_+ = [zI_{sm}(q, z) - z]_+ \quad (\mathrm{mod}\ \mathfrak{a}_\varepsilon),$$

with $\mathfrak{a}_\varepsilon$ the ideal in the Novikov ring generated by $\{q^\beta \mid \beta(L_\theta) > \frac{1}{\varepsilon}\}$.

Corollary **1.3.** — *For any $\varepsilon \geq 0+$, and any $g \geq 1$,*

$$F_g^\varepsilon\big(q, \mathbf{t}(\psi)\big) = F_g^\infty\big(q, \mathbf{t}(\psi) + [zJ_{sm}^\varepsilon(q) - z]_+|_{z=-\psi}\big).$$

Further, in genus $g = 0$ the same relation holds after discarding from $F_0^\infty(q, \mathbf{t}(\psi))$ the terms corresponding to pairs (β, k) for which $Q_{0,k}^\varepsilon(X, \beta)$ is not defined.

Proof. — The ψ-classes at the markings $1, \ldots, k$ pull-back under the maps b_A, c, c_A, and i. Applying the virtual class wall-crossing (1.2.2) in Conjecture 1.1 successively for the walls from 1 to ε (and using the projection formula) gives the equality of the Taylor coefficients of the two sides in the claimed equality. □

Remark **1.4.** — (*i*) The formula in Corollary 1.3 is equivalent to

$$F_g^\varepsilon\big(q, \mathbf{t}(\psi) - [zJ_{sm}^\varepsilon(q) - z]_+|_{z=-\psi}\big) = F_g^\infty\big(q, \mathbf{t}(\psi)\big).$$

(*ii*) Assuming only the formula (1.2.1) from Conjecture 1.1 gives the weaker equality

$$F_g^\varepsilon\big(q, \bar{\mathbf{t}}(\psi)\big) = F_g^\infty\big(q, \bar{\mathbf{t}}(\psi) + [zJ_{sm}^\varepsilon(q) - z]_+|_{z=-\psi}\big),$$

with $\bar{\mathbf{t}}(\psi)$ the restriction of $\mathbf{t}(\psi)$ to the subring $i^*A^*(\mathbf{P}^m)_{\mathbf{Q}} \subset A^*(X)_{\mathbf{Q}}$.

1.3.2. *Semi-positive targets.* — Recall that a triple $(\mathrm{W}, \mathbf{G}, \theta)$ is called semi-positive if

$$\beta(\det \mathrm{T_W}) = \beta(-\mathrm{K}_{[\mathrm{W}/\mathbf{G}]}) \geq 0$$

for every $\beta \in \mathrm{Eff}(\mathrm{W}, \mathbf{G}, \theta)$. For such targets we have

$$\left[z\mathrm{J}^{\varepsilon}_{sm}(q) - z\right]_{+} = \mathrm{J}^{\varepsilon}_{1}(q) + \left(\mathrm{J}^{\varepsilon}_{0}(q) - 1\right)z,$$

since $\deg(\mu_\beta(z)) \leq 1$ for all β. The wall-crossing formula of Corollary 1.3 becomes

$$\mathrm{F}^{\varepsilon}_{g}\big(q, \mathbf{t}(\psi)\big) = \mathrm{F}^{\infty}_{g}\big(q, \mathbf{t}(\psi) + \mathrm{J}^{\varepsilon}_{1}(q) - \left(\mathrm{J}^{\varepsilon}_{0}(q) - 1\right)\psi\big). \tag{1.3.1}$$

In fact, equation (1.3.1) is equivalent to the wall-crossing formula conjectured in [9, Conjecture 1.2.1]:

Corollary **1.5.** — *For a semi-positive triple* $(\mathrm{W}, \mathbf{G}, \theta)$ *we have*

$$\left(\mathrm{J}^{\varepsilon}_{0}\right)^{2g-2}\left(\delta^{1}_{g}\left(\frac{\chi_{\mathrm{top}}(\mathrm{X})}{24}\log \mathrm{J}^{\varepsilon}_{0}(q)\right) + \mathrm{F}^{\varepsilon}_{g}\big(q, \mathbf{t}(\psi)\big)\right) = \mathrm{F}^{\infty}_{g}\left(q, \frac{\mathbf{t}(\psi) + \mathrm{J}^{\varepsilon}_{1}(q)}{\mathrm{J}^{\varepsilon}_{0}(q)}\right), \tag{1.3.2}$$

where $\chi_{\mathrm{top}}(\mathrm{X})$ *denotes the topological Euler characteristic and* δ^{1}_{g} *is the Kronecker delta. (In genus* $g = 0$ *we use the same convention as in Corollary 1.3.)*

Proof. — Using the dilaton equation for Gromov-Witten invariants in the right-hand side of (1.3.1) to remove the insertions $(\mathrm{J}^{\varepsilon}_{0}(q) - 1)\psi$ produces exactly (1.3.2). The additional term $\delta^{1}_{g}(\frac{\chi_{\mathrm{top}}(\mathrm{X})}{24}\log \mathrm{J}^{\varepsilon}_{0}(q))$ appears due to the failure of the dilaton equation for $\overline{\mathrm{M}}_{1,1}(\mathrm{X}, 0) = \overline{\mathrm{M}}_{1,1} \times \mathrm{X}$. Namely, since the virtual class is

$$\left[\overline{\mathrm{M}}_{1,1}(\mathrm{X}, 0)\right]^{\mathrm{vir}} = \left(1 \otimes c_{\dim \mathrm{X}}(\mathrm{T_X}) - \psi \otimes c_{\dim \mathrm{X}-1}(\mathrm{T_X})\right) \cap [\overline{\mathrm{M}}_{1,1} \times \mathrm{X}],$$

we have

$$\langle \psi \rangle^{\infty}_{1,1,0} = \int_{\overline{\mathrm{M}}_{1,1}\times \mathrm{X}} \psi \otimes c_{\dim \mathrm{X}}(\mathrm{T_X}) = \frac{1}{24}\chi_{\mathrm{top}}(\mathrm{X}),$$

while the dilaton equation would formally predict $\langle \psi \rangle^{\infty}_{1,1,0} = 0$. □

1.4. *Complete intersections in projective space.* — The main result of the paper is a proof of Conjecture 1.1 for projective complete intersections. In fact, we will prove the following slightly strengthened version.

Let V be the affine space of dimension $n + 1$ with the standard diagonal $\mathbf{G} := \mathbf{C}^*$-action and linearization $\theta = \mathrm{id}_{\mathbf{C}^*}$. Let W be a complete intersection of $r \leq n$ homogeneous hypersurfaces in V. Then $\mathrm{X} := \mathrm{W}/\!\!/_{\theta}\mathbf{G}$ is the corresponding projective complete intersection in $\mathbf{P}(\mathrm{V})$ (and W is the affine cone over X). Assume that the hypersurfaces are general, so that X is smooth. We take $X \hookrightarrow \mathbf{P}(V)$ as our embedding i. In this case, the induced

$$i : Q^{\varepsilon}_{g,k}(X, d) \longrightarrow Q^{\varepsilon}_{g,k}(\mathbf{P}(V), d)$$

are also embeddings. The maps that replace markings by base-points, as well as the contraction maps, respect these embeddings, i.e., given a wall $\varepsilon = 1/d_a$ and $\varepsilon_+ > \varepsilon \geq \varepsilon_-$ nearby, we have restrictions

$$b_A : Q^{\varepsilon_+}_{g,k+A}(X, d^A_0) \longrightarrow Q^{\varepsilon_+}_{g,k}(X, d),$$

where $d^A_0 = d - |A|d_a$, and

$$c : Q^{\varepsilon_+}_{g,k}(X, d) \longrightarrow Q^{\varepsilon_-}_{g,k}(X, d).$$

Theorem **1.6.** — *There is an equality*

$$\begin{aligned}&[Q^{\varepsilon_-}_{g,k}(X, d)]^{\mathrm{vir}} - c_*[Q^{\varepsilon_+}_{g,k}(X, d)]^{\mathrm{vir}} \\ &= \sum_{|A|} \frac{1}{|A|!}(b_A)_*(c_A)_*\left(\prod_{a\in A} ev_a^*\mu_{d_a}(z)|_{z=-\psi_a} \cap [Q^{\varepsilon_+}_{g,k+A}(X, d^A_0)]^{\mathrm{vir}}\right)\end{aligned}$$

in the Chow group $A_*(Q^{\varepsilon_-}_{g,k}(X, d))_{\mathbf{Q}}$.

Since Theorem 1.6 implies the formula (1.2.2), the relations between ε-quasimap invariants and Gromov-Witten invariants in Corollaries 1.3 and 1.5 hold for nonsingular complete intersections $X \subset \mathbf{P}^n$ of codimension $r \leq n$.

Let $l_1, l_2, \ldots, l_r$ be the degrees of the hypersurfaces whose intersection is X. The small I-function of X is given by the well-known formula (see [17])

$$I(q, z) = \mathbf{1} + \sum_{d\geq 1} q^d \frac{\prod_{i=1}^r \prod_{j=1}^{l_i d}(l_i H + jz)}{\prod_{j=1}^d (H + jz)^{n+1}},$$

where H denotes the restriction to X of the hyperplane class on $\mathbf{P}^n$.

If $\sum_{i=1}^r l_i \geq n + 2$, so that X is a variety of general type, we do not know of any simplification of the wall-crossing formula in Corollary 1.3. Note that even in genus $g = 0$ our result is new.

If X is Fano or Calabi-Yau, more precise statements can be made.

The case $\sum_{i=1}^r l_i \leq n - 1$ of complete intersections which are Fano of index at least two is the simplest, since $J^{\varepsilon}_0(q) = 1$ and $J^{\varepsilon}_1(q) = 0$ for all $\varepsilon \geq 0+$. We conclude the following ε-independence result.

Corollary **1.7.** — *The quasimap invariants of a projective complete intersection with* $\sum_i l_i \leq n - 1$ *are independent of* ε.

In the Fano of index one case, $\sum_{i=1}^r l_i = n$, we have $J_0^\varepsilon(q) = 1$ and $J_1^\varepsilon(q) = q(\prod_{i=1}^r l_i!)\mathbf{1}$ for all $0+ \leq \varepsilon \leq 1$.

Corollary **1.8.** — *For a projective complete intersection with* $\sum_i l_i = n$ *and for* $0+ \leq \varepsilon \leq 1$ *we have*

$$F_g^\varepsilon(\mathbf{t}(\psi)) = F_g^\infty\left(\mathbf{t}(\psi) + q\left(\prod_{i=1}^r l_i!\right)\mathbf{1}\right).$$

In particular, if $(g, n) \neq (0, 1), (0, 2)$, *then the* primary *invariants are again* ε*-independent:*

$$\langle \gamma_1, \dots \gamma_n \rangle_{g,n,\beta}^\varepsilon = \langle \gamma_1, \dots \gamma_n \rangle_{g,n,\beta}^\infty.$$

The second equality in Corollary 1.8 is a consequence of the string equation in Gromov-Witten theory.

The most interesting is the Calabi-Yau case $\sum_{i=1}^r l_i = n + 1$, for which

$$J_0^\varepsilon(q) = \sum_{0 \leq d \leq \frac{1}{\varepsilon}} q^d \frac{\prod_{i=1}^r (l_i d)!}{d!^{n+1}},$$

$$J_1^\varepsilon(q) = H \sum_{1 \leq d \leq \frac{1}{\varepsilon}} q^d \frac{\prod_{i=1}^r (l_i d)!}{d!^{n+1}} \left(\sum_{i=1}^r \sum_{k=1}^{l_i d} \frac{l_i}{k} - (n+1) \sum_{k=1}^d \frac{1}{k}\right).$$

For every ε and every d, the virtual dimension of the moduli space $Q_{g,k}^\varepsilon(X, d)$ is equal to $(\dim X - 3)(1 - g) + k$. We split the discussion according to the genus.

1.4.1. *Genus zero.* — The wall-crossing formula (1.3.2) at $g = 0$ for a Calabi-Yau complete intersection is proved in [9, §3] using Dubrovin-type reconstruction arguments and results from [7]. Here we just note that the new proof in this paper *does not* use the torus action on $\mathbf{P}^n$.

1.4.2. *Genus one.* — When $g = 1$, the virtual dimension is independent of the dimension of X. Consider the unpointed case $k = 0$, i.e. the specialization of (1.3.2) at $g = 1$, and $\mathbf{t}(\psi) = 0$. Separating the $d = 0$ contributions and applying the divisor equation in the Gromov-Witten side gives

Corollary **1.9.** — *For a Calabi-Yau complete intersection* $X \subset \mathbf{P}^n$

$$\frac{1}{24}\chi_{\text{top}}(X)\log J_0^\varepsilon + \sum_{d \geq 1} q^d \langle\ \rangle_{1,0,d}^\varepsilon \tag{1.4.1}$$

$$= -\frac{1}{24}\int_X \frac{J_1^\varepsilon}{J_0^\varepsilon} c_{\dim X - 1}(T_X) + \sum_{d \geq 1} q^d \exp\left(\int_{d[line]} \frac{J_1^\varepsilon}{J_0^\varepsilon}\right)\langle\ \rangle_{1,0,d}^\infty.$$

When $\varepsilon = 0+$, the formula (1.4.1) answers a question raised first in [25, §10.2]. Note that the unpointed genus one $(0+)$-invariants $\langle\ \rangle^{0+}_{1,0,d}$ have been recently calculated by Kim and Lho ([21]) in terms of the small I-function. Combining [21, Theorem 1.1] with Corollary 1.9 gives new proofs for the main results on genus one Gromov-Witten invariants of X from [30] and [27].

1.4.3. *Higher genus.* — If $g \geq 2$ and $\dim X \geq 4$, the virtual classes (hence the invariants) vanish by dimension considerations. We restrict to the case of unpointed invariants of Calabi-Yau threefolds. The invariants for $d = 0$ are the same for all stability conditions and are given by the formula

$$\langle\ \rangle^{\varepsilon}_{g,0,0} = \frac{(-1)^g}{2}\chi_{\mathrm{top}}(X)\frac{|B_{2g}|}{2g}\frac{|B_{2g-2}|}{2g-2}\frac{1}{(2g-2)!},$$

with B_{2g}, B_{2g-2} the Bernoulli numbers, see [16], [14].

Corollary **1.10.** — *For a Calabi-Yau threefold complete intersection in* $\mathbf{P}^n$, $g \geq 2$ *and* $\varepsilon \geq 0+$,

$$J^{\varepsilon}_0(q)^{2g-2}\left(\frac{(-1)^g}{2}\chi_{\mathrm{top}}(X)\frac{|B_{2g}|}{2g}\frac{|B_{2g-2}|}{2g-2}\frac{1}{(2g-2)!} + \sum_{d\geq 1} q^d \langle\ \rangle^{\varepsilon}_{g,0,d}\right)$$
$$= \frac{(-1)^g}{2}\chi_{\mathrm{top}}(X)\frac{|B_{2g}|}{2g}\frac{|B_{2g-2}|}{2g-2}\frac{1}{(2g-2)!} + \sum_{d\geq 1} q^d \exp\left(\int_{d[line]}\frac{J^{\varepsilon}_1}{J^{\varepsilon}_0}\right)\langle\ \rangle^{\infty}_{g,0,d}.$$

1.5. *Relation with Mirror Symmetry.* — In this subsection we let X be the quintic hypersurface in $\mathbf{P}^4$ and consider the asymptotic stability condition $\varepsilon = 0+$. (The same analysis will apply to the $(0+)$-theory of any Calabi-Yau threefold for which Conjecture 1.1 holds.)

Fix a genus $g \geq 1$. In their landmark paper [2], Bershadsky, Cecotti, Ooguri, and Vafa studied the string theory B-model of a Calabi-Yau threefold and in particular they proposed a method to calculate the genus g Gromov-Witten potential of the quintic (with no insertions) via Mirror Symmetry. Namely, let $\mathcal{F}^{B}_{g}(q)$ be the holomorphic limit of the genus g partition function of the B-model associated to the *mirror family* of the quintic, where q is the coordinate around the large complex structure point. Let the mirror map be $Q = q\exp(\frac{1}{H}\frac{I_1(q)}{I_0(q)})$, where

$$I_0(q) = 1 + \sum_{d\geq 1} q^d \frac{(5d)!}{d!^5}, \qquad I_1(q) = H\sum_{d\geq 1} q^d \frac{(5d)!}{(d!)^5}\left(\sum_{j=d+1}^{5d}\frac{1}{j}\right).$$

Then the genus $g \geq 2$ Mirror Conjecture of [2] for the quintic threefold is the equality

$$\text{(1.5.1)} \qquad I_0(q)^{2g-2}\mathcal{F}^{B}_{g}(q) = \sum_{d\geq 0} Q^d \langle\ \rangle^{\infty}_{g,0,d}.$$

Hence Corollary 1.10 says precisely that the quasimap partition function $F_g^{0+}|_{\mathbf{t}=0}(q)$ is *equal* to $\mathcal{F}_g^{\mathrm{B}}(q)$, with no mirror map involved. Similarly, Corollary 1.9 gives the same equality in genus $g = 1$. In other words, our results in this paper can be viewed as giving a mathematically rigorous and geometrically meaningful definition of the holomorphic limit of the B-model partition function.

The B-model partition function of the mirror quintic has been studied extensively in the Physics literature. It is expected to have modular properties and to satisfy a recursion in g, determined up to a holomorphic function $f_g(q)$, the so-called "holomorphic ambiguity". The ambiguity has been fixed up to genus $g = 51$ in [20] and this is by far the most efficient computational method for predicting (via the conjectural mirror formula (1.5.1)) the higher genus Gromov-Witten invariants of the quintic. We speculate that the holomorphic ambiguity $f_g(q)$ has an intrinsic meaning in quasimap theory. It would be very interesting to determine if this is indeed the case.

1.6. *Final remarks.* — While the proof of Theorem 1.6 we give here is quite involved, it turns out to be also robust. For example, it extends easily to the case of complete intersections in products of projective spaces. It also applies to proving a wall-crossing formula for the virtual classes of quasimap moduli spaces (with same stability parameter $\varepsilon = 0+$ and target a complete intersection $X \subset \prod \mathbf{P}^{n_i}$) when one usual marking is changed to an infinitesimally weighted marking. To keep this paper from becoming excessively long, we defer the details of these developments to future writings.

1.7. *Acknowledgments.* — I.C.-F. was partially supported by the NSF grants DMS-1305004 and DMS-1601771. B.K. is supported by the KIAS individual grant MG016403. In addition, I.C.-F. thanks KIAS for financial support, excellent working conditions, and an inspiring research environment during visits when a large part of this project was completed. We deeply thanks the anonymous referee for valuable suggestions to improve the readability of the paper.

2. Virtual classes for moduli of quasimaps

2.1. *Overview.* — In this section we give a concrete description of the virtual class of a moduli space of quasimaps to a complete intersection in projective space. This is accomplished by embedding the moduli space into a smooth stack and intersecting the normal cone for this embedding with the zero section of an appropriate vector bundle. This description will be crucially used in the proof of Theorem 1.6 given in Section 3. The construction is uniform for all discrete parameters g, k, d and ε, but requires the existence of the moduli space of stable curves, so it doesn't apply directly to the unpointed elliptic case $(g, k) = (1, 0)$. An appropriate modification, sufficient for completing the proof of Theorem 1.6 in this case as well, will be discussed in Section 3.7.

2.2. *Set-up and conventions.* — From now on we let $\mathbf{G} = \mathbf{C}^*$. Let V be an $n+1$-dimensional $\mathbf{G}$-representation ($n \geq 1$), with weight vector $(1, \dots, 1)$. Let $\mathbf{C}^r_{\vec{l}}$ be an r-dimensional $\mathbf{G}$-representation with positive weight vector $\vec{l} := (l_1, \dots, l_r)$ ($l_j > 0, \forall j$). Assume we are given a $\mathbf{G}$-equivariant map

$$\varphi = \oplus_{i=1}^r \varphi_i : \mathrm{V} \to \mathbf{C}^r_{\vec{l}}$$

such that the closed subscheme $\mathrm{W} := \varphi^{-1}(0)$ is smooth away from $0 \in \mathrm{V}$ and of dimension $\dim \mathrm{W} = n + 1 - r > 0$. We linearize the $\mathbf{G}$ action on V by the character θ of weight 1. The GIT quotient $\mathrm{X} := \mathrm{W}/\!/_\theta \mathbf{G}$ is a nonsingular complete intersection of type $(l_1, \dots, l_r)$ in $\mathbf{P}^n = \mathrm{V}/\!/_\theta \mathbf{G}$, with φ_i its homogeneous equations.

Recall that the inclusion $i : \mathrm{X} \subset \mathbf{P}(\mathrm{V})$ induces an embedding

$$i : \mathrm{Q}^{\varepsilon}_{g,k}(\mathrm{X}, d) \hookrightarrow \mathrm{Q}^{\varepsilon}_{g,k}\big(\mathbf{P}(\mathrm{V}), d\big)$$

for all $\varepsilon \geq 0+$.

We also make the following conventions:

- $\overline{\mathrm{M}}_{g,k}$ denotes the Deligne-Mumford stack of k-pointed stable curves of genus g, while $\mathfrak{M}_{g,k}$ denotes the Artin stack of prestable k-pointed curves of genus g.
- $\mathfrak{Bun}^{g,k}_{\mathbf{G}}$ denotes the moduli stack of principal $\mathbf{G}$-bundles on k-pointed prestable curves of genus g. It is a smooth Artin stack of pure dimension and decomposes as $\coprod_{d \in \mathbf{Z}} \mathfrak{Bun}^{g,k}_{\mathbf{G},d}$, according to the degrees of the principal bundles. There are natural forgetful morphisms

$$\mathrm{Q}^{\varepsilon}_{g,k}\big(\mathbf{P}(\mathrm{V}), d\big) \longrightarrow \mathfrak{Bun}^{g,k}_{\mathbf{G},d} \longrightarrow \mathfrak{M}_{g,k}.$$

- The universal families of curves on various moduli stacks are denoted by $\mathfrak{C}$, usually with decorations recording the discrete data. For example,

$$\begin{array}{ccccc} \mathfrak{C}^{\varepsilon}_{g,k,d} & \longrightarrow & \mathfrak{C}_{g,k} & \longleftarrow & \mathfrak{C}^{\varepsilon'}_{g,k,d'} \\ \downarrow & & \downarrow & & \downarrow \\ \mathrm{Q}^{\varepsilon}_{g,k}(\mathrm{X}, d) & \longrightarrow & \overline{\mathrm{M}}_{g,k} & \longleftarrow & \mathrm{Q}^{\varepsilon'}_{g,k}(\mathbf{P}(\mathrm{V} \otimes \mathbf{C}^{\mathrm{N}}), d'). \end{array}$$

  We will abuse notation and denote always by π the projection from the universal curve to the base.

We will represent quasimaps to a projective space $\mathbf{P}(\mathrm{V})$ as tuples

$$\big((\mathrm{C}, p_1, \dots, p_k), \mathrm{L}, u\big)$$

with L a line bundle on C and u a section of $L \otimes V$ (as in [6]). Quasimaps to $X \subset \mathbf{P}(V)$ will then be such tuples for which the components $u_1, \dots, u_{\dim V}$ of u (once a basis of V is chosen) satisfy the homogeneous equations of X. The base-points of the quasimap are the points of C where all the u_i's vanish and the length $\ell(x)$ at a point $x \in C$ is the common order of vanishing. Given $\varepsilon \in \mathbf{Q}_{>0}$, recall that the definition of ε-stability requires the following conditions be satisfied:

(1) the base-points are away from nodes and markings;
(2) $\varepsilon\ell(x) \leq 1$ for all $x \in C$;
(3) the line bundle $\omega_C(p_1 + \cdots + p_k) \otimes L^\varepsilon$ is ample.

For $\varepsilon = 0+$ condition (2) is empty and is discarded, while condition (3) translates into the absence of rational tails in C and the strict positivity of $\deg L$ on rational bridges (rational components of C containing exactly two special points).

Finally, recall that the theory of virtual classes was first developed by Li and Tian in [24], and by Behrend and Fantechi in [1]. In this paper we use the formalism of [1].

2.3. *Twisting line bundles.* — Fix $(g, k) \neq (1, 0)$.

For each $\varepsilon \geq 0+$ we construct a line bundle $\mathscr{M}_\varepsilon$ on the universal curve

$$\mathfrak{C}^\varepsilon_{g,k,d} \longrightarrow Q^\varepsilon_{g,k}\big(\mathbf{P}(V), d\big)$$

as follows.

When $g = 0$, we take the trivial line bundle $\mathscr{M}_\varepsilon = \mathcal{O}$.

When $g \geq 1$ and $g + k \geq 2$, the moduli stack $\overline{M}_{g,k}$ exists and we have the diagram

$$\begin{array}{ccc} \mathfrak{C}^\varepsilon_{g,k,d} & \xrightarrow{\widetilde{st}_\varepsilon} & \mathfrak{C}_{g,k} \\ \downarrow \pi & & \Sigma_i \uparrow\downarrow \pi \\ Q^\varepsilon_{g,k}(X, d) & \xrightarrow{st_\varepsilon} & \overline{M}_{g,k} \end{array}$$

with $st_\varepsilon, \widetilde{st}_\varepsilon$ the stabilization morphisms and Σ_i the sections of π corresponding to the k markings. The logarithmic relative dualising sheaf $\omega_{\log} := \omega_\pi(\Sigma_1 + \cdots \Sigma_k)$ on $\mathfrak{C}_{g,k}$ is π-ample and we choose a positive integer p such that $\omega_{\log}^{\otimes p}$ is π-relatively very ample. We also choose a very ample line bundle on the (projective!) coarse moduli of $\overline{M}_{g,k}$ and denote by $\mathscr{H}$ its pull-back to the stack $\overline{M}_{g,k}$. Now set

$$\mathscr{M}_\varepsilon := \widetilde{st}_\varepsilon^{\,*}\big(\pi^*\mathscr{H} \otimes \omega_{\log}^{\otimes p}\big).$$

Lemma **2.1.** — *The line bundles $\mathscr{M}_\varepsilon$ satisfy the following properties:*

(i) *If $\varepsilon > \varepsilon'$, then $\mathscr{M}_\varepsilon = \tilde{c}^*\mathscr{M}_{\varepsilon'}$, where $\tilde{c}$ is the induced contraction morphism on universal curves in the diagram*

$$\begin{array}{ccc} \mathfrak{C}^{\varepsilon}_{g,k,d} & \xrightarrow{\ \tilde{c}\ } & \mathfrak{C}^{\varepsilon'}_{g,k,d} \\ \downarrow & & \downarrow \\ Q^{\varepsilon}_{g,k}(\mathbf{P}(V), d) & \xrightarrow{\ c\ } & Q^{\varepsilon'}_{g,k}(\mathbf{P}(V), d) \end{array}$$

(ii) *For every geometric fiber C of $\mathfrak{C}^{\varepsilon}_{g,k,d} \to Q^{\varepsilon}_{g,k}(\mathbf{P}(V), d)$ we have*

$$H^1(C, \mathscr{L} \otimes \mathscr{M}_\varepsilon|_C) = 0,$$

where $\mathscr{L}$ denotes the universal line bundle associated to the universal principal **G**-*bundle on the universal curve.*

Proof. — Part (*i*) is obvious from the definition, since C and $\tilde{c}$ are compatible with the forgetful stabilization maps. For part (*ii*), notice that $\deg \mathscr{L}$ is nonnegative on every component of every geometric fiber C and by stability it is strictly positive on every rational component with at most two special points. On the other hand, by construction $\mathscr{M}_\varepsilon$ has vanishing H^1 on the stabilization of C and is trivial on rational tails and rational bridges. The required vanishing follows. □

Choose once and for all global sections $\{\tau_1, \ldots, \tau_N\}$ giving a basis of $\Gamma(\mathfrak{C}_{g,k}, \pi^*\mathscr{H} \otimes \omega_{\log}^{\otimes p})$, and hence an embedding

$$h : \mathfrak{C}_{g,k} \longrightarrow \mathbf{P}(\mathbf{C}^N).$$

Let $s_j^\varepsilon := \widetilde{ft}_\varepsilon^{\,*}\tau_j$ of $\mathscr{M}_\varepsilon$ be the induced sections of $\mathscr{M}_\varepsilon$, determining the map $h_\varepsilon := h \circ \widetilde{ft}_\varepsilon$, with $\mathscr{M}_\varepsilon = h_\varepsilon^*\mathcal{O}_{\mathbf{P}(\mathbf{C}^N)}(1)$. When the parameter ε is understood we will drop it from the notation and write simply $\mathscr{M}$ and s_j for the twisting line bundle and its sections. Furthermore, we will use the same notations when considering the restriction of the set-up in this subsection to the moduli spaces $Q^{\varepsilon}_{g,k}(X, d)$ via the embedding i.

Note that the degree of $\mathscr{M}$ on the fibers of the universal curve is a constant positive integer $d_{\mathscr{M}}$ depending only on (g, k), but not on d, or on the dimension of $\mathbf{P}(V)$.

2.4. *Perfect obstruction theory of $Q^{\varepsilon}_{g,k}(X, d)$.* — Fix $(g, k) \neq (1, 0)$ and $\varepsilon \geq 0+$. Consider the line bundle $\mathscr{L}' := \mathscr{L} \otimes \mathscr{M}$ on the universal curve $\mathfrak{C}^{\varepsilon}_{g,k,d}$ over $Q^{\varepsilon}_{g,k}(X, d)$. There is a commuting diagram with exact rows

(**2.4.1**)
$$\begin{array}{ccccccccc}
0 & \longrightarrow & \mathscr{L}\otimes \mathrm{V} & \xrightarrow{\oplus_j s_j} & \oplus_{j=1}^{\mathrm{N}} \mathscr{L}'\otimes \mathrm{V} & \xrightarrow{\alpha_0} & \mathscr{P} & \longrightarrow & 0 \\
 & & \Big\downarrow {\scriptstyle \oplus_i d\varphi_i} & & \Big\downarrow {\scriptstyle \oplus_{i,j} s_j^{l_i-1} d\varphi_i} & & \Big\downarrow {\scriptstyle f} & & \\
0 & \longrightarrow & \oplus_{i=1}^{r} \mathscr{L}^{l_i} & \xrightarrow{\oplus_{i,j} s_j^{l_i}} & \oplus_{i,j} (\mathscr{L}')^{l_i} & \xrightarrow[\alpha_1]{} & \mathscr{Q} & \longrightarrow & 0.
\end{array}$$

The top row is obtained by puling-back the tautological sequence

(**2.4.2**) $$0 \longrightarrow \mathcal{O}_{\mathbf{P}(\mathbf{C}^{\mathrm{N}})}(-1) \longrightarrow \mathcal{O}_{\mathbf{P}(\mathbf{C}^{\mathrm{N}})} \otimes \mathbf{C}^{\mathrm{N}} \longrightarrow \mathrm{Q} \longrightarrow 0$$

via $h_\varepsilon : \mathfrak{C}^{\varepsilon}_{g,k,d} \longrightarrow \mathbf{P}(\mathbf{C}^{\mathrm{N}})$ and tensoring with $\mathscr{L}'\otimes \mathrm{V}$. The bottom row comes from (2.4.2) similarly, by taking the direct sum of its pull-backs via $g_{l_i} \circ h_\varepsilon$, tensored with $(\mathscr{L}')^{l_i}$, where $g_{l_i} : \mathbf{P}(\mathbf{C}^{\mathrm{N}}) \longrightarrow \mathbf{P}(\mathbf{C}^{\mathrm{N}})$ is the degree l_i map $[t_1 : \cdots : t_{\mathrm{N}}] \mapsto [t_1^{l_i} : \cdots : t_{\mathrm{N}}^{l_i}]$. In particular, $\mathscr{P}$ and $\mathscr{Q}$ are vector bundles.

The components $d\varphi_i$ of the vertical homomorphism on the left are given as follows. Let $\Delta \subset \mathfrak{C}^{\varepsilon}_{g,k,d}$ be an open substack. After choosing coordinates $(x_0, \ldots x_n)$ on V, we may write φ_i as a homogeneous polynomial of degree l_i and a local section v of $\mathscr{L}\otimes \mathrm{V}$ on Δ as $v = (v_0, \ldots v_n)$. Then we put

$$d\varphi_i(v) = \nabla\varphi_i(u|_\Delta)\cdot v = \sum_{m=0}^{n} \frac{\partial \varphi_i}{\partial x_m}(u|_\Delta) v_m,$$

where $u = (u_0, \ldots, u_n)$ is the universal section of $\mathscr{L}\otimes \mathrm{V}$ on $\mathfrak{C}^{\varepsilon}_{g,k}$. Similarly, for fixed i and j and a local section $v' = (v'_0, \ldots v'_n)$ of $\mathscr{L}'\otimes \mathrm{V}$,

$$s_j^{l_i-1} d\varphi_i(v') = \sum_{m=0}^{n} \frac{\partial \varphi_i}{\partial x_m}(u\otimes s_j|_\Delta) v'_m = \sum_{m=0}^{n} s_j^{l_i-1}|_\Delta \frac{\partial \varphi_i}{\partial x_m}(u|_\Delta) v'_m.$$

Viewing (2.4.1) as an exact sequence of two-term complexes, it follows that the two-term vertical complex on the left in (2.4.1) is quasi-isomorphic to the shifted mapping cone $\mathrm{A}^\bullet := \mathrm{Cone}(\alpha)[-1]$ of the homomorphism $\alpha = (\alpha_0, \alpha_1)$. Denote

$$\mathscr{R} := \oplus_{i,j} (\mathscr{L}')^{l_i}.$$

Define a coherent sheaf $\mathscr{E}$ (in fact, a vector bundle) by the exact sequence

(**2.4.3**) $$0 \to \mathscr{E} \to \mathscr{P} \oplus \mathscr{R} \to \mathscr{Q} \to 0,$$

where $\mathscr{P} \oplus \mathscr{R} \to \mathscr{Q}$ is given by $(x, y) \mapsto f(x) - \alpha_1(y)$. Then $\mathrm{A}^\bullet$ is quasi-isomorphic to

(**2.4.4**) $$\oplus_{j=1}^{\mathrm{N}} \mathscr{L}'\otimes \mathrm{V} \to \mathscr{E}.$$

On the other hand, if

$$\rho : Prin(\mathscr{L}) \times_{\mathbf{G}} \mathrm{W} \to \mathfrak{C}^{\varepsilon}_{g,k,d}$$

denotes the universal W-fiber bundle with $Prin(\mathscr{L})$ the principal **G**-bundle associated to $\mathscr{L}$ and we view u as the universal section of ρ, then the pull-back $u^*\mathbf{T}_\rho$ of the relative tangent complex of ρ coincides with the two-term complex $\mathscr{L} \otimes \mathrm{V} \to \oplus_{i=1}^{r} \mathscr{L}^{l_i}$ on the left of (2.4.1). We conclude that $u^*\mathbf{T}_\rho$ is quasi-isomorphic to (2.4.4) at amplitude $[0, 1]$.

Part *(ii)* of Lemma 2.1 gives the vanishing $\mathrm{R}^1\pi_*\mathscr{L}' = 0$. This in turn implies that $\mathrm{R}^1\pi_*\mathscr{P} = \mathrm{R}^1\pi_*\mathscr{Q} = 0$. Since the derived push-forward of $u^*\mathbf{T}_\rho$ has amplitude in $[0, 1]$ by [12, Theorem 4.5.2], the same is true for the derived push-forward of the shifted mapping cone $\mathrm{A}^\bullet$. Hence the map $\pi_*(\mathscr{P} \oplus \mathscr{R}) \to \pi_*\mathscr{Q}$ is surjective and then $\mathrm{R}^1\pi_*\mathscr{E} = 0$ from (2.4.3). It follows that

$$\mathrm{E}^{\varepsilon}_d := \pi_*\mathscr{E} \tag{2.4.5}$$

is a locally free sheaf on $\mathrm{Q}^{\varepsilon}_{g,k}(\mathrm{X}, d)$ and we obtain a perfect complex

$$\oplus_{j=1}^{\mathrm{N}} \pi_*\mathscr{L}' \otimes \mathrm{V} \to \mathrm{E}^{\varepsilon}_d, \tag{2.4.6}$$

whose dual represents the canonical perfect obstruction theory

$$\left(\mathrm{R}^\bullet\pi_* u^*\mathbf{T}_\rho\right)^\vee$$

for $\mathrm{Q}^{\varepsilon}_{g,k}(\mathrm{X}, d)$ relative to $\mathfrak{Bun}^{g,k}_{\mathbf{G}}$. We have proved the following result.

Proposition **2.2.** — *The virtual fundamental class of* $\mathrm{Q}^{\varepsilon}_{g,k}(\mathrm{X}, \beta)$ *is*

$$\left[\mathrm{Q}^{\varepsilon}_{g,k}(\mathrm{X}, d)\right]^{\mathrm{vir}} = 0^!_{\mathrm{E}^{\varepsilon}_d}\left([\mathbf{C}_\varepsilon]\right)$$

where $\mathbf{C}_\varepsilon \subset \mathrm{E}^{\varepsilon}_d$ *denotes the Behrend-Fantechi obstruction cone, see [1], associated to the relative perfect obstruction theory given by* (2.4.6).

2.5. *An embedding of* $\mathrm{Q}^{\varepsilon}_{g,k}(\mathrm{X}, d)$ *into a smooth stack.* — Set

$$d' := d + d_{\mathscr{M}} = d + \deg(\mathscr{M}|_{\mathrm{C}}).$$

Consider the moduli stack $\mathrm{Q}^{\varepsilon}_{g,k}(\mathbf{P}(\mathrm{V} \otimes \mathbf{C}^{\mathrm{N}}), d')$, with universal curve $\mathfrak{C}^{\varepsilon}_{g,k,d'}$. By a slight abuse, denote also by $\mathscr{M}$ the twisting line bundle on $\mathfrak{C}^{\varepsilon}_{g,k,d'}$ (defined by the construction in Section 2.3, as the pull-back of $\pi^*\mathscr{H} \otimes \omega_{\log}^{\otimes p}$ on $\mathfrak{C}_{g,k}$ by the stabilization morphism).

Definition **2.3.** — *Define* $\mathrm{U}^{\varepsilon}_{d'} \subset \mathrm{Q}^{\varepsilon}_{g,k}(\mathbf{P}(\mathrm{V} \otimes \mathbf{C}^{\mathrm{N}}), d')$ *as the open substack consisting of the* ε*-stable quasimaps*

$$\left((\mathrm{C}, p_1, \ldots, p_k), \mathrm{L}', u'\right)$$

to $\mathbf{P}(\mathrm{V} \otimes \mathbf{C}^{\mathrm{N}})$ *such that* $\mathrm{H}^1(\mathrm{C}, \mathrm{L}') = 0$.

Note that $U^{\varepsilon}_{d'}$ is the complement of the support of the coherent sheaf $R^1\pi_*\mathscr{L}'$, so it is indeed an open substack.

Lemma **2.4.** — *The stack* $U^{\varepsilon}_{d'}$ *is a separated DM-stack of finite type, smooth and of pure dimension over* $\mathfrak{Bun}^{g,k}_{\mathbf{G}}$*, and hence over* $\mathfrak{M}_{g,k}$*. In particular, fixing a locally-closed substack of* $\mathfrak{Bun}^{g,k}_{\mathbf{G}}$ *parametrizing prestable curves with fixed topological type, together with line bundles of given degrees on the components, produces a corresponding locally-closed substack of* $U^{\varepsilon}_{d'}$ *with the same codimension.*

Proof. — The separatedness and finite type properties follow from the corresponding ones for $Q^{\varepsilon}_{g,k}(\mathbf{P}(V\otimes\mathbf{C}^N), d')$. By definition, the quasimaps in $U^{\varepsilon}_{d'}$ are unobstructed, which gives the smoothness and the pure dimensionality. (In fact, $U^{\varepsilon}_{d'}$ is also irreducible, since it is the smooth locus in the "main component" of $Q^{\varepsilon}_{g,k}(\mathbf{P}(V\otimes\mathbf{C}^N), d')$. Irreducibility of the "main component" follows from the connectedness of $\overline{M}_{g,k}(\mathbf{P}(V\otimes\mathbf{C}^N), d')$, proven in [22].) □

Let $\pi:\mathfrak{C}^{\varepsilon}_{g,k,d'}\to U^{\varepsilon}_{d'}$ be the universal curve and let $\mathscr{L}'$ be the universal line bundle of π-relative degree d' on $\mathfrak{C}^{\varepsilon}_{g,k,d'}$. By the very definition of $U^{\varepsilon}_{d'}$, the sheaf $\pi_*\mathscr{L}'$ is locally free. Put

$$\mathscr{L}:=\mathscr{L}'\otimes\mathscr{M}^{-1},$$

and consider the diagram of vector bundles on $\mathfrak{C}^{\varepsilon}_{g,k,d'}$

$$\begin{array}{ccccccccc} 0 & \longrightarrow & \mathscr{L}\otimes V & \xrightarrow{\oplus_j s_j} & \oplus_{j=1}^{N}\mathscr{L}'\otimes V & \longrightarrow & \mathscr{P}^{\varepsilon}_{d'} & \longrightarrow & 0 \\ & & & & \big\downarrow{\scriptstyle \oplus_j(\oplus_i d\varphi_i)} & & & & \\ 0 & \longrightarrow & \oplus_{i=1}^{r}\mathscr{L}^{l_i} & \xrightarrow{\oplus_{i,j}s_j^{l_i}} & \oplus_{i,j}(\mathscr{L}')^{l_i} & \longrightarrow & \mathscr{Q}^{\varepsilon}_{d'} & \longrightarrow & 0. \end{array} \tag{2.5.1}$$

As before, the exact rows are obtained from the tautological exact sequence (2.4.2) on $\mathbf{P}(\mathbf{C}^N)$ via pull-backs, tensoring with appropriate line bundles, and taking direct sums. The components of the map between the middle terms (for fixed i and j) are given by

$$d\varphi_i(v'_{j0},\dots v'_{jn})=\sum_{m=0}^{n}\frac{\partial\varphi_i}{\partial x_m}((u'_{j0},\dots u'_{jn})|_{\Delta})v'_{jm},$$

where

$$u'=(u'_{10},\dots,u'_{1n},u'_{20},\dots,u'_{2n},\dots,u'_{N0},\dots,u'_{Nn}) \tag{2.5.2}$$

is the universal global section of $\oplus_{j=1}^{N}\mathscr{L}'\otimes V$ on $\mathfrak{C}^{\varepsilon}_{g,k,d'}$ and $(v'_{10},\dots,v'_{1n},\dots,v'_{N0},\dots,v'_{Nn})$ is a local section of $\oplus_{j=1}^{N}\mathscr{L}'\otimes V$ over an open $\Delta\subset\mathfrak{C}^{\varepsilon}_{g,k,d'}$.

Let us denote

$$\mathscr{A}^{\varepsilon}_{d'} := \oplus_{j=1}^{N} \mathscr{L}' \otimes V, \qquad \mathscr{R}^{\varepsilon}_{i,d'} := \oplus_{j=1}^{N} \big(\mathscr{L}'\big)^{l_i}, \qquad \mathscr{R}^{\varepsilon}_{d'} := \oplus_{i=1}^{r} \mathscr{R}^{\varepsilon}_{i,d'}.$$

The tautological section τ^{ε} of $\pi_* \mathscr{A}^{\varepsilon}_{d'}$ induces a natural section σ^{ε}_{P} of the vector bundle

$$P^{\varepsilon}_{d'} := \pi_* \mathscr{P}^{\varepsilon}_{d'}$$

on $U^{\varepsilon}_{d'}$. On the other hand, we also have the section σ^{ε}_{R} of the vector bundle

$$R^{\varepsilon}_{d'} := \pi_* \mathscr{R}^{\varepsilon}_{d'}$$

whose (i,j)-component is given by $\varphi_i(u'_{j0}, \dots, u'_{jn})$. Set

$$\textbf{(2.5.3)} \qquad \sigma^{\varepsilon} := \big(\sigma^{\varepsilon}_{P}, \sigma^{\varepsilon}_{R}\big) \in H^0\big(U^{\varepsilon}_{d'}, P^{\varepsilon}_{d'} \oplus R^{\varepsilon}_{d'}\big).$$

Because the exactness of the rows of (2.5.1) is preserved for any base change, it follows immediately that the zero locus of the section σ^{ε} is identified with the stack $Q^{\varepsilon}_{g,k}(X, d)$. Thus, we have an explicit embedding of $Q^{\varepsilon}_{g,k}(X, d)$ in the smooth stack $U^{\varepsilon}_{d'}$, summarized in the diagram

$$\textbf{(2.5.4)} \qquad \begin{array}{ccc} & & P^{\varepsilon}_{d'} \oplus R^{\varepsilon}_{d'} \\ & & \downarrow \uparrow \sigma^{\varepsilon} \\ Q^{\varepsilon}_{g,k}(X, d) \cong (\sigma^{\varepsilon})^{-1}(0) & \xrightarrow{\text{closed}} & U^{\varepsilon}_{d'} \\ & \searrow & \downarrow \text{smooth} \\ & & \mathfrak{Bun}^{g,k}_{\mathbf{G}}. \end{array}$$

Over $Q^{\varepsilon}_{g,k}(X, d)$, the diagram (2.5.1) restricts to the diagram (2.4.1). Denoting by $\mathscr{I}$ the ideal sheaf of the closed substack $Q^{\varepsilon}_{g,k}(X, d)$ in $U^{\varepsilon}_{d'}$ and setting

$$\textbf{(2.5.5)} \qquad F^{\varepsilon}_{d'} := P^{\varepsilon}_{d'} \oplus R^{\varepsilon}_{d'} = \pi_* \mathscr{P}^{\varepsilon}_{d'} \oplus \pi_* \mathscr{R}^{\varepsilon}_{d'},$$

we obtain the commuting diagram of coherent sheaves

$$\textbf{(2.5.6)} \qquad \begin{array}{ccccc} (F^{\varepsilon}_{d'}|_{Q^{\varepsilon}_{g,k}(X,d)})^{\vee} & \longrightarrow & (E^{\varepsilon}_{d})^{\vee} & \longrightarrow & (\pi_* \mathscr{A}^{\varepsilon}_{d'}|_{Q^{\varepsilon}_{g,k}(X,d)})^{\vee} \\ & \searrow (\sigma^{\varepsilon})^{\vee} & \downarrow & & \| \\ & & \mathscr{I}/\mathscr{I}^2 & \longrightarrow & \Omega_{U^{\varepsilon}_{d'}/\mathfrak{Bun}^{g,k}_{\mathbf{G}}}|_{Q^{\varepsilon}_{g,k}(X,d)} \end{array},$$

where the existence of the surjection $(E^{\varepsilon}_{d})^{\vee} \twoheadrightarrow \mathscr{I}/\mathscr{I}^2$ follows from a standard deformation theory calculation.

The square in the diagram (2.5.6) is precisely the map of complexes from the obstruction theory (2.4.6) to the two-term truncation of the relative cotangent complex $\mathbf{L}_{Q^{\varepsilon}_{g,k}(X,d)/\mathfrak{B}un^{g,k}_{\mathbf{G}}}$. The indicated equality $(\pi_*\mathscr{A}^{\varepsilon}_{d'})^{\vee} = \Omega_{U^{\varepsilon}_{d'}/\mathfrak{B}un^{g,k}_{\mathbf{G}}}$ follows from the definition of $U^{\varepsilon}_{d'}$ and the identification of $(R^{\bullet}\pi_*\mathscr{A}^{\varepsilon}_{d'})^{\vee}$ with the relative obstruction theory over $\mathfrak{B}un^{g,k}_{\mathbf{G}}$ for $Q^{\varepsilon}_{g,k}(\mathbf{P}(V \otimes \mathbf{C}^N), d')$, see [6, §5.3]. Here $\mathscr{L}'$ denotes, by abusing notation, also the universal line bundle on the universal curve on $Q^{\varepsilon}_{g,k}(\mathbf{P}(V \otimes \mathbf{C}^N), d')$.

Lemma **2.5.** — *The relative normal cone* $\mathbf{C}_{Q^{\varepsilon}_{g,k}(X,d)/U^{\varepsilon}_{d'}}$ *for the embedding in* (2.5.4) *coincides with the obstruction cone* $\mathbf{C}_{\varepsilon} \subset E^{\varepsilon}_{d}$.

Proof. — First, we have by definition

$$\mathbf{C}_{\varepsilon} = \mathbf{C}_{in} \times_{[E^{\varepsilon}_{d}/T_{U^{\varepsilon}_{d'}/\mathfrak{B}un^{g,k}_{\mathbf{G}}}]} E^{\varepsilon}_{d},$$

where $\mathbf{C}_{in}$ is the relative intrinsic normal cone of $Q^{\varepsilon}_{g,k}(X, d)$ over $\mathfrak{B}un^{g,k}_{\mathbf{G}}$ (see [1]) and $[E^{\varepsilon}_{d}/T_{U^{\varepsilon}_{d'}/\mathfrak{B}un^{g,k}_{\mathbf{G}}}]$ denotes the stack quotient. Since $\mathbf{C}_{in} = [\mathbf{C}_{Q^{\varepsilon}_{g,k}(X,d)/U^{\varepsilon}_{d'}}/T_{U^{\varepsilon}_{d'}/\mathfrak{B}un^{g,k}_{\mathbf{G}}}]$, the Lemma follows. □

Proposition 2.2 and Lemma 2.5 imply the following concrete description of the virtual classes of moduli spaces of ε-stable quasimaps to X.

Corollary **2.6.**

$$[Q^{\varepsilon}_{g,k}(X, d)]^{\mathrm{vir}} = 0^{!}_{E^{\varepsilon}_{d}}\big([\mathbf{C}_{Q^{\varepsilon}_{g,k}(X,d)/U^{\varepsilon}_{d'}}]\big).$$

Remark **2.7.** — Recall that in genus zero we take a trivial twisting line bundle $\mathscr{M}$, so in this case $U^{\varepsilon}_{d'} = Q^{\varepsilon}_{0,k}(\mathbf{P}(V), d)$ and the construction reduces to the known realization of $Q^{\varepsilon}_{g,k}(X, d)$ as the zero locus of a section of the bundle $\oplus_i \pi_*(\mathscr{L})^{l_i}$ on $Q^{\varepsilon}_{0,k}(\mathbf{P}(V), d)$. This bundle has "correct" rank $d\sum_i l_i + r$, hence its refined top Chern class gives $[Q^{\varepsilon}_{g,k}(X, d)]^{\mathrm{vir}}$. However, for $g \geq 1$ the rank of the bundle $F^{\varepsilon}_{d'} = \pi_*\mathscr{P}^{\varepsilon}_{d'} \oplus \pi_*\mathscr{R}^{\varepsilon}_{d'}$ is larger than the virtual codimension of $Q^{\varepsilon}_{g,k}(X, d)$ in $U^{\varepsilon}_{d'}$, so the virtual class is *not* the refined top Chern class.

3. Proof of Theorem 1.6

3.1. *Overview.* — Adapting an idea of Bertram from [3], we consider a one-parameter degeneration of the diagram (2.5.4) which is obtained via a refinement of MacPherson's Graph Construction. The proof of Theorem 1.6 will then follow by analyzing the central fiber limit of the virtual cycle $[Q^{+}_{g,k}(X, d)]^{\mathrm{vir}}$ in this degeneration.

3.2. *Boundary strata.* — Let ε_0 be a wall, so that $m := 1/\varepsilon_0$ is a positive integer. Let $\varepsilon_+ > \varepsilon_0 \geq \varepsilon_-$ be stability parameters separated only by the single wall ε_0. Fix the numerical data (g, k, d). We will denote by $\mathrm{Q}^{\pm}_{g,k}(\mathrm{X}, d)$, $\mathrm{U}^{\pm}_{d'}$ etc. the moduli spaces corresponding to the stability parameters $\varepsilon_\pm$. The contraction morphisms with the abused notation

$$c : \mathrm{Q}^{+}_{g,k}(\mathrm{X}, d) \longrightarrow \mathrm{Q}^{-}_{g,k}(\mathrm{X}, d), \qquad c : \mathrm{U}^{+}_{d'} \longrightarrow \mathrm{U}^{-}_{d'}$$

contract precisely the rational tails of degree m.

The evaluation maps at the markings will be denoted by $\hat{ev}_j$ for $\mathrm{Q}^{\varepsilon}_{g,k}(\mathbf{P}(\mathrm{V}\otimes\mathbf{C}^{\mathrm{N}}), d')$ and for its open substack $\mathrm{U}^{\varepsilon}_{d'}$, while we reserve the notation ev_j for the evaluation maps on $\mathrm{Q}^{\varepsilon}_{g,k}(\mathbf{P}(\mathrm{V}), d)$ and on $\mathrm{Q}^{\varepsilon}_{g,k}(\mathrm{X}, d)$.

For a finite index set A, with $|\mathrm{A}| = 1, 2, \ldots, [\frac{d}{m}]$ we associate to each $a \in \mathrm{A}$ the integer $d_a = m$ and set

$$d_0 = d_0^{\mathrm{A}} := d - \sum_{a\in\mathrm{A}} d_a = d - |\mathrm{A}|m \geq 0. \tag{3.2.1}$$

Denote

$$\mathrm{D}_{\mathrm{A}} := \mathrm{U}^{+}_{k+\mathrm{A},d_0'} \times_{\mathbf{P}(\mathrm{V}\otimes\mathbf{C}^{\mathrm{N}})^{\mathrm{A}}} \prod_{a\in\mathrm{A}} \mathrm{Q}^{+}_{0,a}(\mathbf{P}(\mathrm{V}\otimes\mathbf{C}^{\mathrm{N}}), d_a),$$
$$\tilde{\mathrm{D}}_{\mathrm{A}} := \mathrm{U}^{+}_{k+\mathrm{A},d_0'} \times_{\mathbf{P}(\mathrm{V}\otimes\mathbf{C}^{\mathrm{N}})^{\mathrm{A}}} \prod_{a\in\mathrm{A}} \mathfrak{C}^{+}_{0,a,d_a},$$

where $\mathfrak{C}^{+}_{0,a,d_a} \to \mathrm{Q}^{+}_{0,a}(\mathbf{P}(\mathrm{V}\otimes\mathbf{C}^{\mathrm{N}}), d_a)$ is the universal curve, the notations $\mathrm{U}^{\pm}_{k+\mathrm{A},d_0'}$ are the obvious ones, and the fiber products are made via $(\hat{ev}_a)_{a\in\mathrm{A}}$ on the left and $\prod_{a\in\mathrm{A}} \hat{ev}_a$ on the right. The easiest way to describe the evaluation map $\hat{ev}_a : \mathfrak{C}^{+}_{0,a,d_a} \to \mathbf{P}(\mathrm{V}\otimes\mathbf{C}^{\mathrm{N}})$ is by identifying $\mathfrak{C}^{+}_{0,a,d_a}$ with the moduli stack $\mathrm{Q}^{+}_{0,a|1}(\mathbf{P}(\mathrm{V}\otimes\mathbf{C}^{\mathrm{N}}), d_a)$ which parametrizes ε_+-stable quasimaps of degree d_a from rational curves with one marking a of weight 1 and one additional marking of weight $0+$, see [8] for more on these moduli stacks.

We will need an alternative description of these boundary strata which takes into account the twisting line bundles $\mathscr{M}$.

Consider the diagram of universal curves

(3.2.2)

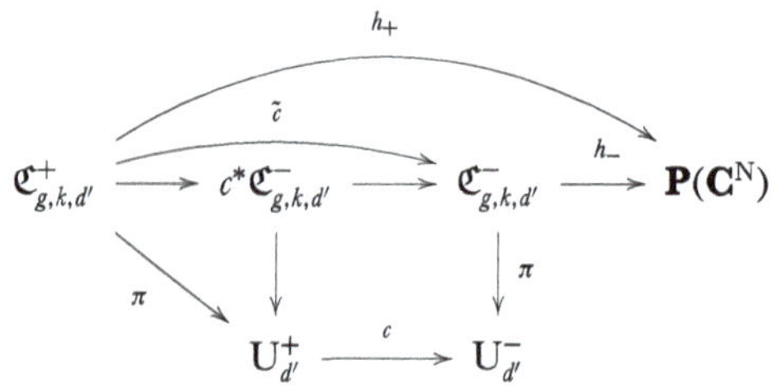

with cartesian square and the maps $h_\pm$ given by the sections $s_1, \dots s_N \in \Gamma(\mathfrak{C}^-_{g,k,d'}, \mathscr{M}_-)$, so that $\mathscr{M}_\pm = (h_\pm)^*(\mathcal{O}_{\mathbf{P}(\mathbf{C}^N)}(1))$. For each $a \in A$ we obtain maps

$$h_a^\pm : U^\pm_{k+A,d_0'} \longrightarrow \mathbf{P}(\mathbf{C}^N) \tag{3.2.3}$$

as the compositions

$$h_a^- : U^-_{k+A,d_0'} \xrightarrow{\Sigma_a} \mathfrak{C}^-_{g,k+A,d_0'} \xrightarrow{\tilde{b}_A} \mathfrak{C}^-_{g,k,d'} \xrightarrow{h_-} \mathbf{P}(\mathbf{C}^N),$$

$$h_a^+ : U^+_{k+A,d_0'} \xrightarrow{c_A} U^-_{k+A,d_0'} \xrightarrow{h_a^-} \mathbf{P}(\mathbf{C}^N).$$

Here Σ_a is the section corresponding to the marking $a \in A$, $\tilde{b}_A$ is the map that trades each marking in A for a base-point of length d_a, and c_A is the contraction of rational tails of degree d_a. There is a natural identification

$$D_A \cong U^+_{k+A,d_0'} \times_{(\mathbf{P}(V\otimes\mathbf{C}^N)\times\mathbf{P}(\mathbf{C}^N))^A} \prod_{a\in A} \left(Q^+_{0,a}(\mathbf{P}(V\otimes\mathbf{C}^N), d_a) \times \mathbf{P}(\mathbf{C}^N)\right), \tag{3.2.4}$$

where the fiber product is now done using $((\hat{ev}_a, h_a^+))_{a\in A}$ on the left and $\prod_{a\in A}(\hat{ev}_a \times \mathrm{id}_{\mathbf{P}(\mathbf{C}^N)})$ on the right. Similarly,

$$\tilde{D}_A \cong U^+_{k+A,d_0'} \times_{(\mathbf{P}(V\otimes\mathbf{C}^N)\times\mathbf{P}(\mathbf{C}^N))^A} \prod_{a\in A} \left(\mathfrak{C}^+_{0,a,d_a} \times \mathbf{P}(\mathbf{C}^N)\right).$$

We have the following commuting diagram of canonical morphisms:

$$\begin{array}{ccccc}
U^+_{d'} & & \xrightarrow{\quad c \quad} & & U^-_{d'} \\
\uparrow{\scriptstyle \nu_A} & & & & \uparrow{\scriptstyle b_A} \\
D_A & \xrightarrow{\mathrm{Pr}_A} & U^+_{k+A,d_0'} & \xrightarrow{c_A} & U^-_{k+A,d_0'} \\
\downarrow{\scriptstyle \mathrm{pr}_A} & & & & \\
\prod_{a\in A}(Q^+_{0,a}(\mathbf{P}(V\otimes\mathbf{C}^N), d_a) \times \mathbf{P}(\mathbf{C}^N)), & & & &
\end{array} \tag{3.2.5}$$

where b_A denotes the morphism which trades the markings A for base points of length d_a. The two projections pr_A and Pr_A are those coming from the fiber product description (3.2.4) of D_A. The map ν_A has degree $|A|!$ and sends D_A onto the boundary stratum of $U^+_{d'}$ generically parametrizing (unobstructed) ε_+-stable quasimaps to $\mathbf{P}(V \otimes \mathbf{C}^N)$ whose domain curves have exactly $|A|$ unordered rational tails of degree d_a. In particular, for $A = \{a\}$ the map $\nu_{\{a\}}$ is an embedding of $D_{\{a\}}$ as a boundary divisor.

The contractions c, c_{A} are isomorphisms over the (nonempty) loci of quasimaps with irreducible domain curves. By Lemma 2.4, the complements of these loci have positive codimension and we conclude that c, c_{A} are birational morphisms and hence degree 1 maps.

We finally introduce one more piece of notation. Let p_a denote the Cartier divisor on the universal curve $\mathfrak{C}^{\pm}_{g,k+\{a\},d_0'}$ of the moduli spaces $\mathrm{U}^{\pm}_{k+\{a\},d_0'}$ which is the image of the section Σ_a corresponding to the marking a. Similarly, we have the Cartier divisor p_a^{tail} on the universal curve $\mathfrak{C}^{+}_{0,a,d_a} \times \mathbf{P}(\mathbf{C}^{\mathrm{N}})$ of $\mathrm{Q}^{+}_{0,a}(\mathbf{P}(\mathrm{V} \otimes \mathbf{C}^{\mathrm{N}}), d_a) \times \mathbf{P}(\mathbf{C}^{\mathrm{N}})$ defined by the image of the section $\Sigma_{tail,a}$ corresponding to the marking a. As usual, $\mathcal{O}(p_a)$, respectively $\mathcal{O}(p_a^{tail})$, will stand for the associated line bundles; and $\mathcal{O}_{p_a}$, respectively $\mathcal{O}_{p_a^{tail}}$ will stand for the coherent sheaves $\Sigma_{a*}\Sigma_a^*\mathcal{O}$, $\Sigma_{tail,a*}\Sigma_{tail,a}^*\mathcal{O}$ on the universal curves. Then $\Sigma_a^*\mathcal{O}(-p_a)$, respectively $\Sigma_{tail,a}^*\mathcal{O}(-p_a^{tail})$, is identified with the line bundle with first Chern class ψ_a on $\mathrm{U}^{\pm}_{k+\mathrm{A},d_0'}$, respectively ψ_a^{tail} on $\mathrm{Q}^{+}_{0,a}(\mathbf{P}(\mathrm{V} \otimes \mathbf{C}^{\mathrm{N}}), d_a) \times \mathbf{P}(\mathbf{C}^{\mathrm{N}})$. Abusing notation, we will write $\mathcal{O}(\psi_a)$ and $\mathcal{O}(\psi_a^{tail})$ for these line bundles, and $\mathcal{O}(-\psi_a)$, $\mathcal{O}(-\psi_a^{tail})$ for their duals.

3.3. *MacPherson's Graph Construction.* — For easy notation, for $\mathrm{A} = \{a\}$ in (3.2.5) we write D_a, Pr_a, c_a, b_a, etc instead of $\mathrm{D}_{\{a\}}$, $\mathrm{Pr}_{\{a\}}$ $c_{\{a\}}$ $b_{\{a\}}$, etc. Let $\pi : \mathfrak{C}^{\pm}_{g,k,d'} \to \mathrm{U}^{\pm}_{d'}$ be the universal curve and denote by $\tilde{c}$ the contraction morphism from $\mathfrak{C}^{+}_{g,k,d'}$ to $\mathfrak{C}^{-}_{g,k,d'}$, which is an isomorphism outside the divisor $\tilde{\mathrm{D}}_a$. Hence $\mathscr{L}'_+ \cong \tilde{c}^*\mathscr{L}'_-(-d_a\tilde{\mathrm{D}}_a)$. Here the coefficient $-d_a$ is obtained by the consideration of $\deg \mathscr{L}'_+|_{\mathrm{C}_a} = d_a$, $\deg \mathcal{O}_{\mathrm{C}_a}(\mathrm{C}_a) = -1$ for the contracted rational tail C_a on the fiber curve of π over a general closed point of D_a. It follows that for every $l \geq 1$ there are homomorphisms

$$(\mathscr{L}'_+)^l \cong \tilde{c}^*(\mathscr{L}'_-)^l(-ld_a\tilde{\mathrm{D}}_a) \to \tilde{c}^*(\mathscr{L}'_-)^l$$

of line bundles on $\mathfrak{C}^{+}_{g,k,d'}$.

In particular, taking $l = 1$ and using the top line of the diagram (2.5.1) gives a map $\mathscr{P}^{+}_{d'} \longrightarrow \tilde{c}^*(\mathscr{P}^{-}_{d'})$. Applying π_* we obtain homomorphisms

$$\Phi_{\mathrm{P}} : \mathrm{P}^{+}_{d'} \longrightarrow c^*\mathrm{P}^{-}_{d'}, \qquad \Phi_{\mathrm{R}} : \mathrm{R}^{+}_{d'} \longrightarrow c^*\mathrm{R}^{-}_{d'},$$
$$\Phi = (\Phi_{\mathrm{P}}, \Phi_{\mathrm{R}}) : \mathrm{F}^{+}_{d'} \longrightarrow c^*\mathrm{F}^{-}_{d'}$$

of vector bundles on $\mathrm{U}^{+}_{d'}$, which are isomorphisms outside D_a. We have used here the canonical isomorphisms $\pi_*\tilde{c}^*\mathscr{R}^{-}_{d'} \cong c^*\pi_*\mathscr{R}^{-}_{d'}$ and $\pi_*\tilde{c}^*\mathscr{P}^{-}_{d'} \cong c^*\pi_*\mathscr{P}^{-}_{d'}$ obtained by applying to (3.2.2) the base-change followed by the projection formula.

Consider the Grassmann bundle over $\mathrm{U}^{+}_{d'}$

$$\mathbf{Gr} := \mathbf{Gr}\big(\mathrm{F}^{+}_{d'} \oplus c^*\mathrm{F}^{-}_{d'}\big) := \mathit{Grass}\big(r_d, \mathrm{F}^{+}_{d'} \oplus c^*\mathrm{F}^{-}_{d'}\big),$$

with $r_d = \mathrm{rank}(\mathrm{F}^{+}_{d'})$. Let $\eta : \mathbf{Gr} \to \mathrm{U}^{+}_{d'}$ be the projection and denote by ζ the tautological subbundle of rank r_d in $\eta^*(\mathrm{F}^{+}_{d'} \oplus c^*\mathrm{F}^{-}_{d'})$.

The map $\eta \times \mathrm{id}$ has a section

$$v : \mathrm{U}^+_{d'} \times \mathbf{A}^1 \longrightarrow \mathbf{Gr} \times \mathbf{A}^1, \qquad v(y, \lambda) = \big(y, \mathrm{graph}\big(\lambda(\Phi)_y\big), \lambda\big).$$

Define the closed substack

$$\Gamma := \overline{\mathrm{Im}(v)} \subset \mathbf{Gr} \times \mathbf{P}^1$$

as the stack-theoretic closure of the image of v. As $\mathrm{U}^+_{d'}$ is nonsingular and irreducible, Γ is also irreducible, of dimension equal to $1 + \dim \mathrm{U}^+_{d'}$.

In fact, if we consider the "component" Grassmann bundles

$$\mathbf{Gr}_{\mathrm{P}} := \mathbf{Gr}\big(\mathrm{P}^+_{d'} \oplus c^*\mathrm{P}^-_{d'}\big) := \mathit{Grass}\big(r_{\mathrm{P}}, \mathrm{P}^+_{d'} \oplus c^*\mathrm{P}^-_{d'}\big),$$
$$\mathbf{Gr}_{\mathrm{R}} := \mathbf{Gr}\big(\mathrm{R}^+_{d'} \oplus c^*\mathrm{R}^-_{d'}\big) := \mathit{Grass}\big(r_{\mathrm{R}}, \mathrm{R}^+_{d'} \oplus c^*\mathrm{R}^-_{d'}\big),$$

with projections $\eta_{\mathrm{P}}, \eta_{\mathrm{R}}$ and tautological subbundles $\zeta_{\mathrm{P}}, \zeta_{\mathrm{R}}$, then there is a natural inclusion

$$\mathbf{Gr}_{\mathrm{P}} \times_{\mathrm{U}^+_{d'}} \mathbf{Gr}_{\mathrm{R}} \subset \mathbf{Gr}$$

such that ζ restricts to $\zeta_{\mathrm{P}} \boxplus \zeta_{\mathrm{R}}$ and the inclusion of Γ in $\mathbf{Gr} \times \mathbf{P}^1$ factors through $(\mathbf{Gr}_{\mathrm{P}} \times_{\mathrm{U}^+_{d'}} \mathbf{Gr}_{\mathrm{R}}) \times \mathbf{P}^1$.

For $\lambda \in \mathbf{P}^1 = \mathbf{A}^1 \cup \{\lambda = \infty\}$ denote by Γ_λ the fiber of the projection $\Gamma \to \mathbf{P}^1$. When $\lambda \in \mathbf{A}^1$, under the identifications $v_\lambda : \mathrm{U}^+_{d'} \xrightarrow{\cong} \Gamma_\lambda$, we have

$$v_\lambda^* \zeta = \mathrm{Im}\big(\mathrm{F}^+_{d'} \xrightarrow{(\mathrm{id}, \lambda\Phi)} \mathrm{F}^+_{d'} \oplus c^*\mathrm{F}^-_{d'}\big).$$

In particular, at $\lambda = 0$ we have $v_0^*\zeta = \mathrm{F}^+_{d'} \oplus \{0\}$.

At $\lambda = \infty$ the fiber breaks into components encoding the degeneracy of the map Φ, as in [15, Example 18.1.6]. First of all, there is a distinguished component $\Gamma_{\infty,dist}$ which has multiplicity one and projects birationally to $\mathrm{U}^+_{d'}$, while $\zeta|_{\Gamma_{\infty,dist}} = \{0\} \oplus c^*\mathrm{F}^-_{d'}$. All other components of Γ_∞ come with some multiplicities and project into D_a under η. Their description is our next task. The analysis is similar to the one in the proof of [3, Lemma 4.4], where a related genus zero case is treated. In our situation there are complications due to the twisting by $\mathscr{M}$, but also slight simplifications, due to the fact that c only contracts rational tails of fixed degree d_a, which therefore do not interfere with each other.

3.3.1. *Description of* Γ_∞. — For each $j_a \geq 1$ consider the $\mathbf{P}^1$-bundle over D_a

$$\mathbf{P}_{j_a} := \mathbf{P}\big(\mathrm{pr}_a^*\mathcal{O}\big(j_a\psi_a^{tail}\big) \oplus \mathrm{Pr}_a^*\mathcal{O}(-j_a\psi_a)\big)$$

and their fiber product

$$\mathbf{P}_{j_A} := \prod_{a \in A} \mathbf{P}_{j_a}|_{D_A}$$

over D_A.

Theorem **3.1.** — *Let j_A be the multi-index $(j_a)_{a\in A}$ with each j_a in the range $1 \le j_a \le \max\{d_a, d_a l_i \mid i = 1, \dots, r\}$ and let $m_{j_A} := \prod_{a\in A} j_a$. For each j_A, there exists a map $\alpha_{j_A} : \mathbf{P}_{j_A} \to \mathbf{Gr}$, described below, satisfying that*

$$(\mathbf{3.3.1}) \qquad [\Gamma_\infty] = [\Gamma_{\infty,dist}] + \sum_{(A,j_A)} m_{j_A}[\Gamma_{\infty,j_A}] = [\Gamma_{\infty,dist}] + \sum_{(A,j_A)} \frac{m_{j_A}}{|A|!}(\alpha_{j_A})_*[\mathbf{P}_{j_A}]$$

in the Chow group $A_*(\mathbf{Gr})_{\mathbf{Q}}$. Here Γ_{∞,j_A} is the image stack of α_{j_A}. Furthermore Γ_{∞,j_A} projects to D_A under the projection map $\eta : \mathbf{Gr} \to U^+_{d'}$.

Defining α_{j_A} amounts to finding a subbundle ξ^{j_A} of $\pi^*_{\mathbf{P}}\nu^*_A(F^+_{d'} \oplus c^*F^-_{d'})$ with its rank equal to the rank of $F^+_{d'}$. Denote by $\pi_{\mathbf{P}} : \mathbf{P}_{j_A} \to D_A$ the projection map. Then the vector bundle ξ^{j_A} will be constructed as an extension of $\boxplus_{a\in A}\mathcal{O}_{\mathbf{P}_{j_a}}(-1) \otimes \pi^*_{\mathbf{P}}F^{j_a}$ by $\pi^*_{\mathbf{P}}(\mathrm{pr}^*_A F^{+j_A+1}_{tail,d_a} \oplus \mathrm{Pr}^*_A c^*_A F^{-j_A-1}_{d'_0})$ for some vector bundles

$$F^{j_a}, F^{+j_A+1}_{tail,d_a}, F^{-j_A-1}_{d'_0} \text{ on } D_a,$$

$$\prod_{a\in A} Q^+_{0,a}(\mathbf{P}(V \otimes \mathbf{C}^N), d_a) \times \mathbf{P}(\mathbf{C}^N), U^-_{k+A,d'_0} \text{ respectively.}$$

The bundles $\mathrm{pr}^*_a F^{+j_a}_{tail,d_a}$ (resp. $\mathrm{Pr}^*_a c^*_a F^{-j_a}_{d'_0}$) for j_a will form a decreasing (resp. increasing) filtration of the kernel sheaf of $\nu^*_a\Phi$ (resp. of the sheaf $\nu^*_a c^* F^-_{d'}$).

3.3.2. *Description of the vector bundle $F^{+j_a+1}_{tail,d_a}$ on $Q^+_{0,a}(\mathbf{P}(V \otimes \mathbf{C}^N), d_a) \times \mathbf{P}(\mathbf{C}^N)$.* — Consider first the case $A = \{a\}$ of the boundary divisor D_a. On the universal curve

$$\pi : \mathfrak{C}^+_{0,a,d_a} \times \mathbf{P}(\mathbf{C}^N) \to Q^+_{0,a}(\mathbf{P}(V \otimes \mathbf{C}^N), d_a) \times \mathbf{P}(\mathbf{C}^N),$$

put $\mathcal{L}_+ := \mathcal{L}'_+ \boxtimes \mathcal{O}_{\mathbf{P}(\mathbf{C}^N)}(-1)$. We have the diagram

$$\begin{array}{ccccccccc} 0 & \longrightarrow & \mathcal{L}_+ \otimes V & \xrightarrow{\oplus_j s_j} & \oplus_{j=1}^N \mathcal{L}'_+ \otimes V & \longrightarrow & \mathcal{P}^+_{tail,d_a} & \longrightarrow & 0 \\ & & & & \downarrow{\scriptstyle \oplus_j(\oplus_i d\varphi_i)} & & & & \\ 0 & \longrightarrow & \oplus_{i=1}^r \mathcal{L}_+^{l_i} & \xrightarrow{\oplus_{i,j} s_j^{l_i}} & \oplus_{i,j}(\mathcal{L}'_+)^{l_i} & \longrightarrow & \mathcal{Q}^+_{tail,d_a} & \longrightarrow & 0, \end{array}$$

whose rows are obtained from the exact sequence

$$0 \to \mathcal{O}_{\mathbf{P}(\mathbf{C}^N)}(-1) \to \oplus_{j=1}^{N} \mathcal{O}_{\mathbf{P}(\mathbf{C}^N)} \to \mathrm{Q} \to 0$$

via pull-backs, tensoring with appropriate line bundles, and taking direct sums, as explained in Section 3. Now define the vector bundles

$$\mathrm{P}^{+}_{tail,d_a} := \pi_* \mathscr{P}^{+}_{tail,d_a}, \qquad \mathscr{R}^{+}_{tail,d_a} := \oplus_{i,j} (\mathscr{L}'_{+})^{l_i},$$
$$\mathrm{R}^{+}_{tail,d_a} := \pi_* \mathscr{R}^{+}_{tail,d_a}, \qquad \mathrm{F}^{+}_{tail,d_a} := \mathrm{P}^{+}_{tail,d_a} \oplus \mathrm{R}^{+}_{tail,d_a}.$$

For integers $j_a = 1, \ldots, \max\{d_a, d_a l_i \mid i = 1, \ldots, r\}$, we have the subbundles

$$\mathrm{P}^{+,j_a}_{tail,d_a} := \pi_*\big(\mathscr{P}^{+}_{tail,d_a}(-j_a p_a^{tail})\big), \tag{3.3.2}$$
$$\mathrm{R}^{+,j_a}_{tail,d_a} := \pi_*\big(\mathscr{R}^{+}_{tail,d_a}(-j_a p_a^{tail})\big) \tag{3.3.3}$$

of vector bundles $\mathrm{P}^{+}_{tail,d_a}$, $\mathrm{R}^{+}_{tail,d_a}$ respectively. They are vector bundles on $\mathrm{Q}^{+}_{0,a}(\mathbf{P}(\mathrm{V} \otimes \mathbf{C}^N), d_a) \times \mathbf{P}(\mathbf{C}^N)$. We also put

$$\mathrm{P}^{+,0}_{tail,d_a} := \mathrm{P}^{+}_{tail,d_a}, \qquad \mathrm{R}^{+,0}_{tail,d_a} := \mathrm{R}^{+}_{tail,d_a}, \qquad \mathrm{F}^{+,0}_{tail,d_a} := \mathrm{F}^{+}_{tail,d_a}.$$

Note that $\mathrm{P}^{+,j_a}_{tail,d_a} = 0$ if $j_a > d_a$, and that $(\mathscr{L}'_{+})^{l_i}$ does not contribute to $\mathrm{R}^{+,j_a}_{tail,d_a}$ if $j_a > l_i d_a$. Hence the quotients of the decreasing filtrations given by (3.3.2) and (3.3.3) are

$$0 \to \mathrm{P}^{+,j_a+1}_{tail,d_a} \to \mathrm{P}^{+,j_a}_{tail,d_a} \to \mathrm{P}^{j_a}_{tail} \otimes \mathcal{O}\big(j_a \psi_a^{tail}\big) \to 0,$$
$$0 \to \mathrm{R}^{+,j_a+1}_{tail,d_a} \to \mathrm{R}^{+,j_a}_{tail,d_a} \to \mathrm{R}^{j_a}_{tail} \otimes \mathcal{O}\big(j_a \psi_a^{tail}\big) \to 0,$$

where we put for each $0 \le j_a \le \max\{d_a, l_i d_a \mid i = 1, \ldots, r\}$

$$\mathrm{P}^{j_a}_{tail} := \begin{cases} (ev_a \times \mathrm{id}_{\mathbf{P}(\mathbf{C}^N)})^*((\mathcal{O}_{\mathbf{P}(\mathrm{V}\otimes\mathbf{C}^N)}(1) \otimes \mathrm{V}) \boxtimes \mathrm{Q}), & \text{if } j_a \le d_a \\ 0, & \text{if } j_a > d_a \end{cases}$$

and

$$\mathrm{R}^{j_a}_{tail} := \oplus_{i=1}^{r} \mathrm{R}^{j_a}_{i,tail},$$
$$\mathrm{R}^{j_a}_{i,tail} := \begin{cases} (ev_a \times \mathrm{id}_{\mathbf{P}(\mathbf{C}^N)})^*(\mathcal{O}_{\mathbf{P}(\mathrm{V}\otimes\mathbf{C}^N)}(l_i) \boxtimes \oplus_{j=1}^{N} \mathcal{O}_{\mathbf{P}(\mathbf{C}^N)}), & \text{if } j_a \le l_i d_a \\ 0, & \text{if } j_a > l_i d_a \end{cases}.$$

Alternatively, when they are not set to zero,

$$\mathrm{P}^{j_a}_{tail} = \pi_*\big(\mathscr{P}^{+}_{tail,d_a} \otimes \mathcal{O}_{p_a^{tail}}\big), \qquad \mathrm{R}^{j_a}_{tail} = \pi_*\big(\mathscr{R}^{+}_{tail,d_a} \otimes \mathcal{O}_{p_a^{tail}}\big).$$

Taking the direct sums

$$\mathrm{F}^{+,j_a}_{tail,d_a} := \mathrm{P}^{+,j_a}_{tail,d_a} \oplus \mathrm{R}^{+,j_a}_{tail,d_a}, \qquad \mathrm{F}^{j_a}_{tail} := \mathrm{P}^{j_a}_{tail} \oplus \mathrm{R}^{j_a}_{tail}$$

gives a filtration of the vector bundle F^{+}_{tail,d_a} on $Q^{+}_{0,a}(\mathbf{P}(V\otimes \mathbf{C}^N), d_a)\times \mathbf{P}(\mathbf{C}^N)$, with quotients $F^{j_a}_{tail}\otimes \mathcal{O}(j_a\psi^{tail}_a)$. The pull-back $\nu_a^*F^+_{d'}$ can be written as the extension

$$0 \longrightarrow \mathrm{pr}_a^*F^{+,1}_{tail,d_a} \longrightarrow \nu_a^*F^+_{d'} \xrightarrow{res} \mathrm{Pr}_a^*F^+_{d_0'} \longrightarrow 0\ . \tag{3.3.4}$$

3.3.3. *Description of the vector bundle* $F^{-,j_a-1}_{d_0'}$ *on* $U^-_{k+\{a\},d_0'}$. — Let $F^{\pm}_{b_A,d_0'}$ denote the vector bundles on $U^{\pm}_{k+A,d_0'}$ defined as in (2.5.5), but using the twisting line bundles $\mathscr{M}^{\pm}$ *induced from* $\mathfrak{C}^-_{g,k,d'}$ (and hence from $\overline{M}_{g,k}$) via pull-back by

$$\tilde{b}_A : \mathfrak{C}^-_{g,k+A,d_0'} \longrightarrow \mathfrak{C}^-_{g,k,d'}.$$

The homomorphism Φ factors when pulled-back to D_A as

$$\nu_A^*F^+_{d'} \xrightarrow{res} \mathrm{Pr}_A^*F^+_{b_A,d_0'} \xrightarrow{\text{generic. isom}} \mathrm{Pr}_A^*c_A^*F^-_{b_A,d_0'} \hookrightarrow \mathrm{Pr}_A^*c_A^*b_A^*F^-_{d'} = \nu_A^*c^*F^-_{d'}\ .$$

Here the first map *res* is given by the restriction of sections to the non-contracted parts of the universal curve. The middle arrow is the pull-back by Pr_A of the map Φ on $U^+_{k+A,d_0'}$ and is therefore an isomorphism generically on D_A. The third map is induced from the canonical injections on the universal curve $\mathscr{L}'_{-,d_0'} \to \mathscr{L}'_{-,d_0'}(\sum_a d_a p_a) = \tilde{b}_A^*\mathscr{L}'_{-,d'}$ and $(\mathscr{L}'_{-,d_0'})^{l_i} \to (\mathscr{L}'_{-,d_0'})^{l_i}(\sum_a l_i d_a p_a) = \tilde{b}_A^*(\mathscr{L}'_{-,d'})^{l_i}$.

Consider the codomain $\mathrm{Pr}_a^*c_a^*b_a^*F^-_{d'}$ of $\Phi|_{D_a}$ and the square diagram of universal curves

$$\begin{array}{ccc}
\mathfrak{C}^-_{g,k+\{a\},d_0'} & \longrightarrow & \mathfrak{C}^-_{g,k,d'} \\
{\scriptstyle \pi}\downarrow & & \downarrow \\
U^-_{k+\{a\},d_0'} & \xrightarrow[b_a]{} & U^-_{k,d'}.
\end{array}$$

In the bundle $b_a^*F^-_{d'}$ on $U^-_{k+\{a\},d_0'}$ we have the increasing filtrations

$$P^{-,0}_{d_0'} \subset P^{-,1}_{d_0'} \subset \cdots \subset P^{-,d_a}_{d_0'} = b_a^*P^-_{d'},$$
$$R^{-,0}_{d_0'} \subset R^{-,1}_{d_0'} \subset \cdots \subset R^{-,\max_i\{d_a l_i\}}_{d_0'} = b_a^*R^-_{d'}$$

induced via the subbundles

$$P^{-,j_a}_{d_0'} := \pi_*\big(\mathscr{P}^-_{d_0'}(j_a p_a)\big) \cap b_a^*P^-_{d'}, \quad j_a = 0, 1, \ldots, d_a,$$

$$\mathrm{R}_{d_0'}^{-j_a} := \pi_*\big(\mathscr{R}_{d_0'}^-(j_a p_a)\big) \cap b_a^* \mathrm{R}_{d'}^-, \quad j_a = 0, 1, \dots, \max_i\{l_i d_a\}.$$

Here we use the natural injections $\mathscr{P}_{d_0'}^-(j_a p_a) \to \mathscr{P}_{d_0'}^-(d_a p_a) \cong \tilde{b}_a^* \mathscr{P}_{d'}^-$ for $j_a \leq d_a$ and $\mathscr{R}_{i,d_0'}^-(j_a p_a) \to \mathscr{R}_{i,d_0'}^-(l_i d_a p_a) \cong \tilde{b}_a^* \mathscr{R}_{i,d'}^-$ for $j_a \leq l_i d_a$. The quotients are

$$0 \to \mathrm{P}_{d_0'}^{-j_a - 1} \to \mathrm{P}_{d_0'}^{-j_a} \to \mathrm{P}^{-j_a} \otimes \mathcal{O}(-j_a \psi_a) \to 0,$$
$$0 \to \mathrm{R}_{d_0'}^{-j_a - 1} \to \mathrm{R}_{d_0'}^{-j_a} \to \mathrm{R}^{-j_a} \otimes \mathcal{O}(-j_a \psi_a) \to 0,$$

where we put for each $0 \leq j_a \leq \max\{d_a, l_i d_a \mid i = 1, \dots, r\}$

(**3.3.5**)
$$\mathrm{P}^{-j_a} := \begin{cases} \pi_*(\mathscr{P}_{d_0'}^- \otimes \mathcal{O}_{p_a}), & \text{if } j_a \leq d_a \\ 0, & \text{if } j_a > d_a \end{cases},$$

and

(**3.3.6**)
$$\mathrm{R}^{-j_a} := \oplus_{i=1}^r \mathrm{R}_i^{-j_a},$$
$$\mathrm{R}_i^{-j_a} := \begin{cases} \pi_*(\oplus_{j=1}^{\mathrm{N}} (\mathscr{L}_-')^{l_i} \otimes \mathcal{O}_{p_a}), & \text{if } j_a \leq l_i d_a \\ 0, & \text{if } j_a > l_i d_a \end{cases}.$$

Setting

$$\mathrm{F}_{d_0'}^{-j_a} := \mathrm{P}_{d_0'}^{-j_a} \oplus \mathrm{R}_{d_0'}^{-j_a}$$

gives an increasing filtration of the vector bundle $b_a^* \mathrm{F}_{d'}^-$ on $\mathrm{U}_{k+\{a\},d_0'}^-$ with quotients $\mathrm{F}^{-j_a} \otimes \mathcal{O}(-j_a \psi_a)$ and $\mathrm{F}^{-j_a} := \mathrm{P}^{-j_a} \oplus \mathrm{R}^{-j_a}$.

3.3.4. *Description of* $\alpha_{j_a} : \mathbf{P}_{j_a} \to \mathbf{Gr}$. — For each $j_a \geq 1$ recall the $\mathbf{P}^1$-bundle over D_a

$$\mathbf{P}_{j_a} := \mathbf{P}\big(\mathrm{pr}_a^* \mathcal{O}\big(j_a \psi_a^{tail}\big) \oplus \mathrm{Pr}_a^* \mathcal{O}(-j_a \psi_a)\big),$$

with projection $\pi_{\mathbf{P}} : \mathbf{P}_{j_a} \longrightarrow \mathrm{D}_a$. Consider the tautological sequence

$$0 \longrightarrow \mathcal{O}_{\mathbf{P}_{j_a}}(-1) \longrightarrow \pi_{\mathbf{P}}^*\big(\mathrm{pr}_a^* \mathcal{O}\big(j_a \psi_a^{tail}\big) \oplus \mathrm{Pr}_a^* \mathcal{O}(-j_a \psi_a)\big)$$
$$\longrightarrow \mathcal{O}_{\mathbf{P}_{j_a}}(1) \longrightarrow 0.$$

Now define the extension $\xi_{\mathrm{P}}^{j_a}$ as the vector bundle uniquely fitting in the commuting diagram with exact columns

$$\begin{array}{ccc}
0 & & 0 \\
\uparrow & & \uparrow \\
\mathcal{O}_{\mathbf{P}_{j_a}}(-1)\otimes \pi_{\mathbf{P}}^*\mathrm{P}^{j_a} & \hookrightarrow & \pi_{\mathbf{P}}^*((\mathrm{pr}_a^*\mathcal{O}(j_a\psi_a^{tail})\oplus \mathrm{Pr}_a^*\mathcal{O}(-j_a\psi_a))\otimes \mathrm{P}^{j_a}) \\
\uparrow & & \uparrow \\
\xi_{\mathrm{P}}^{j_a} & \hookrightarrow & \pi_{\mathbf{P}}^*(\mathrm{pr}_a^*\mathrm{P}^{+j_a}_{tail,d_a}\oplus \mathrm{Pr}_a^*c_a^*\mathrm{P}^{-j_a}_{d_0'}) \\
\uparrow & & \uparrow \\
\pi_{\mathbf{P}}^*(\mathrm{pr}_a^*\mathrm{P}^{+j_a+1}_{tail,d_a}\oplus \mathrm{Pr}_a^*c_a^*\mathrm{P}^{-j_a-1}_{d_0'}) & \xrightarrow{=} & \pi_{\mathbf{P}}^*(\mathrm{pr}_a^*\mathrm{P}^{+j_a+1}_{tail,d_a}\oplus \mathrm{Pr}_a^*c_a^*\mathrm{P}^{-j_a-1}_{d_0'}) \\
\uparrow & & \uparrow \\
0 & & 0
\end{array}$$

where the horizontal arrows are injective as *maps of vector bundles* and

$$\mathrm{P}^{j_a} := \mathrm{pr}_a^*\mathrm{P}^{j_a}_{tail} \cong \mathrm{Pr}_a^*c_a^*\mathrm{P}^{-j_a}.$$

Similarly, we define $\xi_{\mathrm{R}}^{j_a}$ as an extension, via

$$\begin{array}{ccc}
0 & & 0 \\
\uparrow & & \uparrow \\
\mathcal{O}_{\mathbf{P}_{j_a}}(-1)\otimes \pi_{\mathbf{P}}^*\mathrm{R}^{j_a} & \hookrightarrow & \pi_{\mathbf{P}}^*((\mathrm{pr}_a^*\mathcal{O}(j_a\psi_a^{tail})\oplus \mathrm{Pr}_a^*\mathcal{O}(-j_a\psi_a))\otimes \mathrm{R}^{j_a}) \\
\uparrow & & \uparrow \\
\xi_{\mathrm{R}}^{j_a} & \hookrightarrow & \pi_{\mathbf{P}}^*(\mathrm{pr}_a^*\mathrm{R}^{+j_a}_{tail,d_a}\oplus \mathrm{Pr}_a^*c_a^*\mathrm{R}^{-j_a}_{d_0'}) \\
\uparrow & & \uparrow \\
\pi_{\mathbf{P}}^*(\mathrm{pr}_a^*\mathrm{R}^{+j_a+1}_{tail,d_a}\oplus \mathrm{Pr}_a^*c_a^*\mathrm{R}^{-j_a-1}_{d_0'}) & \xrightarrow{=} & \pi_{\mathbf{P}}^*(\mathrm{pr}_a^*\mathrm{R}^{+j_a+1}_{tail,d_a}\oplus \mathrm{Pr}_a^*c_a^*\mathrm{R}^{-j_a-1}_{d_0'}) \\
\uparrow & & \uparrow \\
0 & & 0
\end{array}$$

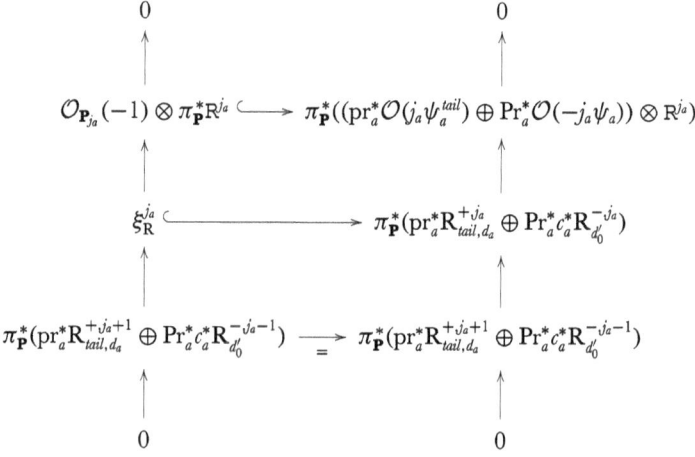

where

$$\mathrm{R}^{j_a} := \mathrm{pr}_a^* \mathrm{R}^{j_a}_{tail} \cong \mathrm{Pr}_a^* c_a^* \mathrm{R}^{-j_a}.$$

Since

$$\xi^{j_a} := \xi^{j_a}_{\mathrm{P}} \oplus \xi^{j_a}_{\mathrm{R}}$$

is canonically a subbundle of $\pi_{\mathbf{P}}^* \nu_a^* (\mathrm{F}^+_{d'} \oplus c^* \mathrm{F}^-_{d'})$ whose rank is equal to the rank of $\mathrm{F}^+_{d'}$, it gives rise to a morphism

$$\alpha_{j_a} : \mathbf{P}_{j_a} \longrightarrow \mathbf{Gr}\big(\mathrm{F}^+_{d'} \oplus c^* \mathrm{F}^-_{d'}\big)$$

which is birational onto its image and such that $\xi^{j_a} = \alpha^*_{j_a} \zeta$ (respecting the decompositions into P and R components). We will show in Section 3.3.6 that the image is a component of the limit fiber Γ_∞ which we denote by Γ_{∞, j_a} and which has multiplicity j_a in the fiber.

3.3.5. *Description of* $\alpha_{j_{\mathrm{A}}} : \mathbf{P}_{j_{\mathrm{A}}} \to \mathbf{Gr}$ *and the vector bundle* $\mathbb{F}^{j_a}$ *on* D_a. — For general A the above analysis extends immediately, as the various rational tails may be treated independently. Specifically, this means that we now consider a collection $j_{\mathrm{A}} := \{j_a \mid a \in \mathrm{A}\}$ of positive integers and define

$$\mathrm{P}^{+j_{\mathrm{A}}+1}_{tail, d_a} := \boxplus_{a\in \mathrm{A}} \pi_*\big(\mathscr{P}^+_{tail, d_a}\big(-(j_a+1)p_a^{tail}\big)\big),$$
$$\mathrm{R}^{+j_{\mathrm{A}}+1}_{tail, d_a} := \boxplus_{a\in \mathrm{A}} \pi_*\big(\mathscr{R}^+_{tail, d_a}\big(-(j_a+1)p_a^{tail}\big)\big)$$

on $\prod_{a\in\mathrm{A}} (\mathrm{Q}^+_{0,a}(\mathbf{P}(\mathrm{V}\to\mathbf{C}^{\mathrm{N}}), d_a) \times \mathbf{P}(\mathbf{C}^{\mathrm{N}}))$ and

$$\mathrm{P}^{-j_{\mathrm{A}}-1}_{d_0'} := \pi_*\Bigg(\mathscr{P}^-_{d_0'}\Big(\sum_{a\in\mathrm{A}}(j_a-1)p_a\Big)\Bigg) \cap b_{\mathrm{A}}^* \mathrm{P}^-_d,$$

$$\mathrm{R}^{-j_{\mathrm{A}}-1}_{d_0'} := \pi_*\Bigg(\mathscr{R}^-_{d_0'}\Big(\sum_{a\in\mathrm{A}}(j_a-1)p_a\Big)\Bigg) \cap b_{\mathrm{A}}^* \mathrm{R}^-_d$$

on $\mathrm{U}^-_{k+\mathrm{A}, d_0'}$. Further, we put

$$\mathrm{F}^{+j_{\mathrm{A}}+1}_{tail, d_a} := \mathrm{P}^{+j_{\mathrm{A}}+1}_{tail, d_a} \oplus \mathrm{R}^{+j_{\mathrm{A}}+1}_{tail, d_a}, \qquad \mathrm{F}^{-j_{\mathrm{A}}-1}_{d_0'} := \mathrm{P}^{-j_{\mathrm{A}}-1}_{d_0'} \oplus \mathrm{R}^{-j_{\mathrm{A}}-1}_{d_0'}.$$

Setting

$$\mathbf{P}_{j_{\mathrm{A}}} := \prod_{a\in\mathrm{A}} \mathbf{P}_{j_a}|_{\mathrm{D}_{\mathrm{A}}},$$

where the product is fiber product over $\mathrm{D_A}$, we have the projection $\pi_{\mathbf{P}} : \mathbf{P}_{j_A} \to \mathrm{D_A}$ and extensions

$$(\mathbf{3.3.7}) \qquad 0 \to \pi_{\mathbf{P}}^*\big(\mathrm{pr}_A^* \mathrm{P}_{tail,d_a}^{+,j_A+1} \oplus \mathrm{Pr}_A^* c_A^* \mathrm{P}_{d_0'}^{-,j_A-1}\big) \to \xi_{\mathrm{P}}^{j_A} \to \boxplus_{a\in A}\big(\mathcal{O}_{\mathbf{P}_{j_a}}(-1) \otimes \pi_{\mathbf{P}}^* \mathrm{P}^{j_a}\big) \to 0,$$

$$(\mathbf{3.3.8}) \qquad 0 \to \pi_{\mathbf{P}}^*\big(\mathrm{pr}_A^* \mathrm{R}_{tail,d_a}^{+,j_A+1} \oplus \mathrm{Pr}_A^* c_A^* \mathrm{R}_{d_0'}^{-,j_A-1}\big) \to \xi_{\mathrm{R}}^{j_A} \to \boxplus_{a\in A}\big(\mathcal{O}_{\mathbf{P}_{j_a}}(-1) \otimes \pi_{\mathbf{P}}^* \mathrm{R}^{j_a}\big) \to 0,$$

$$(\mathbf{3.3.9}) \qquad 0 \to \pi_{\mathbf{P}}^*\big(\mathrm{pr}_A^* \mathrm{F}_{tail,d_a}^{+,j_A+1} \oplus \mathrm{Pr}_A^* c_A^* \mathrm{F}_{d_0'}^{-,j_A-1}\big) \to \xi^{j_A} \to \boxplus_{a\in A}\big(\mathcal{O}_{\mathbf{P}_{j_a}}(-1) \otimes \pi_{\mathbf{P}}^* \mathrm{F}^{j_a}\big) \to 0,$$

with

$$(\mathbf{3.3.10}) \qquad \xi^{j_A} := \xi_{\mathrm{P}}^{j_A} \oplus \xi_{\mathrm{R}}^{j_A}, \qquad \mathrm{F}^{j_a} := \mathrm{P}^{j_a} \oplus \mathrm{R}^{j_a}.$$

As before, this gives a morphism $\alpha_{j_A} : \mathbf{P}_{j_A} \to \mathbf{Gr}$ such that $\xi^{j_A} = \alpha_{j_A}^* \zeta$. We will show in Section 3.3.6 that the image of α_{j_A}, denoted Γ_{∞,j_A}, is a component of the limit fiber, with multiplicity $m_{j_A} := \prod_{a\in A} j_a$.

3.3.6. *Proof of Theorem 3.1.* — The description of the components Γ_{∞,j_A} of Γ_∞ supported over $\mathrm{D_A}$, with their multiplicities, as well as the fact that they exhaust the special fiber, all follow from writing explicitly the map Φ in local coordinates in an analytic (or étale) neighborhood of a general point p of the boundary stratum $\mathrm{D_A}$. An explicit proof is as follows.

Choose an étale open neighborhood U of $\mathrm{U}_{d'}^+$ such that p is a closed point in the scheme U. Let $\hat{\mathcal{O}}_p$ be the completion of $\mathcal{O}_{\mathrm{U},p}$ and let C be the fiber curve of π over p. The curve C has exactly $|A|$-many nodal points q. Let $\mathrm{C}_{tail,q}$ be the rational tail component of C which meets q and let C_{main} be the remained component of C so that $\mathrm{C} = \cup_q \mathrm{C}_{tail,q} \cup \mathrm{C}_{main}$. We may express the completion $\hat{\mathcal{O}}_q$ at the node as

$$\hat{\mathcal{O}}_q \cong \hat{\mathcal{O}}_p[[x_q, y_q]]/(x_q y_q - t_q)$$

with local defining equations $x_q \in \hat{\mathcal{O}}_q$, $t_q \in \hat{\mathcal{O}}_p$ of the divisors $\tilde{\mathrm{D}}_a$, D_a respectively.

Consider a commuting diagram of natural $\hat{\mathcal{O}}_p$-module homomorphisms

$$\begin{array}{ccc}
(\pi_*(\mathscr{P}_{d'}^+ \oplus \bigoplus_i \mathscr{R}_{i,d'}^+))_p \otimes \hat{\mathcal{O}}_p & \overset{\phi_1}{\hookrightarrow} & \oplus_q(\mathscr{P}_{d'}^+ \oplus \bigoplus_i \mathscr{R}_{i,d'}^+)_q \otimes \hat{\mathcal{O}}_q \\
\Phi_p \otimes \mathrm{id} \Big\downarrow =:\Phi_{\hat{p}} & & \Big\downarrow \oplus_q \Psi_q \\
(\pi_*(\mathscr{P}_{d'}^+(d_a\tilde{\mathrm{D}}_a) \oplus \bigoplus_i \mathscr{R}_{i,d'}^+(l_i d_a \tilde{\mathrm{D}}_a)))_p \otimes \hat{\mathcal{O}}_p & \overset{\phi_2}{\hookrightarrow} & \oplus_q(\mathscr{P}_{d'}^+(d_a\tilde{\mathrm{D}}_a) \oplus \bigoplus_i \mathscr{R}_{i,d'}^+(l_i d_a\tilde{\mathrm{D}}_a))_q \otimes \hat{\mathcal{O}}_q
\end{array}$$

where $\mathscr{P}^+_{d'} := \mathscr{L}'_+ \otimes V \otimes Q$, $\mathscr{R}^+_{i,d'} := \oplus_{j=1}^N (\mathscr{L}'_+)^{l_i}$ as in (2.5.1), the horizontal maps ϕ_i are the restriction maps, and Ψ_q are the natural maps.

Since the horizontal restriction maps ϕ_i are injections, we will use $\oplus_q \Psi_q$ to express $\Phi_{\hat{p}}$ explicitly. For this, let us choose a $\hat{\mathcal{O}}_q$-basis $\{e^q_{0,j}\}_{j=1}^{(N-1)\dim V}$ of $\mathscr{P}^+_{d',q} \otimes \hat{\mathcal{O}}_q$ and a $\hat{\mathcal{O}}_q$-basis $\{e^q_{i,j}\}_{i=1,j=1}^{r,N}$ of $\oplus_i \mathscr{R}^+_{i,q} \otimes \hat{\mathcal{O}}_q$. With respect to this basis, we have also a basis $\{e^q_{0,j} \otimes x_q^{-d_a}\}_{j=1}^{(N-1)\dim V}$ of $\mathscr{P}^+_{d',q} \otimes \hat{\mathcal{O}}_q(d_a\tilde{D}_a) \cong \mathscr{P}^+_{d',q}(d_a\tilde{D}_a) \otimes \hat{\mathcal{O}}_q$ and a basis $\{e^q_{i,j} \otimes x_q^{-l_i d_a}\}_{i=1,j=1}^{r,N}$ of $\oplus_i \mathscr{R}^+_{i,q} \otimes \hat{\mathcal{O}}_q(l_i d_a \tilde{D}_a) \cong \oplus_i \mathscr{R}^+_{i,q}(l_i d_a\tilde{D}_a) \otimes \hat{\mathcal{O}}_q$. With respect to these bases, the right vertical map Ψ_q is the component-wise multiplication by $x_q^{d_a}, x_q^{l_1 d_1}, \dots, x_q^{l_r d_r}$.

Let $\mathbf{k}(p)$ be the residue field of $\mathcal{O}_p$ and let $\bar{e}^q_{0,j}$, $\bar{e}^q_{i,j}$ be the restrictions in $(\mathscr{P}^+_{d'} \oplus \bigoplus_i \mathscr{R}^+_{i,d'})_q \otimes \hat{\mathcal{O}}_q \otimes \mathbf{k}(p)$ of $e^q_{0,j}$, $e^q_{i,j}$ respectively. Choose also a $\mathbf{k}(p)$-basis $\mathcal{B}_{main}$ of $H^0(C_{main}, (\mathscr{P}^+_{d'} \oplus \bigoplus_i \mathscr{R}^+_{i,d'})|_{C_{main}}(-\sum_q q))$ by taking the union of some bases of $H^0(C_{main}, \mathscr{P}^+_{d'}|_{C_{main}}(-\sum_q q))$, $H^0(C_{main}, \mathscr{R}^+_{i,d'}|_{C_{main}}(-\sum_q q))$, $\forall i$. Consider the following subset

$$\textbf{(3.3.11)}\quad \{\oplus_q s_q\}_{s\in\mathcal{B}_{main}} \cup \bigcup_q \{\bar{e}^q_{0,j}, y_q\bar{e}^q_{0,j}, \dots, y_q^{d_a}\bar{e}^q_{0,j}\}_{j=1}^{(N-1)\dim V} \cup \bigcup_q \{\bar{e}^q_{i,j}, y_q\bar{e}^q_{i,j}, \dots, y_q^{l_i d_a}\bar{e}^q_{i,j}\}_{i=1,j=1}^{r,N}$$

of $\oplus_q(\mathscr{P}^+_{d'} \oplus \bigoplus_i \mathscr{R}^+_{i,d'})_q \otimes \hat{\mathcal{O}}_q \otimes \mathbf{k}(p)$. Here s_q denotes the stalk of s at $q \in C_{main}$. Note that (3.3.11) is a $\mathbf{k}(p)$-basis of the subspace $H^0(C, (\mathscr{P}^+_{d'} \oplus \bigoplus_i \mathscr{R}^+_{i,d'})|_C)$. Extend this $\mathbf{k}(p)$-basis (3.3.11) to a basis of $(\pi_*(\mathscr{P}^+_{d'} \oplus \bigoplus_i \mathscr{R}^+_{i,d'}))_p \otimes \hat{\mathcal{O}}_p$ as a $\hat{\mathcal{O}}_p$-module,

$$\textbf{(3.3.12)}\quad \{\oplus_q \tilde{s}_q\}_{s\in\mathcal{B}_{main}} \cup \bigcup_q \{e^q_{0,j}, y_q e^q_{0,j}, \dots, y_q^{d_a} e^q_{0,j}\}_{j=1}^{(N-1)\dim V} \cup \bigcup_q \{e^q_{i,j}, y_q e^q_{i,j}, \dots, y_q^{l_i d_a} e^q_{i,j}\}_{i=1,j=1}^{r,N}$$

where $\tilde{s} \in \pi_*(\mathscr{P}^+_{d'} \oplus \bigoplus_i \mathscr{R}^+_{i,d'}) \otimes \hat{\mathcal{O}}_p$ is an extension of s.

Let $l_0 = 1$ and let $l(s) = l_0$ for $s \in \mathcal{B}_{main}$ if s comes from $\mathscr{P}^+_{d'}|_{C_{main}}(-\sum_q q)$, $l(s) = l_i$ if s comes from $\mathscr{R}^+_{i,d'}|_{C_{main}}(-\sum_q q)$. Choose also a basis of $(\pi_*(\mathscr{P}^+_{d'}(d_a\tilde{D}_a) \oplus \bigoplus_i \mathscr{R}^+_{i,d'}(l_i d_a \tilde{D}_a)))_p \otimes \hat{\mathcal{O}}_p$ which is expressed via ϕ_2 as

$$\textbf{(3.3.13)}\quad \begin{aligned} &\{\oplus_q \tilde{s}_q\}_{s\in\mathcal{B}_{main}} \\ &\cup \bigcup\nolimits_q \left\{x_q^{d_a}\left(e^q_{0,j} \otimes x_q^{-d_a}\right), x_q^{d_a-1}\left(e^q_{0,j} \otimes x_q^{-d_a}\right), \dots, e^q_{0,j} \otimes x_q^{-d_a}\right)\right\}_{j=1}^{(N-1)\dim V} \\ &\cup \bigcup\nolimits_q \left\{x_q^{l_i d_a}\left(e^q_{i,j} \otimes x_q^{-d_a}\right), x_q^{l_i d_a-1}\left(e^q_{i,j} \otimes x_q^{-l_i d_a}\right), \dots, e^q_{i,j} \otimes x_q^{-l_i d_a}\right)\right\}_{i=1,j=1}^{r,N}. \end{aligned}$$

The map $\lambda\Phi_{\hat{p}}$ sends

$$\oplus_q \tilde{s}_q \mapsto \oplus_q \lambda \tilde{s}_q, \quad \text{and} \quad y_q^k e^q_{i,j} \mapsto \lambda t_q^k x_q^{l_i d_a - k}\left(e^q_{i,j} \otimes x_q^{-l_i d_a}\right),$$
$$i = 0, 1, \dots, r;\ k = 0, 1, \dots, l_i d_a;\ \forall j$$

so that with respect to the $\hat{\mathcal{O}}_p$-bases (3.3.12) and (3.3.13), $\lambda\Phi_{\hat{p}}$ is a diagonal matrix with entries λ's, λt_q^k, $k = 0, 1, \ldots, \max_{i=0,\ldots,r}\{l_i d_a\}$.

Now according to the fate of λt_q^k, $k = 0, 1, \ldots$, as $\lambda \to \infty$ and $t_q \to 0$ $\forall q$, the cycle class $[\Gamma_\infty]$ can be easily identified yielding the decomposition (3.3.1) for each A. Namely, for the node q corresponding to a, if $\lambda t_q^{j_a}$ goes to a nonzero number $w_a \in \mathbf{C}$ for some j_a, then the limit of graph($\lambda\Phi$) in the region is the point $Point(j_a, w_a)_{a\in \mathrm{A}}$ in $\mathbf{Gr}|_p$ corresponding to the direct sum of the following three subspaces (i), (ii), (iii)

$$
\begin{aligned}
&\text{(i)} && \mathrm{F}^{+j_{\mathrm{A}}+1}_{tail,d_a}|_{\mathrm{pr}_{\mathrm{A}}(p)} = \oplus_{a\in \mathrm{A}} \oplus_{i,j} \langle y_q^{j_a+1}\bar{e}^q_{i,j}, \ldots, y_q^{l_i d_a}\bar{e}^q_{i,j}\rangle \subset \mathrm{F}^+_{d'}|_p;\\
&\text{(ii)} && \mathrm{F}^{-j_{\mathrm{A}}-1}_{d'_0}|_{\mathrm{Pr}_{\mathrm{A}}\circ c_{\mathrm{A}}(p)} = \mathrm{H}^0(\mathrm{C}_{main}, \mathscr{P}^-_{d'}|_{\mathrm{C}_{main}}(-\textstyle\sum_a d_a q))\oplus\\
&&& \qquad \oplus_{i\geq 1}\mathrm{H}^0(\mathrm{C}_{main}, \mathscr{R}^-_{i,d'}|_{\mathrm{C}_{main}}(-\textstyle\sum_a l_i d_a q))\oplus\\
&&& \qquad \oplus_{a\in \mathrm{A}} \oplus_{i,j} \langle x_q^{l_i d_a}(\bar{e}^q_{i,j}\otimes x_q^{-l_i d_a}), \ldots, x_q^{l_i d_a-(j_a-1)}(\bar{e}^q_{i,j}\otimes x_q^{-l_i d_a})\rangle\\
&&& \qquad \subset c^*\mathrm{F}^-_{d'}|_p;\\
&\text{(iii)} && \oplus_{a\in \mathrm{A}} \oplus_{i,j} y_q^{j_a}\bar{e}^q_{i,j} \oplus w_a x_q^{l_i d_a - j_a}(\bar{e}^q_{i,j}\otimes x_q^{-l_i d_a}) \subset \mathrm{pr}^*_{\mathrm{A}}\mathrm{F}^{+j_{\mathrm{A}}}_{tail,d_a}|_p \oplus \mathrm{Pr}^*_{\mathrm{A}} c^*_{\mathrm{A}}\mathrm{F}^{-j_{\mathrm{A}}}_{d'_0}|_p.
\end{aligned}
$$

It is clear that there is a natural correspondence between the irreducible components of Γ_∞ and $Point(j_a, 1)_{a\in \mathrm{A}}$ $\forall j_{\mathrm{A}}$. Denote by $\Gamma_{\infty j_{\mathrm{A}}}$ the component corresponding to $Point(j_a, 1)_{a\in \mathrm{A}}$. The intersection multiplicity of $\Gamma_\infty \cap \{\lambda = \infty\}$ at $\Gamma_{\infty j_{\mathrm{A}}}$ is $m_{j_{\mathrm{A}}} := \prod_{a\in \mathrm{A}} j_a$ according to the equations $t_q^{j_a} = 0$, $\forall a \in \mathrm{A}$ in the open affine coordinate ring of $\mathbf{Gr}$ around $Point(j_a, 1)_{a\in \mathrm{A}}$.

3.3.7. *Remark.* — Denoting by $\mathbf{e}$ the Euler class, [15, Example 18.1.6] gives

$$
(\mathbf{3.3.14}) \qquad \mathbf{e}(\mathrm{F}^+_{d'}) \cap [\mathrm{U}^+_{d'}] - \mathbf{e}(c^*\mathrm{F}^-_{d'}) \cap [\mathrm{U}^+_{d'}] = \sum_{(\mathrm{A}, j_{\mathrm{A}})} \frac{m_{j_{\mathrm{A}}}}{|\mathrm{A}|!}(\eta|_{\Gamma_{\infty j_{\mathrm{A}}}})_*(\mathbf{e}(\zeta) \cap [\Gamma_{\infty j_{\mathrm{A}}}]).
$$

For $g = 0$, when no twisting occurs, $\mathrm{U}^{\pm}_{d'}$ reduces to $\mathrm{Q}^{\pm}_{0,k}(\mathbf{P}(\mathrm{V}), d)$, while $\mathrm{F}^{\pm}_{d'} = \pi_*(\oplus_{i=1}^r \mathscr{L}^{l_i}_{\pm})$. After applying c_*, the left-hand side of (3.3.14) becomes precisely

$$
c_* i_* [\mathrm{Q}^+_{0,k}(\mathrm{X}, d)]^{\mathrm{vir}} - i_*[\mathrm{Q}^-_{0,k}(\mathrm{X}, d)]^{\mathrm{vir}}.
$$

On the other hand, it is not too difficult to show[1] that the right-hand side can be written in the form

$$
\sum_{\mathrm{A}} \frac{1}{|\mathrm{A}|!}(b_{\mathrm{A}})_*(c_{\mathrm{A}})_* i_* \Big(\prod_{a\in \mathrm{A}} ev_a^* \mu_{d_a}(z)|_{z=-\psi_a} \cap [\mathrm{Q}^+_{0,k+\mathrm{A}}(\mathrm{X}, d_0^{\mathrm{A}})]^{\mathrm{vir}}\Big),
$$

for *some* polynomial Chow cohomology class $\mu_{d_a}(z) \in \mathrm{A}^*(\mathrm{X})_{\mathbf{Q}}[z]$. Combined with the identification of μ_{d_a} in Section 3.6 below, this proves for X the weaker equality (1.2.1) in Conjecture 1.1 in genus zero.

[1] The argument is a considerably simplified version of the proof of Theorem 3.8 in Section 3.5 below.

3.4. *A refinement of the graph construction.* — The equality (3.3.14) may be viewed as a degeneration formula for the top Chern class of the vector bundle $F^+_{d'}$ on $U^\pm_{d'}$. As a main step in our proof of Theorem 1.6, we establish in this subsection a refined degeneration formula which relates the Gysin pull-backs $0^!_{E^\varepsilon_d}([\mathbf{C}_{Q^\varepsilon_{g,k}(X,d)/U^\varepsilon_{d'}}])$ of the normal cones from Corollary 2.6.

3.4.1. *Deformation of the embedding (2.5.4).* — The map Φ fits in the following commuting diagram

$$\begin{array}{ccccc} F^+_{d'} & \xrightarrow{\Phi} & c^*F^-_{d'} & \longrightarrow & F^-_{d'} \\ {\scriptstyle\sigma^+}\uparrow & & \uparrow{\scriptstyle c^*(\sigma^-)} & & \uparrow{\scriptstyle\sigma^-} \\ U^+_{d'} & \xrightarrow[=]{} & U^+_{d'} & \xrightarrow[c]{} & U^-_{d'} \end{array}$$

with $\sigma^\pm$ the canonical sections (2.5.3). Recall that the zero locus of $\sigma^\pm$, call it $Y^\pm$, is identified with $Q^\pm_{g,k}(X, d)$. Denote by $Z = Z_{g,k,d}$ the zero locus of $c^*(\sigma^-) = \Phi \circ \sigma^+$; in other words, $Z = c^{-1}(Q^-_{g,k}(X, d))$. Observe that there is a closed embedding $Y^+ \hookrightarrow Z$.

Remark **3.2.** — If we restrict c further to $Y^+ \subset Z$, the resulting map coincides with the contraction $c : Y^+ \to Y^-$ induced from the natural embedding $X \subset \mathbf{P}(V)$ and the contraction $c : Q^+_{g,k}(\mathbf{P}(V), d) \to Q^-_{g,k}(\mathbf{P}(V), d)$. This follows from the fact that the twisting line bundle $\mathscr{M}$ is trivial on the rational tails.

It turns out that it is better to consider the deformation of Z induced by the family $\Gamma \to \mathbf{P}^1$. To this end, consider the *universal quotient bundle* Υ on $\mathbf{Gr}$, so that

$$0 \to \zeta \to \eta^*\left(F^+_{d'} \oplus c^*F^-_{d'}\right) \to \Upsilon \to 0,$$

is exact. As before, we also consider the universal quotient bundles Υ_P on $\mathbf{Gr}_P$ and Υ_R on $\mathbf{Gr}_R$. *We will use the same notations for the induced vector bundles on* Γ.

The section $\eta^*(\sigma^+, c^*\sigma^-)$ of $\eta^*(F^+_{d'} \oplus c^*F^-_{d'})$ induces a section

$$\overline{\sigma} \in H^0(\Gamma, \Upsilon)$$

of Υ on Γ, via composition with the projection.

Let

$$\Gamma^0 := \overline{\sigma}^{-1}(0) \subset \Gamma \subset \mathbf{Gr} \times \mathbf{P}^1,$$

be the zero locus of $\overline{\sigma}$.

As before, let Γ^0_λ denote the fiber of Γ^0 over $\lambda \in \mathbf{P}^1$. For $\lambda \neq 1, \infty$, under the isomorphism $v_\lambda : U^+_{d'} \times \{\lambda\} \to \Gamma_\lambda$, the section $\overline{\sigma}$ corresponds to the section $(1-\lambda)c^*\sigma^-$ of $F^-_{d'}$. Hence, for $\lambda \notin \{1, \infty\}$, we get that Γ^0_λ is isomorphic to Z.

The fiber over $1 \in \mathbf{A}^1$ is the entire $U^+_{d'}$, so from now on we will consider the families Γ and Γ^0 *only over* $\mathbf{P}^1 \setminus \{1\}$ (but will keep the same notation).

The fiber over $\infty \in \mathbf{P}^1$ decomposes in the Chow group as

$$[\Gamma^0_\infty] = [\Gamma^0_{\infty,dist}] + \sum_{(A,j_A)} m_{j_A}[\Gamma^0_{\infty j_A}],$$

with $\Gamma^0_{\infty,dist} := \Gamma_{\infty,dist} \times_\Gamma \Gamma^0$ and $\Gamma^0_{\infty j_A} := \Gamma_{\infty j_A} \times_\Gamma \Gamma^0$.

Note that on $\Gamma_{\infty,dist} = U^+_{d'}$ the quotient bundle Υ is equal to $\eta^* F^+_{d'} \oplus \{0\}$ and $\overline{\sigma} = (\sigma^+, 0)$, hence $\Gamma^0_{\infty,dist}$ is identified with $Q^+_{g,k}(X, d)$, embedded as in (2.5.4).

3.4.2. *Deformation of the obstruction theory.* — The normal cone $\mathbf{C}_{\Gamma^0/\Gamma}$ is a subcone of $\Upsilon|_{\Gamma^0}$. We claim that, possibly after a birational modification of the fiber Γ_∞, it actually sits inside a subbundle Υ^0 of the "correct" rank.

Recall the twisting line bundle $\mathscr{M}$ on the universal curve $\mathfrak{C}^\pm_{g,k,d'}$ of $U^\pm_{d'}$ introduced in the beginning of Section 2.5 and recall s_j the sections $\tilde{\mathit{ft}}^*_\pm \tau_j$ of $\mathscr{M}$ where $\tilde{\mathit{ft}}_\pm : \mathfrak{C}^\pm_{g,k,d'} \to \mathfrak{C}_{g,k}$ is the stabilization map; see Section 2.3 for the definition of τ_j. Here $\mathfrak{C}_{g,k}$ is the universal curve over $\overline{M}_{g,k}$.

On the universal curve $\mathfrak{C}^+_{g,k,d'}$ over $U^+_{d'}$, there is a vector bundle monomorphism

$$\mathscr{P}^+_{d'} \hookrightarrow \mathscr{P}^+_{d',big} := \mathscr{L}'_+ \otimes \mathscr{M} \otimes V \otimes \mathbf{C}^{\binom{N}{2}}$$

induced from the homomorphism

$$\oplus_j \mathscr{L}'_+ \otimes V \to \mathscr{P}^+_{d',big}, \quad (v_j)_{j=1}^N \mapsto \oplus_{j_1 > j_2}(s_{j_1} v_{j_2} - s_{j_2} v_{j_1}).$$

Similarly there are vector bundle monomorphisms

$$\mathscr{P}^-_{d'} \hookrightarrow \mathscr{P}^-_{d',big} := \mathscr{L}'_- \otimes \mathscr{M} \otimes V \otimes \mathbf{C}^{\binom{N}{2}};$$

$$\mathscr{Q}^\pm_{d'} \hookrightarrow \mathscr{Q}^\pm_{d',big} := \oplus_i (\mathscr{L}'_\pm \otimes \mathscr{M})^{l_i} \otimes \mathbf{C}^{\binom{N}{2}}.$$

We replace the stack Γ by the closed substack Γ^{new} of the product $\mathbf{Gr}^{new} \times \mathbf{P}^1$ defined via the MacPherson graph construction, where $\mathbf{Gr}^{new}$ is now the fibered product over $U^+_{d'}$ of the various Grassmann bundles:

$$\begin{aligned}(3.4.1)\qquad \mathbf{Gr}^{new} = {} & \mathbf{Gr}(\pi_* \mathscr{P}^+_{d'} \oplus c^* \pi_* \mathscr{P}^-_{d'}) \times_{U^+_{d'}} \mathbf{Gr}(\pi_* \mathscr{R}^+_{d'} \oplus c^* \pi_* \mathscr{R}^-_{d'}) \\ & \times_{U^+_{d'}} \mathbf{Gr}\Big(\oplus_j \pi_* \mathscr{L}'_+ \otimes V \bigoplus \oplus_j c^* \pi_* \mathscr{L}'_- \otimes V\Big) \times_{U^+_{d'}} \mathbf{Gr}(\pi_* \mathscr{Q}^+_{d'} \oplus c^* \pi_* \mathscr{Q}^-_{d'}) \\ & \times_{U^+_{d'}} \mathbf{Gr}(\pi_* \mathscr{P}^+_{d',big} \oplus c^* \pi_* \mathscr{P}^-_{d',big}) \times_{U^+_{d'}} \mathbf{Gr}(\pi_* \mathscr{Q}^+_{d',big} \oplus c^* \pi_* \mathscr{Q}^-_{d',big}).\end{aligned}$$

The projection onto the first two factors induces a birational morphism $p_{12} : \Gamma^{new} \to \Gamma$, which is an isomorphism outside $\infty \in \mathbf{P}^1$.

Denote by $\Upsilon_{\oplus_j \mathscr{L}' \otimes V}, \Upsilon_{\mathscr{P}_{big}}, \Upsilon_{\mathscr{R}}, \Upsilon_{\mathscr{Q}_{big}}, \dots$ the universal quotient bundles on $\Gamma^{new} \subset \mathbf{Gr}^{new} \times \mathbf{P}^1$ obtained via pull-back from the third, the fifth, the second, the sixth, ... factor of $\mathbf{Gr}^{new}$ respectively. Similarly, denote by $\zeta_{\oplus_j \mathscr{L}' \otimes V}, \dots,$ the universal subbundles on Γ^{new}. Recall that $\Upsilon_{\mathscr{P}}$ and $\Upsilon_{\mathscr{R}}$ come with the sections $\overline{\sigma}_P$ and $\overline{\sigma}_R$, the components of the section $\overline{\sigma}$ of

$$\Upsilon = \Upsilon_{\mathscr{P}} \oplus \Upsilon_{\mathscr{R}}$$

(see Section 3.4.1). We set

$$\Gamma^{new,0} = \overline{\sigma}^{-1}(0).$$

As in the case when we had only the fibered product of the first two relative Grassmannians, for each j_A there is a natural morphism

$$\alpha_{j_A}^{new} : \mathbf{P}_{j_A} \longrightarrow \mathbf{Gr}^{new} \times \{\infty\},$$

which has generic degree $|A|!$ to the image and such that the relation (3.3.1) still holds for the new special fiber (in other words, the birational modification $p_{12} : \Gamma^{new} \to \Gamma$ does *not* introduce additional components over $\infty \in \mathbf{P}^1$). These morphisms are obtained by constructing extensions analogous to (3.3.7) and (3.3.8) for the remaining four factors in (3.4.1). We have $\alpha_{j_A} = p_{12} \circ \alpha_{j_A}^{new}$. Our proof of Theorem 1.6 will eventually reduce to intersection-theoretic computations performed after transfering everything to the $\mathbf{P}_{j_A}$'s. Hence it is harmless to drop from now on the superscript "new" from the notations for $\mathbf{Gr}$, Γ, Γ etc.

We are now ready to construct the required vector bundle Υ. Define two homomorphisms

$$d\varphi_{\pm,big} : \mathscr{P}^{\pm}_{d',big} \to \mathscr{Q}^{\pm}_{d',big}, \quad (v_{j_1,j_2}) \mapsto \oplus_i \oplus_{j_1 > j_2} \nabla \varphi_i \big(s_{j_1} u'_{j_2}\big) \cdot v_{j_1,j_2},$$

where $\oplus_j u'_j$ is the universal sections of $\oplus_j \pi_* \mathscr{L}'_{\pm} \otimes V$ as in (2.5.2).

On Γ, there is a natural diagram

(3.4.2)
$$\begin{array}{ccc} \Upsilon_{\oplus_j \mathscr{L}' \otimes V} & \longrightarrow & \Upsilon_{\mathscr{P}_{big}} \\ \big\downarrow & & \big\downarrow {\scriptstyle \overline{\pi_* d\varphi_{big}}} \\ \Upsilon_{\mathscr{R}} & \longrightarrow & \Upsilon_{\mathscr{Q}_{big}} \end{array}$$

which is not necessarily commutative. Here $\overline{\pi_* d\varphi_{big}}$ is the homomorphism induced from $d\varphi_{\pm,big}$ via push-forward. The remaining three arrows are all constructed by the same procedure. For example, the top horizontal homomorphism is obtained as follows. The composition of natural maps

$$\zeta_{\oplus_j \mathscr{L}' \otimes V} \to \eta^*\big(\pi_* \mathscr{P}^+_{d'} \oplus c^* \pi_* \mathscr{P}^-_{d'}\big) \to \eta^*\big(\pi_* \mathscr{P}^+_{d',big} \oplus c^* \pi_* \mathscr{P}^-_{d',big}\big) \to \Upsilon_{\mathscr{P}_{big}}$$

vanishes on $\Gamma \setminus \Gamma_\infty$ and hence vanishes on the closure Γ.

Let $\tilde{\eta}$ denote the composition of natural maps $\Gamma \to \mathbf{Gr} \times (\mathbf{P}^1 \setminus \{1\}) \to \mathrm{U}^+_{d'}$.

Lemma **3.3.** — *The following hold.*

(1) *The zero locus of the* P*-component* $\overline{\sigma}_{\mathrm{P}}$ *of* $\overline{\sigma}$ *is contained in the zero locus of* $\tilde{\eta}^*\sigma^-_{\mathrm{P}}$ *(see* (2.5.3) *for the definition of* σ^-_{P}*).*

(2) $(\sigma^+_{\mathrm{P}})^{-1}(0) = \mathrm{Q}^+_{g,k}(\mathbf{P}(\mathrm{V}), d) = (c^*\sigma^-_{\mathrm{P}})^{-1}(0)$

(3) *The diagram* (3.4.2) *becomes commutative when it is restricted to* $\overline{\sigma}_{\mathrm{P}}^{-1}(0)$.

Proof. — (1) Consider the homomorphism of locally free sheaves

$$\left(\eta^*\mathrm{P}^+_{d'} \oplus \eta^*c^*\mathrm{P}^-_{d'}\right) \boxplus \mathcal{O}_{\mathbf{P}^1\setminus\{1\}} \to \eta^*c^*\mathrm{P}^-_{d'} \boxplus \left(\mathcal{O}_{\mathbf{P}^1}(1)\right)|_{\mathbf{P}^1\setminus\{1\}},$$

$$\left(v^+, v^-\right) \mapsto \lambda_0\Phi\left(v^+\right) - \lambda_1 v^-,$$

where λ_0, λ_1 denote homogeneous coordinates of $\mathbf{P}^1$. Since $\zeta|_\Gamma$ is contained in the kernel of the above homomorphism, there is a map $\Upsilon_{\mathrm{P}} \to \eta^*c^*\mathrm{P}^-_{d'} \boxplus (\mathcal{O}_{\mathbf{P}^1}(1))|_{\mathbf{P}^1\setminus\{1\}}$, under which the section $\overline{\sigma}_{\mathrm{P}}$ goes to $(\lambda_0 - \lambda_1)c^*\sigma^-_{\mathrm{P}}$. Therefore the zero locus of $\overline{\sigma}_{\mathrm{P}}$ is contained in the zero locus of $\tilde{\eta}^*\sigma^-_{\mathrm{P}}$.

(2) The first equality is clear. The second equality is the claim

$$\mathrm{Q}^+_{g,k}(\mathbf{P}(\mathrm{V}), d) = c^{-1}\left(\mathrm{Q}^-_{g,k}(\mathbf{P}(\mathrm{V}), d)\right).$$

The claim is obvious since for a T-family of ε_+-stable quasimaps to $\mathbf{P}(\mathrm{V} \otimes \mathbf{C}^{\mathrm{N}})$, it is a T-family of ε_+-stable quasimaps to $\mathbf{P}(\mathrm{V})$ if and only if the family restricted to every geometric point of the test scheme T is a ε_+-stable quasimaps to $\mathbf{P}(\mathrm{V})$.

(3) The diagram (3.4.2) is by definition induced, by the pullback $\tilde{\eta}^*$, from the diagram of homomorphisms of locally free sheaves on $\mathrm{U}^+_{d'}$

$$\begin{array}{ccc} \pi_* \oplus_j \mathscr{L}'_+ \otimes \mathrm{V} \oplus c^*\pi_*(\oplus_j \mathscr{L}'_- \otimes \mathrm{V}) & \longrightarrow & \pi_*\mathscr{P}^+_{big} \oplus c^*\pi_*\mathscr{P}^-_{big} \\ \downarrow & & \downarrow {\scriptstyle \pi_* d\varphi_{+,big} \oplus c^*\pi_* d\varphi_{-,big}} \\ \pi_*\mathscr{R}^+_{d'} \oplus c^*\pi_*\mathscr{R}^-_{d'} & \longrightarrow & \pi_*\mathscr{Q}^+_{big} \oplus c^*\pi_*\mathscr{Q}^-_{big}. \end{array} \qquad \textbf{(3.4.3)}$$

The diagram (3.4.3) is commutative on the zero locus $\mathrm{Q}^+_{g,k}(\mathbf{P}(\mathrm{V}), d)$ of the section σ^+_{P} since the difference of the clockwise path and the counterclockwise path in each $\pm$-component

$$\begin{aligned} &\oplus_i\left(\nabla\varphi_i\left(s_{j_1}u'_{j_2}\right) \cdot (s_{j_1}v_{j_2} - s_{j_2}v_{j_1}) - \left(s^{l_i}_{j_1}\nabla\varphi_i\left(u'_{j_2}\right) \cdot v_{j_2} - s^{l_i}_{j_2}\nabla\varphi_i\left(u'_{j_1}\right) \cdot v_{j_1}\right)\right) \\ &= \oplus_i\left(-\nabla\varphi_i\left(s_{j_1}u'_{j_2}\right) \cdot s_{j_2}v_{j_1} + \nabla\varphi_i\left(s_{j_2}u'_{j_1}\right) \cdot s_{j_2}v_{j_1}\right) \end{aligned}$$

vanishes for the universal section $(u'_j)_j$ of $\oplus_j\mathscr{L}'_\pm \otimes \mathrm{V}$ with the vanishing condition $s_{j_1}u'_{j_2} - s_{j_2}u'_{j_1} = 0$. Hence it is enough to show that the zero locus of $\overline{\sigma}_{\mathrm{P}}$ contained in $\Gamma \times_{\mathrm{U}^+_{d'}} (\sigma^+_{\mathrm{P}})^{-1}(0)$. This follows from (1) and (2) above. □

In particular, the diagram (3.4.2) commutes when restricted to Γ^0. Since the horizontal maps factor through $\Upsilon_{\mathscr{P}}$ and $\Upsilon_{\mathscr{Q}}$, it follows that on Γ^0 we have the commuting diagram

$$\begin{array}{ccc} \Upsilon_{\oplus_j \mathscr{L}'\otimes V}|_{\Gamma^0} & \longrightarrow & \Upsilon_{\mathscr{P}}|_{\Gamma^0} \\ \downarrow & & \downarrow {\scriptstyle f_\Upsilon} \\ \Upsilon_{\mathscr{R}}|_{\Gamma^0} & \xrightarrow{\alpha_{1,\Upsilon}} & \Upsilon_{\mathscr{Q}}|_{\Gamma^0}, \end{array}$$

where $f_\Upsilon = \overline{\pi_* d\varphi_{big}}|_{\Upsilon_{\mathscr{P}}|_{\Gamma^0}}$. The map of vector bundles

$$\gamma : (\Upsilon = \Upsilon_{\mathscr{P}} \oplus \Upsilon_{\mathscr{R}})|_{\Gamma^0} \to \Upsilon_{\mathscr{Q}}|_{\Gamma^0}, \qquad \gamma(x, y) = f_\Upsilon(x) - \alpha_{1,\Upsilon}(y)$$

is surjective since it is so at each closed point of Γ (this needs to be checked at points on the special fiber Γ_∞, where it follows by pulling-back to the appropriate $\mathbf{P}_{j_A}$ and using the description of the three universal quotient bundles as extensions, as in e.g. (3.5.11) below). Define the required vector bundle on Γ to be

$$\Upsilon^0 := \ker \gamma.$$

Lemma **3.4.** — *The normal cone* $\mathbf{C}_{\Gamma^0/\Gamma}$ *is a subcone of* Υ^0.

Proof. — Let $\mathscr{I}_{\Gamma^0}$ denote the defining ideal sheaf of the closed substack Γ^0 of Γ. We will check that the induced homomorphism $(\Upsilon_{\mathscr{Q}})|^\vee_{\Gamma^0} \to \mathscr{I}_{\Gamma^0}/\mathscr{I}^2_{\Gamma^0}$ is identically zero. For this consider the commuting diagram

$$\begin{array}{ccccc} (\Upsilon_{\mathscr{Q}_{big}})^\vee & \longrightarrow & \Upsilon^\vee & \longrightarrow & \mathscr{I}_{\Gamma^0} \\ \downarrow & & \downarrow & & \downarrow \\ \tilde{\eta}^*(\pi_* \mathscr{Q}^+_{d',big} \oplus c^* \pi_* \mathscr{Q}^-_{d',big})^\vee & \longrightarrow & \tilde{\eta}^*(\mathrm{F}^+_{d'} \oplus c^* \mathrm{F}^-_{d'})^\vee & \longrightarrow & \mathscr{O}_\Gamma, \end{array}$$

where $\tilde{\eta}$ denotes the composition $\Gamma \to \mathbf{Gr} \times (\mathbf{P}^1 \setminus \{1\}) \to \mathrm{U}^+_{d'}$. By the above commuting diagram and the surjection $(\Upsilon_{\mathscr{Q}_{big}})|^\vee_{\Gamma^0} \twoheadrightarrow (\Upsilon_{\mathscr{Q}})|^\vee_{\Gamma^0}$, it is enough to show that the composition of the bottom arrows lands in $\mathscr{I}^2_{\Gamma^0}$. On the other hand $\tilde{\eta}^* \mathrm{Im}(\sigma_\mathrm{P}^{+\vee}) \subset \mathscr{I}_{\Gamma^0}$ by Lemma 3.3 (1). Here we view the dual $\sigma_\mathrm{P}^{+\vee}$ of σ_P^+ as the cosection $\sigma_\mathrm{P}^{+\vee} : (\mathrm{P}^+_{d'})^\vee \to \mathscr{O}_{\mathrm{U}^+_{d'}}$. Hence by Lemma 3.3 (2) it is enough to check that the composition *comp* of $(\pi_* \mathscr{Q}^\pm_{d',big})^\vee \to (\mathrm{F}^\pm_{d'})^\vee \to \mathscr{O}_{\mathrm{U}^\pm_{d'}}$ lands in $(\mathrm{Im}\sigma_\mathrm{P}^{\pm\vee})^2$. This is easy to check as follows. Recalling the definition of $\sigma_\mathrm{R}^\pm$, $\sigma_\mathrm{P}^\pm$ in (2.5.3), note that, for $\delta \in (\pi_* \mathscr{Q}^\pm_{d',big})^\vee$

$$\begin{aligned} comp(\delta) &= \left\langle \delta, \oplus_i \oplus_{j_1 > j_2} \nabla\varphi_i\left(s_{j_1} u'_{j_2}\right) \cdot \left(s_{j_1} u'_{j_2} - s_{j_2} u'_{j_1}\right) - \left(\varphi_i\left(s_{j_1} u'_{j_2}\right) - \varphi_i\left(s_{j_2} u'_{j_1}\right)\right)\right\rangle \\ &\in \left(\mathrm{Im}\sigma_\mathrm{P}^{\pm\vee}\right)^2. \end{aligned}$$

Here the last line is due to the Taylor expansion of the last term $\varphi_i(s_{j_2}u'_{j_1})$ in the first line:

$$\varphi_i\big(s_{j_2}u'_{j_1}\big) = \varphi_i\big(s_{j_1}u'_{j_2}\big) + \nabla\varphi_i\big(s_{j_1}u'_{j_2}\big)\cdot\big(s_{j_2}u'_{j_1} - s_{j_1}u'_{j_2}\big)$$

modulo the square of the ideal $\mathrm{Im}\sigma_{\mathrm{P}}^{\pm\vee}$ generated by $s_{j_2}u'_{j_1} - s_{j_1}u'_{j_2}$. □

By construction, on the fiber $\Gamma_0^0 := \Gamma^0 \times_\Gamma \Gamma_0$ we have

$$\Upsilon^0|_{\Gamma_0^0} = c^*\mathrm{E}_d^-,$$

while on the distinguished component $\Gamma^0_{\infty,dist} := \Gamma^0 \times_\Gamma \Gamma_{\infty,dist}$ over $\lambda = \infty$,

$$\Upsilon^0|_{\Gamma^0_{\infty,dist}} = \mathrm{E}_d^+,$$

with $\mathrm{E}_d^\pm$ as defined in (2.4.5).

3.4.3. *Refined degeneration formula.* — Consider the diagram, whose squares are all cartesian,

$$\begin{array}{ccccccccc}
\lambda & \longleftarrow & \mathbf{Gr}_{\mathrm{Z}} & \xleftarrow{\iota_\lambda} & \Gamma^0_\lambda & \longrightarrow & \mathbf{C}_{\Gamma^0/\Gamma}|_\lambda & \longrightarrow & \lambda \\
{\scriptstyle\lambda}\downarrow & & \downarrow & & \downarrow & & \downarrow & & \downarrow{\scriptstyle\lambda} \\
\mathbf{P}^1\setminus\{1\} & \longleftarrow & \mathbf{Gr}_{\mathrm{Z}}\times(\mathbf{P}^1\setminus\{1\}) & \xleftarrow[\iota]{} & \Gamma^0 & \longrightarrow & \mathbf{C}_{\Gamma^0/\Gamma} & \longrightarrow & \mathbf{P}^1\setminus\{1\} \\
 & & & & \downarrow & & \downarrow & & \\
 & & & & \Gamma^0 & \xrightarrow{0} & \Upsilon^0 & &
\end{array}$$

where $\mathbf{Gr}_{\mathrm{Z}}$ denotes the relative Grassmannian $\mathbf{Gr}$ restricted to Z, with projection $\eta|_{\mathrm{Z}} : \mathbf{Gr}_{\mathrm{Z}} \to \mathrm{Z}$.

Lemma **3.5.** — *In* $\mathrm{A}_*(\mathrm{Z})_{\mathbf{Q}}$ *we have the equality*

$$(\mathbf{3.4.4})\qquad (\eta|_{\mathrm{Z}})_*(\iota_0)_*0^!_{\Upsilon^0|_{\Gamma^0_0}}\big([\mathbf{C}_{\Gamma^0_0/\Gamma_0}]\big) - (\eta|_{\mathrm{Z}})_*(\iota_\infty)_*\big(0^!_{\Upsilon^0|_{\Gamma^0_{\infty,dist}}}\big([\mathbf{C}_{dist}]\big)\big)$$
$$= \sum_{(\mathrm{A},j_\mathrm{A})} m_{j_\mathrm{A}}(\eta|_{\mathrm{Z}})_*(\iota_\infty)_*\big(0^!_{\Upsilon^0|_{\Gamma^0_{\infty j_\mathrm{A}}}}\big([\mathbf{C}_{j_\mathrm{A}}]\big)\big),$$

where $\mathbf{C}_{dist}$ *is the normal cone* $\mathbf{C}_{\Gamma^0_{\infty,dist}/\Gamma_{\infty,dist}}$ *and* $\mathbf{C}_{j_\mathrm{A}}$ *is the normal cone* $\mathbf{C}_{\Gamma^0_{\infty j_\mathrm{A}}/\Gamma_{\infty j_\mathrm{A}}}$.

Proof. — By Theorem 6.2.(a) and Theorem 6.4 in [15] (as extended to DM-stacks in [29]), we have

$$(\mathbf{3.4.5})\qquad \lambda^!\iota_*0^![\mathbf{C}_{\Gamma^0/\Gamma}] = (\iota_\lambda)_*\lambda^!0^![\mathbf{C}_{\Gamma^0/\Gamma}] = (\iota_\lambda)_*0^!\lambda^![\mathbf{C}_{\Gamma^0/\Gamma}].$$

When $\lambda = 0$,

$$0^!\lambda^![\mathbf{C}_{\Gamma^0/\Gamma}] = 0^!_{\Upsilon^0|_{\Gamma^0_0}}\big([\mathbf{C}_{\Gamma^0_0/\Gamma_0}]\big).$$

By Lemma 3.6 below, when $\lambda = \infty$,

$$0^!\lambda^![\mathbf{C}_{\Gamma^0/\Gamma}] = 0^!_{\Upsilon^0|_{\Gamma^0_{\infty,dist}}}\big([\mathbf{C}_{dist}]\big) + \sum_{(A,j_A)} m_{j_A} 0^!_{\Upsilon^0|_{\Gamma^0_{\infty,j_A}}}\big([\mathbf{C}_{j_A}]\big).$$

The first term in (3.4.5) is independent of λ. Hence

$$\begin{aligned}(\iota_0)_* 0^!_{\Upsilon^0|_{\Gamma^0_0}}\big([\mathbf{C}_{\Gamma^0_0/\Gamma_0}]\big) = {} & (\iota_\infty)_*\big(0^!_{\Upsilon^0|_{\Gamma^0_{\infty,dist}}}\big([\mathbf{C}_{dist}]\big)\big) \\ & + \sum_{(A,j_A)} m_{j_A}(\iota_\infty)_*\big(0^!_{\Upsilon^0|_{\Gamma^0_{\infty,j_A}}}\big([\mathbf{C}_{j_A}]\big)\big)\end{aligned}$$

in $A_*(\mathbf{Gr}_Z)_{\mathbf{Q}}$. Pushing forward to Z we get (3.4.4). □

To state Lemma 3.6 used in the above proof, we set up some notation first. Recall from [23, p. 489] that for a local embedding $\mathcal{X} \to \mathcal{Y}$ of algebraic stacks of finite type over the base field, one has the normal cone $\mathbf{C}_{\mathcal{X}/\mathcal{Y}}$ to $\mathcal{X}$ in $\mathcal{Y}$ and also the deformation of normal cone, denoted $M^\circ_{\mathcal{X}}(\mathcal{Y})$. This is a stack with a morphism to $\mathbf{P}^1$ such that the general fiber is isomorphic to $\mathcal{Y}$ and the special fiber at $t = 0 \in \mathbf{P}^1$ is isomorphic to $\mathbf{C}_{\mathcal{X}/\mathcal{Y}}$. If $\mathcal{X}$ is a closed substack in $\mathcal{Y}$, the deformation can be obtained as in [15, Chapter 5], by constructing

$$M_{\mathcal{X}}(\mathcal{Y}) := Bl_{\mathcal{X}\times\{0\}}\mathcal{Y} \times \mathbf{P}^1$$

and setting

$$M^\circ_{\mathcal{X}}(\mathcal{Y}) := M_{\mathcal{X}}(\mathcal{Y}) \setminus Bl_{\mathcal{X}\times\{0\}}\mathcal{Y} \times \{0\}.$$

Now form the commuting diagram, whose squares are all cartesian

$$\begin{array}{ccccccc}
\mathbf{C}_{\Gamma^0_\infty/\Gamma_\infty} & \underset{j}{\hookrightarrow} & \mathbf{C}_{\Gamma^0/\Gamma}|_{\lambda=\infty} & \longrightarrow & \mathbf{C}_{\Gamma^0/\Gamma} & \longrightarrow & t=0 \\
\downarrow & & \downarrow & & \downarrow & & \downarrow{\scriptstyle v_0} \\
M^\circ_{\Gamma^0_\infty}(\Gamma_\infty) & \underset{\text{closed}}{\overset{i}{\hookrightarrow}} & M^\circ_{\Gamma^0}(\Gamma)|_{\lambda=\infty} & \longrightarrow & M^\circ_{\Gamma^0}(\Gamma) & \longrightarrow & \mathbf{P}^1 \\
 & & \downarrow & & \downarrow & & \\
 & & \lambda=\infty & \longrightarrow & \mathbf{P}^1\setminus\{1\}. & &
\end{array}$$

Lemma **3.6.** — *The equalities*

$$\infty^![\mathbf{C}_{\Gamma^0/\Gamma}] = j_*[\mathbf{C}_{\Gamma^0_\infty/\Gamma_\infty}] = [\mathbf{C}_{dist}] + \sum_{A,j_A} m_{j_A}[\mathbf{C}_{j_A}]$$

hold in $A_*(\mathbf{C}_{\Gamma^0/\Gamma}|_{\lambda=\infty})_{\mathbf{Q}}$.

Proof. — The equality $\infty^![\mathbf{C}_{\Gamma^0/\Gamma}] = j_*[\mathbf{C}_{\Gamma^0_\infty/\Gamma_\infty}]$ is a consequence of the definition of Gysin maps, their commutativity, and their compatibility with proper push-forward, as follows:

$$\begin{aligned}\infty^![\mathbf{C}_{\Gamma^0/\Gamma}] &= \infty^! v_0^![\mathrm{M}^\circ_{\Gamma^0}(\Gamma)] = v_0^!\infty^![\mathrm{M}^\circ_{\Gamma^0}(\Gamma)] = v_0^![\mathrm{M}^\circ_{\Gamma^0}(\Gamma)|_\infty]\\ &= v_0^![\overline{\Gamma_\infty \times (\mathbf{P}^1 - \{t=0\})}] = v_0^! i_*[\mathrm{M}^\circ_{\Gamma^0_\infty}(\Gamma_\infty)] = j_* v_0^![\mathrm{M}^\circ_{\Gamma^0_\infty}(\Gamma_\infty)]\\ &= j_*[\mathbf{C}_{\Gamma^0_\infty/\Gamma_\infty}].\end{aligned}$$

Here some explanation is in order. For the third equality in the above chain, note that $\mathrm{M}^\circ_{\Gamma^0}(\Gamma)$ is irreducible and dominant over $\mathbf{P}^1 \setminus \{1\}$. The closure is taken in $\mathrm{M}^\circ_{\Gamma^0}(\Gamma)|_\infty$. The fifth equality follows by the very definition of proper push-forward.

The decomposition

$$j_*[\mathbf{C}_{\Gamma^0_\infty/\Gamma_\infty}] = [\mathbf{C}_{dist}] + \sum_{A,j_A} m_{j_A}[\mathbf{C}_{j_A}]$$

is a consequence of the decomposition $[\Gamma_\infty] = [\Gamma_{\infty,dist}] + \sum_{A,j_A} m_{j_A}[\Gamma_{\infty,j_A}]$ in $A_*(\Gamma_\infty)_{\mathbf{Q}}$ (Theorem 3.1), via the specialization to the normal cone homomorphism $A_*(\Gamma_\infty)_{\mathbf{Q}} \to A_*(\mathbf{C}_{\Gamma^0_\infty/\Gamma_\infty})_{\mathbf{Q}}$. □

We finish this subsection by recording a basic intersection-theoretic Lemma which will be used several times in the sequel.

Lemma **3.7.** — *Let* $f : \mathcal{Y}' \longrightarrow \mathcal{Y}$ *be a proper morphism between finite type Deligne-Mumford stacks of the same pure dimension. Let* $i : \mathcal{X} \hookrightarrow \mathcal{Y}$ *be a closed embedding and form the fiber square*

$$\begin{array}{ccc}\mathcal{X}' & \longrightarrow & \mathcal{Y}'\\ \downarrow & & \downarrow f\\ \mathcal{X} & \xrightarrow{\ i\ } & \mathcal{Y}.\end{array}$$

Let $\tilde{f} : \mathbf{C}_{\mathcal{X}'/\mathcal{Y}'} \longrightarrow \mathbf{C}_{\mathcal{X}/\mathcal{Y}}$ *be the induced map between normal cones. If* $f_*[\mathcal{Y}'] = m[\mathcal{Y}]$ *for a nonnegative rational number* m, *then* $\tilde{f}_*[\mathbf{C}_{\mathcal{X}'/\mathcal{Y}'}] = m[\mathbf{C}_{\mathcal{X}/\mathcal{Y}}]$.

Proof. — When $\mathcal{Y}$, $\mathcal{X}$, and $\mathcal{Y}'$ are schemes, this is [29, Lemma 3.15]. For the convenience of the reader, we give a short argument. Consider the deformations to the normal cone

$$M_{\mathcal{X}}\mathcal{Y} = Bl_{\mathcal{X}\times\{0\}}\mathcal{Y} \times \mathbf{P}^1, \qquad M_{\mathcal{X}'}\mathcal{Y}' = Bl_{\mathcal{X}'\times\{0\}}\mathcal{Y}' \times \mathbf{P}^1.$$

The map $\phi : M_{\mathcal{X}'}\mathcal{Y}' \longrightarrow M_{\mathcal{X}}\mathcal{Y}$ induced by f is proper and $\phi_*[M_{\mathcal{X}'}\mathcal{Y}'] = m[M_{\mathcal{X}}\mathcal{Y}]$. Let $v_0 : \{0\} \hookrightarrow \mathbf{P}^1$ be the inclusion. Denoting by $\mathbf{1}$ the trivial rank one vector bundle, we have

$$(\mathbf{3.4.6}) \qquad m\big[\mathbf{P}(\mathbf{C}_{\mathcal{X}/\mathcal{Y}} \oplus \mathbf{1})\big] + m[Bl_{\mathcal{X}}\mathcal{Y}] = mv_0^![M_{\mathcal{X}}\mathcal{Y}] = v_0^!\phi_*\big[M_{\mathcal{X}'}\mathcal{Y}'\big]$$
$$= (\phi|_{t=0})_* v_0^!\big[M_{\mathcal{X}'}\mathcal{Y}'\big],$$

where we have used the commutativity of Gysin maps with proper push-forward for the last equality. Since

$$v_0^!\big[M_{\mathcal{X}'}\mathcal{Y}'\big] = \big[\mathbf{P}(\mathbf{C}_{\mathcal{X}'/\mathcal{Y}'} \oplus \mathbf{1})\big] + \big[Bl_{\mathcal{X}'}\mathcal{Y}'\big]$$

and $(\phi|_{t=0})_*[Bl_{\mathcal{X}'}\mathcal{Y}'] = m[Bl_{\mathcal{X}}\mathcal{Y}]$, we conclude from (3.4.6) that

$$(\phi|_{t=0})_*\big[\mathbf{P}(\mathbf{C}_{\mathcal{X}'/\mathcal{Y}'} \oplus \mathbf{1})\big] = m\big[\mathbf{P}(\mathbf{C}_{\mathcal{X}/\mathcal{Y}} \oplus \mathbf{1})\big].$$

The Lemma follows, since $\tilde{f}$ is the restriction to $\mathbf{C}_{\mathcal{X}'/\mathcal{Y}'}$ of $\phi|_{t=0}$. □

3.5. *The correcting classes $\mu^{\mathrm{N}}_{d_a}(z)$.* — Consider the Segre embedding

$$(\mathbf{3.5.1}) \qquad Seg : \mathbf{P}(\mathrm{V}) \times \mathbf{P}(\mathbf{C}^{\mathrm{N}}) \longrightarrow \mathbf{P}(\mathrm{V} \otimes \mathbf{C}^{\mathrm{N}}).$$

Recall the map $h_a^+ : \mathrm{U}^+_{k+\mathrm{A},d_0'} \longrightarrow \mathbf{P}(\mathbf{C}^{\mathrm{N}})$ given by the twisting line bundle $\mathscr{M}_+$ and its sections $s_1, \ldots, s_{\mathrm{N}}$; see (3.2.3). Viewing $\mathrm{Q}^+_{g,k+\mathrm{A}}(\mathrm{X}, d_0)$ as a substack of $\mathrm{U}^+_{k+\mathrm{A},d_0'}$ via the embedding (2.5.4) for the bundle $\mathrm{F}^+_{d_0'}$, we have the restriction $h_a^+ : \mathrm{Q}^+_{g,k+\mathrm{A}}(\mathrm{X}, d_0) \longrightarrow \mathbf{P}(\mathbf{C}^{\mathrm{N}})$; see (3.2.1) for notation $d_0 = d_0^{\mathrm{A}}$. The two evaluation maps on $\mathrm{Q}^+_{g,k+\mathrm{A}}(\mathrm{X}, d_0)$ at markings in A are related by

$$\hat{ev}_a|_{\mathrm{Q}^+_{g,k+\mathrm{A}}(\mathrm{X},d_0)} = Seg \circ \big(ev_a, h_a^+\big);$$

see Section 3.2 for notations $\hat{ev}_a$ and ev_a.

In this subsection we prove the following weaker version of the main theorem.

Theorem **3.8.** — *Let z be a formal variable. There exists a Chow cohomology class $\mu^{\mathrm{N}}_{d_a}(z) \in \mathrm{A}^*(\mathrm{X} \times \mathbf{P}(\mathbf{C}^{\mathrm{N}}))_{\mathbf{Q}}[z]$, dependent on g and k only through the dependence on N, such that after push-forward to $\mathrm{A}_*(\mathrm{Q}^-_{g,k}(\mathrm{X}, d))_{\mathbf{Q}}$ by $c|_{\mathrm{Z}}$, the equality of Lemma 3.5 becomes*

$$(\mathbf{3.5.2}) \qquad \big[\mathrm{Q}^-_{g,k}(\mathrm{X}, d)\big]^{\mathrm{vir}} - c_*\big[\mathrm{Q}^+_{g,k}(\mathrm{X}, d)\big]^{\mathrm{vir}}$$
$$= \sum_{\mathrm{A}} \frac{1}{|\mathrm{A}|!}(b_{\mathrm{A}})_*(c_{\mathrm{A}})_*\Big(\prod_{a\in\mathrm{A}}\big(ev_a, h_a^+\big)^*\mu^{\mathrm{N}}_{d_a}(z)|_{z=-\psi_a} \cap \big[\mathrm{Q}^+_{g,k+\mathrm{A}}\big(\mathrm{X}, d_0^{\mathrm{A}}\big)\big]^{\mathrm{vir}}\Big).$$

Proof. — We analyze the push-forward to $A_*(Q^-_{g,k}(X,d))_{\mathbf{Q}}$ of each term in (3.4.4) by $c|_Z$ which will be also denoted by c for easy notation. We have also induced maps

$$\mathbf{C}_{\Gamma^0_0/\Gamma_0} \to c^*\mathbf{C}_{Q^-_{g,k}(X,d)/U^-_{d'}} \to \mathbf{C}_{Q^-_{g,k}(X,d)/U^-_{d'}},$$

whose composition will be denoted by $c_{\mathbf{c}}$.

The terms on the left-hand side are very easy. First, by the identifications $(\Gamma^0_0 \subset \Gamma_0) = (Z \subset U^+_{d'})$ and $\Upsilon^0|_{\Gamma^0_0} = c^*E^-_d$ we have

$$\begin{aligned} c_*(\eta|_Z)_*(\iota_0)_*\left(0^!_{\Upsilon^0|_{\Gamma^0_0}}\left([\mathbf{C}_{\Gamma^0_0/\Gamma_0}]\right)\right) &= 0^!_{E^-_d}\left(c_{\mathbf{c}*}[\mathbf{C}_{\Gamma^0_0/\Gamma_0}]\right) \\ &= 0^!_{E^-_d}\left([\mathbf{C}_{Q^-_{g,k}(X,d)/U^-_{d'}}]\right) \\ &= \left[Q^-_{g,k}(X,d)\right]^{\mathrm{vir}}, \end{aligned}$$

where we have used standard properties of the Gysin map for the first equality, Lemma 3.7 for the second equality, and Corollary 2.6 for the third equality.

Second,

$$c_*(\eta|_Z)_*(\iota_\infty)_*0^!_{\Upsilon^0|_{\Gamma_{dist}}}\left([\mathbf{C}_{dist}]\right) = c_*\left(\left[Q^+_{g,k}(X,d)\right]^{\mathrm{vir}}\right),$$

again by the identifications $(\Gamma^0_{\infty,dist} \subset \Gamma_{\infty,dist}) = (Q^+_{g,k}(X,d) \subset U^+_{d'})$ and $\Upsilon^0|_{\Gamma_{\infty,dist}} = E^+_d$, together with Corollary 2.6.

The analysis of the right-hand side of (3.4.4) is significantly more subtle, so we divide it into several steps for clarity.

Step 1: Transferring the computation to $\mathbf{P}_{j_A}$. The Segre embedding (3.5.1), together with the inclusion $i : X \hookrightarrow \mathbf{P}(V)$, induces the embedding

(3.5.3)
$$\begin{aligned} i_{Seg} : X \times \mathbf{P}(\mathbf{C}^N) &\hookrightarrow \mathbf{P}(V \otimes \mathbf{C}^N) \times \mathbf{P}(\mathbf{C}^N), \\ (x,y) &\mapsto \left(Seg(i(x),y),y\right). \end{aligned}$$

We identify $X \times \mathbf{P}(\mathbf{C}^N)$ with its image under i_{Seg}. Set

$$Q^+_{tail,a} := (\hat{ev}_a \times \mathrm{id}_{\mathbf{P}(\mathbf{C}^N)})^{-1}\left(X \times \mathbf{P}(\mathbf{C}^N)\right),$$

a closed substack in $Q^+_{0,a}(\mathbf{P}(V \otimes \mathbf{C}^N), d_a) \times \mathbf{P}(\mathbf{C}^N)$, and

$$Q^+_{tail,A} := \prod_{a \in A} Q^+_{tail,a},$$

so that we have the cartesian square

$$\begin{array}{ccc} Q^+_{tail,A} & \longrightarrow & \prod_{a\in A}(Q^+_{0,a}(\mathbf{P}(V\otimes\mathbf{C}^N),d_a)\times\mathbf{P}(\mathbf{C}^N)) \\ \downarrow & & \downarrow {\scriptstyle \prod_a(\hat{ev}_a\times\mathrm{id}_{\mathbf{P}(\mathbf{C}^N)})} \\ (X\times\mathbf{P}(\mathbf{C}^N))^A & \xrightarrow{\prod_a i_{Seg}} & (\mathbf{P}(V\otimes\mathbf{C}^N)\times\mathbf{P}(\mathbf{C}^N))^A. \end{array}$$

Further, define the closed substack $D_{X,A} \subset D_A$ by the cartesian square

$$\textbf{(3.5.4)} \qquad \begin{array}{ccc} D_{X,A} & \xrightarrow{\ \mathrm{Pr}_A\ } & Q^+_{g,k+A}(X, d_0) \\ {\scriptstyle \mathrm{pr}_A}\big\downarrow & & \big\downarrow{\scriptstyle ((ev_a, h_a^+))_{a\in A}} \\ Q^+_{tail,A} & \xrightarrow[\prod_a(\hat{ev}_a \times \mathrm{id}_{\mathbf{P}(\mathbf{C}^N)})]{} & (X \times \mathbf{P}(\mathbf{C}^N))^A, \end{array}$$

where by abusing notation Pr_A, pr_A denote $\mathrm{Pr}_A|_{D_{X,A}}$, $\mathrm{pr}|_{D_{X,A}}$ respectively. Note that $\prod_a(\hat{ev}_a \times \mathrm{id}_{\mathbf{P}(\mathbf{C}^N)})$ is a flat map (in fact, smooth) and therefore so is Pr_A.

Now fix the pair (A, j_A) and define $Z_{j_A} \subset \mathbf{P}_{j_A}$ by the cartesian square

$$\textbf{(3.5.5)} \qquad \begin{array}{ccc} Z_{j_A} & \longrightarrow & \mathbf{P}_{j_A} \\ {\scriptstyle \alpha_{j_A}}\big\downarrow & & \big\downarrow{\scriptstyle \alpha_{j_A}} \\ \Gamma^0_{\infty,j_A} & \longrightarrow & \Gamma_{\infty,j_A}. \end{array}$$

Z_{j_A} is the zero locus of the section $\alpha^*_{j_A}\overline{\sigma} \in H^0(\mathbf{P}_{j_A}, \alpha^*_{j_A}\Upsilon)$. The restriction to Z_{j_A} of the projection $\pi_{\mathbf{P}} : \mathbf{P}_{j_A} \longrightarrow D_A$ factors through $D_{X,A}$.

We assemble everything in the commuting diagram

$$\textbf{(3.5.6)} \qquad \begin{array}{ccccccc} \Gamma^0_{\infty,j_A} & \xrightarrow{(\eta|_Z)\circ\iota_\infty} & Z & \xrightarrow{\ c\ } & \longrightarrow & \longrightarrow & Q^-_{g,k}(X, d) \\ {\scriptstyle \alpha_{j_A}}\big\uparrow & & \big\uparrow{\scriptstyle \nu_A} & & & & \big\uparrow{\scriptstyle b_A} \\ Z_{j_A} & \xrightarrow{\ \pi_{\mathbf{P}}\ } & D_{X,A} & \xrightarrow{\ \mathrm{Pr}_A\ } & Q^+_{g,k+A}(X, d_0) & \xrightarrow{\ c_A\ } & Q^-_{g,k+A}(X, d_0) \\ & & {\scriptstyle \mathrm{pr}_A}\big\downarrow & & \big\downarrow{\scriptstyle ((ev_a, h_a^+))_{a\in A}} & & \\ & & Q^+_{tail,A} & \xrightarrow[\prod(\hat{ev}_a \times \mathrm{id})]{} & (X \times \mathbf{P}(\mathbf{C}^N))^A & & \end{array}$$

with abusing notation again $c = c|_Z$, $c_A = c_A|_{Q^+_{g,k+A}(X,d_0)}$ (this notation is justified by Remark 3.2 in Section 3.4.1) and $\nu_A = \nu_A|_{D_{X,A}}$ etc.

Let

$$\textbf{(3.5.7)} \qquad \widetilde{\mathbf{C}_{j_A}} := \mathbf{C}_{Z_{j_A}/\mathbf{P}_{j_A}}.$$

By Lemma 3.7 applied to (3.5.5) and the commutativity of the Gysin map with push-forward,

$$0^!_{\Upsilon^0|_{\Gamma^0_{\infty,j_A}}}([\mathbf{C}_{j_A}]) = \frac{1}{|A|!}(\alpha_{j_A})_* 0^!_{\alpha^*_{j_A}(\Upsilon^0|_{\Gamma^0_{\infty,j_A}})}([\widetilde{\mathbf{C}_{j_A}}]),$$

where $\mathbf{C}_{j_A} := \mathbf{C}_{\Gamma^0_{\infty j_A}/\Gamma_{\infty j_A}}$ as defined in Lemma 3.5. From the diagram (3.5.6),

$$\begin{aligned}&\frac{1}{|A|!}c_*(\eta|z)_*(\iota_\infty)_*(\alpha_{j_A})_*\big(0^!_{\alpha^*_{j_A}(\Upsilon^0|_{\Gamma^0_{\infty j_A}})}([\widetilde{\mathbf{C}_{j_A}}])\big)\\ &=\frac{1}{|A|!}(b_A)_*(c_A)_*(\mathrm{Pr}_A)_*(\pi_{\mathbf{P}})_*\big(0^!_{\alpha^*_{j_A}(\Upsilon^0|_{\Gamma^0_{\infty j_A}})}([\widetilde{\mathbf{C}_{j_A}}])\big).\end{aligned}$$

Letting $\Upsilon^0_{j_A}$ denote $\alpha^*_{j_A}\Upsilon^0|_{\Gamma^0_{\infty j_A}}$, it remains to show that

$$\sum_{j_A} m_{j_A}(\mathrm{Pr}_A)_*(\pi_{\mathbf{P}})_*\big(0^!_{\Upsilon^0_{j_A}}([\widetilde{\mathbf{C}_{j_A}}])\big) \tag{3.5.8}$$

has the form

$$\Big(\prod_{a\in A}\big(ev_a,h_a^+\big)^*\mu_{d_a}(z)|_{z=-\psi_a}\Big)\cap\big[Q^+_{g,k+A}(X,d_0^A)\big]^{\mathrm{vir}},$$

as claimed in Theorem 3.8.

Step 2: Description of Υ_{j_A}. We start by describing first

$$\Upsilon_{j_A} := \alpha^*_{j_A}\Upsilon|_{\Gamma_{\infty j_A}} \tag{3.5.9}$$

on $\mathbf{P}_{j_A}$. Define vector bundles $G^{+,j_A}_{d'}$ and $G^{-,j_A}_{d'_0}$ on D_A via exact sequences

$$\begin{aligned}&0\to \mathrm{pr}^*_A F^{+,j_A}_{tail,d_a}\to \nu^*_A F^+_{d'}\to G^{+,j_A}_{d'}\to 0,\\ &0\to \mathrm{Pr}^*_A c^*_A F^{-,j_A}_{d'_0}\to \nu^*_A c^* F^-_{d'}\to G^{-,j_A}_{d'_0}\to 0.\end{aligned} \tag{3.5.10}$$

By (3.3.9), we have an extension

$$0\to \boxplus_{a\in A}\big(\mathcal{O}_{\mathbf{P}_{j_a}}(1)\otimes\pi^*_{\mathbf{P}}F^{j_a}\big)\to\Upsilon_{j_A}\to\pi^*_{\mathbf{P}}\big(G^{+,j_A}_{d'}\oplus G^{-,j_A}_{d'_0}\big)\to 0. \tag{3.5.11}$$

Further, if we let

$$G^{+,j_A}_{tail,d_a} := \big(\boxplus_{a\in A}\mathrm{pr}^*_a F^{+,1}_{tail,d_a}\big)/\mathrm{pr}^*_A F^{+,j_A}_{tail,d_a},$$

then from (3.3.4) and (3.5.10) it follows that $G^{+,j_A}_{d'}$ fits into an extension

$$0\to G^{+,j_A}_{tail,d_a}\to G^{+,j_A}_{d'}\to \mathrm{Pr}^*_A F^+_{d'_0}\to 0. \tag{3.5.12}$$

Note that we may write alternatively

$$\begin{aligned}G^{-,j_A}_{d'_0} =&\mathrm{Pr}^*_A c^*_A\big(\oplus_{a\in A; j_a\le d_a}\pi_*\big(\mathscr{P}^-_{d'_0}\otimes\mathcal{O}_{(d_a-j_a)p_a}(d_a p_a)\big)\big)\oplus\\ &\mathrm{Pr}^*_A c^*_A\big(\oplus_{a\in A}\pi_*\big(\oplus_{i:j_a\le l_i d_a}\mathscr{R}^-_{i,d'_0}\otimes\mathcal{O}_{(l_i d_a-j_a)p_a}(l_i d_a p_a)\big)\big),\end{aligned}$$

and

$$\mathrm{G}^{+,j_{\mathrm{A}}}_{tail,d_a} = \begin{cases} \mathrm{pr}^*_{\mathrm{A}}(\boxplus_{a\in\mathrm{A}}\pi_*((\mathscr{P}^+_{d_a}\oplus\oplus_i\mathscr{R}^+_{i,d_a})\otimes\mathcal{O}_{(j_a-1)p^{tail}_a}(-p^{tail}_a))), & \text{if } j_a\le d_a,\\ \mathrm{pr}^*_{\mathrm{A}}(\boxplus_{a\in\mathrm{A}}\pi_*(\oplus_{i:j_a\le l_i d_a}\mathscr{R}^+_{i,d_a}\otimes\mathcal{O}_{(j_a-1)p^{tail}_a}(-p^{tail}_a))), & \text{if } j_a> d_a,\end{cases}$$

from which it follows that in the K-group of vector bundles on D_{A}

$$\mathrm{G}^{-,j_{\mathrm{A}}}_{d'_0}\sim\left(\boxplus_{a\in\mathrm{A}}\oplus_{m=j_a+1}^{d_a}\left(\mathrm{Pr}^*_a c^*_a\mathcal{O}(-m\psi_a)\otimes \mathrm{P}^{d_a-m}\right)\right)\oplus$$
$$\left(\boxplus_{a\in\mathrm{A}}\oplus_{i=1}^{r}\oplus_{m=j_a+1}^{l_i d_a}\left(\mathrm{Pr}^*_a c^*_a\mathcal{O}(-m\psi_a)\otimes \mathrm{R}_i^{l_i d_a-m}\right)\right),$$

and[2]

$$\mathrm{G}^{+,j_{\mathrm{A}}}_{tail,d_a}\sim\boxplus_{a\in\mathrm{A}}\left(\oplus_{m=1}^{j_a-1}\left(\mathrm{pr}^*_a\mathcal{O}\left(m\psi^{tail}_a\right)\otimes \mathrm{F}^m\right)\right)$$

where $\mathrm{P}^{d_a-m}:=\mathrm{Pr}^*_a c^*_a\mathrm{P}^{-,d_a-m}$, $\mathrm{R}_i^{l_i d_a-m}:=\mathrm{Pr}^*_a c^*_a\mathrm{R}_i^{-,l_i d_a-m}$ (see (3.3.5), (3.3.6), (3.3.10) for the definition of P^{-,d_a-m}, $\mathrm{R}_i^{-,l_i d_a-m}$, F^m respectively).

To summarize, the outer terms of the exact sequences (3.5.11) and (3.5.12) give four pieces that combine to make $\Upsilon_{j_{\mathrm{A}}}$.

We now move to the description of the subbundle $\Upsilon^0_{j_{\mathrm{A}}}\subset\Upsilon_{j_{\mathrm{A}}}|_{\mathrm{Z}_{j_{\mathrm{A}}}}$ (see (3.5.8) for the notation $\Upsilon^0_{j_{\mathrm{A}}}$). For each $1\le i\le r$ and $0\le j_a$, introduce the bundles

$$\mathrm{R}^{j_a}_{i,small}:=\begin{cases}\mathrm{pr}^*_a(\hat{ev}_a\times \mathrm{id}_{\mathbf{P}(\mathbf{C}^{\mathrm{N}})})^*(\mathcal{O}_{\mathbf{P}(\mathrm{V}\otimes\mathbf{C}^{\mathrm{N}})}(l_i)\boxtimes\mathcal{O}_{\mathbf{P}(\mathbf{C}^{\mathrm{N}})}(-l_i)), & \text{if } j_a\le l_i d_a,\\ 0, & \text{if } j_a> l_i d_a,\end{cases}$$

on D_a. We use the same notation for the restrictions of $\mathrm{R}^{j_a}_{i,small}$ to the substacks D_{A} and $\mathrm{D}_{\mathrm{X,A}}$ of D_a. Further, we set

$$\mathrm{R}^{j_a}_{small}:=\oplus_{i=1}^r\mathrm{R}^{j_a}_{i,small}.$$

Note that, alternatively, we may write on $\mathrm{D}_{\mathrm{X,A}}$

$$\begin{aligned}\mathrm{R}^{j_a}_{small}&=\mathrm{pr}^*_{\mathrm{A}}\left(\pi_*\left(\oplus_i(\mathscr{L}_{+,d_a})^{l_i}\otimes\mathcal{O}_{p^{tail}_a}\right)\right)\\&=\mathrm{Pr}^*_{\mathrm{A}}\left(\pi_*\left(\oplus_i(\mathscr{L}_{+,d_0})^{l_i}\otimes\mathcal{O}_{p_a}\right)\right),\end{aligned}$$

for $j_a\le l_i d_a$. Finally, put

$$\mathrm{F}^{j_a}_{small}:=\mathrm{P}^{j_a}\oplus\mathrm{R}^{j_a}_{small}.$$

---

[2] The notation F^m is a little ambiguous, since the dependence on the marking a is not apparent anymore. The same will happen later, e.g., with the bundles F^0 in (3.5.17) below. Hopefully this will not cause any confusion.

The surjection $\Upsilon_{j_A} \twoheadrightarrow \pi_{\mathbf{P}}^*\mathrm{Pr}_A^*\mathrm{F}_{d_0'}^+$ on $\mathbf{P}_{j_A}$ (coming from (3.5.11) and (3.5.12)) induces a surjection $\Upsilon_{j_A}^0 \twoheadrightarrow \pi_{\mathbf{P}}^*\mathrm{Pr}_A^*\mathrm{E}_{d_0}^+$ on Z_{j_A}. Define the *excess bundles* $\Upsilon_{j_A,ex}$ and $\Upsilon_{j_A,ex}^0$ as the corresponding kernels:

$$(3.5.13)\qquad \begin{aligned} 0 \longrightarrow \Upsilon_{j_A,ex} \longrightarrow \Upsilon_{j_A} \longrightarrow \pi_{\mathbf{P}}^*\mathrm{Pr}_A^*\mathrm{F}_{d_0'}^+ \longrightarrow 0,\\ 0 \longrightarrow \Upsilon_{j_A,ex}^0 \longrightarrow \Upsilon_{j_A}^0 \longrightarrow \pi_{\mathbf{P}}^*\mathrm{Pr}_A^*\mathrm{E}_{d_0}^+ \longrightarrow 0. \end{aligned}$$

To complete the description of $\Upsilon_{j_A}^0$, we note that the excess bundle in turn fits into an extension

$$(3.5.14)\qquad \begin{aligned} 0 \to \boxplus_{a\in A}\big(\mathcal{O}_{\mathbf{P}_{j_a}}(1)|_{Z_{j_A}} \otimes \pi_{\mathbf{P}}^*\mathrm{F}_{small}^{j_a}\big) \to \Upsilon_{j_A,ex}^0 \to \pi_{\mathbf{P}}^*\big(\mathrm{G}_{tail,d_a,small}^{+j_A} \oplus \mathrm{G}_{d_0,small}^{-j_A}\big)\\ \to 0, \end{aligned}$$

with

$$\begin{aligned} \mathrm{G}_{tail,d_a,small}^{+j_A} &\sim \oplus_{a\in A}\big(\oplus_{m=1}^{j_a-1}(\mathrm{pr}_A^*\mathcal{O}(m\psi_a^{tail}) \otimes \mathrm{F}_{small}^m)\big),\\ \mathrm{G}_{d_0,small}^{-j_A} &\sim \big(\oplus_{a\in A}\oplus_{m=j_a+1}^{d_a}(\mathrm{Pr}_A^*c_A^*\mathcal{O}(-m\psi_a)\otimes \mathrm{P}^{d_a-m})\big)\oplus\\ &\qquad \big(\oplus_{a\in A}\oplus_{i=1}^{r}\oplus_{m=j_a+1}^{l_id_a}(\mathrm{Pr}_A^*c_A^*\mathcal{O}(-m\psi_a)\otimes \mathrm{R}_{i,small}^{l_id_a-m})\big) \end{aligned}$$

in the K-group of D_A. For later use, we note that from the above K-group expressions it follows that the Euler classes of these bundles have the form

$$(3.5.15)\qquad \mathbf{e}\big(\mathrm{G}_{tail,d_a,small}^{+j_A}\big) = \mathrm{pr}_A^*\prod_{a\in A}(\hat{ev}_a \times \mathrm{id}_{\mathbf{P}(\mathbf{C}^N)})^* f_{d_a}^{+j_a}(z)|_{z=\psi_a^{tail}},$$

$$(3.5.16)\qquad \mathbf{e}\big(\mathrm{G}_{d_0,small}^{-j_A}\big) = \mathrm{Pr}_A^*\prod_{a\in A}\big(ev_a, h_a^+\big)^* f_{d_a}^{-j_a}(z)|_{z=-\psi_a},$$

where the Chow cohomology classes

$$f_{d_a}^{+j_a}(z), f_{d_a}^{-j_a}(z) \in \mathrm{A}^*\big(\mathrm{X}\times\mathbf{P}(\mathbf{C}^N)\big)_{\mathbf{Q}}[z] = \big(\mathrm{A}^*(\mathrm{X})_{\mathbf{Q}}\otimes \mathrm{A}^*(\mathbf{P}(\mathbf{C}^N))_{\mathbf{Q}}\big)[z]$$

are polynomials in z with coefficients which are *universal* expressions in Chern classes of various tautological bundles $\mathcal{O}_X(l)$ on X, and $\mathcal{O}_{\mathbf{P}(\mathbf{C}^N)}(m)$ and the tautological quotient bundle Q on $\mathbf{P}(\mathbf{C}^N)$.

In the formula (3.5.16) we have used that the ψ-classes at markings in A on $Q_{g,k+A}^-(X, d_0)$ and $Q_{g,k+A}^+(X, d_0)$ pull-back under c_A, that is, $c_A^*\psi_a = \psi_a$.

Step 3: Deformation. The idea for computing (3.5.8) is to deform the bundle $\Upsilon_{j_A}^0$, together with its closed subcone $\widetilde{\mathbf{C}}_{j_A}$ (see (3.5.7) for the notation $\widetilde{\mathbf{C}}_{j_A}$), to the bundle $\Upsilon_{j_A,ex}^0 \oplus \pi_{\mathbf{P}}^*\mathrm{Pr}_A^*\mathrm{E}_{d_0}^+$ with the closed cone $\pi_{\mathbf{P}}^*\mathrm{Pr}_A^*\mathbf{C}_{Q_{g,k+A}^+(X,d_0)/U_{k+A,(d_0,d_0')}^+}$ (see (3.5.13) for the notation $\Upsilon_{j_A,ex}^0$).

To begin with, consider on D_A the vector bundle homomorphisms

$$\mathrm{pr}_A^*(\oplus_{a\in A}F^+_{tail,d_a}) \xrightarrow{\oplus_a r_a^{tail}} \oplus_{a\in A}F^0,$$

$$\mathrm{Pr}_A^*F^+_{d_0'} \xrightarrow{\oplus_a r_a} \oplus_{a\in A}F^0,$$

where r_a^{tail} and r_a are given by "restricting sections at the marking a". The resulting surjective gluing map

$$\mathrm{Pr}_A^*F^+_{d_0'} \oplus \mathrm{pr}_A^*(\oplus_{a\in A}F^+_{tail,d_a}) \xrightarrow{\oplus_a(r_a - r_a^{tail})} \oplus_{a\in A}F^0 \longrightarrow 0$$

has kernel $\nu_A^*F^+_{d'}$.

Via its embedding in $\pi_{\mathbf{P}}^*(\nu_A^*F^+_{d'} \oplus \mathrm{Pr}_A^*c_A^*F^-_{d'})$, we may view $\alpha_{j_A}^*(\zeta_{\mathscr{P}\oplus\mathscr{R}}|_{\Gamma_{\infty j_A}})$ as a subbundle

$$\alpha_{j_A}^*(\zeta_{\mathscr{P}\oplus\mathscr{R}}|_{\Gamma_{\infty j_A}}) \subset \pi_{\mathbf{P}}^*\big(\mathrm{Pr}_A^*F^+_{d_0'} \oplus \mathrm{pr}_A^*\big(\oplus_{a\in A}F^+_{tail,d_a}\big) \oplus \mathrm{Pr}_A^*c_A^*F^-_{d'}\big).$$

The quotient is an "unglued" version of Υ_{j_A}. Precisely, it splits as $\pi_{\mathbf{P}}^*\mathrm{Pr}_A^*(F^+_{d_0'}) \oplus \Upsilon_{j_A,ex,\hat{0}}$, and there are exact sequences

$$0 \longrightarrow \Upsilon_{j_A,ex} \longrightarrow \Upsilon_{j_A,ex,\hat{0}} \xrightarrow{\oplus_a r_a^{tail}} \pi_{\mathbf{P}}^*(\oplus_a F^0) \longrightarrow 0$$

and

(**3.5.17**)
$$0 \longrightarrow \Upsilon_{j_A} \longrightarrow \pi_{\mathbf{P}}^*\mathrm{Pr}_A^*F^+_{d_0'} \oplus \Upsilon_{j_A,ex,\hat{0}} \xrightarrow{\oplus_a(r_a - r_a^{tail})} \pi_{\mathbf{P}}^*(\oplus_a F^0) \longrightarrow 0$$

on $\mathbf{P}_{j_A} \xrightarrow{\pi_{\mathbf{P}}} D_A$. Composing the section $\overline{\sigma} : \mathcal{O}_{\mathbf{P}_{j_A}} \longrightarrow \Upsilon_{j_A}$ with the monomorphism in (3.5.17) gives the section

$$\big(\pi_{\mathbf{P}}^*\mathrm{Pr}_A^*\sigma^+_{d_0'}, \overline{\sigma}_{ex}\big) : \mathcal{O}_{\mathbf{P}_{j_A}} \longrightarrow \pi_{\mathbf{P}}^*\mathrm{Pr}_A^*F^+_{d_0'} \oplus \Upsilon_{j_A,ex,\hat{0}}.$$

The base of our deformation will be $\mathbf{A}^1$ with coordinate t. Denote $\varrho : \mathbf{P}_{j_A} \times \mathbf{A}^1 \longrightarrow \mathbf{P}_{j_A}$ the projection. Define on $\mathbf{P}_{j_A} \times \mathbf{A}^1$ the vector bundle ker via the exact sequence

$$0 \longrightarrow \ker \longrightarrow \varrho^*(\pi_{\mathbf{P}}^*\mathrm{Pr}_A^*F^+_{d_0'} \oplus \Upsilon_{j_A,ex,\hat{0}}) \xrightarrow{\oplus_a(tr_a - r_a^{tail})} \varrho^*\pi_{\mathbf{P}}^*(\oplus_a F^0) \longrightarrow 0$$

deforming (3.5.17). The section

$$\widetilde{\sigma} := \left(\varrho^*\pi_{\mathbf{P}}^*\mathrm{Pr}_{\mathrm{A}}^*\sigma^+_{d_0'},\, t\varrho^*\overline{\sigma}_{ex}\right)$$

of $\varrho^*(\pi_{\mathbf{P}}^*\mathrm{Pr}_{\mathrm{A}}^*\mathrm{F}^+_{d_0'} \oplus \Upsilon_{j_{\mathrm{A}},ex,\hat{0}})$ factors through ker, so we will view it from now on as a section of ker. We have the identifications

$$(\ker|_{t=1}, \widetilde{\sigma}|_{t=1}) = (\Upsilon_{j_{\mathrm{A}}}, \overline{\sigma})$$

and

$$(\ker|_{t=0}, \widetilde{\sigma}|_{t=0}) = \left(\pi_{\mathbf{P}}^*\mathrm{Pr}_{\mathrm{A}}^*\mathrm{F}^+_{d_0'} \oplus \Upsilon_{j_{\mathrm{A}},ex}, \left(\pi_{\mathbf{P}}^*\mathrm{Pr}_{\mathrm{A}}^*\sigma^+_{d_0'}, 0\right)\right).$$

Let

$$\widetilde{\mathrm{Z}} := \widetilde{\sigma}^{-1}(0) \subset \mathbf{P}_{j_{\mathrm{A}}} \times \mathbf{A}^1$$

be the zero locus and observe that we have in fact

$$\widetilde{\mathrm{Z}} \subset \mathbf{P}_{j_{\mathrm{A}}}|_{\mathrm{D}_{\mathrm{X,A}}} \times \mathbf{A}^1,$$

where $\mathbf{P}_{j_{\mathrm{A}}}|_{\mathrm{D}_{\mathrm{A}}}$ is the fibered product

$$\begin{array}{ccccc}
\mathbf{P}_{j_{\mathrm{A}}}|_{\mathrm{D}_{\mathrm{X,A}}} & \xrightarrow{\pi_{\mathbf{P}}} & \mathrm{D}_{\mathrm{X,A}} & \xrightarrow{\mathrm{Pr}_{\mathrm{A}}} & \mathrm{Q}^+_{g,k+\mathrm{A}}(\mathrm{X}, d_0) \\
\downarrow & & \downarrow & & \downarrow \\
\mathbf{P}_{j_{\mathrm{A}}} & \xrightarrow[\pi_{\mathbf{P}}]{} & \mathrm{D}_{\mathrm{A}} & \xrightarrow[\mathrm{Pr}_{\mathrm{A}}]{} & \mathrm{U}^+_{k+\mathrm{A},d_0'}.
\end{array}$$

The fibers of the $\mathbf{A}^1$-family $\widetilde{\mathrm{Z}}$ at $t = 1$ and at $t = 0$ are

$$\widetilde{\mathrm{Z}}|_{t=1} = \mathrm{Z}_{j_{\mathrm{A}}}, \qquad \widetilde{\mathrm{Z}}|_{t=0} = \mathbf{P}_{j_{\mathrm{A}}}|_{\mathrm{D}_{\mathrm{X,A}}}.$$

Notice that the normal cones satisfy

$$(\mathbf{3.5.18}) \qquad [\mathbf{C}_{\widetilde{\mathrm{Z}}/(\mathbf{P}_{j_{\mathrm{A}}}\times\mathbf{A}^1)}|_{t=0}] = [\mathbf{C}_{(\mathbf{P}_{j_{\mathrm{A}}}|_{\mathrm{D}_{\mathrm{X,A}}})/\mathbf{P}_{j_{\mathrm{A}}}}] = \pi_{\mathbf{P}}^*\mathrm{Pr}_{\mathrm{A}}^*[\mathbf{C}_{\mathrm{Q}^+_{g,k+\mathrm{A}}(\mathrm{X},d_0)/\mathrm{U}^+_{k+\mathrm{A},d_0'}}],$$

and

$$(\mathbf{3.5.19}) \qquad [\mathbf{C}_{\widetilde{\mathrm{Z}}/(\mathbf{P}_{j_{\mathrm{A}}}\times\mathbf{A}^1)}|_{t=1}] = [\widetilde{\mathbf{C}_{j_{\mathrm{A}}}}],$$

as desired.

The "correct" obstruction bundle $\Upsilon^0_{j\mathrm{A}}$ also deforms. Namely, if we repeat the construction in this step, but with the bundles $\mathscr{P}^{\pm} \oplus \mathscr{R}^{\pm}$, $\mathrm{F}^{\pm}_{d'}$ replaced by $\mathscr{Q}^{\pm}$, $\mathrm{Q}^{\pm}_{d'} := \pi_* \mathscr{Q}^{\pm}_{d'}$ respectively, we obtain an unglued version of $\Upsilon_{\mathscr{Q}\, j\mathrm{A}} := \alpha^*_{j\mathrm{A}} \Upsilon_{\mathscr{Q}}|_{\Gamma_{\infty j\mathrm{A}}}$ given as the extension

$$0 \longrightarrow \Upsilon_{\mathscr{Q}\, j\mathrm{A}} \longrightarrow \pi^*_{\mathbf{P}} \mathrm{Pr}^*_{\mathrm{A}} \mathrm{Q}^+_{d'_0} \oplus \Upsilon_{\mathscr{Q}\, j\mathrm{A}, ex, \hat{0}} \xrightarrow{\oplus_a (r_a - r_a^{tail})} \pi^*_{\mathbf{P}}(\oplus_a \mathrm{F}^0_{\mathscr{Q}}) \longrightarrow 0,$$

and a vector bundle $\ker_{\mathscr{Q}}$ on $\mathbf{P}_{j\mathrm{A}} \times \mathbf{A}^1$ defined via the deformation

$$0 \longrightarrow \ker_{\mathscr{Q}} \longrightarrow \varrho^*(\pi^*_{\mathbf{P}} \mathrm{Pr}^*_{\mathrm{A}} \mathrm{Q}^+_{d'_0} \oplus \Upsilon_{\mathscr{Q}\, j\mathrm{A}, ex, \hat{0}}) \xrightarrow{\oplus_a (\mathrm{tr}_a - r_a^{tail})} \varrho^* \pi^*_{\mathbf{P}}(\oplus_a \mathrm{F}^0_{\mathscr{Q}}) \longrightarrow 0.$$

Here $\mathrm{F}^0_{\mathscr{Q}}$ "at the marking a" is the cokernel of $0 \to \mathrm{F}^0_{small} \to \mathrm{F}^0$; alternatively,

$$\mathrm{F}^0_{\mathscr{Q}} = \mathrm{pr}^*_{\mathrm{A}}\big(\pi_*\big(\mathscr{Q}^+_{d'_a} \otimes \mathcal{O}_{p_a^{tail}}\big)\big) = \mathrm{Pr}^*_{\mathrm{A}}\big(\pi_*\big(\mathscr{Q}^+_{d'_0} \otimes \mathcal{O}_{p_a}\big)\big).$$

After restricting to $\widetilde{\mathrm{Z}}$, there is a surjection

$$\varrho^*\big(\pi^*_{\mathbf{P}} \mathrm{Pr}^*_{\mathrm{A}} \mathrm{F}^+_{d'_0} \oplus \Upsilon_{j\mathrm{A}, ex, \hat{0}}\big) \longrightarrow \varrho^*\big(\pi^*_{\mathbf{P}} \mathrm{Pr}^*_{\mathrm{A}} \mathrm{Q}^+_{d'_0} \oplus \Upsilon_{\mathscr{Q}\, j\mathrm{A}, ex, \hat{0}}\big) \longrightarrow 0,$$

(just as in Section 3.4.2), making the diagram

$$\begin{array}{ccccc}
\varrho^*(\pi^*_{\mathbf{P}} \mathrm{Pr}^*_{\mathrm{A}} \mathrm{F}^+_{d'_0} \oplus \Upsilon_{j\mathrm{A}, ex, \hat{0}}) & \xrightarrow{\oplus_a (\mathrm{tr}_a - r_a^{tail})} & \varrho^* \pi^*_{\mathbf{P}}(\oplus_a \mathrm{F}^0) & \longrightarrow & 0 \\
\downarrow & & \downarrow & & \\
\varrho^*(\pi^*_{\mathbf{P}} \mathrm{Pr}^*_{\mathrm{A}} \mathrm{Q}^+_{d'_0} \oplus \Upsilon_{\mathscr{Q}\, j\mathrm{A}, ex, \hat{0}}) & \xrightarrow{\oplus_a (\mathrm{tr}_a - r_a^{tail})} & \varrho^* \pi^*_{\mathbf{P}}(\oplus_a \mathrm{F}^0_{\mathscr{Q}}) & \longrightarrow & 0 \\
\downarrow & & \downarrow & & \\
0 & & 0 & &
\end{array}$$

commutative. We conclude that there is an induced map of vector bundles

$$\ker \longrightarrow \ker_{\mathscr{Q}},$$

which is easily seen to be surjective at all closed points, and hence surjective. Now define the correct obstruction bundle $\widetilde{\Upsilon}$ on $\widetilde{\mathrm{Z}}$ as the kernel:

$$0 \longrightarrow \widetilde{\Upsilon} \longrightarrow \ker \longrightarrow \ker_{\mathscr{Q}} \longrightarrow 0.$$

At $t = 1$ we have

$$\widetilde{\Upsilon}|_{t=1} = \Upsilon^0_{j\mathrm{A}}, \qquad \textbf{(3.5.20)}$$

while at $t = 0$

$$\widetilde{\Upsilon}|_{t=0} = \pi_{\mathbf{P}}^* \mathrm{Pr}_{\mathrm{A}}^* \mathrm{E}_{d_0}^+ \oplus \Upsilon_{j\mathrm{A},ex}^0. \tag{3.5.21}$$

Here $\Upsilon_{j\mathrm{A},ex}^0$ on $\mathbf{P}_{j\mathrm{A}}|_{\mathrm{D_{X,A}}}$ is given by the same extension as in (3.5.14):

$$0 \to \boxplus_{a\in\mathrm{A}}\left(\mathcal{O}_{\mathbf{P}_{j_a}}(1) \otimes \pi_{\mathbf{P}}^* \mathrm{F}_{small}^{j_a}\right) \to \Upsilon_{j\mathrm{A},ex}^0 \to \pi_{\mathbf{P}}^*\left(\mathrm{G}_{tail,d_a,small}^{+,j\mathrm{A}} \oplus \mathrm{G}_{d_0,small}^{-,j\mathrm{A}}\right) \to 0. \tag{3.5.22}$$

By a calculation similar to the one used to prove Lemma 3.4, one checks that the normal cone $\mathbf{C}_{\widetilde{\mathrm{Z}}/(\mathbf{P}_{j\mathrm{A}} \times \mathbf{A}^1)}$ is a subcone of $\widetilde{\Upsilon}$.

Let $\iota : \widetilde{\mathrm{Z}} \hookrightarrow \mathbf{P}_{j\mathrm{A}}|_{\mathrm{D_{X,A}}} \times \mathbf{A}^1$ denote the inclusion and consider the diagram

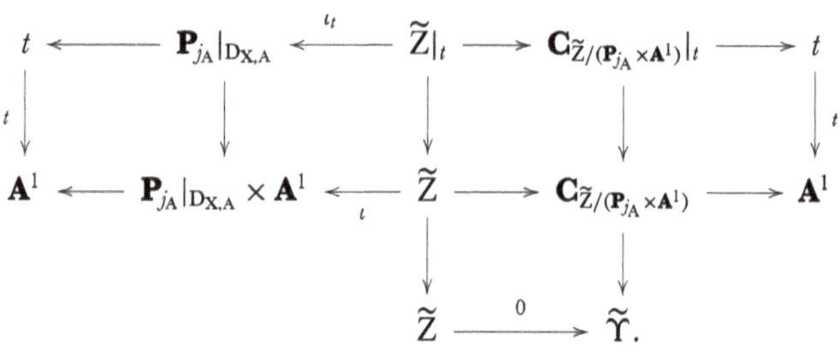

The proof of Lemma 3.5 shows the equality

$$(\iota_1)_* 0^!_{\widetilde{\Upsilon}|_{t=1}}\left([\mathbf{C}_{\widetilde{\mathrm{Z}}/(\mathbf{P}_{j\mathrm{A}}\times\mathbf{A}^1)}|_{t=1}]\right) = 0^!_{\widetilde{\Upsilon}|_{t=0}}\left([\mathbf{C}_{\widetilde{\mathrm{Z}}/(\mathbf{P}_{j\mathrm{A}}\times\mathbf{A}^1)}|_{t=0}]\right)$$

in the Chow group of $\mathbf{P}_{j\mathrm{A}}|_{\mathrm{D_{X,A}}}$. By (3.5.18), (3.5.19), (3.5.20), (3.5.21), the Excess Intersection Formula ([15, Theorem 6.3]), the compatibility of Gysin maps with flat pull-back, and Corollary 2.6, this can be rewritten as

$$(\iota_1)_* 0^!_{\Upsilon^0_{j\mathrm{A}}}([\widetilde{\mathbf{C}}_{j\mathrm{A}}]) = \mathbf{e}\left(\Upsilon^0_{j\mathrm{A},ex}\right) \cap \pi_{\mathbf{P}}^* \mathrm{Pr}_{\mathrm{A}}^* [\mathrm{Q}^+_{g,k+\mathrm{A}}(\mathrm{X}, d_0)]^{\mathrm{vir}}, \tag{3.5.23}$$

where $\mathbf{e}$ denotes the Euler class and $\pi_{\mathbf{P}}^*$, $\mathrm{Pr}_{\mathrm{A}}^*$ are the flat pull-backs.

Step 4: Final calculation. Recall the diagram from (3.5.6)

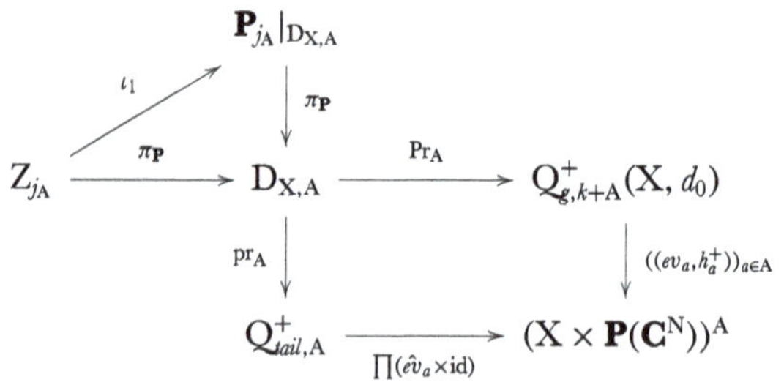

and that we want to compute (3.5.8). From (3.5.23) this is the same as computing

$$\textbf{(3.5.24)} \qquad \sum_{j_A} m_{j_A} (\mathrm{Pr_A})_* (\pi_{\mathbf{P}})_* \big(\mathbf{e}\big(\Upsilon^0_{j_A,ex}\big) \cap \pi^*_{\mathbf{P}} \mathrm{Pr}^*_A \big[\mathrm{Q}^+_{g,k+A}(\mathrm{X}, d_0)\big]^{\mathrm{vir}}\big).$$

By (3.5.22),

$$\mathbf{e}\big(\Upsilon^0_{j_A,ex}\big) = \mathbf{e}\big(\boxplus_{a\in A}\big(\mathcal{O}_{\mathbf{P}_{j_a}}(1) \otimes \pi^*_{\mathbf{P}} \mathrm{F}^{j_a}_{small}\big)\big)\mathbf{e}\big(\pi^*_{\mathbf{P}}\big(\mathrm{G}^{+j_A}_{tail,d_a,small}\big)\big) \times \mathbf{e}\big((\pi_{\mathbf{P}})^*\big(\mathrm{G}^{-j_A}_{d_0,small}\big)\big).$$

Set $\alpha := \mathbf{e}(\mathrm{G}^{+j_A}_{tail,d_a,small})\mathbf{e}(\mathrm{G}^{-j_A}_{d_0,small}) \cap \mathrm{Pr}^*_A[\mathrm{Q}^+_{g,k+A}(\mathrm{X}, d_0)]^{\mathrm{vir}}$. Then (3.5.24) can be successively rewritten as

$$\sum_{j_A} m_{j_A} (\mathrm{Pr_A})_* \big\{(\pi_{\mathbf{P}})_*\big(\mathbf{e}\big(\boxplus_{a\in A}\big(\mathcal{O}_{\mathbf{P}_{j_a}}(1) \otimes \pi^*_{\mathbf{P}} \mathrm{F}^{j_a}_{small}\big)\big) \cap \pi^*_{\mathbf{P}}\alpha\big)\big\}$$

$$= \sum_{j_A} m_{j_A} (\mathrm{Pr_A})_* \prod_{a\in A} (\pi_{\mathbf{P}})_* \Bigg(\sum_{m=0}^{\mathrm{rk}\,(\mathrm{F}^{j_a}_{small})} c_1\big(\mathcal{O}_{\mathbf{P}_{j_a}}(1)\big)^m \cap \pi^*_{\mathbf{P}}\big(c_{\mathrm{rk}\,(\mathrm{F}^{j_a}_{small})-m}\big(\mathrm{F}^{j_a}_{small}\big) \cap \alpha\big)\Bigg)$$

$$= \sum_{j_A} m_{j_A} (\mathrm{Pr_A})_* \prod_{a\in A} \Bigg(\sum_{m=0}^{\mathrm{rk}\,(\mathrm{F}^{j_a}_{small})} s_{m-1}\big(\mathrm{pr}^*_A \mathcal{O}\big(j_a \psi^{tail}_a\big) \oplus \mathrm{Pr}^*_A \mathcal{O}(-j_a\psi_a)\big) c_{\mathrm{rk}\,(\mathrm{F}^{j_a}_{small})-m}\big(\mathrm{F}^{j_a}_{small}\big) \cap \alpha \Bigg),$$

where s_{m-1} denote the Segre classes.

The Chow cohomology class

$$\sum_{m=0}^{\mathrm{rk}\,(\mathrm{F}^{j_a}_{small})} s_{m-1}\big(\mathrm{pr}^*_A \mathcal{O}\big(j_a \psi^{tail}_a\big) \oplus \mathrm{Pr}^*_A \mathcal{O}(-j_a\psi_a)\big) c_{\mathrm{rk}\,(\mathrm{F}^{j_a}_{small})-m}\big(\mathrm{F}^{j_a}_{small}\big)$$

is a polynomial in $\mathrm{Pr}^*_A \psi_a$, of the form

$$\sum_b \mathrm{pr}^*_A\big((\hat{ev}_a \times \mathrm{id})^* \delta_b(z)|_{z=\psi^{tail}_a}\big) \mathrm{Pr}^*_A \psi^b_a,$$

where the δ_b's are themselves polynomials with coefficients given by universal expressions in Chern classes of various tautological bundles $\mathcal{O}_{\mathrm{X}}(l)$ on X, and $\mathcal{O}_{\mathbf{P}(\mathbf{C}^{\mathrm{N}})}(m)$ and Q on $\mathbf{P}(\mathbf{C}^{\mathrm{N}})$. Further, by (3.5.15), (3.5.16), the Euler classes $\mathbf{e}(\mathrm{G}^{+j_A}_{tail,d_a,small})$ and

$\mathbf{e}(G^{-j_A}_{d_0,small})$ appearing in α are given respectively by the universal expressions $\prod_a \mathrm{pr}^*_A(\hat{ev}_a \times \mathrm{id})^* f^{+j_a}_{d_a}(\psi^{tail}_a)$ and $\prod_a \mathrm{Pr}^*_A(ev_a, h^+_a)^* f^{-j_a}_{d_a}(-\psi_a)$.

Setting

$$\gamma_b := (\hat{ev}_a \times \mathrm{id})^*\left(\delta_b f^{+j_a}_{d_a}\right)\left(\psi^{tail}_a\right) \in A^*\left(Q^+_{tail,a}\right)_{\mathbf{Q}}$$

and recalling that $m_{j_A} = \prod_{a\in A} j_a$, we conclude that (3.5.24) has the form

$$\textbf{(3.5.25)}\qquad \prod_{a\in A}\Bigg(\sum_{j_a=1}^{\max_i\{l_i d_i\}} j_a(\mathrm{Pr}_A)_*\Big\{\sum_b \mathrm{pr}^*_A(\gamma_b)\mathrm{Pr}^*_A\big(\psi^b_a\big(ev_a, h^+_a\big)^* f^{-j_a}_{d_a}(-\psi_a)\big)\Big\}$$
$$\times\left(\left[Q^+_{g,k+A}(X, d_0)\right]^{\mathrm{vir}}\right)\Bigg).$$

Here $(\mathrm{Pr}_A)_* : A^*(D_{X,A})_{\mathbf{Q}} \longrightarrow A^*(Q^+_{g,k+A}(X, d_0))_{\mathbf{Q}}$ denotes the Gysin map induced by the bivariant class $[\mathrm{Pr}_A]$ corresponding to the canonical orientation of the flat proper morphism Pr_A, see equation (G_2) in [15, §17.4]. Applying [15, Example 17.4.1(b)] to the cartesian square (3.5.4) and using the projection formula for bivariant classes, equation (3.5.25) proves Theorem 3.8, with

$$\textbf{(3.5.26)}\qquad \mu^N_{d_a}(z) := \sum_{j_a=1}^{\max_i\{l_i d_i\}} j_a \sum_b (-z)^b f^{-j_a}_{d_a}(z)(\hat{ev}_a \times \mathrm{id})_*(\gamma_b) \in A^*\left(X \times \mathbf{P}(\mathbf{C}^N)\right)_{\mathbf{Q}}[z].$$

□

We stress again that our argument shows that the formula (3.5.26) for the correcting class μ^N_d is universal in the following sense: it depends on g and k only through the dependence on N of the polynomials $f^{+j_a}_{d_a}(z), f^{-j_a}_{d_a}(z), \delta_b(z) \in A^*(X \times \mathbf{P}(\mathbf{C}^N))_{\mathbf{Q}}[z]$. This will be used in the next subsection.

3.6. *Identification of the correcting class.* — In this subsection we finish the proof of Theorem 1.6 (for $(g, k) \neq (1, 0)$) by showing that the class (3.5.26) satisfies

$$\textbf{(3.6.1)}\qquad \mu^N_{d_a}(z) = \text{coefficient of } q^{d_a} \text{ in } z\big(J^{\varepsilon_-}_{sm}(z) - J^{\varepsilon_+}_{sm}(z)\big) \otimes \mathbf{1}_{\mathbf{P}(\mathbf{C}^N)}.$$

Indeed, assuming (3.6.1), it follows first that the coefficient of q^{d_a} in $z(J^{\varepsilon_-}_{sm}(z) - J^{\varepsilon_+}_{sm}(z))$ is a polynomial in z (because the left-hand side is such) and then by the general asymptotic properties of the small J^{ε}-functions it coincides with the coefficient of q^{d_a} in $[zI_{sm}(q, z) - z]_+$. Second, (3.6.1) also shows that the class $(ev_a, h^+_a)^* \mu^N_{d_a}(z)$ is independent of N, so that we may replace it by $ev^*_a \mu_{d_a}(z)$ in the formula (3.5.2). Hence Theorem 3.8 together with (3.6.1) imply Theorem 1.6.

To prove (3.6.1), we take $d = d_a$ (so that $d_0 = 0$) and consider the graph spaces $QG^{\pm}_{0,0,d_a}(X)$. These are the moduli stacks of $\varepsilon_{\pm}$-stable quasimaps of degree d_a to X, whose domains are genus zero unpointed curves with a component which is a parametrized $\mathbf{P}^1$, see [7, 12]. Similarly, we have the moduli stacks $QG^{\pm}_{0,0,d_a}(\mathbf{P}(V))$ and $QG^{\pm}_{0,0,d_a}(\mathbf{P}(V \otimes \mathbf{C}^N))$, which are smooth. The ε_--stability condition implies that the domain curve must be an irreducible parametrized $\mathbf{P}^1$, while ε_+-stability allows in addition quasimaps with domain consisting of one rational tail and the parametrized $\mathbf{P}^1$. These quasimaps have degree d_a on the rational tail and are constant maps on the parametrized $\mathbf{P}^1$. In particular, there are identifications

$$QG^{-}_{0,0,d_a}(\mathbf{P}(V)) \cong \mathbf{P}(\mathrm{Sym}^{d_a}(\mathbf{C}^2) \otimes V),$$
$$QG^{-}_{0,0,d_a}(\mathbf{P}(V \otimes \mathbf{C}^N)) \cong \mathbf{P}(\mathrm{Sym}^{d_a}(\mathbf{C}^2) \otimes V \otimes \mathbf{C}^N).$$

Recall that we have the embeddings

$$X \times \mathbf{P}(\mathbf{C}^N) \hookrightarrow \mathbf{P}(V) \times \mathbf{P}(\mathbf{C}^N) \hookrightarrow \mathbf{P}(V \otimes \mathbf{C}^N) \times \mathbf{P}(\mathbf{C}^N),$$

whose composition is the map i_{Seg} from (3.5.3). The induced embeddings of graph spaces commute with the contraction maps:

(3.6.2)
$$\begin{array}{ccc} QG^{+}_{0,0,d_a}(X) \times \mathbf{P}(\mathbf{C}^N) & \hookrightarrow & QG^{+}_{0,0,d_a}(\mathbf{P}(V \otimes \mathbf{C}^N)) \times \mathbf{P}(\mathbf{C}^N) \\ \downarrow{\scriptstyle c\times \mathrm{id}} & & \downarrow{\scriptstyle c\times \mathrm{id}} \\ QG^{-}_{0,0,d_a}(X) \times \mathbf{P}(\mathbf{C}^N) & \hookrightarrow & QG^{-}_{0,0,d_a}(\mathbf{P}(V \otimes \mathbf{C}^N)) \times \mathbf{P}(\mathbf{C}^N). \end{array}$$

The right contraction map $c \times \mathrm{id}$ is an isomorphism outside the boundary divisor

$$\begin{aligned} D_a &\cong (Q^{+}_{0,\{a\}}(\mathbf{P}(V \otimes \mathbf{C}^N), d_a) \times \mathbf{P}(\mathbf{C}^N)) \\ &\quad \times_{\mathbf{P}(V\otimes\mathbf{C}^N)\times\mathbf{P}(\mathbf{C}^N)} (QG^{+}_{0,\{a\},0}(\mathbf{P}(V \otimes \mathbf{C}^N)) \times \mathbf{P}(\mathbf{C}^N)) \\ &\cong (Q^{+}_{0,\{a\}}(\mathbf{P}(V \otimes \mathbf{C}^N), d_a) \times \mathbf{P}(\mathbf{C}^N)) \\ &\quad \times_{\mathbf{P}(V\otimes\mathbf{C}^N)\times\mathbf{P}(\mathbf{C}^N)} (\mathbf{P}(V \otimes \mathbf{C}^N) \times \mathbf{P}^1 \times \mathbf{P}(\mathbf{C}^N)), \end{aligned}$$

where $QG^{+}_{0,\{a\},0}(\mathbf{P}(V \otimes \mathbf{C}^N)) \cong \mathbf{P}(V \otimes \mathbf{C}^N) \times \mathbf{P}^1$ is the moduli stack of ε_+-stable quasimaps of degree 0 to $\mathbf{P}(V \otimes \mathbf{C}^N)$, whose domains are genus zero *one-pointed* curves with a component which is a parametrized $\mathbf{P}^1$, see [7, 12]. Let $\mathscr{L}_{\pm}$ denote the universal line bundles of degree d_a on the fibers of the universal curves over the various $QG^{\pm} \times \mathbf{P}(\mathbf{C}^N)$. Let $\mathscr{M}$ denote the pull-back of $\mathcal{O}_{\mathbf{P}(\mathbf{C}^N)}(1)$ to $QG^{\pm} \times \mathbf{P}(\mathbf{C}^N)$, with the basis $\{t_1, \ldots, t_N\}$ of global sections, and set $\mathscr{L}'_{\pm} = \mathscr{L}_{\pm} \otimes \mathscr{M}$. With these notations (which are justified, since the line bundles are compatible with the above embeddings), the construction of Section 2.4 produces the obstruction theory (2.4.6) of $QG^{\pm}_{0,0,d_a}(X) \times \mathbf{P}(\mathbf{C}^N)$ relative to the smooth, pure dimensional stack $\mathfrak{Bun}^{\mathbf{P}^1}_{\mathbf{G}} \times \mathbf{P}(\mathbf{C}^N)$. Here $\mathfrak{Bun}^{\mathbf{P}^1}_{\mathbf{G}} \longrightarrow \widetilde{\mathbf{P}^1[0]}$

is the relative Picard stack over the Fulton-MacPherson stack $\widetilde{\mathbf{P}^1[0]}$ of unpointed rational curves with one parametrized component. The corresponding virtual class is $[QG^{\pm}_{0,0,d_a}(X)]^{\mathrm{vir}} \times [\mathbf{P}(\mathbf{C}^N)]$. Note that for all universal curves, the map h to $\mathbf{P}(\mathbf{C}^N)$ is just the projection.

Further, if we put

$$U^{\pm} := QG^{\pm}_{0,0,d_a}(\mathbf{P}(V \otimes \mathbf{C}^N)) \times \mathbf{P}(\mathbf{C}^N),$$

then the construction of Section 2.5 also applies to produce the vector bundles $F^{\pm}$ on $U^{\pm}$, with sections $\sigma^{\pm}$ such that $(\sigma^{\pm})^{-1}(0) \cong QG^{+}_{0,0,d_a}(X) \times \mathbf{P}(\mathbf{C}^N)$. This embedding of $QG^{+}_{0,0,d_a}(X) \times \mathbf{P}(\mathbf{C}^N)$ in $U^{\pm}$ is precisely the one in (3.6.2). The diagram (2.5.6) holds as well, hence we have the concrete description

$$[QG^{\pm}_{0,0,d_a}(X)]^{\mathrm{vir}} \times [\mathbf{P}(\mathbf{C}^N)] = 0^{!}_{E^{\pm}}(\mathbf{C}_{QG^{\pm}_{0,0,d_a}(X)\times \mathbf{P}(\mathbf{C}^N)/U^{\pm}})$$

as in Corollary 2.6.

From the degeneration analysis in Section 3.2 – Section 3.5, it follows that Theorem 3.8 holds in the situation considered in this section, giving the equality

$$\begin{aligned} (\mathbf{3.6.3}) \qquad & [QG^{-}_{0,0,d_a}(X)]^{\mathrm{vir}} \times [\mathbf{P}(\mathbf{C}^N)] - (c \times \mathrm{id})_*([QG^{+}_{0,0,d_a}(X)]^{\mathrm{vir}} \times [\mathbf{P}(\mathbf{C}^N)]) \\ &= (b_a \times \mathrm{id})_*((ev_a, h^{+}_a)^* \mu^{N}_{d_a}(-\psi_a) \cap ([QG^{+}_{0,\{a\},0}(X)]^{\mathrm{vir}} \times [\mathbf{P}(\mathbf{C}^N)]), \end{aligned}$$

with $\mu^{N}_{d_a}$ the universal class in (3.5.26). Notice that the one-pointed, *degree zero* graph space is identified with $X \times \mathbf{P}^1$, with virtual class the usual fundamental class (for any stability parameter ε), while the maps

$$ev_a : X \times \mathbf{P}^1 \times \mathbf{P}(\mathbf{C}^N) \longrightarrow X, \qquad h^{+}_a : X \times \mathbf{P}^1 \times \mathbf{P}(\mathbf{C}^N) \longrightarrow \mathbf{P}(\mathbf{C}^N)$$

are respectively the first and third projections. The class ψ_a is the pull-back of $c_1(\omega_{\mathbf{P}^1})$ via the second projection.

Now recall that graph spaces carry a $\mathbf{C}^*$-action (induced by the standard action on the parametrized domain component) for which the maps c and b_a are equivariant. It is customary to denote by z the equivariant parameter for this action. In each graph space there is a distinguished part of the $\mathbf{C}^*$-fixed locus corresponding to quasimaps for which the entire nontrivial data is concentrated over the point 0 in the parametrized domain component. The restrictions of the maps c and b_a to the fixed point locus respect the decomposition into distinguished and non-distinguished parts. It follows that if we apply the virtual localization formula of [19] to (3.6.3) (using the trivial action on the $\mathbf{P}(\mathbf{C}^N)$ factors) and discard from both sides the localization residues at all non-distinguished fixed-point loci, we still have an equality between the remaining distinguished residues.

In our particular case, the distinguished fixed locus in $QG^{-}_{0,0,d_a}(X) \times \mathbf{P}(\mathbf{C}^N)$ is identified with $X \times \mathbf{P}(\mathbf{C}^N)$, the distinguished fixed locus in $QG^{+}_{0,0,d_a}(X) \times \mathbf{P}(\mathbf{C}^N)$ is identified with $Q^{+}_{0,1}(X, d_a) \times \mathbf{P}(\mathbf{C}^N)$, and the distinguished fixed locus in $QG^{+}_{0,\{a\},0}(X) \times \mathbf{P}(\mathbf{C}^N) = X \times \mathbf{P}^1 \times \mathbf{P}(\mathbf{C}^N)$ is $X \times \{0\} \times \mathbf{P}(\mathbf{C}^N)$. Moreover, the restriction of $c \times \mathrm{id}$ to the distinguished fixed locus is $ev_1 \times \mathrm{id}$, while $b_a \times \mathrm{id}$, (ev_a, h^+_a) are the identity map on the distinguished fixed locus. The equality of distinguished residues of (3.6.3) becomes

$$\textbf{(3.6.4)} \qquad \text{coefficient of } q^{d_a} \text{ in } \left(J^{\varepsilon-}_{sm}(z) - J^{\varepsilon+}_{sm}(z)\right) \otimes \mathbf{1}_{\mathbf{P}(\mathbf{C}^N)} = \frac{\mu^N_{d_a}(z)}{z}$$

in $A^*(X \times \mathbf{P}(\mathbf{C}^N))_{\mathbf{Q}}[z, z^{-1}]$, proving (3.6.1). Indeed, the left-hand side is as stated by the very definition of the small J-functions in (5.1.1) of [7], while for the right-hand side we used that, in the right-hand side of (3.6.3), $\psi_a|_{X\times\{0\}\times\mathbf{P}(\mathbf{C}^N)} = -z$, and that the equivariant normal bundle of $\{0\} \subset \mathbf{P}^1$ has first Chern class z, i.e., the denominator z in the right-hand side of (3.6.4) so that $\frac{1}{z}$ is the distinguished residue of $[QG^{+}_{0,\{a\},0}(X)]^{\mathrm{vir}} \times [\mathbf{P}(\mathbf{C}^N)]$.

3.7. *The unpointed genus 1 case.* — Since $\overline{M}_{1,0}$ is empty, we do not have the twisting line bundles $\mathscr{M}$ satisfying Lemma 2.1 and which are all compatible. However, it turns out that an appropriate modification of the set-up in Section 2 allows for an application of the arguments in Section 3 to establish Theorem 1.6 in this case as well.

3.7.1. *Set-up.* — By an unpointed semistable genus 1 curve we mean an unpointed prestable genus 1 curve with no rational tails. Let $\mathfrak{M}^{ss}_{1,0}$ denote the moduli stack of semistable genus 1 curves.

Fix positive integers d and e. Let M_N denote the moduli stack of degree e unpointed genus 1 *stable maps* to $\mathbf{P}(\mathbf{C}^N)$ with *semistable* domain curves. Since all line bundles of degree e on semistable genus 1 curves are non-special, M_N is a smooth (non-proper) Deligne-Mumford stack. Denote by $\mathfrak{C}^{ss}_{1,0} \longrightarrow M_N$ the universal curve and by

$$h : \mathfrak{C}^{ss}_{1,0} \longrightarrow \mathbf{P}(\mathbf{C}^N)$$

the universal map.

Let $d' = d + e$ and let $Q^{\varepsilon,unob}_{1,0}(\mathbf{P}(V \otimes \mathbf{C}^N), d')$ be the open substack of $Q^{\varepsilon}_{1,0}(\mathbf{P}(V \otimes \mathbf{C}^N), d')$ consisting of ε-stable quasimaps (C, L', u') with vanishing $H^1(C, L')$. Define $U^{\varepsilon,N}_{d'}$ as the fiber product

$$Q^{\varepsilon,unob}_{1,0}(\mathbf{P}(V \otimes \mathbf{C}^N), d') \times_{\mathfrak{M}^{ss}_{1,0}} M_N.$$

Here the morphism $Q^{\varepsilon,unob}_{1,0}(\mathbf{P}(V \otimes \mathbf{C}^N), d') \to \mathfrak{M}^{ss}_{1,0}$ is the composite of the contraction map $Q^{\varepsilon,unob}_{1,0}(\mathbf{P}(V \otimes \mathbf{C}^N), d') \to Q^{0+}_{1,0}(\mathbf{P}(V \otimes \mathbf{C}^N), d')$ and the forgetful map $Q^{0+}_{1,0}(\mathbf{P}(V \otimes \mathbf{C}^N), d') \to \mathfrak{M}^{ss}_{1,0}$.

Since M_N is smooth over $\mathfrak{M}^{ss}_{1,0}$ and $Q^{\varepsilon,unob}_{1,0}(\mathbf{P}(V \otimes \mathbf{C}^N), d')$ is smooth over $\mathfrak{Bun}^{1,0}_{\mathbf{G}}$, the stack $U^{\varepsilon,N}_{d'}$ is smooth over $\mathfrak{Bun}^{1,0}_{\mathbf{G}}$.

The universal curve $\mathfrak{C}^{\varepsilon}_{1,0,d'}$ over $U^{\varepsilon,N}_{d'}$ has a semistabilization morphism ss_{ε} : $\mathfrak{C}^{\varepsilon}_{1,0,d'} \to \mathfrak{C}^{ss}_{1,0}$ (the contraction of rational tails of universal curves), fitting into the commuting diagram

$$\begin{array}{ccccc} \mathfrak{C}^{\varepsilon}_{1,0,d'} & \xrightarrow[ss_{\varepsilon}]{} & \mathfrak{C}^{ss}_{1,0} & \xrightarrow[h]{} & \mathbf{P}(\mathbf{C}^N) \\ \downarrow \pi & & \downarrow & & \\ U^{\varepsilon,N}_{d'} & \xrightarrow[\text{proj}]{} & M_N. & & \end{array}$$

We set $h_{\varepsilon} = h \circ ss_{\varepsilon} : \mathfrak{C}^{\varepsilon}_{1,0,d'} \to \mathbf{P}(\mathbf{C}^N)$ and $\mathscr{M}_{\varepsilon} = h^*_{\varepsilon}\mathcal{O}_{\mathbf{P}(\mathbf{C}^N)}(1)$. Further, the sections t_j of $\mathcal{O}_{\mathbf{P}(\mathbf{C}^N)}(1)$ associated to the homogeneous coordinates of $\mathbf{P}(\mathbf{C}^N)$ give the sections $s_j := h^*_{\varepsilon}t_j \in H^0(\mathfrak{C}^{\varepsilon}_{1,0,d'}, \mathscr{M}_{\varepsilon}), j = 1, \dots, N$.

3.7.2. *Obstruction theory for* $Q^{\varepsilon}_{1,0}(X, d) \times_{\mathfrak{M}^{ss}_{1,0}} M_N$ *relative to* $\mathfrak{Bun}^{1,0}_{\mathbf{G}}$. — Denote by $\mathscr{L}'_{\varepsilon}$ the universal line bundle on the universal curve $\mathfrak{C}^{\varepsilon}_{1,0,d'}$ of $U^{\varepsilon,N}_{d'}$ and put $\mathscr{L}_{\varepsilon} := \mathscr{L}'_{\varepsilon} \otimes \mathscr{M}^{-1}_{\varepsilon}$.

Consider the diagram of vector bundles and $\mathcal{O}_{\mathfrak{C}^{\varepsilon}_{1,0,d'}}$-linear maps, corresponding to (2.5.1),

$$\begin{array}{ccccccccc} 0 & \to & \mathscr{L}_{\varepsilon} \otimes V \oplus h^*_{\varepsilon}T_{\mathbf{P}(\mathbf{C}^N)} & \xrightarrow{(\oplus_j s_j, \mathrm{id})} & \oplus_{j=1}^{N} \mathscr{L}'_{\varepsilon} \otimes V \oplus h^*_{\varepsilon}T_{\mathbf{P}(\mathbf{C}^N)} & \to & \mathscr{P}^{\varepsilon}_{d'} & \to & 0 \\ & & & & \downarrow (\oplus_j(\oplus_i d\varphi_i), 0) & & & & \\ 0 & \to & \oplus_{i=1}^{r} \mathscr{L}^{l_i}_{\varepsilon} & \xrightarrow{\oplus_{i,j} s_j^{l_i}} & \oplus_{i,j}(\mathscr{L}'_{\varepsilon})^{l_i} & \to & \mathscr{Q}^{\varepsilon}_{d'} & \to & 0. \end{array}$$

Let $Q^{\varepsilon}_X := Q^{\varepsilon}_{1,0}(X, d)$. As before, there is a vector bundle

$$P^{\varepsilon}_{d'} \oplus R^{\varepsilon}_{d'} := \pi_* \mathscr{P}^{\varepsilon}_{d'} \oplus \pi_*\big(\oplus_{i,j}(\mathscr{L}'_{\varepsilon})^{l_i}\big)$$

on $U^{\varepsilon,N}_{d'}$, with a section σ^{ε} whose zero locus is naturally isomorphic to the product stack

$$Q^{\varepsilon}_X \times_{\mathfrak{M}^{ss}_{1,0}} M_N.$$

On the universal curve $\mathfrak{C}^{\varepsilon}_X$ over $Q^{\varepsilon}_X \times_{\mathfrak{M}^{ss}_{1,0}} M_N$ (associated to the universal curve of Q^{ε}_X), we may complete the diagram above to a homomorphism of short exact sequences. In particular, we obtain a natural homomorphism

$$\mathscr{L}_{\varepsilon} \otimes V \oplus h^*_{\varepsilon}T_{\mathbf{P}(\mathbf{C}^N)} \to \oplus_{i=1}^{r} \mathscr{L}^{l_i}_{\varepsilon}$$

and an exact sequence

$$0 \to \mathscr{E}^{\varepsilon}_{d} \to \mathscr{P}^{\varepsilon}_{d'} \oplus \big(\oplus_{i,j}(\mathscr{L}'_{\varepsilon})^{l_i}\big) \to \mathscr{Q}^{\varepsilon}_{d'} \to 0,$$

defining a vector bundle $\mathscr{E}^{\varepsilon}_{d}$ on $\mathfrak{C}^{\varepsilon}_X$, with $\pi_*\mathscr{E}^{\varepsilon}_{d}$ also locally-free.

Denote by $\mathbf{C}_{\sigma^\varepsilon}$ the normal cone to $Q_X^\varepsilon \times_{\mathfrak{M}_{1,0}^{ss}} M_N$ in $U_{d'}^{\varepsilon,N}$. As before, $\mathbf{C}_{\sigma^\varepsilon}$ is a closed subcone of the vector bundle $\pi_* \mathscr{E}_d^\varepsilon$, with the embedding induced by a surjection $\pi_* \mathscr{E}_d^\varepsilon \twoheadrightarrow \mathscr{I}/\mathscr{I}^2$, where $\mathscr{I}$ is the ideal sheaf of the closed substack $Q_X^\varepsilon \times_{\mathfrak{M}_{1,0}^{ss}} M_N$.

Consider the following commuting diagram

$$
\begin{array}{ccc}
Q_X^\varepsilon \times_{\mathfrak{M}_{1,0}^{ss}} M_N & \xhookrightarrow{\ \ \text{closed}\ \ } U_{d'}^{\varepsilon,N} \longrightarrow M_N \\
\downarrow & \qquad\qquad \downarrow \scriptstyle{\text{smooth}} \\
& Q_{1,0}^{\varepsilon,unob}(\mathbf{P}(V \otimes \mathbf{C}^N), d') \longrightarrow \mathfrak{M}_{1,0}^{ss} \\
& \downarrow \scriptstyle{\text{smooth}} \\
& \mathfrak{B}un_{\mathbf{G}}^{1,0}
\end{array}
$$

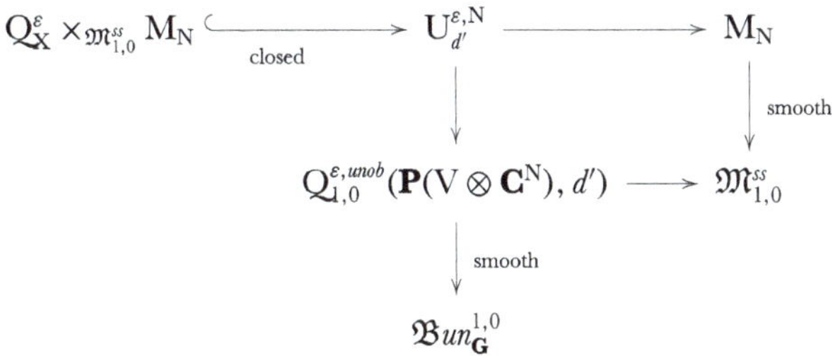

and *define* a perfect obstruction theory $\mathbf{E}$ for $Q_X^\varepsilon \times_{\mathfrak{M}_{1,0}^{ss}} M_N$ relative to $\mathfrak{B}un_{\mathbf{G}}^{1,0}$ by

$$
\left[R^\bullet \pi_* \big(\mathscr{L}_\varepsilon \otimes V \oplus h_\varepsilon^* T_{\mathbf{P}(\mathbf{C}^N)} \to \oplus_{i=1}^r \mathscr{L}_\varepsilon^{l_i} \big) \right]^\vee
$$
$$
\overset{\text{qiso}}{\sim} \left[\big(\pi_* \mathscr{E}_d^\varepsilon \big)^\vee \to \big(\oplus_{j=1}^N \pi_* \mathscr{L}_\varepsilon' \otimes V \oplus \pi_* h_\varepsilon^* T_{\mathbf{P}(\mathbf{C}^N)} \big)^\vee \right] =: \mathbf{E}
$$
$$
\downarrow \qquad\qquad \downarrow \cong
$$
$$
\left[\mathscr{I}/\mathscr{I}^2 \to \Omega_{U_{d'}^{\varepsilon,N}/\mathfrak{B}un_{\mathbf{G}}^{1,0}} |_{Q_X^\varepsilon \times_{\mathfrak{M}_{1,0}^{ss}} M_N} \right].
$$

The associated virtual class is, by definition,

$$
\left[Q_X^\varepsilon \times_{\mathfrak{M}_{1,0}^{ss}} M_N \right]^{\text{vir}} := 0^!_{\pi_* \mathscr{W}_{\varepsilon,d}} [\mathbf{C}_{\sigma^\varepsilon}].
$$

3.7.3. *Wall-crossing.* — We will compare the virtual classes $[Q_X^\pm \times_{\mathfrak{M}_{1,0}^{ss}} M_N]^{\text{vir}}$ under the contraction map $c : Q_X^+ \times_{\mathfrak{M}_{1,0}^{ss}} M_N \to Q_X^- \times_{\mathfrak{M}_{1,0}^{ss}} M_N$, where the contraction map does not do anything on the M_N factor.

The comparison can be carried out as before. Similar to (3.2.2), there is a commuting diagram

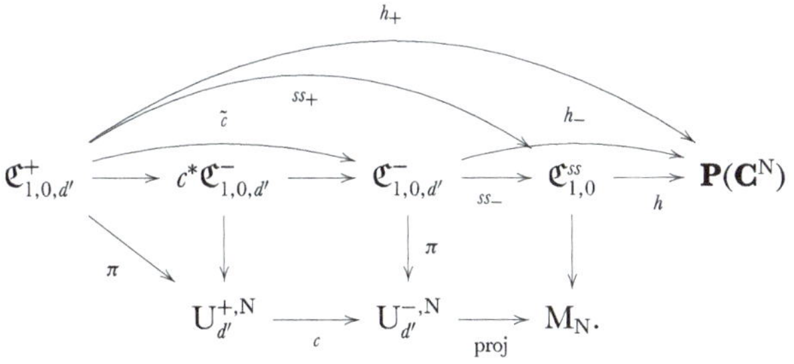

First use the homomorphism $\Phi : P^+_{d'} \oplus R^+_{d'} \to c^*P^-_{d'} \oplus c^*R^-_{d'}$ induced from the contraction map to perform the MacPherson graph construction. Second, deform the obstruction normal cone of $c^{-1}(Q^-_X \times_{\mathfrak{M}^{ss}_{1,0}} M_N)$ in $U^{+,N}_{d'}$ using the induced section of the universal quotient bundle of $\mathbf{Gr}(P^+_{d'} \oplus R^+_{d'} \oplus c^*P^-_{d'} \oplus c^*R^-_{d'})$.

Repeating word for word the arguments of Sections 3.3–3.6, we obtain the following analogue of Theorem 3.8. Let z be a formal variable. Let the Chow cohomology class $\mu^N_{d_a}(z) \in A^*(X \times \mathbf{P}(\mathbf{C}^N))_{\mathbf{Q}}[z]$ be given by the universal formula (3.5.26). The equality

$$\textbf{(3.7.1)}\qquad [Q^-_{1,0}(X,d) \times_{\mathfrak{M}^{ss}_{1,0}} M_N]^{vir} - c_*[Q^+_{1,0}(X,d) \times_{\mathfrak{M}^{ss}_{1,0}} M_N]^{vir} =$$
$$\sum_A \frac{1}{|A|!}(b_A)_*(c_A)_*\Big(\prod_{a\in A}(ev_a, h^+_a)^*\mu^N_{d_a}(z)|_{z=-\psi_a} \cap [Q^+_{1,A}(X,d^A_0) \times_{\mathfrak{M}^{ss}_{1,0}} M_N]^{vir}\Big)$$

holds in the Chow group $A_*(Q^-_{1,0}(X,d) \times_{\mathfrak{M}^{ss}_{1,0}} M_N)_{\mathbf{Q}}$, where

- c_A is the contraction map

$$Q^+_{1,A}(X,d^A_0) \times_{\mathfrak{M}^{ss}_{1,0}} M_N \to Q^-_{1,A}(X,d^A_0) \times_{\mathfrak{M}^{ss}_{1,0}} M_N,$$

- b_A is the morphism

$$Q^-_{1,A}(X,d^A_0) \times_{\mathfrak{M}^{ss}_{1,0}} M_N \to Q^-_{1,0}(X,d) \times_{\mathfrak{M}^{ss}_{1,0}} M_N$$

  which trades the markings A for base points of length d_a,
- the morphism $h^+_a : Q^+_{1,A}(X,d^A_0) \times_{\mathfrak{M}^{ss}_{1,0}} M_N \to \mathbf{P}(\mathbf{C}^N)$ is the composite of the contraction

$$Q^+_{1,A}(X,d^A_0) \times_{\mathfrak{M}^{ss}_{1,0}} M_N \to Q^-_{1,A}(X,d^A_0) \times_{\mathfrak{M}^{ss}_{1,0}} M_N,$$

  the marking section

$$\Sigma_a : Q^-_{1,A}(X,d^A_0) \times_{\mathfrak{M}^{ss}_{1,0}} M_N \to \mathfrak{C}^-_{A,X}$$

  of the universal curve over $Q^-_{1,A}(X,d^A_0) \times_{\mathfrak{M}^{ss}_{1,0}} M_N$ (associated to the universal curve of $Q^-_{1,A}(X,d^A_0)$), the morphism

$$\mathfrak{C}^-_{A,X} \to \mathfrak{C}^-_X$$

  induced from b_A, and finally $h_-|_{\mathfrak{C}^-_X} : \mathfrak{C}^-_X \to \mathbf{P}(\mathbf{C}^N)$.

3.7.4. *Relation between* $[\mathrm{Q}^{\varepsilon}_{\mathrm{X}} \times_{\mathfrak{M}^{ss}_{1,0}} \mathrm{M_N}]^{\mathrm{vir}}$ *and* $[\mathrm{Q}^{\varepsilon}_{\mathrm{X}}]^{\mathrm{vir}}$. — By a result of Cooper, [13], the stack $\mathrm{Q}^{0+}_{1,0}(\mathbf{P}(\mathrm{V}), d)$ has projective coarse moduli and hence there is a morphism from the universal curve of $\mathrm{Q}^{0+}_{1,0}(\mathbf{P}(\mathrm{V}), d)$ to $\mathbf{P}(\mathbf{C}^{\mathrm{N}})$ for some N such that the morphism does not contract any irreducible component of any fiber of the universal curve. Fix such a morphism ϕ and let e be the degree of a fiber curve under ϕ. The degree e is independent of the choice of fiber since $\mathrm{Q}^{0+}_{1,0}(\mathbf{P}(\mathrm{V}), d)$ is connected. (In fact, $\mathrm{Q}^{0+}_{1,0}(\mathbf{P}(\mathrm{V}), d)$ is irreducible; this follows from the connectedness of $\overline{\mathrm{M}}_{1,0}(\mathbf{P}(\mathrm{V}), d)$ (see [22]), the surjectivity of the contraction map $\overline{\mathrm{M}}_{1,0}(\mathbf{P}(\mathrm{V}), d) \to \mathrm{Q}^{0+}_{1,0}(\mathbf{P}(\mathrm{V}), d)$, and the smoothness of $\mathrm{Q}^{0+}_{1,0}(\mathbf{P}(\mathrm{V}), d)$ (see [25]).) From now on *we work with the stack* $\mathrm{M_N}$ *corresponding to these particular choices of* N *and* e.

By the universal property of $\mathrm{M_N}$, upon restricting ϕ to the universal curve over $\mathrm{Q}^{0+}_{\mathrm{X}}$, we obtain a morphism $\underline{h}_{1,0} : \mathrm{Q}^{0+}_{\mathrm{X}} \to \mathrm{M_N}$ fitting in the diagram with the cartesian square

$$\begin{array}{ccccc} \mathfrak{C}^{0+}_{1,0} & \longrightarrow & \mathfrak{C}^{ss}_{1,0} & \xrightarrow{h} & \mathbf{P}(\mathbf{C}^{\mathrm{N}}) \\ \downarrow & & \downarrow & & \\ \mathrm{Q}^{0+}_{\mathrm{X}} & \longrightarrow \; \mathrm{Q}^{0+}_{1,0}(\mathbf{P}(\mathrm{V}), d) \; \longrightarrow & \mathrm{M_N}. & & \end{array}$$

(with $\phi : \mathfrak{C}^{0+}_{1,0} \to \mathbf{P}(\mathbf{C}^{\mathrm{N}})$ the composite along the top and $\underline{h}_{1,0} : \mathrm{Q}^{0+}_{\mathrm{X}} \to \mathrm{M_N}$ the composite along the bottom)

We also let

$$\underline{h}^{\varepsilon}_{1,0} : \mathrm{Q}^{\varepsilon}_{\mathrm{X}} \to \mathrm{Q}^{0+}_{\mathrm{X}} \xrightarrow{\underline{h}_{1,0}} \mathrm{M_N}$$

denote the composition of $\underline{h}_{1,0}$ and the contraction $\mathrm{Q}^{\varepsilon}_{\mathrm{X}} \to \mathrm{Q}^{0+}_{\mathrm{X}}$.

One checks directly that there is a natural cartesian square

$$\begin{array}{ccc} \mathrm{Q}^{\varepsilon}_{\mathrm{X}} & \xrightarrow{(\mathrm{id}, \underline{h}^{\varepsilon}_{1,0})} & \mathrm{Q}^{\varepsilon}_{\mathrm{X}} \times_{\mathfrak{M}^{ss}_{1,0}} \mathrm{M_N} \\ {\scriptstyle \underline{h}^{\varepsilon}_{1,0}} \downarrow & & \downarrow {\scriptstyle (\underline{h}^{\varepsilon}_{1,0}, \mathrm{id})} \\ \mathrm{M_N} & \xrightarrow{\Delta} & \mathrm{M_N} \times_{\mathfrak{M}^{ss}_{1,0}} \mathrm{M_N}. \end{array}$$

In the derived category of coherent sheaves on $\mathrm{Q}^{\varepsilon}_{\mathrm{X}}$ there is a commuting diagram

$$\begin{array}{ccc} (\underline{h}^{\varepsilon}_{1,0})^*(\mathrm{L}_{\Delta}[-1] \cong (\pi_* h^* \mathrm{T}_{\mathbf{P}(\mathbf{C}^{\mathrm{N}})})^{\vee}) & \longrightarrow & (\mathrm{id}, \underline{h}^{\varepsilon}_{1,0})^* \mathbf{E} \\ \downarrow & & \downarrow \\ \mathbf{L}_{\mathrm{Q}^{\varepsilon}_{\mathrm{X}}/\mathrm{Q}^{\varepsilon}_{\mathrm{X}} \times_{\mathfrak{M}^{ss}_{1,0}} \mathrm{M_N}}[-1] & \longrightarrow & (\mathrm{id}, \underline{h}^{\varepsilon}_{1,0})^* \mathbf{L}_{\mathrm{Q}^{\varepsilon}_{\mathrm{X}} \times_{\mathfrak{M}^{ss}_{1,0}} \mathrm{M_N}/\mathfrak{B}un^{1,0}_{\mathbf{G}}} \end{array}$$

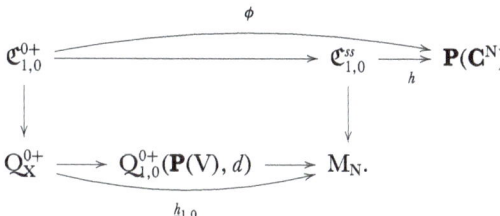

whose mapping cone is the obstruction theory for Q^{ε}_{X} relative to $\mathfrak{Bun}^{1,0}_{\mathbf{G}}$, as in Section 2.4. The functoriality result of [1, Proposition 5.10] implies the relation

(3.7.2) $$\Delta^![Q^{\varepsilon}_{X} \times_{\mathfrak{M}^{ss}_{1,0}} M_N]^{\mathrm{vir}} = [Q^{\varepsilon}_{X}]^{\mathrm{vir}}.$$

Now apply $\Delta^!$ to (3.7.1). Using the compatibility of the Gysin homomorphism for proper push-forward, the commutativity of Chern classes with Gysin homomorphism, the relation (3.7.2), and the identification of $\mu^{N}_{d_a}(z)$ from Section 3.6, we conclude the proof of Theorem 1.6 in the remaining case $(g, k) = (1, 0)$.

Publisher's Note

Springer Nature remains neutral with regard to jurisdictional claims in published maps and institutional affiliations.

REFERENCES

1. K. Behrend and B. Fantechi, The intrinsic normal cone, *Invent. Math.*, **128** (1997), 45–88.
2. M. Bershadsky, S. Cecotti, H. Ooguri and C. Vafa, Holomorphic anomalies in topological field theories, *Nucl. Phys. B*, **405** (1993), 279–304.
3. A. Bertram, Another way to enumerate rational curves with torus actions, *Invent. Math.*, **142** (2000), 487–512.
4. A. Bertram, I. Ciocan-Fontanine and B. Kim, Two proofs of a conjecture of Hori and Vafa, *Duke Math. J.*, **126** (2005), 101–136.
5. A. Bertram, I. Ciocan-Fontanine and B. Kim, Gromov-Witten invariants for nonabelian and abelian quotients, *J. Algebraic Geom.*, **17** (2008), 275–294.
6. I. Ciocan-Fontanine and B. Kim, Moduli stacks of stable toric quasimaps, *Adv. Math.*, **225** (2010), 3022–3051.
7. I. Ciocan-Fontanine and B. Kim, Wall-crossing in genus zero quasimap theory and mirror maps, *Algebraic Geom.*, **1** (2014), 400–448.
8. I. Ciocan-Fontanine and B. Kim, Big I-functions, in *Development of Moduli Theory*, Advanced Studies in Pure Mathematics, vol. 69, *Kyoto 2013*, pp. 323–347, (2016) (volume in honor of S. Mukai's 60th birthday).
9. I. Ciocan-Fontanine and B. Kim, Higher genus quasimap wall-crossing for semi-positive targets, *J. Eur. Math. Soc.*, **19** (2017), 2051–2102.
10. I. Ciocan-Fontanine, B. Kim and C. Sabbah, The abelian/nonabelian correspondence and Frobenius manifolds, *Invent. Math.*, **171** (2008), 301–343.
11. I. Ciocan-Fontanine, M. Konvalinka and I. Pak, Quantum cohomology of $Hilb_n(\mathbf{C}^2)$ and the weighted hook walk on Young diagrams, *J. Algebra*, **349** (2012), 268–283.
12. I. Ciocan-Fontanine, B. Kim and D. Maulik, Stable quasimaps to GIT quotients, *J. Geom. Phys.*, **75** (2014), 17–47.
13. Y. Cooper, The geometry of stable quotients in genus one, *Math. Ann.*, **361** (2015), 943–979.
14. C. Faber and R. Pandharipande, Hodge integrals and Gromov-Witten theory, *Invent. Math.*, **139** (2000), 173–199.
15. W. Fulton, *Intersection theory*, Ergebnisse der Mathematik und ihrer Grenzgebiete, Springer, Berlin, 1984.
16. E. Getzler and R. Pandharipande, Virasoro constraints and the Chern classes of the Hodge bundle, *Nucl. Phys. B*, **530** (1998), 701–714.
17. A. Givental, Equivariant Gromov-Witten invariants, *Int. Math. Res. Not.*, **13** (1996), 613–663.
18. A. Givental, A mirror theorem for toric complete intersections, in *Topological Field Theory, Primitive Forms and Related Topics*, Progr. Math., vol. 160, *Kyoto, 1996*, pp. 141–175, Birkhäuser Boston, Boston, 1998.
19. T. Graber and R. Pandharipande, Localization of virtual classes, *Invent. Math.*, **135** (1999), 487–518.
20. M-x. Huang, A. Klemm and S. Quackenbush, Topological string theory on compact Calabi-Yau: modularity and boundary conditions, in A. Kapustin, M. Kreuzer and K.-G. Schelsinger (eds.) *Homological Mirror Symmetry – New Developments and Perspectives*, Lecture Notes in Physics, vol. 757, pp. 45–102, 2009.

21. B. Kim and H. Lho, Mirror theorem for elliptic quasimap invariants, *Geom. Topol.*, **22** (2018), 1459–1481.
22. B. Kim and R. Pandharipande, The Connectedness of the moduli space of maps to homogeneous spaces, in K. Fukaya, Y.-G. Oh, K. Ono and G. Tian (eds.) *Symplectic Geometry and Mirror Symmetry: Proceedings of the 4th KIAS Annual International Conference*, pp. 187–201, 2001.
23. A. Kresch, Canonical rational equivalence of intersections of divisors, *Invent. Math.*, **136** (1999), 483–496.
24. J. Li and G. Tian, Virtual moduli cycles and Gromov-Witten invariants of algebraic varieties, *J. Am. Math. Soc.*, **11** (1998), 119–174.
25. A. Marian, D. Oprea and R. Pandharipande, The moduli space of stable quotients, *Geom. Topol.*, **15** (2011), 1651–1706.
26. A. Mustaţă and A. Mustaţă, Intermediate moduli spaces of stable maps, *Invent. Math.*, **167** (2007), 47–90.
27. A. Popa, The genus one Gromov-Witten invariants of Calabi-Yau complete intersections, *Trans. Am. Math. Soc.*, **365** (2013), 1149–1181.
28. Y. Toda, Moduli spaces of stable quotients and wall-crossing phenomena, *Compos. Math.*, **147** (2011), 1479–1518.
29. A. Vistoli, Intersection theory on algebraic stacks and on their moduli spaces, *Invent. Math.*, **97** (1989), 613–670.
30. A. Zinger, The reduced genus 1 Gromov-Witten invariants of Calabi-Yau hypersurfaces, *J. Am. Math. Soc.*, **22** (2009), 691–737.

I. C.-F.
School of Mathematics, University of Minnesota, 206 Church St. SE, Minneapolis, MN 55455, USA
and
School of Mathematics, Korea Institute for Advanced Study, 85 Hoegiro, Dongdaemun-gu, Seoul, 02455, Korea
ciocan@math.umn.edu

B. K.
School of Mathematics, Korea Institute for Advanced Study, 85 Hoegiro, Dongdaemun-gu, Seoul, 02455, Korea
bumsig@kias.re.kr

Manuscrit reçu le 1 août 2018
Version révisée le 3 décembre 2019
Manuscrit accepté le 24 janvier 2020
publié en ligne le 7 février 2020.

GPSR Compliance
The European Union's (EU) General Product Safety Regulation (GPSR) is a set of rules that requires consumer products to be safe and our obligations to ensure this.

If you have any concerns about our products, you can contact us on

ProductSafety@springernature.com

In case Publisher is established outside the EU, the EU authorized representative is:

Springer Nature Customer Service Center GmbH
Europaplatz 3
69115 Heidelberg, Germany

www.ingramcontent.com/pod-product-compliance
Ingram Content Group UK Ltd.
Pitfield, Milton Keynes, MK11 3LW, UK
UKHW021829080526
470863UK00003B/53

9789819503841